Interdisciplinary Applied Mathematics

Volume 22

Problems in engineering, computational science, and the physical and biological sciences are using increasingly sophisticated mathematical techniques. Thus, the bridge between the mathematical sciences and other disciplines is heavily traveled. The correspondingly increased dialog between the disciplines has led to the establishment of the series: *Interdisciplinary Applied Mathematics.*

The purpose of this series is to meet the current and future needs for the interaction between various science and technology areas on the one hand and mathematics on the other. This is done, firstly, by encouraging the ways that mathematics may be applied in traditional areas, and well as point towards new and innovative areas of applications; and, secondly, by encouraging other scientific disciplines to engage in a dialog with mathematicians outlining their problems to both access new methods and suggest innovative developments within mathematics itself.

The series will consist of monographs and high-level texts from researchers working on the interplay between mathematics and other fields of science and technology.

Interdisciplinary Applied Mathematics

Volumes published are listed at the end of this book.

Springer Science+Business Media, LLC

Muhammad Sahimi

Heterogeneous Materials I

Linear Transport and Optical Properties

With 145 Illustrations

Springer

Muhammad Sahimi
Department of Chemical Engineering
University of Southern California
Los Angeles, CA 90089-1211
USA
and
Institute for Advanced Studies in Basic Sciences
Gava Zang, Zanjan 45195-159
Iran
moe@iran.usc.edu

Mathematics Subject Classification (2000): 82-02, 65M

Library of Congress Cataloging-in-Publication Data
Sahimi, Muhammad.
Heterogeneous materials / Muhammad Sahimi.
p. cm. — (Interdisciplinary applied mathematics ; 22-23)
Includes bibliographical references and indexes.
Contents: [1] Linear transport and optical properties — [2] Nonlinear and breakdown properties and atomistic modeling.

ISBN 978-0-387-00167-8 ISBN 978-0-387-21705-5 (eBook)
DOI 10.1007/978-0-387-21705-5

1. Inhomogeneous materials. 2. Composite materials. I. Title. II. Interdisciplinary applied mathematiccs ; v. 22-23.
TA418.9.I53 S24 2003
620.1′1—dc21 2002042744

Printed on acid-free paper.

9 8 7 6 5 4 3 2 1 SPIN 10900956

www.springer-ny.com

To children of the third world

who have the talent but not the means to succeed

and to

the memory of my father, Habibollah Sahimi,

who instilled in me, a third world child, the love of reading

Preface

Disorder plays a fundamental role in many natural and man-made systems that are of industrial and scientific importance. Of all the disordered systems, heterogeneous materials are perhaps the most heavily utilized in all aspects of our daily lives, and hence have been studied for a long time. With the advent of new experimental techniques, it is now possible to study the morphology of disordered materials and gain a much deeper understanding of their properties. Novel techniques have also allowed us to design materials of morphologies with the properties that are suitable for intended applications.

With the development of a class of powerful theoretical methods, we now have the ability for interpreting the experimental data and predicting many properties of disordered materials at many length scales. Included in this class are renormalization group theory, various versions of effective-medium approximation, percolation theory, variational principles that lead to rigorous bounds to the effective properties, and Green function formulations and perturbation expansions. The theoretical developments have been accompanied by a tremendous increase in the computational power and the emergence of massively parallel computational strategies. Hence, we are now able to model many materials at molecular scales and predict many of their properties based on first-principle computations.

In this two-volume book we describe and discuss various theoretical and computational approaches for understanding and predicting the effective macroscopic properties of heterogeneous materials. Most of the book is devoted to comparing and contrasting the two main classes of, and approaches to, disordered materials, namely, the continuum models and the discrete models. Predicting the effective properties of composite materials based on the continuum models, which are based on solving the classical continuum equations of transport, has a long history and goes back to at least the middle of the nineteenth century. Even a glance at the literature on the subject of heterogeneous materials will reveal the tremendous amount of work that has been carried out in the area of continuum modeling. Rarely, however, can such continuum models provide accurate predictions of the effective macroscopic properties of *strongly* disordered multiphase materials. In particular, if the contrast between the properties of a material's phases is large, and the phases form large clusters, most continuum models break down. At the same time, due to their very nature, the discrete models, which are based on a lattice representation of a material's morphology, have the ability for providing accurate predictions for the effective properties of heterogeneous materials, even when the heterogeneities are strong, while another class of discrete models, that represent a material as a collection of its constituent atoms and molecules, provides accurate predictions of the material's properties at mesoscopic scales, and thus, in this sense, the discrete models are complementary to the continuum models. The last three decades of the twentieth century witnessed great advances in discrete modeling of materials and predicting their macroscopic properties, and one main goal of this book is to describe these advances and compare their predictions with those of the continuum models. In Volume I we consider characterization and modeling of the morphology of disordered materials, and describe theoretical and computational approaches for predicting their *linear* transport and optical properties, while Volume II focuses on nonlinear properties, and fracture and breakdown of disordered materials, in addition to describing their atomistic modeling. Some of the theoretical and computational approaches are rather old, while others are very new, and therefore we attempt to take the reader through a journey to see the history of the development of the subjects that are discussed in this book. Most importantly, we always compare the predictions with the relevant experimental data in order to gain a better understanding of the strengths and/or shortcomings of the two classes of models.

A large number of people have helped me gain deeper understanding of the topics discussed in this book, and hence have helped me to write about them. Not being able to name them all, I limit myself to a few of them who, directly or indirectly, influenced the style and contents of this book. Dietrich Stauffer has greatly contributed to my understanding of percolation theory, disordered media, and critical phenomena, some of the main themes of this book; I am deeply grateful to him. For their tireless help in the preparation of various portions of this book, I would like to thank two of my graduate students, Sushma Dhulipala and Alberto Schroth. Although they may not be aware of it, Professors Pedro Ponte Castañeda of the University of Pennsylvania and Salvatore Torquato of Princeton University provided great help by guiding me through their excellent work, which is described in this book; I would like to thank them both. Some of my own work described in this book has been carried out in collaboration with many people; I am pleased to acknowledge their great contributions, especially those of Dr. Sepehr Arbabi, my former doctoral student. The constant encouragement and support offered by many of my colleagues, a list of whom is too long to be given here, are also gratefully acknowledged. I would like particularly to express my deep gratitude to my former doctoral student Dr. Jaleh Ghassemzadeh, who provided me with critical help at all stages of preparation of this book. Several chapters of this book have been used, in their preliminary versions, in some of the courses that I teach, and I would like to acknowledge the comments that I received from my students.

My wife, Mahnoush, and son, Ali, put up with the countless hours, days, weeks, and months that I spent in preparing this book and my almost complete absence during the time that I was writing, but never denied me their love and support without which this book would have never been completed; I love and cherish them both.

Muhammad Sahimi
Los Angeles, California, USA
May 2002

Contents

Preface vii

Abbreviated Contents for Volume II xx

1 Introduction 1
1.0 Historical Perspective . . . 1
1.1 Heterogeneous Materials . . . 2
1.2 Effective Properties of Heterogeneous Materials . . . 4
1.3 Linear Transport Properties . . . 5
1.3.1 The Effective Conductivity . . . 5
1.3.2 The Effective Dielectric Constant . . . 6
1.3.3 The Effective Elastic Moduli . . . 6
1.4 Nonlinear Transport Properties . . . 7
1.4.1 Constitutive Nonlinearity . . . 7
1.4.2 Threshold Nonlinearity . . . 8
1.5 Predicting the Effective Properties of Heterogeneous Materials . . . 8
1.5.1 The Continuum Models . . . 9
1.5.2 The Discrete Models . . . 10
1.6 The Organization of the Book . . . 11

I Characterization and Modeling of the Morphology 13

2 Characterization of Connectivity and Clustering 15
2.0 Introduction . . . 15
2.1 Characterization of the Geometry: Self-Similar Fractal Microstructures . . . 17
2.2 Statistical Self-Similarity . . . 19
2.3 Measurement of the Fractal Dimension . . . 20
2.3.1 The Correlation Function Method . . . 21
2.3.2 Small-Angle Scattering . . . 21
2.4 Self-Affine Fractals . . . 24
2.5 Characterization of Connectivity and Clustering . . . 26
2.5.1 Random Bond and Site Percolation . . . 27
2.5.2 Percolation Thresholds . . . 29

2.5.3 Bicontinuous Materials, Phase-Inversion Symmetry, and Percolation 30
2.5.4 Computer Generation of a Single Cluster 30
2.6 Percolation Properties 31
2.6.1 Morphological Properties 32
2.6.2 Transport Properties 32
2.6.3 The Structure of the Sample-Spanning Cluster 34
2.7 Universal Scaling Properties of Percolation 35
2.7.1 Morphological Properties 35
2.7.2 Transport Properties 36
2.7.3 Practical Significance of the Critical Exponents 37
2.8 Scale-Dependent Properties of Percolation Composites 38
2.9 Finite-Size Scaling 40
2.10 Percolation in Random Networks and Continua 41
2.10.1 Percolation Thresholds: Materials with Very Low or High Thresholds 42
2.10.2 The Ornstein–Zernike Formulation 44
2.11 Differences Between Lattice and Continuum Percolation 45
2.12 Correlated Percolation 48
2.12.1 Short-Range Correlations 49
2.12.2 Long-Range Correlations 50
2.13 Experimental Measurement of Percolation Properties 52
Summary 56

3 Characterization and Modeling of the Morphology **57**
3.0 Introduction 57
3.1 Models of Heterogeneous Materials 58
3.2 One-Dimensional Models 59
3.3 Spatially Periodic Models 61
3.4 Continuum Models 62
3.4.1 Dispersion of Spheres 63
3.4.1.1 Equilibrium Hard-Sphere Model 65
3.4.1.2 Random Close Packing Versus Maximally Random Jamming 66
3.4.1.3 Particle Distribution and Correlation Functions 66
3.4.1.4 The n-Particle Probability Density 73
3.4.2 Distribution of Equal-Size Particles 75
3.4.2.1 Fully Penetrable Spheres 75
3.4.2.2 Fully Impenetrable Spheres 77
3.4.2.3 Interpenetrable Spheres 80
3.4.3 Distribution of Polydispersed Spheres 80
3.4.3.1 Fully Penetrable Spheres 81
3.4.3.2 Fully Impenetrable Spheres 82
3.4.4 Simulation of Dispersion of Spheres 83
3.4.5 Models of Anisotropic Materials 84

3.4.6 Tessellation Models of Cellular Materials 85
3.4.7 Gaussian Random Field Models of Amorphous Materials 91
3.5 Discrete Models 95
3.5.1 Network Models 95
3.5.2 Bethe Lattice Models 96
3.6 Reconstruction of Heterogeneous Materials: Simulated Annealing 97
Summary 101

II Linear Transport and Optical Properties 103

4 Effective Conductivity, Dielectric Constant, and Optical Properties: The Continuum Approach 105
4.0 Introduction 105
4.1 Symmetry Properties of the Conductivity Tensor 108
4.2 General Results 108
4.3 Effective Conductivity of Dispersion of Spheres: Exact Results 110
4.3.1 Three-Dimensional Regular Arrays of Spheres 111
4.3.1.1 Simple-Cubic arrays 111
4.3.1.2 Body-Centered and Face-Centered Cubic Arrays 118
4.3.2 Two-Dimensional Arrays of Cylinders 120
4.3.2.1 Hexagonal Arrays 121
4.3.2.2 Square Arrays 123
4.4 Exact Results for Coated Spheres and Laminates 125
4.5 Perturbation Expansion for the Effective Conductivity 127
4.5.1 Isotropic Materials: Strong-Contrast Expansion 128
4.5.2 Approximations 132
4.5.3 The Microstructural Parameter ζ_2 133
4.5.4 Anisotropic Materials 136
4.6 Bounds on Effective Conductivity 138
4.6.1 Isotropic Materials 139
4.6.1.1 Two-Point Bounds 139
4.6.1.2 Cluster Bounds 141
4.6.1.3 Three- and Four-Point Bounds 146
4.6.1.4 Cluster Expansions for the Effective Conductivity 151
4.6.2 Anisotropic Materials 152
4.6.2.1 Two-Point Bounds 152
4.6.2.2 Three- and Four-Point Bounds 153
4.6.2.3 Simplification of the Bounds 154
4.6.2.4 Cluster Expansions for the Effective Conductivity 158
4.7 The Effect of the Interface on the Effective Conductivity 160

4.8 Exact Duality Relations 161
4.9 Effective-Medium Approximation 162
4.9.1 Isotropic Materials 163
4.9.2 Anisotropic Materials 164
4.9.3 Critique of the Effective-Medium Approximation 167
4.9.4 The Maxwell–Garnett Approximation 168
4.10 The Random Walk Method 169
4.11 The Effective Dielectric Constant 176
4.11.1 Spectral Representation 177
4.11.2 Perturbation Expansion 179
4.11.3 Rigorous Bounds 180
4.12 Optical Properties 182
4.12.1 Conductor–Insulator Composites 183
4.12.2 Conductor–Superconductor Composites 186
4.12.3 Anisotropic Materials 187
4.12.4 The Cole–Cole Representation 188
4.13 Beyond the Quasi-static Approximation: Mie Scattering 189
4.14 Dynamical Effective-Medium Approximation 192
4.15 The Effect of Large-Scale Morphology 194
4.16 Multiple-Scattering Approach 194
Summary 196

5 Effective Conductivity and Dielectric Constant: The Discrete Approach **197**
5.0 Introduction 197
5.1 Experimental Data for Conduction in Heterogeneous Materials 198
5.1.1 Powders 198
5.1.2 Polymer Composites 199
5.1.3 Conductor–Insulator Composites 200
5.2 Conductivity of a Random Resistor Network 202
5.3 Exact Solution for Bethe Lattices 203
5.3.1 The Microscopic Conductivity 204
5.3.2 Effective-Medium Approximation 206
5.3.3 Conductor–Insulator Composites 206
5.3.4 Conductor–Superconductor Composites 207
5.4 Exact Results for Two-Dimensional Composites 207
5.4.1 Exact Duality Relations 207
5.4.2 Log-Normal Conductance Distribution 209
5.5 Green Function Formulation and Perturbation Expansion 209
5.5.1 Properties of the Green Functions 211
5.6 Effective-Medium Approximation 213
5.6.1 Conductor–Insulator Composites 216
5.6.2 Conductor–Superconductor Composites 218

5.6.3 Resistor Networks with Multiple Coordination Numbers 218
5.6.4 Materials with Zero Percolation Threshold 221
5.6.5 Comparison with the Experimental Data 223
5.6.6 Accuracy of the Effective-Medium Approximation 225
5.6.7 Cluster Effective-Medium Approximation 229
5.6.8 Coherent-Potential Approximation 230
5.7 Effective-Medium Approximation for Site Percolation 231
5.8 Effective-Medium Approximation for Correlated Composites 234
5.9 Effective-Medium Approximation for Anisotropic Materials 236
5.9.1 The Green Functions 237
5.9.2 Conductivity Anisotropy Near the Percolation Threshold 240
5.9.3 Comparison with the Experimental Data 241
5.10 Cumulant Approximation 242
5.10.1 The Lorentz Field 244
5.10.2 Perturbation Expansion 244
5.10.3 Computation of the Lowest-Order Terms 245
5.10.4 Bond Percolation 247
5.11 Position-Space Renormalization Group Methods 248
5.12 Renormalized Effective-Medium Approximation 253
5.13 The Critical Path Method 255
5.14 Numerical Computation of the Effective Conductivity 258
5.14.1 The Conjugate-Gradient Method 258
5.14.2 Transfer-Matrix Method 259
5.14.3 Network Reduction: The Lobb–Frank–Fogelholm Methods 261
5.14.4 Random Walk Method 263
5.15 Estimation of the Critical Exponents of the Conductivity 264
5.15.1 Finite-Size Scaling 264
5.15.2 Position-Space Renormalization Group Method 265
5.15.3 Series Expansion 266
5.15.4 Field-Theoretic Approach 267
5.15.5 Comparison with the Experimental Measurements . . . 268
5.16 Resistance Fluctuations, Moments of the Current Distribution, and Flicker Noise 270
5.16.1 Tellegen's Theorems 271
5.16.2 Cohn's Theorem 271
5.16.3 Scaling Properties 272
5.16.4 Comparison with the Experimental Data 274
5.17 Hall Conductivity 275
5.17.1 Effective-Medium Approximation 276
5.17.2 Network Model 279

5.17.3 Exact Duality Relations 280
5.17.4 Scaling Properties 281
5.17.5 Comparison with Experimental Data 282
5.18 Classical Aspects of Superconductivity 284
5.18.1 Magnetoconductivity 284
5.18.2 Magnetic Properties 285
5.18.3 Comparison with the Experimental Data 286
5.18.3.1 The London Penetration Depth 286
5.18.3.2 The Specific Heat 287
5.18.3.3 The Critical Current 287
5.18.3.4 The Critical Fields 288
5.18.3.5 Differential Diamagnetic Susceptibility . . . 289
Summary 290

6 Frequency-Dependent Properties: The Discrete Approach 291
6.0 Introduction 291
6.1 Diffusion in Heterogeneous Materials 292
6.1.1 Green Function Formulation and Perturbation Expansion 293
6.1.2 Self-Consistent Approach 294
6.1.3 Self-Consistent Approximation, Generalized Master Equation and Continuous-Time Random Walks 295
6.1.4 The Green Functions 295
6.1.5 Effective-Medium Approximation 297
6.1.6 The Mean Square Displacement 298
6.1.7 Difference Between Transport in Low- and High-Dimensional Materials 299
6.1.8 Predictions of the Effective-Medium Approximation 300
6.1.8.1 One-Dimensional Materials 300
6.1.8.2 Two- and Three-Dimensional Materials 302
6.1.9 Anomalous Diffusion 305
6.1.10 Scaling Theory of Anomalous Diffusion 306
6.1.11 Comparison with the Experimental Data 308
6.1.12 The Governing Equation for Anomalous Diffusion 309
6.2 Hopping Conductivity 309
6.2.1 The Miller–Abrahams Network Model 311
6.2.2 The Symmetric Hopping Model 314
6.2.2.1 Exact Solution for One-Dimensional Materials 314
6.2.2.2 Exact Solution for Bethe Lattices 316
6.2.2.3 Perturbation Expansion and Effective-Medium Approximation 317

6.2.3 The Asymmetric Hopping Model: Perturbation Expansion 319
6.2.3.1 Exact Solution for Bethe Lattices 321
6.2.3.2 Two-Site Self-Consistent Approximation 322
6.2.3.3 Two-Site Effective-Medium Approximation 323
6.2.3.4 Energy-Dependent Effective-Medium Approximation 324
6.2.4 Variable-Range Hopping: The Critical Path Method 325
6.2.4.1 Effect of a Variable Density of States 329
6.2.4.2 Effect of Coulomb Interactions 330
6.2.4.3 Comparison with the Experimental Data 330
6.2.4.4 Fractal Morphology and Superlocalization 335
6.2.5 Continuous-Time Random Walk Model 337
6.3 AC Conductivity 339
6.3.1 Universality of AC Conductivity 340
6.3.2 Resistor–Capacitor Model 345
6.3.3 Universal AC Conductivity: Effective-Medium Approximation 346
6.3.4 Universal AC Conductivity: Symmetric Hopping Model 349
6.3.5 Role of Percolation in Universality of AC Conductivity 350
6.4 Dielectric Constant and Optical Properties 352
6.4.1 Resistor–Capacitor Model 352
6.4.2 Resistor–Capacitor–Inductor Model 354
6.4.3 Position-Space Renormalization Group Approach . . . 356
6.4.4 Effective-Medium Approximation 360
6.4.5 Random Walk Model 362
6.5 Scaling Properties of AC Conductivity and Dielectric Constant 365
6.5.1 Comparison with the Experimental Data 369
6.6 Vibrational Density of States: The Scalar Approximation 375
6.6.1 Numerical Computation 377
6.6.2 Effective-Medium Approximation 378
6.6.3 Cluster Effective-Medium Approximation 382
6.6.4 Scaling Theory: Phonons Versus Fractons 383
6.6.5 Characteristics of Fractons 384
6.6.5.1 Localization 384
6.6.5.2 Dispersion Relation 385
6.6.5.3 Crossover from Phonons to Fractons 385

6.6.6 Large-Scale Computer Simulations 386
6.6.7 Missing Modes 389
6.6.8 Localization Properties of Fractons 393
6.6.8.1 Mode Patterns of Fractons 393
6.6.8.2 Ensemble-Averaged Fractons 393
6.6.9 Comparison with the Experimental Data 395
6.7 The Dynamical Structure Factor 398
6.7.1 Theoretical Analysis 399
6.7.2 Scaling Analysis 400
6.7.3 Numerical Computation 402
6.8 Fractons and Thermal Transport in Heterogeneous Materials 404
6.8.1 Anharmonicity 405
6.8.2 Phonon-Assisted Fracton Hopping 405
6.8.3 Dependence of Sound Velocity on Temperature 407
Summary 409

7 Rigidity and Elastic Properties: The Continuum Approach 410
7.0 Introduction 410
7.1 The Stress and Strain Tensors 411
7.1.1 Symmetry Properties of the Stiffness Tensor 412
7.1.2 Theorems of Minimum Energy 415
7.1.3 The Strain Energy of a Composite Material 416
7.1.4 Volume Averaging 417
7.2 Exact Results 419
7.2.1 Interrelations Between Two- and Three-Dimensional Moduli 420
7.2.2 Exact Results for Regular Arrays of Spheres 421
7.2.3 Exact Results for Coated Spheres and Laminates 422
7.2.4 Connection to Two-Dimensional Conductivity 425
7.2.5 Exact Duality Relations 425
7.2.6 The Cherkaev–Lurie–Milton Theorem and Transformation 426
7.2.7 Universality of Poisson's Ratio in Percolation Composites 427
7.2.8 Composite Materials with Equal Shear Moduli 429
7.2.9 Dundurs Constants 429
7.2.10 Relations Between Elastic Moduli and Thermoelastic Properties 430
7.3 Dispersion of Spherical Inclusions 431
7.3.1 The Dilute Limit: A Single Sphere 431
7.3.2 Nondilute Dispersions 433
7.3.3 Two Spherical Inclusions 437
7.4 Exact Strong-Contrast Expansions 447
7.4.1 Integral Equation for the Cavity Strain Field 447
7.4.2 Exact Series Expansions 452

7.4.3 Exact Series Expansions for Isotropic Materials 455
7.4.4 Macroscopically Anisotropic Materials 460
7.4.5 The Microstructural Parameter η_2 460
7.4.6 Comparison with Numerical Simulation 461
7.4.6.1 Two-Dimensional materials 461
7.4.6.2 Three-Dimensional materials 466
7.5 Rigorous Bounds 471
7.5.1 Isotropic Materials 472
7.5.1.1 One-Point Bounds 472
7.5.1.2 Two-Point Bounds 472
7.5.1.3 Cluster Bounds 475
7.5.1.4 Three- and Four-Point Bounds 485
7.5.2 Anisotropic Materials 488
7.6 Multiple Scattering Method 493
7.6.1 The Dilute Limit 497
7.6.2 Nondilute Systems 499
7.6.3 Comparison with the Experimental Data 501
7.7 Effective-Medium Approximations 504
7.7.1 Fundamental Tensors and Invariant Properties 505
7.7.2 Symmetric Effective-Medium Approximation 508
7.7.3 Asymmetric Effective-Medium Approximation 510
7.7.4 The Maxwell–Garnett Approximations 513
7.8 Numerical Simulation 514
7.8.1 Finite-Difference Methods 514
7.8.2 Boundary-Element and Finite-Element Methods 514
7.9 Links Between the Conductivity and Elastic Moduli 518
7.9.1 Two-Dimensional Materials 518
7.9.1.1 Conductivity-Bulk Modulus Bounds 520
7.9.1.2 Conductivity-Shear Modulus Bounds 521
7.9.1.3 Applications 525
7.9.2 Three-Dimensional Materials 526
7.9.2.1 Conductivity-Bulk Modulus Bounds 527
7.9.2.2 Applications 530
Summary 531

8 Rigidity and Elastic Properties: The Discrete Approach 532
8.0 Introduction 532
8.1 Elastic Networks in Biological Materials 534
8.2 Number of Elastic Moduli of a Lattice 536
8.3 Numerical Simulation and Finite-Size Scaling 537
8.4 Derivation of Elastic Networks from Continuum Elasticity 539
8.4.1 The Born Model 541
8.4.2 Shortcomings of the Born Model 541
8.5 The Central-Force Network 543

8.6 Rigidity Percolation 545
8.6.1 Static and Dynamic Rigidity and Floppiness of Networks 546
8.6.2 The Correlation Length of Rigidity Percolation 548
8.6.3 The Force Distribution 549
8.6.4 Determination of the Percolation Threshold 550
8.6.4.1 Moments of the Force Distribution . . . 550
8.6.4.2 The Pebble Game 551
8.6.4.3 Constraint-Counting Method 554
8.6.5 Mapping Between Rigidity Percolation and Resistor Networks 555
8.6.6 Nature of the Phase Transition 556
8.6.7 Scaling Properties of the Elastic Moduli 557
8.7 Green Function Formulation and Perturbation Expansion . . . 560
8.7.1 Effective-Medium Approximation 561
8.7.2 The Born Model 561
8.7.3 Rigidity Percolation 563
8.8 The Critical Path Method 565
8.9 Central-Force Networks at Nonzero Temperatures and Under Stress 565
8.10 Shortcomings of the Central-Force Networks 568
8.11 Elastic Percolation Networks with Bond-Bending Forces . . . 569
8.11.1 The Kirkwood–Keating Model 570
8.11.2 The Bond-Bending Model 570
8.11.3 The Percolation Thresholds 571
8.11.4 The Force Distribution 572
8.11.5 Comparison of the Central-Force and Bond-Bending Networks 573
8.11.6 Scaling Properties 574
8.11.7 Relation with Scalar Percolation 576
8.11.8 Fixed Points of Vector Percolation: University of the Poisson's Ratio 577
8.11.9 Position-Space Renormalization Group Method 578
8.11.10 Effective-Medium Approximation 581
8.12 Transfer-Matrix Method 581
8.13 The Beam Model 582
8.14 The Granular Model 583
8.15 Entropic Networks 584
Summary 585

9 Rigidity and Elastic Properties of Network Glasses, Polymers, and Composite Solids: The Discrete Approach **587**
9.0 Introduction 587
9.1 Network Glasses 589
9.1.1 Rigidity Transition 591
9.1.2 Comparison with the Experimental Data 592

9.1.3 Rigidity Transition at High Coordination Numbers 595
9.1.4 Effect of Onefold-Coordinated Atoms 597
9.1.5 Stress-Free Versus Stressed Transition 598
9.2 Branched Polymers and Gels 599
9.2.1 Percolation Model of Polymerization and Gelation 602
9.2.2 Morphological Properties of Branched Polymers and Gels 602
9.2.2.1 Gel Polymers 603
9.2.2.2 Comparison with the Experimental Data 604
9.2.2.3 Branched Polymers 606
9.2.2.4 Comparison with the Experimental Data 608
9.2.3 Rheology of Critical Gels: Dynamic-Mechanical Experiments 609
9.2.4 The Relaxation Time Spectrum 611
9.2.5 Comparison with the Experimental Data 614
9.2.5.1 Physical Gels 614
9.2.5.2 Chemical Gels 614
9.2.5.3 Enthalpic Versus Entropic Elasticity 616
9.2.5.4 Viscosity of Near-Critical Gelling Solutions 618
9.3 Mechanical Properties of Foams 621
9.4 Mechanical Properties of Composite Solids 624
9.4.1 Porous Materials 624
9.4.2 Superrigid Materials 627
9.5 Wave Speeds in Porous Materials 627
9.6 Elastic Properties of Composite Materials with Length Mismatch 628
9.7 Materials with Negative Poisson's Ratio 631
9.8 Vibrational Density of States: Vector Percolation Model 634
9.8.1 Scaling Theory 634
9.8.2 Crossover Between Scalar Approximation and Vector Density of States 637
9.8.3 Large-Scale Computer Simulation 637
9.8.4 Comparison with the Experimental Data 640
Summary 642

References **643**

Index **685**

Abbreviated Contents for Volume II

Preface

Abbreviated Contents for Volume I

Introduction to Volume II

1 Characterization of Surface Morphology

I Effective Properties of Heterogeneous Materials with Constitutive Nonlinearities

2 Nonlinear Conductivity and Dielectric Constant: The Continuum Approach

3 Nonlinear Conductivity, Dielectric Constant, and Optical Properties: The Discrete Approach

4 Nonlinear Rigidity and Elastic Moduli: The Continuum Approach

II Fracture and Breakdown of Heterogeneous Materials

5 Electrical and Dielectric Breakdown: The Discrete Approach

6 Fracture: Basic Concepts and Experimental Techniques

7 Brittle Fracture: The Continuum Approach

8 Brittle Fracture: The Discrete Approach

III Atomistic and Multiscale Modeling of Materials

9 Atomistic Modeling of Materials

10 Multiscale Modeling of Materials: Joining Atomistic Models with Continuum Mechanics

References

Index

1 Introduction

1.0 Historical Perspective

This two-volume book was written at the dawn of the 21st century. Although development of materials science was actually commenced about 2 million years ago, at the beginning of the Stone Age, human civilization would not have been where it is now, had it not been for the astonishing progress that materials science has made since 5000 years ago, when the Bronze Age started in the Far East, and especially since about 3000 years ago when the Iron Age began in Asia Minor. The last 100 years have, however, witnessed stunning advances in materials science which far exceed those of all prior centuries combined. Whereas at the beginning of the 20th century the atomic structure of any material was not known, today we are able

(1) to make materials in which atoms are placed at specific locations;
(2) to manufacture composites with properties that Nature cannot produce;
(3) to make materials that can change some of their specific features as the environment around them also varies—the so-called *smart materials*, and
(4) to even manipulate individual atoms in a material so as to investigate the most fundamental issues in science.

Despite the huge progress, the goal in developing and advancing the science of materials has not changed over the past 2 million years: Elevating the quality of people's lives and making efficient and better use of the resources that are available on our planet.

The advances of materials research can perhaps be best illustrated by three classes of materials that are of prime interest today, namely, structural, polymeric, and electronic materials. Modern materials science owes its development to structural materials, and in particular to metallurgy and metal physics. It has been a long journey from the Bronze Age to sending the Pioneer 1 spacecraft outside the Solar System, but after this long and tortuous journey we can now claim that we have mastered air travel, surface transportation, housing, and to some extent space flights, owing all of them to structural materials. Even before we discovered and understood them, polymeric materials were being used by Nature to perform extremely complex functions, some of which we do not quite understand yet. In fact, human beings, and more generally every living being as we know it, are made of carbon-based materials and are probably the most complex polymers that can be made. However, over the last 100 years, the science of polymeric materials has progressed so much that we now use polymers in fabrics, food packaging, structural applications, and many other instances. In fact, polymers are playing an increasingly important role in a new branch of materials, namely, biomaterials. Electronic materials, on the other hand, could not have been conceived before quantum mechanics was discovered in the first few decades of the 20th century. Our understanding of such materials has progressed so much that it is extremely difficult to even imagine our world today without telecommunication, computers, radio, and television, all of which use electronic materials. In fact, the astonishing advances in increasing the speed of computers are largely due to the progress that has been made in understanding the properties of electronic materials, which in turn has helped us in their design and manufacturing.

1.1 Heterogeneous Materials

The advances in materials science and research have not, however, come easily. The fact is that *Nature is disordered*, and so also are most materials that she produces. Over the past few decades, it has become increasingly clear that what Nature produces often has an optimized structure. Therefore, heterogeneous materials that Nature produces must also have this property. This fact has been recognized by human beings, since throughout history the use of composites and heterogeneous materials has been widespread. However, this recognition was at first only empirical; only the last few decades have witnessed further development of heterogeneous materials in a scientific manner. We now recognize that only in our imagination can we buy clean, pure, completely characterized and geometrically perfect materials. Engineers and materials scientists work in a world of composites and mixtures, and biologists do so even more. Even an experimentalist who focuses on the purest of materials, such as carefully grown crystals, can seldom escape the effects of defects, trace impurities, and finite boundaries. There are few concepts in science more elegant to contemplate than an infinite, perfectly periodic crystal lattice, and few systems as remote from reality. We must therefore come to terms with heterogeneous materials. The heterogeneities manifest themselves in the apparent randomness in the morphology of materials. The morphology of a material has two major aspects: *topology*, the interconnectiveness of its individual elements; and *geometry*, the shape and size of these individual elements.

However, a heterogeneous morphology is only half of the story. We know that however random the stage upon which the drama of Nature is played out, it is also at times very difficult to follow the script. We believe, at least above the quantum mechanical level, in the doctrine of determinism, yet important *continuous* systems exist in which deterministic descriptions are beyond hope. Typical examples are diffusion and Brownian motion where, over certain length scales, we observe an apparent random process, or *disordered dynamics*. Hence, Nature is disordered both in her structure and the processes she supports. Indeed, the two types of disorder are often coupled. An example is fluid flow through a porous material where the interplay between the disordered morphology of the pore space and the dynamics of fluid motion gives rise to a rich variety of phenomena. Therefore, to understand and predict the properties of a wide variety of materials, and to develop the capability for manipulating their morphology to suite a particular application, we must understand the role of heterogeneity in the materials' structure and the processes that take place therein.

Although the existence and role of the heterogeneities seem rather obvious, these topics were, for several decades, familiar to most physicists, engineers, and materials scientists either in empirical manners, or in the form of statistical mechanics, and the application of such equations as the Boltzmann's equation. Researchers in these fields made remarkable progress by taking advantage of periodic structures. Despite their impressive progress though, it was always felt that something was missing because, as always, one still had to face the real materials which are usually disordered. Therefore, it became clear that a statistical physics of disordered media must be devised to provide methods for deriving macroscopic properties of such materials from the laws that govern the microscopic world or, alternatively, for deducing their microscopic properties from the macroscopic data that can be obtained by experimental techniques. To be successful and accurate, such a statistical physics of disordered materials must take into account the effect of the materials' topology *and* geometry. While the role of the latter was already appreciated in the early years of the 20th century, the effect of the former was ignored for many decades, or at best was treated in an unrealistic or empirical manner, simply because it was thought to be too difficult to be taken into account rigorously.

The effort for devising such a statistical physics of disordered materials was begun in earnest in the 1960s and has been, ever since, in a rapid stage of progress. The reason for this progress is fourfold.

(1) Advanced theoretical methods for predicting the macroscopic properties of such materials have been established.
(2) The vast increase in computational power has enabled us to study by computer simulations very complex materials at several disparate length scales, ranging from molecular to macroscopic scales.
(3) The advent of novel experimental techniques has allowed us to collect highly precise data.
(4) The crucial role of interconnectivity of the microscopic elements of heterogeneous materials and its effect on their macroscopic properties have been recognized and understood.

The net result of these developments is that we now have a much better understanding of many properties of a wide variety of heterogeneous materials. The goal of this two-volume book is to describe and discuss these developments for several classes of such materials.

1.2 Effective Properties of Heterogeneous Materials

Transport processes in disordered materials, the main focus of this book, constitute an important class of problems, in view of their relevance to understanding and modeling of a wide variety of phenomena in natural and industrial processes. A partial list of examples of a transport process in a disordered material includes hopping transport in amorphous semiconductors, frequency-dependent conduction in superionic conductors, diffusion in microporous materials and through biological tissues, conduction in, and deformation, electrical and dielectric breakdown and fracture of, composite solids, optical processes in heterogeneous materials, and many more.

What are the effective properties that are of interest to us in this book? Some of the effective properties of heterogeneous materials that will be studied in this two-volume book are as follows.

(1) The effective (electrical, thermal, hopping, and Hall) conductivity
(2) The effective dielectric constant
(3) The effective elastic moduli
(4) The electrical and dielectric breakdown fields
(5) The effective yield strength

In addition, we also consider some aspects of the classical (as opposed to quantum-mechanical) superconductivity of composite materials. Because of the exact mathematical analogy between the formulations of the thermal and electrical conduction in heterogeneous materials, and due to the fact that their mechanisms in solid materials are the same, in most cases that are discussed in this book, we do not distinguish between thermal and electrical conductivities. Moreover, the effective diffusivity tensor $\mathbf{D}_e$ is directly related to the effective conductivity tensor $\mathbf{g}_e$, and therefore from knowledge of $\mathbf{g}_e$ one obtains an estimate of $\mathbf{D}_e$, and vice versa. In addition, we also consider closely-related properties, such as the vibrational density of states of heterogeneous materials (which are related to other properties, such as the thermal conductivity), as well as the optical properties that are directly linked to the effective frequency-dependent dielectric constant.

To define such effective macroscopic properties, one must start from *local* transport equations that govern the transport processes at the *microscopic* length scales. If, for example, a heterogeneous material is composed of domains of different materials, then the relevant microscopic length scale is the typical or average domain size. If the length scale over which a material can be considered as heterogeneous is much smaller than the macroscopic length of the sample, then the average or effective properties of the heterogeneous material are well-defined and independent of the sample size. Otherwise, the material is macroscopically heterogeneous, and its effective properties depend on the sample size. To make the discussion more specific, we consider linear and nonlinear transport properties separately. Volume I describes prediction of the effective linear properties, while nonlinear properties will be taken up in Volume II.

1.3 Linear Transport Properties

Suppose that a heterogeneous material is a domain of space with volume Ω which is composed of two regions or phases: phase 1 of volume fraction ϕ_1 and phase 2 of volume fraction ϕ_2. For clarity and simplicity of our discussion we assume that the material is statistically homogeneous, but contains microscopic heterogeneities (although in at least some instances that will be considered in this book this condition may be violated). We now consider some of the properties of interest.

1.3.1 The Effective Conductivity

Consider the thermal conductivity of a two-phase heterogeneous linear material, such that phases 1 and 2 have isotropic thermal conductivities g_1 and g_2, respectively (the extension to a material with more than two phases is obvious). The governing equation for steady-state conduction in the material at a point $\mathbf{x}$ is given by

$$\nabla \cdot \mathbf{q}(\mathbf{x}) = 0, \quad \text{for both phases,} \tag{1}$$

where $\mathbf{q}$ is the local heat flux. We also need a *local* constitutive relation that relates the flux $\mathbf{q}$ to the thermal conductivities, namely, the Fourier's law,

$$\mathbf{q}(\mathbf{x}) = -g(\mathbf{x})\nabla T(\mathbf{x}), \quad \text{for both phases,} \tag{2}$$

where $T(\mathbf{x})$ and $g(\mathbf{x})$ are the local temperature and thermal conductivity at point $\mathbf{x}$, respectively. We assume that T and the normal component of $\mathbf{q}$ across the interface (or surface) between phases 1 and 2 are continuous.

We now define an *ensemble-averaged* constitutive relation that defines the symmetric, second-rank effective conductivity tensor $\mathbf{g}_e$:

$$\langle \mathbf{q}(\mathbf{x}) \rangle = -\mathbf{g}_e \cdot \langle \nabla T(\mathbf{x}) \rangle \tag{3}$$

where $\langle \cdot \rangle$ denotes an *ensemble average*, i.e., the average over a large number of samples which are identical in their macroscopic details but are different at the microscopic length scale. Such an ensemble-averaged effective conductivity $\mathbf{g}_e$ exists in the limit that the ratio of the microscopic and macroscopic length scales approaches 0. Note that, although we assumed that phases 1 and 2 have *isotropic* conductivities, the effective conductivity $\mathbf{g}_e$ is in general a tensor, and hence the system behaves anisotropically. The anisotropy can be due to a variety of factors, such as layering, embedding of oriented inclusions in a matrix, etc.

1.3.2 *The Effective Dielectric Constant*

For simplicity we restrict our discussion here to the zero frequency (static) limit (in this book and Volume II we will, however, consider frequency-dependent dielectric constant). In this limit the electric field $\mathbf{E}$ is curl-free, i.e., one has $\nabla \times \mathbf{E} = \mathbf{0}$, and satisfies the following equation,

$$\nabla \cdot (\epsilon \mathbf{E}) = 0, \tag{4}$$

where $\epsilon(\mathbf{x})$ is the local value of the dielectric constant which varies from ϵ_1 for phase 1 to ϵ_2 for phase 2.

To define the effective dielectric constant of the material we consider the electrostatic energy stored in the system if a given potential drop is applied across the material. The average electric field $\langle \mathbf{E} \rangle$ in the system is given by

$$\langle \mathbf{E} \rangle = \frac{1}{\Omega} \int \mathbf{E}(\mathbf{x}) d\Omega, \tag{5}$$

where Ω is the volume of the material. The average electric field depends only on the applied potential. The effective dielectric constant ϵ_e of the heterogeneous material is then defined by

$$\epsilon_e \langle \mathbf{E} \rangle^2 = \frac{1}{\Omega} \int \epsilon(\mathbf{x}) E^2(\mathbf{x}) d\Omega. \tag{6}$$

One can also calculate ϵ_e by an alternative method based on the local and average electric displacement fields, $\mathbf{D}(\mathbf{x})$ and $\langle \mathbf{D} \rangle$, respectively, which are defined by

$$\mathbf{D}(\mathbf{x}) = \epsilon \mathbf{E}(\mathbf{x}), \quad \langle \mathbf{D} \rangle = \epsilon_e \langle \mathbf{E} \rangle. \tag{7}$$

In terms of these two quantities, the effective dielectric constant ϵ_e is given by

$$\frac{\langle \mathbf{D} \rangle^2}{\epsilon_e} = \frac{1}{\Omega} \int \frac{D^2(\mathbf{x})}{\epsilon(\mathbf{x})} d\Omega. \tag{8}$$

1.3.3 *The Effective Elastic Moduli*

To formally define the effective stiffness tensor $\mathbf{C}_e$ one proceeds in a similar manner. Suppose, for example, that a linear heterogeneous material is composed of isotropic phases 1 and 2 with elastic bulk moduli K_1 and K_2 and shear moduli μ_1 and μ_2, respectively. At equilibrium we must have

$$\nabla \cdot \boldsymbol{\sigma}(\mathbf{x}) = 0 \quad \text{for both phases}, \tag{9}$$

where $\boldsymbol{\sigma}(\mathbf{x})$ is the symmetric, second-rank local stress tensor. The local constitutive equation is given by

$$\boldsymbol{\sigma}(\mathbf{x}) = \lambda(\mathbf{x}) \text{tr}[\boldsymbol{\epsilon}(\mathbf{x})]\mathbf{U} + 2\mu(\mathbf{x})\boldsymbol{\epsilon}(\mathbf{x}), \quad \text{for both phases}, \tag{10}$$

with

$$\boldsymbol{\epsilon}(\mathbf{x}) = \frac{1}{2}\left[\nabla \mathbf{u}(\mathbf{x}) + \nabla \mathbf{u}(\mathbf{x})^{\mathrm{T}}\right]. \tag{11}$$

Here $\epsilon(\mathbf{x})$ is the symmetric, second-rank local strain stensor, $\mathbf{u}(\mathbf{x})$ is the local displacement, $\lambda(\mathbf{x})$ and $\mu(\mathbf{x})$ are the local Lamé constants, $\mathbf{U}$ is the unit dyadic, and superscript T denotes transpose operation on the tensor.

Given the above local constitutive relation, the symmetric, *fourth-rank* effective stiffness tensor $\mathbf{C}_e$ is defined by the following ensemble-averaged relation:

$$\langle\sigma(\mathbf{x})\rangle = \mathbf{C}_e : \langle\epsilon(\mathbf{x})\rangle. \tag{12}$$

The conditions under which the above equation holds are similar to those for the conduction problem discussed above. For macroscopically-isotropic materials $\mathbf{C}_e$ can be expressed in terms of only two independent effective elastic moduli K_e and μ_e, since,

$$\langle\sigma(\mathbf{x})\rangle = \left(K_e - 2\frac{\mu_e}{d}\right)\mathrm{tr}\langle\epsilon\rangle\mathbf{U} + 2\mu_e\langle\epsilon\rangle, \tag{13}$$

where d is the dimensionality of the system. To characterize anisotropic materials one needs more than two effective elastic moduli. For example, it can be shown that there are five effective elastic moduli for transversely isotropic fiber-reinforced two-phase materials, but only three of them are independent, and hence the remaining two moduli can be obtained from a combination of the three independent moduli which usually are the effective axial shear modulus μ_e^a, the effective bulk modulus K_e, and the effective transverse shear modulus μ_e^t.

1.4 Nonlinear Transport Properties

For several decades most of the transport phenomena in heterogeneous materials that were studied were linear processes. In practice, however, there are many natural and industrial processes in which a *nonlinear* transport phenomenon takes place. Volume II will consider such nonlinear transport processes, where we divide them into two groups.

1.4.1 Constitutive Nonlinearity

Materials and media of this type *always* behave nonlinearly. For example, if in a composite material the relation between the current I and voltage V is given by

$$I = gV^n \tag{14}$$

where g is a generalized conductance of the material, then, as far as the electrical conductivity is concerned, for $n \neq 1$ the material *always* behaves nonlinearly. Another example is provided by flow of polymers in a porous medium which is always a nonlinear process, as the relation between the volumetric flow rate of the polymer and the pressure gradient applied to the medium is *always* nonlinear. We will discuss such nonlinear phenomena in Chapters 2–4 of Volume II.

1.4.2 *Threshold Nonlinearity*

In this class of materials are those for which the nonlinearity arises as a result of imposing on them an external field of sufficient intensity. Brittle fracture and dielectric breakdown of composite solids are two important examples of such nonlinear transport processes. In brittle fracture, for example, the elastic response of a solid material is governed by the equations of linear elasticity until the external stress or strain that has been imposed on the material exceeds a critical value, at which time the material breaks down and microcracks begin to emerge. In many doped polycrystalline semiconductors (for example, ZnO) the electrical conductivity is highly nonlinear above a threshold voltage. Nonlinear conduction has also been observed in many ceramic conductors (Einziger, 1987; Niklasson, 1989b). A list of all possible nonlinear transport processes of this type is very long. One important point to remember is that, the interplay between a nonlinear transport process and the disordered morphology of a composite material gives rise to a rich variety of phenomena that are usually far more complex than what one usually must deal with in linear processes. Over the past 15 years, an increasing number of investigations have been devoted to such nonlinear transport processes, and deeper insight into their properties has been acquired. A major goal of Volume II is to describe this progress and compare various properties of nonlinear transport processes in heterogeneous materials with their linear counterparts.

Note that while for linear materials the nature of the boundary conditions does not play a critical role, for nonlinear materials, at least those that are strongly disordered, the boundary conditions must be precisely defined; otherwise, one may not even be able to define the effective (ensemble-averaged) macroscopic properties. For example, Chapters 2 and 3 of Volume II will make it clear that the effective nonlinear conductivity of strongly disordered materials that follow Eq. (14) can be defined only for *two-terminal* systems, i.e., one for which a current is injected into the material at one point and extracted at a second point (or one in which a potential gradient is established between two points). However, if there are more than one injection and/or extraction points, then it is not even clear how the effective nonlinear conductivity of the material should be defined.

1.5 Predicting the Effective Properties of Heterogeneous Materials

Now that we have stated the types of effective properties of heterogeneous materials that we are interested in, it is also essential to consider the types of models that have been developed over the past several decades for describing transport phenomena in composite materials. The analysis of transport processes in solid materials has a long history. However, it is only in the past three decades that this analysis has been extended to include detailed structural properties of the materials, and in particular the distribution of their heterogeneities.

Making *exact* predictions for the effective properties of composite materials with anything but the simplest morphologies is extremely difficult, if not impossible. With the advent of powerful computers, efficient computational algorithms have been developed for estimating various properties of heterogeneous materials. Large-scale computer simulations that are based on such algorithms either numerically solve the governing equations, or use the analogy between various phenomena (for example, the analogy between random walk processes and diffusion and conduction) in order to estimate the effective transport properties. While the applications of such methods and techniques to linear transport processes have proven fruitful, devising similar algorithms for nonlinear transport processes has been considerably more difficult. Despite this difficulty, considerable progress has also been made in this arena which will be described in this book.

To describe the theoretical approaches for estimating the effective properties of composite materials, we divide them into two classes. In the first class of models are what we refer to as the *continuum models*, while the second class is made of the *discrete models*. Both types of models are described and analyzed in this Volume and Volume II, and what follows is a brief description of the general features of each class of models.

1.5.1 The Continuum Models

Continuum models represent the classical approach to describing and analyzing transport processes in materials of complex and irregular morphology. The physical laws that govern the transport processes at the microscopic level are well understood. One can, in principle, write down the differential equations that describe transport of energy, charge, or stress in a material and specify the associated initial and boundary conditions. However, as the morphology of most real composite materials is very irregular, practical and economically feasible computations for exact estimation of the effective properties are still very difficult—even in the event that one knows the detailed morphology of the material. Moreover, even if the solution of the problem could be obtained in such great detail, it would contain much more information than would be useful in any practical sense. Thus, it becomes essential to adopt a macroscopic description at a length scale much larger than the dimension of the individual phases of a composite material.

As described above, the effective properties of a material are defined as averages of the corresponding microscopic quantities. The averages must be taken over a volume that is small enough compared with the volume of the material, yet large enough for the equations of change to hold when applied to that volume. At every point in the material one uses the smallest such volume, and thereby generates macroscopic field variables that follow equations such as Eq. (14) for electrical conduction, or Fourier's law of thermal conduction. Even when the averaging is theoretically sound, due to the complex morphology of composite materials, predicting the effective properties, without resorting to numerical simulations, is often very difficult. As mentioned above, many of such past theoretical attempts to derive expressions for the effective transport coefficients of disordered materials from their microstructure entailed a simplified representation of the morphology. Having derived the macroscopic governing equations, one has the classical description of a composite material as a continuum. These equations are then discretized and solved numerically, provided that the effective properties that appear in the transport equations are either supplied as the inputs (through, for example, experimental measurements), or else a model for the morphology of the material is assumed so that the effective transport properties can be somehow estimated, so that the numerical solution yields other quantities of interest, such as the potential distribution in the material. We refer to various models associated with this classical description as the *continuum models*. These models have been widely used because of their convenience and familiarity to the engineers and materials scientists. They do have some limitations, one of which was noted above in the discussion concerning scales and averaging. They are also not well-suited for describing those phenomena in which the interconnectivity of different parts of a material plays a major role. Such models also break down if there are correlations in the morphology of the composite material with an extent that is comparable with the material's linear size.

In addition to deriving the effective macroscopic equations and obtaining their solution by numerical calculations, one may also derive exact results in terms of rigorous upper and lower bounds to the properties of interest. Hence, powerful tools have been developed for deriving accurate upper and lower bounds and estimates. Finally, various approximations, such as the mean-field and effective-medium approximations, have also been developed in the context of the continuum models. We will describe most of these theoretical approaches throughout both this book and Volume II.

1.5.2 The Discrete Models

The second class of models, the *discrete models*, are free of many limitations of the continuum models. They themselves are divided into two groups.

(1) In the first class of discrete models, a material is represented by a discrete set of atoms and molecules that interact with each other through interatomic potentials. In a solid material, the distance between the atoms is fixed. One then carries out atomistic simulations of the materials' behavior under a variety of conditions. Several types of such simulations have been developed over the past few decades. With the advent of massively-parallel computational algorithms, atomistic simulations have increasingly become a viable and *quantitative* method of predicting the effective properties of materials.

(2) In the second class are lattice models of composite materials. The bonds of the lattices represent microscopic elements of the material. For example, they represent a conducting or insulating elements, or an elastic or a plastic region. They do *not* represent molecular bonds, and therefore such lattice models are appropriate for length scales that are much larger than molecular scales. These models have been advanced to describe various phenomena at the microscopic

level and have been extended in the last several years to also describe them at the macroscopic length scales. We will describe both classes of discrete models in this Volume and Volume II.

The main shortcoming of both groups of the discrete models, from a practical point of view, is the large computational effort required for a realistic discrete representation of the material and simulating its behavior, although the ever-increasing computational power is addressing this difficulty. Although the original idea of lattice representation of a material is rather old and goes back to the early 1950s, but it was only in the early 1970s that systematic and rigorous procedures were developed to map, in principle, any disordered material onto an equivalent lattice model. Moreover, it was recognized in the late 1960s that percolation theory is the natural language for describing the effect of interconnectivity of the microscopic elements of a disordered material on its effective macroscopic properties. Describing application of percolation theory to estimating the effective properties of materials constitutes an important aspect of both Volume I and II.

1.6 The Organization of the Book

What we intend to do in this book and Volume II is describing the most important developments in predicting the effective properties of linear and nonlinear composite materials, and comparing the predictions with the relevant experimental data. To accomplish our goal, we describe and discuss, for each effective property, the continuum and discrete models separately. Then, in the last chapter of Volume II we will describe recent advances in *multiscale* modeling of materials' properties—a methodology that combines a discrete approach with a continuum model.

The structure of each chapter in both Volume I and Volume II is as follows.

(1) The main problem(s) of interest is (are) introduced.
(2) The problem(s) is (are) then analyzed by several methods, each of which provide valuable insight into the solution of the problem(s) and the physical phenomena that it (they) represent. Typically, each chapter starts with exact and rigorous results, then describes analytical approximations, and finally discusses the numerical and computer simulation methods. The weakness and strengths of each method are also pointed out. In this way, the most important progress in understanding the physical phenomena of interest is described and discussed.
(3) When possible (which is almost always the case), we compare the theoretical predictions with the experimental data and/or high-resolution computer simulation results.
(4) We also emphasize an important aspect of the phenomena that are studied, namely, their universal properties, i.e., those that are independent of the microscopic features and the structure of the distribution of the heterogeneities and depend only on the dimensionality of the materials. The existence of such

universal properties is particularly important because, if these phenomena do possess universal properties, then one may employ the *simplest* models that can produce such universal properties in order to study the phenomena of interest in composite materials. In addition, such universal properties provide a guide for developing either accurate correlations for the experimental data, or accurate approximate expressions.

We divide the present Volume into two major parts. In Part I (Chapters 2 and 3) we describe characterization of the microstructure of composite materials and the various models that have been developed for representing them, while Part II (Chapters 4–9) contains the description and discussion of the methods for predicting the effective linear properties of heterogeneous materials, and their comparison with the experimental data.

Some aspects of characterization of materials' morphology, which are more relevant to what is discussed in Volume II, will be described in its Chapter 1. Aside from this chapter, Volume II is divided into three parts. In Part I (Chapters 2–4) we study transport processes in heterogeneous materials that are characterized by constitutive nonlinearities. Part II (Chapters 5–8) contains the description and discussion of transport processes with threshold nonlinearity, including electrical and dielectric breakdown, and brittle fracture of disordered materials. Finally, in Part III we will describe (in Chapters 9 and 10) advances in atomistic modeling of materials, and how a powerful new approach that combines atomistic simulations with the continuum description—in effect a combination of a discrete approach with a continuum model—promises to provide much deeper understanding of materials, and deliver quantitative predictions for their effective properties.

Let us emphasize that, although every attempt has been made to discuss and cite the relevant literature on every subject that we consider, what we do cite and bring to the attention of the reader represents what was known to us at the time of writing this book, and/or what we considered to be the most relevant. As such, this two-volume book represents the author's biased view of the subject of composite materials.

Part I

Characterization and Modeling of the Morphology

2
Characterization of Connectivity and Clustering

2.0 Introduction

As pointed out in Chapter 1, Nature produces a wide variety of materials with enormous variations in their morphology. The same is true about man-made materials. The morphology of materials' microstructure consists of its geometry, topology and surface structure. The geometry describes the shapes and sizes of the micro- and mesoscale elements of the material. The shapes can vary anywhere from completely ordered and Euclidean to complex and seemingly chaotic patterns. In general, Euclidean shapes are formed under close-to-equilibrium conditions, although even in such cases equilibrium thermodynamics is often inadequate for describing the process that gives rise to such shapes. The topology of materials' structure describes how the micro- or mesoscale elements are connected to one another. In addition, many materials, especially those that are made under far-from-equilibrium conditions, have very complex surface structure. In particular, many materials have very rough surface with the roughness seemingly following a very complex pattern which, however, is amenable to precise characterization. In recent years, it has become clear that characterizing surface roughness will go a long way toward providing us with a much better understanding of materials' microstructure and hence their effective properties. In particular, materials' rough surfaces are often directly linked with the nonlinear phenomena that will be described in Volume II, and therefore we will describe their characterization and modeling in Volume II. However, when we speak of the surface roughness, it is important to specify the length scales over which the roughness is observed. Even the most rugged mountains look perfectly smooth when viewed from the outer space! Therefore, surface roughness, and more generally the characteristics of the morphology, *depend on the length scale of observations or measurements*.

Examples of such disordered morphologies, and in particular rough surfaces, are abundant. Some of the well-known examples are microporous materials, such as membranes, catalysts and adsorbent. The geometry of these materials is disordered, as their pores have a wide variety of shapes and sizes. Their topology is also disordered, since the connectivity of the pores varies in the pore space, with some pores being connected to a larger number of pores than others. The surface of the pores is, over certain length scales which depend on the type of the material, also rough. One of the best examples of a material with a rough surface is the thin films that are made by molecular beam epitaxy and are used for manufacturing of semiconductors and computer chips. Such thin films are made of silicon and other elements, and are formed by deposition of atoms on a very clean surface. Thin films with rough surfaces are also made by sputtering in which an energized beam of particles is sent toward the bulk of a material. Collision of the beam particles with the material causes ejection of some particles from the material's surface, which then deposit on another surface and start to grow a thin film of the original material.

The enormous variations in the morphology of the natural, or even man-made, materials are such that, up until a few decades ago, the task of describing and quantifying such morphologies seemed hopeless. This task was complicated further by the fact that in many instances the materials' morphology undergoes a type of phase transition, i.e., it changes from one distinct structure to another, as the conditions that give rise to the morphology change. A well-known example is electrochemical deposition where the electrochemical patterns change as the deposition conditions are varied. This morphological phase transition is reminiscent of ordinary or thermal phase transitions which are now well-understood. Hence, similar to thermal phase transitions, one may have a *morphological phase diagram* in terms of the parameters that control the material's morphology. Such a diagram maps out all the possible morphologies that a given material made by a given process can possess. Unlike thermal phase transitions, however, and despite extensive studies, the nature of the transitions between various morphological phases is not well-understood yet.

However, several developments have considerably changed the bleak outlook for characterizing the morphology of materials that had persisted for several decades, and have brightened the prospects for deeper understanding of materials' microstructures. Some of these developments were already discussed in Chapter 1. Among them are the advent of powerful computers and novel experimental techniques that allow highly sophisticated computations of materials' properties and their measurement. Two other developments that have advanced our understanding of materials' microstructure and properties are as follows.

(1) It has been widely recognized that the complex microstructure and behavior of a wide variety of materials can be quantitatively characterized by using the ideas of fractal distributions, which correspond in a unique way to the geometrical and dynamical properties of the systems under study. As we discuss in this chapter (and in Chapter 1 of Volume II), fractal concepts provide us with an important tool for characterizing the geometry and surface structure of heterogeneous materials, and long-range correlations that often exist in their morphology. Even if a material's morphology does not possess fractal properties at significant length scales, the concepts of fractal geometry often provide useful means of obtaining deeper insights into the structure of the material.

(2) Another important tool for characterizing the effect of the topology, i.e., connectivity, of a disordered material on its transport properties is the concepts of percolation theory. For example, how the conducting parts of a composite material that consists of conducting and insulating phases cluster together and form sample-spanning paths for transport of current, heat, or stress across the

material is precisely quantified by percolation theory. Moreover, as we discuss in the subsequent chapters, percolation concepts are essential to correct interpretation of the experimental data for various properties of disordered materials.

Thus, the purpose of this chapter is to review and discuss the essential concepts and ideas of fractal structures and percolation processes. We begin by describing and discussing the basic principles of fractal geometry, after which percolation processes and their essential features will be described.

2.1 Characterization of the Geometry: Self-Similar Fractal Microstructures

Although it is possible to give a formal mathematical definition of a fractal system or set (Mandelbrot, 1982), an intuitive and less formal definition of a fractal system may be more useful: In a fractal structure the part is reminiscent of the whole. This implies that the system is *self-similar* and possesses scale-invariant properties, i.e., its morphology repeats itself at different length scales. That is, there exist pieces of the system above a certain length scale—the lower cutoff scale for fractality—that can be magnified to recover the structure of the system at larger length scales up to another length scale—the upper cutoff for its fractality. Below the lower cutoff and above the upper cutoff scales the system loses its self-similarity. There are disordered media that are self-similar at *any* length scale, while natural materials and media that exhibit fractal properties typically lose their fractal characteristics at sufficiently small or large length scales. Moreover, it should be pointed out that natural systems are generally not, *in the strict mathematical sense*, fractal; rather their behavior *approaches* what is envisioned in fractal geometry, which explains why it is useful to use fractals for describing them.

One of the simplest characteristics of a fractal is its fractal dimension D_f, which is defined as follows. We cover the fractal system by non-overlapping d-dimensional spheres of Euclidean radius r, or boxes of linear size R, and count the number $N(r)$ of such spheres that is required for complete coverage. The fractal dimension of the system is then defined by

$$D_f = \lim_{r \to 0} \frac{\ln N}{\ln(1/r)}. \tag{1}$$

Estimating D_f based on Eq. (1) is called the *box-counting method*. Non-fractal objects, such as a straight line, a square, or a sphere, are Euclidean and their effective dimensionality is d, the dimensionality of the space in which they are embedded. Note that, in order to be able to write down Eq. (1), we have implicitly assumed the existence of a lower and an upper cutoff length scale for the fractality of the system which are, respectively, the radius r of the spheres and the linear size L of the system.

One may also define the fractal dimension D_f through the relation between the system's mass M and its characteristic length scale L. If the system is composed of particles of radius r and mass m, then

$$M = cm(L/r)^{D_f}, \tag{2}$$

where c is a geometrical constant of order 1. Since we can fix the dependence of M on m and r, we can write

$$M(L) \sim L^{D_f}. \tag{3}$$

Equation (3) is consistent with the corresponding scaling relationship $M \sim L^d$ for Euclidean objects. Often, measuring M entails using an ensemble of samples with similar structures, rather than a single sample. In this case

$$\langle M \rangle = cm(L/r)^{D_f}, \tag{4}$$

where $\langle \cdot \rangle$ indicates an average over the mass of a large number of samples with linear sizes in the range $L \pm \delta L$, centered on L. It is clear that Eqs. (2)–(4) suggest a method for estimating the fractal dimension from a logarithmic plot of the mass M versus the length scale L. Equations (2) and (4) both imply that the mean density $\rho = M/\Omega$, where Ω is the volume of the system that contains the fractal in the embedding space, is *not* a constant, but *decreases* with the length scale L according to

$$\rho \sim L^{D_f - d} \sim M^{(D_f - d)/D_f}. \tag{5}$$

Equation (5) is one of the most important characteristics of self-similar fractal objects. A fractal dimension less than the Euclidean dimension of the space and a decreasing density with increasing length scales both imply that the fractal object cannot fill the space and has a sparse structure.

In Figure 2.1 we show the construction of a classical self-similar fractal called the *Sierpinski gasket*. Every triangle in each generation is replaced by $N = 3$ smaller triangles that have been scaled down by a factor $r = 1/2$, and therefore using

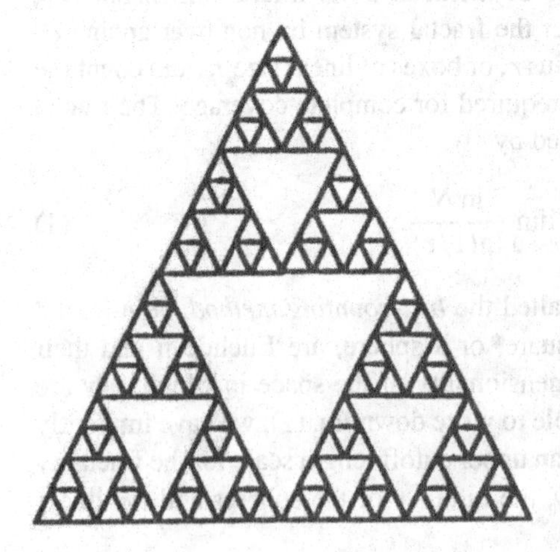

FIGURE 2.1. A two-dimensional Sierpinski gasket.

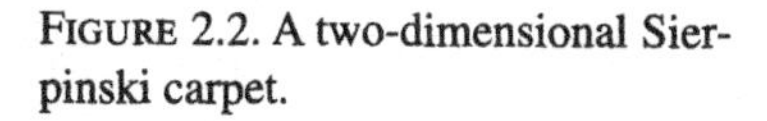

FIGURE 2.2. A two-dimensional Sierpinski carpet.

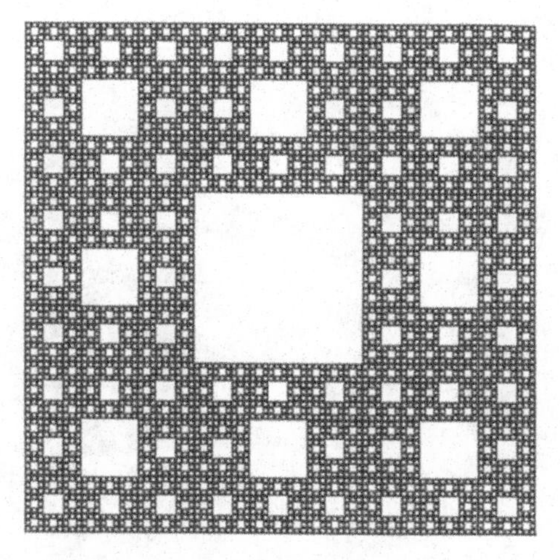

Eq. (1) we obtain $D_f = \ln 3/\ln 2 \simeq 1.58$, while for the three-dimensional (3D) version of the gasket we obtain $D_f = \ln 4/\ln 2 = 2$. Another example is provided by the *Sierpinski carpet*; see Figure 2.2. To construct this fractal, we start with a black square and subdivide it into nine equal squares. We then erase the central square and apply the same procedure to each of the remaining eight black squares and continue this *ad infinitum*. Then, $D_f = \ln 8/\ln 3 \simeq 1.893$. More generally, we can subdivide the squares into $(2n+1)^2$ identical parts and generate a carpet with $D_f = \ln[(2n+1)^2 - 1]/\ln(2n+1)$, where $n = 1$ corresponds to Figure 2.2.

2.2 Statistical Self-Similarity

The Sierpinski gasket and carpet are examples of what are usually referred to as *exact* fractals, because their self-similarity is exact at any length scale. Most natural or even man-made fractals are, however, *statistically self-similar* fractals, because their self-similarity is only in an average sense. One of the most important examples of such fractals is one which is generated by the *diffusion-limited aggregation* model (Witten and Sander, 1981). In this model the site at the center of a lattice is occupied by a stationary particle. A new particle is then injected into the lattice, far from the center, which diffuses on the lattice until it reaches a surface site, i.e., an empty site which is a nearest neighbor of the stationary particle. The diffusing particle then sticks to surface site and remains there permanently. Another diffusing particle is injected into the lattice to reach another surface (empty) site and stick to it, and so on. If this process is continued for a long time, a large aggregate is formed. The most important property of diffusion-limited aggregates is that they are self-similar and fractal. Extensive computer simulations (for a review see, for example, Meakin, 1998) indicate that $D_f \simeq 1.7$ and 2.45 for 2D and 3D aggregates, respectively. A 2D example of such aggregates is shown in Figure 2.3.

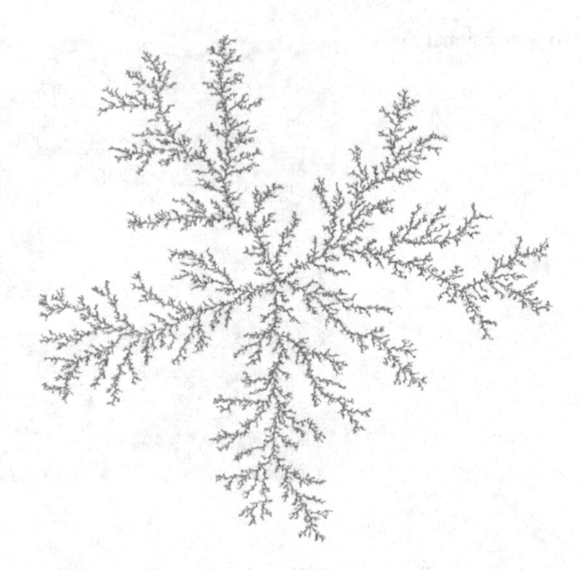

FIGURE 2.3. A two-dimensional diffusion-limited aggregate.

Diffusion-limited aggregates have found wide applications, ranging from colloidal systems to miscible displacement processes in porous media. We will come back to this model in Chapters 5 and 8 of Volume II, where we describe models of dielectric breakdown and fracture of composite materials.

2.3 Measurement of the Fractal Dimension

If a fractal morphology is to be formed at small length scales, it will require formation of a large surface that separates the material from the embedding space. Since formation of a large surface requires a large amount of energy, fractal materials do not usually form under equilibrium conditions. However, constrained materials, such as polymers, membranes, and froths, are exceptions to this general rule. A polymer, for example, is a constrained material because, although it is in equilibrium, its structure is subject to a constraint which is the way its monomers are joined. Therefore, a large surface is formed without forcing the polymer to collapse. But in any case, regardless of whether a fractal structure has been formed under equilibrium or non-equilibrium conditions, it is usually highly disordered, because of which its self-similarity is only in a statistical sense. Therefore, it may be more appropriate to refer to disordered fractals as scale-invariant rather than self-similar structures. A visual check of self-similarity of disordered materials is clearly impossible, and thus one must estimate their fractal dimension in order to ascertain their fractality. In addition to the box-counting method described above, there are several other methods of estimating the fractal dimension, two of which are now described.

2.3.1 The Correlation Function Method

A powerful method for testing self-similarity of disordered materials is to construct a correlation function $C_n(\mathbf{r}^n)$ defined by

$$C_n(\mathbf{r}^n) = \langle \rho(\mathbf{r}_0)\rho(\mathbf{r}_0 + \mathbf{r}_1) + \cdots + \rho(\mathbf{r}_0 + \mathbf{r}_n) \rangle, \tag{6}$$

where $\rho(\mathbf{r})$ is the density at position $\mathbf{r}$, and the average is taken over all possible values of $\mathbf{r}_0$. Here $\mathbf{r}^n$ denotes the set of points, $\{\mathbf{r}_1, \cdots, \mathbf{r}_n\}$. If an object is fractal, then its correlation function defined by Eq. (6) should remain the same, up to a constant factor, if all the length scales of the system are rescaled by a constant factor b. Thus, one must have

$$C_n(b\mathbf{r}_1, b\mathbf{r}_2, \cdots, b\mathbf{r}_n) = b^{-nx} C_n(\mathbf{r}_1, \cdots, \mathbf{r}_n). \tag{7}$$

It is not difficult to see that only a *power-law* correlation function can satisfy Eq. (7). Moreover, it can be shown that one must have $x = d - D_f$, where the quantity x is called the *co-dimensionality*. However, in most cases only the two-point, or the direct, correlation function can be computed or measured with high precisions, and therefore we focus on this quantity. In practice, to construct the direct correlation function for use in analyzing a fractal structure, one typically employs a digitized image of the system. The correlation function is then written as

$$C(\mathbf{r}) = \frac{1}{\Omega} \sum_{\mathbf{r}'} s(\mathbf{r}')s(\mathbf{r} + \mathbf{r}'), \tag{8}$$

where $s(\mathbf{r})$ is a function such that $s(\mathbf{r}) = 1$ if a point at $\mathbf{r}$ belongs to the system, $s(\mathbf{r}) = 0$ otherwise, and $r = |\mathbf{r}|$. There are also indirect methods of measuring the correlation function; see below. Because of self-similarity of the fractal structure, the direct correlation function $C(r)$ *decays* as

$$C(r) \sim r^{D_f - d}. \tag{9}$$

Such power-law decay of $C(r)$ not only provides a test of self-similarity of a disordered medium or material, but also provides us with a means of estimating its fractal dimension since, according to Eq. (9), if one prepares a logarithmic plot of $C(r)$ versus r, then for a fractal object one should obtain a straight line with a slope $D_f - d$. Estimating the fractal dimension based on the direct correlation function has proven to be a very robust and reliable method. Note that Eq. (9) has an important implication: In a fractal system *there are long-range correlations*, since $C(r) \to 0$ only as $r \to \infty$. The existence of such correlations has important implications for estimating the effective transport properties of disordered materials which will be discussed in the subsequent chapters.

2.3.2 Small-Angle Scattering

Small-angle scattering provides a measure of fractal behavior at length scales between 0.5 and 50 mm. In a scattering experiment the observed scattering density

$I(\mathbf{q})$ is given by the Fourier transform of the direct correlation function $C(\mathbf{r})$:

$$I(\mathbf{q}) = \int_0^{\infty} C(\mathbf{r}) \exp(i\mathbf{q} \cdot \mathbf{r}) d^3\mathbf{r}, \tag{10}$$

where $\mathbf{q}$ is the scattering vector, the magnitude of which is given by

$$q = \frac{4\pi}{\lambda} \sin\left(\frac{1}{2}\theta\right) \tag{11}$$

where λ is the wavelength of the radiation scattered by the sample through an angle θ. In a scattering experiment, $C(\mathbf{r})$ refers to spatial variations in scattering amplitude per unit volume, rather than its physical density. In many cases it is not unreasonable to assume that, to a good approximation, there is no interference scattering, and therefore the total scattering intensity is the sum of the scattering from all portions of the system. For an isotropic medium, $C(\mathbf{r}) = C(r)$, where $r = |\mathbf{r}|$, and Eq. (10) becomes

$$I(q) = \int_0^{\infty} 4\pi r^2 \frac{\sin(qr)}{qr} C(r) dr. \tag{12}$$

Since the correlation function for a 3D system is given by [see Eq. (9)], $C(r) \sim r^{D_f - 3}$, we obtain

$$I(q) \sim q^{-D_f} \Gamma(D_f - 1) \sin[(D_f - 1)\pi/2], \tag{13}$$

where Γ is the gamma function. Equation (13) is valid if $qd_p \ll 1$ and $qD_m \gg 1$, where d_p and D_m are, respectively, the effective diameters of an individual scattering particle and the material. Both light scattering and small-angle X-ray scattering from silica aggregation clusters confirmed Eq. (13) (Schaefer *et al.*, 1984). Note that a small-angle scattering experiment also provides an indirect way of measuring the direct correlation function $C(r)$.

As discussed above, the range of scale-invariance and fractal behavior in real materials may be limited by the lower and upper cutoff length scales ℓ_l and ℓ_u. Finite size of a system also limits the range of self-similarity. Under such conditions, the assumption that the total scattering intensity is simply the sum of scatterings by the individual portions may break down, leading to interference scattering. To remedy this situation, Sinha *et al.* (1984) included in $C(r)$ an exponentially decaying term, incorporating a scattering correlation length ξ_s which reflects the upper limit of self-similarity, namely,

$$C(r) \sim r^{D_f - 3} \exp(-r/\xi_s), \tag{14}$$

which, when used in Eq. (12), yields

$$I(q) \sim q^{-1} \Gamma(D_f - 1) \xi_s^{D_f - 1} \sin\left[(D_f - 1) \tan^{-1}(q\xi_s)\right] [1 + (q\xi_s)^2]^{(1-D_f)/2}. \tag{15}$$

The validity of Eq. (15) was confirmed by the scattering experiments of Sinha *et al.* (1984) on silica particle aggregates. Note that in the limit $\xi_s \to \infty$, we recover Eq. (13), and for small values of $q\xi_s$ and $D_f = 3$ (homogeneous materials) we obtain

$$I(q) = \frac{8\pi\xi_s^2}{1+(q\xi_s)^2}, \tag{16}$$

which is the classical result of Debye *et al.* (1957).

If r is small, i.e., scattering is at larger values of q but still within the small-angle approximation, then the scattering reflects the nature of the boundaries between, for example, the micropores of a material and their surfaces. When the surface is rough and the roughness pattern is self-similar, one may characterize the surface by a fractal dimension D_s which can be estimated by the scattering technique. If S is the surface area and ℓ is a basic characteristic length scale of the rough, fractal surface, then D_s is defined by $S \sim \ell^{D_s}$, implying that the actual area of a fractal surface is *larger* than what is expected of a smooth surface, and hence one usually has $D_s > 2$. The surface fractal dimension D_s may or may not be the same as the fractal dimension D_f of the bulk of the material. One may also have a non-fractal material which, however, possesses a fractal surface, and vice-versa. Bale and Schmidt (1984) showed that for rough surfaces described by a surface fractal dimension $D_s > 2$, the direct correlation function takes on the following form

$$C(r) \sim 1 - ar^{3-D_s}, \tag{17}$$

in which $a = S_0[4\phi(1-\phi)\Omega]^{-1}$, where Ω is the sample volume, ϕ the porosity, and S_0 a material-dependent constant having the dimensions of area, which is equal to the pore surface area when the surface is smooth and non-fractal. Substituting Eq. (17) into (10) yields

$$I(q) \sim q^{D_s-6}\Gamma(5-D_s)\sin[(D_s-1)\pi/2], \tag{18}$$

which reduces to $I(q) \sim q^{-4}$, the classical Porod law for smooth surfaces for which $D_s = 2$, which is valid at the *shortest* length scales. Bale and Schmidt (1984) were able to confirm Eq. (18) for pores in lignites and sub-bituminous coals using small-angle X-ray scattering. If both the bulk material and the surface are fractal and $D_f \neq D_s$, then it is not difficult to show that

$$I(q) \sim q^{D_s-2D_f}. \tag{19}$$

Therefore, one has a crossover from the q^{-D_f} behavior of $I(q)$ to the $q^{D_s-2D_f}$ behavior, and to the q^{D_s-6} law. The crossover between the q^{-D_f} and q^{D_s-6} laws occurs at $\xi_s \sim q^{-1}$. If D_s is close to 3, the crossover between the q^{-D_f} and q^{D_s-6} regimes may be difficult to discern. Finally, if the direct correlation function is not isotropic, but only possesses rotational symmetry around a unique axis, one would obtain a scattering law which has an elliptically symmetric dependence on the azimuthal orientation of $\mathbf{q}$. This elliptical dependence can be removed by averaging the scattering law in terms of a reduced scattering vector. Typically, small-angle scattering methods encounter difficulty at the smallest length scales.

In practice, estimating the fractal dimension by an scattering experiment may not be as straightforward as making a logarithmic plot of $I(q)$ versus q. One reason for this is that the range of scaling is limited, and hence one must be careful about the possible crossover effects, such as those discussed above. Moreover, it is difficult sometimes to distinguish scattering data for a fractal material from those for Euclidean materials that are made of particles of different sizes with a power-law size distribution. It is also clear that in all the methods of estimating the fractal dimension of a heterogeneous material one always deals with power laws and hence ends up making a logarithmic plot of one quantity, e.g., the scattering intensity, versus another quantity. However, if the power law, and hence the resulting straight line, are observed over less than one order of magnitude variations in the independent parameter (for example, the length scale L or the magnitude of the scattering vector q), then interpreting the straight line as an indication of the existence of a fractal structure should be done with caution. Crossover and finite-size effects may significantly affect the quality of the data, and thus their effect should be taken into account. Often, a material is complex enough that a simple fractal analysis of the type discussed above may not be sufficient for analyzing its structure, and more sophisticated methods may be needed.

One major problem with equations such as (13) and (15) is that, they cannot distinguish different types of fractal materials which have similar fractal dimensions. For example, while 3D diffusion-limited aggregates and percolation composites (see below) both have a fractal dimension of about 2.5, their morphologies are completely different. Various methods have been suggested for distinguishing such materials, none of which is definitive. For example, Chen *et al.* (1987) suggested that *second-order* light scattering (SOLS) can distinguish between materials that have roughly the same fractal dimension D_f. In SOLS, each electric field amplitude contributing to the detected signal involves polarization of *two* particles (as opposed to the usual small-angle scattering that involves contributions from individual particles). Chen *et al.* (1987) showed that adding these amplitudes, and squaring and ensemble averaging them yield the observed intensity $I^{(2)}$ of SOLS. Therefore, two-, three-, *and* four-body correlation effects all contribute to SOLS. For self-similar materials, one has, $I^{(2)} \sim M^x$, where M is the mass of the material, and x is a material-dependent exponent. For example, for 3D diffusion-limited aggregates, $x \simeq 1.1$, while for 3D percolation composites, $x \simeq 1.5$, and therefore the two types of materials exhibit different scaling of the intensity of SOLS.

2.4 Self-Affine Fractals

The self-similarity of a fractal structure implies that its microstructure is invariant under an isotropic rescaling of lengths, i.e., if all the lengths in all the directions are rescaled by the same scale factor. However, there are many fractals that preserve their scale-invariance only if all the lengths in different directions are rescaled by direction-dependent scale factors. In other words, the scale-invariance of such systems is preserved only if lengths in the $x-$, $y-$, and z-directions are scaled by scale factors b_x, b_y, and b_z, where in general these scale factors are not equal. This type of scale-invariance—under a direction-dependent rescaling—implies

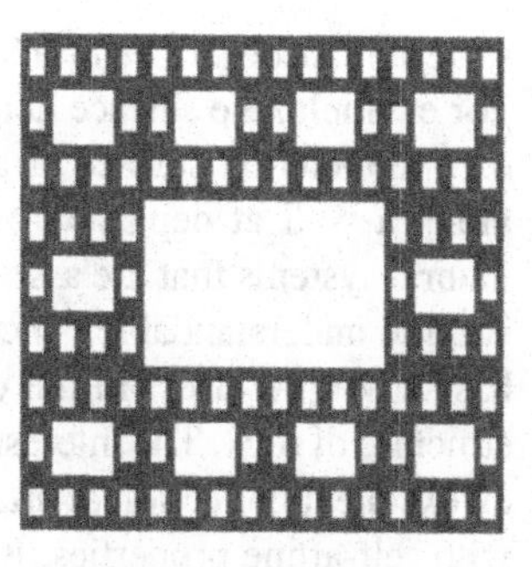

FIGURE 2.4. An example of a two-dimensional self-affine fractal.

that the fractal system is *anisotropic*. Such fractal systems are called *self-affine*, a term that was used first by Mandelbrot (1985). An example is shown in Figure 2.4, which can be thought of as the self-affine version of the Sierpinski carpet of Figure 2.2.

Unlike self-similar system, a self-affine fractal cannot be described by a single fractal dimension D_f, and in fact if one utilizes any of the methods of estimating D_f that were described above, then the resulting fractal dimension would depend on the length scales over which the method is utilized. Despite this difficulty, Mandelbrot (1986) showed that for a self-affine fractal that has scaling factors b_1, $b_2, \cdots, b_n$ for a subdivision into N parts, one may define a useful quantity, the *gap dimension* D_g, given by

$$D_g = \frac{\log N}{\log(b_1 b_2 \cdots b_n)^{1/n}}. \tag{20}$$

In general, D_g and D_f are not equal.

A simple example of a self-affine random fractal is the curve in the (x, t) plane, generated by the equation, $x = F(t)$, where $F(t)$ is the displacement of a particle undergoing 1D diffusion (Brownian motion). In this example, for any $b > 0$, $F(t)$ and $b^{-1/2}F(bt)$ are statistically identical. To see this, recall that the probability density function $P(x, t)$ for finding a diffusing particle at position x at time t satisfies the usual diffusion equation,

$$\frac{\partial P}{\partial t} = D\frac{\partial^2 P}{\partial x^2}, \tag{21}$$

where D is the diffusivity. It is straightforward to show that Eq. (21) is invariant under the rescaling, $(x, t) \to (b^{1/2}x, bt)$, i.e., the invariance is preserved by rescaling the x and t axes by different scale factors.

Another well-known example of a process that gives rise to a self-affine fractal is the marginally stable growth of an interface between, for example, two fluids. For example, if water displaces oil in a porous medium, the interface between water and oil is a self-affine fractal. Well-known examples of man-made materials with self-affine surfaces include thin films that are formed by molecular beam epitaxy, and fracture surface of composite materials (see Volume II). Among natural surfaces that have self-affine properties are bacterial colonies, and pores and fractures of rock and other types of porous media. Many properties of such materials are described by a function $f(\mathbf{x})$ that also possesses a self-affine structure. For example, the surface height $h(x, y)$ at a lateral position $\mathbf{x}$ of a rough surface, e.g., the internal surface of a rock fracture, and the porosity distribution of rock along a well at depths $\mathbf{x}$, both have self-affine property. Self-affinity of many natural systems that are associated with Earth, such as various properties of rock, is quite understandable, since gravity plays a dominant role in one direction but has very little effect in the other directions, hence generating anisotropy in the structure of rock. The interested reader is referred to Family and Vicsek (1991) for an excellent collection of articles which describe a wide variety of rough surfaces with self-affine properties. It should be clear to the reader that, similar to exactly self-similar fractals, exactly self-affine fractals are not of great practical interest. Indeed, natural exactly self-affine fractals are very rare. Instead, self-affine fractals that one encounters in practical situations are typically disordered, and thus their self-affinity is only in a statistical sense.

2.5 Characterization of Connectivity and Clustering

As already pointed out in the Introduction, the effect of connectivity of microscopic elements of a heterogeneous material on its macroscopic properties is quantified by percolation theory. In this section, we describe the essential concepts of this theory, and outline their application to not only characterizing the morphology of a material, but also estimating its effective properties. In subsequent chapters, such applications will be described in considerable details.

Percolation processes were first developed by Flory (1941) and Stockmayer (1943) for describing polymerization by which small branching molecules react and form large macromolecules. It also appeared in the work of Good (1949) on the theory of branching processes. However, they did not use the terminology of percolation theory, and in particular Flory and Stockmayer developed their theory of polymerization using a special lattice, namely, the Bethe lattice, an endlessly-branching structure without any closed loops, an example of which is shown in Figure 2.5. We will come back to their theory in Chapter 9.

In its present form, percolation theory first appeared in the mathematics literature, and was introduced by Broadbent and Hammersley (1957), who dealt with the problem of spread of a hypothetical fluid through a random medium. This process usually involves some random elements. However, the underlying randomness might be of two very different types. In one type, the randomness is imposed by the *fluid*, in that it is the fluid that decides what path to take in the medium. In this case the motion of the fluid is described by the classical diffusion process. On the other hand, the randomness may also be imposed by the *medium* itself; this was the new situation that was studied by Broadbent and Hammersley (1957). They dubbed the new phenomenon a percolation process, since they thought that the spread of a fluid through the random medium resembled flow of coffee in a percolator!

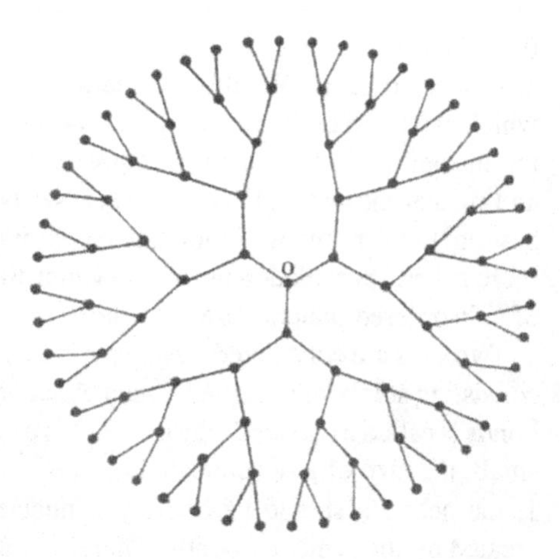

FIGURE 2.5. A Bethe lattice of coordination number 3.

2.5.1 Random Bond and Site Percolation

Since percolation deals with the effect of the connectivity and clustering of microscopic elements of a disordered medium, such as a heterogeneous material, on its effective properties, to make our discussions concrete we need a model of the material. Although models of microstructure of heterogeneous materials will be described in detail in Chapter 3, for now we assume that we can represent a composite material by a network or lattice in which each bond represents a microscopic piece of the material. In their original paper, Broadbent and Hammersley studied two different problems. One was the *bond percolation problem*, in which the bonds of the network are either randomly and independently occupied or intact with probability p, or they are *vacant* or removed with probability $1 - p$. For a large lattice, this assignment is equivalent to removing a fraction $1 - p$ of all the bonds at random. Figure 2.6 shows a random square network in which a fraction

FIGURE 2.6. Bond percolation on the square network in which a fraction $p = 0.6$ of the bonds are intact.

0.4 of the bonds has been removed at random, i.e., $p = 0.6$. The intact bonds represent, for example, the conducting regions of a composite material through which electrical or thermal current flows, in which case the vacant bonds represent the insulating portion of the composite. In a heterogeneous solid the conducting and insulating materials are not necessarily distributed randomly throughout the system, but for now we ignore such complications. In Section 2.12 we consider a more general percolation process in which the correlations between various regions of a disordered material are not ignored.

Two sites are considered *connected* if there exists at least one path between them consisting solely of occupied bonds. A set of connected sites surrounded by vacant bonds is called a *cluster*. If the network is of very large extent and if p is sufficiently small, the size of any connected cluster is likely to be small. But if p is close to 1, the network should be entirely connected, except for occasional small holes created by the removed bonds. Therefore, at some well-defined value of p, there should be a transition in the topological (connectivity) structure of the network; this value is called the *bond percolation threshold* p_{cb} which is the largest fraction of occupied bonds such that for $p < p_{cb}$ there is no sample-spanning cluster of occupied bonds, whereas for $p > p_{cb}$ a sample-spanning cluster of occupied bonds does form.

The second problem that was studied by Broadbent and Hammersley was a *site percolation process* in which sites of the network are occupied with probability p and vacant with probability $1 - p$. Two nearest-neighbor sites are considered connected if they are both occupied, and connected clusters on the network are defined in a manner similar to that for bond percolation. There also exists a *site percolation threshold* p_{cs} such that for $p > p_{cs}$ there is an infinite (sample-spanning) cluster of occupied sites, whereas for $p < p_{cs}$ the network is macroscopically disconnected. Figure 2.7 shows site percolation clusters on a square network.

We point out that the percolation processes as defined here are *static* processes; that is, once a percolation network is generated, its configuration does not change with time. Many variants of the classical random bond or site percolation processes have been developed in order to explain a wide variety of phenomena in disordered media. We refer the interested reader to Stauffer *et al.* (1982) and Sahimi (1994a) for discussions of such variants of the classical percolation problem and their applications.

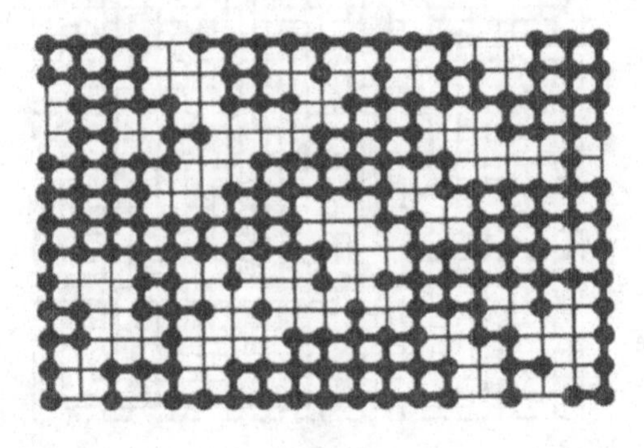

FIGURE 2.7. Site percolation on the square network in which a fraction $p = 0.59$ of the sites are intact.

2.5.2 *Percolation Thresholds*

The derivation of the exact values of p_{cb} and p_{cs} is an extremely difficult problem. In fact, such a derivation has been possible to date only for certain lattices related to the Bethe lattice and for a few 2D networks. For the Bethe lattice Fisher and Essam (1961) showed that

$$p_{cb} = p_{cs} = \frac{1}{Z-1} \tag{22}$$

where Z is the coordination number of the lattice, i.e., the number of bonds connected to the same site. We compile the current estimates of p_{cb} and p_{cs} (or their exact values if they are known) for three common 2D lattices in Table 2.1, while the most accurate numerical values of p_{cb} and p_{cs} for four common 3D lattices are listed in Table 2.2. The bond percolation thresholds of d-dimensional lattices is well-approximated by

$$p_{cb} \simeq \frac{d}{Z(d-1)}. \tag{23}$$

Also shown in these tables is the product $B_c = Zp_{cb}$ which, as can be seen, is an almost invariant of percolation networks with

$$B_c \simeq \frac{d}{d-1}. \tag{24}$$

The significance of B_c is discussed below. Generally speaking, since in any lattice a bond has more nearest-neighbor bonds than a site has nearest-neighbor sites, we have $p_{cs} \geq p_{cb}$. Moreover, because lattices in higher dimensions are better connected than those in lower dimensions, percolation thresholds decrease with increasing d. This is also clearly manifested by Eq. (24).

TABLE 2.1. Values of bond percolation threshold p_{cb}, site percolation threshold p_{cs}, and $B_c = Zp_{cb}$ for three common two-dimensional networks.

Network	Z	p_{cb}	B_c	p_{cs}
Honeycomb	3	$1 - 2\sin(\pi/18) \simeq 0.6527$*	1.96	0.6962
Square	4	1/2*	2	0.5927
Triangular	6	$2\sin(\pi/18) \simeq 0.3473$*	2.08	1/2*

*Exact result

TABLE 2.2. Numerical estimates of bond percolation threshold p_{cb}, site percolation threshold p_{cs}, and $B_c = Zp_{cb}$ for four common three-dimensional networks.

Network	Z	p_{cb}	B_c	p_{cs}
Diamond	4	0.3886	1.55	0.4299
Simple-Cubic	6	0.2488	1.49	0.3116
BCC	8	0.1795	1.44	0.2464
FCC	12	0.119	1.43	0.199

2.5.3 Bicontinuous Materials, Phase-Inversion Symmetry, and Percolation

An important problem is whether both phases of a two-phase material can simultaneously form sample-spanning clusters. If this is possible, then the material is called *bicontinuous*. If the present and vacant bonds or sites of a lattice model of a disordered material are thought of as constituting its two phases, then percolation theory provides a clear answer to the question of bicontinuity of the material. Random percolation systems possess *phase-inversion symmetry*, i.e., the morphology of phase 1 at volume fraction p is statistically identical to that of phase 2 with the same volume fraction. Clearly, then, no random 1D material can be bicontinuous, as its (bond or site) percolation threshold p_c is 1 (since removal of even a single bond or site breaks the 1D system into two pieces). For $d \geq 2$, any system possessing phase-inversion symmetry will be bicontinuous for $p_c < p < 1 - p_c$ (where p_c is either the site or bond percolation threshold), provided that $p_c < 1/2$. Thus, all the 3D lattices listed in Table 2.2 can be bicontinuous in either bond or site percolation. However, for 3D systems, neither $p_c < 1/2$ nor phase-inversion symmetry is a necessary condition for bicontinuity.

Two-dimensional materials are much more difficult to be made bicontinuous. In many respects, randomly disordered two-phase materials correspond to site percolation systems. If so, then since there is no 2D lattice with a site percolation threshold $p_{cs} < 1/2$ (see Table 2.1), we may conclude that no 2D randomly-disordered material can be bicontinuous. We will come back to this issue in Section 2.10.1.

2.5.4 Computer Generation of a Single Cluster

Random (or correlated) removal of sites or bonds, i.e., random assignment of bonds or sites as insulating, may not be suitable for many applications, because such a procedure generates, in addition to the sample-spanning cluster, isolated finite clusters as well. In some applications only the sample-spanning cluster is of interest, or at least the process of interest begins with a single cluster, and therefore we must first delete all the isolated clusters from the system. A method that was originally developed by Leath (1976) and Alexandrowicz (1980) generates only the sample-spanning cluster for $p > p_c$, or the largest cluster for $p \leq p_c$. In their method one begins with a single occupied site which is usually selected to be the center of the network. One then identifies the nearest-neighbor sites of the occupied site and considers them occupied and adds them to the cluster if random numbers R, uniformly distributed in (0,1) and assigned to the sites, are less than the fixed value p. The perimeter (the nearest-neighbor empty sites) of these sites are then identified and the process of occupying the sites continues in the same way. If a selected perimeter site is not designated as occupied, then it remains unoccupied forever. The generalization of this method for generating a cluster of occupied bonds is obvious. The main advantage of this method is its efficiency which makes

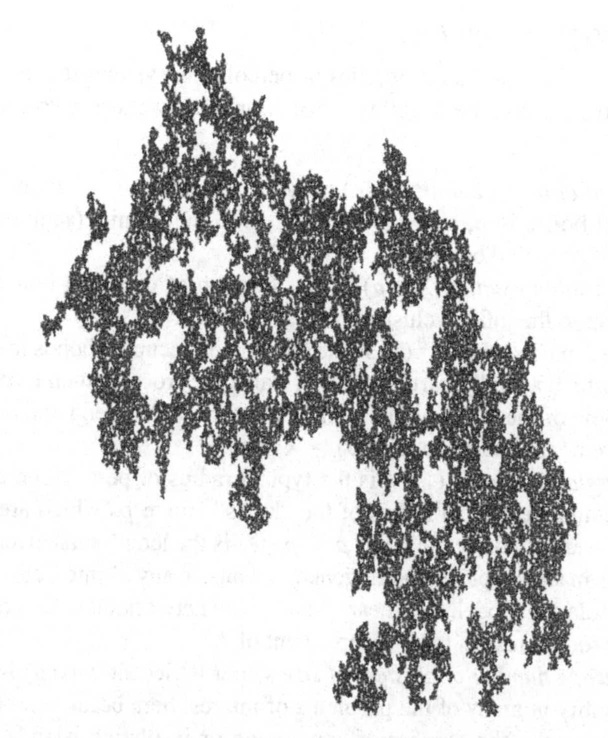

FIGURE 2.8. The largest percolation cluster at the site percolation threshold $p_c = 1/2$ of a $10^4 \times 10^4$ triangular lattice.

it possible to generate very large clusters. Figure 2.8 presents the largest percolation cluster at the site percolation threshold $p_{cs} = 1/2$ of a $10^4 \times 10^4$ triangular lattice.

An important quantity in percolation systems is the number of clusters of a given size. For example, in a composite solid which is a random mixture of conducting and insulating materials, one may be interested in the number of insulating or conducting islands in the material. Counting the number of such clusters in computer simulation of percolation systems is a difficult task. An algorithm due to Hoshen and Kopelman (1976), and its modification by Newman and Ziff (2000), perform this task very efficiently.

2.6 Percolation Properties

In addition to the percolation thresholds p_{cb} and p_{cs}, the behavior of percolation systems is quantified by several other important quantities. Some of these quantities describe the topology and geometry of percolation systems, while others characterize their effective transport properties. We now describe these properties.

2.6.1 Morphological Properties

Some of the most important properties of percolation systems that describe their topology and geometry are as follows. For simplicity, we use p_c to denote p_{cs} or p_{cb}.

(1) The *percolation probability* $P(p)$ is the probability that, when the fraction of occupied bonds is p, a given site belongs to the infinite (sample-spanning) cluster of occupied bonds.

(2) The *accessible fraction* $X^A(p)$ is that fraction of occupied bonds (or sites) belonging to the infinite cluster.

(3) The *backbone fraction* $X^B(p)$ is the fraction of occupied bonds in the infinite cluster which actually participate in a transport process, such as conduction, since some of the bonds in the infinite cluster are dead-end and do not carry any current, and therefore, $X^A(p) \geq X^B(p)$.

(4) The *correlation length* $\xi_p(p)$ is the typical radius of percolation clusters for $p < p_c$, and the typical radius of the "holes" above p_c which are generated by the vacant bonds or sites. For $p > p_c$, ξ_p is the length scale over which the system is macroscopically homogeneous. Thus, in any Monte Carlo simulation of a percolation model, the linear size L of the network must be larger than ξ_p in order for the results to be independent of L.

(5) The *average number of clusters of size* s (per lattice site) $n_s(p)$ is an important quantity in many of the problems of interest here because it corresponds to, for example, the number of conducting or insulating islands of a given size in a conductor-insulator composite solid. Figure 2.9 shows the typical p-dependence of some of these quantities in site percolation on the simple-cubic lattice.

(6) The probability that two sites, one at the origin and another one at a distance $\mathbf{r}$, are both occupied and belong to the same cluster of occupied sites, is $p^2 P_2(\mathbf{r})$, where $P_2(\mathbf{r})$ is called the *pair-connectedness function*.

(7) The *mean cluster size* S (also called the *site-averaged* cluster number) is the average number of sites in the cluster that contains a randomly-selected site, and is given by,

$$S = \frac{\sum_s s^2 n_s(p)}{\sum_s s n_s(p)}. \tag{25}$$

Essam (1972) showed that S and the pair-connectedness function $P_2(\mathbf{r})$ are related through a simple relation:

$$S = 1 + p \sum_{\mathbf{r}} P_2(\mathbf{r}). \tag{26}$$

2.6.2 Transport Properties

Because a major application of percolation theory has been modeling of transport in disordered materials, and in particular composite solids, we must also consider the effective transport properties of percolation system, namely, their conductivity,

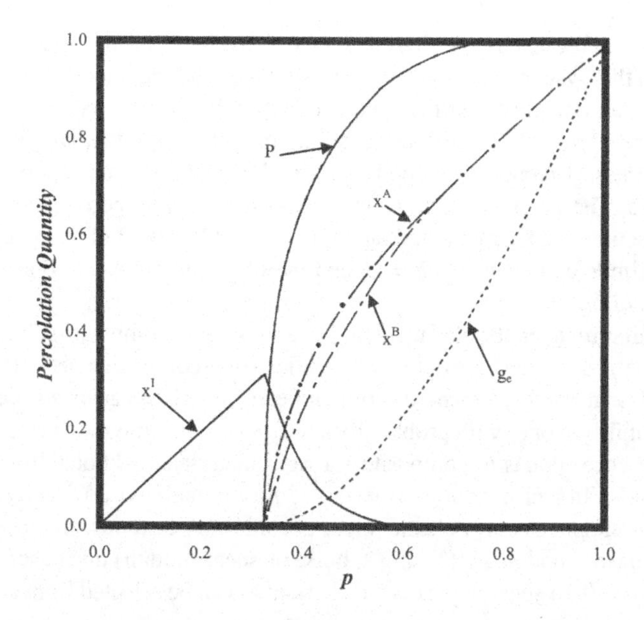

FIGURE 2.9. Typical dependence of several percolation quantities on the fraction p of the active sites in a simple-cubic lattice. X^I is the fraction of the isolated active sites; that is, $X^I = p - X^A$.

diffusivity, elastic moduli, and dielectric constant. We first consider the conductivity of a two-phase composite material modeled as a two-component network in which each (randomly-selected) bond has a conductance g_1 with probability p or g_2 with probability $q = 1 - p$. It is straightforward to show that the effective electrical conductivity g_e (or the thermal conductivity k_e) of the network is a *homogeneous function* and takes on the following form,

$$g_e(p, g_1, g_2) = g_1 F(p, h), \tag{27}$$

where $h = g_2/g_1$. Due to the assumption of randomness of the material's morphology, g_e is invariant under the interchange of g_1 and g_2 (phase-inversion symmetry), and therefore we must have

$$g_e(p, g_1, g_2) = g_e(q, g_2, g_1), \quad \text{and} \quad F(p, h) = hF(q, 1/h). \tag{28}$$

The limit in which $g_2 = 0$ and g_1 is finite corresponds to a conductor-insulator mixture. In this case, as $p \to p_c$, more and more bonds are insulating, the conduction paths become very tortuous, and therefore g_e decreases; at p_c one has $g_e(p_c) = 0$, since no sample-spanning conduction path exists any more. More generally, the conductance g_1 can be selected from a statistical distribution, which is in fact the case in most practical systems, such as porous materials and composite solids.

The limit in which $g_1 = \infty$ and g_2 is finite represents a conductor-superconductor mixture. All quantum-mechanical aspects of real superconductors are ignored in this definition, and we are concerned only with the effect of the local connectivity of the material on this conductivity. It is clear that the effective conductivity g_e of this system is dominated by the superconducting bonds. If $p < p_c$, then a sample-spanning cluster of the superconducting bonds does not exist, and g_e is finite. As $p \to p_c^-$, the effective conductivity g_e increases until a sample-spanning cluster of the superconducting bonds is formed for the first time at $p = p_c$, where g_e *diverges*. The practical significance of superconducting percolation networks will be discussed in Chapters 5, 6, and 9 (see also Chapter 3 of Volume II). Note that both limits correspond to $h = 0$, and therefore the point $h = 0$ at $p = p_c$ is particularly important.

In a similar manner, the elastic moduli of a two-phase composite solid, modeled by a percolation network, are defined. Consider a two-component network in which each bond is an elastic element (a spring or beam) which has an elastic constant e_1 with probability p or e_2 with probability $q = 1 - p$. The limit in which $e_2 = 0$ and e_1 is finite corresponds to composites of rigid materials and holes (for example, porous solids). In such networks, as $p \to p_c$, more bonds have no rigidity, the paths for transmission of stress or elastic forces become very tortuous, and therefore the effective elastic moduli E_e (Young's, bulk, or shear moduli) decrease; at p_c one has $E_e(p_c) = 0$. In general, the elastic constant e_1 can be selected from a statistical distribution.

The limit in which $e_1 = \infty$ and e_2 is finite represents mixtures of rigid-superrigid materials. In this case the effective elastic moduli E_e of the system are dominated by the superrigid bonds. If $p < p_c$, then a sample-spanning cluster of the superrigid bonds cannot form, and E_e is finite. As $p \to p_c^-$, the effective elastic moduli increase until the percolation threshold p_c of the rigid phase is reached at which a sample-spanning cluster of the superrigid bonds is formed for the first time, and the effective elastic moduli *diverge*. The significance of superrigid percolation networks will be discussed in Chapters 8 and 9.

Similarly, the effective dielectric constant ϵ_e of a two-phase composite (and insulating) material, modeled by a percolation network, may be defined and in fact, as shown in Chapters 5 and 6, ϵ_e is closely related to the conductor-superconductor model described above. Finally, the effective diffusivity D_e of a porous material can also be defined in a similar manner.

2.6.3 *The Structure of the Sample-Spanning Cluster*

The sample-spanning cluster, through which transport of current or stress takes place, can be divided into two parts: The dead-end part that carries no current or stress, and the backbone which is the multiply-connected part of the cluster defined above. Near p_c the bonds in the backbone are also divided into two groups: Those that are in the *blobs* which are multiply connected and make the transport paths very tortuous, and the *red bonds* which are those that, if cut, would split the backbone into two parts. Such bonds are called red because in transport of heat or electrical currents they carry the largest current fluxes and therefore, in analogy with a real network, become very hot and hence red. Figure 2.10 shows a sample-spanning percolation cluster and its backbone on the square network. Currently, the most

FIGURE 2.10. The sample-spanning cluster (top) and its backbone (bottom) in bond percolation in the square network, in which the fraction of conducting bonds is $p = 0.55$.

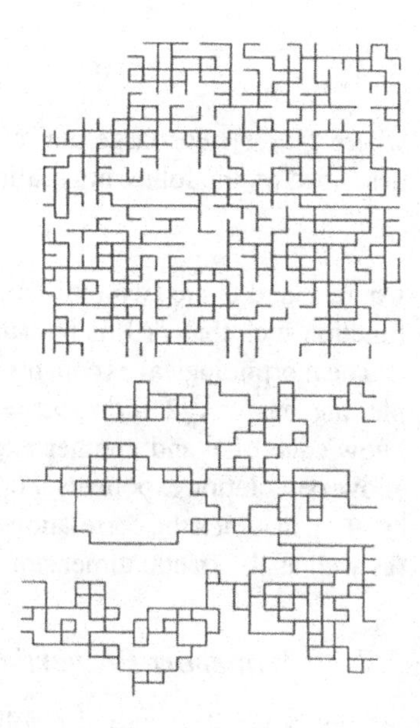

efficient algorithm for identifying the backbone of a percolation network is due to Knackstedt *et al.* (2000).

2.7 Universal Scaling Properties of Percolation

One of the most important characteristics of percolation systems is their *universal* properties. The behavior of many percolation quantities near p_c is insensitive to the microstructure (for example, the coordination number) of the network, and to whether the percolation process is a site or a bond problem. The quantitative statement of this universality is that many percolation properties follow power laws near p_c, and the *critical exponents* that characterize such power laws are universal and depend only on the Euclidean dimensionality d of the system. We first describe the universal properties of the quantities that characterize the morphology of percolation systems, and then present and discuss those of transport properties.

2.7.1 *Morphological Properties*

In general, the following power laws hold near p_c,

$$P(p) \sim (p - p_c)^{\beta}, \tag{29}$$

$$X^A(p) \sim (p - p_c)^{\beta}, \tag{30}$$

$$X^B(p) \sim (p - p_c)^{\beta_{bb}}, \tag{31}$$

$$\xi_p(p) \sim |p - p_c|^{-\nu}, \tag{32}$$

$$S(p) \sim |p - p_c|^{-\gamma}, \tag{33}$$

$$P_2(\mathbf{r}) \sim \begin{cases} r^{2-d-\eta}, & p = p_c, \\ \exp(-r/\xi_p), & \text{otherwise}, \end{cases} \tag{34}$$

where $r = |\mathbf{r}|$. For large clusters near p_c, the cluster size distribution $n_s(p)$ is described by the following scaling law,

$$n_s \sim s^{-\tau} f[(p - p_c)s^{\sigma}], \tag{35}$$

where τ and σ are two more universal critical exponents, and $f(x)$ is a scaling function such that $f(0)$ is not singular.

The morphological exponents defined above are not all independent. For example, one has, $\tau = 2 + \beta\sigma$, $\nu d = \beta + 1/\sigma = 2\beta + \gamma$, and $\gamma = (2 - \eta)\nu$. In fact, knowledge of ν and another exponent is sufficient for determining most of the above percolation exponents. For random percolation models, or for correlated percolation in which the correlations are short-ranged, all the above critical exponents (as well as the fractal dimensions described below) are completely universal.

2.7.2 *Transport Properties*

Similar power laws are also followed by the transport properties of percolation composites. In particular,

$$g_e(p) \sim (p - p_c)^{\mu}, \quad \text{conductor-insulator composites} \tag{36}$$

$$g_e(p) \sim (p_c - p)^{-s}, \quad \text{conductor-superconductor composites} \tag{37}$$

$$E_e(p) \sim (p - p_c)^{f}, \quad \text{rigid-soft composites} \tag{38}$$

$$E_e(p) \sim (p_c - p)^{-\chi}, \quad \text{rigid-superrigid composites.} \tag{39}$$

The power law that characterizes the behavior of the effective diffusivity $D_e(p)$ near p_c is derived from that of $g_e(p)$. According to Einstein's relation, $g_e \sim \rho D_e$, where ρ is the density of the carriers. Although a diffusing species can move on all the clusters, above p_c only diffusion on the sample-spanning cluster contributes significantly to D_e, so that $\rho \sim X^A(p)$. Hence, $g_e(p) \sim X^A(p) D_e(p)$, and therefore

$$D_e(p) \sim (p - p_c)^{\mu - \beta}. \tag{40}$$

For length scales $L < \xi_p$, the resistance R between two end points of a box of linear size L scales with L as $R \sim L^{\tilde{\zeta}}$. It is not difficult to show that

$$\mu = (d - 2)\nu + \zeta, \tag{41}$$

where, $\zeta = \tilde{\zeta}\nu$. It has been shown (Straley, 1977b) that in 2D, $\mu = s$ (see also Chapters 4 and 5).

Equations (36) and (37) can be unified by using the two-component resistor network described above. In the critical region, i.e., the region near p_c, where both $|p - p_c|$ and $h = g_2/g_1$ are small, the effective conductivity g_e follows the following scaling law (Efros and Shklovskii, 1976; Straley, 1976)

$$g_e \sim g_1 |p - p_c|^{\mu} \Phi_{\pm}(h|p - p_c|^{-\mu - s}). \tag{42}$$

where Φ_+ and Φ_- are two homogeneous functions corresponding, respectively, to the regions above and below p_c, and are, similar to μ and s, universal. For any fixed and non-zero h, g_e has a smooth dependence on $p - p_c$. This becomes clearer if we rewrite Eq. (42) as

$$g_e \sim g_1 h^{\mu/(\mu+s)} \Psi \left[|p - p_c| h^{-1/(\mu+s)} \right], \tag{43}$$

where $\Psi(x) = x^{\mu}\Phi_+(x^{-\mu-s}) = (-x)^{\mu}\Phi_-[(-x)^{-\mu-s}]$. Since the function $\Psi(x)$ is universal, the implication of Eq. (43) is that if one plots $g_e/[g_1 h^{\mu/(\mu+s)}]$ versus $|p - p_c|h^{-1/(\mu+s)}$ for all networks (or randomly-disordered materials) which have the same Euclidean dimensionality, all the results (or measurements) should collapse onto a single universal curve. This provides a powerful tool for estimating the conductivity of a composite for any value of h given the conductivities for two other values of h (so that the universal curve can be constructed). Somewhat similar, but more complex, scaling equations can be developed for the elastic moduli, dielectric constant and other properties of percolation composites, but we defer their description to future chapters. The implied prefactors in all the above power laws depend on the type of lattice and are not universal.

No exact relation is known between the transport and morphological exponents. This is perhaps because the transport exponents describe *dynamical* properties of percolation systems, whereas the morphological exponents describe the *static* properties, and in general there is no reason to expect a direct relation between the two.

The exponents μ, s, f and χ are universal if two conditions are satisfied. (1) There should not be any long-range correlations in the system. (2) The distribution $\psi(x)$ of the conductance or elastic constant of the bonds should be such that (Kogut and Straley, 1979; Straley, 1982; Sahimi *et al.*, 1983a)

$$\psi_{-1} = \int_0^{\infty} \frac{1}{x} \psi(x) dx < \infty. \tag{44}$$

Some correlated or continuous systems violate one or both of these conditions. Later in this chapter, we will discuss the implications of this violation. In Table 2.3 the current estimates of the critical exponents in 2D and 3D are compiled.

2.7.3 *Practical Significance of the Critical Exponents*

If two physical phenomena in heterogeneous materials that contain percolation disorder are described by two different sets of critical exponents, then the physical laws governing the two phenomena must be fundamentally different. Thus, critical exponents help one to distinguish between different classes of problems and the physical laws that govern them. Moreover, since the numerical values of the percolation properties are not universal and vary from one system to another, but the scaling and power laws that they follow near p_c are universal and do not depend on the details of the system, estimates of the critical exponents for a certain phenomenon are used for establishing the relevance of a particular percolation model to that phe-

TABLE 2.3. Values of the critical exponents and fractal dimensions of percolation. The geometrical exponents and the fractal dimension D_f for 2D systems, as well as all the quantities for Bethe lattices, are exact. Values of μ and s for 2D and 3D systems represent the exponents for disordered composites with a conductance distribution that does not violate inequality (44). The elasticity exponents f and χ are described and tabulated in Chapter 8.

Exponent	$d = 2$	$d = 3$	Bethe Lattice
β	5/36	0.41	1
β_B	0.48	1.05	2
τ	187/91	2.18	5/2
σ	36/91	0.45	1/2
ν	4/3	0.88	1/2
η	5/24	-0.068	0
D_f	91/48	2.53	4
D_{bb}	1.64	1.87	2
D_r	3/4	1.36	2
D_{min}	1.13	1.34	2
μ	1.3	2.0	3
s	1.3	0.73	0

nomenon in disordered materials. Finally, in many conductor-insulator composite materials, power laws such as (36), (37), (42), and (43) have been found to be valid over a wide range of the volume fraction of the conducting component of the composite, and therefore such power laws are very useful for correlating the data and interpreting them in terms of a well-understood theory, i.e., percolation theory.

2.8 Scale-Dependent Properties of Percolation Composites

As mentioned above, the correlation length ξ_p has the physical significance that for length scales $L > \xi_p$ the percolation system is macroscopically homogeneous. However, for $L < \xi_p$ the system is *not* homogeneous and its macroscopic properties do depend on L. In this regime, the sample-spanning cluster is statistically self-similar at all length scales less than ξ_p, and its mass M (the total number of occupied bonds or sites that it contains) scales with L as

$$M \propto L^{D_f}, \quad L < \xi_p \tag{45}$$

where D_f is the fractal dimension of the cluster. For $L > \xi_p$ one has, $M \propto L^d$ and thus $D_f = d$, where d is the Euclidean dimension of the system. The crossover between the fractal and Euclidean regimes takes place at $L \simeq \xi_p$. Thus, we can combine the two regimes and write down a unified scaling equation for $M(L, \xi_p)$:

$$M(L, \xi_p) = L^{D_f} h(L/\xi_p), \tag{46}$$

where $h(x)$ is a scaling function. For $L \gg \xi_p$ ($x \gg 1$) we expect to have $M \sim L^d$, and hence $h(x) \sim x^{D_f-d}$, whereas for $x \ll 1$ we must have $h(x) \simeq$ constant. The fractal dimension D_f of the sample-spanning percolation cluster is related to the percolation exponents defined above. The mass M of the cluster is proportional to $X^A(p)\xi_p^d$, since only a fraction $X^A(p)$ of all the occupied bonds or sites are in the cluster. Therefore, we must have $M \propto \xi_p^{d-\beta/\nu}$. For $L < \xi_p$ we must replace ξ_p with L, since in this regime L is the dominant (and the only relevant) length scale of the system. Thus, $M \propto L^{d-\beta/\nu}$, which, when compared with Eq. (45), yields

$$D_f = d - \beta/\nu, \tag{47}$$

so that $D_f(d = 2) = 91/48 \simeq 1.9$ and $D_f(d = 3) \simeq 2.53$. Similarly, for $L < \xi_p$ the backbone is also a fractal object and its fractal dimension D_{bb} is given by

$$D_{bb} = d - \beta_{bb}/\nu. \tag{48}$$

Note that at $p = p_c$ the correlation length is divergent, so that at this point the sample-spanning cluster and its backbone are fractal objects at *any* length scale.

The number M_r of the red bonds of the backbone that are in a box of linear size $L < \xi_p$ scales with L as $M_r \propto L^{D_r}$, where D_r is the fractal dimension of the set of the red bonds. It has been shown (Coniglio, 1981) that

$$D_r = \nu^{-1}. \tag{49}$$

Another important concept is the *minimum* or *chemical* path between two points of a percolation cluster, which is the shortest path between the two points. For $L < \xi_p$, the length L_{min} of the path scales with L as

$$L_{min} \sim L^{D_{min}}, \tag{50}$$

so that D_{min} is the fractal dimension of the minimal path.

If $L < \xi_p$, i.e., if the percolation system is in the fractal regime, all the percolation properties become scale dependent. Since $|p - p_c| \sim \xi_p^{-1/\nu}$, one can rewrite all the above power laws in terms of ξ_p. For example, Eq. (29) is rewritten as, $P(p) \sim \xi_p^{-\beta/\nu}$. For $L < \xi_p$ one replaces ξ_p by L, since in this regime L is the dominant length scale in the system (for $L > \xi_p$ the correlation length ξ_p is the dominant length scale), and therefore

$$P(L) \sim L^{-\beta/\nu}, \tag{51}$$

$$X^A(L) \sim L^{-\beta/\nu}, \tag{52}$$

$$X^B(L) \sim L^{-\beta_{bb}/\nu}, \tag{53}$$

$$g_e(L) \sim L^{-\mu/\nu}, \quad \text{conductor-insulator composites} \tag{54}$$

$$g_e(L) \sim L^{s/\nu}, \quad \text{conductor-superconductor composites} \tag{55}$$

$$E_e(L) \sim L^{-f/\nu}, \quad \text{rigid-soft composites} \tag{56}$$

$$E_e(L) \sim L^{\chi/\nu}, \quad \text{rigid-superrigid composites} \tag{57}$$

$$D_e(L) \sim L^{-\theta}, \tag{58}$$

where $\theta = (\mu - \beta)/\nu$. Thus, *scale-dependent properties are a signature of a fractal morphology*. Similar to the mass of the sample-spanning cluster for which we could write down scaling Eq. (46), we can also write down scaling equations for length-scale dependence of all other percolation properties. For example, Eqs. (36) and (54) are combined to yield

$$g_e(L, \xi_p) = \xi_p^{-\mu/\nu} G(L/\xi_p), \tag{59}$$

where $G(x)$ is a (universal) scaling function. Once it is established that a material possesses a fractal morphology, many classical laws of physics must be significantly modified. For example, Fick's law of diffusion with a constant diffusivity can no longer describe diffusion processes in fractal materials, a subject that will be described and discussed in Chapter 6.

2.9 Finite-Size Scaling

So far we have described percolation in disordered materials that are of infinite extent. However, percolation in finite systems deserves discussion since, both in practical applications and in computer simulations, one usually deals with a system of a finite extent. In such systems, as p_c is approached ξ_p eventually exceeds the linear size L of the material, in which case, as discussed above, L becomes the dominant length scale of the system. Fisher (1971) developed a theory for the scaling properties of a finite system near a critical point, such as a percolation threshold, which is usually called *finite-size scaling*. According to Fisher's theory, the variation of *any* property P_L of a system of linear size L is written as

$$P_L \sim L^{-\zeta} f(u) \tag{60}$$

with $u = L^{1/\nu}(p - p_c) \sim (L/\xi_p)^{1/\nu}$, where $f(u)$ is a non-singular function. If, in the limit $L \to \infty$, P_∞ follows a power law such as, $P_\infty \sim (p - p_c)^\delta$, then one must have $\zeta = \delta/\nu$. Therefore, the variations with L of $P_L(p)$ in a *finite network* at p_c of the *infinite network* can be used to obtain information about the quantities of interest for an *infinite network near* p_c.

Finite-size scaling theory has been used successfully for obtaining accurate estimates of the critical exponents, and even the percolation thresholds, from simulation of finite systems. The finite size of a network causes a shift in its percolation threshold (Levinshtein *et al.*, 1976; Reynolds *et al.*, 1980):

$$p_c - p_c(L) \sim L^{-1/\nu}. \tag{61}$$

In this equation p_c is the percolation threshold of the infinite system, and $p_c(L)$ is an *effective percolation threshold* for a finite system of linear dimension L. However, we should point out that Eqs. (60) and (61) are valid for very large network size L, whereas in practice very large systems cannot easily be simulated. To remedy this situation, Eq. (60) is modified to

$$P_L \sim L^{-\zeta}[a_1 + a_2 h_1(L) + a_2 h_2(L) + \cdots], \tag{62}$$

where a_1, a_2 and a_3 are three fitting parameters, and h_1 and h_2 are called the *correction-to-scaling* functions. These functions are particularly important when small and moderate system size L are used in the simulations. For *transport properties*, $h_1 = (\ln L)^{-1}$ and $h_2 = L^{-1}$ often provide accurate estimates of ζ (Sahimi and Arbabi, 1991). Equation (62) provides us with a means of estimating a critical exponent: Calculate P_L at $p = p_c$ for several system sizes L and fit the results to Eq. (62) to estimate ζ and thus $\delta = \zeta\nu$.

2.10 Percolation in Random Networks and Continua

Although percolation in regular networks—those in which the coordination number Z is the same everywhere—has been extensively invoked for investigating transport in disordered materials, percolation in continua and in topologically-random networks—those in which the coordination number varies from site to site—are of great interest, since in most practical situations one must deal with such irregular and continuous systems. For example, continuum percolation is directly applicable to characterization and modeling of the morphology and effective transport properties of microemulsions (see Chapter 6), polymer blends (see Chapter 9), sintered materials (see Chapter 5), sol-gel transitions (see Chapter 9) and many more, although discrete percolation models that take into account the fine details of the morphology of such materials are equally applicable, and in fact are easier to use.

There are at least three ways of realizing a percolating continuum. In the first method, one inserts a random (or correlated) distribution of inclusions, such as circles, spheres or ellipses, in an otherwise uniform background. The correlations arise when one imposes certain constraints on the system. For example, if the particles are not allowed to overlap, or if the extent of their penetrability or overlap is fixed, then the resulting morphology contains correlations the extent of which depends on the type of the constraints imposed. In such models percolation is defined either as the formation of a sample-spanning cluster of the channels *between* the inclusions, or as the formation of a sample-spanning cluster of touching or overlapping inclusions. In the second method, the percolating composite is generated by tessellating the space into regular or random polyhedra (Winterfeld *et al.*, 1981). A (volume) fraction of the polygons or polyhedra are designated as one phase of the material, while the rest constitutes the second phase, with each phase having its own distinct effective properties. Figure 2.11 presents an example of such a model. The characterization of the microstructure of these models, and how they are utilized for modeling heterogeneous materials, will be described in Chapter 3. Finally, the third method consists of constructing a distribution of conducting sticks of a given aspect ratio, or plates of a given extent, inserted in a uniform background. Such models have been proposed for representing fibrous materials. An example is shown in Figure 2.12.

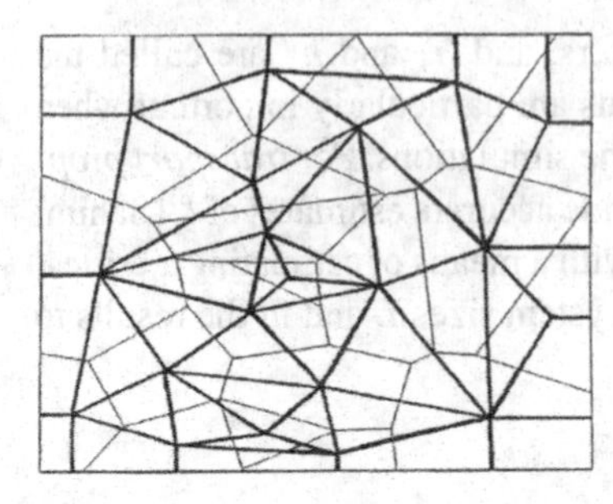

FIGURE 2.11. Two-dimensional Voronoi tessellation of space (thin lines) and the corresponding Voronoi network (thick lines).

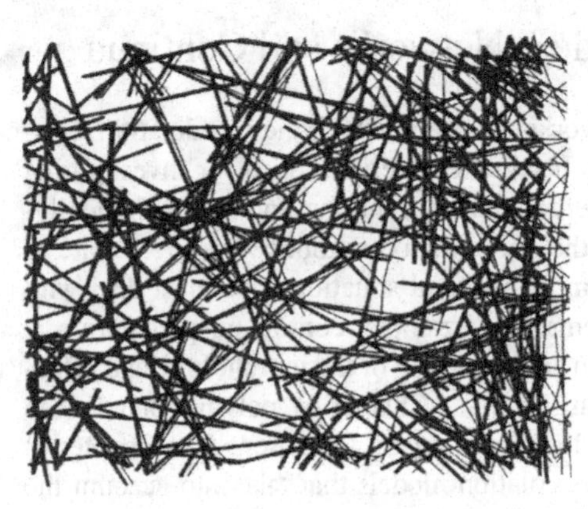

FIGURE 2.12. A two-dimensional model of fibrous materials.

2.10.1 Percolation Thresholds: Materials with Very Low or High Thresholds

There are several techniques by which one can estimate the percolation threshold of a continuum percolation system. Somewhat similar to lattice percolation, the percolation thresholds of continuum d-dimensional systems with $d \geq 2$ are not known exactly, except in one special case. The exceptions are for certain 2D symmetric-cell models. Two-phase symmetric-cell models are constructed by partitioning space into cells of arbitrary shapes and sizes, with the cells being randomly designated as phases 1 and 2 with probabilities ϕ_1 and ϕ_2. A simple example is a 2D system which is tessellated into square cells, where each cell belongs either to phase 1 or phase 2 with probabilities ϕ_1 and ϕ_2. The Voronoi tessellation described above provides another example of such models. We will come back to these models in Chapter 3 where we consider them as models of composite materials. Unlike distributions of particles in a matrix, the symmetric-cell models possess phase-inversion symmetry defined in Section 2.5.3.

It is generally believed that for any 2D symmetric-cell model in which the centers of the cells are the sites of a fully triangulated lattice one has, $\phi_{1c} = \phi_{2c} = 1/2$, although we are not aware of any rigorous proof of this prediction. In this case, the continuum percolation threshold and the site percolation threshold of the lattice are identical. Examples include hexagonal and Voronoi tessellations, both of which have site percolation thresholds of 1/2.

Suppose that particles that are distributed in a uniform matrix constitute phase 2 of a continuum percolation model of a composite. Pike and Seager (1974) and Haan and Zwanzig (1977) were among the first to obtain accurate estimates of the percolation threshold ϕ_{2c} for randomly-distributed overlapping disks and spheres. The most accurate estimate of ϕ_{2c} for 2D (disk) systems that is currently available is due to Quintanilla *et al.* (2000) who reported that $\phi_{2c} \simeq 0.67637 \pm 0.00005$, while for 3D (spherical particles) systems, Rintoul and Torquato (1997) provided the estimate, $\phi_{2c} \simeq 0.2895 \pm 0.00005$. The corresponding values for overlapping oriented squares and cubes are (Pike and Seager, 1974; Haan and Zwanzig, 1977), $\phi_{2c} \simeq 0.67 \pm 0.01$ and 0.28 ± 0.01, respectively. In these models, all the particles had the same size. In general, if the particles do not all have the same size, then ϕ_{2c} depends continuously on the size distribution of the particles. Numerical simulations indicate (see, for example, Pike and Seager, 1974, Lorenz *et al.*, 1993), however, that ϕ_{2c} depends only weakly on the particle size distribution, although certain pathological exceptions to this general result can also be constructed. Meester *et al.* (1994) conjectured that the monodisperse distribution *minimizes* the percolation threshold.

One may prepare materials that have exceedingly small percolation thresholds. A well-known example is an *aerogel*, an ultra-light material which is formed by a sort of a gelation process (see Chapter 9 for description and discussion of gelation). Aerogels have a fractal structure and the volume fraction of solid materials in them is typically 0.5%, indicating that the percolation threshold of the solid phase is even smaller than 0.5%. In Chapter 9 we will describe more some fundamental properties of aerogels and their characterization and modeling. One may also prepare two-phase composite materials by compacting binary mixtures of particles of widely disparate sizes in such a way that the smaller particles form a sample-spanning cluster that resides essentially on the surfaces of the larger particles (see, for example, Malliaris and Turner, 1971; Kusy, 1977).

In the opposite limit, one may also prepare composite materials with percolation thresholds that approach unity. This can be achieved if the particle size distribution is very broad, and each particle possesses a "soft" repulsive interparticle potential with a range which is larger than the size of the particle. Unlike monodisperse particles, polydispersivity causes the particles to fill the space, but the repulsive interactions prevent formation of a sample-spanning cluster until the system is essentially completely filled by the particles. An example of this type of materials is colloidal dispersions.

One of the most important discoveries for continuum percolation is (Scher and Zallen, 1970) that a critical occupied volume fraction ϕ_c, which is defined as

$$\phi_{2c} = p_{cs} f_l, \tag{63}$$

where f_l is the filling factor of a lattice when each site of the lattice is occupied by a sphere in such a way that two nearest-neighbor impermeable spheres touch one another at the midpoint, appears to be an almost invariant of the system with a value of about 0.17 for 3D systems. Shante and Kirkpatrick (1971) generalized this idea to permeable spheres and showed that the average number B_c of bonds per sites at p_c is related to ϕ_c by

$$\phi_{2c} = 1 - \exp(-\frac{1}{8}B_c), \tag{64}$$

and that for continuous systems, $B_c \to p_{cs}Z$, as Z becomes large. It is clear from Table 2.3 and Eq. (24) that in 3D, $B_c \simeq 1.5$. It has been shown (see, for example, Balberg and Binenbaum, 1985, and references therein) that the morphological exponents, defined by Eqs. (29)–(35), are equal for lattice and continuous systems.

Similar to lattice systems described earlier, one may also speak of two-phase continuum composites that are bicontinuous. Hence, extending the criteria of bicontinuity for lattice systems, described above, to continuum systems, we say that a d-dimensional two-phase continuum system that possesses phase-inversion symmetry is bicontinuous for $\phi_{2c} < \phi_2 < 1 - \phi_{2c}$ for $d \geq 2$. However, these are not necessary conditions. For example, a bicontinuous structure without phase-inversion symmetry but with $\phi_{2c} < 1/2$ is a 3D distribution of identical overlapping spheres, where for $0.29 < \phi_2 < 0.97$ both the particle phase and the space between the particles are percolating. An example of a bicontinuous composite without phase-inversion symmetry is a close-packed face-centered-cubic lattice. For example, at the percolation threshold of the particle (sphere) phase, $\phi_{2c} = \pi/\sqrt{18}$, the space between the particles also percolates.

2.10.2 *The Ornstein–Zernike Formulation*

Continuum percolation can be described by an Ornstein–Zernike (OZ) formulation. In words, the OZ formulation states that the total correlation between two particles at $\mathbf{r}_1$ and $\mathbf{r}_2$ can be written as the sum of two contributions: (1) A *direct* effect of the particle at $\mathbf{r}_1$ on the particle at $\mathbf{r}_2$, which is generally short-ranged and is characterized by a type of correlation function $D_2(\mathbf{r})$, and (2) an *indirect* effect in which a particle at $\mathbf{r}_1$ affects a particle at $\mathbf{r}_3$, which in turn affects a particle at $\mathbf{r}_2$, directly or indirectly through other particles. The indirect effect is weighted by the density and averaged over all possible $\mathbf{r}_3$. It should be clear to the reader that the indirect effect has a longer range than $D_2(\mathbf{r})$. We do not give the details of the OZ formulation (see Section 3.4.1.4 for more details), but note that Coniglio *et al.* (1977) utilized this formulation to derive the following equation for the pair-connectedness function $P_2(\mathbf{r})$ defined above:

$$P_2(\mathbf{r}) = D_2(\mathbf{r}) + \rho D_2(\mathbf{r}) \otimes P_2(\mathbf{r}), \tag{65}$$

where $D_2(\mathbf{r})$ is the *direct connectedness function* [$D_2(\mathbf{r})$ is in fact the analogue of the correlation function $C(r)$ defined in Section 2.3.1], ρ is the number density of the particles, and $\otimes$ denotes a convolution integral. Taking the Fourier transform of both sides of Eq. (65) yields

$$\tilde{P}_2(\omega) = \frac{\tilde{D}_2(\omega)}{1 - \rho\tilde{D}_2(\omega)}. \tag{66}$$

On the other hand, the mean cluster size S is related to $P_2(\mathbf{r})$ through the relation

$$S = 1 + \rho \int P_2(\mathbf{r}) d\mathbf{r} \tag{67}$$

which is in fact the continuum analogue of Eq. (26). Equations (65)–(67) are the basis for studying continuum percolation system using a variety of models and approximations.

Note that Eq. (67) implies that

$$S = 1 + \rho \tilde{P}_2(0) = [1 - \rho \tilde{D}_2(0)]^{-1}. \tag{68}$$

Since at the percolation threshold the mean cluster size S diverges, Eq. (68) immediately yields an estimate for the critical value ρ_c of the number density ρ:

$$\rho_c = [\tilde{D}_2(0)]^{-1} = \left[\int D_2(\mathbf{r}) d\mathbf{r} \right]^{-1}. \tag{69}$$

In the OZ formulation, the average coordination number $\langle Z \rangle$ of the system is given by

$$\langle Z \rangle = \rho \Omega_s(d) \int_0^D r^{d-1} P_2(r) dr, \tag{70}$$

where $\Omega_s(d)$ is the total solid angle contained in a d-dimensional sphere, $\Omega_s(d) = 2\pi^{d/2}/\Gamma(d/2)$, with Γ being the gamma function, and D is the diameter of the (possibly overlapping) spheres.

2.11 Differences Between Lattice and Continuum Percolation

Transport in percolating continua can be quite different from that in percolation networks. Consider, for example, a model in which one inserts at random circular or spherical inclusions in an otherwise homogeneous medium. For the obvious reason, this is called the *Swiss-cheese model* (see Figure 2.13). If transport takes place through the matrix (through the channels between the non-overlapping spheres), then the problem is mapped onto an equivalent problem on the *edges* of the Voronoi polygons or polyhedra (Kerstein, 1983). Such a network has an exact coordination number of 4 in 3D. To generate a Voronoi tessellation of space one distributes Poisson (random) points in the space, each of which is the basis for a Voronoi polygon or polyhedron, which is that part of the space that is closer to its Poisson point than to any other Poisson point. A Voronoi *network* is constructed by connecting to each other the Poisson points in the neighboring polygons or polyhedra (see Figure 2.11). One may also construct the *inverted Swiss-cheese model* in which the roles of the two phases in the Swiss-cheese model are switched, i.e., transport takes place through the circular or spherical inclusions, in which case the transport problem can be mapped onto an equivalent one in the Voronoi network. In both models it is assumed that all the inclusions have the same size. van der Marck

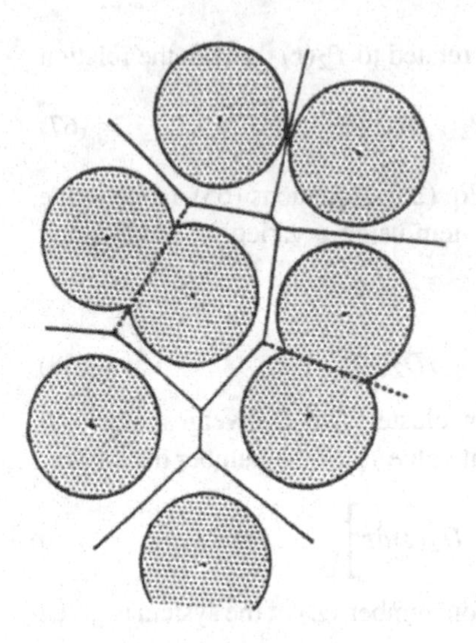

FIGURE 2.13. Two-dimensional Swiss-cheese model.

(1996) proposed a more general method for mapping transport in continua that consists of a distribution of inclusions of various sizes in a uniform background, onto an equivalent problem in an equivalent network model. We should, however, point out that if one utilizes the equivalent network of a given continuum, one can no longer assign the transport properties of the network's bonds from an arbitrary distribution, because there is a *natural* distribution of the conductances (or other transport properties) of the transport channels in the continuum which must be constructed based on the shapes and sizes that the channels take on (see below).

The Voronoi network was utilized by Jerauld *et al.* (1984a,b) and Sahimi and Tostsis (1997) as a prototype of irregular networks to study transport in disordered composites. The *average* coordination number of the Voronoi network is about 6 and 15.5 in 2D and 3D, respectively. Jerauld *et al.* (1984a,b) showed that the geometrical critical exponents of percolation in such random networks are the same as those for regular networks. Moreover, they established that as long as the average coordination number of a regular network and a topologically-random one (for example, the 2D Voronoi and triangular networks) are about the same, many transport properties of the two systems are, for all practical purposes, identical.

However, certain continuous percolation models violate the condition expressed by inequality (44) for the universality of the transport exponents (Halperin *et al.*, 1985; Feng *et al.*, 1987), in which case the transport exponents for the continuous models are not necessarily the same as those in random networks, and thus they must be estimated separately. The differences between the critical exponents of lattice and continuum percolation are caused by the natural distributions of the transport properties that give rise to new scaling laws which cannot be predicted by the network models, unless the same natural distributions are also utilized in the network models as well.

Consider, as an example, the Swiss-cheese model. Figure 2.14 shows a typical channel between two spheres (disks in 2D) of radius a. We imagine that a rectangu-

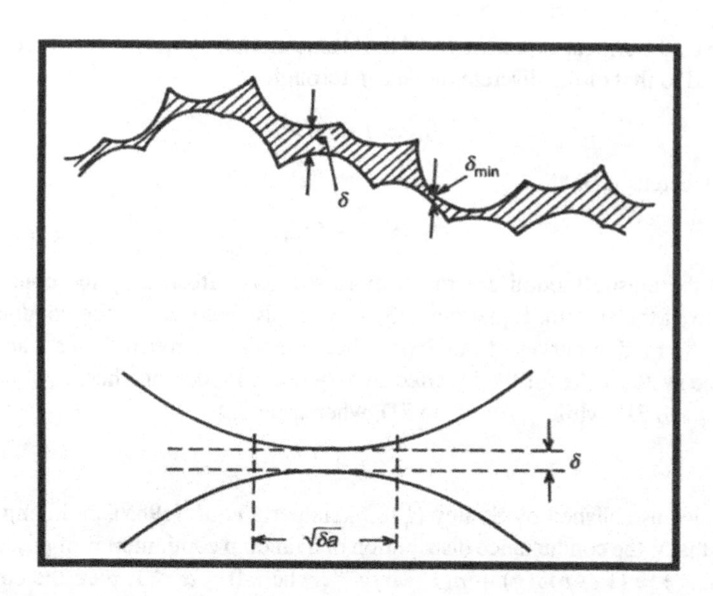

FIGURE 2.14. The conducting necks (upper part of the figure) in the Swiss- cheese model, and determining the approximate channels (the dashed lines) that represent them.

lar strip of width δ is formed, in which case the length $2l$ of the strip is of the order of $\sqrt{\delta a}$. This is due to the fact that the length l forms the base of a right-angled triangle with the apex at the center of a sphere, the sphere's center being the adjacent site, and the hypotenuse being of length $a + \delta$, so that $a^2 + l^2 = (a + \delta)^2$, hence yielding $l \sim \sqrt{\delta a}$. The conductivity g of any such bond in d dimensions is then, $g \sim \delta^{d-1}/l \sim \delta^{d-3/2}/\sqrt{l}$. The distribution function $\psi(g)$ of such conductivities is then finite as $\delta \to \delta_{\min}$. The node-link model described above can then be utilized for estimating the effective conductivity of such a model (Halperin *et al.*, 1985), which is then given by

$$g_e \sim \xi_p^{-(d-1)} \xi_p \Sigma_L \sim \xi_p^{2-d} \Sigma_L, \tag{71}$$

where

$$\Sigma_L^{-1} = \sum_i g_i^{-1} \sim \int_{\delta_{\min}}^{\infty} \psi(g) g^{-x} dg, \tag{72}$$

so that the effective conductivity depends on the structure of $\psi(g)$.

Based on such considerations, Halperin *et al.* (1985) and Feng *et al.* (1987) showed that the critical exponents that characterize the power laws for the transport properties of the Swiss-cheese and the inverted Swiss-cheese models near their percolation thresholds can be very different from those of a random network. In particular, if μ_{sc} and μ_{isc} denote, respectively, the critical exponents of the conductivity of the Swiss-cheese and the inverted Swiss-cheese models, then in 2D, $\mu_{sc} = \mu_{isc} = \mu$, while in 3D, $\mu_{isc} = \mu$, and

$$\mu_{sc} \simeq \mu + 1/2. \tag{73}$$

Moreover, the critical exponent f_{sc} of the elastic moduli of the Swiss-cheese model in related to that of the discrete models f through

$$f_{sc} \simeq f + 3/2, \tag{74}$$

in 2D, whereas in 3D

$$f_{sc} \simeq f + 5/2, \tag{75}$$

so that the elastic moduli are much more strongly affected by the continuous structure of the system. Equations (73)–(75) imply that near p_c the conductivity and elastic moduli curves of the Swiss-cheese model are much flatter than those of lattice systems. As for the inverted Swiss-cheese model, one has, $\mu_{isc} = \mu$, in both 2D and 3D, while $f_{isc} = f$ in 2D, whereas in 3D

$$f_{isc} \simeq f + 1/2. \tag{76}$$

It was also established by Straley (1982), Halperin *et al.* (1985), and Feng *et al.* (1987) that if the conductance distribution of a random continuum is of power-law form, $\psi(g) = (1-p)\delta(g) + p(1-\alpha)g^{-\alpha}$, where $0 \leq \alpha \leq 1$, then the conductivity critical exponent $\mu(\alpha)$ of the continuum is *not* universal, but is given by

$$\mu_l(\alpha) = 1 + (d-2)\nu + \frac{\alpha}{1-\alpha} \leq \mu(\alpha) \leq \mu_u(\alpha) = \mu + \frac{1}{1-\alpha}. \tag{77}$$

Such power-law conductance distributions are actually very common among disordered continua. For example, for the Swiss-cheese models in d-dimensions one has, $\alpha = (2d-5)/(2d-3)$. In 2D, $\alpha = -1$, and the conductance distribution is not singular at $g = 0$, and therefore $\mu_{sc} = \mu$. In 3D, $\alpha = 1/3$, and the conductance distribution is singular at $g = 0$, hence giving rise to Eq. (73). Moreover, Straley (1982) argued that

$$\mu(\alpha) = \min[\mu, \mu_l(\alpha)]. \tag{78}$$

These results have not only been confirmed by computer simulations, but have also received experimental verification (see, for example, Garfunkel and Weissman, 1985; Koch *et al.*, 1985; Chen and Chou, 1985; Rudman *et al.*, 1986), some of which will be discussed later in this book. A more detailed discussion of continuum percolation is given by Balberg (1987).

2.12 Correlated Percolation

In the percolation phenomena described so far, no correlations between various segments of the system were assumed. However, disorder in many important heterogeneous materials is not completely random, as there usually are correlations with a range that may be finite but large. For example, in packing of solid particles, the microstructure of which will be described in Chapter 3, there are short-range correlations. The universal scaling properties of percolation systems with finite-range correlations are the same as those of random percolation, if the length scale of interest is larger than the correlation length. Moreover, if the correlation function $C(r)$ decays as r^{-d} or faster, where d is the Euclidean dimensionality of the system, then the scaling properties of the system are identical with those of random percolation. This is not totally unexpected because even in random percolation, as p decreases toward p_c, correlations start to build up (recall that, as discussed above, for length scales $L < \xi_p$ the sample-spanning cluster is a fractal object), and hence the introduction of any type of correlations with a range shorter than ξ_p cannot change its scaling properties. In many other cases, e.g., in some disordered elastic materials, there are long-range correlations (see Chapter 8 and 9). We now describe and discuss a few models of correlated percolation and point out their relevance to the physical phenomena of interest to us in this book.

2.12.1 Short-Range Correlations

The earliest correlated percolation model that we are aware of is a correlated bond percolation mode due to Kirkpatrick (1973). In his model, which was intended for hopping conduction in disordered solids (see Chapters 5 and 6), to each site of a lattice is assigned a random number s_i, uniformly distributed between -1 and $+1$. Each bond ij is assigned a number b_{ij} which is calculated by

$$b_{ij} = \frac{1}{2}(|s_i| + |s_j| + |s_i - s_j|), \tag{79}$$

and all bonds with $b_{ij} > \epsilon$ are removed, where ϵ is some selected value. Thus, a bond remains intact only if b_{ij} is sufficiently small, implying that $|s_i|$ and $|s_j|$ must both be small. Therefore, if a bond is present or conducting, its neighbors are also likely to be conducting, and hence the conducting bonds are clustered together. Many other percolation models with short-range correlations have been developed, most of which are described elsewhere (Sahimi, 1998).

Positive correlations (that is, correlations that cluster the conducting bonds together, as opposed to negative correlations that make it more likely that a conducting bond is next to an insulating one, and vice versa), whether short- or long-ranged, usually *reduce* the percolation threshold of the system. For example, in Kirkpatrick's model, $p_{cb} \simeq 0.1$ for the simple-cubic lattice, sharply lower than $p_{cb} \simeq 0.249$ for random bond percolation (see Table 2.2). This is due to the clustering of the conducting bonds caused by the positive correlations, as a result of which formation of sample-spanning transport paths is possible even at low values of the fraction of the conducting phase. On the other hand, Bug *et al.* (1985) presented evidence based on computer simulations that indicate that positive correlations may sometimes increase the percolation threshold and at other times decrease it, depending on the distance between two occupied sites that are considered connected, the dimensionality of the system, and its temperature.

2.12.2 Long-Range Correlations

Interesting percolation models with long-range correlations, in which the correlations decay with the distance between 2 points on the sample-spanning percolation cluster, were developed by Weinrib (1984; see also Weinrib and Halperin, 1983) and Prakash *et al.* (1992). In the latter case, the correlation function $C(r)$ in a d-dimensional system was assumed to be given by

$$C(r) \sim r^{-(d-\lambda)}, \tag{80}$$

where $-2 \leq \lambda \leq 2$ is a parameter of the model, such that $0 \leq \lambda \leq 2$ represents positive correlations, while $-2 \leq \lambda \leq 0$ corresponds to negative correlations. Percolation models in which the correlations increase with the distance between two sites or bonds on the sample-spanning cluster were first studied by Sahimi (1994b, 1995a) and Sahimi and Mukhopadhyay (1996). However, their model was motivated by its applications to flow in geological formations, and thus is not described here.

The first percolation model with long-range correlations that we are aware of, which is also relevant to the problems we discuss in this book, was proposed by Pollak and Riess (1975) and Chalupa *et al.* (1979), and is known as the *bootstrap percolation*. In this problem sites of a lattice are initially randomly occupied. Then, those sites that do not have at least Z_c nearest-neighbor occupied sites are removed (note that $Z_c = 0$ is the usual random percolation). The *interactions* between the sites are short-ranged, but the correlations between them increase as the distance between two occupied sites also increases. Figure 2.15 shows a 2D example of a bootstrap percolation cluster. The original motivation for developing this model was to explain the behavior of some disordered materials in which magnetic impurities are randomly distributed in a host of non-magnetic metals. It is believed that in some of such materials an impurity atom cannot sustain a localized

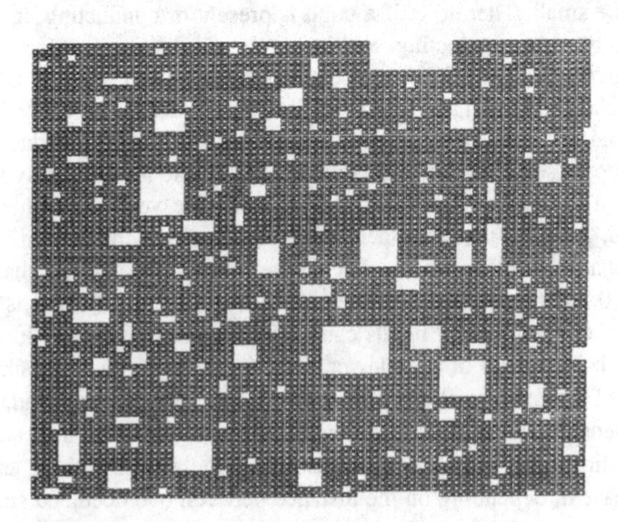

FIGURE 2.15. Two-dimensional bootstrap percolation on the square network. White areas indicate the absent or cut bonds.

magnetic moment unless it is surrounded by a minimum number of magnetic neighbors (see, for example, Jaccarino and Walker, 1965). Bootstrap percolation has also been used for studying orientational order in orthoparahydrogen mixtures (Adler *et al.*, 1987) and dynamics of glass transitions (Ertel *et al.*, 1988).

Extensive numerical studies of bootstrap percolation were begun by Kogut and Leath (1981). Since then, bootstrap percolation has been studied extensively (for a review see Adler, 1991; see also Stauffer and de Arcangelis, 1996, for more recent developments), and has proven to contain a rich variety of behavior which is a strong function of the parameter Z_c. For example, an important question is the nature of the percolation transition in this model. It now appears that for sufficiently high values of $Z_c \leq Z$ (where Z is the coordination number of the lattice), the percolation transition is *first-order*, i.e., discontinuous, whereas for low values of Z_c the transition is continuous and second-order. A first-order phase transition is characterized by sharp discontinuities at p_c, namely, percolation quantities vary smoothly with p anywhere from $p = 1$ or $p = 0$ to just above p_c, but at p_c abruptly (discontinuously) changes their value to zero. If the phase transition is first-order, then the percolation threshold of the system is in fact $p_c = 1$, the sample-spanning cluster is compact (i.e., its fractal dimension $D_f = d$), and power laws (29)–(35) are no longer valid. For example, bootstrap percolation transition is first-order in the simple-cubic lattice with $Z_c \geq 4$, in the triangular lattice with $Z_c \geq 4$, and in the square lattice with $Z_c \geq 3$.

However, we must point out that, although when bootstrap percolation transition is first-order, the percolation threshold is $p_c = 1$, extensive simulations have shown that sample-size effects are very large and strong, so that the approach to the asymptotic value of $p_c = 1$ is very slow, so that in small or even moderately large lattices one may obtain an *effective percolation threshold* p_c which is less than 1. Sample-size effects also smear the first-order phase transition. The reason for the peculiar behavior of this model seems to be the long-range nature of the correlations between two occupied sites, and the fact that for large enough Z_c these correlations do not decay with the distance between the two sites, rather they *increase* with the distance.

The richness of bootstrap percolation is not, however, limited to the nature of the percolation transition at p_c. In the first simulation of this model, Kogut and Leath (1981) found that in the triangular and simple-cubic lattices with $Z_c = 3$ the value of the critical exponent β, defined by Eqs. (29) and (30), seems to be different from that of random percolation. Adler and Stauffer (1990) carried out an extensive study of this model in the simple-cubic lattice and found that, $\beta(Z_c = 0) = \beta(Z_c = 1) = \beta(Z_c = 2) \simeq 0.41$, the same as that of random percolation (see Table 2.3), but that, $\beta(Z_c = 3) \simeq 0.6$, whereas the corresponding value of ν appeared to be the same as in random percolation for $0 \leq Z_c \leq 3$. Conduction in bootstrap percolation systems was studied by Sahimi and Ray (1991) in both 2D and 3D, who showed that in the simple-cubic lattice the critical exponent μ of the conductivity for $Z_c =$ 0, 1, 2, and 3 is the same as that of random percolation listed in Table 2.3. Figure 2.16 shows the behavior of the conductivity, which also demonstrates the first-order phase transition when Z_c is large enough. Sahimi and Ray also speculated that

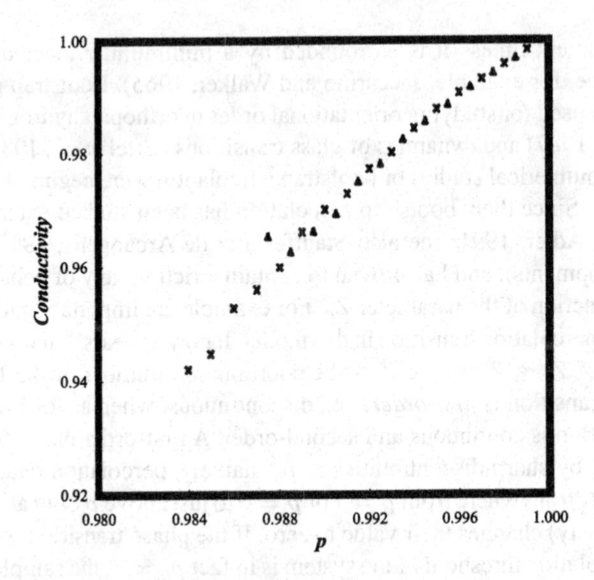

FIGURE 2.16. Conductivity of bootstrap percolation in a $L \times L \times L$ simple-cubix network, versus the fraction p of occupied sites, with $Z_c = 5$, where $(\times)$ are the results for $L = 25$, while (Δ) are for $L = 45$.

bootstrap percolation with large enough Z_c (i.e., when the percolation transition is first-order) may be relevant to modeling of mechanical (see Chapter 8) and fracture properties (see Chapter 8 of Volume II) of solids since, similar to bootstrap percolation with large enough Z_c, one may also have first-order phase transitions in such phenomena. Their speculation has in fact turned out to be true.

2.13 Experimental Measurement of Percolation Properties

Let us now discuss experimental measurements of percolation quantities in composite materials in order to establish their relevance. To do this, we consider conductor-insulator composites that have wide applications, and describe and discuss measurement of their morphological properties. Measurement of their transport properties will be described and analyzed in the subsequent chapters.

Thin metal films are two-phase mixtures of metal and non-metal components that have interesting morphological and transport properties. Their electrical conductivity varies continuously with the volume fraction ϕ of the metallic phase. If ϕ is large enough, the metal phase forms a sample-spanning cluster, while the non-metallic phase is dispersed in the composite in the form of isolated islands. In this regime, the electrical conductivity of the system is large and its temperature coefficient of resistance is positive. If the volume fraction of the metallic phase is close to its percolation threshold ϕ_c, one has a complex composite of isolated metallic islands, a tortuous and sample-spanning cluster of the metallic component, as well as islands of the non-metallic component. If the non-metallic phase forms a sample-spanning cluster, the system is in the dielectric regime, and its electrical conductivity is very small. Transmission electron microscopy (TEM) indicates that thin metal films have a discontinuous structure in which the metallic phase constitutes only a fraction of the total volume of the system. Such films have a nearly 2D structure and are formed in the early stages of film growth by a variety of techniques, such as evaporation or sputtering. The metallic phase first forms isolated islands that at later stages of the process join together and form a continuous film. Generally speaking, a metal film is considered thin if its thickness ℓ is less than the percolation correlation length ξ_p of the 3D composite.

Important studies of the morphological properties of thin metallic films were carried out by Voss *et al.* (1982) and Kapitulnik and Deutscher (1982), the results of which were published in back-to-back papers in the *Physical Review Letters*. Voss *et al.* prepared Au films by electron-beam evaporation onto 30 nm-thick amorphous Si_3N_4 windows grown on a Si wafer frame. A series of samples were prepared simultaneously which ranged from 6 to 10 nm thick and varied from electrically insulating to conducting. Transmission electron micrographs were digitized, and with the use of threshold detection and a connectivity-checking algorithm the individual Au clusters were isolated and their statistical properties were analyzed. Voss *et al.* found that the Au clusters were irregularly shaped and ramified. At large scales most of the film properties were uniform, but at small scales they were not. Since the deposited Au atoms had some initial mobility, the Au clusters were not totally random, as a result of which the percolation threshold of the system, $\phi_c \simeq 0.74$, was larger than 0.5, the expected value for 2D continuum percolation (see above). However, Voss *et al.* found that at ϕ_c the largest Au cluster is a fractal object with a fractal dimension, $D_f \simeq 1.9$, in excellent agreement with that of 2D percolation, $D_f = 91/48 \simeq 1.896$.

Kapitulnik and Deutscher (1982) prepared samples by successive deposition of amorphous Ge as the substrate, and of thin Pb films as the metal. This allowed them to obtain deposition with only short-range correlations. The size of the Pb crystallites, defined as the average Pb thickness, was about 200 Å. Since the continuity of the metallic cluster is controlled by its thickness, varying the sample thickness is equivalent to generating percolation networks with varying fractions of conducting components. Kapitulnik and Deutscher (1982) deposited the thin films on TEM grids and photographed it with a very large magnification. The pictures were then analyzed for various percolation properties. Figure 2.17 presents their results for the density $\rho = M/L^2$ of the sample-spanning cluster and its backbone. Over about one order of magnitude variations in L the sample-spanning cluster and its backbone exhibit fractal behavior, with $D_f \simeq 1.9$ for the cluster and $D_{bb} \simeq 1.65$ for the backbone, in excellent agreement with the fractal dimensions for 2D percolation (see Table 2.3). Kapitulnik and Deutscher (1982) also measured n_s, the number of metallic clusters of mass s below the percolation threshold. According to Eq. (35), one must have $n_s \sim s^{-\tau}$, and measurements of Kapitulnik and Deutscher yielded, $\tau \simeq 2.1 \pm 0.2$, completely consistent with $\tau = 187/91 \simeq 2.054$ for 2D percolation (see Table 2.3).

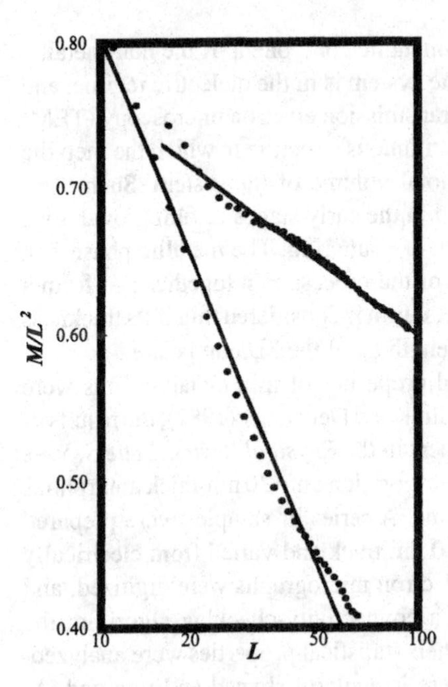

FIGURE 2.17. The mass M of the sample-spanning cluster and its backbone (upper data) for Pb films versus the linear size L of the material (after Kapitulnik and Deutscher, 1982).

More recently, Viswanathan and Heaney (1995) succeeded in direct imaging of a percolation network in a 3D conductor-insulator composite, and obtaining accurate estimates of its percolation properties. They used electric force microscopy (EFM) which is a type of scanning probe microscopy that measures electric field gradients near the surface of a sample using a sharp conductive tip. The gradients are measured by the grounded tip which is oscillated near its resonant frequency. By altering the effective spring constant of the tip as it encounters a force gradient from the electric field, the cantilever's resonant frequency is varied. The resulting change is then monitored, producing a map of the strength of the electric field gradients. The EFM mode generates two images, one of topography and the other of electric field gradients. These maps are generated by the "lift mode" technique by which a line is scanned to give topography; a second pass is made over the same line a prescribed distance above the topography to image the electric field gradients. A commercial Digital Instruments Nanoscope Multimode Atomic Force Microscopy (for EFM imaging) was used for imaging the composites. Tips were metal coated and made of single-crystal silicon with a 5 nm radius curvature. The composite samples were attached on top of the metal substrates using conductive silver paint. The samples were composed of commercial carbon-black and polymer. The carbon-black resistivity was of order of 10^{-2} Ωcm. It consisted of 200 nm mean diameter aggregates composed of smaller fused semi-spherical particles of 80 nm mean diameter. The polymer was high density polyethylene with a resistivity of 10^{18} Ωcm.

To quantify the morphology of the sample, the EFM image was binarized in order to highlight the conductive features. The resulting image was then analyzed

FIGURE 2.18. Logarithmic plot of the average density $\rho(L)$ of the conducting regions versus the linear size L of the sample. The minimum conducting unit is a square of size $L = 0.037\ \mu$m (after Viswanathan and Heaney, 1995).

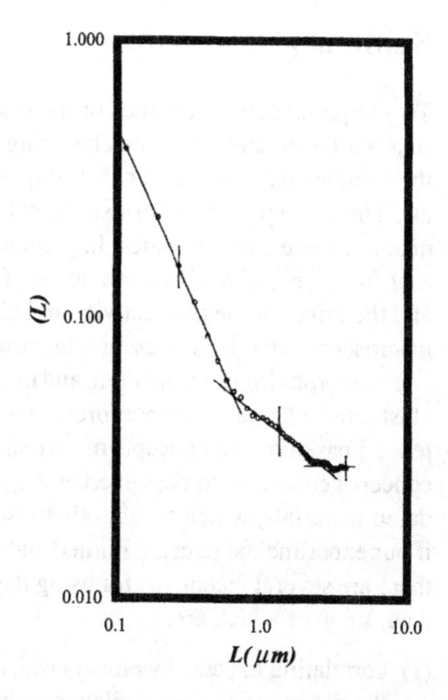

in terms of the scaling theory of percolation described above. A point that was part of the conductive area or "mass" M in squares centered on this origin was counted as a function of the square linear size L. The density $\rho(L) = M(L)/L^2$ of the conductive area was then computed. The procedure was repeated for 25 non-overlapping areas on the binarized image, and the results were averaged. Figure 2.18 shows the results and indicates the existence of three regimes. For $L < 0.6\ \mu$m, the density satisfies a power law corresponding to $M(L) \propto L^{D_{\text{eff}}}$, where D_{eff} is the effective fractal dimension of the conductive areas. The data yield, $D_{\text{eff}} \simeq 0.9 \pm 0.1$. Since the fractal dimension D_{eff} of an object embedded in a d-dimensional space and cut by a surface of dimensionality D_s is given by

$$D_{\text{eff}} = D_f - (d - D_s), \tag{81}$$

the fractal dimension D_f of the carbon-black aggregate is estimated to be $D_f \simeq 1.9 \pm 0.1$, in excellent agreement with that of 2D percolation.

For $0.8\ \mu\text{m} < L < 2\ \mu$m, the density follows the same type of power law that corresponds to $D_f \simeq 2.6 \pm 0.1$, in good agreement with the fractal dimension of 3D percolation clusters at the percolation threshold, $D_f \simeq 2.53$ (see Table 2.3), while for $L > 3\ \mu$m the density is roughly constant, corresponding to $D_f = 3$ for a homogeneous material. The crossover from the power law behavior occurs at $L \simeq \xi_p$, where ξ_p is the correlation length of percolation, and hence for these materials $\xi_p \simeq 3\ \mu$m. These measurements all agree with the predictions of percolation theory, hence firmly establishing the quantitative relevance of percolation to describing the morphology (and transport properties) of heterogeneous materials.

Summary

Two important characteristics of the morphology of disordered multiphase materials are the connectivity and clustering of their phases, and the shape and size of the sample-spanning cluster that allows a transport process occur in the materials. The concepts of modern statistical physics of disordered materials can now quantify these characteristics. In particular, percolation theory is an indispensable tool for quantitative characterization of the morphology of composite materials, and the effect of the connectivity and clustering of their phases on their effective macroscopic characteristics, and in particular transport properties.

The morphology of a material, and in particular the shape of the sample-spanning cluster that allows a transport process to occur in the material, can also be characterized based on the concepts of self-similarity, self-affinity, and fractality. These concepts enable us to characterize the geometry and surface roughness of disordered materials, which would otherwise seem very difficult to understand. Even if our experimental data are limited and have been measured over a limited range, there are several incentives for using the ideas of fractal geometry to interpret the data, some of which are,

(1) correlating the data by power laws, a main tool of fractal geometry, condenses the description of a complex morphology.
(2) Self-similarity and self-affinity allow us to correlate in a simple way properties and performance of a material with its morphology and, possibly, dynamics of its formation, as with the case of rough surfaces that will be described in Volume II.
(3) If we choose *not* to use the concepts of fractal geometry, then, in many instances our only choice will be to discard the data altogether, whereas fractals provide us with at least an approximate description and correlation of the data.

Moreover, both percolation theory and fractal geometry point to the fundamental role of the length scale and long-range correlations in the macroscopic homogeneity of a heterogeneous material. If the largest relevant length scale of a material, e.g., its linear size, is less than the length scale at which it can be considered homogeneous, which for percolation disorder is the correlation length ξ_p, then the classical equations that describe transport processes in the material must be fundamentally modified. These issues will be discussed in detail in the subsequent chapters, where the relevance of the concepts that were described in the present chapter will be demonstrated.

3
Characterization and Modeling of the Morphology

3.0 Introduction

In Chapter 2 we learned how to characterize two important aspects of the morphology of a disordered multiphase material, namely, the connectivity and clustering of its phases, and pointed out their deep effect on the material's macroscopic properties. We also described how the correlations, especially long-range correlations, in the morphology of a material are characterized in terms of fractal geometry. The next step is to complete the task of characterization of the morphology, and develop models that can realistically represent the connectivity and correlation properties of a heterogeneous material, as investigation of transport phenomena in such materials must include as the first, and perhaps the most important, ingredient, a realistic model of the material itself. However, such models must logically depend on the type of material that one wishes to deal with, *and* on the computational resources that are available to one, since there is no "universal" model that can represent every type of material. Of course, as more mictrostructural details are added to a model of a heterogeneous material, it becomes more realistic but also, by necessity, more complex. However, the process of adding more mictrostructural details to a model of a material cannot go on *ad infinitum*, because our computational power is not unlimited and thus cannot handle a model with "infinite" complexity. Hence, we face the task of deciding which aspects of a material's microstructure is crucial to its macroscopic properties, and in particular to its transport properties which are the focus of this book. Before making this decision, we must answer the following key question: *do we wish to only gain qualitative insight into a given phenomenon that occurs in a heterogeneous material, or is our model intended for quantitative modeling of the phenomenon?* If the former is of prime interest to us, then we should develop the *simplest* model that allows us to investigate the transport process of interest in the material with a reasonable computation time and can predict the essential features of the phenomenon of interest. If the predictions of the model are intended to be quantitative, then two key ingredients must be included in the model:

(1) The material's morphology must be accurately represented by the model, and
(2) the mechanism(s) of the transport process must also be accurately described.

In this chapter we consider the first ingredient—the morphology of a material—and describe and discuss various models in which the fundamental aspects of materials' heterogeneities have been, to various degrees of details, incorporated.

Before describing the models, let us discuss a few important concepts that will be referred to throughout this chapter. If a material is statistically homogeneous, we may invoke an *ergodic hypothesis* to compute its volume-averaged properties. According to this hypothesis, averaging a property over *all* realizations of an ensemble is equivalent to averaging over the volume of *one* realization in the limit of infinite volume. In addition, a material is considered *strictly isotropic* if the joint distributions that describe some stochastic process in the material are *rotationally invariant*, i.e., they are invariant under rigid-body rotation of the spatial coordinates. Finally, as described in Chapter 2 (see Section 2.5.3), a two-phase material is said to possess *phase-inversion symmetry* if the morphology of phase 1 at volume fraction ϕ_1 is statistically identical with that of phase 2 when its volume fraction is ϕ_1.

3.1 Models of Heterogeneous Materials

Accurate models are needed for evaluating the transport coefficients and other important dynamical properties of heterogeneous materials. The simplest of such properties is perhaps the effective thermal or electrical conductivity g_e. Thus, one major goal of modeling the morphology of a material has always been predicting such properties, given some information or data about the material's morphology. If, for example, a material consists of two different constituents or phases, each having its own distinct thermal or electrical conductivity, then, one would like to be able to predict the overall effective conductivity of the material as a function of the volume fractions of its constituents. Over the years, many models of heterogeneous materials have been developed, most of which have been motivated by a certain application. It may happen that a model is useful for studying a particular phenomenon and can provide reasonable predictions for some of its properties. However, if the model does not realistically represent the material's morphology, then it will not be useful for investigating other types of transport phenomena in the same material and providing accurate predictions for their properties. Such models typically contain parameters that are either defined very vaguely, or have no physical meaning whatsoever; their sole purpose is to make the models' predictions agree with the experimental data for a particular phenomenon. What we intend to do in this chapter is not describing such models, but those that are useful for predicting several transport properties of heterogeneous materials.

Aside from one-dimensional (1D) models, one can divide useful models of heterogeneous materials into two groups. In one group are what we referred to, in Chapter 1, as the continuum models, while the second group consists of the discrete or lattice models of heterogeneous materials. Models in each group have their own strengths as well as shortcomings which will be pointed out throughout this chapter.

3.2 One-Dimensional Models

If a set of atoms lie along a line, then we have a 1D model, or a linear chain. If the distance r_{ij} between two neighboring atoms i and j is the same everywhere in the system, then the 1D model is usually called an *ordered chain*. In practice, this cannot be achieved exactly. At best, the variable r_{ij} is a stochastically distributed quantity with a very small variance. If, however, the r_{ij} are random variables with a significant variance, then the resulting linear chain is sometimes called a 1D *liquid* or a 1D *glass*. If each r_{ij} is an independent variable, distributed with uniform probability along the entire length of the chain, then the system is called a *linear gas*. It is clear that a 1D system cannot be topologically disordered, as every atom is a neighbor to exactly two other atoms everywhere in the chain. Therefore, the chain's disorder must be compositional (i.e., the disorder must be in the local transport properties). Consider, as an example, diffusion of a particle on a linear chain which is usually described by the following master equation

$$\frac{\partial P_i}{\partial t} = W_{i,i-1}(P_{i-1} - P_i) + W_{i,i+1}(P_{i+1} - P_i), \tag{1}$$

where P_i is the probability of finding the particle at site (or atom) i at time t, and W_{ij} is the transition rate for making a jump from site i to site j. If the transition probabilities are independent random variables, then Eq. (1) describes diffusion in a compositionally or substitutionally disordered linear chain, a problem of considerable interest in the late 1970s and early 1980s. We will come back to this problem in Chapter 6.

Another well-studied 1D model is the so-called 1D hard rod model, in which rods of length ℓ are randomly, sequentially, and irreversibly placed on a line of length L at a rate J of per unit length and per unit time. A problem of interest would then be the expected value for coverage at the saturation limit as $L \to \infty$. This problem belongs to the more general class of problems known as *random sequential additional* (RSA) processes (see, for example, Reńyi, 1963; Widom, 1966; Feder, 1980; Cooper, 1988; Viot *et al.*, 1993). In these problems, non-overlapping objects (rods, disks, spheres, etc.) are randomly, irreversibly, and sequentially placed into a system, a problem that was apparently first studied by Widom (1966), when the objects are d-dimensional identical spheres. The particles do not sample the configuration space uniformly, as their locations are frozen after they are placed into the system. If the objects are d-dimensional identical spheres, then the filling process ends at the saturation, or *jamming*, limit, beyond which no addition will be possible. For the 1D hard rod model, which is also known as the *car-parking problem*, the saturation limit is reached when the coverage is about 0.7476. We will come back to the RSA problem later in this chapter.

There has been considerable interest over the past few years in the fabrication and characterization of nanoscale structures, such as quantum dots and wires. Electronic, magnetic, and mechanical properties of such materials are of great interest, since they are believed to play important roles in high-storage capacity, and high-performance transport devices. With the aid of advanced experimental techniques, materials of this type have been fabricated which exhibit 1D properties. However, quantum effects play a prominent role in most of such materials, and therefore we do not consider them in this book. The interested reader should consult *Materials Research Society Bulletin*, volume 24 (No. 8) (1999) for a current review of this subject.

Aside from nanoscale structures, materials that are truly 1D are very rare. What are usually referred to as 1D materials are in fact *quasi*-1D systems. For example, one may have a magnetic chain if one can find a crystalline material in which the magnetic ions interact with each other along chains that are isolated from each other by, for example, large non-magnetic ions, or radicals, or molecular groupings. Well-known examples include copper benzoate [$Cu(C_6H_5COO)_2{\cdot}3H_2O$], copper dipyridine dichloride ($CuCl_2{\cdot}2NC_5H_5$), superionic conductors such as β-eucryptite ($LiAlSO_4$) in which fast ionic motion takes place mostly along 1D channels, organic materials such as tetrathiafulvalene-tetracyanoquinodimethne (TTF-TCNQ) which displays anisotropic, essentially 1D conductivity, and many polymeric chains. In copper benzoate, for example, the packing of large flat benzoate groups allows the magnetic Cu^{++} ions to come within exchange distance along the *c* axis (see Figure 3.1), but keeps these chains well apart in the *a* and *b* directions of the monoclinic lattice. This material is known to be a linear Heisenberg antiferromagnet, a subject of many studies.

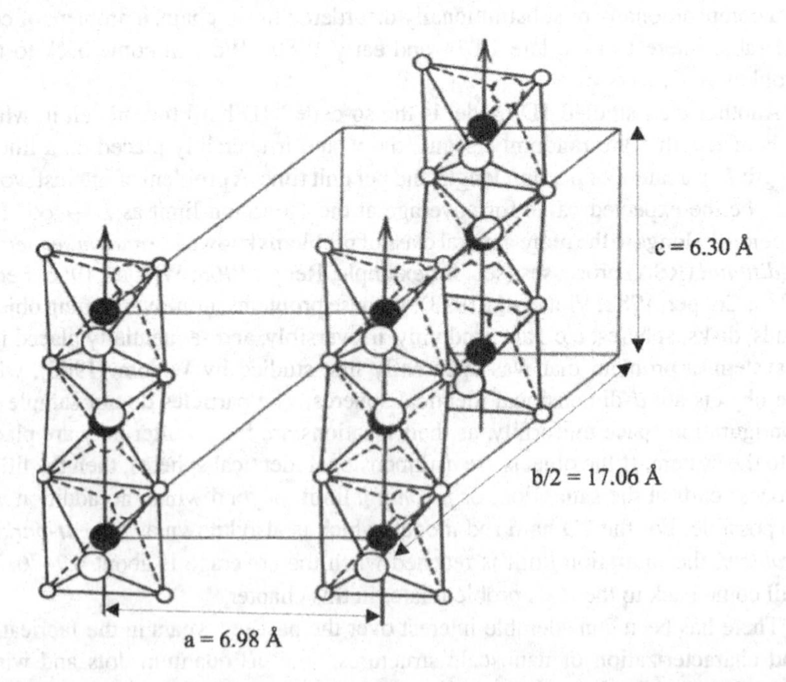

FIGURE 3.1. Model of copper benzoate that shows the magnetic chains. Black circles represent the Cu atoms, large circles the $O(C_6H_5COO^-)$ complex, while small circles show the $O(H_2O)$ molecules.

Steady-state diffusion and conduction, and mechanical properties of such linear chains are simple problems that can be solved exactly. However, unsteady-state diffusion and frequency-dependent AC conductivity of disordered linear chains are non-trivial problems that are described by the master equation (1). We will come back to these problems in Chapter 6.

3.3 Spatially Periodic Models

In this class of models the material's microstructure is represented by a periodic arrangement of microscopic elements in repetitive units called unit cells. The unit cell can have an ordered structure or contain a small degree of heterogeneity. An example is the so-called Wigner–Seitz cell shown in Figure 3.2 which exhibits translational invariance. A spatially-periodic model is also characterized by an associated lattice that contains the translational symmetries of the material for which the model is intended. Because of its periodic structure, the lattice is of infinite extent and is generated from any one lattice point by discrete displacements of the form,

$$\mathbf{R} = i_1\mathbf{e}_1 + i_2\mathbf{e}_2 + i_3\mathbf{e}_3, \tag{2}$$

where $\mathbf{I} = (i_1, i_2, i_3)$ is a triplet of integers, and $\{\mathbf{e}_1, \mathbf{e}_2, \mathbf{e}_3\}$ is a triad of basic lattice vectors. This triad is *not* unique because, by applying any unimodular 3×3 matrix with entries that are integer to the basis $\{\mathbf{e}_1, \mathbf{e}_2, \mathbf{e}_3\}$, one can obtain another equally valid basis. It is often convenient to use cells that are parallelepipeds built on a given choice of basic lattice vectors. Given the flexibility that one has with the choice of the unit cells, their shapes are often ambiguously defined.

Spatially-periodic systems are also characterized by two basic length scales. One is the microscopic length scale of the lattice l_m defined as, $l_m = \max[d_{min}(\mathbf{r})]$,

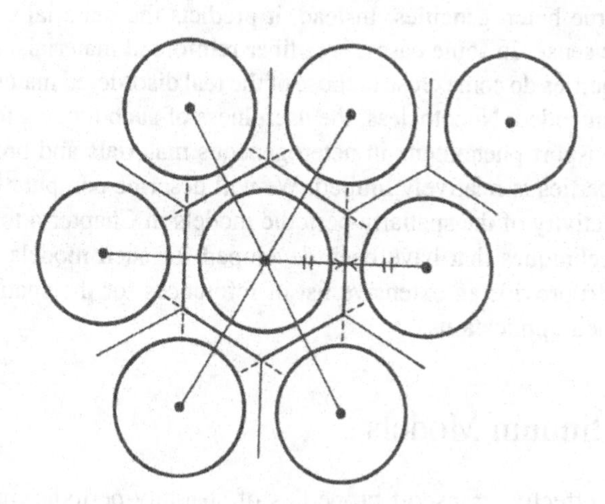

FIGURE 3.2. The Wigner–Seitz cell, which can be thought of as the *ordered* version of a Voronoi cell.

where $d_{min}(\mathbf{r})$ is the distance between a point at $\mathbf{r}$ and the nearest lattice points. For example, for a cubic lattice of size a, $l_m = \sqrt{3}a/2$. The second scale is the length scale L over which the averages of some physical fields, e.g., an electric or stress field, vary in a reasonably smooth manner. This length scale is typically of the order of the linear size of the material for which the model is intended. For the material to be macroscopically homogeneous, one must have $L \gg l_m$.

The simplest spatially-periodic model consists of a 2D array of infinitely-long circular cylinders which is in fact a reasonable model for fiber-reinforced materials, and as such was first introduced by Hashin and Rosen (1964). Each individual fiber is represented by a cylinder of radius R_c. Associated with each cylinder is an annulus of matrix material of radius R_s, and each of such combination is called a composite cylinder. Due to its anisotropy, the system is characterized by five elastic moduli, and it can be shown that the analysis of a composite cylinder will suffice for determining four of the five elastic moduli. This model is in fact the 2D analog of a 3D spatially-periodic model—a composite sphere model—in which the spheres sit on the sites of a regular periodic lattice and are surrounded by the matrix (the remaining of the lattice). The spheres can be rigid, and it is well-known that such rigid inclusions can degrade the strength of the matrix. However, the stiffening effect is usually not large, and the (spherical) inclusions improve other important parameters of the material, such as its cost, or the characteristics of its dynamic response. We will come back to these models in Chapters 4 and 7.

The analysis of transport processes in such models is a relatively simple task, when numerical or analytical calculations are confined to a unit cell. In principle, the unit cell can have an arbitrary shape, but if one is to analyze a disordered unit cell of arbitrary shape, the analysis would be no easier than that of other models of heterogeneous materials, described below, that are in fact more realistic than the spatially-periodic models. In a sense, a spatially-periodic model represents a sort of mean-field approximation to truly disordered materials because it does not contain any true heterogeneities. Instead, it predicts the material's properties in some average sense. In some cases, e.g., fiber reinforced materials, the predicted effective properties do come close to those of the real disordered material for which the model is intended. Nonetheless, the usefulness of such models for simulation of various transport phenomena in heterogeneous materials and predicting their effective properties is relatively limited. We will describe computation of the effective conductivity of the spatially-periodic models in Chapter 4 to demonstrate the elegant techniques that have been developed for such models. Nitsche and Brenner (1989) provide an extensive list of references for the spatially-periodic models and their applications.

3.4 Continuum Models

Although the effective transport properties of spatially-periodic models of heterogeneous materials have been extensively studied (see Chapter 4), transport processes in models in which some type of disorder is explicitly included have also been of great interest, since in most practical situations one must deal with materials that are, at least over certain length scales, heterogeneous. There are several such models which, as discussed above, can be classified into two major groups—the continuum and the discrete or lattice models.

Generally speaking, there are at least three different classes of continuum models. In the first class of models, one has a distribution of inclusions, such as circles, ellipses, cylinders, spheres, or ellipsoids, in an otherwise uniform matrix. The inclusions and the matrix represent the two phases of the material. By a phase we mean a portion of the material which has effective properties (conductivity, elastic constants, etc.) that are distinct from those of other portions of the material. In the second class of continuum models, one generates a model composite material by tessellating the space into regular or random polygons or polyhedra and designating at random (or in prescribed manner) some of the polygons or polyhedra as representing one phase of the composite, with the rest representing the second phase. Finally, the third class of models consists of those in which one distributes conducting (or elastic) sticks of a given aspect ratio, or plates of given extents, and studies transport in such systems. Each of these classes of models has been designed for representing a particular type of heterogeneous materials. For example, the models in the third class have been used for representing fibrous materials, with the sticks representing the fibers. In what follows we describe these models in detail, and discuss their properties.

However, before describing the continuum models let us point out that the main attractive feature of such models is that, with the appropriate choice of the parameters (to be described below), they can be used to represent many real heterogeneous materials. The main disadvantage of these models is the complexities that are involved if one is to utilize them for studying transport processes in real heterogeneous materials. Thus, although the effective transport properties of such continuum models can be and have been calculated using a variety of techniques (see Chapters 4 and 7), they are still too complex for routine use in the investigation of many important phenomena in heterogeneous materials

3.4.1 *Dispersion of Spheres*

Consider a statistical distribution of N identical d-dimensional spheres of radius R, which we call it phase 2, distributed in an otherwise uniform matrix that represents phase 1 of a two-phase disordered material. The total volume of the system is Ω. This model is not as simple or restricted as it may seem. For example, if the particles do overlap and form clusters, one can generate a wide variety of models with complex microstructures. Its 1D version, i.e., a system of rods mentioned earlier, can represent certain laminates. Its 2D version, i.e., a system of disks or, equivalently, a system of infinitely-long cylinders, distributed in the matrix, can be utilized for modeling fiber-reinforced materials and thin films. An example is shown in Figure 3.3, where the disks' sizes are distributed according to a uniform probability density function. The 3D version can be used for modeling unconsolidated media (for example, suspensions of particles), if the spheres do not overlap,

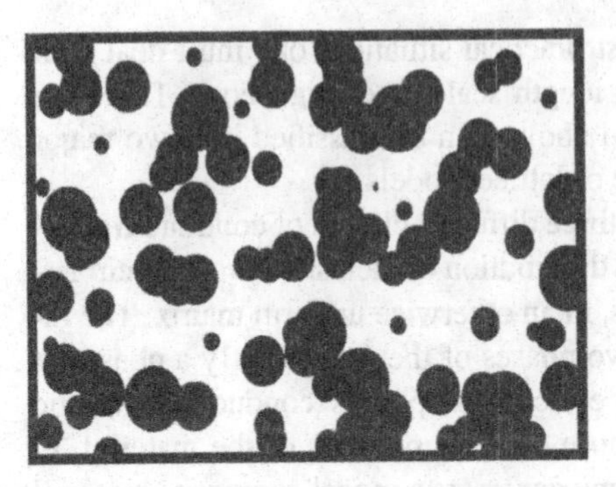

FIGURE 3.3. Two-dimensional model of overlapping disks (or infinitely-long cylinders). The radii of the disks are distributed uniformly.

as well as consolidated systems such as sintered materials and cements and a variety of other composite materials, if the spheres are allowed to overlap.

If the spheres are not allowed to overlap at all, then one obtains what is popularly referred to as *fully-impenetrable* or *hard-particle* model. This model has been used for studying liquids (see, for example, Reiss *et al.*, 1959; Hansen and McDonald, 1986), glasses (see, for example, Zallen, 1983), powders (Shahinpour, 1980), cell membranes (Cornell *et al.*, 1981), thin films (Quickenden and Tan, 1974), particulate composites, to be discussed in this book, colloidal dispersions (Russel *et al.*, 1989), granular materials, and many other types of material. However, the model becomes quite general if the spheres are allowed to overlap (Weissberg, 1963). The intersection of the spheres does not have to represent a true physical entity, but can be only a way of generating a heterogeneous material with a certain microstructure. An example is the *penetrable-concentric shell* model or the *cherry-pit* model (Torquato, 1984), in which each d-dimensional sphere of diameter $2R$ is composed of a hard impenetrable core of diameter $2\lambda R$, encompassed by a perfectly penetrable shell of thickness $(1-\lambda)R$, an example of which is shown in Figure 3.4. The limits $\lambda = 0$ and 1 correspond, respectively, to the cases of fully-penetrable sphere model, also called the Swiss-cheese model (see Section

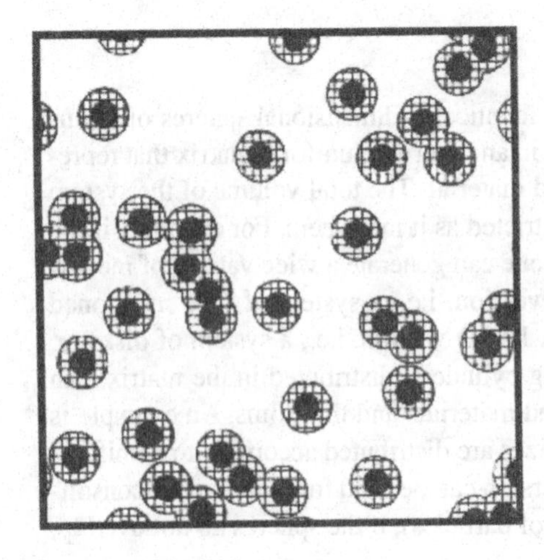

FIGURE 3.4. The cherry-pit, or the penetrable-concentric shell, model. The diameter of the hard particles (the black circles) is $2R$, while the diameter of their impenetrable hard core (black circles plus the core around them) is $2\lambda R$.

2.11), and the totally-impenetrable sphere model. Thus, varying λ between 0 and 1 allows one to tune the connectivity of the particles (i.e., the number of particles that are in contact with a given particle) and obtain a wide variety of models with desired morphology.

In the fully-penetrable sphere model (the limit $\lambda = 0$) there is no correlation between the particles' positions. If ϕ_2 is the volume fraction of the particle phase, then this phase becomes sample spanning (i.e., it percolates) for $\phi_{2c} \simeq 0.67$ and 0.29, for $d = 2$ and 3, respectively (see Section 2.10.1 for the precise values of ϕ_{2c}). Note that, as discussed in Chapter 2, no bicontinuous structure can exist in 2D, and therefore when the particle phase is sample-spanning, the matrix is discontinuous. On the other hand, for $d = 3$ the system can be bicontinuous. In particular, both the particle phase and the matrix are sample spanning for $0.3 \leq \phi_2 \leq 0.97$.

3.4.1.1 Equilibrium Hard-Sphere Model

In this model, the particles do not interact if the interparticle separation is larger than the sphere diameter, but there is an infinite repulsive force between them if the distance is less than or equal to the sphere diameter. An important property of this model is that the impenetrability constraint does not uniquely specify the statistical ensemble. That is, the system can be either in thermal equilibrium, or in one of the infinitely many non-equilibrium states, such as the random sequential adsorption mentioned earlier.

Although the equilibrium hard-sphere model may appear to be simple, it exhibits surprisingly rich behavior. For example, extensive experimental data and large-scale computer simulations indicate that the model in $d \geq 2$ undergoes a *disorder-order* phase transition, if the particle density is high enough (see, for example, Reiss and Hammerich, 1986). Moreover, although the existence of a first-order liquid-solid phase transition in 3D has not been proven rigorously, computer simulations which started with the work of Hoover and Ree (1968), as well as experiments on colloidal hard-sphere systems (Pusey and van Megan, 1986; Rutgers *et al.*, 1996), do provide strong evidence for the existence of this phase transition. The situation for the 2D systems, i.e., systems of hard disks, is not yet completely clear.

We should note a fundamental difference between equilibrium 2D hard-sphere model and its 3D counterpart. The difference is due to the fact that in 2D the densest *global* packing is consistent with densest local packing. The maximum number of d-dimensional spheres that can be packed in such a way that each sphere touches the others is $d + 1$ (which is why the coordination number of the 3D network that results from mapping the channels between the touching spheres onto a lattice is exactly 4). The d-dimensional polyhedron that results by taking the sphere centers as vertices is a *simplex*, resulting in line segments, equilateral triangles, and regular tetrahedra for $d = 1$, 2, and 3, respectively. Whereas identical simplices can fill 1D and 2D systems with overlapping, the same is not possible for 3D systems, i.e., one cannot fill them with identical non-overlapping regular tetrahedra. Due to this inconsistency between the local packing rules and the global packing constraints, the 3D systems are said to be *geometrically frustrated.*

3.4.1.2 Random Close Packing Versus Maximally Random Jamming

The notion of random close packing (RCP) has been used for decades. This state is usually defined as the maximum volume fraction ϕ_{cp} that a large, random collection of spheres can attain. It has been *assumed* that this state is well-defined and unique. However, in practice, there has been some difficulty testing this hypothesis. For example, in the now-classic experiments of Scott and Kilgour (1969), an estimate, $\phi_{cp} \simeq 0.637$, was obtained by pouring ball bearings into a large container, vibrating the system vertically for a long time to achieve maximum densification, and then, in order to eliminate the finite-size effects, extrapolating the measured volume fractions. However, the estimate of ϕ_{cp} obtained this way may depend on the pouring rate and the frequency and amplitude of the vibrations. Indeed, Pouliquen *et al.* (1997) showed that one can obtain denser, partially crystalline packings, if one pours the ball bearings at low rates in horizontally-shaken containers! Computer simulations of the same problem have not been of much help either, yielding estimates of ϕ_{cp} that vary anywhere from 0.6 (Visscher and Bolsterli, 1972) to 0.68 (Tobochnik and Chapin, 1988), presumably because different algorithms have been used in such simulations.

Torquato *et al.* (2000) have argued that these inconsistencies are due to RCP being an ill-defined state, pointing out that the words "random" and "close packed" are in fact inconsistent with each other, since close packed implies that the spheres are in contact with each other such that the average coordination number is maximum, but as the coordination number increases, the system's disorder decreases in some sense. Torquato *et al.* (2000) have instead suggested the notion of *maximally random jamming* (MRJ). A particle is considered jammed if it cannot be translated while fixing the positions of all other particles in the system. A system of particles is jammed if each of its particles is jammed, and there can be no collective motion of any contacting subset of its particles that leads to unjamming. The advantage of a MRJ process over a RCP one is that, the former can be defined in a mathematically rigorous manner.

Torquato *et al.* (2000; see also Truskett *et al.*, 2000), using Molecular Dynamics simulations (see Chapter 9 of Volume II for detailed discussion of this simulation method) and an algorithm due to Lubachevsky and Stillinger (1990), obtained an estimate $\phi_{rj} \simeq 0.64$ for the volume fraction at MRJ, which is in fact very close to the estimate of Scott and Kilgour (1969). There are still many unresolved issues concerning a maximally jammed state. Therefore, while throughout this book we may use the notion of RCP, we should keep in mind its ambiguity and the alternative packing or jamming states that have been proposed by Torquato *et al.* (2000).

3.4.1.3 Particle Distribution and Correlation Functions

A heterogeneous material and the model that represents it, including a dispersion of spheres, can be characterized by a number of fundamental mictrostructural properties, including several n-point correlation and distribution functions (Torquato, 2002). The most important of such distribution, correlation functions, and mictrostructural properties, are as follows.

(1) The most fundamental distribution function is perhaps $\rho_n(\mathbf{r}^n)$, where $\rho_n(\mathbf{r}^n)d\mathbf{r}_1 d\mathbf{r}_2 \cdots d\mathbf{r}_n$ is the probability of simultaneously finding a particle centered in volume element around $d\mathbf{r}_1$, another particle centered in volume element around $d\mathbf{r}_2$, etc, where $\mathbf{r}^n$ represents the set $\{\mathbf{r}_1 \cdots \mathbf{r}_n\}$. For statistically homogeneous media the probability density $\rho_n(\mathbf{r}^n)$ depends only upon the relative distances $\mathbf{r}_{12}, \cdots, \mathbf{r}_{1n}$, where $\mathbf{r}_{1i} = \mathbf{r}_i - \mathbf{r}_1$. Thus, for example, $\rho_2(\mathbf{r}_1, \mathbf{r}_2) = \rho_2(r_{12})$. Note also that $\rho_1(\mathbf{r}_1) = \rho$, where ρ is the density of the particles

(2) An important probability distribution function is $S_n(\mathbf{x}^n)$, which is defined as the probability of simultaneously finding n points at positions $\mathbf{x}_1, \mathbf{x}_2, \cdots, \mathbf{x}_n$ in one of the phases. Similarly, $S_n^{(i)}(\mathbf{x}^n)$ is the probability that the n points at $\mathbf{x}_1$, $\mathbf{x}_2, \cdots, \mathbf{x}_n$ are in phase i of a multiphase material. If the material is statistically homogeneous, then $S_1^{(i)}$ is simply the volume fraction ϕ_i of phase i. Moreover, $S_2(\mathbf{r}_1, \mathbf{r}_2) = S_2(r_{12})$. If the material is statistically heterogeneous, then the correlation functions S_n can be systematically represented in terms of the n-particle probability density function $\rho_n(\mathbf{r}^n)$ described above (see below). Such n-point probability functions were originally introduced by Brown (1955) in the context of determining the effective transport properties of composite materials. Chapters 4 and 7, as well as Chapters 2 and 4 of Volume II, will demonstrate the significance of such n-point probability functions.

If a material is statistically heterogeneous, then the probability function $S_n^{(i)}$ depends on the absolute positions $\mathbf{x}_1, \mathbf{x}_2, \cdots, \mathbf{x}_n$. In this case, $S_1^{(i)}$ varies in space and denotes the *local* volume fraction of phase i. For statistically homogeneous (or strictly stationary) materials, on the other hand, the joint probability distributions are translationally invariant, so that, $S_n^{(i)}(\mathbf{x}^n) = S_n^{(i)}(\mathbf{x}^n + \mathbf{y})$. If a two-phase material possesses phase-inversion symmetry, then $S_n^{(1)}(\mathbf{x}^n; \phi_1, \phi_2) = S_n^{(2)}(\mathbf{x}^n; \phi_2, \phi_1)$.

(3) One may also define the point/q-particle distribution functions $G_n^{(i)}(\mathbf{x}_1; \mathbf{r}^q)$ which characterize the probability of finding a point at $\mathbf{x}_1$ in phase i and a configuration of q spheres with their centers at $\mathbf{r}^n$, respectively, with $n = q + 1$. More general functions, such as $G^{(s)}(\mathbf{x}_s; \mathbf{r}^n)$ with obvious meanings, can also be defined.

(4) An important aspect of the microstructure of disordered two-phase materials is the properties of the interface between the two phases, which can be characterized by surface correlation functions. For example, one can define surface-surface, surface-matrix, and surface-void correlation functions $F_{ss}(\mathbf{x}_1, \mathbf{x}_2)$, $F_{sm}(\mathbf{x}_1, \mathbf{x}_2)$, and $F_{sv}(\mathbf{x}_1, \mathbf{x}_2)$, associated with finding a point on the interface between the two phases and another point in the matrix phase, or on the interface, or in the pore space (for porous materials), respectively. In similar fashions, higher-order correlation functions, such as F_{ssm} and F_{smm} can be defined. Other valuable information may be gained from a surface-particle correlation function F_{sp} which yields the correlations associated with finding a point on the particle-matrix interface and the center of a sphere in some volume element.

The functions F_{ss} and F_{sv} can be obtained from any plane cut through an isotropic material. For homogeneous and isotropic materials, these functions depend only on $|r| = |\mathbf{x}_2 - \mathbf{x}_1|$. If the two points are widely separated in materials without long-range order, then $F_{ss}(\mathbf{x}_1, \mathbf{x}_2) \to s(\mathbf{x}_1)s(\mathbf{x}_2)$, and $F_{sv}(\mathbf{x}_1, \mathbf{x}_2) \to s(\mathbf{x}_1)S_1(\mathbf{x}_2)$, where $s(\mathbf{x}_1)$ is the specific surface area (surface area per unit volume) at $\mathbf{x}_1$. Thus, for homogeneous materials, as $|\mathbf{r}| \to \infty$, one obtains, $F_{sv}(\mathbf{r}) \to s\phi_1$ and $F_{ss}(\mathbf{r}) \to s^2$.

(5) Another important property of heterogeneous materials are the *exclusion* probabilities $E_V(r)$ and $E_P(r)$. The former is the probability of finding a region Ω_V—taken to be a d-dimensional spherical cavity of radius r, centered at some arbitrary point—to be empty of inclusion centers. Physically, $E_V(r)$ is the expected fraction of space available to a test sphere of radius $r - R$, inserted into the system. $E_P(r)$, on the other hand, is the probability of finding a region Ω_P—taken to be a d-dimensional spherical cavity of radius r, centered at some arbitrary particle center—to be empty of other particle centers.

(6) One can also define *void nearest-neighbor* probability density functions $H_V(r)$ and $H_P(r)$. $H_V(r)dr$ is the probability that at an arbitrary point in the system the center of the nearest particle lies at a distance in the interval $(r, r + dr)$, while $H_P(r)dr$ is the probability that at an arbitrary particle center in the system the center of the nearest particle lies at a distance in the interval $(r, r + dr)$.

From the definitions of the void nearest-neighbor and exclusion probabilities, it should be clear that

$$E_i(r) = 1 - \int_0^r H_i(x)dx, \quad i = V, P \tag{3}$$

or, equivalently, $H_i(r) = -\partial E_i/\partial r$. One can also define a mean nearest-neighbor distance $\langle \ell_P \rangle$ by

$$\langle \ell_P \rangle = \int_0^\infty rH(r)dr = 2R + \int_0^\infty E_P(r)dr, \tag{4}$$

which is an important characteristic of packing of spheres. Utilizing $\langle \ell_P \rangle$, one can precisely define the random-close packing fraction ϕ_{cp} in the fully-impenetrable spheres model. Specifically, ϕ_{cp} can be defined as the maximum of packing fraction ϕ_2 over all the homogeneous and isotropic ensembles of the packings at which $\langle \ell_P \rangle = 2R$, where R is the spheres' radius. If the spheres are polydispersed with a mean radius $\langle R \rangle$, then one may define ϕ_{cp} in the same manner by $\langle \ell_P \rangle = 2\langle R \rangle$. Practically, ϕ_{cp} is the packing fraction at which randomly-arranged hard (impenetrable) spheres cannot further be compressed if a hydrostatic pressure is applied to the system. At ϕ_c the exclusion and the void nearest-neighbor probabilities are both zero for $r > 2R$. The numerical estimates for ϕ_{cp}, as mentioned earlier, are $\phi_{cp} \simeq 0.82$ and 0.64 for $d = 2$ and 3, respectively. Note that ϕ_{cp} is different from the *closest packing fraction* which are $\pi/(2\sqrt{3}) \simeq 0.907$ and $\pi/(3\sqrt{2}) \simeq 0.74$ for $d = 2$ and 3, respectively. It is also different from the *freezing packing fraction* ϕ_f for which the numerical estimates are $\phi_f \simeq 0.69$ and 0.49 for $d = 2$ and 3, respectively.

(7) A particularly important property is the pore-size probability density function $P(\delta)$, where $P(\delta)d\delta$ yields the probability that a point in phase 1 lies at a distance between δ and $\delta + d\delta$ from the nearest point on the pore-solid interface. However, we should keep in mind that for porous materials, $P(\delta)$ is *not* the standard pore-size distribution (Sahimi, 1993b, 1995b) which is measured by many experimental techniques. Two limiting values of P are $P(0) = s/\phi_1$ and $P(\infty) = 0$, where s/ϕ_1 is the interfacial area per unit volume.

The pore-size probability density function $P(\delta)$ is actually related to the void nearest-neighbor probability density function H_V defined above. As an example, we consider a system of interacting identical spheres of radius R. Using the definition of $H_V(r)$, it is clear that $\delta = r - R$, and therefore, $P(\delta) = \phi_1^{-1} H_V(\delta + R)$, for $\delta \geq 0$.

(8) Two other important characteristic properties of multiphase heterogeneous materials are the *lineal-path function* $L_p^{(i)}(z)$ (Lu and Torquato, 1992a) and the *chord-length distribution function* $L_c^{(i)}(z)$ (Torquato and Lu, 1993). $L_p^{(i)}(z)$ is the probability that a line segment of length z is entirely in phase i, when randomly thrown into the sample. For 3D systems, $L_p^{(i)}(z)$ is equivalent to the area fraction of phase i measured from the projected image onto a plane, a highly important problem in stereology (Underwood, 1970). Moreover, $L_p^{(i)}(0) = \phi_i$ and $L_p^{(i)}(\infty) = 0$. $L_p^{(i)}(z)$ contains information on the microscopic connectivity of the material, and thus it is a useful quantity. In stochastic geometry, the quantity $\phi_i[1 - L_p^{(i)}(z)]$ is sometimes referred to as the *linear contact distribution function* (Stoyan *et al.*, 1995).

$L_c^{(i)}(z)dz$, on the other hand, is the probability of finding a chord of length between z and $z + dz$ in phase i [thus, $L_c^{(i)}(z)$ is actually a probability density]. Chords are distributions of lengths between intersections of lines with the interface between the phases. An example is shown in Figure 3.5. Chord-length distributions are relevant to transport processes in heterogeneous media, such as diffusion, radiative heat transfer (Ho and Strieder, 1979; Tassopoulos and Rosner, 1992), and flow and conduction in porous materials (Krohn and Thompson, 1986). The quantities $L_p^{(i)}(z)$ and $L_c^{(i)}(z)$ are in fact closely related; their exact mathematical relation will be derived below.

(9) A very general distribution function for characterizing the microstructure of a heterogeneous material is $G_n(\mathbf{x}^p; \mathbf{r}^q)$, which is the n-point distribution function that characterizes the correlations associated with finding p particles centered at positions $\mathbf{x}^p = \{\mathbf{x}_1, \cdots, \mathbf{x}_p\}$ and q particles centered at positions $\mathbf{r}^q = \{\mathbf{r}_1, \cdots, \mathbf{r}_q\}$, with $n = p + q$. Clearly, for $q = 0$ we have $G_n(\mathbf{x}^n; \emptyset) = S_n(\mathbf{x}_n)$, where $\emptyset$ denotes the null set, and in the limit $p = 0$, $G_n(\emptyset; \mathbf{r}^q) = \rho_n(\mathbf{r}^n)$. Moreover, if $p = 1$ and the radius of the p-particles is taken to be zero, then, $G_n(\mathbf{x}_1; \mathbf{r}^q) = G^{(1)}(\mathbf{x}_1; \mathbf{r}^q)$, already defined above.

Note that $G_n(\mathbf{x}^p; \mathbf{r}^q)$ is *not* a pure probability function, but, rather, is a hybrid quantity. It is a probability function with respect to the positions $\mathbf{x}^p$ and, if properly scaled, is a probability density function with respect to the positions $\mathbf{r}^q$.

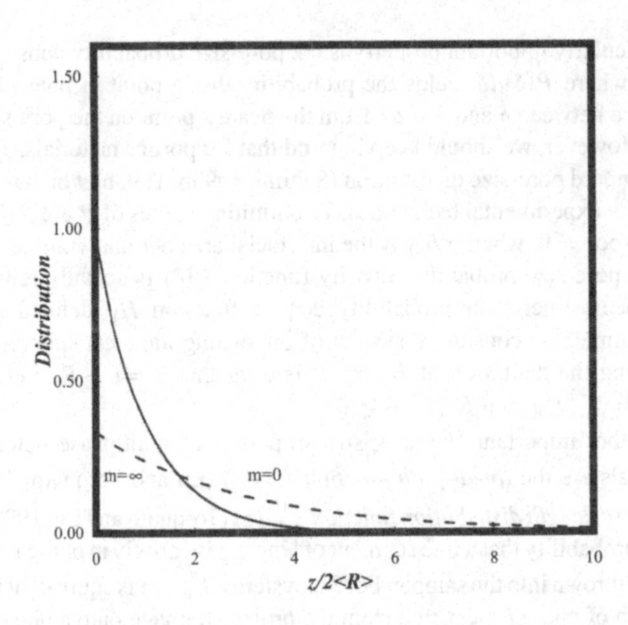

FIGURE 3.5. Chord-length distribution function L_C vs dimensionless distance $z/2\langle R\rangle$ for a 3D fully-impenetrable polydispersed system. The particles' sizes follow a Schulz distribution, $f(R) \sim R^m \exp[-(m+1)R/\langle R\rangle]$ $(m > -1)$. The volume fraction of the spheres is $\phi_2 = 0.4$ (after Torquato and Lu, 1993).

(10) One can define a most general n-point distribution function $H_n(\mathbf{x}^m; \mathbf{x}^{p-m}; \mathbf{r}^q)$ (Torquato, 1986b), defined to be the correlation function associated with finding m points with positions $\mathbf{x}^m$ on certain surfaces within the material, $p-m$ with positions $\mathbf{x}^{p-m}$ in certain spaces *exterior* to the spheres, and q sphere centers with positions $\mathbf{r}^q$ with $n = p + q$, in a statistically *heterogeneous* material of N identical d-dimensional spheres. As discussed below, most of the n-point correlation and distribution functions that were discussed above represent some limiting cases of $H_n(\mathbf{x}^m; \mathbf{x}^{p-m}; \mathbf{r}^q)$.

By the way of its construction, the function H_n, unlike the less general function G_n, contains interfacial information about the available surfaces associated with the first m test particles. However, this part of H_n does not have a probabilistic interpretation, but it can be interpreted probabilistically by considering infinitesimal dilations of the exclusion spheres. That is, $H_n(\mathbf{x}^n; \emptyset; \emptyset)$ is *proportional to* the probability that the points $\mathbf{x}^n$ lie in the dilated regions surrounding the interfaces $A_1, A_2, \cdots, A_m$, respectively. Finally, note that in the cherry-pit model, $H_n(\mathbf{x}^m; \mathbf{x}^{p-m}; \mathbf{r}^q)$ is identically zero for certain $\mathbf{r}^q$. In particular, for any value of the impenetrability λ (see above), we have, $H_n(\mathbf{x}^m; \mathbf{x}^{p-m}; \mathbf{r}^q) = 0$, if $|\mathbf{r}_i - \mathbf{r}_j| < 2\lambda R$, where R is the radius of the spheres.

These functions are not all independent and are in fact related to one another. To obtain their interrelationships, suppose that the d-dimensional spheres are spatially distributed according to a specific N-particle probability density $P_N(\mathbf{r}^N)$. Then, the ensemble average of any many-body function $F(\mathbf{r}^N)$ is defined by

$$\langle F(\mathbf{r}^N)\rangle = \int F(\mathbf{r}^N) P_N(\mathbf{r}^N) d\mathbf{r}^N. \tag{5}$$

$\rho_n(\mathbf{r}^n)$, which is sometimes called the n-particle *generic* probability density, is related to $P_N(\mathbf{r}^N)$ by

$$\rho_n(\mathbf{r}^n) = \frac{N!}{(N-n)!} \int P_N(\mathbf{r}^N) d\mathbf{r}_{n+1} \cdots d\mathbf{r}_N. \tag{6}$$

Torquato (1986b) developed an exact series representation of H_n which is given by

$$H_n(\mathbf{x}^m; \mathbf{x}^{p-m}; \mathbf{r}^q) = (-1)^m \frac{\partial}{\partial a_1} \cdots \frac{\partial}{\partial a_m} G_n(\mathbf{x}^p; \mathbf{r}^q), \tag{7}$$

with

$$G_n(\mathbf{x}^p; \mathbf{r}^q) = \sum_{s=0}^{\infty} (-1)^s G_n^{(s)}(\mathbf{x}^p; \mathbf{r}^q), \tag{8}$$

$$G_n^{(s)}(\mathbf{x}^p; \mathbf{r}^q) = \frac{1}{s!} \prod_{i=1}^{q} \prod_{k=1}^{p} e(y_{ki}; a_k) \int \rho_{q+s}(\mathbf{r}^{q+s}) \prod_{j=q+1}^{q+s} m^{(p)}(\mathbf{x}^p; \mathbf{r}_j) d\mathbf{r}_j, \tag{9}$$

$$m^{(p)}(\mathbf{x}^p; \mathbf{r}_j) = 1 - \prod_{i=1}^{p} [1 - m(y_{ij}; a_i)]. \tag{10}$$

Here $m(y_{ij}; a)$ is an *exclusion indicator function* defined by

$$m(y_{ij}; a) = \begin{cases} 1, & y_{ij} < a \\ 0, & \text{otherwise} \end{cases} \tag{11}$$

with

$$e(y_{ij}; a) = 1 - m(y_{ij}; a), \quad y_{ij} = |\mathbf{x}_i - \mathbf{r}_j| \tag{12}$$

Note that, in the limit $q = 0$, the factor multiplying the integral in Eq. (9) is by definition equal to unity.

In order to derive Eq. (7), Torquato (1986b) developed the following key idea. One must consider adding p test particles of radii $b_1, \cdots\cdots, b_p$ to the system of N spherical inclusions of radius R, with $p \ll N$. Because the ith test particle is capable of excluding the centers of the actual inclusions from spheres of radius a_i (where, for $b_i > 0$, $a_i = R + b_i$, and, for $b_i = 0$, $a_i = R - c_i$, $0 \le c_i \le R$), then it is natural to associate with each test particle a subdivision of space into two regions: Ω_i, the space available to the ith test particle, i.e., the space outside N spheres of radius a_i centered at $\mathbf{r}_i$, and the complement space Ω_i^*. Suppose that $\mathcal{S}_i$ is the surface between Ω_i and Ω_i^*. Then, $H_n(\mathbf{x}^m; \mathbf{x}^{p-m}; \mathbf{r}^q)$ yields the correlations associated with finding the center of a test particle of radius b_1 at $\mathbf{x}_1$ on $\mathcal{S}_1, \cdots\cdots$, the center of a test particle of radius b_m at $\mathbf{x}_m$ on $\mathcal{S}_m$, the center of a test particle of radius b_{m+1} at $\mathbf{x}_{m+1}, \cdots$, the center of a test particle of radius b_p at $\mathbf{x}_p$ in Ω_p, and of finding any q inclusions with configuration $\mathbf{r}^q$, where $\mathbf{x}^{p-m} \equiv \{\mathbf{x}_{m+1}, \cdots, \mathbf{x}_p\}$, with $n = p + q$. Torquato (1986b) has given upper and lower bounds for H_n for various limits of the dispersion of spheres model.

Given the n-point distribution function H_n, the other correlation and distribution functions defined above can be computed. Hence

$$S_n(\mathbf{x}^n) = \lim_{a_i \to R} H_n(\emptyset; \mathbf{x}^n; \emptyset), \quad \forall\, i, \tag{13}$$

$$G_n(\mathbf{x}_1; \mathbf{r}^q) = \lim_{a_1 \to R} H_n(\emptyset; \mathbf{x}_1; \mathbf{r}^q), \tag{14}$$

$$F_{ss}(\mathbf{x}_1; \mathbf{x}_2) = \lim_{a_i \to R} H_2(\mathbf{x}_1, \mathbf{x}_2; \emptyset, \emptyset), \quad \forall\, i. \tag{15}$$

$$H_V(r) = H_1(\mathbf{x}_1; \emptyset, \emptyset), \quad E_V(r) = H_1(\emptyset; \mathbf{x}_1; \emptyset). \tag{16}$$

One can also show that $E_P(r)$ is the limiting value of $H_2(\emptyset; \mathbf{x}_1; \mathbf{r}_1)/\rho_1(\mathbf{r}_1)$, as $|\mathbf{x}_1 - \mathbf{r}_1| \to \mathbf{0}$, from which $H_P(r) = -\partial E_P/\partial r$ is obtained easily.

To derive a general equation for the lineal-path function $L_p^{(i)}(z)$ for a multiphase material, we focus on the corresponding quantity $L_p(z)$ for a two-phase heterogeneous material; the generalization of the result to a material with more than two phases will then be obvious. The key idea is that $L_p(z)$, which is a sort of exclusion probability function, yields the probability of inserting a test particle—a line of length z—into the system, which is equal to the probability of finding an exclusion zone $\Omega_E(z)$ between a line of length z and a sphere of radius R. The region Ω_E is a d-dimensional spherocylinder of cylindrical length z and radius R with hemispherical caps of radius R on either end; see Figure 3.6. Therefore, if we define the exclusion indicator function $m(\mathbf{y}; z)$ by

$$m(\mathbf{y}; z) = \begin{cases} 1, & \mathbf{y} \in \Omega_E(z) \\ 0, & \text{otherwise} \end{cases} \tag{17}$$

then, $L_p(z)$ is given by (Lu and Torquato, 1992a)

$$L_p(z) = 1 + \sum_{k=1}^{\infty} \frac{(-1)^k}{k!} \int \rho_k(\mathbf{r}^k) \prod_{j=1}^{k} m_j(\mathbf{x} - \mathbf{r}_j; z) d\mathbf{r}_j. \tag{18}$$

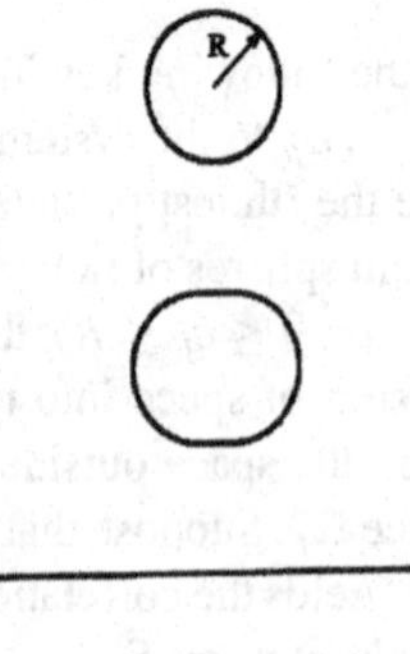

FIGURE 3.6. The exclusion zone Ω_E, showing the expansion of a spherical cavity of radius R (top) into an intermediate spherocylinder (middle), and the final spherocylinder of length z and hemispherical caps of the same radius R (after Lu and Torquato, 1992a).

Lu and Torquato (1993a,b) showed that, for statistically isotropic two-phase materials of arbitrary microgeometry, the chord-length distribution function is related to the lineal-path function $L_p(z)$ by

$$L_c(z) = \frac{\ell_C}{\phi_1}\frac{d^2 L_p(z)}{dz^2}, \tag{19}$$

where ℓ_C is the average of $L_c(z)$ and is given by

$$\ell_C = \int_0^\infty z L_c(z) dz. \tag{20}$$

Lu and Torquato (1991) generalized the above formulation, Eqs. (7)–(18), to a polydispersed composite material in which the spheres' radii follow a probability density function. The most important results obtained from such a formulation will be described below.

3.4.1.4 The n-Particle Probability Density

The above formulation of the various mictrostructural properties, the correlation and distribution functions, and the relationships between them imply that, given the n-particle probability density ρ_n, one can compute the correlation function H_n, from which most other properties follow. The advantage of this formulation is that, the function ρ_n has been studied in great details in the context of the statistical mechanics of liquids (see, for example, Hansen and McDonald, 1986), and therefore the extensive results obtained in those studies can be immediately employed for modeling composite materials. For example, in the fully-penetrable sphere model (i.e., the limit $\lambda = 0$ of the impenetrability parameter λ) with a particle density (number of particles per unit volume) ρ, there is no spatial correlation between the particles. Therefore, one has the *exact* relation

$$\rho_n(\mathbf{r}^n) = \rho^n, \quad \forall\, n. \tag{21}$$

In this case, it is easy to see that

$$S_n(\mathbf{r}^n) = \exp(-\rho V_n), \tag{22}$$

where V_n is the union volume of the n spheres (see below).

Determining $\rho_n(\mathbf{r}^n)$ for other types of the dispersion of spheres is considerably more difficult. For example, for fully-impenetrable spheres the impenetrability condition cannot by itself uniquely determine the ensemble, and one must supply more information. One must, for example, state that the spheres are distributed in the most random fashion which, together with the impenetrability condition, can determine the state of the system. Among all the $\rho_n(\mathbf{r}^n)$, the two-particle probability density $\rho_2(\mathbf{r}_1, \mathbf{r}_2)$ is one of the most important properties. For isotropic equilibrium distribution of the particles, the Ornstein–Zernike equation (Ornstein and Zernike, 1914)

$$h(r_{12}) = C(r_{12}) + \rho \int C(|\mathbf{r}_{23} - \mathbf{r}_{12}|) h(r_{23}) d\mathbf{r}_3\,, \tag{23}$$

[see also Section 2.10.2 where, for percolation, we used $P_2(r)$ and $D_2(r)$ instead of $h(r)$ and $C(r)$, respectively] is an important working tool, where $h(r)$ is the total correlation function, given by

$$h(r) = \frac{1}{\rho}\rho_2(r) - 1, \tag{24}$$

with ρ being the particle density, $\mathbf{r}_{ij} \equiv \mathbf{r}_j - \mathbf{r}_i$, and $r_{ij} \equiv |\mathbf{r}_{ij}|$. Here, $C(r)$ is the usual direct correlation function, as it measures the direct effect of particle 1 on particle 2, whereas $h(r_{12})$ is a measure of the total effect, direct and indirect, of particle 1 on 2. In general, $C(r) \to 0$ as $r \to \infty$. Therefore, if $C(r)$ can somehow be computed or guessed, then Eq. (23) is solved numerically in order to determine $h(r)$ and hence $\rho_2(r)$. For hard spheres (i.e., the limit $\lambda = 1$ in which no overlap between the spherical particles is allowed) a widely-used and very accurate approximation to $C(r)$ is the Percus–Yevick approximation (Percus and Yevick, 1958):

$$C(r) = \begin{cases} c_0 + c_1(r/2R) + c_3(r/2R)^3, & r < 2R \\ 0 & r > 2R \end{cases} \tag{25}$$

where the three constants are given by

$$c_0 = -\frac{1+2\phi_2}{(1-\phi_2)^4}, \tag{26}$$

$$c_1 = 6\phi_2\frac{1+\phi_2/2}{(1-\phi_2)^4}, \quad c_3 = \frac{\phi_2 c_1}{2}, \tag{27}$$

with ϕ_2 being the volume fraction of the spherical particles.

Integral equations similar to Eq. (23) have also been derived for the higher order ρ_n (see, for example, Hansen and McDonald, 1986), but as n increases they become very difficult to solve. For low particle densities, the following approximation has proven to be very useful:

$$\rho_3(r_{12}, r_{13}, r_{23}) = \frac{1}{\rho^3}\rho_2(r_{12})\rho_2(r_{13})\rho_2(r_{23}). \tag{28}$$

For the fully-impenetrable spheres model, the following low-density expansions have been derived (Widom, 1966),

$$\rho_2(r_{12}) = \rho^2\Theta(r_{12} - 2R)[1 + \rho v_2^i(r_{12}; 2R)] + O(\rho^4), \tag{29}$$

$$\rho_3(r_{12}, r_{13}, r_{23}) = \rho^3\Theta(r_{12} - 2R)\Theta(r_{13} - 2R)\Theta(r_{23} - 2R) + O(\rho^4), \tag{30}$$

where

$$\Theta(r) = \begin{cases} 1, & r > 0 \\ 0, & r < 0 \end{cases} \tag{31}$$

Here, $v_2^i(r; R, R)$, which is the intersection volume of two identical d-dimensional spheres of radius R with their centers separated by a distance r, is given by

$$v_2^i(r; R, R) = (2R - r)\Theta(2R - r), \quad d = 1 \tag{32}$$

$$v_2^i(r; R, R) = 2R^2 \left[\cos^{-1}\left(\frac{r}{2R}\right) - \frac{r}{2R}\left(1 - \frac{r^2}{4R^2}\right)^{1/2}\right]\Theta(2R - r), \quad d = 2 \tag{33}$$

$$v_2^i(r; R, R) = \frac{4\pi R^3}{3}\left(1 - \frac{3r}{4R} + \frac{r^3}{16R^3}\right)\Theta(2R - r), \quad d = 3 \tag{34}$$

Determining the intersection volume of three or more spheres is non-trivial, especially if the spheres' radii are not the same. Powell (1964) carried out the computations for three spheres of equal volume, Helte (1994) presented the results for four spheres, while Roberts and Knackstedt (1996) gave the corresponding expression for three unequal spheres.

3.4.2 *Distribution of Equal-Size Particles*

Let us now describe and summarize some mictrostructural properties of the dispersion of spheres model. Its transport properties will be described in Chapters 4 and 7. We first consider the case in which all the spheres have the same radius. In the next section we consider a polydisperse system in which the spheres' radii are distributed according to a statistical distribution. Torquato (2002) reviewed many other properties of these models.

Consider first a system of identical particles of *arbitrary* shapes with number density ρ. We define a dimensionless density $\eta = \rho v$, where v is the volume of a particle which, for example, for a d-dimensional sphere of radius R is given by

$$v(R) = \frac{\pi^{d/2} R^d}{\Gamma(1 + d/2)}, \tag{35}$$

where $\Gamma(x)$ is the gamma function. For fully-impenetrable particles, the reduced density η is *exactly* the particle volume fraction ϕ_2, i.e.,

$$\eta = \phi_2 = 1 - \phi_1. \tag{36}$$

This equality is not obeyed if the particles can overlap. In particular, for the penetrable-concentric shell model one has

$$\eta(\lambda) \geq \phi_2(\lambda), \tag{37}$$

where λ is the impenetrability index defined above, so that the equality applies when $\lambda = 1$. Having defined these essential quantities, we can now summarize various mictrostructural properties for the dispersion of spheres model.

3.4.2.1 Fully Penetrable Spheres

For this model, which represents the limit $\lambda = 0$, the first two functions S_1 and S_2 are given by

$$S_n = \begin{cases} \phi_1 = 1 - \phi_2 = \exp(-\eta), & n = 1, \\ \exp\left[-\eta \dfrac{V_2(r; R)}{v}\right], & n = 2, \end{cases} \tag{38}$$

where V_2 is the *union* volume of two spheres, to be defined shortly. Since for this model the n-particle probability density function is given by Eq. (21), calculation of $H_n(\mathbf{x}^m; \mathbf{x}^{p-m}; \mathbf{r}^q)$ is straightforward. In fact, according to Eqs. (7) and (21) H_n is expressible in terms of purely geometrical properties of the system. In particular, if we let $m = q = 0$, then $H_n(\mathbf{x}^n)$, the probability of inserting n spheres of radii $a_1, \cdot\cdot, a_n$ [see Eq. (7)] into a system of N spheres of radius R at positions $\mathbf{x}_1, \cdots\cdots, \mathbf{x}_n$, i.e., inserting the particles into the available space of the composite, is given by

$$H_n(\mathbf{x}^n) = \exp\left[-\rho V_n(\mathbf{x}^n; a_1, \cdots, a_n)\right]. \tag{39}$$

Here, $V_n(\mathbf{x}^n; a_1, \cdots, a_n)$ is the union volume of n d-dimensional spheres of radii $a_1, \cdots, a_n$, centered at $\mathbf{x}_1, \cdots, \mathbf{x}_n$, respectively. In the limit that $a_i \to R$ for $\forall\, i$, one recovers Eq. (22). Note that the union volume $V_2(r_{12}; a_1, a_2)$ of two spheres is given by, $V_2(r_{12}; a_1, a_2) = v(a_1) + v(a_2) - v_2^i(r_{12}; a_1, a_2)$, and

$$\begin{aligned} V_3(r_{12}, r_{13}, r_{23}; a_1, a_2, a_3) &= v(a_1) + v(a_2) + v(a_3) - v_2^i(r_{12}; a_1, a_2) \\ &\quad - v_2^i(r_{13}; a_1, a_3) - v_2^i(r_{23}; a_2, a_3) + v_3^i(r_{12}, r_{13}, r_{23}; a_1, a_2, a_3), \end{aligned} \tag{40}$$

with $r_{ij} = |\mathbf{x}_i - \mathbf{x}_j|$, and v being the volume of one sphere given by Eq. (35).

Other properties of this model can also be computed. The specific surface area s of the model is given by

$$s = \lim_{a_1 \to R} H_1(\mathbf{x}_1; \emptyset, \emptyset) = \rho\phi_1 \frac{dv}{dR} = \frac{\eta\phi_1 d}{R}, \tag{41}$$

where v and ϕ_1 are given by Eqs. (35) and (38), respectively. Equation (41) must be compared with the corresponding results for the impenetrable sphere model: $s = \rho dv/dR$. The surface correlation functions defined above can also be computed. For example, the surface-surface correlation function F_{ss} [that is, the limit $m = 2$, $p = 2$, and $q = 0$ of $H_n(\mathbf{x}^n; \mathbf{x}^{p-m}; \mathbf{r}^q)$] is given by

$$F_{ss}(\mathbf{x}_1, \mathbf{x}_2) = -\lim_{a_i \to R} \frac{\partial}{\partial a_1}\frac{\partial}{\partial a_2} \exp[-\rho V_2(\mathbf{x}_1, \mathbf{x}_2; a_1, a_2)], \quad \forall\, i \tag{42}$$

and the surface-void correlation function F_{sv} [that is, the limit $m = 1$, $p = 2$, and $q = 0$ of $H_n(\mathbf{x}^n; \mathbf{x}^{p-m}; \mathbf{r}^q)$] is given by

$$F_{sv}(\mathbf{x}_1, \mathbf{x}_2) = -\lim_{a_1 \to R} \frac{\partial}{\partial a_1} \exp[-\rho V_2(\mathbf{x}_1, \mathbf{x}_2; a_1, R)], \tag{43}$$

both of which were first derived by Doi (1976). Thus, in general, we need the union volume V_2 of two spheres of radii a_1 and a_2. The non-trivial part of this computation is determining the intersection volume $v_2^i(r; a_1, a_2)$. For example, for $d = 3$ one has $v_2^i(r; a_1, a_2) = v(a_1)$ for $0 \le r \le a_2 - a_1$; $v_2^i(r; a_1, a_2) = 4\pi f(r; a_1, a_2)/3$ for $a_2 - a_1 \le r \le a_2 + a_1$, and $v_2^i(r; a_1, a_2) = 0$ for $r \ge a_2 + a_1$, where

$$f(r; a_1, a_2) = -\frac{3}{16r}(a_2^2 - a_1^2)^2 + \frac{1}{2}(a_2^3 + a_1^3) - \frac{3r}{8}(a_2^2 + a_1^2) + \frac{1}{16}r^3. \tag{44}$$

Since there are no correlations in this model, there is no difference between the void and particle nearest-neighbor functions, $H_P(r) = H_V(r) = H(r)$. We can easily see that,

$$H(r) = \rho \frac{dv(r)}{dr} \exp[-\rho v(r)], \tag{45}$$

where $v(r)$ is given by Eq. (35). Using Eqs. (3) and (45), we then obtain

$$E(r) = \exp\left[-\eta \frac{v_d(x)}{v_d(1)}\right], \tag{46}$$

where

$$v_d(x) = \frac{1}{2^d} \frac{\pi^{d/2} x^d}{\Gamma(1 + d/2)}. \tag{47}$$

We can also compute the lineal-path function $L_p(z) = L_p^{(1)}(z)$ for this model. Using Eqs. (18) and (21), it is not difficult to see that

$$L_p(z) = \exp[-\rho v_E(z)], \tag{48}$$

where $v_E(z)$ is the d-dimensional volume of the exclusion zone Ω_E defined by Eq. (17), which in d dimensions is given by

$$v_E(z) = \frac{\pi^{d/2} R^d}{\Gamma(1 + d/2)} + \frac{\pi^{(d-1)/2} R^{d-1}}{\Gamma[1 + (d-1)/2]} z. \tag{49}$$

Once $L_p(z)$ is known, the chord-length distribution $L_c(z)$ can be immediately computed using Eq. (19). Utilizing Eq. (38), we can rewrite the results for $L_p(z)$ in terms of the volume fraction ϕ_1 of the matrix with the results being,

$$L_p(x) = \begin{cases} \phi_1^{1+x}, & d = 1, \\ \phi_1^{1+(4/\pi)x}, & d = 2, \\ \phi_1^{1+(3/2)x}, & d = 3, \end{cases} \tag{50}$$

where $x = z/(2R)$. Equations (50) indicate that with increasing z the lineal-path function decreases with the matrix volume fraction, since it becomes increasingly more difficult to insert a line segment of length z in the matrix. It will not be difficult to show that, the pore-size distribution $P(\delta)$ is given by, $P(\delta) = \rho\phi_1^{-1} dv/dr \exp[-\rho V_(R + \delta)]$, where the derivative is evaluated at $r = R + \delta$.

3.4.2.2 Fully Impenetrable Spheres

Unlike the case of fully-penetrable spheres, the totally-impenetrable sphere model is much more difficult to analyze, because there are significant correlations in the system imposed by the impenetrability condition. Despite this difficulty, significant progress has been made.

Torquato (1986b) proved that for fully-impenetrable spheres, and more generally particles of any shape, many of the n-point correlation functions truncate exactly after n-body terms. For example,

$$S_2(r) = 1 - 2\phi_2 + \rho m \otimes m + \rho_2 \otimes m \otimes m, \tag{51}$$

$$G_2(r) = e(r; R)[\rho - \rho_2 \otimes m], \tag{52}$$

$$F_{ss}(r) = \rho\delta(r - R) \otimes \delta(r - R) + \rho_2 \otimes \delta(r - R) \otimes \delta(r - R), \tag{53}$$

$$F_{sm}(r) = s - \rho\delta(r - R) \otimes m - \rho_2 \otimes \delta(r - R) \otimes m, \tag{54}$$

where ϕ_2 is the sphere volume fraction, s is the specific surface area given by, $s = \rho dv/dR = d\eta/R$, $\rho_2(r)$ is the two-particle probability density, $m(r; R)$ is the exclusion indicator function defined by Eq. (11) with $a = R$, $e(r; R)$ is defined by Eq. (12), and $\otimes$ denotes a convolution integral such that, for example, $\rho_2 \otimes m \equiv \int \rho_2(r')m(r - r'; R)dr'$. For example, for an equilibrium system of hard rods of length D, an exact expression for $S_2(x)$ has been obtained (Torquato and Lado, 1985; Quintanilla and Torquato, 1996):

$$S_2(x) = (1 - \eta)\sum_{k=0}^{M} \frac{1}{k!}\left[\frac{(x - k)\eta}{1 - \eta}\right]^k \exp\left[-\frac{(x - k)\eta}{1 - \eta}\right], \tag{55}$$

where $x = r_{12}/D$ and $M \le x \le M + 1$.

Torquato and Stell (1985) evaluated the above convolution integrals for $d = 2$ and 3 using a variety of approximations to $\rho_2(r)$, such as the Percus–Yevick approximation described above, and Seaton and Glandt (1986) did the same using computer simulations. Figure 3.7 compares the results for $S_2(r)$ for fully-penetrable and completely-impenetrable sphere models for $d = 3$ and a sphere

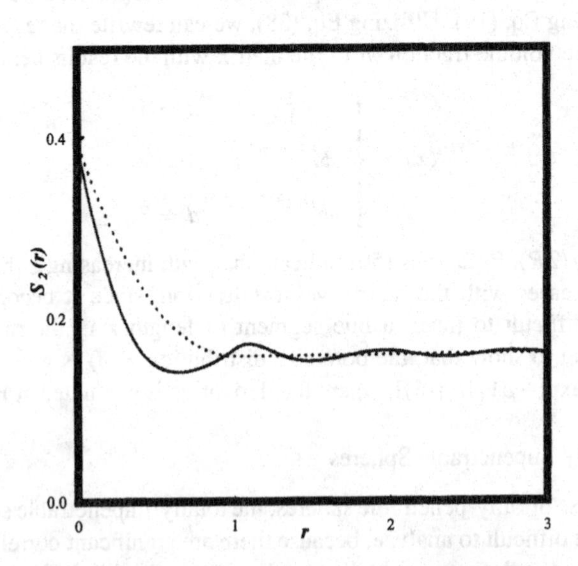

FIGURE 3.7. The two-point probability function $S_2(r)$ vs the distance r between the two points, at a sphere volume fraction of $\phi_2 = 0.6$. The solid and dashed curves represent, respectively, models of hard and overlapping spheres. The diameter of the spheres is 1 (after Torquato and Stell, 1985).

volume fraction $\phi_2 = 0.6$. The diameter of the spheres is unity. As can be seen, in the case of fully-penetrable spheres $S_2(r)$ decreases exponentially until it approaches its asymptotic value $S_2(r = 2R) = \phi_1^2$. However, in the case of fully-impenetrable spheres and for small r, $S_2(r)$ oscillates about its long-range value. The oscillations are indicative of short-range order which is caused by the excluded volume effect. Three-point and higher-order correlation functions for such models were also evaluated by Torquato (1986b).

Similarly, it is difficult to derive analytical, easy-to-use, expressions for the nearest-neighbor functions for fully-impenetrable sphere model. Note that one must have $E_P(r) = 1$ and $H_P(r) = 0$ for $0 \leq r < 2R$. It can also be shown that $E_V(r) = 1 - \rho v(r)$ and $H_V(r) = \rho s$ for $0 \leq r \leq R$, where v is the volume of a cavity of radius r. In general, the functions $H_P(r)$ and $H_V(r)$ are not truncated series, but are expressed as infinite series. Since for $d = 1$, i.e., for hard rods, the n-particle probability density $\rho_n(r)$ is known exactly for an equilibrium ensemble, $H_P(r)$ and $H_V(r)$ can be calculated exactly as well. In fact, Torquato *et al.* (1990) showed that for rods

$$H_i(x) = \begin{cases} \dfrac{\eta}{R} \exp\left[-\dfrac{\eta(2x-1)}{1-\eta}\right], & x \geq 1/2 \text{ and } i = V, \\[2ex] \dfrac{2\eta}{D(1-\eta)} \exp\left[-\dfrac{2\eta(x-1)}{1-\eta}\right], & x \geq 1 \text{ and } i = P, \end{cases} \tag{56}$$

where $\eta = \rho v$ is the dimensionless density (volume fraction of the particles) defined above, and $x = r/(2R)$. For $d = 2$ an accurate approximation to $H_P(x)$ is given by (Torquato *et al.*, 1990)

$$H_P(x) = \frac{4\eta(2r-\eta)}{(1-\eta)^2} \exp\left\{-\frac{4\eta}{(1-\eta)^2}[x^2 - 1 + \eta(x-1)]\right\}, \quad x > 1, \tag{57}$$

while for $d = 3$ it is given by

$$H_P(x) = 24\eta(f_1 + f_2 x + f_3 x^3) \exp\left\{-\eta[24 f_1(x-1) + 12 f_2(x^2-1) + 8 f_3(x^3-1)]\right\}, \quad x > 1, \tag{58}$$

with

$$f_1 = \frac{1}{2}\eta^2 f(\eta), \quad f_2 = -\frac{1}{2}\eta(3+\eta) f(\eta), \quad f_3 = (1+\eta) f(\eta), \tag{59}$$

where $f(\eta) = (1-\eta)^{-3}$. Other accurate expressions for $H_P(x)$ and $H_V(x)$ are given by Torquato (1995). Given expressions such as Eqs. (56)–(58) for $H_P(r)$ and $H_V(r)$, the corresponding equations for $E_P(r)$ and $E_V(r)$ can be immediately computed using Eq. (3). Accurate approximations for $L_p(z) = L_p^{(1)}(z)$ are given by (Lu and Torquato, 1992a)

$$L_p(z) = (1-\eta) \exp\left[-\frac{\eta z}{2R(1-\eta)}\right], \quad d = 1, \tag{60}$$

$$L_p(z) = (1-\eta) \exp\left[-\frac{2\eta z}{\pi R(1-\eta)}\right], \quad d = 2, \tag{61}$$

$$L_p(z) = (1-\eta)\exp\left[-\frac{3\eta z}{4R(1-\eta)}\right], \quad d=3, \tag{62}$$

from which the corresponding expressions for the chord-length distribution $L_c(z)$ can be obtained using Eq. (19). Note that Eq. (60) is exact.

Another important property of fully-impenetrable spheres, that can be measured indirectly by X-ray scattering, is the radial distribution function $g(r)$, where $g(r)d\mathbf{r}$ is proportional to the probability of finding a particle in the volume element $d\mathbf{r}$ at a distance r from a given particle [in 3D $g(r)r^2\Delta r$ is proportional to the mean number of particles in a shell of radius r and thickness Δr surrounding the particle]. Near the random close-packing fraction ϕ_{cp}, the radial distribution function $g(r)$ follows the following power law (Song *et al.*, 1988; Tobochnik and Chapin, 1988)

$$g(r) \sim (\phi_{cp} - \phi)^{-\zeta}, \quad r = 2R, \quad \phi \to \phi_{cp}, \tag{63}$$

where it appears that $\zeta = 1$ in both 2D and 3D.

Finally, it can be shown that, for any ergodic ensemble of isotropic packings of identical d-dimensional hard spheres, the mean nearest-neighbor distance $\langle \ell_P \rangle$, defined by Eq. (4), is bounded from above: $\langle \ell_P \rangle \leq 1 + (2^d \eta d)^{-1}$.

3.4.2.3 Interpenetrable Spheres

Lee and Torquato (1988) calculated by computer simulations the matrix volume fraction ϕ_1 in the penetrable concentric shell model as a function of the parameter λ. The following approximate, but very accurate, formulae for ϕ_1 were derived by Rikvold and Stell (1985) which are in excellent agreement with the numerical results of Lee and Torquato (1988):

$$\phi_1(\eta, \lambda) = (1 - \lambda^d \eta)\exp\left[-\frac{(1-\lambda^d)\eta}{(1-\lambda^d\eta)^d}\right]\Psi_d(\eta, \lambda), \tag{64}$$

where $\Psi_d(\eta, \lambda)$ is a d-dependent function such that $\Psi_1 = 1$, and

$$\Psi_2(\eta, \lambda) = \exp\left[-\frac{\lambda^2\eta^2(1-\lambda)^2}{(1-\lambda^2\eta)^2}\right], \tag{65}$$

$$\Psi_3(\eta, \lambda) = \exp\left[-\frac{3\lambda^3\eta^2}{2(1-\lambda^3\eta)^3}(2 - 3\lambda + \lambda^3 - 3\lambda^4\eta + 6\lambda^5\eta - 3\lambda^6\eta)\right]. \tag{66}$$

Note that $\eta\lambda^d$ represents the hard-core volume fraction in d dimensions.

3.4.3 *Distribution of Polydispersed Spheres*

A more realistic version of model of dispersion of spheres is one in which the radii of the spheres are distributed according to a normalized probability density function $f(R)$. Polydispersivity leads to more flexibility in the model and hence can be exploited for a variety of purposes. It also results in a wider choice of possible definitions for the various distribution functions discussed above. For example, in studying the nearest-neighbor distribution, one can specify the nearest sphere *surface* to a reference point, or the nearest sphere *center* to a reference point. These matters have been discussed by Lu and Torquato (1992a,b). What follows is a summary of the most important results for the microstructure of this class of models in which every averaged property is defined with respect to the distribution $f(R)$.

3.4.3.1 Fully Penetrable Spheres

For this class of models the volume fraction of the matrix can be obtained from a modification of Eq. (38) (Chiew and Glandt, 1984),

$$S_1 = \phi_1 = \exp[-\rho\langle v(R)\rangle], \tag{67}$$

and the specific surface s area is given by

$$s = \rho \frac{\partial \langle v(R)\rangle}{\partial R} \exp[-\rho\langle v(R)\rangle]. \tag{68}$$

Stell and Rikvold (1987) showed that

$$S_n(\mathbf{x}^n) = \exp[-\rho\langle V_n(\mathbf{x}^n; R, \cdots, R)\rangle], \tag{69}$$

where V_n is the union volume of n spheres of radius R defined and discussed above. Moreover,

$$G_2(r) = \rho\phi_1\langle\Theta(R-r)\rangle. \tag{70}$$

The surface correlation functions have also been computed for this class of models. In particular, for $d = 3$ one has (Torquato and Lu, 1990)

$$F_{ss}(r) = \left[16\pi^2\rho^2\langle I(R)^2\rangle + \frac{2\pi\rho}{r}\langle R^2\Theta(2R-r)\rangle\right] S_2(r), \tag{71}$$

$$F_{sv}(r) = 4\pi\rho\langle I(R)\rangle S_2(r), \tag{72}$$

where

$$I(R) = R^2 - \left(\frac{R^2}{2} - \frac{rR}{4}\right)\Theta(2R-r), \tag{73}$$

and $S_2(r)$ is obtained from Eq. (69).

The lineal-path function has also been computed for this class of models. In order to compute this function, the exclusion indicator function defined by Eq. (17) must be generalized for a polydispersed system. In this case one defines an exclusion *region* indicator function by

$$m_j(\mathbf{x}; z) = \begin{cases} 1, & \mathbf{x} \in \Omega_E(z, R_j) \\ 0, & \text{otherwise,} \end{cases} \tag{74}$$

where R_j is the radius of the jth sphere. The lineal-path function $L_p(z) = L_p^{(i)}(z)$ is then given by (Lu and Torquato, 1993a,b)

$$L_p(z) = 1 + \sum_{k=1}^{\infty} \frac{(-1)^k}{k!} \int dR_1 \cdots dR_k f(R_k) \rho_k(\mathbf{r}^k; R_1, \cdots, R_k) \prod_{j=1}^{k} m_j(\mathbf{x}; z) d\mathbf{r}_j, \tag{75}$$

which is a generalization of Eq. (18) to polydispersed systems. Then, using Eqs. (21) and (75), one obtains

$$L_p(z) = \exp[-\rho \langle v_E(z, R) \rangle], \tag{76}$$

where $v_E(z, R)$ is given by Eq. (49). Combining Eqs. (38), (49) and (76), and proceeding in the same manner as that for the equal-sized particles, we obtain the analogs of Eqs. (50) for polydispersed systems:

$$L_p(z) = \begin{cases} \phi_1^{1+2\langle R\rangle z/(\pi \langle R^2 \rangle)}, & d = 2, \\ \phi_1^{1+3\langle R^2\rangle z/(4\langle R^3 \rangle)}, & d = 3. \end{cases} \tag{77}$$

The corresponding equation for $d = 1$ is obvious. Using Eqs. (19) and (77), the corresponding results for the chord-length distribution $L_c(z)$ can be immediately computed.

3.4.3.2 Fully Impenetrable Spheres

These models have many intriguing properties, much of which is not well-understood yet. For example, Henderson *et al.* (1996) showed that even a relatively small degree of polydispersivity can completely suppress the order-disorder phase transition seen in monodisperse hard-sphere model, mentioned in Section 3.4.1.1. In addition, equilibrium mixtures of small and large hard spheres can *phase separate*, i.e., the small and large hard spheres *demix*, at sufficiently high densities. However, the nature of such phase transitions (first-order versus second-order) is not understood yet.

Very little is also known about the packing characteristics of polydisperse hard spheres. For example, even if we consider one of simplest of such polydisperse packings, namely, a binary mixture of hard spheres of arbitrary radii R_1 and R_2, its largest achievable packing fraction is not known. Despite such difficulties, some progress has been made which is now summarized.

For polydisperse impenetrable spheres, the following results are known:

$$S_1 = \phi_1 = 1 - \rho \langle v(R) \rangle, \tag{78}$$

and the specific surface area s is given by

$$s = \rho \frac{d\langle v(R) \rangle}{dR} = d\eta \frac{\langle R^{d-1} \rangle}{\langle R^d \rangle}. \tag{79}$$

Accurate approximations for the lineal-path function $L_p(z) = L_p^{(1)}(z)$ have also been derived for this class of models, with the result for a d-dimensional system being,

$$L_p(z) = \phi_1 \exp\left[-\frac{\Gamma(1+\frac{1}{2}d)\eta\langle R^{d-1}\rangle}{\pi^{d/2}\phi_1\langle R^d\rangle} \right]. \tag{80}$$

If we compare Eq. (80) with Eq. (62), then generalizations of Eqs. (60) and (61) to polydisperse systems become obvious. Given $L_p(z)$, the chord-length distribution $L_c(z)$ can be immediately computed using Eq. (19).

3.4.4 *Simulation of Dispersion of Spheres*

How does one carry out computer simulation of a system of spherical particles? If the spheres are allowed to overlap, then the simulation is rather straightforward. The problem is more difficult if one wishes to model the fully-impenetrable spheres model, or to take into account the effect of the hard cores that the spheres may have (represented by the impenetrability parameter λ).

The classical method of computer generation of a packing of spherical particles in the fully-impenetrable spheres model is that of Visscher and Bolsterli (1972). In their algorithm, a sphere is dropped into the simulation box from the top. If the particle hits the "floor," it stops. If it hits another particle, say p_1, it rolls down on p_1 until it hit another particle p_2. Then it rolls on both p_1 and p_2 until it hits particle p_3. If its contact with p_1, p_2, and p_3 is stable, then it stops. Otherwise, it rolls on the double contacts, and so on. Since the particle's motion is always downward, the effect of the gravity is automatically taken into account.

As a simple method that takes into account the effect of the hard core of spherical particles, consider, for example, a system of d-dimensional spheres. The number N of the particles and the volume Ω of the system are fixed. The particles are initially placed in a cubical cell (with volume $\Omega = L^d$) on the sites of a regular lattice, e.g., the face-centered or body-centered lattice. No hard core overlap is assumed initially. The particles are then moved randomly by a small distance to new positions. These new positions are either accepted or rejected according to whether or not the hard cores overlap. One usually uses *periodic boundary conditions* which means that if a particle exits from an external face of the system, an identical particle enters the system from the opposite face of the system. This sort of boundary condition eliminates the boundary effects, hence allowing one to essentially simulate an infinitely-large system. After the particles have been moved a sufficiently large number of times, the system reaches equilibrium, and its configuration no longer changes.

Using such algorithms, Smith and Torquato (1989), Lee and Torquato (1989) and Miller and Torquato (1990) calculated various mictrostructural and transport properties of the sphere model, and compared the results with the theoretical results discussed above.

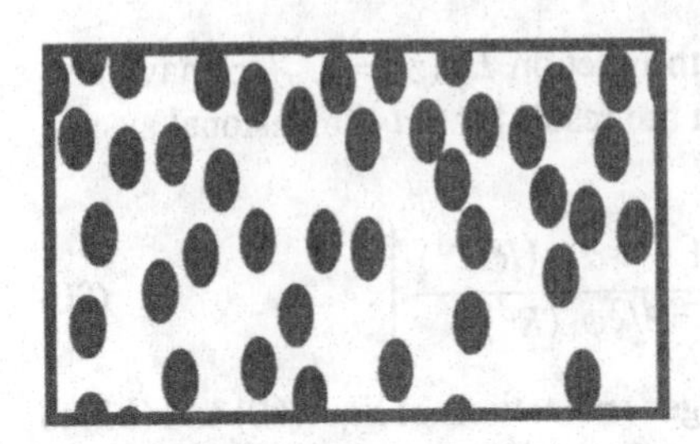

FIGURE 3.8. A 2D model of anisotropic materials that consists of elliptical particles, distributed in a uniform matrix.

3.4.5 *Models of Anisotropic Materials*

The models based on dispersion of spheres can be significantly generalized to anisotropic media, either particulate or fibrous, that are composed of particles of arbitrary shapes inserted in a matrix. For example, a two-phase anisotropic material may be composed of a background matrix and ellipsoidal inclusions inserted in the matrix, an example of which is shown in Figure 3.8. Alternatively, the material may be composed of a background matrix with long fibers, represented by sticks of fixed aspect ratios, inserted in the matrix. Many of the results for dispersion of spheres that were discussed above can be generalized to such models. The key idea, developed by Torquato and Sen (1990), is to generalize the inclusion indicator function $m(\mathbf{x})$, defined by Eq. (11), to non-spherical inclusions in which each configurational coordinate is fully specified by its center-of-mass position. Consider a point $\mathbf{x} = (x_1, \cdots, x_d)$, where x_i denotes the component of $\mathbf{x}$ in the principal axes coordinate system with $i = 1, \cdots, d$. Then, for a rectangle with sides of lengths $2a$ and $2b$ one has

$$m(\mathbf{x}) = \begin{cases} 1, & |x_1| \le a \text{ and } |x_2| \le b, \\ 0, & \text{otherwise.} \end{cases} \tag{81}$$

For a rectangular parallelepiped with sides $2a$, $2b$, and $2c$, we have

$$m(\mathbf{x}) = \begin{cases} 1, & |x_1| \le a \text{ and } |x_2| \le b \text{ and } |x_3| \le c, \\ 0, & \text{otherwise.} \end{cases} \tag{82}$$

For an ellipse with axes of lengths $2a$ and $2b$, one has

$$m(\mathbf{x}) = \begin{cases} 1, & x_1^2/a^2 + x_2^2/b^2 \le 1, \\ 0, & \text{otherwise.} \end{cases} \tag{83}$$

For an ellipsoidal inclusion with axes of lengths $2a$, $2b$ and $2c$, the inclusion indicator function is given by

$$m(\mathbf{x}) = \begin{cases} 1, & x_1^2/a^2 + x_2^2/b^2 + x_3^2/c^2 \le 1, \\ 0, & \text{otherwise.} \end{cases} \tag{84}$$

Finally, the inclusion indicator function for a circular cylinder of diameter $2R$ and length 2ℓ is given by

$$m(\mathbf{x}) = \begin{cases} 1, & x_1^2 + x_2^2 \le R^2 \text{ and } |x_3| \le \ell, \\ 0, & \text{otherwise.} \end{cases} \tag{85}$$

In these examples, the material's microstructure is similar to that of an idealized *nematic liquid crystal* in which there are long-range orientational order and short-range positional order. Clearly, one may also consider cylindrical inclusions with elliptical or rectangular cross sections.

The n-point matrix probability functions $S_n(\mathbf{x}^n)$ for fully-penetrable inclusions are still given by Eq. (22), except that V_n, the union volume of n oriented inclusions centered at $\mathbf{x}^n$, must be computed. For $n = 2$ or 3 this can be done analytically. For example, for rectangular inclusions and oriented parallelepipeds, one has

$$V_2(\mathbf{x}) = \begin{cases} 8ab - (2a - x)(2b - y)\Theta(2a - x)\Theta(2b - y), & d = 2, \\ 16abc - (2a - x)(2b - y)(2c - z)\Theta(2a - x)\Theta(2b - y)\Theta(2c - z), & d = 3, \end{cases} \quad (86)$$

where, for example, $x = |\mathbf{x}|$, and x, y and z are the distances between the centroids of the two rectangular or parallelepiped regions in the x_1, x_2 and x_3 directions, respectively. For spheroidal inclusions in which $a = c$, and their symmetry axis is aligned with the z-axis, we have $V_2(\mathbf{r}) = \frac{8}{3}\pi a^2 b - v_2^i(\mathbf{r})$, where

$$v_2^i = \frac{4}{3}\pi a^2 b \left\{1 - \frac{3}{4}\frac{r}{R(\theta)} + \frac{1}{16}\left[\frac{r}{R(\theta)}\right]^3\right\}\Theta[2R(\theta) - r], \quad (87)$$

where $R(\theta) = a[1 - (1 - a^2/b^2)\cos^2\theta]^{-1/2}$. For circular cylindrical inclusions we have

$$V_2(\mathbf{x}) = 4\pi a^2 b - (2b - |x\cos\theta|)I(|x\sin\theta|)\Theta(2a - |x\sin\theta|)\Theta(2b - |x\sin\theta|), \quad (88)$$

where θ is the polar angle that $\mathbf{x}$ makes with the x_3 axis, and

$$I(x) = 2a^2\left(\cos^{-1}\frac{x}{2a} - \frac{x}{2a}\sqrt{1 - \frac{x^2}{4a^2}}\right)\Theta(2a - r) \quad (89)$$

Clearly, one may still generalize the above methodology to the cases where the inclusions are, for example, long sticks or fibers, which are reasonable models of fibrous materials, such as printing papers.

3.4.6 Tessellation Models of Cellular Materials

There is a large class of natural or man-made materials that have cellular microstructure. Examples include corks, balsa, sponge, cancellous bone, wood, foods such as chocolate bar, foams, emulsions, a wide variety of biological materials, cracked mud, surfaces of shrunken gels, and many more. Materials can also be *fabricated* to have a cellular microstructure. Examples include metals, ceramics, glasses, and composites. Such materials have wide applications in thermal insulation, packaging, structural materials (for example, wood, cancellous bone, and coral), floating materials (for example, plastic foams used for flotation in boats), and many other applications. Moreover, foams are used (Weaire and Hutzler, 1999) as filters, carriers for inks, dyes, lubricants and enzymes, and as water-repellent

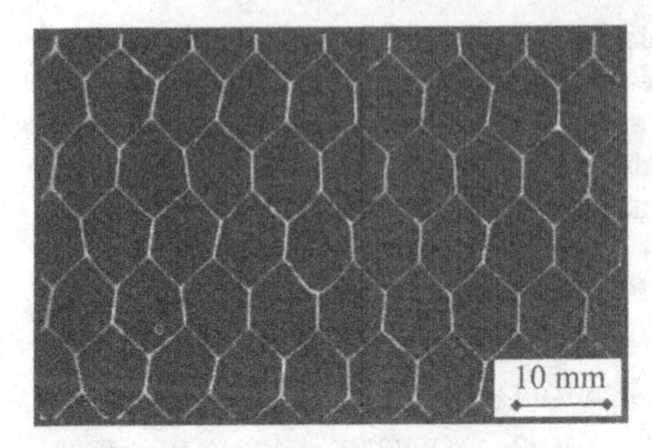

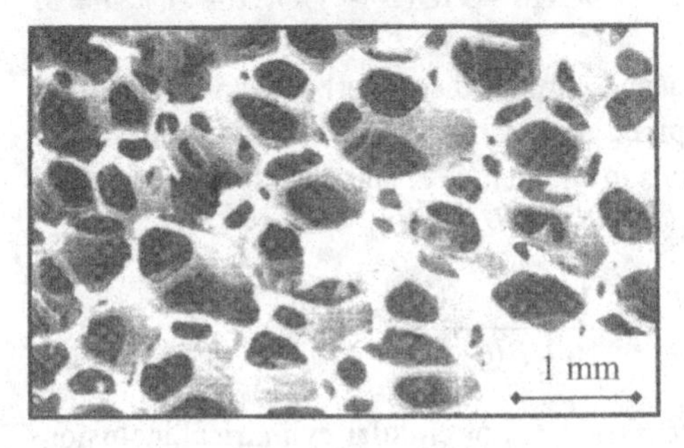

FIGURE 3.9. Examples of cellular solids: A 2D aluminum honeycomb (top), and a 3D polymeric foam with both open and closed cells (bottom) (after Gibson and Ashby, 1997).

membranes. Gibson and Ashby (1997) provide a detailed discussion of cellular solids and their properties. We will discuss in the subsequent chapters the transport properties of some of such materials.

A cellular solid is composed of an interconnected network of solid struts or plates which form the edges and faces of the cell. Figure 3.9 shows two well-known examples, a 2D honeycomb and a 3D foam. Honeycomb represents the simplest type of such materials as it is made of regular and equal-size polygons which pack to fill a plane. More generally, one may consider 3D structures in which polyhedra pack to fill the space, an example of which is the foam shown in Figure 3.9. An important characteristic of a cellular solid is its *relative density*, $\rho_r = \rho_c/\rho_s$, where ρ_c is the density of the cellular material, and ρ_s is the density of the material from which the cells are made. There are foams with ρ_r as low as 10^{-3}. Polymeric foams, which have wide applications in cushioning, packaging and insulation, have $0.05 < \rho_r < 0.2$.

The ubiquity of cellular materials has inspired development of a class of models for their microstructure which we call the *tessellation models*. Such models are constructed by partitioning or tessellating the space into cells—polygons or polyhedra—and designating, either randomly or according to a given correlation function, some of the cells as phase 1, some as phase 2, etc. The morphology of the model and the corresponding cellular material, and hence their overall properties, depend on the size, shape and connectivity of the cells. However, while the cells' sizes do have an effect on the overall properties of the model and material, their shapes are more important to their macroscopic features. Depending on the intended application, the cells can have a variety of shapes. In some cases they have regular shapes, while in other cases they may have very irregular structures. If the cells are isotropic, i.e., if they are equiaxed, the material's macroscopic properties are also isotropic. However, if the cells are even slightly anisotropic, e.g., if they are slightly elongated, then the material has direction-dependent properties and behaves anisotropically. Each cell has two connectivity characteristics. One is the connectivity Z_f of cell *faces* which is the number of faces that meet at an edge. Its value can be as high as six, although it is typically three. The second one is Z_e, the connectivity of the cell *edges*, which is the number of edges that meet at a node or vertex. For example, in a honeycomb made of hexagons the edge connectivity is $Z_e = 3$, while in foams it is usually 4, but can be higher. In more general multiphase materials the local connectivity of cells of the same type plays a crucial role in the materials' macroscopic properties. For example, if cells of phase 1 form sample-spanning pathways between two opposite faces of the material, then the material's properties are dominated by those of phase 1. The effect of the connectivity is quantified by percolation theory which was the subject of Chapter 2. Due to the flexibility that one is afforded in developing a tessellation models of cellular materials, such models can be tailored to closely mimic the materials' microstructure.

Compared with 2D cellular materials and their models, the 3D materials are much more complex. Although the Greek mathematician Plateau identified a rhombic dodecahedron, a twelve-faced, garnet shaped figure, as the basic building block of 3D cellular solids which can fill the space, this does not provide the most efficient way of doing so, in the sense of minimizing the surface area per unit volume. Up until recently, it was believed that the space-filling 3D cell that has this property is Kelvin's tetrakaidecahedron with slightly curved faces; see Figure 3.10. However, using computer simulations Weaire and Phelan (1994) identified a unit cell with a surface area per unit volume which is lower than that of Kelvin's cell by about 0.3%; this is also shown in Figure 3.10. The Weaire–Phelan unit cell is made of six 14-sided cells, with 12 pentagonal and 2 hexagonal faces, and two pentagonal dodecahedra, all of equal volume. While the hexagonal faces are planar, the pentagonal faces are curved. The 14-sided cells are arranged in three orthogonal axes with the 12-sided cells lying in the interstices between them. This generates an overall simple-cubic lattice structure.

It may be useful to think of the tessellation models as consisting of *vertices*, joined by *edges*, which surround *faces*, which enclose *cells* (in 2D faces and cells become the same). If V, E, F and C are, respectively, the number of vertices, edges, faces and cells, then according to *Euler's law*

$$F - E + V = 1, \quad 2D, \tag{90}$$

$$-C + F - E + V = 1, \quad 3D. \tag{91}$$

Euler's law has several interesting consequences. For example, a honeycomb with regular hexagonal cells has six edges surrounding each face. Euler's law dictates that an irregular three-connected honeycomb also has, on average, six sides per face. Hence, if, for example, a four-sided face is introduced into the system, then

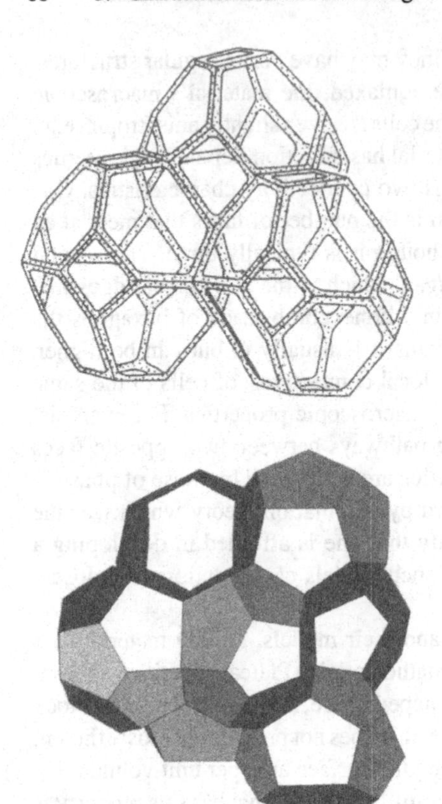

FIGURE 3.10. (a) Kelvin's tetrakaidecahedron unit cell, and (b) the Weaire–Phelan unit call that consists of 6 fourteen-sided polyhedra and 2 twelve-sided polyhedra.

an eight-sided face should also be inserted somewhere in the system; insertion of a five-sided face requires a complementary insertion of a seven-sided face, and so on. The practical implication of Euler's law is that, most cells have six sides, and those which do not are paired. To see how Euler's law dictates this, note that since $Z_e = 3$ and each edge is shared between two vertices, we have $E/V = 3/2$. If f_n is the number of faces with n sides, since an edge separates 2 faces, we must have, $\sum_n nf_n/2 = E$. We then have, from Euler's law, $6 - \sum_n nf_n/F = 6/F$. As F increases, $6/F$ becomes negligibly small. Moreover, $\sum_n nf_n/F$ is just the average number of sides per face, $\langle n \rangle$, and therefore $\langle n \rangle = 6$ in 2D. More generally, with an edge-coordination number Z_e we have

$$\langle n \rangle = \frac{2Z_e}{Z_e - 2}, \tag{92}$$

so that with $Z_e = 3$ one obtains $\langle n \rangle = 6$. It is not easy to generalize Eq. (92) to 3D. However, for an isolated cell ($C = 1$) with an edge connectivity of Z_e and face-connectivity of Z_f we have

$$\langle n \rangle = \frac{Z_e Z_f}{Z_e - 2}\left(1 - \frac{2}{f}\right), \tag{93}$$

where f is the number of faces. Typically, of course, $Z_e = 3$ and $Z_f = 2$, so that

$$\langle n \rangle = 6\left(1 - \frac{2}{f}\right). \tag{94}$$

Observe that, Eq. (94) predicts that if f varies from 12 (for dodecahedra cells) to 20 (for icosdahedra), $\langle n \rangle$ varies only between 5 and 5.4. This explains why most foams have faces with five edges, regardless of the shape of their basic cell. Many grain boundaries in metals and ceramics have $f \approx 14$ and $\langle n \rangle \approx 5.1$.

As mentioned above, it is generally true that a cell with more sides than the average has neighbors which, taken together, have less than the average number. This observation is described for honeycombs by the *Aboav–Weaire rule* (Aboav, 1970; Weaire, 1974):

$$\langle m \rangle = 5 + \frac{6}{n}, \tag{95}$$

where n is the number of edges of the candidate cell and $\langle m \rangle$ is the average number of edges of its n neighbors. Equation (95) is sometimes written in a more general form, $\langle m \rangle = 6 - a + n^{-1}(6a + \mu_2)$, where a is a parameter that is of the order of unity for many natural structures, and μ_2 is the second moment of the probability distribution of finding a cell with n edges. On the other hand, Lewis (1923) (see Lewis, 1946, for details) observed that in a variety of 2D cellular patterns the area $A(n)$ of a cell with n sides is given by

$$\frac{A(n)}{A(\langle n \rangle)} = \frac{n-2}{\langle n \rangle - 2}, \tag{96}$$

where $A(\langle n \rangle)$ is the area of a cell with the average number of sides $\langle n \rangle$. This rule was actually proven by Rivier and Lessowski (1982). Moreover, Rivier (1982) generalized Eq. (96) to 3D to show that the volume $\Omega(f)$ of a polyhedron cell with f faces satisfies the following equation

$$\frac{\Omega(f)}{\Omega(\langle f \rangle)} = \frac{f-3}{\langle f \rangle - 3}. \tag{97}$$

A particularly simple model is constructed by tessellating a d-dimensional cubical subspace into M^d identical d-dimensional cubical cells, with the cells randomly and independently designated as phase 1 (say, white cells) or phase 2 (say, black cells) with probabilities (volume fractions) ϕ_1 and ϕ_2, respectively. The black cells correspond to the particles in the fully-impenetrable spheres model. Suppose that N is the total number of the black cells, so that the black cells' volume fraction is $\phi_2 = N/M^d$. Lu and Torquato (1990) showed that for this model

$$\rho_n(\mathbf{r}^n) = \frac{N!}{(N-n)!}\frac{(M^d-n)!}{M^d!}\prod_{i=1}^{n}\left[\sum_{j=1}^{M^d}\delta(\mathbf{r}_i - \mathbf{R}_j)\right]\prod_{i,j}\theta(i,j), \tag{98}$$

where

$$\theta(i,j) = \begin{cases} 0, & \mathbf{r}_i = \mathbf{r}_j, \\ 1, & \text{otherwise}, \end{cases} \tag{99}$$

and $\mathbf{R}_j$ denotes the position of the center of the jth cell. Given $\rho_n(\mathbf{r}^n)$, one can calculate other important properties of the model. For example, an important property of this model is that, although $S_1(\mathbf{x}) = \phi_1 = 1 - N/M^d$, for points within the system the higher-order quantities depend on the absolute positions $\mathbf{r}^n$, indicating that the system is statistically inhomogeneous. Suppose, for example, that $\mathbf{x}_1$ and $\mathbf{x}_2$ lie anywhere in the *same* white cell. Then,

$$S_2(\mathbf{x}_1, \mathbf{x}_2) = \phi_1, \tag{100}$$

whereas if $\mathbf{x}_1$ and $\mathbf{x}_2$ lie anywhere in *different* white cells, then

$$S_2(\mathbf{x}_1, \mathbf{x}_2) = 1 - 2\phi_2 + \phi_2 \frac{\phi_2 - 1/M}{1 - 1/M}. \tag{101}$$

So far, we have discussed tessellation models in which all the cells have the same size. However, natural cellular materials usually have a range of cell sizes. The size distribution can be very narrow or very broad, with over two orders of magnitude difference between the smallest and largest cell sizes. One particularly well-known model for such cases is the *Voronoi tessellation* (Voronoi, 1908), mentioned in Chapter 2, which is also known as *Dirichlet tessellation*, and is sometimes referred to as the *Wigner–Seitz cell* (see Figure 3.2). In this model one first creates a 2D or 3D simulation box and tessellate it by inserting a number of random (Poisson) points in it, each of which is the basis for a Voronoi polygon or polyhedron, which is that part of the space which is nearer to its Poisson point than to any other Poisson point. In practice, the Voronoi tessellation of space is done by first joining the Poisson points by straight lines and then bisecting these by plane surfaces. The envelope of the surface which surrounds a point is its Voronoi polyhedron or cell. In 2D, the boundary of the Voronoi polygon is composed of segments of the perpendicular bisectors of each edge that connects a Poisson point to its *nearest-neighbor* sites, which are the points that share a Voronoi edge. In 3D, the boundary of a Voronoi polyhedron consists of planes that perpendicularly bisect each edge that connects a Poisson point to its nearest-neighbor sites, which are those points that share a Voronoi face. A 2D example is shown in Figure 3.11.

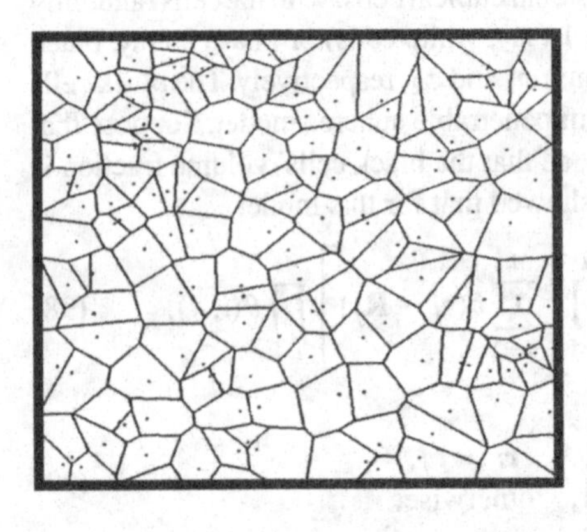

FIGURE 3.11. A 2D Voronoi tessellation of space. The dots represent the initial Poisson points.

Although this model may seem somewhat artificial, it in fact arises naturally in several important problems. For example, if in the fully-impenetrable spheres model the spheres are non-conducting and transport in the composite is through the matrix in between the non-overlapping spheres, then the transport problem can be mapped *exactly* onto an equivalent problem on the *edges* of the Voronoi polyhedra. The model also has applications in theory of liquids, flow through packed-bed reactors, biology, meteorology, metallurgy, crystallography, forestry, ecology, and several other important problems and phenomena. Note that for the Voronoi model in 2D, $\langle n \rangle = 6$ (which is dictated by Euler's law), while in 3D each Voronoi polyhedron has on average $f \approx 15.54$ faces not 14. The surface (in 2D) and volume (in 3D) of the Voronoi cells also follow Eq. (96) and Eq. (97), respectively. Moreover, Gilbert (1962) showed that for the Voronoi model the two-point probability function $S_2(r)$ defined above is given by a double quadrature. Many other properties of this model have been determined either analytically (see, for example, Meijering, 1953) or by computer simulations (see, for example, Winterfeld *et al.*, 1981).

3.4.7 Gaussian Random Field Models of Amorphous Materials

Neither dispersion of spheres nor the tessellation models can be accurate representation of amorphous materials which arise in alloys, microemulsions, some polymers, and many other materials. Although the overlapping spheres model can capture certain features of such materials, it also suffers from several shortcomings. For example, the inclusion and matrix phases are not topologically equivalent; due to overlap between the spheres there are not long-range correlations in the system, and the interface between the matrix and the inclusion phase is spherical. For these reasons, the overlapping spheres model has not been used extensively for modeling amorphous materials.

To remedy these shortcomings, a new class of models has been developed in which the interface between the phases is taken to be a level cut of a random field. The original idea was developed by Cahn (1965) who was studying the influence of spinodal decomposition on the morphology of isotropic two-phase systems. Berk (1987, 1991) generalized the model in order to study scattering properties of microemulsions, while Roberts and Teubner (1995) and Roberts and Knackstedt (1996) made further refinements in order to study transport properties of amorphous heterogeneous materials. We follow the last two papers and describe the essentials of this interesting class of models.

Consider a Gaussian random field (GRF) $y(\mathbf{r})$ and let the level sets $y(\mathbf{r}) = \alpha$ define the interface between the two phases of a two-phase material, with $y > \alpha$ being phase 1. The probability distribution function $S_n(\mathbf{r}^n)$ is given by the volume average

$$S_n(\mathbf{r}^n) = \langle H(y_1 - \alpha) \cdots H(y_n - \alpha) \rangle, \tag{102}$$

where $H(y)$ is the Heaviside function and $y_i = y(\mathbf{r}_i)$. We assume that the material is macroscopically homogeneous and isotropic, so that volume and ensemble averages are equivalent. The joint probability density function of y_i is assumed to be given by

$$P_n(y_1, y_2, \cdots, y_n) = \frac{1}{\sqrt{(2\pi)^n|\mathbf{G}|}} \exp\left(-\frac{1}{2}\mathbf{y}^T\mathbf{G}^{-1}\mathbf{y}\right), \quad (103)$$

where the elements of $\mathbf{G}$, $g_{ij} = \langle y_i y_j \rangle$, usually referred to as the field-field correlation functions, are given by

$$g_{ij} = 4\pi \int_0^\infty \omega^2 S(\omega) \frac{\sin \omega|\mathbf{r}_i - \mathbf{r}_j|}{\omega|\mathbf{r}_i - \mathbf{r}_j|} d\omega, \quad (104)$$

where $S(\omega)$ is the spectral density of the random field, i.e., the Fourier transformation of its covariance. However, in general, $P_n(\mathbf{y})$ does not have to be Gaussian. The lower-order n-point probability functions $S_n(\mathbf{x}^n)$, are defined by

$$\begin{aligned} S_n(\mathbf{x}^n) &= \langle \Pi_{i=1}^n \Theta[y(\mathbf{x}_i) - y_0] \rangle \\ &= \int_{-\infty}^\infty \cdots \int_{-\infty}^\infty \left\{ \Pi_{i=1}^n \Theta[y(\mathbf{x}_i) - y_0] \right\} P_n(\mathbf{y}) d\mathbf{y}, \end{aligned} \quad (105)$$

from which we can immediately obtain

$$S_n = \begin{cases} \phi_1 = \dfrac{1}{\sqrt{2\pi}} \displaystyle\int_\alpha^\infty \exp\left(-\frac{1}{2}t^2\right) dt, & n = 1, \\ \dfrac{1}{2\pi} \displaystyle\int_0^{g_{ij}} \exp\left(-\frac{y_0^2}{1+t}\right) \frac{1}{\sqrt{1-t^2}} dt + \phi_1^2, & n = 2, \end{cases} \quad (106)$$

so that if we specify ϕ_1, the corresponding value of α can be computed (for example, $\alpha = 0$ for $\phi_1 = 0.5$). Therefore, if the spectral density $S(\omega)$ is specified, the field-field correlation function and hence the GRF can be generated. Then, phase 1 is identified by the region $y > \alpha$ and the interface between the two phases is defined by the level sets $y(\mathbf{r}) = \alpha$.

The above model has proven to be very useful for describing many materials. However, it is not completely appropriate for describing a class of two-phase materials in which one of the phases remains sample-spanning down to very low values of its volume fraction. An example is polystyrene foam, a highly porous material shown in Figure 3.12. The solid phase has a sheet-like structure which is quite different from those found in cellular, particulate, or even the one-level cut GRF models described above. As a result, the one-level cut GRF model is not capable of producing the mictrostructure of this material, and in particular a sample-spanning solid phase at very low volume fractions. To address this problem, Roberts and Knackstedt (1996) defined phase 1 to be the region in space where $\alpha \leq y(\mathbf{r}) \leq \beta$, with the remaining space being phase 2, and therefore the interface between the two phases is defined by a two-level cut. In the limit $\beta \to \infty$ one recovers the one-level cut model described above. This model is much more flexible than the one-level cut model. Roberts and Knackstedt (1996) considered a field-field correlation function similar to Eq. (104), except that the upper limit of the integral

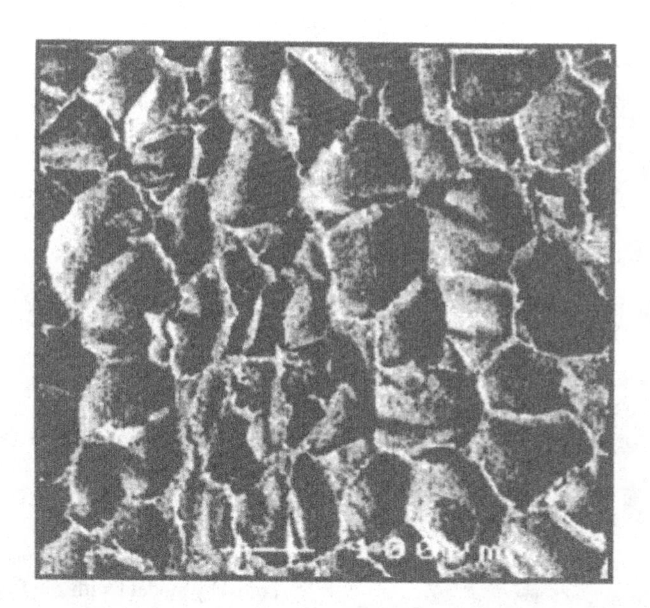

FIGURE 3.12. Morphology of polystyrene foam (after Roberts and Knackstedt, 1996).

was cut off at a value K, so that

$$g_K(r) = 4\pi \int_0^K \omega^2 S(\omega) \frac{\sin \omega r}{\omega r} d\omega, \tag{107}$$

where K is a parameter of the model. Given the volume fraction ϕ_1 one must now specify the parameters α and β. There are several ways of doing so. For example, one requires that an equivalent fraction of phase 1 lie on either side of a particular level cut $y(\mathbf{r}) = \gamma$, so that the model is in some sense "symmetric." Then, a parameter s defined by

$$s = \frac{1}{\sqrt{2\pi}} \int_\gamma^\infty \exp\left(-\frac{1}{2}t^2\right) dt, \tag{108}$$

is introduced; clearly $s \in [0, 1]$. For a given ϕ_1, the parameters α and β are defined through the relations

$$\frac{1}{2}\phi_1 = \frac{1}{\sqrt{2\pi}} \int_\alpha^\gamma \exp\left(-\frac{1}{2}t^2\right) dt = \frac{1}{\sqrt{2\pi}} \int_\gamma^\beta \exp\left(-\frac{1}{2}t^2\right) dt. \tag{109}$$

Roberts and Knackstedt (1996) considered two classes of models. One, model I, was defined by

$$S(\omega) = \frac{1}{N\pi^2[(1 - \nu^2 + \omega^2)^2 + 4\nu^2]}, \tag{110}$$

where N is a renormalization constant to ensure that $g_K(0) = 1$, ν is a model parameter, and

$$\lim_{K\to\infty} g_K(r) = e^{-r} \frac{\sin \nu r}{\nu r}. \tag{111}$$

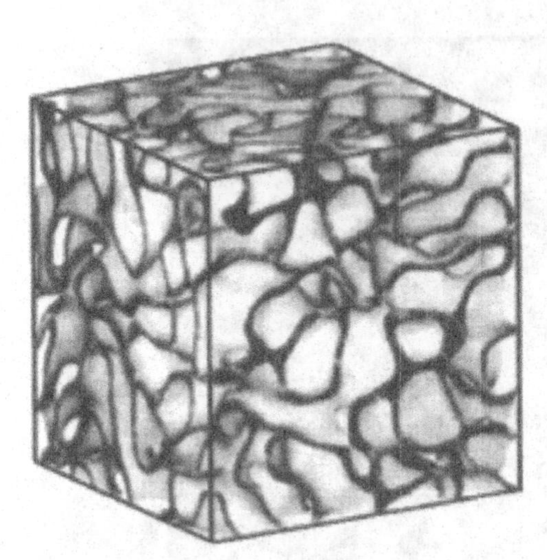

FIGURE 3.13. Interface of a material generated by model II with $s = 0.5$ and $\mu = 1.5$, at a pore volume fraction $\phi_1 = 0.2$. The dark region is represented by the cut $-0.253 < y(\mathbf{r}) < 0.253$ (after Roberts and Knackstedt, 1996).

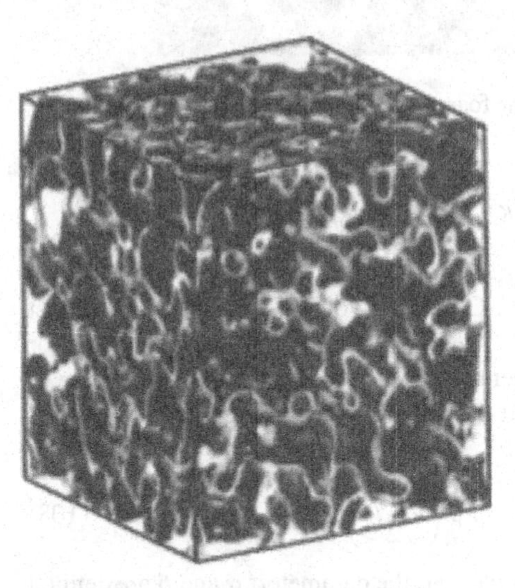

FIGURE 3.14. Interface of a material generated by model I with $s = 0.5$, $\nu = 0.0$ and a volume fraction $\phi_1 = 0.2$. The light region is represented by the cut $-0.253 < y(\mathbf{r}) < 0.253$ (after Roberts and Knackstedt, 1996).

Model II was defined by

$$S(\omega) = \frac{3}{4\pi(\mu^3 - 1)}[H(\mu) - H(1)], \quad \mu > 1, \tag{112}$$

$$g_K(r) = \frac{3(\sin \mu r - \mu r \cos \mu r - \sin r + r \cos r)}{r^3(\mu^3 - 1)}, \quad \mu \leq K. \tag{113}$$

Figure 3.13 shows the resulting model material using model II with $s = 0.5, \mu = 1.5$ and $\phi_1 = 0.2$. The dark region is defined by $-0.253 < y(\mathbf{r}) < 0.253$. In this case the model contains highly connected structures. Figure 3.14 presents the result using model I with $s = 0.5, \phi_1 = 0.2$ and $\nu = 0$. The light region corresponds to $-0.253 < y(\mathbf{r}) < 0.253$. Finally, Figure 3.15 depicts the model material using model II and the same parameters as in Figure 3.12 but with $s = 0.2$. In this case, the dark region is defined by $-1.28 < y(\mathbf{r}) < -0.253$. As can be seen, by varying

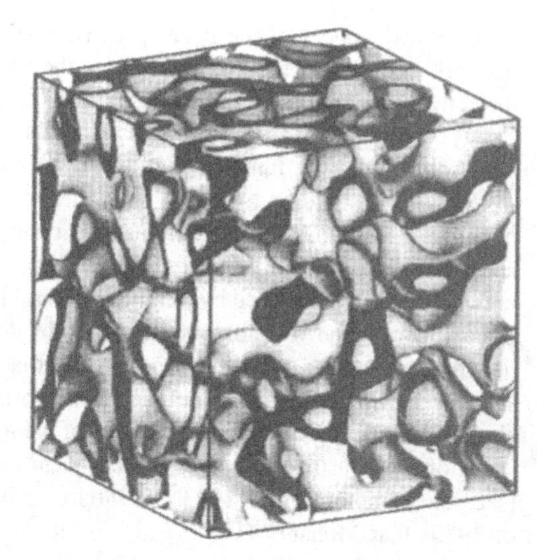

FIGURE 3.15. Interface of a material generated by model II with $s = 0.2$ at a volume fraction $\phi_1 = 0.2$. The dark region is represented by the cut $-1.28 < y(\mathbf{r}) < 1.28$ (after Roberts and Knackstedt, 1996).

a few parameters one can generate a wide variety of interesting microstructures which, in many cases, appear very realistic. In addition, transport properties of such model materials can be computed and compared with the relevant experimental data. This comparison can be used for tuning the parameters of the model.

3.5 Discrete Models

So far, we have discussed models of microstructure of heterogeneous materials that are considered as a sort of disordered continua. We now describe the discrete models that have been used extensively in the development of statistical physics of disordered media. These models have proven to be particularly useful when the heterogeneous system or material contains several phases, one or more of which are sample-spanning. Then, the behavior of the material in the vicinity of the percolation threshold of a given phase, i.e., the point at which the phase becomes sample-spanning, is particularly interesting and is described by percolation theory, the essential elements of which were described in Chapter 2. What follows is a brief description of the discrete models. In the subsequent chapters, where we study the transport properties of heterogeneous materials using such models, more properties of the discrete models will be described.

3.5.1 Network Models

The idea of representing a disordered material by a network is particularly self-evident and intuitively appealing for at least one class of such materials, namely, microporous materials, since it is clear to everyone that fluid paths in such materials may branch and, later on, join one another. This prompted many people to think of a network model of the pore space of a microporous material in which the network's bonds represent in some sense the flow paths or channels that connect the network's sites where the flow paths meet. To each bond is assigned an effective radius or, equivalently, an effective flow conductance, which can be selected from a probability density function that can, in principle, be measured. The network may have a disordered topology, such that the coordination number Z of each node, i.e., the number of bonds that are connected to the node, is distributed according to a probability density function. An example is provided by the Voronoi network, which is obtained by joining by straight lines the centers of the polygons or polyhedra in the Voronoi tessellation model described above. Alternatively, the network can have an ordered topology, so that the coordination number is the same everywhere. Examples include the networks that one can construct by using simple-cubic lattices. Although the idea of using a network to represent the pore space of a porous medium is intuitively clear and has been used for a long time (see, for example, Sahimi, 1993b, 1995b for reviews), it was only in the beginning of 1980s that Mohanty (1981) and Lin and Cohen (1982) provided a rigorous mathematical foundation to such modeling approaches.

More generally, one may consider the bonds of a network to be resistors with conductances that are selected from a probability distribution function. Then, a fraction p_1 of the bonds are designated as phase 1, a fraction p_2 as phase 2, and so on. Each phase may have its own probability density function for the its bonds' conductances. In this way, one obtains a discrete model of a multiphase material that can be used for studying such transport properties as the effective conductivity of the material. If vector transport properties, such as the elastic moduli of the material, are of interest, then each bond represents an elastic element, such as a spring or beam, to which effective elastic constants are assigned.

Such network models can also be obtained from discretization of the governing equations for transport processes in heterogeneous media. Depending on whether one uses a finite-difference or a finite-element approach for discretizing the transport equations, a variety of networks can be obtained. These matters will be discussed in the subsequent chapters where we study various transport processes.

3.5.2 *Bethe Lattice Models*

Aside from the discrete one-dimensional models described in 3.2, the simplest discrete models are branching networks which are based on the Bethe lattices of a given coordination number, an example of which is shown in Figure 2.5. Such models have been used routinely in the statistical mechanics literature for investigating critical phenomena in the mean-field approximation. As far as their applicability to modeling heterogeneous media is concerned, branching networks suffer from two major shortcomings. First, although they contain interconnected bonds that can mimic the interconnectivity of the microscopic elements of a heterogeneous material, they lack closed loops of bonds which restricts the interactions between the microscopic elements. Secondly, for a Bethe lattice of coordination number Z, the ratio $\mathcal{R}$ of the number of sites on the external surface of the network and the total number of sites in the lattice is given by

$$\mathcal{R} = \frac{Z-2}{Z-1}, \tag{114}$$

which takes on finite values for any $Z \neq 2$, whereas for large 3D networks, and indeed most real materials, this ratio is essentially zero. Thus, surface effects strongly affect the properties of a Bethe lattice, leading to anomalous phenomena such as those discussed by Hughes and Sahimi (1982) who investigated diffusion processes on such lattices. The advantage of the Bethe lattice models is that it is often possible to derive analytical equations for the properties of interest. Surprisingly, some of the predictions of these equations agree quite well with those of 3D systems. An example is conduction in disordered Bethe lattices that will be described and discussed in Chapter 5 and 6, and also Chapter 3 of Volume II.

To our knowledge, Liao and Scheidegger (1969) and Torelli and Scheidegger (1972) were the first to use the Bethe lattices for modeling transport in a disordered system which, in their case, was a porous medium. They studied the phenomenon of hydrodynamic dispersion, i.e., mixing of two miscible fluids flowing in a porous medium, where the medium was modeled by a Bethe lattice. Others have used Bethe lattices to model transport and chemical reactions in catalytic and non-catalytic materials (for a review see Sahimi *et al.*, 1990). Stinchcombe (1973, 1974) appears to be the first to use this model for calculating the electrical conductivity of disordered materials (see Chapters 5 and 6, and also Chapter 3 of Volume II).

3.6 Reconstruction of Heterogeneous Materials: Simulated Annealing

The models discussed so far can be accurate representation of a large class of materials. However, a challenging problem is generating a realization of a heterogeneous material given specific but, by necessity, limited mictrostructural information on the material. In this context, limited mictrostructural information implies lower-order statistical correlation functions, such as those discussed above. One hurdle to this difficult problem is that, often the information that is available is only from 2D cross sections of the material, whereas the goal is to infer the 3D microstructure from the 2D information. Moreover, the degree with which the structure of a real material can be mimicked by a model which is constructed based on low-order statistical information provides us with important insight about the usefulness of the correlation and distribution functions described above. In particular, the solution of this problem yields insight into the usefulness of *two-point* correlation functions that can be measured experimentally by small-angle scattering (see Section 2.3.2).

A heavily-used method for constructing a model of a heterogeneous material is based on the Gaussian random fields (GRF) that were described above (for a review see, for example, Levitz, 1998). However, a fundamental problem may limit, at least in some cases, the usefulness of model construction methods based on the GRFs. This limitation is the fact that the material's microstructure may be correlated (in fact, in practice that is usually the case) with a relatively large correlation length, and therefore its reconstruction cannot be done with models such as the GRF model. In addition, as discussed above, the GRF approach assumes that in order to construct a model it suffices to specify the volume fractions of the two phases and the standard two-point correlation functions (such as the power spectra). However, imposing constraints only on the volume fractions, which represent the first moments of the distribution functions, and the two-point correlation functions, may not be adequate as GRFs have the property that their higher-order cumulants are all zero. On the other hand, imposing constraints on the higher-order cumulants of the distribution and correlation functions is not straightforward and, as a result, it is not clear to what extent the generated model can match the actual material, even if the material can in fact be modeled by a GRF.

A natural approach to the reconstruction problem, i.e., one in which one tries to develop a model that can mimic certain amount and type of experimental information, is to formulate it as an optimization problem. Then, the most appropriate model is one that minimizes the differences between the properties of the model and the data measured for these properties. In this method one starts with a reference system—one for which one has some mictrostructural information which has been either measured directly, or has been "guessed" based on intuition and insight. Then, an optimization method is used to search for the best model that can reproduce those aspects of the reference model for which experimental information is available. One of the most efficient optimization algorithms is simulated annealing (SA) developed by Kirkpatrick *et al.* (1983). It is particularly useful for those systems where a global minimum is hidden among many local extrema. It is based on a well-known physical fact which states that if a system is heated to a high temperature T and then slowly cooled down to absolute zero, the system equilibrates to its ground state. At a given T, the probability of finding the system in a state with energy $\mathcal{H}$ is given by the Boltzmann distribution, $P(\mathcal{H}) \sim \exp(-\mathcal{H}/k_B T)$, where k_B is the Boltzmann's constant. Given an optimization problem, SA can be used if one has,

(1) a set of possible configurations of the system;
(2) a method for systematically changing the configuration;
(3) an "energy" function, in analogy with the work of Kirkpatrick *et al.* (1983), to minimize, and
(4) an annealing schedule of changing a temperature-like variable, so that the system can reach its minimum at which the model takes on its optimal configuration that honors the experimental data.

The precise procedure according to which this method is implemented is as follows. One starts with a model of the reference system. Then, SA is used to find an appropriate pattern of, say, the spatial distribution of various phases of the reference system, given the experimental information about material. From our point of view, the most important issue to be resolved is to find out how the two (or more) phases are distributed, so that the transport properties of the model can mimic those of the reference system, and hence honor the data.

For the starting model of microstructure of a heterogeneous material the first requirement depends on the type of the material that one is dealing with. For example, if the material is known to be particulate, then clearly the appropriate starting model is a dispersion of inclusions, distributed in a background matrix. Suppose that $\{C\}$ denotes the set of all possible configurations of such a model that we are to develop. We need one or more criteria according to which we can select the most appropriate model from the set $\{C\}$. This requirement (3) does in fact provide a method according to which a given configuration of the model is changed until the most appropriate is reached. To make the discussion concrete, imagine that we wish to develop a model for a two-phase composite material consisting of conducting and insulating phases. Then $\{C\}$ is the set of all configurations of the system in which some portions of the system are "on," i.e., they allow conduction, and some are "off," i.e., they are insulators. Suppose that one assigns a probability function for randomly selecting a microscopic portion of the system. Then, if the selected portion is on, it is turned off and vice-versa. This generates a new configuration of the system. One now defines the "neighborhood" N_C of C as the set of all configurations that are very close to C or, loosely speaking, they are one step away from C. Annealing the system means picking a configuration from N_C and comparing it with C. In order to make the comparison precise, one needs to define an "energy" $\mathcal{H}$—the third ingredient of SA. Usually $\mathcal{H}$ is defined by

$$\mathcal{H} = \sum_j |f(\mathbf{M}_j) - f(\mathbf{S}_j)|^\gamma, \tag{115}$$

where $\mathbf{M}_j$ and $\mathbf{S}_j$ are vectors of *measured* and *simulated* responses, respectively, γ a constant and $f(\mathbf{x})$ a real monotonic function. The measured data could supply morphological information or provide data about the transport or other properties of the system. A probability distribution of the configurations is then assumed to be expressible as a Boltzmann distribution,

$$P(C) = a \exp[\mathcal{H}(C)/kT], \tag{116}$$

where a is the renormalization constant which is very difficult to estimate, because one must know the energy of *all* configurations, which is impossible, and T is a temperature-like variable to be defined below.

Because $P(C)$ is a Boltzmann distribution, C, the current configuration, can be modeled as a Markov random field, which means that the transition probability for moving from C to C' depends only on C and C' and not on the previous configurations from the set $\{C\}$. Thus, the transition probability can change from configuration to configuration, but it does not depend on the previous configurations that were examined. Therefore, given C and N_C, the transition probability for moving from C to C' (given our current configuration C) is equal to the probability that we select C', times the probability that the system would make the transition to a given configuration C'. Therefore,

$$P\{C \to C'|C\} = \begin{cases} 0 & C' \notin N_C \\ P(C'|C) \times 1 & C' \in N_C, \\ & \mathcal{H}(C') - \mathcal{H}(C) \leq 0, \\ P(C'|C) \exp\{[\mathcal{H}(C) - \mathcal{H}(C')]/T\} & C' \in N_C, \\ & \mathcal{H}(C') - \mathcal{H}(C) > 0, \end{cases} \tag{117}$$

where for the last equation we must have $C \neq C'$. The final ingredient of the model is a schedule for lowering the temperature as annealing proceeds. This means that as annealing continues, one is less likely to keep those configurations that increase $\mathcal{H}$. For example, an *ad hoc*, but effective, schedule was suggested by Press *et al.* (1992) who proposed that the temperature should be changed after a number of configuration iterations which is sufficient to produce a fixed number of acceptable changes. At the end of each iteration i, the temperature T_i is decreased using a geometric series

$$T_{i+1} = T_i R_i^i, \tag{118}$$

where $0 < R_i < 1$. Thus, one only needs to select the initial temperature. This is selected such that it is of the same order of magnitude as the energy difference between the first two configurations, so that the energy difference between successive configurations remains (most of the time) between zero and 1. This schedule does not, however, guarantee convergence to the true optimal configuration, but it is usually effective. Alternatively, one may use the Metropolis algorithm as the acceptance criterion, in which case the logarithmic annealing schedule which decreases the temperature according to

$$T_i \sim \frac{1}{\ln i}, \tag{119}$$

would evolve the system to its ground state. Of course, such a logarithmic schedule may also cause very slow convergence. However, the convergence of the system to the true optimal configuration is ensured.

What does one do if, for example, quantitative information, such as the range of possible responses, is available, but there are no actual measurements? For example, suppose one wishes to predict the electrical current I at a point in the system far from a point at which a measurement was done, and has the information that the currents were observed to be between a and b, $I \in [a, b]$. In this case, each time a configuration is changed, the following steps are also taken:

(1) The new point is added to the configuration and I is calculated at that point;
(2) a new energy function $\mathcal{H}'$ is calculated such that $\mathcal{H}' = 0$, if the calculated I is in $[a, b]$, $\mathcal{H}' = (I - a)^\gamma$, if $Q < a$, and $\mathcal{H}' = (I - b)^\gamma$ if $Q > b$;
(3) if $\mathcal{H}(C') + \mathcal{H}'(C') < \mathcal{H}(C) + \mathcal{H}'(C)$, then C' is kept. Otherwise, the usual annealing probability is used to keep or reject C', and
(4) finally, the new point is removed and the process continues.

This completes the SA method for selecting a configuration of a two-phase material that mimics its important features and reproduces its measured properties.

Summary

Several classes of realistic models for the morphology of a wide variety of heterogeneous materials are now available. Depending on the available computational power, and the extent of accurate experimental data for the morphological characteristics of a material, one can generate a model of a material in considerable detail, and employ the techniques that will be described in the subsequent chapters to estimate its transport properties. Each class of models has its own strengths and weaknesses. At the same time, there is no "universal" model that can be used for every type of material. Therefore, depending on the complexity of a material's microstructure, the amount of information available for the microstructure, and the computational power that is available, one may choose the appropriate model to represent a material.

Part II

Linear Transport and Optical Properties

4
Effective Conductivity, Dielectric Constant, and Optical Properties: The Continuum Approach

4.0 Introduction

In Chapters 2 and 3 we described and discussed the main concepts and methods for characterization and modeling of morphology of heterogeneous materials. Beginning with the present chapter, we describe and discuss the relation between the effective (macroscopic) transport and optical properties of a disordered material and its morphology. If by morphology one implies the arrangement and motion of atoms and/or molecules, then, depending on the specific phenomenon under study, the problem of estimating the effective transport properties has been called, over the last several decades, kinetic theory, liquid state theory, solid state physics, condensed matter physics, metal physics, or crystallography. However, in the present chapter we are not interested in these phenomena at the atomic or molecular length scales; these phenomena will be taken up in Chapters 9 and 10 of Volume II. In many applications, the term microscopic refers to a length scale large compared with molecular dimensions but small compared with the overall dimensions of a given heterogeneous material. Under such circumstances the material can still be considered as a continuum when viewed on the microscopic length scale, but the problem of estimating its effective properties is no longer studied as a molecular-level phenomenon treated by statistical mechanics and theoretical physics, but one at larger length scales which is usually studied by applied physicists and engineers. It is this domain that is of interest to us in this chapter.

How can one estimate the effective transport properties of a heterogeneous material? Because of disorder in a material's microstructure, statistical description of its morphology is appropriate and indeed unavoidable. In Chapters 2 and 3 we described various statistical methods and models for analyzing and representing heterogeneous materials' morphology. In the same spirit, statistical methods play an indispensable role in estimating the effective transport properties of heterogeneous materials. However, although characterization of the morphology of *ordered* materials—those that do not contain disorder of *any* sort—is a straightforward task, the same is not true about estimating their effective transport properties. For example, the problem of estimating the effective (electrical or thermal) conductivity of a regular d-dimensional array of conducting (or insulating) spheres inserted in a uniform matrix of a different conductivity was studied for almost a *century* before its solution was derived, even though the formulation of the problem and deriving its solution seemed straightforward. At the same time, transport properties of ordered multiphase materials often provide key insights into those of disordered materials. For this reason, throughout this book, and whenever appropriate, we describe and discuss methods for estimating effective transport properties of ordered materials and use the results as a guide for understanding those of heterogeneous materials.

Experience with homogeneous materials, and also general theory of transport processes, shows that any *small* departure from (thermal, mechanical, or electrical) equilibrium, which is represented by the existence of a gradient of intensity ∇G in a material, gives rise to an associated flux $\mathbf{J}$ given by a linear relation

$$\mathbf{J} = -\mathcal{T} \cdot \nabla G, \tag{1}$$

where the proportionality factor $\mathcal{T}$ is a *local* transport coefficient that characterizes the material at the *local* level. $\mathcal{T}$ is in general a second-rank tensor when a scalar quantity is transported in the medium, such as heat or mass, and a fourth-rank tensor when ∇G represents the gradient of a vector quantity, e.g., stress. If the material's morphology is isotropic, the resulting symmetry conditions on $\mathcal{T}$ simplify its form greatly, in which case a single transport coefficient may suffice for the description of, say, thermal conductivity. If the material is statistically homogeneous (but microscopically heterogeneous), we expect a linear relation such as (1) to also exist between an imposed *small* average gradient of intensity $\langle \nabla G \rangle$ and the resulting *average* flux $\langle \mathbf{J} \rangle$:

$$\langle \mathbf{J} \rangle = -\mathcal{T}_e \langle \nabla G \rangle, \tag{2}$$

where $\mathcal{T}_e$ is now the effective transport coefficient that characterizes the material *macroscopically*, and the averages are over the ensemble of all possible configurations of the disordered material. In general, $\mathcal{T}_e$ differs from a simple average of the statistical distribution of the local transport coefficient $\mathcal{T}$. However, in many cases such simple and linear equations as (1) and (2) do not exist for a material, in which case the material is said to behave nonlinearly. We will consider such materials in Volume II. In this and the next few chapters we focus on estimating the effective linear transport properties of heterogeneous materials. Clearly, the microstructure of a material controls its effective transport properties. If detailed information is available on the materials' morphology, then it should be exploited to the full extent for estimating their transport properties, which is what we try to accomplish throughout this book. If, on the other hand, little is known about the morphology, then one is forced to fall back on observations, experience or hypothesis, or purely phenomenological theories which in essence are speculations.

If we restrict ourselves to linear transport properties of heterogeneous materials, then we discover an important fact: A remarkably large number of apparently different transport problems can be treated by a common theoretical framework. For example, (1) thermal conductors; (2) electrical conductors; (3) electrical insulators, and (4) dia- or para-magnetic materials are all described by Eq. (2), where the transport coefficient is called, respectively, (1) thermal conductivity; (2) electrical conductivity; (3) dielectric constant, and (4) magnetic permeability. At steady state, the flux $\mathbf{J}$ in all the four phenomena satisfies a conservation equation:

$$\nabla \cdot \mathbf{J} = 0. \tag{3}$$

TABLE 4.1. Mathematically equivalent transport problems with the flux **J** and transport coefficient $\mathcal{T}$.

Phenomenon	Potential	Driving force	**J**	$\mathcal{T}$
Heat conduction	Temperature	Temperature gradient	Heat flux	k
Electrical conduction	Electric potential	Electric field	Current density	g
Diffusion	Density	Density gradient	Diffusant current density	$\mathcal{D}$
Electrostatics	Electric potential	Electric field	Electric displacement	ϵ
Magnetostatics	Magnetic potential	Magnetic field	Magnetic induction	$\mathcal{P}$

Table 4.1 lists several transport problems that are mathematically equivalent. It is precisely their mathematical equivalency that makes it possible to treat these problems within a common theoretical framework. However, perhaps even more remarkable is the fact that considerable common ground exists, both in the methods of analysis and in the nature of the results, between transport problems with mathematical formulations that are *not* identical. For example, if we consider a two-phase heterogeneous material, the differences between the formulations of two transport problems in the material may arise from different boundary conditions that must be satisfied at the interface between the two phases, or from different constitutive equations that describe transport in different phases. However, such differences are, in many cases, matters of analytical details. Although such similarities may seem obvious to us now, they were not noted for many decades, as a result of which research on various transport processes in heterogeneous materials proceeded independently which we now recognize, with the benefit of hindsight, as unnecessary. As we discuss in this and the next few chapters, the results for one type of transport process often have their analogues for several other transport problems.

In this chapter we describe and discuss the continuum methods for estimating the effective transport and optical properties of disordered materials. By continuum methods we mean those that rely on the continuum models of the morphology of disordered materials that were described in Chapter 3. In particular, we describe and discuss various theoretical and computer simulation methods for estimating the effective conductivity, diffusivity and dielectric constant. Estimating vector transport properties, such as the effective elastic moduli of disordered materials based on the continuum models, will be considered in Chapter 7. After describing some symmetry properties of the conductivity tensor, our discussions will focus on describing some general results that are applicable to all the transport processes that are discussed in this book. These results are general in the sense that they are applicable to continuum models of transport processes *regardless* of the volume fraction of the inclusion phase. We then describe methods of estimating the (thermal or electrical) conductivity and dielectric constant of heterogeneous materials with a microstructure that can be described by a dispersion of inclusions inserted in a uniform matrix. Both isotropic and anisotropic materials will be considered. We are always interested in ergodic materials (that is, those for which the ensemble-averaged properties are equal to the volume-averaged properties, when the volume of the system tends to infinity), for which the effective conductivity tensor $\mathbf{g}_e$

exists and is independent of the boundary conditions imposed on the surface of the materials.

4.1 Symmetry Properties of the Conductivity Tensor

It is not difficult to show that, in general, the conductivity tensor **g** must be symmetric, i.e., $g_{ij} = g_{ji}$, from which it follows that the corresponding resistivity tensor must also be symmetric. This symmetry implies that the number of independent components of **g** reduces from d^2 (for a d-dimensional system) to $\frac{1}{2}d(d+1)$. Thus, in 3D, the conductivity tensor has only six independent components:

$$\mathbf{g} = \begin{bmatrix} g_{11} & g_{12} & g_{13} \\ g_{12} & g_{22} & g_{23} \\ g_{13} & g_{23} & g_{33} \end{bmatrix}. \tag{4}$$

For symmetry with respect to a plane, say the $x_1 - x_2$ plane, **g** has four independent components as $g_{13} = g_{23} = 0$. This is called *monoclinic symmetry.* If **g** is symmetric with respect to three orthogonal planes, then it has three independent components which are the g_{ii} for $i = 1, 2$ and 3. The other three components vanish. This is referred to as *orthotropic symmetry.* If **g** is symmetric with respect to a 90° rotation about, say the x_3 axis, then **g** has only two non-zero and independent components which are $g_{11} = g_{22}$ and g_{33}. The rest of the components are zero. This is referred to as *transversely isotropic symmetry.* In general, a crystal has n-fold rotational symmetry ($n > 0$) if it appears unchanged when it is rotated about an axis through $2\pi/n$ radians. Thus, for example, crystals with 6-fold rotational symmetry axis (i.e., hexagonal), a 4-fold rotational symmetry axis (i.e., tetragonal), or a 3-fold rotational symmetry axis (i.e., trigonal) are all transversely isotropic with respect to the conductivity, although neither a tetragonal nor a trigonal crystal has elastic transverse isotropy; see Chapter 7. Finally, when the conductivity is independent of the orientation of the coordinate system, then **g** has only a single independent component, $g_{ii} = g$, for $i = 1, 2$ and 3. The rest of the components are zero. This is the standard *isotropic symmetry.*

4.2 General Results

For two-phase materials, general results that are applicable over the entire range of the volume fraction ϕ_2 of the inclusion phase are rare, and it is precisely for this reason that they are also important. In what follows we describe a few of such results.

(1) For any statistically homogeneous (but microscopically disordered) material the mean flux is a maximum when the local transport properties of the material are *uniform* (Batchelor, 1974). The proof is straightforward.

(2) One can calculate the effective transport properties *exactly* if the local transport properties of the material vary in one direction only. For example, this

corresponds, in the case of dispersion of inclusions, to particles in the form of infinite plane slabs parallel to each other. This reduces the problem to an essentially 1D problem. As an example, consider thermal or electrical conduction in a composite material in which the conductivity of the matrix is g_m with volume fraction ϕ_1, and that of the inclusions is αg_m with a volume fraction ϕ_2. If the inclusions are arranged such that x is the direction along which the conductivities vary (so that x is an axis of symmetry), then the principal axes of the effective conductivity tensor $\mathbf{g}_e$ are parallel and perpendicular to x, with its diagonal elements given by $\langle 1/g \rangle$, $\langle g \rangle$ and $\langle g \rangle$, i.e., thermal resistances in series and parallel, respectively, where $\langle g \rangle = g_m(1 - \phi_2 + \alpha\phi_2)$, and $\langle 1/g \rangle = \alpha g_m/(\alpha + \phi_2 - \alpha\phi_2)$. In addition, for regular arrays of d-dimensional spheres embedded in a uniform matrix, one can derive exact solutions for the effective transport properties of the composites. These solutions will be presented and discussed below.

(3) The effective conductivity tensor $\mathbf{g}_e$ is positive definite and symmetric if the *local* conductivity tensor $\mathbf{g}$ is also positive definite and symmetric.

(4) Heterogeneous materials, that consist of isotropic phases, possess the following properties. (i) Statistically isotropic composite materials are always *macroscopically* isotropic. (ii) Statistical anisotropy, as measured by the correlation functions described in Chapter 3, does *not* necessarily imply a *macroscopically anisotropic* material. For example, composite materials with cubic symmetry are statistically (microscopically) anisotropic but are macroscopically isotropic. Cubic lattices of spheres that are studied later in this chapter are examples of such composite materials. (iii) Macroscopically anisotropic composite materials are necessarily microscopically anisotropic.

(5) The effective conductivity tensor $\mathbf{g}_e$ of a macroscopically anisotropic composite material with n isotropic phases is a *homogeneous function of degree one* in its n scalar phase conductivities $g_1, \cdots, g_n$. This means, for $n = 2$, that, $\mathbf{g}_e(\alpha g_1, \alpha g_2) = \alpha \mathbf{g}_e(g_1, g_2)$, for all values of α. Thus, if we set $\alpha = 1/g_1$, we immediately obtain

$$\mathbf{g}_e = g_1 \mathbf{g}_e(1, g_2/g_1), \tag{5}$$

implying that the homogeneity relation reduces, without loss of generality, the number of independent variables from two to one.

(6) If the volume fraction of the inclusions is small, or if the fluctuations in the relevant transport property of the material are small compared with their mean value, then, many rigorous results can be derived. Consider, for example, thermal conduction again. If θ represents small perturbations in the temperature field, then Eq. (3) implies that, $\nabla^2\theta \approx -\langle \mathbf{G} \rangle \cdot \nabla(g - \langle g \rangle)/\langle g \rangle$, if we assume that $|g - \langle g \rangle|/\langle g \rangle$ is small everywhere (which implies that $|\nabla\theta|/\langle \mathbf{G} \rangle$ is also small everywhere). If $\mathcal{S}(\boldsymbol{\omega})$ is the power spectrum of $g - \langle g \rangle$, then

$$\langle (g - \langle g \rangle) \rangle \nabla\theta = -\int \boldsymbol{\omega} \left(\frac{\langle \mathbf{G} \rangle \cdot \boldsymbol{\omega}}{\omega^2} \right) \left[\frac{\mathcal{S}(\boldsymbol{\omega})}{\langle g \rangle} \right] d\boldsymbol{\omega}. \tag{6}$$

It then follows that

$$\langle \mathbf{J} \rangle = \langle g(\langle \mathbf{G} \rangle + \nabla\theta) \rangle \approx \langle g \rangle \langle \mathbf{G} \rangle - \langle (g - \langle g \rangle) \nabla\theta \rangle, \tag{7}$$

and therefore the effective conductivity tensor is given by

$$\mathbf{g}_e = \langle g \rangle \mathbf{U} - \int \left(\frac{\boldsymbol{\omega}\boldsymbol{\omega}}{\omega^2} \right) \left[\frac{S(\boldsymbol{\omega})}{\langle g \rangle} \right] d\boldsymbol{\omega}, \tag{8}$$

where $\mathbf{U}$ is the unit tensor. For example, for isotropic materials $S(\boldsymbol{\omega})$ is independent of direction of $\boldsymbol{\omega}$, and therefore

$$\mathbf{g}_e = \langle g \rangle - \frac{\langle (g - \langle g \rangle)^2 \rangle}{3 \langle g \rangle}. \tag{9}$$

Equation (9) was first given by Beran (1968); the derivation presented here was given by Batchelor (1974). For a two-phase isotropic and dispersed material of the type discussed above, Eq. (9) simplifies further to

$$g_e \approx \langle g \rangle - \frac{1}{3} g_m \phi_2 (1 - \phi_2)(\alpha - 1)^2. \tag{10}$$

We should emphasize that these results are valid for the case in which the fluctuations in the local conductivity are weak. For a 2D system (i.e., a system of long cylinders distributed in the matrix) the fractor 1/3 should be replaced by 1/2. In this case, if the temperature gradient is parallel to the axes of the cylinders, then the effective conductivity is simply $\langle g \rangle$.

(7) One can derive rigorous upper and lower bounds to the effective transport properties of a heterogeneous material. The usefulness of the bounds depends on the complexity of the material's microstructure and the amount of microstructural information that is included in the bounds. In some cases the bounds provide very accurate estimates of the effective transport properties, while in other cases they are practically useless. We will describe and discuss the derivation of such bounds and provide the expressions for the most accurate bounds currently available.

Having stated the above general results, we now begin describing the estimation of scalar transport properties of heterogeneous materials. We begin by describing the results for dispersion of spheres.

4.3 Effective Conductivity of Dispersion of Spheres: Exact Results

We first consider the cases for which an exact solution can be derived. The best-known of such solutions are for regular arrays of d-dimensional identical spheres of radius R inserted in a uniform matrix. The conductivities of the spheres and the matrix are different, and the goal is to obtain the effective conductivity of the composite material. This problem, as described here, was first studied by Lord

Rayleigh (1872). Since then it has been studied by many authors, including Runge (1925) who noticed an error in Lord Rayleigh's results, and Kharadly and Jackson (1952) who, despite the error, found good agreement between Rayleigh's results and their own experimental data (see below). The problem has been studied in both 2D and 3D, but we first consider the 3D microstructures which are more general, and then summarize the results for the 2D cases.

4.3.1 Three-Dimensional Regular Arrays of Spheres

In addition to the early investigators of this problem mentioned above, Meredith and Tobias (1960) also analyzed this problem, but their solution suffers from a fundamental error resulting from use of an axially symmetric potential (with axis in the direction of the macroscopic potential gradient), rather than a potential possessing the four-fold symmetry of the cubic array. Other papers reporting a study of this problem include those of Keller (1963), Zuzovsky and Brenner (1977), Doyle (1978), McPhedran and McKenzie (1978), McKenzie *et al.* (1978), and Sangani and Acrivos (1983), using a variety of techniques. Here we describe the method due to McPhedran and McKenzie (1978) and McKenzie *et al.* (1978), as their work presents a unified approach to both the 2D and 3D problems. We first consider a simple-cubic array of spheres, after which face-centered cubic (FCC) and body-centered cubic (BCC) lattices are discussed.

4.3.1.1 Simple-Cubic arrays

Consider a simple-cubic lattice of identical spheres inserted in a uniform matrix. The ratio of the conductivities of the spheres and the matrix is σ, and we wish to determine the effective conductivity g_e of the composite normalized by that of the matrix. The sphere radius is R, measured in units of the lattice spacing, and thus the sphere volume fraction is $\phi_2 = 4\pi R^3/3$. Due to the symmetry of the lattice, the effective conductivity of the composite is independent of the direction along which an external field $\mathbf{E}_0$ is applied, and therefore we assume that $\mathbf{E}_0$ is applied in, say, the x-direction. The most convenient coordinate system, relative to the center O of a sphere, is the spherical coordinates (r, θ, φ), where θ is measured from the x-axis and φ is an azimuthal angle measured from the xy plane. The voltage distribution in the composite satisfies the Laplace equation, $\nabla^2 V = 0$. To compute g_e one must find the voltage distribution both inside and outside the spheres. A general expansion about the point O, appropriate inside the spheres, i.e., for $r < R$, is given by

$$V_s(r, \theta, \varphi) = C_0 + \sum_{l=1}^{\infty} \sum_{m=-l}^{l} C_{lm}\, r^l Y_{lm}(\theta, \varphi), \tag{11}$$

while the corresponding expansion for the matrix region is

$$V_m(r, \theta, \varphi) = A_0 + \sum_{l=1}^{\infty} \sum_{m=-l}^{l} (A_{lm} r^l + B_{lm} r^{-l-1}) Y_{lm}(\theta, \varphi). \tag{12}$$

Here Y_{lm} is the spherical harmonic of order (l, m), i.e., the solution of the Laplace equation in spherical coordinates, which can be represented in terms of the Legendre functions, $Y_{lm}(\theta, \varphi) = [(l-|m|)!/(l+|m|)!]^{1/2} P_l^{|m|}(\cos\theta)\exp(im\varphi)$.

The boundary conditions at the interface between the spheres and the matrix, i.e., at $r = R$, are

$$V_m = V_s, \quad \sigma\frac{\partial V_s}{\partial n} = \frac{\partial V_m}{\partial n}, \tag{13}$$

where n is normal to the surface of the spheres. One finds on applying these boundary conditions that

$$A_{lm} = \frac{B_{lm}[\sigma + (l+1)/l]}{R^{2l+1}(1-\sigma)}. \tag{14}$$

Due to the symmetry of the lattice the voltage V must be an anti-symmetric function of θ about $\theta = \pi/2$, which implies that only odd values of l in Eqs. (11) and (12) must be allowed. Moreover, V must also be a symmetric function of φ about $\varphi = \pi/4$, so that only those values of m that are multiples of 4 must be allowed in the expansions.

The expansion for V_m contains the term r^l which increases with r, and thus represents the contributions to the solution from sources other than those on the central sphere centered at O (for sources on this sphere the contributions decrease with r). These sources are at infinity and on all the spheres other than the central one. The term r^{-l-1}, which decreases with increasing r and is singular at $r = 0$, represents the contributions of the sources on the central sphere. Thus, the voltage expansion not singular at O is due to terms originating at infinity and at the other lattice sites, which means that

$$A_0 + \sum_{l=1}^{\infty}\sum_{m=-2l+1}^{2l-1} A_{2l-1,m} r^{2l-1} Y_{2l-1,m}(\theta, \varphi)$$
$$\equiv \sum_{l=1}^{\infty}\sum_{m=-2l+1}^{2l-1}\sum_{i=1}^{\infty} B_{2l-1,m}\rho_i^{-2l} Y_{2l-1,m}(\theta_i, \varphi_i) + E_0 x. \tag{15}$$

The sum over i is over all spheres other than the central one, while the coordinates $(\rho_i, \theta_i, \varphi_i)$ are measured relative to the center of the ith sphere. We group together terms of Eq. (15) having positive and negative values of m, rescale the coefficients, and express the spherical harmonics in terms of the Legendre functions $P_l^m(\cos\theta)$. This yields

$$A_0 + \sum_{l=1}^{\infty}\sum_{m=0}^{2l-1} A_{2l-1,m} r^{2l-1} P_{2l-1}^m(\cos\theta)\cos(m\varphi) - E_0 x$$
$$\equiv \sum_{l=1}^{\infty}\sum_{m=0}^{2l-1}\sum_{i=0}^{\infty} B_{2l-1,m}\rho_i^{-2l} P_{2l-1}^m(\cos\theta_i)\cos(m\varphi_i). \tag{16}$$

In his treatment of this problem, Lord Rayleigh truncated terms higher than those involving r^3 from the Legendre polynomials.

It remains to determine the unknown coefficients $B_{2l-1,m}$. They are obtained by equating the partial derivatives of all orders with respect to x of the two sides of identity (16) and evaluating them at a point $Q(r_0, \theta_0, \varphi_0)$ within a unit cell that contains O. Hence, if we equate the $(2n+1)$th partial derivatives of both sides of (16) at the point Q, we obtain

$$\sum_{l=n+1}^{\infty} \sum_{m=0}^{2l-2n-2} \binom{m+2l-1}{2n+1} P_{2l-2n-2}^{m}(\cos\theta_0)\cos(m\varphi_0)A_{2l-1,m}$$

$$+\sum_{l=1}^{\infty}\sum_{m=0}^{2l-1}\sum_{i=1}^{\infty}\binom{2l+2n-m}{2n+1}\rho_i^{-2l-2n-1}P_{2l+2n}^{m}(\cos\theta_i)\cos(m\varphi_i)B_{2l-1,m} = E_0\delta_{n,0}, \tag{17}$$

where

$$\binom{n}{r} = \frac{n(n-1)\cdots(n-r+1)}{r!}.$$

In the sum over i we run over all positive or negative values of the lattice points (u, v, w) excluding $(0, 0, 0)$. Thus, if

$$U_l^m(Q) = \sum_{i=1}^{\infty} \rho_i^{-l-1} P_l^m(\cos\theta_i)\cos(m\varphi_i), \tag{18}$$

where ρ_i is the distance between the points (x_0, y_0, z_0) and (u, v, w), $\cos\theta_i = (x_0 - u)/\rho_i$, and $\cos\varphi_i = (y_0 - v)/(z_0 - w)$, then Eq. (17) becomes

$$\sum_{l=n+1}^{\infty}\sum_{m=0}^{2l-2n-2}\binom{m+2l-1}{2n+1}P_{2l-2n-2}^{m}(\cos\theta_0)r_0^{2l-2n-2}\cos(m\varphi_0)\frac{B_{2l-1,m}[\sigma+2l/(2l-1)]}{R^{4l-1}(1-\sigma)}$$

$$+\sum_{l=1}^{\infty}\sum_{m=0}^{2l-1}\binom{2l+2n-m}{2n+1}U_{2l+2n}^{m}(Q)B_{2l-1,m} = E_0\delta_{n,0}. \tag{19}$$

In general, one must determine M_θ coefficients of the form $B_{2l-1,0}$ and M_φ coefficients of the form $B_{2l-1,m}$, representing the effect of the azimuthal terms, where m is a multiple of 4 and does not take on a zero value. For the M_θ coefficients the point Q is taken to coincide with O, hence yielding

$$\frac{B_{2n+1,0}[\sigma+(2n+2)/(2n+1)]}{R^{4n+3}(1-\sigma)} + \sum_{l=1}^{\infty}\sum_{m=0}^{2l-1}\binom{2l+2n-m}{2n+1}U_{2l+2n}^{m}(\mathrm{O})B_{2l-1,m}$$

$$= E_0\delta_{n,0}, \tag{20}$$

which yields M_θ coefficients. For the M_φ coefficients one takes the point Q to be away from O and on the line $x_0 = y_0 = z_0$, where $x_0 = (4\sqrt{3})^{-1}$. However, the sums in Eq. (19), and in particular the term $U_{2l}^m(Q)$, depend on the assumed form of the surface at infinity which encloses the lattice. This dependence had prompted some researchers (for example, Jeffrey, 1973) to cast doubt on the validity of Lord Rayleigh's approach. However, an elegant argument by McPhedran and McKenzie (1978) established that the shape dependence of these sums can be removed if one replaces the field E_0 in Eq. (19) with the *total* field $E = E_0 + E_p$, where E_p the polarization field that represents the effect of polarization charges on the exterior boundary of the cubic array, and if one sets $U_2^0(\mathrm{O}) = U_2^0(Q) = 2\pi/3$. Then, the resulting modified Eq. (19) can be used for determining the coefficients $B_{2l-1,m}$.

To make further progress, one needs to evaluate the terms $U_{2l}^m(\mathrm{O})$ and $U_{2l}^m(Q)$. McPhedran and McKenzie (1978) showed that

$$U_{2l}^0(\mathrm{O}) = 2 + 4P_{2l}(\mathrm{O}), \quad l > 10, \tag{21}$$

$$U_{2l}^m(\mathrm{O}) = 4P_{2l}^m(\mathrm{O}), \quad l \geq 13, \tag{22}$$

provide very accurate estimates of $U_{2l}^m(\mathrm{O})$. The numerical values for the first few terms of $U_{2l}^m(\mathrm{O})$ are, $U_4^0(\mathrm{O}) \simeq 3.108$, $U_6^0(\mathrm{O}) \simeq 0.57333$, and $U_8^0(\mathrm{O}) \simeq 3.2593$. McPhedran and McKenzie (1978) list the values of $U_{2l}^m(\mathrm{O})$ for $m > 0$ (with m being multiples of 4). In addition, if $r_0 = 1/4$, $r_1 = \sqrt{1 + r_0^2 - 2x_0}$, $r_2 = \sqrt{1 + r_0^2 + 2x_0}$, $\tan\varphi_1 = (x_0 - 1)/x_0$, and $\tan\varphi_2 = (x_0 + 1)/x_0$, then

$$U_{2l}^0(Q) = \{2P_{2l}(x_0/r_1) + P_{2l}[(x_0 - 1)/r_1]\}r_1^{-2l-1}, \quad l > 25, \tag{23}$$

and

$$U_{2l}^m(Q) = \{2P_{2l}^m(x_0/r_1)\cos(m\varphi_1) + P_{2l}^m[(x_0 - 1)/r_1]\cos(m\pi/4)\}r_1^{-2l-1}$$
$$+ \{2P_{2l}^m(x_0/r_2)\cos(m\varphi_2) + P_{2l}^m[(x_0 + 1)/r_2]\cos(m\pi/4)\}r_2^{-2l-1}, \quad l \geq 13. \tag{24}$$

Values of the first few $U_{2l}^m(Q)$, where $Q = (x_0, y_0, z_0)$ and $x_0 = (4\sqrt{3})^{-1}$, are $U_4^0(Q) \simeq 2.752$, $U_6^0(Q) \simeq 0.41415$, and $U_8^0(Q) \simeq 1.3989$. The corresponding values for $m > 0$ (and multiples of 4) are listed by McPhedran and McKenzie (1978). However, numerical calculations of McPhedran and McKenzie (1978) showed that the effect of the azimuthal terms is in fact negligible. Having determined values of $U_{2l}^m(\mathrm{O})$ and $U_{2l}^m(Q)$, one can now determine the coefficients $B_{2l-1,m}$ to any desired accuracy by truncating Eqs. (19) and (20) and solving them (in which E_0 is replaced by $E = E_0 + E_p$).

Application of the Green theorem shows that the effective conductivity of the cubic array is given by

$$g_e = 1 - 4\pi\frac{B_{1,0}}{E}, \tag{25}$$

which is an important result since it indicates that one only needs to compute $B_{1,0}$. Figure 4.1 presents the predicted effective conductivity of the composite versus the volume fraction ϕ_2 of the spheres. The spheres are perfectly conducting (i.e., the conductivity ratio $\sigma = \infty$), and the conductivity of the composite has been normalized by that of the matrix. Note that this is a mixture of conducting and superconducting materials, as problem that was described in Chapter 2 (see Section 2.6.2) in the context of a percolation model of conducting and superconducting resistors. The results shown are for various values of $M = M_\theta$, as the azimuthal contributions have been neglected (i.e., $M_\varphi = 0$). The lowest value, $M = 1$, corresponds to a sort of a mean-field approximation first derived by Maxwell (1873):

$$g_e = 1 - 3\phi_2 \left(\frac{2+\sigma}{1-\sigma} + \phi_2 \right)^{-1}, \tag{26}$$

so that in the limit $\sigma \to \infty$ one has $g_e = (1 + 2\phi_2)/(1 - \phi_2)$. Although this approximation is accurate for low values of ϕ_2, it becomes increasingly inaccurate for larger values of ϕ_2 and breaks down completely for $\phi_2 > 0.6$. In particular, it predicts that the conductivity of the composite diverges only when $\phi_2 \to 1$, which, as percolation theory predicts, is wrong. In reality, the conductivity diverges when $\phi_2 = \phi_c = \pi/6 \simeq 0.523$, which happens when the spheres begin to touch. This is actually predicted very accurately by the theory of McPhedran and McKenzie (1978) when M becomes very large; see Figure 4.1. However, compared to the $M = 1$ case, the solution for $M = 2$, i.e., when octupoles are taken into account, indicates a dramatic improvement, and multipoles of order 2^7, which correspond to $M = 4$, provide very substantial improvements. For the $M = 4$ case and neglecting the contributions of the azimuthal terms, the approximate solution is given by

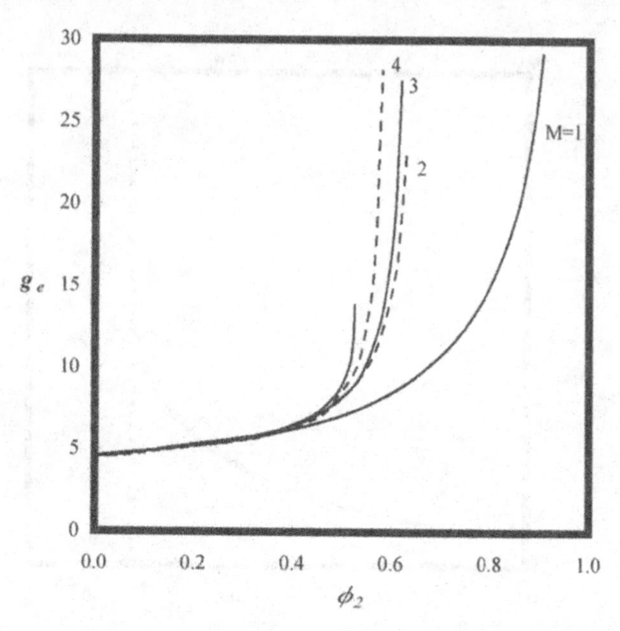

FIGURE 4.1. Effective conductivity g_e of a simple cubic array of perfectly conducting spheres as a function of their volume fraction ϕ_2. M is the order of the theory, with $M = 1, 2$, and 3 corresponding to the theories of Maxwell (1973), Rayleigh (1892), and Meredith and Tobias (1960). $M = \infty$ (the first curve from the left) is the exact solution (after McPhedran and McKenzie, 1978).

(McPhedran and McKenzie, 1978)

$$g_e = 1 - 3\frac{\phi_2}{D}, \tag{27}$$

where

$$D = \Re_1^{-1} + \phi_2 - b_1\Re_5\phi_2^{14/3} - c_1\Re_7\phi_2^6 - a_1\phi_2^{10/3}\frac{1 - c_2\Re_5\phi_2^{11/3} + c_3\Re_5^2\phi_2^{22/3}}{1/\Re_3 + b_2\phi_2^{7/3} - c_4\Re_5\phi_2^6} \tag{28}$$

with

$$\Re_n = \frac{1-\sigma}{\sigma + 1 + n^{-1}}, \tag{29}$$

and $a_1 = 1.305$, $b_1 = 0.01479$, $b_2 = 0.4054$, $c_1 = 0.1259$, $c_2 = 0.5289$, $c_3 = 0.06993$, and $c_4 = 6.1673$. If the coefficients $c_1 - c_4$ are set to zero, then one obtains the formula suggested by Meredith and Tobias (1960). Figure 4.2 compares the theoretical predictions with the experimental data of Meredith and Tobias (1960), and the agreement is excellent.

Let us mention some of the important results that have been obtained by various groups and compare them with those of McPhedran and McKenzie (1978) discussed above. Zuzovsky and Brenner (1977) presented the following expression

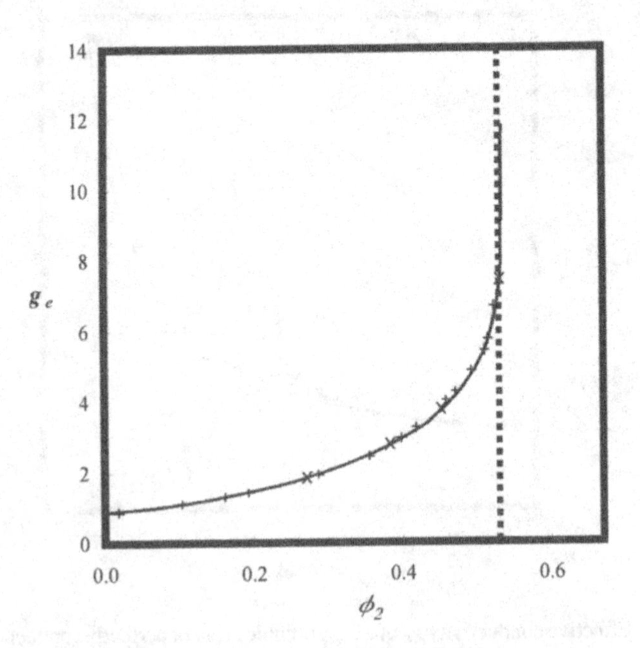

FIGURE 4.2. Comparison of the experimental data of Kharadly and Jackson (+) and Meredith and Tobias (×) for the effective conductivity of a simple cubic array of perfectly conducting spheres with the exact solution ($M = \infty$). The vertical dashed lines corresponds to $\phi_2 = \frac{1}{6}\pi$ (after McPhedran and McKenzie, 1978).

for the effective conductivity of the simple-cubic array of spheres:

$$g_e = 1 - 3\phi_2 \left[\frac{2+\sigma}{1-\sigma} + \phi_2 - \frac{16 d_1^2 \phi_2^{10/3}}{1/\Re_3 + 20 d_2 \phi_2^{7/3}} - 53.5 d_2^2 \Re_5 \phi_2^{14/3} + O(\phi_2^6) \right]^{-1}, \tag{30}$$

where $d_1 = 0.2857$ and $d_2 = 0.02036$. The predictions of this equation agree closely with those of McPhedran and McKenzie (1978). Note that, unlike McPhedran and McKenzie (1978), Zuzovsky and Brenner did not neglect the contributions of the azimuthal terms. In the limit $\sigma \to 0$, i.e., when the spheres become insulating, Eq. (30) reduces to

$$g_e = 1 - 3\phi_2 \left[2 + \phi_2 - \frac{1.306 \phi_2^{10/3}}{4/3 + 0.407 \phi_2^{7/3}} - 0.018 \phi_2^{14/3} + O(\phi_2^6) \right]^{-1}, \tag{31}$$

the predictions of which also agree with the experimental data of Meredith and Tobias (1960). In the opposite limit of $\sigma \to \infty$, i.e., when the spheres become perfectly conducting, Eq. (30) reduces to

$$g_e = 1 + 3\phi_2 \left[1 - \phi_2 - \frac{1.306 \phi_2^{10/3}}{1 - 0.407 \phi_2^{7/3}} - 0.022 \phi_2^{14/3} + O(\phi_2^6) \right]^{-1}, \tag{32}$$

the predictions of which also agree with the experimental data of Meredith and Tobias (1960) for ϕ_2 as large as 0.45. Sangani and Acrivos (1983) presented expressions similar to those of Zuzovsky and Brenner (1977), except that theirs were accurate to order $O(\phi_2^{25/3})$. Batchelor and O'Brien (1977) considered the problem of calculating the conductivity of a lattice of touching, finitely conducting spheres, and derived the following expression for the conductivity of a simple-cubic array of touching spheres,

$$g_e \approx \pi \ln \sigma, \quad \sigma \gg 1. \tag{33}$$

The predictions of McPhedran and McKenzie (1978) agree with those of Eq. (33) for large values of σ. Furthermore, Keller (1963) suggested that for the case of perfectly conducting spheres and for ϕ_2 close to ϕ_c one has

$$g_e \approx -\frac{\pi}{2} \ln(\phi_c - \phi_2), \tag{34}$$

with $\phi_c = \pi/6$, as mentioned above. The predictions of Keller's equation also agree with the results of McPhedran and McKenzie (1978). Since ϕ_c represents a sort of percolation threshold, Eq. (34) represents a mean-field approximation to the problem of conductivity of a percolation system of conducting-superconducting mixture discussed in Chapter 2, because Equation (34) predicts that the conductivity of the mixture diverges logarithmically as ϕ_c is approached, which is precisely what the mean-field theory predicts. Moreover, the logarithmic divergence of the conductivity implies that the critical exponent s defined by Eq. (2.37) is zero (see Table 2.3).

4.3.1.2 Body-Centered and Face-Centered Cubic Arrays

For these cases we follow the treatment of McKenzie *et al.* (1978). The potential expansions are still given by Eqs. (11) and (12), where the coefficients A_{lm} are related to B_{lm} through Eq. (14), and

$$C_{lm} = \frac{B_{lm}(1/\mathfrak{R}_l + 1)}{R^{2l+1}}, \tag{35}$$

where $\mathfrak{R}_l$ is given by Eq. (29). For the FCC and BCC lattices the axes can be selected in such a way that l is restricted to odd values, and that the potential is an even function of φ. Equations (16) and (19) still hold for both lattices. Actually, these equations, with a most minor modification, hold for many other Bravais lattices.

For the BCC lattice, the volume fraction of the spheres is $\phi_2 = 8\pi R^3/3$, with its critical value, i.e., when the spheres touch (i.e., at $R = \sqrt{3}/4$), being $\phi_c = \sqrt{3}\pi/8$. The x-direction, along which the external field is applied, is taken to lie along the (111) direction. The reason for this choice is that, for the case of perfectly conducting spheres, as ϕ_2 approaches ϕ_c a divergent current occurs along the line $\theta = 0$, which can cause convergence problems for the series in Eqs. (11) and (12). Since $V(r, \theta, \varphi) = V(r, \theta, \varphi + \frac{2}{3}\pi k)$, where k is an integer, we must restrict m in the series to multiples of 3. To apply the Green theorem and relate the conductivity to the coefficients $B_{2l-1,m}$, a unit cell must be selected. McKenzie *et al.* (1978) chose a hexagonal prism with its end faces on the planes perpendicular to the x-axis at $x = \pm\sqrt{3}/4$, and its cross section being of external radius $\sqrt{2/3}$. A sphere lies at the center of the unit cell and each side edge of the prism passes through a sphere center. There are 3 spheres contained within this cell which has volume 3/2.

Unlike the case of the simple-cubic lattice, the contributions of the azimuthal terms for the BCC lattice cannot be neglected. Thus, M_θ equations are obtained from (19) by taking Q to be at the origin O, while M_φ equations are obtained by taking Q to be at (x_0, x_0, x_0) with $x_0 = \sqrt{3}/12$. Thus, a total of $M = M_\theta + M_\varphi$ terms are used. Computations of McKenzie *et al.* (1978) showed that a very accurate solution is obtained even with $M_\theta = M_\varphi = 4$. To calculate the coefficients $B_{2l-1,m}$ one must compute $U_{2l}^m(\mathrm{O})$ and $U_{2l}^m(Q)$. To remove the shape-dependence of the solution, one must have, $U_2^0(\mathrm{O}) = U_2^0(Q) = 4\pi/3$. Values of the first few terms are, $U_4^0(\mathrm{O}) \simeq 2.071$, $U_6^0(\mathrm{O}) \simeq 9.68277$, and $U_8^0(\mathrm{O}) \simeq 2.26619$. With Q at (x_0, y_0, z_0) and $x_0 = \sqrt{3}/12$, values of the first few terms of $U_{2l}^m(Q)$ are, $U_4^0(Q) \simeq 2.5322$, $U_6^0(Q) \simeq 5.50568$, and $U_8^0(Q) \simeq -4.97972$. Other relevant values of U_{2l}^m (with m being multiples of 3) both at O and at Q are listed by McKenzie *et al.* (1978). We should also keep in mind that, due to the invariance of the lattice under the cubic point group, and the transformation properties of the spherical harmonics under rotations, U_{2l}^m are not all independent. For example, one can show that, $U_4^3(\mathrm{O}) + 60\sqrt{2}U_4^0(\mathrm{O}) = 0$. Knowledge of U_{2l}^m enables one to compute the coefficients $B_{2l-1,m}$ to any desired accuracy by truncating and solving Eqs. (19) and (20).

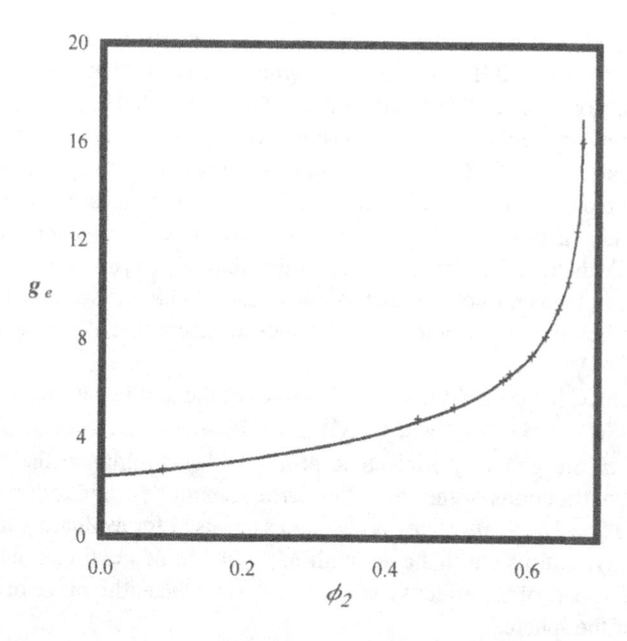

FIGURE 4.3. Comparison of the experimental data for the effective conductivity g_e of a BCC array of perfectly conducting spheres (symbols) with the theoretical predictions (after McKenzie *et al.*, 1978).

Application of the Green theorem to the unit cell shows that the effective conductivity of the BCC composite is given by

$$g_e = 1 - 8\pi \frac{B_{1,0}}{E}, \tag{36}$$

where E is the total field defined above. Figure 4.3 compares the predictions of the theory with the experimental data of McKenzie *et al.* (1978) for the case of perfectly conducting spheres; the agreement between the theory and the experimental data is excellent. In particular, both the theory and the data correctly indicate that the conductivity of the mixture diverges at a volume fraction $\phi_c \simeq 0.68$. An equation similar to (28) also gives accurate estimates of the conductivity, where for the BCC lattice, $a_1 = 0.057467$, $b_1 = 0.166117$, $b_2 = 1.35858$, $c_1 = 0.000950738$, $c_2 = 0.733934$, $c_3 = 0.134665$, and $c_4 = 0.0465862$. In this case ϕ_2 must be relatively small, as the analogue of Eq. (28) for the BCC lattice was obtained with $M_\theta = 4$ and $M_\varphi = 0$, i.e., the contributions of the azimuthal terms were neglected, whereas, as discussed above, for the BCC lattice such contributions are not negligible if the sphere volume fraction ϕ_2 is relatively large. The Zuzovsky–Brenner expression, Eq. (30), also provides accurate estimates of the effective conductivity with $d_1 = -0.0897$ and $d_2 = 0.03811$.

In a similar manner, the conductivity of a dispersion of equal-sized spheres inserted at the nodes of a FCC lattice in a uniform matrix are computed. In this case the volume fraction of the spheres is $\phi_2 = 16\pi R^3/3$, so that the critical volume fraction at which the spheres begin to touch [i.e., at $R = 1/(2\sqrt{2})$] is $\phi_c = \sqrt{2}\pi/6 \simeq 0.740$. The x-axis, along which the external potential is applied, is taken to lie along the (110) lattice direction. The cell selected for applying the Green theorem is a rectangular prism with its end faces being the planes perpendicular to the x-axis at $x = \pm\sqrt{2}/8$. The cross section of the prism in the yz plane is defined by the four points $(y, z) = (\pm\sqrt{2}/4, \pm 1/2)$. The side faces of the prism lie in mirror planes of symmetry in the lattice. Moreover, with this choice of the axes, the potential distribution satisfies the condition that, $V(r, \theta, \varphi) = V(r, \theta, \varphi + k\pi)$, where k is an integer, implying that we must restrict m in the series to even integer values. As before, l is restricted to odd values to ensure that V is an even function of φ.

For the FCC lattice, $U_2^0(\mathrm{O}) = 8\pi/3$. Values of the first few terms are, $U_4^0(\mathrm{O}) \simeq 1.8815$, $U_6^0(\mathrm{O}) \simeq 43.2817$, and $U_8^0(\mathrm{O}) \simeq 45.6674$. Other values of U_{2l}^m for even values of m are given by McKenzie *et al.* (1978). Similar to the BCC lattice case, the contributions of the azimuthal terms cannot be neglected, and therefore a total of $M = M_\theta + M_\varphi$ terms of the series are used for evaluating the effective conductivity. It turns out that even with $M_\theta = 4$ and $M_\varphi = 6$ one obtains a very accurate estimate of the effective conductivity over the entire range of the volume fraction of the spheres.

Application of the Green theorem to the unit cell of the composite yields the result

$$g_e = 1 - 16\pi \frac{B_{1,0}}{E}, \tag{37}$$

where, as before, E is the total field. Equation (28) provides accurate estimates of the conductivity for low values of the sphere volume fraction ϕ_2 with $a_1 = 0.0047058, b_1 = 0.130683, b_2 = 1.20500, c_1 = 0.00603255, c_2 = 5.73021, c_3 = 8.20884$, and $c_4 = 0.295595$. These coefficients were estimated with $M_\theta = 4$ and $M_\varphi = 0$. The Zuzovsky–Brenner expression, Eq. (30), also provides accurate estimates of the effective conductivity with $d_1 = -0.0685$ and $d_2 = -0.03767$.

4.3.2 *Two-Dimensional Arrays of Cylinders*

In 2D, the problem is deriving the effective conductivity of a regular array of infinitely long cylinders with circular cross sections, inserted in a uniform matrix. As in the 3D case, the conductivities of the cylinders (circles) and the matrix are different. The radius of the cylinders is R. Such 2D regular arrays of cylinders inserted in a uniform matrix of a different conductivity are of practical interest. For example, thin films made by sputtering by ion bombardment often have a columnar structure, similar to what we consider here. One may also wish to model thin fibrous materials in which the role of the fibers is played by the long cylinders. Certain cermets—composites of metals and ceramic materials—also have such a structure. According to the *International Tables for X-Ray Crystallography*, there are five ways of packing cylinders in regular 2D arrays. Here we consider two of such arrays, namely, square and hexagonal arrays.

4.3.2.1 Hexagonal Arrays

This problem has been studied by Lord Rayleigh (1892), Runge (1925), Keller (1963), Keller and Sachs (1964), Ninham and Sammut (1976), Doyle (1978), and Perrins *et al.* (1979b). Keller and Sachs (1964) considered the case in which the cylinders were perfectly conducting. We discuss the more general problem in which both the cylinders and the matrix have finite conductivities, and summarize the solution obtained by Perrins *et al.* (1979b) whose method is similar to that of McPhedran and McKenzie (1978) and McKenzie *et al.* (1978) that we described in the last section.

Consider a hexagonal array of cylinders of radius R, as shown in Figure 4.4. Without loss of generality, one may assume a unit separation between the cylinders' centers. The volume fraction ϕ_2 of the cylinders is given by $\phi_2 = 2\pi R^2/\sqrt{3}$. The cylinders touch when $R = 1/2$, and therefore the critical volume fraction is $\phi_c = \pi/(2\sqrt{3})$. The ratio of the conductivities of the cylinders and the matrix is σ, and the goal is to determine the effective conductivity g_e of the composite normalized by the conductivity of the matrix. We use polar coordinates (r, θ) with r being measured from a cylinder center and θ from the horizontal line emanating from the center; see Figure 4.4. An electric field $\mathbf{E}_0$ is applied to the system parallel to the horizontal line. Similar to the 3D case, one must consider the voltage distributions both outside and inside the cylinders. For outside the cylinders, i.e., for $r > R$, Perrins *et al.* (1979b) expanded the voltage about the center as

$$V_m(r, \theta) = A_0 + \sum_{l=1}^{\infty}(A_l r^l + B_l r^{-l}) \cos l\theta, \tag{38}$$

whereas for $r < R$, i.e., inside the cylinders, one has

$$V_c(r, \theta) = C_0 + \sum_{l=1}^{\infty} C_l r^l \cos l\theta. \tag{39}$$

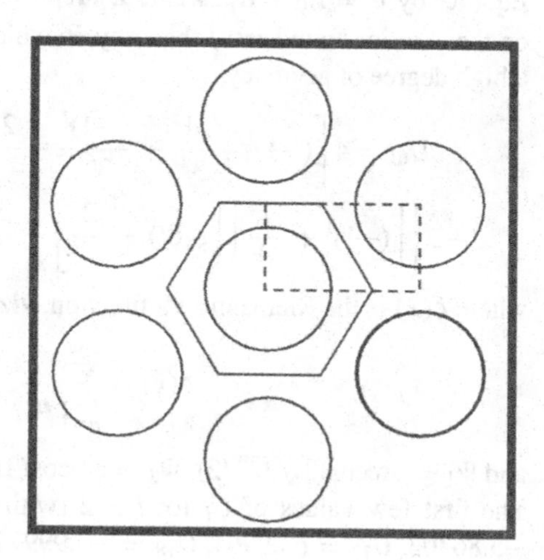

FIGURE 4.4. A portion of the hexagonal array of disks (or cylinders) used in the computations. The basic hexagonal cell, as well as a subunit (dashed outlines), are shown (after Perrins *et al.*, 1979b).

The boundary conditions at the interface between the cylinders and the matrix are again given by Eq. (13). If we set $V_c(0,\theta)=0$ and use the boundary conditions (13), we find that $A_0=C_0=0$. This implies immediately that only odd values of l must be allowed in the expansions for V_m and V_c. Moreover, suppose that, in addition to E_0, we apply fields $E_0(-1/2,-\sqrt{3}/2)$ and $E_0(-1/2,\sqrt{3}/2)$, so that the total applied field is zero. If $V_m(r,\theta)$ is the potential distribution due to E_0, then the symmetry of the system yields $V_m(r,\theta+\frac{2}{3}\pi)$ and $V_m(r,\theta+\frac{4}{3}\pi)$ for the potentials at (r,θ) due to the other two external fields. This implies that, $V_m(r,\theta)+V_m(r,\theta+\frac{2}{3}\pi)+V_m(r,\theta+\frac{4}{3}\pi)=0$. Substituting Eq. (39) into this expression, we can see immediately that l cannot be a multiple of 3, and therefore $A_l=B_l=0$, when l is a multiple of 3.

The analogue of Eq. (19) for the hexagonal array is

$$E_0\delta_{n,1}-(2n-1)!\frac{B_{2n-1}(1+\sigma)}{R^{4n-2}(1-\sigma)}=\sum_{m=1,2,\cdots}\frac{(2n+2m-3)!}{(2m-2)!}U_{2n+2m-3}B_{2m-1}. \tag{40}$$

Here, the quantity U_l is given by

$$U_l=\sum_{j=1}^{\infty}(x_j+iy_j)^{-l}=\sum_{(r,\theta)}r^{-l}\exp(-il\theta), \tag{41}$$

where $i=\sqrt{-1}$, and (x_j,y_j) represents the cylinders' centers, excluding the central one. Despite seemingly being a complex number, all the non-zero U_l are real. Note that, if we assume that the array remains unchanged by an angular rotation through an angle φ, then we must have $U_l=U_l\exp(-il\varphi)$, for all integer values $l>2$. Since U_l is non-zero, we must have $l=2m\pi/\varphi$. For a hexagonal array, $\varphi=\pi/3$, and thus, for $l>2$, the only non-zero sums have an order l which is a multiple of 6. However, similar to the 3D systems, U_2 is conditionally convergent as its value depends upon the shape of the exterior boundary of the array at infinity. But, as in the 3D case, this shape dependence is removed by replacing E_0 in Eq. (40) by $E=E_0+E_p$, where E_p represents the effect of polarization charges on the exterior boundary of the array, in which case, $U_2=2\pi/\sqrt{3}$. In general, to a high degree of accuracy,

$$U_{6l}=4\left[(-1)^l+\frac{1}{3^{3l}}+\frac{(-1)^l}{3^{6l}}+\frac{2\cos(6l\Psi_1)}{7^{3l}}+\frac{2\cos(6l\Psi_2)}{13^{3l}}\right]$$
$$+2\left\{\left[(-1)^l+\frac{1}{3^{3l}}\right]\left[\zeta(6l)+\frac{2}{4^{3l}}\right]+\frac{2\cos(6l\Psi_3)}{7^{3l}}\right\},\quad l=1,2\cdots \tag{42}$$

where $\zeta(x)$ is the Riemann zeta function, viz.

$$\zeta(x)=\sum_{n=1}^{\infty}\frac{1}{n^x},\quad x>1,$$

and $\Psi_1=\arccos(3\sqrt{3/7}/2)$, $\Psi_2=\arccos(3\sqrt{3/13}/2)$, and $\Psi_3=\arccos(\sqrt{3/7})$. The first few values of U_l for $l>2$ (with l being a multiple of 6) are $U_6=-5.86303$, $U_{12}=6.00964$, $U_{18}=-5.99972$, and $U_{24}=6.00001$.

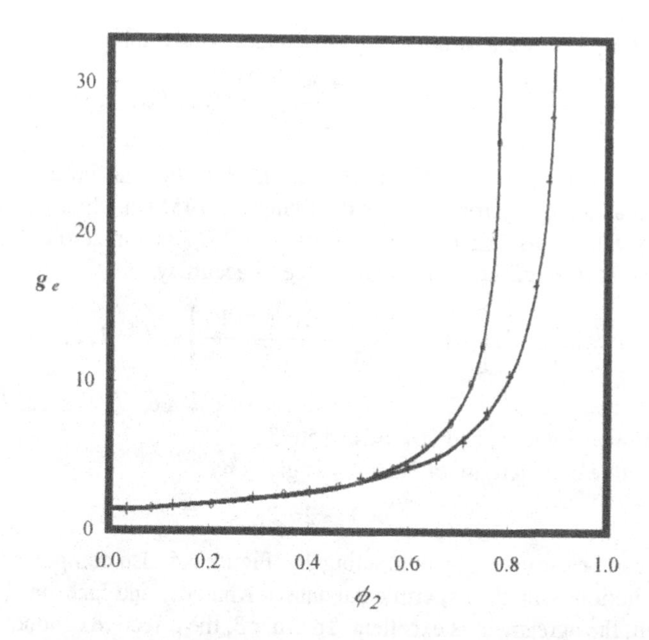

FIGURE 4.5. Comparison of the experimental data (symbols) for the effective conductivity g_e of 2D arrays of perfectly conducting spheres with the theoretical predictions. The curves on the left and right correspond, respectively, to the square and hexagonal arrays (after Perrins *et al.*, 1979a).

Application of the Green theorem then yields

$$g_e = 1 - \frac{4\pi}{\sqrt{3}} \frac{B_1}{E}, \tag{43}$$

so that the effective conductivity is obtained by estimating B_1 which itself is determined by truncating and solving Eq. (40) (in which E_0 is replaced by E). The degree of truncation depends upon the desired accuracy of g_e. Figure 4.5 compares the theoretical predictions with the experimental data; the agreement is excellent. If the conductivity ratio σ is small, or if the cylinders are not too close to each other, then the following approximate solution, obtained with the first 4 terms of the infinite series (40), provides accurate predictions:

$$g_e = 1 - 2\phi_2 \left(\Re + \phi_2 - \frac{0.075422\Re\phi_2^6}{\Re^2 - 1.060283\phi_2^{12}} - \frac{0.000076\phi_2^{12}}{\Re} \right)^{-1}, \tag{44}$$

where $\Re = (1+\sigma)/(1-\sigma)$.

4.3.2.2 Square Arrays

The problem in this case is completely analogous to the hexagonal array (Perrins *et al.*, 1979a). The critical volume fraction at which the cylinders start to touch is $\phi_c = \pi/4$. Equations (38) and (39) still express the voltage distributions outside and inside the cylinders, respectively, in which l is restricted to odd values. The analogue of Eq. (40) for a square array of cylinders is given by

$$E_0\delta_{n,1} - n!\frac{1+\sigma}{1-\sigma}\frac{B_n}{R^{2n}} = \sum_{m=1,3,\cdots} \frac{(n+m-1)!}{(m-1)!}U_{n+m}B_m, \quad n = 1, 3, \cdots \tag{45}$$

The notation is the same as before. The coefficients B_m can be determined to any desired accuracy by truncating and solving Eq. (45) (in which E_0 has been replaced by E). In this case $U_2 = \pi$ and for $m > 2$ U_m is non-zero only when m is a multiple of 4. In general, to a high degree of accuracy,

$$U_{4l} \simeq 4\left\{\zeta(4l)\left[1 + \frac{(-1)^l}{2^{2l}}\right] + \frac{2\cos(4l\Psi)}{5^{2l}}\right\}, \quad l = 1, 2, \cdots, \tag{46}$$

where $\Psi = \arccos(2/\sqrt{5})$. The first few values of U_m are, $U_4 = 3.15121$, $U_8 = 4.25577$, $U_{12} = 3.93885$, and $U_{16} = 4.015695$.

The effective conductivity of the array is given by

$$g_e = 1 - 2\pi B_1, \tag{47}$$

so that g_e can be estimated by computing B_1. Figure 4.5 also compares the theoretical predictions with the experimental data of Kharadly and Jackson (1952). As can be seen, the agreement is excellent. To order 3, the effective conductivity can be approximated by

$$g_e = 1 - 2\phi_2\left(\Re + \phi_2 - \frac{0.305827\Re\phi_2^4}{\Re^2 - 1.402948\phi_2^8} - \frac{0.013362\phi_2^8}{\Re}\right)^{-1}, \tag{48}$$

where, as before, $\Re = (1+\sigma)/(1-\sigma)$.

The limit in which the cylinders are in contact is of interest. Asymptotic analysis of O'Brien (1979) indicates that, for a pair of cylinders of conductivity σ which are separated by a small gap h, the current $I(\sigma, h)$ flowing in response to a unit applied field is given by

$$I(\sigma, h) \simeq \left(\frac{2}{\pi}\frac{\ln\sigma}{\sigma} + \frac{1}{\pi}\sqrt{\frac{h}{R}}\right)^{-1}, \quad \sigma \gg 1. \tag{49}$$

For a square array of cylinders, I is also the total current flowing in the unit cell, if the contribution to g_e from the region around the touching line is much larger than that from other regions. For a hexagonal array of cylinders, I, as given above, must be multiplied by $\sqrt{3}$ in order to obtain g_e. In the limit of touching cylinders ($h = 0$) we obtain

$$g_e \simeq \begin{cases} \dfrac{\pi}{2}\dfrac{\sigma}{\ln\sigma}, & \text{square array,} \\ \dfrac{\sqrt{3}\pi}{2}\dfrac{\sigma}{\ln\sigma}, & \text{hexagonal array,} \end{cases} \tag{50}$$

which agree with the calculations of Perrins *et al.* (1979a,b) if $\sigma > 80$. See also Milton *et al.* (1981) for arrays of intersecting square prisms and a comparison with experimental data.

Keller (1963) derived asymptotic expressions for the effective conductivity of a square array of identical and perfectly conducting circular cylinders in a matrix of conductivity g_1 near the inclusion-phase threshold $\phi_c = \phi_{2c} = \pi/4$, and for a square array of perfectly insulating inclusions near ϕ_c. His results are given by

$$\frac{g_e}{g_1} = \begin{cases} \dfrac{\pi^{3/2}}{2}(\phi_c - \phi_2)^{-1/2}, & \text{perfectly conducting,} \\ \dfrac{2}{\pi^{3/2}}(\phi_c - \phi_2)^{1/2}, & \text{perfectly insulating.} \end{cases} \tag{51}$$

Observe that the right sides of the two equations are the reciprocal of each other, which is a general property of 2D systems. We will come back to this point later in this chapter.

4.4 Exact Results for Coated Spheres and Laminates

In addition to the exact results derived in Section 4.3, one can also obtain an exact expression for the effective conductivity of a particular model of two-phase materials, first suggested by Hashin and Shtrikman (1962a,b). The material in this model consists of composite spheres that are composed of a spherical core of conductivity g_2 and radius a, surrounded by a concentric shell of conductivity g_1 with an outer radius $b > a$. The ratio a/b is fixed and in d-dimensions, $\phi_2 = (a/b)^d$. The composite spheres fill the space, implying that there is a sphere size distribution that extends to infinitesimally-small particles. This model is referred to as the *coated-spheres model.*

To derive an expression for g_e, it is sufficient to obtain the local fields for a single d-dimensional composite sphere. Hence, we consider a uniform material with (as yet) unknown conductivity g_e on which a uniform field $\mathbf{E}_0$ is imposed. We then replace a d-dimensional sphere of radius b of the material with a composite sphere, and attempt to determine g_e in such a way that the field outside the composite sphere remains unchanged, which implies that there would be no change in the material's conductivity due to the replacement. Repeating the procedure of replacing spheres with composite spheres *ad infinitum* will yield the coated-spheres model with the effective conductivity g_e.

Thus, consider a d-dimensional composite sphere in an infinite matrix of conductivity g_e, and let $\mathbf{r}$ be the radial vector emanating from the sphere center. The voltage (or temperature) distribution satisfies the Laplace equation, subject to continuity of the voltage and the radial component of its flux at $r = a$ and $r = b$. The general solution for the voltage V is then given by

$$V = \begin{cases} c_1 E_0 r \cos\theta, & r \le a, \\ c_2 E_0 r \cos\theta + c_3 \dfrac{E_0 \cos\theta}{r^{d-1}}, & a \le r \le b, \\ -E_0 r \cos\theta, & r \ge b, \end{cases} \tag{52}$$

where $E_0 = |\mathbf{E}_0|$, θ is the angle between $\mathbf{r}$ and $\mathbf{E}_0$ (aligned along the z-axis), and c_1, c_2 and c_3 are three constants that are determined from continuity of V at $r = a$ and $r = b$ and that of its radial flux at $r = a$. In particular,

$$c_2 = \frac{(1-d)g_1 - g_2}{g_2 + (d-1)g_1 - \phi_2(g_2 - g_1)}, \tag{53}$$

and

$$c_3 = \frac{a^d(g_2 - g_1)}{g_2 + (d-1)g_1 - \phi_2(g_2 - g_1)}, \tag{54}$$

where $\phi_2 = (a/b)^d$, as mentioned earlier. Continuity of the radial flux at $r = b$ (the interface between the composite sphere and the uniform matrix) then yields

$$g_e = \frac{c_3 g_1(d-1)}{b^2} - c_2 g_1 = \phi_1 g_1 + \phi_2 g_2 - \frac{\phi_1\phi_2(g_2 - g_1)^2}{\phi_2 g_1 + \phi_1 g_2 + g_1(d-1)}, \tag{55}$$

which is an exact result.

Note that, except for the trivial limit of $\phi_2 = 1$, the particle phase never percolates in the composite material. Later in this chapter (see Section 4.6) we will discuss the connection between Eq. (55) and the Hashin–Shtrikman rigorous bounds to the effective conductivity of composite materials.

Another class of models of heterogeneous materials for which exact results have been derived consists of hierarchical laminates—layered materials with n different structural levels. For example, a random first-rank laminate is composed of alternating layers of phases 1 and 2, according to some random process such that, for example, the thicknesses of the layers vary in space. A random second-rank laminate is constructed in two stages. Assume, for simplicity, that the first stage is a series of parallel slabs of *fixed* width W_1 oriented in a given direction, say the y-direction, and generated by some 1D process. Let $\phi_1^{(1)}$ and $\phi_2^{(1)}$ be the volume fractions of the matrix and the inclusion phases, respectively. In the second stage one adds ***perpendicular*** slabs of width W_2 in the gaps of the first stage. Let $\phi_1^{(2)}$ and $\phi_2^{(2)}$ be the volume fractions of phases 1 and 2, respectively, for the second-stage process. Clearly, $\phi_1^{(1)} + \phi_2^{(1)} = \phi_1^{(2)} + \phi_2^{(2)}$. Moreover, a point is in phase 1 of the entire laminate exactly when its x-coordinate is in phase 1 of the first stage and also in phase 1 of the second stage of lamination. However, since these events are independent, the volume fraction of phase 1 of the entire laminate is given by, $\phi_1 = 1 - \phi_2 = \phi_1^{(1)}\phi_1^{(2)}$. Clearly, one can generalize this construction to nth-rank laminates. It should also be clear to the reader that laminates are macroscopically anisotropic.

As an example, consider a laminate of rank 1 that consists of alternating layers of isotropic phases 1 and 2 of random thicknesses. Phase i has conductivity g_i and volume fraction ϕ_i. We assume that the laminate is of an infinite extent. It is not difficult to show that the effective conductivity tensor of this composite is given by

$$\mathbf{g}_e = \begin{bmatrix} \langle g^{-1}\rangle^{-1} & 0 & 0 \\ 0 & \langle g\rangle & 0 \\ 0 & 0 & \langle g\rangle \end{bmatrix}, \tag{56}$$

where $\langle g\rangle = \phi_1 g_1 + \phi_2 g_2$, and $\langle g^{-1}\rangle^{-1} = (\phi_1/g_1 + \phi_2/g_2)^{-1}$, and therefore the components of $\mathbf{g}_e$ are simply the arithmetic and harmonic averages of the phase conductivities. If the laminate is of rank 1, but consists of n phases, then one must use $\langle g\rangle = \sum_{i=1}^{n} \phi_i g_i$, and $\langle g^{-1}\rangle^{-1} = (\sum_{i=1}^{n} \phi_i/g_i)^{-1}$.

Although laminates may seem to have a morphology which is too simple for modeling any real disordered material, they have proven to be very useful models. In fact, Schulgasser (1976) showed that the effective conductivity of a large class of composites can be evaluated *exactly* by a lamination construction of high orders. The effective conductivity of such laminates can still be computed analytically, because, similar to laminates of rank 1, the local fields are piecewise constant under the assumption that the length scales between the different structural levels are infinitely separated. If so, one can repeatedly use the effective conductivity tensor of a first-rank laminate successively at each level of the hierarchy to obtain the effective conductivity of the entire laminate. Moreover, Schulgasser (1976), who was interested in finding the maximum range of the effective conductivity that an isotropic polycrystal, composed of grains with principal conductivities g_a, g_b, and g_c, can take on for all possible orientations, shapes, and configurations of the grains, showed that the well-known upper bound,

$$\left[\frac{1}{3}\left(\frac{1}{g_a} + \frac{1}{g_b} + \frac{1}{g_c}\right)\right]^{-1} \leq g_e \leq \frac{1}{3}(g_a + g_b + g_c), \tag{57}$$

derived by Molyneux (1970) for the effective conductivity of macroscopically isotropic polycrystals, is attained by a laminate of rank 3. Milton (1981d) showed that many bounds on the complex dielectric constant of two-phase composites, to be described later in this chapter and Chapters 5 and 6, can be realized by laminates. It can also be shown that laminates of rank d achieve the d-dimensional Hashin–Shtrikman bounds, to be derived later in this chapter. For example, it can be shown that Eq. (55) for coated-spheres model with $d = 2$, which is equal to the Hashin–Shtrikman lower bound for $g_2 \geq g_1$, is attained by a laminate of rank 2.

4.5 Perturbation Expansion for the Effective Conductivity

Regular arrays of conducting spheres embedded in a uniform matrix of a different conductivity belong to a small class of systems for which exact computations can be carried out. Another approach for deriving exact, albeit implicit, solution of the conduction problem in heterogeneous materials is based on perturbation expansions. There are several versions of such expansions which are not necessarily equivalent. In this section we discuss one such expansion for which the original idea belongs to Brown (1955). The formulation of the problem as discussed below is due to Torquato (1985a). One advantage of this perturbation expansion is that it enables one to establish links with rigorous upper and lower bounds to the effective conductivity and dielectric constant of heterogeneous materials which will be derived later in this chapter. Both isotropic and anisotropic materials can be treated, and thus we discuss them separately.

4.5.1 *Isotropic Materials: Strong-Contrast Expansion*

We assume that the material is a domain D of space of volume Ω (or surface S in 2D) which consists of two phases; one with volume (area) fraction ϕ_1 and conductivity g_1 and another (inclusion) phase characterized by a volume (area) fraction ϕ_2 and conductivity g_2. The local conductivity $g(\mathbf{r})$ at position $\mathbf{r} \in D$ is given by

$$g(\mathbf{r}) = g_j + (g_i - g_j)m_i(\mathbf{r}), \quad i \neq j, \tag{58}$$

with

$$m_i(\mathbf{r}) = \begin{cases} 1, & \mathbf{r} \in D_i \\ 0, & \text{otherwise} \end{cases} \tag{59}$$

where D_i represents the domain of phase i. The two-phase material is now subjected to the electric field $\mathbf{E}_0(\mathbf{r})$ which generates a *Lorentz electric field* $\mathbf{E}_L(\mathbf{r})$ given by (Jackson, 1998)

$$\mathbf{E}_L(\mathbf{r}) = \mathbf{E}_0(\mathbf{r}) + \int_{D_i} \mathbf{T}(\mathbf{r} - \mathbf{r}') \cdot \mathbf{P}(\mathbf{r}')d\mathbf{r}', \tag{60}$$

where D_i denotes the domain of the integration (Ω or S) with the exclusion of an infinitesimally small sphere or circular disk centered at $\mathbf{r}$. $\mathbf{E}_L(\mathbf{r})$ is related to the electric field $\mathbf{E}(\mathbf{r})$ in the material through

$$\mathbf{E}_L(\mathbf{r}) = \left[1 + \frac{g(\mathbf{r}) - g_j}{g_j d}\right] \mathbf{E}(\mathbf{r}), \tag{61}$$

where d is the dimensionality of the system, and $\mathbf{P}(\mathbf{r})$ is the induced polarization field, relative to the medium in the absence of phase i which is related to $\mathbf{E}(\mathbf{r})$ by

$$\mathbf{P}(\mathbf{r}) = \frac{g(\mathbf{r}) - g_j}{2(d-1)\pi} \mathbf{E}(\mathbf{r}). \tag{62}$$

The quantity $\mathbf{T}(\mathbf{r})$ is a dipole-dipole interaction tensor given by

$$\mathbf{T}(\mathbf{r}) = \frac{d\mathbf{r}\mathbf{r} - r^2\mathbf{U}}{g_j r^{d+2}}, \tag{63}$$

where $r \equiv |\mathbf{r}|$ and $\mathbf{U}$ is the unit dyadic. If we combine Eqs. (61) and (62), we obtain

$$\mathbf{P}(\mathbf{r}) = \frac{g_j d}{2(d-1)\pi}\left[\frac{g(\mathbf{r}) - g_j}{g(\mathbf{r}) + (d-1)g_j}\right]\mathbf{E}_L(\mathbf{r})$$
$$= \frac{g_j d}{2(d-1)\pi} g_{ij} m_i(\mathbf{r})\mathbf{E}_L(\mathbf{r}), \quad i \neq j, \tag{64}$$

where g_{ij}, given by

$$g_{ij} = \frac{g_i - g_j}{g_i + (d-1)g_j} \tag{65}$$

is proportional to the dipole polarizability of a sphere or a circular disk of conductivity g_i, embedded in a matrix of conductivity g_j. The effective conductivity g_e of the material is then defined through the following relation:

$$\langle \mathbf{P}(\mathbf{r})\rangle = \frac{g_j d}{2(d-1)\pi}\left[\frac{g_e - g_j}{g_e + (d-1)g_j}\right]\langle \mathbf{E}_L(\mathbf{r})\rangle, \tag{66}$$

where $\langle\cdot\rangle$ denotes an ensemble average.

The procedure developed by Brown (1955) (and extended by Torquato, 1985a) for calculating g_e is actually quite involved. If we combine Eqs. (60) and (64) for the Lorentz and polarization fields, we obtain an integral equation that governs $\mathbf{P}(\mathbf{r})$. This equation is solved by successive substitution which yields an expansion in powers of g_{ij}, which is then formally re-expressed in terms of an operator acting on $\mathbf{E}_0(\mathbf{r})$. One then averages the relation between $\mathbf{P}(\mathbf{r})$ and $\mathbf{E}_0(\mathbf{r})$. As discussed for regular arrays of spheres, the relation between the average field and $\mathbf{E}_0(\mathbf{r})$ is shape-dependent. To remove the shape dependence, one inverts the series for $\langle \mathbf{P}(\mathbf{r})\rangle$ in terms of $\mathbf{E}_0(\mathbf{r})$ and eliminates $\mathbf{E}_0(\mathbf{r})$ by averaging Eq. (60) for $\mathbf{E}_L(\mathbf{r})$. This procedure also eliminates the need for excluding, from the domain of integration D_i in Eq. (60), the infinitesimally small sphere or circular disk centered at $\mathbf{r}$. In the limit $\Omega \to \infty$ (or $S \to \infty$) we obtain a perturbation expansion for g_e which is given by

$$(g_{ij}\phi_i)^2\left[\frac{g_e + (d-1)g_j}{g_e - g_j}\right] = \phi_i g_{ij} - \sum_{n=3}^{\infty} A_n^{(i)} g_{ij}^n, \quad i \neq j, \tag{67}$$

where $A_n^{(i)}$ are integrals over a set of n-point probability functions. For $n \geq 3$ one has

$$A_n^{(i)} = \frac{(-1)^n \phi_i^{2-n}}{d}\left[\frac{dg_j}{2\pi(d-1)}\right]^{n-1}\int\cdots$$
$$\int \mathbf{T}(1,2)\cdots\mathbf{T}(n-1,n) : \mathbf{U}C_n^{(i)}(1,2,\cdots,n)d\mathbf{r}_2\cdots d\mathbf{r}_n, \tag{68}$$

while for $n = 2$,

$$A_2^{(i)} = \frac{d\Gamma(d/2)}{2\pi^{d/2}}\int \mathbf{T}(1,2)\left[S_2^{(i)}(\mathbf{r}_1,\mathbf{r}_2) - \phi_2^2\right]d\mathbf{r}_2, \tag{69}$$

with $\mathbf{T}(i,j) = \mathbf{T}(\mathbf{r}_i - \mathbf{r}_j)$, and $C_n^{(i)}$ is the determinant of the following matrix:

$$\begin{bmatrix} S_2^{(i)}(1,2) & S_1^{(i)}(2) & 0 & \cdots & 0 & 0 \\ S_3^{(i)}(1,2,3) & S_2^{(i)}(2,3) & S_1^{(i)}(3) & \cdots & 0 & 0 \\ \cdot & \cdot & \cdot & \cdots & \cdot & \cdot \\ \cdot & \cdot & \cdot & \cdots & \cdot & \cdot \\ S_{n-1}^{(i)}(1,\cdots,n-1) & S_{n-2}^{(i)}(2,\cdots,n-1) & \cdots & \cdots & S_2^{(i)}(n-2,n-1) & S_1^{(i)}(n-1) \\ S_n^{(i)}(1,2,\cdots,n) & S_{n-1}^{(i)}(2,3,\cdots,n) & \cdots & \cdots & S_3^{(i)}(n-2,n-1,n) & S_2^{(i)}(n-1,n) \end{bmatrix} \tag{70}$$

Here $S_n^{(i)}(1,2,\cdots,n) = S_n^{(i)}(\mathbf{r}_{12},\cdots,\mathbf{r}_{1n}) = \langle m_i(\mathbf{r}_1)m_i \cdots m_i(\mathbf{r}_n)\rangle$, is the probability of finding n points with positions $\mathbf{r}_1, \mathbf{r}_2, \cdots, \mathbf{r}_n$, all in phase i, defined in Chapter 3, and $\mathbf{r}_{ij} = \mathbf{r}_i - \mathbf{r}_j$.

Several points are worth mentioning here:

(1) Equation (67) actually represents two series expansions, one for $i = 1$ and $j = 2$, and another one for $i = 2$ and $j = 1$. Because of their dependence on $S_1^{(i)}, S_2^{(i)}, \cdots, S_n^{(i)}$, the parameters $A_n^{(i)}$ are usually referred to as the n-point microstructural parameters. Since determination of $S_n^{(i)}$ for an arbitrary microstructure is not practical, g_e cannot be determined exactly, although Eq. (67) is an exact (but implicit) equation for g_e.

(2) The n-point parameters $A_n^{(i)}$ are independent of the phase properties, and hence represent purely microstructural properties that depend only on the correlation functions $S_n^{(i)}$, and not their derivatives, as in other theories (see, for example, Beran, 1968).

(3) The presence of the tensor $\mathbf{T}(\mathbf{r})$, which decays as r^{-d} for large r, does not make the integrals in Eqs. (68) and (69) conditionally convergent since, due to the asymptotic properties of $S_n^{(i)}$ described in Chapter 3, the determinant $C_n^{(i)}$ in Eq. (68) and the quantity in the brackets in Eq. (69) both identically vanish at the boundary of the sample, and therefore the integrals in Eqs. (68) and (69) are both absolutely convergent. In addition, for $n \geq 3$, the limiting process of excluding an infinitesimally small cavity about $r_{ij} = 0$ in the integral (68) is no longer necessary, since $C_n^{(i)}$ is again identically zero for such values of n.

(4) Classical perturbation expansions involve small parameters that are simple differences in the phase conductivities (see, for example, Beran, 1968). In contrast, the series expansions (67) are *non-classical* because they involve the polarizability g_{ij} given by Eq. (65), a rational function, and therefore their convergence properties are superior to those of the classical expansions.

(5) The expansion parameter g_{ij} arises because of our choice of excluding a spherical cavity at $\mathbf{r}' = \mathbf{r}$ in the integral (60) and integrating by parts. Had we chosen a non-spherical cavity, we would have obtained a different series expansion. However, the implication of using a non-spherical cavity has not been fully understood yet. Moreover, a free parameter of this formulation is the reference material, which does not have to be phase 1 or 2. In fact, as shown by Eyre and Milton (1999), the convergence of the series expansions (67) can be ac-

celerated by judicious choice of the exclusion-cavity shape and the reference material.

The fact that g_e remains invariant under interchange of the phases in Eq. (67) implies that $A_n^{(1)}$ and $A_n^{(2)}$ are dependent upon each other. For isotropic materials $A_2^{(i)} = 0$. For $n = 3$ and $n = 4$ one has

$$\zeta_1 + \zeta_2 = 1, \tag{71}$$

where

$$\zeta_i = \frac{A_3^{(i)}}{\phi_i \phi_j (d-1)}, \quad i \neq j, \tag{72}$$

and

$$\gamma_1 - \gamma_2 = (d-2)(\zeta_2 - \zeta_1), \tag{73}$$

with

$$\gamma_i = \frac{A_4^{(i)}}{\phi_i \phi_j (d-1)}. \tag{74}$$

It can be shown that $0 \le \zeta_i \le 1$. Moreover,

$$\zeta_i = \frac{4}{\pi \phi_1 \phi_2} \int_0^\infty \frac{dr}{r} \int_0^\infty \frac{ds}{s} \int_0^\pi d\theta \cos(2\theta) \left[S_3^{(i)}(r,s,\theta) - \frac{S_2^{(i)}(r) S_2^{(i)}(s)}{\phi_2} \right], \quad d = 2, \tag{75}$$

$$\zeta_i = \frac{9}{2\phi_1 \phi_2} \int_0^\infty \frac{dr}{r} \int_0^\infty \frac{ds}{s} \int_{-1}^1 d(\cos\theta) P_2(\cos\theta) \left[S_3^{(i)}(r,s,\theta) - \frac{S_2^{(i)}(r) S_2^{(i)}(s)}{\phi_2} \right], \quad d = 3, \tag{76}$$

where P_2 is the Legendre polynomial of order 2, and θ is the angle opposite the side of the triangle of length $|\mathbf{r} - \mathbf{s}|$. Note that, due to its relation with $A_3^{(i)}$, ζ_i represents a three-point parameter, whereas γ_i is a four-point parameter.

It is useful to consider microstructures for which the parameters $A_n^{(i)}$ vanish to all orders for all values of the phase conductivities. Hence, setting $A_n^{(i)} = 0$ in Eq. (67) yields

$$\frac{g_e - g_j}{g_e + (d-1) g_j} = \frac{g_i - g_j}{g_i + (d-1) g_j} \phi_i. \tag{77}$$

For 2D and 3D systems, Eq. (77) is identical with the Hashin–Shtrikman bounds, to be described below, on the effective conductivity of two-phase isotropic materials, and therefore, as discussed in Section 4.4, is *exact* for the assemblages of coated spheres and finite-rank laminates. Thus, one may view Eq. (67) as an expansion that perturbs around the Hashin–Shtrikman microstructures, i.e., the coated-spheres model. Since, as discussed in Section 4.4, the inclusion phase in the coated-spheres model is non-percolating, we may expect the first few terms of the series (67) to provide excellent approximation for g_e for any values of the phase conductivities for materials in which the inclusions, taken to be the polarized phase, are prevented from forming large clusters.

Another interesting feature of the series (67) is as follows. If we write the expansion for $i = 1$ and $j = 2$ and add it to the expansion for $i = 2$ and $j = 1$, we obtain

$$\frac{g_e + (d-1)g_2}{g_e - g_2}\phi_1 + \frac{g_e + (d-1)g_1}{g_e - g_1}\phi_2 = 2 - d - \sum_{n=3}^{\infty}\left[\frac{A_n^{(2)}}{\phi_2}g_{21}^{n-2} + \frac{A_n^{(1)}}{\phi_1}g_{12}^{n-2}\right]. \tag{78}$$

If the sum on the right-hand side of Eq. (78) is zero, then Eq. (78) reduces to the effective-medium approximation, to be derived later in this chapter, for two-phase materials with spherical grains. This fact, together with the realization that, $g_{12}/g_{21} = -[1 + (d-2)g_{21}]^{-1}$, enable us to conclude that Eq. (78) is a series expansion of g_e in powers of g_{21} that perturbs about microstructures that realize the two-phase effective-medium approximation.

4.5.2 *Approximations*

Because determining $S_n^{(i)}$ for large values of n is not currently practical, the series (67) cannot be summed exactly, and one must resort to approximate techniques. A particularly useful method for doing so is using Padé approximation (Baker, 1975) which provides an accurate means of approximately summing a series while using only a limited number of its terms. In general, a Padé approximant of a function $f(x)$ is written as a ratio of two truncated power series in x and denoted by $[n_1/n_2]$, where n_1 and n_2 indicate the order of the series in the numerator and denominator, respectively. The Taylor series of $f(x)$ and the Padé approximant agree up to order $n_1 + n_2$. For example, the $[1/1]$ and $[2/2]$ approximants are given, respectively, by

$$\frac{g_e}{g_j} = \frac{1 - (d-1)\phi_i g_{ij}}{1 - \phi_i g_{ij}}, \quad i \neq j \tag{79}$$

$$\frac{g_e}{g_j} = \frac{1 + [(d-1)\phi_i - \gamma_i/\zeta_i]g_{ij} + (1-d)[\phi_j\zeta_i + \phi_i\gamma_i/\zeta_i]g_{ij}^2}{1 - [\phi_i + \gamma_i/\zeta_i]g_{ij} + [\phi_j(1-d)\zeta_i + \phi_i\gamma_i/\zeta_i]g_{ij}^2}, \quad i \neq j. \tag{80}$$

As was discussed by Torquato (1985a), such Padé approximants can produce rigorous upper and lower bounds to the effective conductivity of the material. We will discuss this link between the Padé approximants and the bounds to g_e in the next sections. Moreover, employing the expressions for the effective conductivity of various regular arrays of spheres, e.g., Eqs. (28) and (30), one can show that the first two terms of series (67) for $j = 1$ (i.e., terms through order g_{ij}^2) provide a very good approximation to its left-hand side for the simple-cubic, BCC and FCC lattices for broad ranges of the spheres' volume fraction ϕ_2 and the conductivity ratio σ. Thus, for such microstructures, $\sum_{n=4}^{\infty} A_n^{(2)} g_{21}^n \simeq 0$, which implies that if this remainder is indeed assumed to be zero, and the remaining equation is solved for g_e, one obtains (Torquato, 1985a)

$$\frac{g_e}{g_1} = \frac{1 + 2\phi_2 g_{21} - 2\phi_1\zeta_2 g_{21}^2}{1 - \phi_2 g_{21} - 2\phi_1\zeta_2 g_{21}^2}. \tag{81}$$

Torquato (1985a) showed that Eq. (81) yields accurate estimates for the effective conductivity of 3D dispersions, provided that the mean cluster size of the dispersed phase (the spheres) is much smaller than the macroscopic length scale, i.e., the volume fraction of the dispersed phase is not close to its percolation threshold (see below).

To utilize the series expansion (67), the Padé approximants, Eq. (81), or various bounds that will be described in Section 4.6, one needs to determine the three-point microstructural parameter ζ_2. Therefore, let us summarize and describe the most important results that have been obtained for this parameter.

4.5.3 The Microstructural Parameter ζ_2

Let us write a low-density expansion for ζ_2 given by

$$\zeta_2 = \zeta_2^{(0)} + \zeta_2^{(1)}\phi_2 + \cdots \tag{82}$$

Given Eq. (82), the following results have been obtained for the first term of the above expansion. For a d-dimensional dilute array of randomly oriented ellipsoids one has

$$\zeta_2^{(0)} = \frac{1}{d-1}\left[d\sum_{i=1}^{d} A_i^2 - 1\right], \tag{83}$$

where A_i are the depolarization factors. In particular, for 3D systems,

$$\zeta_2^{(0)} = \begin{cases} 0 & \text{spheres,} \\ \frac{1}{4} & \text{needles,} \\ 1 & \text{disks,} \end{cases} \tag{84}$$

while for 2D systems,

$$\zeta_2^{(0)} = \begin{cases} 0 & \text{disks,} \\ 1 & \text{needles.} \end{cases} \tag{85}$$

In addition, from the work of Helsing (1994) and Hetherington and Thorpe (1992) one can deduce the following results,

$$\zeta_2^{(0)} = \begin{cases} 0.11882 & \text{cubes,} \\ 0.20428 & \text{triangles,} \\ 0.08079 & \text{squares,} \\ 0.02301 & \text{hexagons,} \end{cases} \tag{86}$$

where the first equation is for random stacks of cubes, while the last three are for regular polygonal inclusions in the plane.

One of the simplest models for which the parameter ζ_2 has been estimated is the so-called symmetric cell model (see, for example, Miller, 1969, for one of the earliest work on such models; see also Section 3.4.2). In such models, one partitions the space into cells of arbitrary shapes and sizes. The cells are designated randomly and independently as phase 1 or 2, with probabilities ϕ_1 and ϕ_2, their corresponding volume fractions. Such models possess, in contrast to dispersions of inclusions, topological equivalence in the sense that their morphology with volume fractions ϕ_i is identical with another morphology with volume fractions $1 - \phi_i$ (i.e., they possess phase-inversion symmetry). They are useful models because one can compute the first few terms of series expansion (67) without any need for computing the associated correlation functions $S_n^{(i)}$. Milton (1982a,b) showed that for a d-dimensional symmetric-cell model

$$\zeta_2 = \phi_2 + \frac{(\phi_1 - \phi_2)(\mathcal{S}d^2 - 1)}{d - 1}, \tag{87}$$

where $\mathcal{S}$ is a factor that depends only on the shape of the cells. For spherical or circular cells, $\zeta_2 = \phi_2$, while for plate-like or ribbon-like cells, $\zeta_2 = \phi_1$. These are useful results because, for example, it can be shown that for cylinders (disks for $d = 2$), be they overlapping, non-overlapping, possessing polydispersivity in size, etc., ζ_2 is always bounded from above by the symmetric-cell result, $\zeta_2 = \phi_2$, and this appears to be also true for $d = 3$. Moreover, we observe that for *any* symmetric-cell material at $\phi_1 = \phi_2 = 1/2$, one must have $\zeta_2 = 1/2$.

Consider now an isotropic symmetric-cell material that consists of d-dimensional ellipsoidal cells. These space-filling cells must be randomly oriented and possess a size distribution down to infinitesimally small sizes. It can be shown that for 3D needle-like cells, $\zeta_2 = \frac{1}{4}\phi_1 + \frac{3}{4}\phi_2$, and for 3D disk-like cells or 2D needle-like cells, $\zeta_2 = \phi_1$. For symmetric-cell materials one has

$$\zeta_2 = \begin{cases} 0.11882\phi_1 + 0.88118\phi_2 & \text{cubical cells,} \\ 0.20428\phi_1 + 0.79572\phi_2 & \text{triangular cells,} \\ 0.08079\phi_1 + 0.91921\phi_2 & \text{square cells,} \\ 0.02301\phi_1 + 0.97699\phi_2 & \text{hexagonal cells.} \end{cases} \tag{88}$$

The microstructural parameter ζ_2 has also been computed for several other models of heterogeneous materials. Stell and Rikvold (1987) showed, for a distribution of overlapping spheres embedded in a uniform matrix (the Swiss-cheese model; see Section 2.11), or its 2D versions (overlapping cylinders), that the parameter ζ_2 is insensitive to polydispersivity effect. Moreover, Berryman (1985) and Torquato (1986a) showed that the first term in an expansion of ζ_2 in terms of ϕ_2, such as Eq. (82), is a good approximation to ζ_2 over almost the entire range of ϕ_2 with $\zeta_2 \simeq 0.5615\phi_2$.

However, estimation of ζ_2 for microstructures in which the inclusions do not overlap is quite difficult, since the integrals in Eqs. (75) and (76) involve the probability distribution function $S_n^{(i)}$, the evaluation of which is by itself a formidable task. For an isotropic, equilibrium distribution of hard (impenetrable) spheres, Beasley and Torquato (1986) calculated ζ_2 exactly through third order in ϕ_2 with the result being

$$\zeta_2 = 0.21068\phi_2 - 0.04693\phi_2^2 + 0.00247\phi_2^3, \tag{89}$$

which appears to be very accurate. Even if we truncate Eq. (79) after the quadratic term, it would still be very accurate for a wide range of ϕ_2, which is not really surprising since ζ_2 incorporates primarily up to three-body effects to lowest order, even when ϕ_2 is large. For a random distribution of infinitely long, parallel, identical, circular and non-overlapping cylinders, Torquato and Lado (1988) calculated ζ_2 exactly through second order in ϕ_2:

$$\zeta_2 = \frac{1}{3}\phi_2 - 0.05707\phi_2^2, \tag{90}$$

which is highly accurate in the range $0 \le \phi_2 \le 0.7$, where $\phi_2 = 0.7$ is relatively close to $\phi_c \simeq 0.81$ for random close packing. Thovert *et al.* (1990) computed exactly through first order in ϕ_2 for impenetrable spheres with a polydispersivity in size. For a bidispersed suspension with widely separated particles' sizes, they found

$$\zeta_2 = 0.35534\phi_2, \tag{91}$$

while for a polydispersed suspension containing n different and widely separated particle sizes (as $n \to \infty$), they obtained

$$\zeta_2 = 0.5\phi_2. \tag{92}$$

In addition, numerical values of ζ_2 for the three regular arrays of spheres described in Section 4.2 have also been computed. They are listed in Table 4.2 along with those for random impenetrable spheres, fully penetrable spheres model (the Swiss-cheese model), described in Chapters 2 and 3, and polydispersed impenetrable spheres. Using values of ζ_2 listed in Table 4.2, it can be easily shown that Eq. (81) provides excellent estimates of the effective conductivity of three 3D regular arrays of spheres embedded in a uniform matrix that were discussed in Section 4.2. Finally, Torquato *et al.* (1999) computed ζ_2 for 2D checkerboard geometry—tessellation of 2D space into square cells—with the result being, $\zeta_2 = 0.08079 + 0.83842\phi_2$, which is identical with what is given in Eq. (88), as it should be.

TABLE 4.2. Values of the microstructural parameter ζ_2 for simple-cubic (SC), body-centered cubic (BCC), and face-centered cubic (FCC) lattices (all computed by McPhedran and Milton, 1981), as well as fully-impenetrable spheres (FIS), fully-penetrable spheres (FPS), and polydispersed impenetrable spheres (PIS) (all tabulated by Torquato, 2002). ϕ_2 is the volume fraction of the spheres.

ϕ_2	SC	BCC	FCC	FIS	FPS	PIS
0.10	0.0003	0.0000	0.0000	0.0200	0.0564	0.0500
0.20	0.0050	0.0007	0.0004	0.0410	0.1140	0.1000
0.30	0.0220	0.0031	0.0021	0.0600	0.1712	0.1500
0.40	0.0678	0.0107	0.0078	0.0770	0.2300	0.2000
0.50	0.1738	0.0307	0.0232	0.0940	0.2900	0.2500
0.60	–	0.0796	0.0619	0.1340	0.3510	0.3000

4.5.4 *Anisotropic Materials*

The perturbation expansion derived above is now extended to anisotropic materials. Equations (60)–(65) are still applicable, but, due to the system's anisotropy, Eq. (66) is modified to (Sen and Torquato, 1989)

$$\langle \mathbf{P}(\mathbf{r})\rangle = \frac{g_j d}{2(d-1)\pi}\left[\mathbf{g}_e + (d-1)g_j\mathbf{U}\right]^{-1}(\mathbf{g}_e - g_j\mathbf{U})\langle \mathbf{E}_L(\mathbf{r})\rangle, \qquad (93)$$

where $\mathbf{g}_e$ is now the effective conductivity tensor of the anisotropic material. Then, the analogue of Eq. (67) for anisotropic materials is

$$(g_{ij}\phi_i)^2(\mathbf{g}_e - g_j\mathbf{U})^{-1}[\mathbf{g}_e + (d-1)g_j\mathbf{U}] = \phi_i g_{ij}\mathbf{U} - \sum_{n=2}^{\infty}\mathbf{A}_n^{(i)}g_{ij}^n, \quad i \neq j, \qquad (94)$$

where $\mathbf{A}_n^{(i)}$ are now tensors, the expressions for which will be given shortly. As before, we refer to $\mathbf{A}_n^{(i)}$ as the n-point microstructural parameter.

As in the case of isotropic materials, Eq. (94) for anisotropic materials actually represents two series expansions, one for $i = 1$ and $j = 2$, and a second one for $i = 2$ and $j = 1$. This means that in one case phase 1 is the matrix and phase 2 is not inserted in it, while the reverse is true in the second case.

A few properties of $\mathbf{A}_n^{(i)}$ are worth mentioning here.

(1) $\mathbf{A}_n^{(i)}$ is a symmetric tensor and for $n = 2, 3, \cdots$ will not, in general, have common principal axes, implying that, in general, the principal axes of the conductivity tensor $\mathbf{g}_e$ will rotate as the phase conductivity ratio σ changes. An example is a heterogeneous material with *chirality*, i.e., one with some degree of left- or right-handed asymmetry. Nonetheless, for at least a certain class of model materials, which includes random distributions of oriented and identical ellipsoids or cylinders in a matrix, there are enough symmetries for all the $\mathbf{A}_n^{(i)}$ to possess common principal axes, in which case n-point tensor multiplication is commutative.

(2) For macroscopically-anisotropic materials, the series expansions (94) can be considered as expansions that perturb about the optimal structures that realize the generalized Hashin–Shtrikman bounds, first derived by Willis (1977) and again by Sen and Torquato (1989).

(3) Consider microstructures for which $\mathbf{A}_n^{(i)} = \mathbf{0}$ for $n \geq 3$. Equation (94) reduces to

$$(g_{ij}\phi_i)^2[\mathbf{g}_e - g_j\mathbf{U}]^{-1}\cdot[\mathbf{g}_e + (d-1)g_j\mathbf{U}] = \phi_i g_{ij}\mathbf{U}, \quad i \neq j. \qquad (95)$$

The series (95) are equivalent to the generalized Hashin–Shtrikman bounds, to be derived below, for d-dimensional anisotropic materials, derived by Sen and Torquato (1989) using the method of Padé approximants. Willis (1977) was the first to derive (95) for $d = 3$. The bounds that are embodied in (95) are optimal, since they are achieved by certain oriented singly-coated ellipsoidal assemblages (see, for example, Milton, 1981d), as well as finite-rank laminates

(Tartar, 1985). Note that, as pointed out earlier, in such models one of the phases is always disconnected.

(4) The lowest-order parameter $\mathbf{A}_2^{(i)}$ does not generally vanish for microscopically-anisotropic materials, since $S_2^{(i)}(\mathbf{r})$ depends on $r = |\mathbf{r}|$ as well as the orientation vector $\mathbf{r}$. For macroscopically-isotropic materials, $\mathbf{A}_2^{(i)} = \mathbf{0}$, since it is traceless. However, as pointed out earlier, we should keep in mind that microscopic anisotropy does not necessarily imply macroscopic anisotropy. An example is cubic lattices of spheres which are microscopically anisotropic, but are macroscopically isotropic with respect to their effective conductivity (but not their elastic moduli; see Chapter 7).

If $\delta_{ij} = (g_i - g_j)/g_j$, then Eq. (94) through fourth-order term becomes

$$\frac{\mathbf{g}_e}{g_j} = \mathbf{U} + \mathbf{a}_1^{(i)}\delta_{ij} + \mathbf{a}_2^{(i)}\delta_{ij}^2 + \mathbf{a}_3^{(i)}\delta_{ij}^3 + \mathbf{a}_4^{(i)}\delta_{ij}^4 + O(\delta_{ij}^5), \quad i \neq j, \tag{96}$$

with

$$\mathbf{a}_1^{(i)} = \phi_i \mathbf{U}, \tag{97}$$

$$\mathbf{a}_2^{(i)} = \frac{1}{d}[\mathbf{A}_2^{(i)} - \phi_i\phi_j\mathbf{U}], \tag{98}$$

$$\mathbf{a}_3^{(i)} = \frac{1}{d^2}\left\{\phi_j^2\phi_i\left[\mathbf{U} - \frac{\mathbf{A}_2^{(i)}}{\phi_i\phi_j}\right]^2 + \mathbf{A}_3^{(i)}\right\}, \tag{99}$$

$$\mathbf{a}_4^{(i)} = \frac{1}{d^3}\left[-\phi_i\phi_j^3\mathbf{B}^3 - \phi_j\mathbf{B}\cdot\mathbf{A}_3^{(i)} - \phi_j\mathbf{A}_3^{(i)}\cdot\mathbf{B} - \mathbf{A}_3^{(i)} + \mathbf{A}_4^{(i)}\right], \tag{100}$$

where $\mathbf{B} = \mathbf{U} - \mathbf{A}_2^{(i)}/(\phi_i\phi_j)$. The first few n-point microstructural parameters are as follows,

$$\mathbf{A}_2^{(i)} = \frac{d}{2\pi(d-1)}\int_{D_i}\mathbf{T}(\mathbf{x})[S_2^{(i)}(\mathbf{x}) - \phi_i^2]d\mathbf{x}, \tag{101}$$

$$\mathbf{A}_3^{(i)} = \left[\frac{d}{2\pi(d-1)}\right]^2\int\mathbf{T}(\mathbf{x}_{12})\cdot\mathbf{T}(\mathbf{x}_{23})\left[S_3^{(i)}(\mathbf{x}_{12},\mathbf{x}_{13}) - \frac{S_2^{(i)}(\mathbf{x}_{12})S_2^{(i)}(\mathbf{x}_{23})}{\phi_i}\right]d\mathbf{x}_2 d\mathbf{x}_3, \tag{102}$$

$$\mathbf{A}_4^{(i)} = \left[\frac{d}{2\pi(d-1)}\right]^3\int\mathbf{T}(\mathbf{x}_{12})\cdot\mathbf{T}(\mathbf{x}_{23})\cdot\mathbf{T}(\mathbf{x}_{34})\left[S_4^{(i)}(\mathbf{x}_{12},\mathbf{x}_{13},\mathbf{x}_{14})\right.$$
$$\left. - \frac{S_3^{(i)}(\mathbf{x}_{12},\mathbf{x}_{13})S_2^{(i)}(\mathbf{x}_{34})}{\phi_i} - \frac{S_3^{(i)}(\mathbf{x}_{23},\mathbf{x}_{24})S_2^{(i)}(\mathbf{x}_{12})}{\phi_i} + \frac{S_2^{(i)}(\mathbf{x}_{12})S_2^{(i)}(\mathbf{x}_{23})S_2^{(i)}(\mathbf{x}_{34})}{\phi_i^2}\right]d\mathbf{x}_2 d\mathbf{x}_3 d\mathbf{x}_4, \tag{103}$$

where $\mathbf{x}_{ij} = \mathbf{x}_i - \mathbf{x}_j$, and the rest of the notation is the same as before. Note that, as mentioned earlier (see also Chapter 3), for both isotropic and anisotropic materials one has

$$S_n^{(i)}(\mathbf{x}^n) \equiv \left\langle \prod_{j=1}^{n} m_i(\mathbf{x}_i) \right\rangle. \tag{104}$$

where $m_i(\mathbf{x}_i)$ is defined by Eq. (51), and $\mathbf{x}^n = \{\mathbf{x}_1, \mathbf{x}_2 \cdots, \mathbf{x}_n\}$.

Using the perturbation expansions for the effective conductivity of isotropic as well as anisotropic materials, one can also develop rigorous upper and lower bounds to $\mathbf{g}_e$. We describe and discuss this in the next section.

4.6 Bounds on Effective Conductivity

Derivation of rigorous bounds to the effective transport properties of heterogeneous materials has a long history which goes back to at least Wiener (1912), who showed that the effective conductivity g_e of a composite material is bounded by

$$\left\langle \frac{1}{g(\mathbf{r})} \right\rangle < g_e < \langle g(\mathbf{r}) \rangle, \tag{105}$$

where the indicated averages are taken with respect to the spatial distribution of the local conductivity $g(\mathbf{r})$ at point $\mathbf{r}$. The bounds (105) state that g_e is bounded from below and above by the conductivity of hypothetical systems in which the local conductivities $\mathbf{g}(\mathbf{r})$ are in series or in parallel, respectively. As such, these bounds are not very sharp and hence are of very limited utility. We refer to (105) as the *one-point* bounds.

Although there are several methods for bounding the effective transport properties of a composite material, we discuss in detail one such method, namely, the variational principles, which has not been used very heavily for nearly four decades. The use of this method was pioneered by Hashin and Shtrikman (1962a,b,1963), Prager (1963) and Beran (1965). Their work was based on a variational principle which is based on the following key theorem.

Theorem: The functional integral defined by

$$\mathcal{H} = \frac{1}{8\pi} \int \left[\epsilon_0 \mathbf{E}_0^2 - \frac{\mathbf{P}^2}{\epsilon - \epsilon_0} + \mathbf{P} \cdot (2\mathbf{E}_0 + \mathbf{E}') \right] d\mathbf{r}, \tag{106}$$

is stationary for arbitrary variations of the polarization $\mathbf{P}$ when

$$\mathbf{P} = (\epsilon - \epsilon_0)\mathbf{E} = \mathbf{D} - \epsilon_0 \mathbf{E}, \tag{107}$$

with $\mathbf{E} = \mathbf{E}_0 + \mathbf{E}'$, subject to the conditions that, $\mathbf{E}' = -\nabla\psi$, and $\epsilon_0 \nabla \cdot \mathbf{E}' = -\nabla \cdot \mathbf{P}$. The boundary condition is that on all the surfaces one must have, $\psi = 0$. Here, $\mathbf{E}(\mathbf{r})$ and $\mathbf{D}(\mathbf{r})$ are the local electric and displacement fields, respectively, $\epsilon(\mathbf{r})$ is the spatially-varying permittivity of the material, ϵ_0 is an arbitrary constant, and $\mathbf{E}_0(\mathbf{r})$ is the electric field if the material were homogeneous. It is straightforward to show that the stationary value of (106) is exactly the electrostatic energy stored in the system, i.e.,

$$\mathcal{H}_s = \frac{1}{8\pi} \int \mathbf{E} \cdot \mathbf{D} \, d\mathbf{r}. \tag{108}$$

4.6.1 *Isotropic Materials*

We first derive the upper and lower bounds for the effective conductivity of isotropic materials, after which anisotropic materials will be considered.

4.6.1.1 Two-Point Bounds

Although the Hashin–Shtrikman (1962a,b) method involved polarization fields, and as such is somewhat different from those of others (see below), we describe their method here both because of its historical significance and the role that it has played in the research field of bounding the effective transport properties of a composite, and also the fact that their work illustrates the essence of the bounding principles. Hashin and Shtrikman studied the permittivity of a heterogeneous material, and considered suitable *trial* choices for $\mathbf{P}$ to be inserted in (106). They used the *ansatz*

$$\mathbf{P} \equiv P_i \mathbf{E}_0, \tag{109}$$

for an m-component composite, where P_i is the polarization for phase i of the material. Then, Eq. (106) yields

$$\frac{8\pi\mathcal{H}}{\Omega} = \epsilon_0 \mathbf{E}_0^2 - \left\langle \frac{P_i^2}{\epsilon - \epsilon_0} \right\rangle + 2\langle P_i \rangle E_0 + \frac{1}{\Omega}\int \mathbf{P}\cdot\mathbf{E}' d\mathbf{r}, \tag{110}$$

where, $\langle P_i \rangle = \sum_i^m \phi_i P_i$, $\langle P_i^2 \rangle = \sum_i^m \phi_i P_i^2$, and Ω is the volume of the system. Thus, one needs only to determine the integral on the right-hand side of Eq. (110). After considerable and tedious algebra (see, for example, Hashin and Shtrikman, 1962a,b; Bergman, 1978) one obtains

$$\frac{8\pi\mathcal{H}}{\Omega} = \epsilon_0 \mathbf{E}_0^2 - \left\langle \frac{P_i^2}{\epsilon - \epsilon_0} \right\rangle + 2\langle P_i \rangle \mathbf{E}_0 - \frac{1}{3\epsilon_0}\left(\langle P_i^2 \rangle - \langle P_i \rangle^2 \right). \tag{111}$$

Equation (111) is now extremised with respect to P_i, choosing a suitable value for ϵ_0. If, for a two-component system ($m = 2$), we choose ϵ_0 to be the minimum or maximum value of ϵ, i.e., ϵ_1 or ϵ_2 (if $\epsilon_1 < \epsilon_2$), respectively, then we find the upper and lower bounds to the effective permittivity, e.g., the dielectric constant ϵ_e, of the material, namely,

$$\epsilon_1 + \frac{\phi_2\epsilon_1}{\epsilon_1/\Delta\epsilon + \phi_1/3} < \epsilon_e < \epsilon_2 + \frac{\phi_1\epsilon_2}{\epsilon_2/\Delta\epsilon + \phi_2/3}, \tag{112}$$

which are the celebrated Hashin–Shtrikman bounds, with $\Delta\epsilon = \epsilon_2 - \epsilon_1$. For a d-dimensional systems, the factor 3 in (112) must be replaced by d.

Are these bounds an inherent property of the model, or are they due to the limitations of the ansatz (109)? To answer this question, Hashin and Shtrikman (1963) considered the so-called *concentric-shell model* (see Chapter 3), which is equivalent to singly-coated spheres model described in Section 4.4. Consider a homogeneous material with permittivity ϵ_0 where the surface potential is prescribed to be ψ_0 so as to create a uniform electric field $\mathbf{E}_0$ within the uniform system. The energy stored in this original system is $\epsilon_0 E_0^2/(8\pi)$. Suppose now that a sphere of radius b in this material is replaced by a composite sphere with an inner radius a and permittivity ϵ_a while the outer concentric shell is of radius b and ϵ_b. The question is: Under what conditions is there no change in the stored energy, i.e., the composite material behaves the same as the original system? The energy will remain unchanged if E_0 outside the composite sphere remains the same by the replacement. We know from elementary electrostatics (see, for example, Jackson, 1998) that for $0 < r < a$, i.e., in the inner sphere, we have

$$\mathbf{E}_a = -c_1 \cos\theta\ \mathbf{e}_r + c_1 \sin\theta\ \mathbf{e}_\theta, \tag{113}$$

where $\mathbf{e}_r$ and $\mathbf{e}_\theta$ are the unit vectors, and c_1 is a constant to be determined. For the shell area, $a < r < b$, we must have

$$\mathbf{E}_b = \left(-c_2 + \frac{c_3}{r^3}\right) \cos\theta\ \mathbf{e}_r + \left(c_2 + \frac{c_3}{r^3}\right) \sin\theta\ \mathbf{e}_\theta, \tag{114}$$

where c_2 and c_3 are also constants to be determined. Finally, outside the sphere and its shell, $b < r < \infty$, we must have no change in the energy, and therefore

$$\mathbf{E} = E_0 \cos\theta\ \mathbf{e}_r - E_0 \sin\theta\ \mathbf{e}_\theta. \tag{115}$$

To determine the constants c_1, c_2 and c_3 we must use the boundary conditions which are the continuity of displacement $D_\perp$ and the electric field $E_\parallel$ across the interfaces at $r = a$ and $r = b$. These four boundary conditions lead us to

$$\begin{cases} E_0 b - c_2 b - c_3/b^2 &= 0 \\ \epsilon_0 E_0 - \epsilon_b c_2 + 2\epsilon_b c_3/b^3 &= 0, \\ c_2 a + c_3/a^2 - c_1 a &= 0, \\ \epsilon_b c_2 - 2\epsilon_b c_2/a^3 - \epsilon_a c_1 &= 0. \end{cases} \tag{116}$$

Since there are four equations and only three unknowns, the determinant of Eqs. (116) must vanish. This leads us to

$$\epsilon_0 = \epsilon_b + \frac{(a/b)^3}{1/(\epsilon_a - \epsilon_b) + [1 - (a/b)^3]/(3\epsilon_b)}. \tag{117}$$

If we carry out the replacement process for every infinitesimal particle in the system, keeping $(a/b)^3$ (i.e., the ratio of the volumes) constant, then we have $\phi_a = (a/b)^3$ and $\phi_b = 1 - (a/b)^3$. Then, by choosing $\epsilon_1 = \epsilon_a, \phi_1 = \phi_a, \epsilon_2 = \epsilon_b$, and $\phi_2 = \phi_b$ we recover the Hashin–Shtrikman lower bound, while selecting $\epsilon_1 = \epsilon_b, \phi_1 = \phi_b, \epsilon_2 = \epsilon_a$ and $\phi_2 = \phi_a$ yields the Hashin–Shtrikman upper bound. This implies that, given only the volume fractions of the two phases and ϵ_1 and ϵ_2, the Hashin–Shtrikman bounds are the *best* possible bounds that one can obtain.

The corresponding Hashin–Shtrikman bounds for the effective conductivity of a two-phase material, consisting of phases 1 and 2 with $g_2 \geq g_1$, are give by

$$\phi_1 g_1 + \phi_2 g_2 - \frac{\phi_1\phi_2(g_1 - g_2)^2}{\phi_2 g_1 + \phi_1 g_2 + g_1(d-1)} \leq g_e \leq$$

$$\phi_1 g_1 + \phi_2 g_2 - \frac{\phi_1\phi_2(g_2 - g_1)^2}{\phi_2 g_1 + \phi_1 g_2 + (d-1)g_2}. \tag{118}$$

The Hashin–Shtrikman bounds are *exact* through second-order in the difference $(g_i - g_2)$. Observe that the lower bound is identical with Eq. (55). These bounds are realized by the singly-coated d-dimensional sphere assemblages and the second-rank laminates described in Section 4.4, which is why they represent the best possible bounds when the only information available is the volume fractions of the phases and g_1 and g_2. If one wishes to obtain sharper bounds to the effective transport properties, then the perturbation expansions derived above must be used which, as discussed above, depend on an infinite number of the n-point microstructural parameters $\mathbf{A}_n^{(i)}$. This will be discussed below.

The corresponding bounds for the effective conductivity g_e of d-dimensional, macroscopically-isotropic materials that consist of n isotropic phases are also known. Suppose that g_M and g_m are, respectively, the maximum and minimum phase conductivities. Then, with $g_m^* = (d-1)g_m$ and $g_M^* = (d-1)g_M$, one has

$$\left[\sum_{i=1}^{n} \phi_i (g_m^* + g_i)^{-1}\right]^{-1} - g_m^* \le g_e \le \left[\sum_{i=1}^{n} \phi_1 (g_M^* + g_i)^{-1}\right]^{-1} - g_M^*. \quad (119)$$

Milton (1981b) showed that these bounds for multiphase materials are realized by certain multicoated spheres for a wide range of cases. The bounds for any dimension $d \ge 2$ are realized by the corresponding d-dimensional multicoated spheres.

4.6.1.2 Cluster Bounds

If use of polarization fields is to be avoided, then the classical variational principles for the effective conductivity tensor $\mathbf{g}_e$ are as follows (Beran, 1968; McPhedran and Milton, 1981):

Minimum potential energy criterion for the upper bound:

$$\langle \nabla T \rangle \cdot \mathbf{g}_e \cdot \langle \nabla T \rangle \le \langle \mathbf{E} \cdot g\mathbf{E} \rangle, \quad \forall \mathbf{E} \in S_U, \quad (120)$$

$$S_U = \{\text{stationary } \phi(\mathbf{r});\ \nabla \times \mathbf{E} = \mathbf{0},\ \langle \mathbf{E} \rangle = \langle \nabla T \rangle\}. \quad (121)$$

Constraints (121) imply that the trial field $\mathbf{E}$ is irrotational and its ensemble average must equal the ensemble average of the true temperature gradient ∇T or voltage gradient ∇V. *Any* trial field $\mathbf{E}$ that satisfies these conditions will yield an upper bound on $\mathbf{g}_e$ given by

$$\mathbf{g}_e \le \frac{\langle g\mathbf{E} \cdot \mathbf{E} \rangle}{\langle \mathbf{E} \rangle \cdot \langle \mathbf{E} \rangle}. \quad (122)$$

Minimum complementary energy criterion for the lower bound:

$$\langle \mathbf{q} \rangle \cdot \mathbf{g}_e^{-1} \cdot \langle \mathbf{q} \rangle \ge \langle \mathbf{J} \cdot g^{-1} \mathbf{J} \rangle, \quad \forall \mathbf{J} \in S_L, \quad (123)$$

$$S_L = \{\text{stationary } \mathbf{J}(\mathbf{r});\ \nabla \cdot \mathbf{J} = 0,\ \langle \mathbf{J} \rangle = \langle \mathbf{q} \rangle\}. \quad (124)$$

As in the case of the upper bound, as long as the trial flux $\mathbf{J}$ satisfies constraints (124), and in particular its ensemble average is equal to the ensemble average of the true flux **q**, it yields a lower bound to the effective conductivity $\mathbf{g}_e$ given by

$$\mathbf{g}_e \geq \frac{\langle \mathbf{J} \rangle \cdot \langle \mathbf{J} \rangle}{\langle g^{-1} \mathbf{J} \cdot \mathbf{J} \rangle}. \tag{125}$$

As an example, if we take the admissible temperature gradient and heat flux to be constant vectors and correspond exactly to the eigenvalues of anisotropic composites composed of alternating slabs parallel or perpendicular to the applied field, we obtain the one-point bounds (105), first derived by Wiener (1912). Having set up the general concepts and ideas of bounding the effective conductivity, we now discuss more specific cases for well-defined geometries.

The perturbation expansions derived in Section 4.5 and the bounding principles discussed above can both be used for deriving upper and lower bounds to the effective conductivity (and also the dielectric constant) of heterogeneous materials. Before proceeding with describing how this is done, let us demonstrate how the variational principles are used for deriving the bounds for a relatively simple model of heterogeneous materials, namely, the fully-penetrable (the Swiss-cheese) model of N equisized spheres of radius R and conductivity g_2 inserted in a uniform matrix of conductivity g_1. The electric field at point **r** can be expanded in a cluster expansion form (Torquato, 1986a):

$$\mathbf{E}(\mathbf{r}) = \langle \mathbf{E}(\mathbf{r}) \rangle + \sum_{k=1}^{n} c_k \mathbf{E}^{(k)}(\mathbf{r}), \tag{126}$$

where the kth-order term, $\mathbf{E}^{(k)}$, is the contribution to **E** by the intrinsic k-body interactions, with the property that for all $k \geq 1$, $\langle \mathbf{E}^{(k)} \rangle = 0$. The constants c_k are selected so as to minimize the upper bound to g_e. Thus, if we substitute (126) into (122) and set $\partial g_e / \partial \mathbf{c} = 0$, where **c** is the vector with components c_k, we obtain

$$g_e \leq \langle g \rangle - \frac{\mathbf{V}^T \mathbf{W}^{-1} \mathbf{V}}{\langle \mathbf{E} \rangle \cdot \langle \mathbf{E} \rangle}, \tag{127}$$

where T denotes the transpose operation, **V** is a vector with components

$$V_k = \langle \mathbf{E} \rangle \cdot \langle g \mathbf{E}^{(k)} \rangle, \tag{128}$$

and the entries of the matrix **W** are given by

$$W_{kl} = \left\langle g \mathbf{E}^{(k)} \cdot \mathbf{E}^{(l)} \right\rangle. \tag{129}$$

Similarly, if the trial flux **J** is taken to be

$$\mathbf{J}(\mathbf{r}) = \langle \mathbf{J} \rangle + \sum_{k=1}^{n} \omega_k \mathbf{J}^{(k)}(\mathbf{r}), \tag{130}$$

and if we proceed as in the case of the upper bound, the following lower bound is obtained

$$g_e \geq \left[\langle 1/g \rangle - \frac{\mathbf{X}^T \mathbf{Y}^{-1} \mathbf{X}}{\langle \mathbf{J} \rangle \cdot \langle \mathbf{J} \rangle} \right]^{-1}, \tag{131}$$

with

$$X_k = \langle \mathbf{J} \rangle \cdot \langle \mathbf{J}^{(k)}/g \rangle, \tag{132}$$

$$Y_{kl} = \langle \mathbf{J}^{(k)} \cdot \mathbf{J}^{(l)}/g \rangle, \tag{133}$$

As pointed out by Torquato (1986a), bounds (127) and (131) are exact through nth order in the sphere volume fraction ϕ_2, because they are based on the exact solution of the electrostatic field equations for n interacting spheres (see below). We refer to them as the nth-order cluster bounds.

It is now possible to derive explicit expressions for the cluster bounds. If the trial fields are taken to be constant vectors added to the sum of the contributions from individual isolated spheres, then for the simplest case, i.e., $n = 1$, one obtains

$$g_e \leq \langle g \rangle - \frac{\langle g\mathbf{E}^{(1)} \rangle \cdot \langle g\mathbf{E}^{(1)} \rangle}{\langle g\mathbf{E}^{(1)} \cdot \mathbf{E}^{(1)} \rangle}, \tag{134}$$

$$g_e \geq \left[\langle 1/g \rangle - \frac{\langle \mathbf{J}^{(1)}/g \rangle \cdot \langle \mathbf{J}^{(1)}/g \rangle}{\langle \mathbf{J}^{(1)} \cdot \mathbf{J}^{(1)}/g \rangle} \right]^{-1}, \tag{135}$$

where $\mathbf{E}^{(1)}$ is the trial field electric field fluctuation, $\mathbf{J}^{(1)}$ is the trial current field fluctuation, and $\langle \cdot \rangle$ denotes an ensemble average. The trial fields $\mathbf{E}^{(1)}$ and $\mathbf{J}^{(1)}$ are obtained from the one-sphere electrostatic boundary value problem (Jackson, 1998):

$$\mathbf{E}^{(1)}(\mathbf{r}; \mathbf{r}^N) = \sum_{i=1}^{N} \bar{\mathbf{K}}(\mathbf{x}_i) \cdot \langle \mathbf{E} \rangle - \int \rho_1(\mathbf{r}_1) \bar{\mathbf{K}}(\mathbf{x}_1) \cdot \langle \mathbf{E} \rangle d\mathbf{r}_1, \tag{136}$$

$$\mathbf{J}^{(1)}(\mathbf{r}; \mathbf{r}^N) = \sum_{i=1}^{N} \bar{\mathbf{M}}(\mathbf{x}_i) \cdot \langle \mathbf{E} \rangle - \int \rho_1(\mathbf{r}_1) \bar{\mathbf{M}}(\mathbf{x}_1) \cdot \langle \mathbf{E} \rangle d\mathbf{r}_1, \tag{137}$$

where $\mathbf{x}_i = \mathbf{r} - \mathbf{r}_i$, $\mathbf{r}^N = \{\mathbf{r}_1, \cdots, \mathbf{r}_n\}$, and

$$\bar{\mathbf{K}}(\mathbf{r}) = \begin{cases} g_{21} R^3 (3\hat{\mathbf{r}}\hat{\mathbf{r}} - \mathbf{U})/r^3, & r > R, \\ -g_{21}\mathbf{U}, & r < R, \end{cases} \tag{138}$$

$$\bar{\mathbf{M}}(\mathbf{r}) = \begin{cases} g_1 g_{21} R^3 (3\hat{\mathbf{r}}\hat{\mathbf{r}} - \mathbf{U})/r^3, & r > R, \\ 2g_1 g_{21}\mathbf{U}, & r < R. \end{cases} \tag{139}$$

Here, $r = |\mathbf{r}|$, $\hat{\mathbf{r}} = \mathbf{r}/r$, $\mathbf{U}$ is the unit dyadic, and $g_{ij} = (g_i - g_j)/(g_j + 2g_i)$. $\rho_1(\mathbf{r})d\mathbf{r}$ is the one-particle limit of $\rho_n(\mathbf{r}^n)d\mathbf{r}^n$, introduced and discussed in detail in Chapter 3 (see Section 3.4.1.3), which gives the probability that the center of exactly one unspecified particle is in volume $d\mathbf{r}_1$ about $\mathbf{r}_1$, the center of exactly one other unspecified particle is in volume $d\mathbf{r}_2$ about $\mathbf{r}_2$, etc. It is now possible to calculate the various needed averages. These are given by (Torquato, 1986a)

$$\langle g \rangle = g_1 + \phi_2(g_2 - g_1), \tag{140}$$

$$\langle 1/g \rangle = 1/g_1 + \phi_2(g_1 - g_2)/(g_1 g_2), \tag{141}$$

$$\langle g\mathbf{E}^{(1)}\rangle = -4\pi g_{21}(g_2 - g_1)\langle \mathbf{E}\rangle \int_0^1 z^2 H_1^{(2)}(z)dz, \tag{142}$$

$$\langle \mathbf{J}^{(1)}/g\rangle = 8\pi g_{21}\left(\frac{g_1 - g_2}{g_2}\right)\langle \mathbf{E}\rangle \int_0^1 z^2 H_1^{(2)}(z)dz, \tag{143}$$

$$\frac{\langle g\mathbf{E}^{(1)} \cdot \mathbf{E}^{(1)}\rangle}{g_{21}^2\langle \mathbf{E}\rangle \cdot \langle \mathbf{E}\rangle} = g_1 A + (g_2 - g_1)B, \tag{144}$$

$$\frac{\langle \mathbf{J}^{(1)} \cdot \mathbf{J}^{(1)}/g\rangle}{g_{21}^2\langle \mathbf{E}\rangle \cdot \langle \mathbf{E}\rangle} = \frac{C}{g_1} + \frac{g_1 - g_2}{g_1 g_2} D, \tag{145}$$

with $A = \sum_{i=1}^{3} A_i$, $B = \sum_{i=1}^{4} B_i$, $C = 2A_1 + 4A_2 + A_3$, $D = 4B_1 + B_2 + 4B_3 + B_4$, where

$$A_1 = 3\eta, \tag{146}$$

$$A_2 = \frac{9}{2}\eta^2 \int_0^1 dz\, z^2 \int_0^1 dy\, y^2 \int_{-1}^1 du\, h(x), \tag{147}$$

$$A_3 = 9\eta^2 \int_1^\infty \frac{dz}{z} \int_1^\infty \frac{dy}{y} \int_{-1}^1 du\, h(x) P_2(u), \tag{148}$$

$$B_1 = 3\eta \int_0^1 dz\, z^2 G_1^{(2)}(z)/\rho, \tag{149}$$

$$B_2 = 6\eta \int_1^\infty \frac{dz}{z^4} G_1^{(2)}(z)/\rho, \tag{150}$$

$$B_3 = \frac{9}{2}\eta^2 \int_0^1 dz\, z^2 \int_0^1 dy\, y^2 \int_{-1}^1 du\, Q(\mathbf{y}, \mathbf{z})/\rho^2, \tag{151}$$

$$B_4 = 9\eta^2 \int_1^\infty \frac{dz}{z} \int_1^\infty \frac{dy}{y} \int_{-1}^1 du\, Q(\mathbf{y}, \mathbf{z}) P_2(u)/\rho^2, \tag{152}$$

and

$$Q(\mathbf{y}, \mathbf{z}) = G_2^{(2)}(\mathbf{y}, \mathbf{z}) - \rho G_1^{(2)}(\mathbf{y}) - \rho G_1^{(2)}(\mathbf{z}) + \rho^2 \phi_2. \tag{153}$$

Here the sphere radius R is taken to be unity, $\rho = N/\Omega$ is the number density, $\eta = \rho v_1$ is a reduced density, v_1 is the volume of one sphere, $P_2(u)$ is the second-order Legendre polynomial, $u = (y^2 + z^2 - x^2)/(2yz)$, and $x = |\mathbf{y} - \mathbf{z}|$.

The quantity $G_n^{(i)}(\mathbf{r}^{n+1})d\mathbf{r}_2 d\mathbf{r}_3 \cdots d\mathbf{r}_n$, the point/$n$-particle distribution function which was introduced and discussed in Chapter 3 (see Section 3.4.1.3), yields the probability of finding phase i at $\mathbf{r}_1$, the center of one unspecified particle in volume $d\mathbf{r}_2$ about $\mathbf{r}_2$, the center of another unspecified particle in volume $d\mathbf{r}_3$ about $\mathbf{r}_3$, etc. As discussed in Chapter 3, for statistically-homogeneous materials, $G_1^{(i)}(\mathbf{r}_1, \mathbf{r}_2) = G_1^{(i)}(r_{12})$, $G_2^{(i)}(\mathbf{r}_1, \mathbf{r}_2, \mathbf{r}_3) = G_2^{(i)}(r_{12}, r_{13}, \hat{\mathbf{r}}_{12} \cdot \hat{\mathbf{r}}_{13})$, etc., where $\mathbf{r}_{ij} = \mathbf{r}_i - \mathbf{r}_j$ and $\hat{\mathbf{r}}_{ij} = \mathbf{r}_{ij}/r_{ij}$. An important property of this function for a two-phase material is that (Torquato, 1986a)

$$G_n^{(1)}(\mathbf{r}; \mathbf{r}^n) + G_n^{(2)}(\mathbf{r}; \mathbf{r}^n) = \rho_n(\mathbf{r}^n). \tag{154}$$

It is also not difficult to show that for $x < 1$ we have $G_1^{(2)}(x) = \rho$. The function $H_1^{(i)}$ that appears in Eqs. (142) and (143) is given by

$$H_n^{(i)}(\mathbf{r}; \mathbf{r}^n) \equiv G_n^{(i)}(\mathbf{r}; \mathbf{r}^n) - G_0^{(i)}(\mathbf{r})\rho_n(\mathbf{r}^n). \tag{155}$$

Thus, it is straightforward to show that $H_1^{(2)}(x) = \rho\phi_1$, provided that $x < 1$. Many other properties of $G_n^{(i)}$ and $H_n^{(i)}$ were given by Torquato (1986a). The function $h(x)$ appearing in Eqs. (147) and (148) is the total correlation function discussed in Chapter 3 (see Section 3.4.1.4) and is given by, $h(x) = \rho_2(x)/\rho$. Thus, the following bounds on the effective conductivity of an isotropic composite consisting of equisized spheres of variable penetrability dispersed throughout a matrix are obtained,

$$g_e \leq \langle g \rangle - \frac{\eta^2\phi_1^2(g_2 - g_1)^2}{g_1 A + (g_2 - g_1)B}, \tag{156}$$

$$g_e \geq \left[\langle 1/g \rangle - \frac{4\eta^2\phi_1^2(g_2 - g_1)^2/(g_1 g_2)}{g_2 C + (g_1 - g_2)D} \right]^{-1}. \tag{157}$$

In both cases an averaged quantity $\langle X \rangle$ is defined as $\langle X \rangle = \phi_1 X_1 + \phi_2 X_2$. As discussed in Chapter 3, for partially-penetrable spheres the inclusion volume fraction ϕ_2 can be related to the reduced density η. For example, for fully-penetrable spheres and totally-impenetrable spheres one has, $\phi_2 = 1 - e^{-\eta}$ and $\phi_2 = \eta$, respectively. Note also that the difference between the upper and lower bounds diverges when $g_1/g_2 \to 0$ or $g_2/g_1 \to \infty$, and in particular the upper bound usually diverges, so that in these limits the upper bound is not useful.

We can make further progress if we specify the microstructure of the model. Consider, as an example, the penetrable-concentric shell (PCS) model (see Chapter 3) in which each d-dimensional sphere of diameter $2R$ is composed of a hard impenetrable core of diameter $2\lambda R$, encompassed by a perfectly penetrable shell of thickness $(1 - \lambda)R$ (see Figure 3.4). The limits $\lambda = 0$ and $\lambda = 1$ correspond, respectively, to the cases of fully-penetrable sphere (the Swiss-cheese) model, and the totally-impenetrable sphere model. For this model, one can develop low-density expansions of the various properties of interest through second order in the density ρ. The result for the total correlation function $h(x)$ is given by (Torquato, 1986b)

$$h(x; \lambda) = \begin{cases} -1, & x < 2\lambda R, \\ 0, & x > 2\lambda R, \end{cases} \tag{158}$$

and

$$G_1^{(2)}(x) = m(x)\rho + [1 - m(x)][v_1 - v_i(x; a, b)]\rho^2 \tag{159}$$

where $m(x) = 1$ if $x < R$ and $m(x) = 0$ otherwise, v_1 is the volume of one sphere, $a = \min(R, 2\lambda R)$, $b = \max(R, 2\lambda R)$, and $v_i(x; a, b)$ is the volume common to two spheres, one of radius $a < b$ and the other of radius b, the centers of which are separated by a distance x (see also Section 3.4.1.4). For $0 \le x \le R_2 - R_1$, v_i is simply the volume of the smaller sphere, $v_i = 4\pi R_1^3/3$, while for $x \ge R_1 + R_2$ the common volume is zero, $v_i = 0$. In between, for $R_2 - R_1 \le x \le R_2 + R_1$ one has

$$v_i(x; R_1, R_2) = \frac{4\pi}{3}\left[-\frac{3}{16x}(R_2^2 - R_1^2)^2 + \frac{1}{2}(R_1^3 + R_2^3) - \frac{3x}{8}(R_1^2 + R_2^2) + \frac{x^3}{16}\right]. \tag{160}$$

We also need $Q(\mathbf{r}_{12}, \mathbf{r}_{13})$ which is given by

$$Q(\mathbf{r}_{12}, \mathbf{r}_{13}) = \begin{cases} 0, & r_{12} > R \text{ and } r_{13} > R, \\ h(r_{23})\rho^2, & r_{12} < R \text{ and } r_{13} > R, \\ h(r_{23})\rho^2, & r_{12} > R \text{ and } r_{13} < R, \\ (h(r_{23}) - 1)\rho^2, & r_{12} < R \text{ and } r_{13} < R. \end{cases} \tag{161}$$

Given Eqs. (158)–(161), the quantities A, B, C, and D are computed using Eqs. (146)–(152). The results through order η^2 are given by

$$A = 3\eta - 3\eta^2(8\lambda^3 - 9\lambda^4 + 2\lambda^6), \tag{162}$$

$$B = \eta + \eta^2\left[1 - 2\lambda^6 + \frac{27\lambda^4}{2} - 16\lambda^3 + \frac{9\lambda^2}{4} + \frac{3\lambda}{4(1+2\lambda)^2} - \frac{3}{8}\ln(1+2\lambda)\right], \tag{163}$$

$$D = 4\eta - \eta^2\left[2 + 8\lambda^6 - \frac{81\lambda^4}{2} + 40\lambda^3 - \frac{9\lambda^2}{4} - \frac{3\lambda}{4(1+2\lambda)^2} + \frac{3}{8}\ln(1+2\lambda)\right], \tag{164}$$

and $C = 2A$. We may also eliminate η in favor of ϕ_2 using the relation

$$\eta = \phi_2 + \phi_2^2\left[4(1-\lambda^3) - \frac{9}{2}(1-\lambda^4) + 1 - \lambda^6\right], \tag{165}$$

which is exact through order ϕ_2^2. The resulting bounds are always sharper than the Hashin–Shtrikman bounds because they contain more microstructural information.

4.6.1.3 Three- and Four-Point Bounds

The above bounds are for a specific model of microstructure of heterogeneous materials. Beran (1965) derived upper and lower bounds on the effective conductivity of *any* statistically isotropic two-phase material which are, through ϕ_2^2, exact. To do this, he utilized trial fields based upon the first two terms of a perturbation expansions for the fields. His bounds are given by

$$\frac{g_e}{g_1} \leq 1 + \frac{g_2 - g_1}{g_1}\left[1 - \frac{g_{12}}{1 + 2g_{12}(1 - f_0)}\right]\phi_2$$

$$+ \frac{g_{12}(g_2 - g_1)}{g_1}\left[\frac{1}{1 + 2g_{12}(1 - f_0)} - g_{12}(1 + 2f_1)\right]\phi_2^2, \tag{166}$$

$$\frac{g_e}{g_1} \geq 1 + \frac{g_2 - g_1}{g_2}\left[1 + \frac{2g_{12}}{1 + g_{12}(f_0 - 1)}\right]\phi_2$$

$$+ \left\{\frac{(g_2 - g_1)^2}{g_2^2}\left[1 + \frac{2g_{12}}{1 + g_{12}(f_0 - 1)}\right]^2\right.$$

$$\left. + \frac{2g_{12}(g_1 - g_2)}{g_2}\left[\frac{1}{1 + g_{12}(f_0 - 1)} + g_{12}(2 + f_1)\right]\right\}\phi_2^2. \tag{167}$$

Here $g_{21} = (g_2 - g_1)/(g_2 + 2g_1)$. The factors f_0 and f_1 depend on the microstructure of the system. If

$$J = \frac{9}{2\phi_1\phi_2}\lim_{L\to\infty}\lim_{\Delta\to 0}\int_\Delta^L \frac{dr}{r}\int_\Delta^L \frac{ds}{s}\int_{-1}^1 d\mu S_3(r, s, \mu)P_2(\mu), \tag{168}$$

where $S_3(r, s, \mu)$ is the three-point probability function which gives the probability of finding the vertices of a triangle with sides of length r and s and angle $\cos^{-1}(\mu)$ in phase 1 (the matrix), and $P_2(x)$ is the second-order Legendre polynomial, and assuming that J can be expanded in powers of ϕ_2, then f_0 and f_1 are defined through the relation

$$J = f_0 + f_1\phi_2 + O(\phi_2^2). \tag{169}$$

In general f_0 depends on one-body information, whereas f_1 depends on one-body and two-body information. It can be shown that for spherical particles $f_0 = 1$. The Beran bounds are third-order bounds in the sense that they are exact through third order in $(g_2 - g_1)$. These bounds, as given by (166)–(169), are difficult to compute. They were simplified by Milton (1981b) and are given by

$$g_e \geq \left[\langle 1/g\rangle - \frac{4\phi_1\phi_2(1/g_2 - 1/g_1)^2}{6/g_1 + (4\phi_1 + 2\zeta_2)(1/g_2 - 1/g_1)}\right]^{-1}, \tag{170}$$

$$g_e \leq \langle g\rangle - \frac{\phi_1\phi_2(g_2 - g_1)^2}{3g_1 + (\phi_1 + 2\zeta_2)(g_2 - g_1)}, \tag{171}$$

where the ensemble average of any quantity X is defined as, $\langle X\rangle = \phi_1 X_1 + \phi_2 X_2$.

The accuracy of the Beran bounds is essentially comparable with that of (156) and (157). They are in fact identical with (156) and (157) for dispersions of identical hard spheres. This fact can be used for developing a numerical procedure for estimating the microstructural parameter ζ_2. A comparison of (156) and (157) with (170) and (171) shows that $A = 3\phi_1\phi_2$, and $B = \phi_1^2\phi_2 + 2\zeta_2\phi_1\phi_2$. In light of Eq. (144) one can then show that

$$\zeta_2 = \frac{\langle m_2 \mathbf{E}^{(1)} \cdot \mathbf{E}^{(1)} \rangle}{2\phi_1\phi_2 g_{12}^2 \langle \mathbf{E} \rangle \cdot \langle \mathbf{E} \rangle} - \frac{1}{2}\phi_1, \tag{172}$$

where m_2 is defined by Eq. (59). Equation (172) provides a means of computing the microstructural parameter ζ_2. This was exploited by Miller and Torquato (1990) who obtained, through computer simulation, accurate estimates of ζ_2.

In general, d-dimensional three-point bounds on the effective conductivity of two-phase isotropic materials of *any* microstructure are given by

$$\phi_1 g_1 + \phi_2 g_2 - \frac{\phi_1\phi_2(g_2 - g_1)^2}{\phi_2 g_1 + \phi_1 g_2 + (d-1)\langle g^{-1}\rangle_\zeta} \le g_e$$
$$\le \phi_1 g_1 + \phi_2 g_2 - \frac{\phi_1\phi_2(g_2 - g_1)^2}{\phi_2 g_1 + \phi_1 g_2 + \langle g \rangle_\zeta}, \tag{173}$$

where for any property p one has, $\langle p \rangle_\zeta = p_1\zeta_1 + p_2\zeta_2$ with $\zeta_1 + \zeta_2 = 1$. Since $\zeta_i \in [0, 1]$, the bounds (173) are always more accurate than the Hashin–Shtrikman bounds. When $\zeta_2 = 0$, the bounds (173) coincide with the Hashin–Shtrikman lower bound (for $g_2 \ge g_1$), whereas when $\zeta_1 = 1$, they coincide with the Hashin–Shtrikman upper bounds (for $g_2 \ge g_1$). The bounds (173) are exact through third order in the difference in the phase conductivities.

Milton (1982a,b) derived a three-point lower bound on g_e for $d = 3$ and $g_2 > g_1$, which improves upon the Beran lower bound, but only slightly:

$$\frac{g_e}{g_1} \ge \frac{1 + (1 + 2\phi_2)g_{21} - 2(\phi_1\zeta_2 - \phi_2)g_{21}^2}{1 + \phi_1 g_{21} - (2\phi_1\zeta_2 + \phi_2)g_{21}^2}, \tag{174}$$

where g_{21} is the same as before, and ζ_2 is given by Eq. (76), or any of its approximations discussed in Section 4.5.3. Milton (1981a) also derived n-point bounds on g_e for $d = 2$ and $d = 3$ which are exact through the nth order in $(g_i - g_j)$. For even values of n, his n-point bounds are exactly realized for space-filling multicoated cylinders (disks in 2D) where each multicoated cylinder or disk has $n/2$ coatings and is similar, to within a scale factor, to any other multicoated cylinder in the composite; see Figure 4.6. In the case of $n = 4$ and for $g_2 \ge g_1$, his lower bound is given by Eq. (80) with $i = 2$ and $j = 1$, while his upper bound is obtained from Eq. (80) with $i = 1$ and $j = 2$. Thus, the Padé approximation that was used to derive Eq. (80) actually yields rigorous upper and lower bounds to the effective conductivity.

Figure 4.7 compares the three-point bounds of Milton and Beran with the simulation results of Kim and Torquato (1992) for an equilibrium distribution of overlapping spheres in a matrix. The ratio of the conductivities of the spheres and the matrix is $\sigma = 10$. As can be seen, for this value of σ, the two bounds provide very sharp estimates of the conductivity. Figure 4.8 makes the same comparison but for $\sigma = 50$, and it is clear that as the conductivity of the spheres becomes much larger than that of the matrix, the bounds begin to become relatively poor. Figure 4.9 compares the results for the case of superconducting spheres ($\sigma = \infty$) with the lower bound of Milton and Beran which, up to about a sphere volume

FIGURE 4.6. The double-coated spheres model in two dimensions.

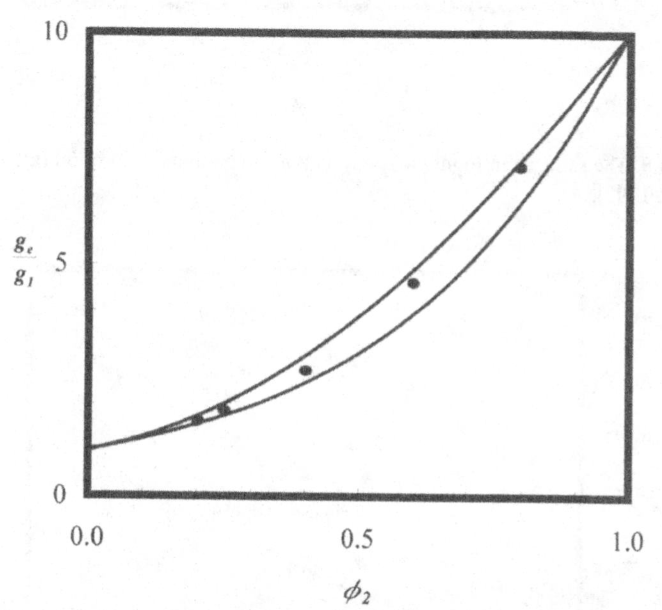

FIGURE 4.7. Comparison of the rigorous three-point bounds for effective conductivity of an equilibrium distribution of overlapping spheres (curves), in a matrix of conductivity g_1, with the simulation results (circles). The ratio of the conductivities of the spheres and the matrix is 10 (after Kim and Torquato, 1992).

fraction of $\phi_2 \simeq 0.2$, provides reasonable estimates of the effective conductivity. Also shown in this figure are the predictions of Eq. (81) which in fact performs better than the lower bound. One can see that, as long as the superconducting spheres are not close to forming a sample-spanning cluster, the three-point bound and Eq. (81) both perform well, but as the sphere volume fraction ϕ_2 approaches the percolation threshold, the lower bound and Eq. (81) both fail.

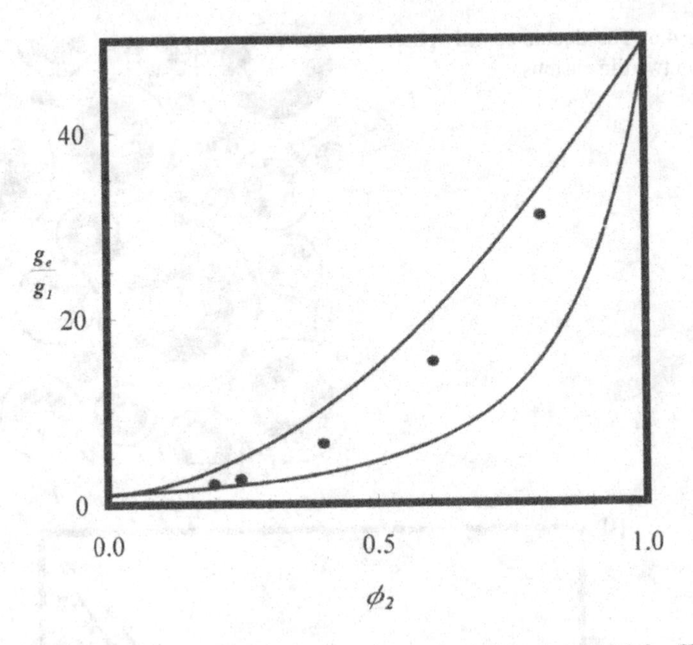

FIGURE 4.8. The same as in Figure 4.7, but with a conductivity ratio of 50 (after Kim and Torquato, 1992).

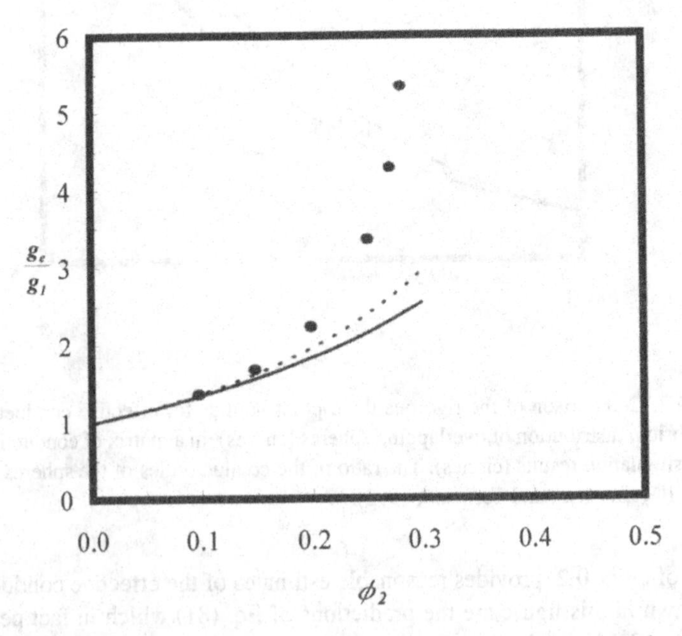

FIGURE 4.9. Effective conductivity of an equilibrium distribution of superconducting overlapping spheres in a matrix of conductivity g_1. The solid curve is the rigorous three-point lower bound, the dotted curve is the approximation given by Eq. (81), while the circles are the numerical simulation results (after Kim and Torquato, 1992).

Four-point bounds on the effective conductivity of two-phase isotropic materials were first derived by Milton (1981b). They are given by

$$\frac{g_L^{(4)}}{g_1} = \frac{1 + [(d-1)\phi_2 - \gamma_2/\zeta_2]g_{21} + (1-d)[\phi_1\zeta_2 + \phi_2\gamma_2/\zeta_2]g_{21}^2}{1 - (\phi_2 + \gamma_2/\zeta_2)g_{21} + [\phi_1(1-d)\zeta_2 + \phi_2\gamma_2/\zeta_2]g_{21}^2}, \tag{175}$$

$$\frac{g_U^{(4)}}{g_1} = \frac{1 + [(d-1)\phi_1 - \gamma_1/\zeta_1]g_{12} + (1-d)[\phi_2\zeta_1 + \phi_1\gamma_1/\zeta_1]g_{12}^2}{1 - (\phi_1 + \gamma_1/\zeta_1)g_{12} + [\phi_2(1-d)\zeta_1 + \phi_1\gamma_1/\zeta_1]g_{12}^2}, \tag{176}$$

where all the parameters were already defined above. Note that for $d = 2$ the four-point bounds are exactly zero [since $\gamma_1 = \gamma_2$; see Eq. (73)], and hence in this case the four-point bounds depend only on the phase conductivities and volume fractions, and the parameter $\zeta_1 = 1 - \zeta_2$.

4.6.1.4 Cluster Expansions for the Effective Conductivity

The cluster expansion derived in Section 4.6.1.2 can also be used for obtaining simple expressions for the effective conductivity of composite materials, since the bounds (127) and (131) are exact through nth order in the sphere volume fraction ϕ_2. We do not give further details of these expansions, and only summarize the results.

Through first order in ϕ_2, the effective conductivity of a dilute dispersion of d-dimensional spheres is given by

$$g_e = g_1 + dg_1g_{21}\phi_2 + O(\phi_2^2). \tag{177}$$

The result for $d = 3$ is usually attributed to Maxwell (1873). It can be shown that the polarization coefficient dg_1g_{21} is not changed if the spheres' sizes are distributed according to a statistical distribution. In fact, the size-independence of the polarization coefficient is independent of the particles' shape. Moreover, if the Hashin–Shtrikman bounds are expanded through first order in ϕ_2, the result would agree with Eq. (177). For the two limiting cases of interest to us, Eq. (177) reduces to

$$\frac{g_e}{g_1} = \begin{cases} 1 - \frac{d}{d-1}\phi_2 + O(\phi_2^2), & \text{insulating spheres,} \\ 1 + d\phi_2 + O(\phi_2^2), & \text{superconducting spheres.} \end{cases} \tag{178}$$

Consider now the non-dilute dispersion of spheres. For the most random distribution of insulating spheres, subject to impenetrability constraint, Jeffrey (1973) derived the following expression:

$$\frac{g_e}{g_1} = 1 - \frac{3}{2}\phi_2 + 0.588\phi_2^2 + O(\phi_2^3), \tag{179}$$

whereas for superconducting d-dimensional spheres one has

$$\frac{g_e}{g_1} = \begin{cases} 1 + 3\phi_2 + 4.51\phi_2^2 + O(\phi_2^3), & d = 3, \\ 1 + 2\phi_2 + 2.74\phi_2^2 + O(\phi_2^3), & d = 2. \end{cases} \tag{180}$$

The 2D result was derived by Peterson and Hermans (1969).

Finally, for the fully-penetrable spheres model, the following results have been derived (Torquato, 1985b)

$$\frac{g_e}{g_1} \simeq \begin{cases} 1 - \frac{3}{2}\phi_2 + 0.345\phi_2^2 + O(\phi_2^3), & \text{insulating spheres,} \\ 1 + 3\phi_2 + 7.56\phi_2^2 + O(\phi_2^3), & \text{superconducting spheres.} \end{cases} \tag{181}$$

4.6.2 Anisotropic Materials

Derivation of upper and lower bounds on the effective conductivity of anisotropic materials is similar to that for the isotropic case, except that for anisotropic materials one must take into account the tensorial nature of the effective conductivity. As usual, we consider a two-phase material. Recall that Eq. (94) actually represents two series expansions, one for $i = 1$ and $j = 2$, and a second one for $i = 2$ and $j = 1$, implying that $\mathbf{g}_e$ remains invariant to interchange of the phases, and thus the $\mathbf{A}_n^{(1)}$ are related to the $\mathbf{A}_n^{(2)}$. Sen and Torquato (1989) showed, in fact, that

$$\mathbf{A}_2^{(1)} = \mathbf{A}_2^{(2)}, \tag{182}$$

$$\mathbf{A}_3^{(1)} + \mathbf{A}_3^{(2)} = (d-1)\phi_1\phi_2\mathbf{U} - (d-2)\mathbf{A}_2^{(1)} - \frac{1}{\phi_1\phi_2}\mathbf{A}_2^{(1)} \cdot \mathbf{A}_2^{(1)}, \tag{183}$$

and

$$\mathbf{A}_4^{(1)} - \mathbf{A}_4^{(2)} = (d-2)\left[\mathbf{A}_3^{(2)} - \mathbf{A}_3^{(1)}\right] + \frac{1}{\phi_1\phi_2}\left[\mathbf{A}_2^{(1)} \cdot \mathbf{A}_3^{(2)} - \mathbf{A}_3^{(1)} \cdot \mathbf{A}_2^{(1)}\right]. \tag{184}$$

Moreover, in 2D all the $\mathbf{A}_{2n+1}^{(i)}$ are scalar multiplier of the unit tensor, and all the $\mathbf{A}_{2n}^{(i)}$ are traceless. For macroscopically-isotropic materials, $\mathbf{A}_2 = \mathbf{A}_2^{(1)} = \mathbf{A}_2^{(2)} = \mathbf{0}$.

To derive the bounds for anisotropic materials one follows the prescription for the isotropic case, namely use of Padé approximation (Sen and Torquato, 1989; Torquato and Sen, 1990), variational principles, or any method that is used for deriving the bounds for isotropic materials, suitably modified for anisotropic materials. The nth-order bounds are exact through order $\delta_{ij} = (g_i - g_j)/g_j$. We summarize the results here.

4.6.2.1 Two-Point Bounds

The two-point anisotropic generalizations of the Hashin–Shtrikman bounds on $\mathbf{g}_e$ (with $g_2 \geq g_1$) are given by

$$\mathbf{g}_L^{(2)} = \langle \mathbf{g} \rangle + (g_1 - g_2)^2 \mathbf{a}_2 \cdot \left[g_1\mathbf{U} + \frac{(g_1 - g_2)^2}{\phi_2}\mathbf{a}_2 \right]^{-1}, \tag{185}$$

$$\mathbf{g}_U^{(2)} = \langle \mathbf{g} \rangle + (g_1 - g_2)^2 \mathbf{a}_2 \cdot \left[g_2\mathbf{U} + \frac{(g_2 - g_1)^2}{\phi_1}\mathbf{a}_2 \right]^{-1}, \tag{186}$$

where [see Eq. (98)] $\mathbf{a}_2 \equiv \mathbf{a}_2^{(1)} = \mathbf{a}_2^{(2)} - (\phi_1\phi_2/d)\mathbf{U}$, and thus the trace of $\mathbf{a}_2$ is just $-\phi_1\phi_2$. This property can be exploited for obtaining simpler bounds on the eigenvalues of $\mathbf{g}_e$. Therefore, these bounds incorporate only volume-fraction information. These bounds were derived by Willis (1977) using the generalization of the Hashin–Shtrikman method described earlier, and by Torquato and Sen (1990) using the method of Padé approximants. The two bounds may be written in terms of a single expression:

$$\frac{\mathbf{g}^{(2)}}{g_1} = \left[\mathbf{U} + \left(\phi_i \mathbf{U} - \frac{1}{\phi_i}\mathbf{a}_2^{(i)}\right)\delta_{ij}\right]\left(\mathbf{U} - \frac{1}{\phi_i}\mathbf{a}_2^{(i)}\delta_{ij}\right)^{-1}, \tag{187}$$

where $\delta_{ij} = (g_i - g_j)/g_j$. For $g_1 \leq g_2$ the lower bound is obtained by setting $i = 2$ and $j = 1$, while the upper bound is obtained with $i = 1$ and $j = 2$. If the material is isotropic, then these bounds reduce to the Hashin–Shtrikman bounds.

Willis (1977) also derived two-point bounds for 3D multiphase anisotropic materials with n anisotropic phases of conductivities $\mathbf{g}_1, \cdots, \mathbf{g}_n$ and a comparison phase 0 of conductivity $\mathbf{g}_0$. They are given by

$$\mathbf{g}_e \begin{cases} \geq \bar{\mathbf{g}}, & \mathbf{g}_i \geq \mathbf{g}_0, \\ \leq \bar{\mathbf{g}}, & \mathbf{g}_i \geq \mathbf{g}_0, \end{cases} \tag{188}$$

for $i = 1, 2, \cdots, n$, where

$$\bar{\mathbf{g}} = \left[\sum_{i=1}^{n} \phi_i \mathbf{g}_i \mathbf{R}_i^{(0)}\right] \cdot \left[\sum_{i=1}^{n} \phi_i \mathbf{R}_i^{(0)}\right]^{-1}, \tag{189}$$

$$\mathbf{R}_i^{(0)} = [\mathbf{U} + \mathbf{P}_0 \cdot (\mathbf{g}_i - \mathbf{g}_0)]^{-1}, \tag{190}$$

$$\mathbf{P}_0 \left[S_2^{(ij)} - \phi_i\phi_j\right] = -\int \mathbf{G}_0(\mathbf{x} - \mathbf{x}')\left[S_2^{(ij)}(\mathbf{x} - \mathbf{x}') - \phi_i\phi_j\right] d\mathbf{x}'. \tag{191}$$

Here $S_2^{(ij)}(\mathbf{x})$ is the probability of finding one point in phase i and another in phase j with a relative separation of $\mathbf{x}$, which, in the absence of long-range order, has the asymptotic properties that, $S_2^{(ij)}(\infty) = \phi_i\phi_j$, and

$$S_2^{(ij)} = \begin{cases} \phi_i, & i = j, \\ 0, & i \neq j. \end{cases} \tag{192}$$

The second-rank tensor $\mathbf{G}_0(\mathbf{r})$ is the double gradient of the Green function for a system with conductivity tensor $\mathbf{g}_0$ [see, for example, Eq. (247) given below]. For two-phase materials ($n = 2$), $\mathbf{P}_0 = -\mathbf{a}_2/(\phi_1\phi_2 g_0)$, and the bounds (188) reduce to (187).

4.6.2.2 Three- and Four-Point Bounds

These bounds are given by (Torquato and Sen, 1990)

$$\left(\frac{\mathbf{g}_e^L}{g_1}\right)^{-1} = (1 + \phi_1\delta_{21})\mathbf{U} + \delta_{21}^2 \mathbf{c}_2 \cdot [(1 - \phi_1\delta_{21})\mathbf{c}_2 + (\mathbf{b}_3 + \phi_1\mathbf{b}_2)\delta_{21}]^{-1} \cdot \mathbf{c}_2, \tag{193}$$

$$\frac{\mathbf{g}_e^U}{g_1} = (1 + \phi_2\delta_{21})\mathbf{U} + \delta_{21}^2 \mathbf{a}_2^{(2)} \cdot (\mathbf{a}_2^{(2)} - \mathbf{a}_3^{(2)}\delta_{21})^{-1} \cdot \mathbf{a}_2^{(2)}, \tag{194}$$

with $\mathbf{b}_2 = \phi_1\mathbf{U} + \mathbf{a}_2^{(1)}$, $\mathbf{c}_2 = \phi_1^2\mathbf{U} - \mathbf{b}_2$, and $\mathbf{b}_3 = -[\phi_1\mathbf{U} + 2\mathbf{a}_2^{(1)} + \mathbf{a}_3^{(1)}]$, where $\mathbf{a}_3^{(1)}$ is given by Eq. (99). For isotropic materials, these bounds reduce to those of Beran described above. Milton (1987), using an elegant continued-fraction formulation of the problem, derived n-point bounds for anisotropic materials which, however, are formal in the sense that the microstructural parameters are not given explicitly in terms of the n-point probability functions that were described in Chapter 3.

The fourth-order bounds are given by

$$\frac{\mathbf{g}_e}{g_j} = \frac{1}{2}\left[\mathbf{AB}^{-1} + (\mathbf{AB}^{-1})^T\right], \tag{195}$$

where

$$\mathbf{A} = \mathbf{U} + \mathbf{p}_1^{(i)}\delta_{ij} + \mathbf{p}_2^{(i)}\delta_{ij}^2, \tag{196}$$

$$\mathbf{B} = [\mathbf{U} + \mathbf{q}_1^{(i)}\delta_{ij} + \mathbf{q}_2^{(i)}\delta_{ij}^2]^{-1}, \tag{197}$$

with

$$\mathbf{q}_1^{(i)} = [\mathbf{a}_2^{(i)}\cdot\mathbf{a}_2^{(i)} - \phi_i\mathbf{a}_3^{(i)}]^{-1}\cdot[\mathbf{a}_1^{(i)}\cdot\mathbf{a}_4^{(i)} - \mathbf{a}_2^{(i)}\cdot\mathbf{a}_3^{(i)}], \tag{198}$$

$$\mathbf{q}_2^{(i)} = -\phi_i^{-1}[\mathbf{a}_3^{(i)} + \mathbf{a}_2^{(i)}\cdot\mathbf{q}_1^{(i)}], \tag{199}$$

where $\mathbf{p}_1^{(i)} = \mathbf{a}_1^{(i)} + \mathbf{q}_1^{(i)}$, and $\mathbf{p}_2^{(i)} = \mathbf{a}_2^{(i)} - \phi_i^{-1}\mathbf{a}_3^{(i)} + [\mathbf{a}_1^{(i)} - \phi_i^{-1}\mathbf{a}_2^{(i)}]\cdot\mathbf{q}_1^{(i)}$, with $\mathbf{a}_4^{(i)}$ being given by Eq. (100). Equation (195), for $g_1 \le g_2, i = 1$ and $j = 2$, yields an upper bound for $\mathbf{g}_e$, while for $i = 2$ and $j = 1$ one obtains a lower bound.

4.6.2.3 Simplification of the Bounds

As Eqs. (101)–(103) indicate, one needs the n-point probability function $S_n^{(i)}(\mathbf{x}^n)$, and in particular $S_n^{(1)}(\mathbf{x}^n)$, in order to estimate the above bounds. In this case

$$S_n^{(1)}(\mathbf{x}^n) = 1 + \sum_{k=1}^{N}\frac{(-1)^k}{k!}\int \rho_k(\mathbf{r}^k)\prod_{j=1}^{k}\left[1 - \prod_{j=1}^{n}[1 - m(\mathbf{x}_i - \mathbf{r}_j)]\right] d\mathbf{r}_j, \tag{200}$$

where, as usual, the inclusion indicator function $m(\mathbf{x})$ depends on the shape of the inclusions. Equation (200) is valid for general heterogeneous arrays of identical, oriented inclusions of arbitrary shapes distributed throughout a matrix. Although $S_n^{(1)}(\mathbf{x}^n)$, as given by Eq. (200), appears to be formidable to compute, it can, for certain limits, be simplified, two of which are as follows (Torquato and Sen, 1990).

Dilute distributions of oriented inclusions

In this limit, even if the inclusions interact with each other with an arbitrary interparticle potential, Eq. (200) can be explicitly evaluated. Through order ρ one has

$$S_n^{(1)}(\mathbf{x}_{12}, \cdots, \mathbf{x}_{1n}) = 1 - \rho \int \left[1 - \prod_{i=1}^{n} [1 - m(\mathbf{x}_i - \mathbf{r}_1)] \right] d\mathbf{r}_1$$
$$= 1 - \rho v_i(\mathbf{x}_{12}, \cdots, \mathbf{x}_{1n}), \tag{201}$$

where v_i, the union or intersection volume of n regions, all with the same size, shape and orientation of an inclusion, with centroids that are separated by the vectors $\mathbf{x}_{12}, \cdots, \mathbf{x}_{1n}$. For $n = 1$, v_i is simply the volume of an inclusion. For $n = 2$, for example, the union volume of two rectangular or two cylindrical inclusions are given, respectively, by Eqs. (3.86) and (3.88).

Overlapping oriented inclusions

As discussed in Chapter 3, for overlapping inclusions $\rho_n(\mathbf{x}^n) = \rho^n$, and therefore Eq. (200) becomes

$$S_n^{(1)}(\mathbf{x}_{12}, \cdots, \mathbf{x}_{1n}) = \exp[-\rho v_i(\mathbf{x}_{12}, \cdots, \mathbf{x}_{1n})]. \tag{202}$$

As usual, the problem is more difficult when the overlapping inclusions form a sample-spanning cluster, in which case the bounds are not usually very useful.

Another limiting case of interest, non-overlapping oriented inclusions, is, as discussed in Chapter 3, a very difficult problem. Computation of even low-order statistics, such as $S_2^{(1)}$, is quite involved. Lado and Torquato (1990) considered the case in which the inclusion was an axially symmetric spheroid aligned with its symmetry axis in the x_3-direction, so that its inclusion indicator function is given by

$$m(\mathbf{r}) = \begin{cases} 1, & (x_1^2 + x_2^2)/a^2 + x_3^2/b^2 < 1, \\ 0, & \text{otherwise} \end{cases} \tag{203}$$

The eccentricity of the spheroid is thus, $\epsilon = b/a$, and therefore the model includes oriented disks ($\epsilon = 0$), spheres ($\epsilon = 1$), and oriented needles ($\epsilon = \infty$). Figure 4.10 presents $S_2(r, \theta) = S_2^{(1)}(\mathbf{x}_{12}) - \phi_1^2$ versus r/b for $\phi_2 = 0.6$ and $\epsilon = 2.0$. Here θ is the polar angle between the x_3-axis and $\mathbf{r} = (x_1, x_2, x_3)$. The results are for three values of θ, and as can be seen, the behavior of $S_2(r, \theta)$ is quite complex.

Further progress in evaluating these bounds can be made if the structure of the inclusions is specified. For example, for any 3D distribution of inclusions aligned in the x_3 direction which possesses transverse isotropy and azimuthal symmetry (for example, circular cylinders or spheroids), the quantity $\mathbf{a}_2 = \mathbf{a}_2^{(i)}$ that appears in the bounds is given by

$$\mathbf{a}_2 = -\phi_1\phi_2 \begin{bmatrix} Q & 0 & 0 \\ 0 & Q & 0 \\ 0 & 0 & 1 - 2Q \end{bmatrix} \tag{204}$$

where

$$Q = \frac{1}{3} - \lim_{\delta \to 0} \frac{1}{4\phi_1\phi_2} \int_\delta^\infty \frac{dx}{x} \int_0^\pi d\theta \sin\theta (1 - 3\cos^2\theta) \left[S_2^{(1)}(\mathbf{x}) - \phi_1^2 \right]. \tag{205}$$

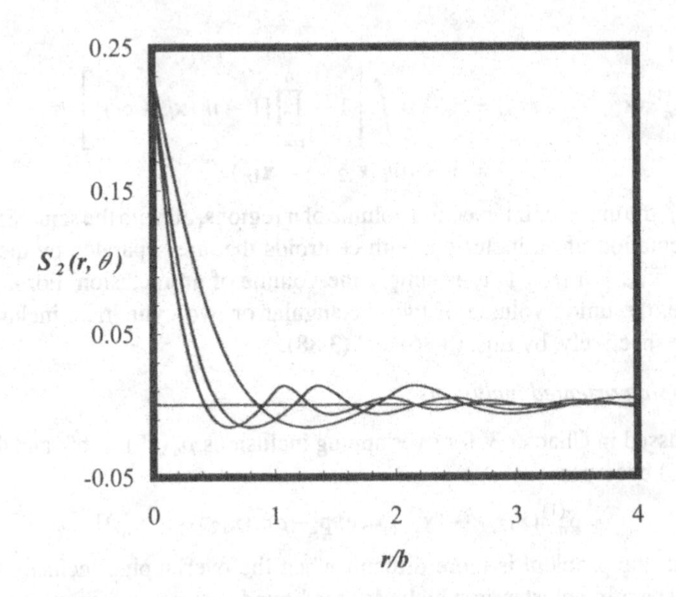

FIGURE 4.10. Cross sections through the two-point probability function $S_2(r, \theta)$ for hard, oriented prolate spheroids at a volume fraction $\phi_2 = 0.6$ of the particles and eccentricity $b/a = 2$. The curves, from right to left on the main peak, are for $\theta = 0°$, 45° and 90° (after Lado and Torquato, 1990).

Here, $x = |\mathbf{x}|$, and θ is the polar angle measured with respect to the x_3-axis. Note that, due to the symmetry in the x_1x_2 plane, the diagonal elements $(a_2)_{11}$ and $(a_2)_{22}$ are equal. It can be shown that for *possibly overlapping* spheroidal inclusions aligned parallel to the x_3-axis with length $2b$ and maximum diameter $2a$ [see Eq. (204)], one has (Willis, 1977)

$$Q = \frac{1}{2}\left\{1 + \frac{1}{(b/a)^2 - 1}\left[1 - \frac{1}{2\chi_1}\ln\left(\frac{1+\chi_1}{1-\chi_1}\right)\right]\right\}, \quad b/a > 1, \tag{206}$$

$$Q = \frac{1}{2}\left\{1 + \frac{1}{(b/a)^2 - 1}\left[1 - \frac{1}{\chi_2}\tan^{-1}(\chi_2)\right]\right\}, \quad b/a < 1, \tag{207}$$

where $\chi_1^2 = -\chi_2^2 = 1 - a^2/b^2$. Numerical values of Q for overlapping cylinders and various values of b/a and ϕ_2 are listed in Table 4.3. Given Eq. (204), the second-order bounds (187) simplify to

$$\frac{(g_e)_{11}}{g_j} = \frac{(g_e)_{22}}{g_j} = \frac{1 + (\phi_i + \phi_j Q)\delta_{ij}}{1 + \phi_j Q\delta_{ij}}, \tag{208}$$

$$\frac{(g_e)_{33}}{g_j} = \frac{1 + [\phi_i + \phi_j(1 - 2Q)]\delta_{ij}}{1 + \phi_j(1 - 2Q)\delta_{ij}}, \tag{209}$$

where, as mentioned earlier, for $g_2 > g_1$ the upper bound is obtained with $i = 1$ and $j = 2$, while $i = 2$ and $j = 1$ yield the lower bound.

Finally, the important case of a model of aligned, infinitely long, equisized, rigid, circular cylinders (or disks in 2D), distributed randomly throughout a matrix,

TABLE 4.3. Dependence of the microstructural parameter Q on the aspect ratio b/a for aligned, overlapping cylinders (Torquato and Sen, 1990) and aligned, non-overlapping spheroids (Willis, 1977). ϕ_2 is the volume fraction of the cylinders. The results for the spheroids are valid for any volume fraction.

b/a	$\phi_2 = 0.1$	$\phi_2 = 0.5$	$\phi_2 = 0.9$	Spheroids
0.01	0.0171	0.0179	0.0199	0.009
0.05	0.0617	0.0640	0.0693	0.037
0.10	0.102	0.105	0.111	0.070
0.20	0.160	0.163	0.169	0.125
0.50	0.263	0.264	0.266	0.236
1.00	0.344	0.343	0.340	0.333
2.00	0.409	0.407	0.402	0.413
5.00	0.460	0.458	0.454	0.472
10.0	0.480	0.478	0.475	0.490
20.0	0.490	0.489	0.487	0.497
∞	0.500	0.500	0.500	0.500

a very useful model of fiber-reinforced materials, should be mentioned. Silnutzer (1972) derived three-point bounds for the effective conductivity of such a material which, in their original forms, were quite complex. Milton (1982a,b) simplified these bounds to

$$g_e \geq \left[\langle 1/g \rangle - \frac{\phi_1\phi_2(1/g_2 - 1/g_1)^2}{\langle 1/\hat{g} \rangle + \langle 1/\tilde{g} \rangle} \right]^{-1}, \tag{210}$$

$$g_e \leq \langle g \rangle - \frac{\phi_1\phi_2(g_2 - g_1)^2}{\langle \hat{g} \rangle + \langle \tilde{g} \rangle}, \tag{211}$$

where $\langle X \rangle = \phi_1 X_1 + \phi_2 X_2$, $\langle \hat{X} \rangle = \phi_2 X_1 + \phi_1 X_2$, and $\langle \tilde{X} \rangle = \zeta_1 X_1 + \zeta_2 X_2$. The microstructural parameters ζ_1 and ζ_2 are given by Eqs. (75) and (77). Milton (1982a,b) also derived fourth-order bounds for the conductivity of such materials which are given by

$$g_e \geq g_1 \left[\frac{(g_1 + g_2)(g_2 + \langle g \rangle) - \phi_1\zeta_2(g_2 - g_1)^2}{(g_1 + g_2)(g_1 + \langle \hat{g} \rangle) - \phi_1\zeta_2(g_2 - g_1)^2} \right], \tag{212}$$

$$g_e \leq g_2 \left[\frac{(g_1 + g_2)(g_1 + \langle g \rangle) - \phi_2\zeta_1(g_2 - g_1)^2}{(g_1 + g_2)(g_2 + \langle \hat{g} \rangle) - \phi_2\zeta_1(g_2 - g_1)^2} \right]. \tag{213}$$

The microstructural parameters ζ_1 and $\zeta_2 = 1 - \zeta_1$ were estimated very accurately by Torquato and Lado (1988); their results are represented by Eq. (90).

Figure 4.11 compares the predictions of the second-order bounds with the computer simulations of Kim and Torquato (1993). Shown in this figure are the diagonal elements of the conductivity tensor $\mathbf{g}_e/g_1$ for a composite that consists of spheroidal inclusions inserted in a uniform matrix, with $b/a = 1/2$ [see Eq. (203)], and a conductivity ratio $\sigma = g_2/g_1 = 10$. It is clear that the bounds provide very

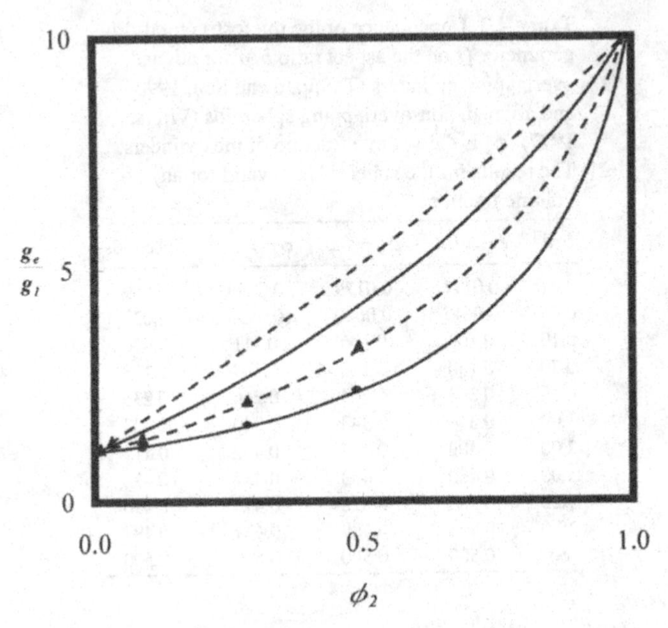

FIGURE 4.11. The diagonal components $[g_e = (g_e)_{ii}]$ of the effective conductivity of a composite material that consists of aligned spheroids of volume fraction ϕ_2, aspect ratio $b/a = 1/2$, and phase conductivity ratio $g_2/g_1 = 10$, in a matrix of conductivity g_1. The dashed curves are the two-point bounds for $(g_e)_{11}/g_1 = (g_e)_{22}/g_1$, the triangles are the corresponding simulation results, while the solid curves and the circles represent the same for $(g_e)_{33}/g_1$ (after Kim and Torquato, 1993).

reasonable estimates of the conductivities. Figure 4.12 makes the same comparison but for $\sigma = \infty$ and $b/a = 0.1$. As in the case of the isotropic case, when the conductivity ratio is large and the inclusions start to form large clusters, the bounds lose their accuracy. In fact, the upper bound diverges, rendering it useless, but, at least over a limited range of the inclusions' volume fraction, the lower bound still provides a useful estimates of the conductivities. The computer simulation results shown in Figures 4.11 and 4.12 were obtained by the random walk method that will be described later in this chapter.

4.6.2.4 Cluster Expansions for the Effective Conductivity

Similar to isotropic materials discussed in Section 4.6.1.4, one may also develop cluster expansions for the effective conductivity of anisotropic materials. Consider, as an example, a dilute suspension of d-dimensional ellipsoidal inclusions with semi-axes $a_1, \cdots, a_d$ and specified orientations. The conductivities of the matrix and inclusions are g_1 and g_2, respectively. Suppose that the ellipsoids are aligned along the x_d-axis. Then,

$$\mathbf{g}_e = \mathbf{g}_1 + \mathbf{M}\phi_2 + O(\phi_2^2), \tag{214}$$

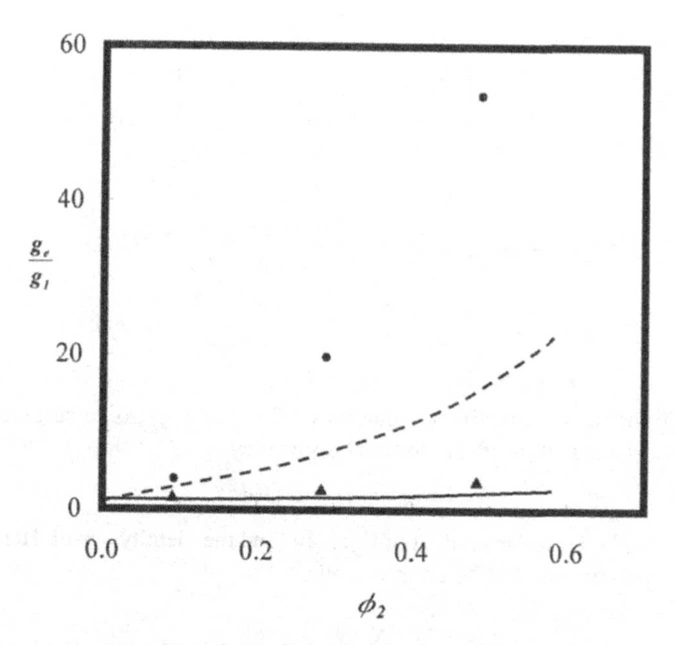

FIGURE 4.12. The diagonal components $[g_e = (g_e)_{ii}]$ of the effective conductivity of a composite that consists of superconducting aligned spheroids of volume fraction ϕ_2, and aspect ratio $b/a = 0.5$. The dashed curve and the triangles are, respectively, the two-point lower bound and the numerical simulation results for $(g_e)_{11} = (g_e)_{22}$, while the solid curve and the circles represent the same for $(g_e)_{33}$. Note that, in this case, the upper bounds diverge (after Kim and Torquato, 1993).

where

$$\mathbf{M} = (g_2 - g_1)\left(\mathbf{U} + \frac{g_2 - g_1}{g_1}\mathbf{A}\right)^{-1}. \tag{215}$$

Here, **M** and **A** are, respectively, the polarization concentration and depolarization tensors for a d-dimensional ellipsoidal inclusion. In particular, the symmetric tensor **A**, in the principal axes frame, has diagonal components A_i, $i = 1, \cdots, d$, given by

$$A_i = \left(\Pi_{j=1}^d \frac{1}{2} a_i\right) \int_0^\infty \frac{1}{(x + a_i^2)\sqrt{\Pi_{j=1}^d (x + a_j)^2}}\, dx, \quad i = 1, \cdots, d. \tag{216}$$

The eigenvalues of **A**, usually referred to as the *depolarization factors*, are all positive. The trace of **A** is unity, and therefore, $0 \le A_i \le 1$. For a d-dimensional sphere, $A_i = 1/d$; for needle-shaped inclusions, $A_1 = A_2 = 1/2$ and $A_3 = 0$, while for disk-shaped inclusions, $A_1 = A_2 = 0$ and $A_3 = 1$.

If the inclusions are aligned 3D spheroids in which $a_1 = a_2 = a$ and $a_3 = b$, then the effective conductivity tensor for this composite, which is *transversely isotropic*, is given by

$$\mathbf{g}_e = \begin{bmatrix} (g_e)_{11} & 0 & 0 \\ 0 & (g_e)_{22} & 0 \\ 0 & 0 & (g_e)_{33} \end{bmatrix}, \tag{217}$$

with

$$(g_e)_{11} = (g_e)_{22} = g_1 + \frac{\phi_2 g_1(g_2 - g_1)}{g_1 + Q(g_2 - g_1)} + O(\phi_2^2), \tag{218}$$

$$(g_e)_{33} = g_1 + \frac{\phi_2 g_1(g_2 - g_1)}{g_1 + (1 - 2Q)(g_2 - g_1)} + O(\phi_2^2), \tag{219}$$

where Q is given by Eqs. (206) and (207).

Finally, the scalar effective conductivity of a dilute isotropic suspension of randomly oriented ellipsoidal inclusions is given by

$$g_e = g_1 + M\phi_2 + O(\phi_2^2), \tag{220}$$

which is simply the scalar product of Eq. (214) and the identity tensor. Here, M is the *scalar* polarization coefficient given by

$$M = \frac{g_2 - g_1}{d} \sum_{i=1}^{d} \frac{g_1}{g_1 + A_i(g_2 - g_1)}, \tag{221}$$

and the A_i are the depolarization factors given above. The 3D version of this result was first given by Polder and Van Santen (1946). With the depolarization factors for needles and disks that we mentioned above, Eqs. (220) and (221) immediately yield

$$\frac{g_e}{g_1} = \begin{cases} 1 + \dfrac{(5g_1 + g_2)(g_2 - g_1)}{3g_1(g_1 + g_2)}\phi_2 + O(\phi_2^2), & \text{needles}, \\ 1 + \dfrac{(g_1 + 2g_2)(g_2 - g_1)}{3g_1 g_2}\phi_2 + O(\phi_2^2), & \text{disks}. \end{cases} \tag{222}$$

4.7 The Effect of the Interface on the Effective Conductivity

So far we have discussed the effective conductivity of composite materials in which the interface between the two phases is "perfect," i.e., it does not offer any resistance to the transport process. However, this may not be the case in many practical applications. For example, thermal or electrical resistance at the interface caused by, e.g., rough surfaces of the two phases in contact, can decrease significantly the effective conductivity of the composite material.

This problem, at least for dispersions of spheres inserted in a uniform matrix, can be solved by the methods that we have described so far in this chapter. Consider, for example, the case in which one has a regular array of d-dimensional spheres inserted in a uniform matrix. The methods that we described in Section 4.3 can be extended to the case in which the interface between the two phases (the inclusions and the matrix) offers resistance to the transport process. One must treat the composite material as a *three-phase* system of similar dispersion in which the spherical inclusions possess a concentric coating of thickness δ and conductivity g_i. The strength of the interface resistance to the transport process may be characterized by a dimensionless parameter $R_h = \hat{R}g_2/R$, where, as usual, R is the radius of the spheres and g_2 their conductivity. Here, $\hat{R} = \delta/g_i$, and of particular interest is the limit in which δ and g_i both approach zero. This would then be equivalent to introducing a temperature or voltage jump across the interface. As an example, consider the electrical conduction case. Then, boundary conditions (13) at the interface should be replaced by, $g_2(V^+ - V^-)/(R_h R) = g_2\partial V^-/\partial n = g_1\partial V^+/\partial n$, where n is the unit outward normal vector, and V^+ and V^- are the voltages on the two sides of the interface. The solution of this problem can be obtained by exactly the same methods that we described in Section 4.3, since the only difference between the perfect and imperfect interfaces is in the boundary conditions at the interface. Cheng and Torquato (1997) actually applied the methods of Section 4.3 to the case of imperfect interfaces. As discussed in Section 4.3, the conductivity of the regular dispersions depends only on the coefficient $B_{1,0}$ [see, for example, Eq. (25) for a simple-cubic array], and thus the difference between the conductivities of such regular composites with perfect or imperfect interfaces appears in the value of the coefficient $B_{0,1}$.

Similarly, one may also derive rigorous upper and lower bounds for the effective conductivity of composite materials with imperfect interfaces (Torquato and Rintoul, 1995). An interesting result that emerges from these works is that there is a critical value of R_h at which the conductivity of the material equals the matrix conductivity g_1, implying that the interface can act in a way that the second (inclusion) phase is *hidden*. This phenomenon has actually been observed in measuring the thermal conductivity of ZnS/diamond composites (Every *et al.*, 1992), and thus it is of practical importance.

4.8 Exact Duality Relations

A reciprocity theorem proved by Keller (1964) and Dykhne (1971) provides an important exact relation for the effective transport properties of 2D materials. Keller (1964) considered a two-phase heterogeneous material consisting of a rectangular lattice of parallel cylinders, each of which is symmetric in the x- and y-axes, with generators parallel to the z-axis, a 2D problem. He proved that the effective conductivity g_e^x in the x-direction is related to the effective conductivity g_e^y of the phase-interchanged composite (i.e., one in which the roles of phases 1 and 2 have been interchanged) in the y-direction by

$$g_e^x(g_1, g_2)g_e^y(g_2, g_1) = g_1 g_2, \tag{223}$$

where g_1 and g_2 are, respectively, the conductivities of the matrix and the inclusions. Hence, if the material is macroscopically isotropic (that is, if the effective conductivity tensor $\mathbf{g}_e$ is rotationally invariant), then Eq. (223) reduces to

$$g_e(g_1, g_2)g_e(g_2, g_1) = g_1 g_2. \tag{224}$$

In an apparently independent work, Dykhne (1971) proved that the effective conductivity of 2D two-phase composite materials that possess phase distributions that are statistically equivalent is given by

$$g_e = \sqrt{g_1 g_2}. \tag{225}$$

Subsequently, Mendelson (1975b) generalized the Keller–Dykhne theorem to *any* 2D two-phase composite, disordered or not, so long as x and y are the principal axes of the effective conductivity tensor, implying that no geometrical symmetries are required. More precisely, composite materials that satisfy Eq. (225) possess phase-inversion symmetry at $\phi_1 = \phi_2 = 1/2$. Equation (225) has another interesting implication: The percolation threshold of a continuous system for which Eq. (225) is valid is *exactly* 1/2.

The proof of the Keller–Dykhne–Mendelson result is based on a theorem that states that a 90° rotation of the effective conductivity tensor of a 2D composite material with local conductivity tensor **g** determines the effective conductivity of the same material, but one in which the local conductivity is the reciprocal tensor $\mathbf{g}^{-1}$ rotated by 90°.

Note that the Keller–Dykhne–Mendelson theorem is independent of the details of morphology of the material, a rare result. Due to this independence, this result provides a very useful test of theoretical and computer simulation estimates of the effective conductivity of 2D composites. For example, Perrins *et al.* (1979b) showed that Eq. (224) is valid for an *arbitrary* order of their exact solutions of the conduction problem for the square and hexagonal arrays, which were derived in Section 4.3.2.

Similar equations hold for the dielectric constant of the material. For example, for the dielectric constant of a 2D anisotropic material in one direction, say x, one has

$$\epsilon_e^{(xx)}(\epsilon_1, \epsilon_2)\epsilon_e^{(yy)}(\epsilon_2, \epsilon_1) = \epsilon_1\epsilon_2, \tag{226}$$

with a similar relation holding for the other direction.

No such reciprocal relation is known for the effective transport properties of 3D materials.

4.9 Effective-Medium Approximation

One of the most successful analytical treatments of the scalar and vector transport properties of composite materials, such as their effective conductivity and elastic moduli, as well as their effective dielectric constant, is the effective-medium approximation (EMA). This approximation has a long history, going back to at least Bruggeman (1935), which was beautifully reviewed by Landauer (1978). The essence of EMA is the replacement of the actual heterogeneous material by an effective homogeneous medium with a conductivity that somehow mimics the behavior of the true effective conductivity g_e of the material. The approximation is very versatile: It can be developed for both isotropic and anisotropic materials, as well as discrete and continuum models of disordered materials. It is *not* restricted to linear transport processes; Chapters 2 and 3 of Volume II will present the derivation of EMAs for several types of nonlinear transport and optical properties. It can also be extended to *vector* transport properties, such as the effective elastic moduli; see Chapters 7–9. As shown in Chapter 6, an EMA can also be developed for time- and frequency-dependent properties of composite materials. The discrete versions of the EMA for scalar and linear transport properties will be derived and discussed in Chapter 5 and 6. In what follows we describe the derivation of an EMA for disordered continua.

4.9.1 Isotropic Materials

As mentioned above, in the EMA approach one replaces the actual inhomogeneous material with one that is homogeneous with an effective conductivity g_e, which is an approximation to the true effective conductivity. The procedure to compute g_e is as follows. For isotropic materials we insert in the effective medium a spherical inclusion of radius R and conductivity g, surrounded by the effective medium with conductivity g_e. For steady-state transport, the governing equations are given by

$$\nabla^2 V_i = 0, \quad 0 < r < R, \tag{227}$$

$$\nabla^2 V_e = 0, \quad r > R, \tag{228}$$

where V_i and V_e are, respectively, the voltage inside the inclusion and in the effective medium. At the interface between the inclusion and the effective medium, we must have the continuity of the currents:

$$g\frac{\partial V_i}{\partial r} = g_e\frac{\partial V_e}{\partial r}, \quad r = R. \tag{229}$$

Moreover, the voltage is a static quantity, and therefore,

$$\oint \nabla V \cdot d\mathbf{r} = 0, \tag{230}$$

which implies that

$$\frac{\partial V_i}{\partial \theta} = \frac{\partial V_e}{\partial \theta}, \quad r = R. \tag{231}$$

Finally, one must impose the physical requirement that, $V_i(r = 0) \neq \infty$.

The general solutions of Eqs. (227) and (228) are given by

$$V_i = A_i + \frac{B_i}{r} + C_i r\cos\theta + D_i\frac{\cos\theta}{r^2}, \tag{232}$$

$$V_e = A_e + \frac{B_e}{r} + C_e r\cos\theta + D_e\frac{\cos\theta}{r^2}. \tag{233}$$

Clearly, one must have $B_i = 0$. The application of the other two boundary conditions, Eqs. (229) and (231), leads us to

$$V_i = A_i + C_e\frac{3g_e}{g + 2g_e}\, r\cos\theta, \tag{234}$$

$$V_e = C_e r \cos\theta - C_e R^3 \frac{g - g_e}{g + 2g_e} \frac{\cos\theta}{r^2}. \tag{235}$$

Suppose now that a constant field $\nabla V_0 = \mathbf{b}$ is applied along the z-axis (where $z = r\cos\theta$). Then, far from the sphere one must have $\nabla V_e = \nabla V_0$, implying that

$$\nabla V_e = br\cos\theta - bR^3 \frac{g - g_e}{g + 2g_e} \frac{\cos\theta}{r^2}. \tag{236}$$

The second term on the right-hand side of Eq. (236) represents the local fluctuations in the field due to the inhomogeneity that insertion of the inclusion causes in the effective medium. The effective-medium conductivity g_e is then computed by requiring that

$$\int f(g) \nabla V_e \, dg = \nabla V_0, \tag{237}$$

which is equivalent to

$$\int \frac{g - g_e}{g + 2g_e} f(g) dg = 0, \tag{238}$$

where $f(g)$ is the distribution of the local conductances of the composite material. More generally, for a d-dimensional isotropic continuum, Eq. (238) generalizes to

$$\int \frac{g - g_e}{g + (d-1)g_e} f(g) dg = 0, \tag{239}$$

which is the equation that was derived by Bruggeman (1935), and independently by Landauer (1952).

It is straightforward to show, upon inserting a conductance distribution $f(g) = (1-\phi)\delta_+(g) + \phi h(g)$ into Eq. (239) [where $h(g)$ is any normalized probability density function], that the EMA predicts a percolation threshold $\phi_c = 1/d$ for a d-dimensional continuous, isotropic material.

If the composite consists of n phases, each with constant conductivity g_i and volume fraction ϕ_i, Eq. (239) becomes

$$\sum_{i=1}^{n} \phi_i \frac{g_i - g_e}{g_i + (d-1)g_e} = 0. \tag{240}$$

4.9.2 *Anisotropic Materials*

The above EMA was generalized by Stroud (1975) to materials that consist of crystallites of arbitrary shape and conductivity tensors of arbitrary symmetry. Suppose that a material is characterized by a conductivity tensor $\mathbf{g(r)}$. If this conductivity field is expanded around a constant reference conductivity $\mathbf{g}_0$ by writing, $\mathbf{g(r)} = \mathbf{g}_0 + \delta\mathbf{g(r)}$, then the average current density $\langle \mathbf{I} \rangle$ in the material is given by

$$\langle \mathbf{I} \rangle = \mathbf{g}_0 \langle \mathbf{E} \rangle + \langle \delta \mathbf{g} \mathbf{E} \rangle, \tag{241}$$

where $\mathbf{E}$ is the electric field in the material. If we define the scalar potential $\Phi(\mathbf{r}) = -\mathbf{E} \cdot \mathbf{r}$, then, using the usual equations of continuity and electrostatics, we obtain, $\nabla \cdot [\mathbf{g}(\mathbf{r})\nabla\Phi(\mathbf{r})] = 0$, which, using the perturbation for the conductivity field, is written as

$$\nabla \cdot (\mathbf{g}\nabla\Phi) = -\nabla \cdot (\delta\mathbf{g}\nabla\Phi) \quad \text{in } \Omega, \tag{242}$$

$$\Phi(\mathbf{r}) = \Phi_0(\mathbf{r}) = -\mathbf{E}_0 \cdot \mathbf{r} \quad \text{on } \partial\Omega. \tag{243}$$

Here, $\mathbf{E}_0$ is the constant far-field electric field imposed on the material, and Ω and $\partial\Omega$ are, respectively, the volume and external surface of the material. If we now introduce a Green function $G(\mathbf{r}, \mathbf{r}')$ such that $G(\mathbf{r}, \mathbf{r}') = 0$ on $\partial\Omega$ and

$$\nabla \cdot [\mathbf{g}_0 \nabla G(\mathbf{r}, \mathbf{r}')] = -\delta(\mathbf{r} - \mathbf{r}') \quad \text{in } \Omega, \tag{244}$$

we can write $\Phi(\mathbf{r})$ in terms of this Green function as

$$\Phi(\mathbf{r}) = \Phi_0(\mathbf{r}) + \int_\Omega G(\mathbf{r}, \mathbf{r}')\nabla' \cdot [\delta\mathbf{g}(\mathbf{r}')\nabla'\Phi(\mathbf{r}')]d\mathbf{r}'. \tag{245}$$

Using integration by parts, taking the gradient of both sides of Eq. (245), writing $\delta\mathbf{I} = \delta\mathbf{g}\mathbf{E}$, and using the fact that $G(\mathbf{r}, \mathbf{r}') = G(\mathbf{r}', \mathbf{r})$, we obtain

$$\mathbf{E}(\mathbf{r}) = \mathbf{E}_0 + \int \mathcal{G}(\mathbf{r}, \mathbf{r}')\delta\mathbf{g}(\mathbf{r}')\mathbf{E}(\mathbf{r}')d\mathbf{r}' \tag{246}$$

where the tensor $\mathcal{G}(\mathbf{r}, \mathbf{r}')$ is defined by

$$\mathcal{G}_{ij}(\mathbf{r}, \mathbf{r}') = \frac{\partial}{\partial r_i'}\frac{\partial}{\partial r_j}G(\mathbf{r}, \mathbf{r}'). \tag{247}$$

One can now obtain an exact, but implicit, expression for the effective medium conductivity tensor $\mathbf{g}_e$. From Eq. (246), we can write

$$\delta\mathbf{g}(\mathbf{r})\mathbf{E}(\mathbf{r}) = \delta\mathbf{g}\left[\mathbf{E}_0 + \int \mathcal{G}(\mathbf{r}, \mathbf{r}')\delta\mathbf{g}(\mathbf{r}')\mathbf{E}(\mathbf{r}')d\mathbf{r}'\right], \tag{248}$$

so that if we define a tensor $\chi(\mathbf{r})$ by

$$\chi(\mathbf{r})\mathbf{E}_0 = \delta\mathbf{g}(\mathbf{r})\mathbf{E}(\mathbf{r}), \tag{249}$$

we obtain

$$\chi(\mathbf{r}) = \delta\mathbf{g}(\mathbf{r})\left[1 + \int \mathcal{G}(\mathbf{r}, \mathbf{r}')\chi(\mathbf{r}')d\mathbf{r}'\right], \tag{250}$$

and therefore

$$\mathbf{g}_e = \mathbf{g}_0 + \langle\chi\rangle. \tag{251}$$

Therefore, the main task is to develop a suitable expression for $\langle\chi\rangle$ which, for a general microstructure, cannot be evaluated exactly, and therefore one must resort to approximate schemes.

Suppose that the material consists of a random assembly of cells or crystallites, each of which is pure, i.e., has a constant conductivity, and that point $\mathbf{r}$ lies in the ith cell which has a volume Ω_i. From Eq. (250) one obtains

$$\chi(\mathbf{r}) = \delta\mathbf{g}_i\left[1 + \int_{\Omega_i} \mathcal{G}(\mathbf{r},\mathbf{r}')\chi(\mathbf{r}')d\mathbf{r}' + \int_{\Omega-\Omega_i} \mathcal{G}(\mathbf{r},\mathbf{r}')\chi(\mathbf{r}')d\mathbf{r}'\right], \tag{252}$$

where $\delta\mathbf{g}_i = \mathbf{g}_i - \mathbf{g}_0$, with $\mathbf{g}_i$ being the conductivity tensor of the ith phase. One approximation consists of

$$\int_{\Omega-\Omega_i} \mathcal{G}(\mathbf{r},\mathbf{r}')\chi(\mathbf{r}')d\mathbf{r}' \simeq \int_{\Omega-\Omega_i} \mathcal{G}(\mathbf{r},\mathbf{r}')\langle\chi\rangle d\mathbf{r}'. \tag{253}$$

However, this approximation is *not* unique, since it depends on the choice of the reference conductivity $\mathbf{g}_0$. The analog of the EMA described above, also referred to as the *self-consistent approximation* (SCA), is obtained by insisting that

$$\langle\chi\rangle = \mathbf{0}, \tag{254}$$

which would imply that $\mathbf{g}_e = \mathbf{g}_0$.

Consider, as an example, a composite material that consists of spherical crystallites of identical composition but anisotropic conductivity tensor. We assume that the principal axes of the crystallites are randomly oriented. It is then easy to see that Eq. (254), the SCA or EMA, becomes

$$\left\langle[1 + (1/3g_e)(\mathbf{g} - g_e\mathbf{U})]^{-1}(\mathbf{g} - g_e\mathbf{U})\right\rangle = 0, \tag{255}$$

where $\mathbf{U}$ is now the identity tensor. If the crystallites are uniaxial, and the coordinate axes are parallel to their principal axes, then

$$\mathbf{g} = g_0\begin{pmatrix} 1 & 0 & 0 \\ 0 & 1 & 0 \\ 0 & 0 & \alpha \end{pmatrix} \tag{256}$$

where α is a constant. Then, the EMA or SCA yields (in 3D)

$$\frac{g_e}{g_0} = 1 + \frac{1}{4}\left\{-3 + [9 + 8(\alpha - 1)]^{1/2}\right\}. \tag{257}$$

As another example, consider the effective conductivity of a polycrystalline metal in a magnetic field $\mathbf{H}$ applied in the z direction. In this case,

$$\mathbf{g}_e = \begin{bmatrix} g_e^{xx} & g_e^{xy} & 0 \\ -g_e^{xy} & g_e^{xx} & 0 \\ 0 & 0 & g_e^{zz} \end{bmatrix} \tag{258}$$

so that the three components of the conductivity tensor, as well as the Green function must be evaluated. With $\mathbf{r} = (x, y, z)$ and $\mathbf{r}' = (x', y', z')$, one has

$$G(\mathbf{r},\mathbf{r}') = \frac{1}{4\pi g_e^{xx}(g_e^{zz})^{1/2}}\left[\frac{(x-x')^2}{g_e^{xx}} + \frac{(y-y')^2}{g_e^{xx}} + \frac{(z-z')^2}{g_e^{zz}}\right]^{-1/2}. \tag{259}$$

Given Eq. (259), the SCA or EMA yields three equations for the three non-zero components of the effective conductivity tensor $\mathbf{g}_e$.

Let us mention that Granqvist and Hunderi (1978) generalized the above EMA to the case where there are dipole-dipole interactions in the material. In this case, the resulting EMA is somewhat similar to Eq. (240) [or (239)] and is given by

$$\sum_{i=1}^{n}\sum_{j=1}^{3}\phi_i\frac{g_i-g_e}{A_jg_i+(1-A_j)g_e}=0, \tag{260}$$

where A_1, A_2 and A_3 are the triplet of depolarization factors which depend on the shape of the inclusions, and were already given in Section 4.6.2.4.

4.9.3 Critique of the Effective-Medium Approximation

Due to its simplicity, and the fact that, under a variety of conditions, it provides reasonable estimates of the effective transport properties, the EMA or SCA has been a popular method. However, there are also many systems for which the EMA provides rather inaccurate estimates of the transport properties. We now discuss the reason for the inaccuracy of the EMA.

Since the continuum EMA derived above does not contain any information about a material's morphology, and because it neglects all the spatial correlations that typically exist in real materials' morphology, its application to estimating the effective transport properties of many materials is questionable. Moreover, as its derivation makes it clear, the EMA assumes that the effective medium exists just outside the test inclusion inserted in the material—a completely wrong assumption for many systems, such as packings of identical spheres which contain gaps between the particles. Acrivos and Chang (1986) addressed this problem by first surrounding the test sphere with a shell of matrix materials, and then inserting it in the effective medium. In addition, the EMA generally violates bounds that improve upon the Hashin–Shtrikman bounds. Finally, although the EMA provides non-trivial predictions for the percolation threshold of various materials, its predictions, except for 2D systems, are generally not accurate. This particular shortcoming of the EMA, and a few others, will be demonstrated in Chapter 5, where we describe and discuss the discrete version of Eq. (239) or (240).

Milton (1985) showed that there is a particular class of morphologies that correspond *exactly* to Eq. (240). For a two-phase system, this class consists of granular aggregates such that spherical grains of the two phases of comparable sizes are well-separated, and the system is self-similar on many length scales. At a particular length scale, large well-separated spheres of the two phases, which correspond to dilute proportions, are surrounded by a matrix that consists of much smaller well-separated spheres of the two phases. They, in turn, are surrounded by a matrix that consists of even much smaller well-separated spheres of the two phases, with the construction continuing *as infinitum*. Thus, the different-sized spheres of the two phases fill up the space in accordance with the overall phase volume fractions. Milton (1985) proved that such a fractal-like model leads to the EMA, Eq. (240).

More recently, Torquato and Hyun (2001) discovered single-scale *periodic* dispersions that realize the two-phase version of the 2D EMA for all volume fractions at a given phase conductivity contrast. Moreover, to a very high degree of accuracy (but not exactly) the same models realize the EMA for almost the entire range of phase conductivities and volume fractions. The Torquato–Hyun model involves a simple inclusion shape at a single length scale, namely, the *generalized hypcylcoid*, define by the equation

$$x^{2/b} + y^{2/b} = a^{2/b}, \tag{261}$$

where a and b are volume fraction-dependent parameters, and all distances are measured in units of the cell length. Except for $b = 1$, the interface between the inclusion and the matrix is not smooth. For example, for $\phi_2 \ll 1$ ($a \to 0$ and $b \to 1$), the model consists of identical circular inclusions of phase 2 on the sites of a square lattice in a matrix of phase 1, while for $\phi_2 = 1/2$ ($a = 1$ and $b = 2$), the model is the periodic checkerboard. Note that both the Milton and Torquato–Hyun models possess phase-inversion symmetry. Note also that, for $\phi_2 = 1/2$, the 2D EMA predicts that $g_e = \sqrt{g_1 g_2}$, and therefore satisfies the exact duality relation (225).

The foregoing discussions lead to the following conclusion. The reason for failure of the EMA in predicting the effective transport properties of many model composites is that, all the phases are treated symmetrically [note that Eq. (240) is symmetric under the interchange $g_1 \longleftrightarrow g_2$ and $\phi_1 \longleftrightarrow \phi_2$].

4.9.4 The Maxwell–Garnett Approximation

Another very famous approximation to the effective conductivity or dielectric constant of disordered materials is due to Maxwell–Garnett (1904), which is also called the Clausius–Mossotti approximation, so named after the closely related works of Mossotti (1850) and Clausius (1879). For a composite material that consists of a matrix of conductivity g_1 and spherical inclusions of conductivity g_2 and volume fraction ϕ_2, the MG approximation is given by

$$\frac{g_e - g_1}{g_e + (d-1)g_1} = \phi_2 \frac{g_2 - g_1}{g_2 + (d-1)g_1}. \tag{262}$$

Note that, unlike the EMA, this approximation is *not* symmetric with respect to interchanging the roles of the matrix and the inclusions. As a result, the slope $dg_e/d\phi_1$ takes on different values in the limits $\phi = 0$ and 1, implying that the MG approximation *cannot* provide a non-trivial (not equal to 0 or 1) estimate for the conductivity or percolation threshold, since this approximation is accurate *only* when the minority component appears as a skin completely surrounding and separating the second phase. Thus, according to the MG approximation, a material remains conducting until every portion of it is replaced by an insulator. Moreover, derivation of Eq. (262) is *independent* of the spheres' sizes, i.e., polydispersivity does not make any difference to the MG approximation, whereas the effective conductivity does, in general, depend on the degree of polydispersivity.

Another shortcoming of the MG approximation is that it *cannot* be generalized to a material with more than two components. To see this, suppose that a material consists of a matrix of conductivity g_0 and two other phases with conductivities

g_1 and g_2 dispersed in the matrix. Then, the generalization of Eq. (262) to this three-component composite is given by

$$\frac{g_e - g_0}{g_e + (d-1)g_0} = \sum_{i=1}^{2} \phi_i \frac{g_i - g_0}{g_i + (d-1)g_0}. \tag{263}$$

However, Eq. (263) predicts that g_e depends on g_0 even when the matrix has completely disappeared, i.e., when $\phi_1 + \phi_2 = 1$!

Extensions of the MG approximation which are capable of predicting a non-trivial percolation threshold have been proposed (Choy *et al.*, 1998). However, these extensions are still inaccurate (albeit more accurate than the original MG approximation), and their predictions are in some cases not completely physical. For example, for a conductor-insulator mixture, the extended MG approximation predicts a percolation threshold, $\phi_c = 0.9103$, while for a conductor-superconductor composite it predicts $\phi_c = 0.7313$, whereas we know that the two percolation thresholds must in fact be identical. Hence, the problem of a correct extension of the MG approximation that can predict non-trivial percolation thresholds, and satisfies all the required symmetries, remains unsolved.

The MG approximation is not completely useless. It yields accurate estimates of g_e at *non-dilute* conditions, if the spheres are well separated from each other, as it is identical with the result for coated-sphere model described in Section 4.4. This would be the case if the volume fraction of the inclusions is small. Moreover, Eq. (262) coincides with the *optimal* Hashin–Shtrikman bounds, i.e., it equals the lower bound in (118) when $g_2 \geq g_1$ and the corresponding upper bound when $g_2 \leq g_1$. When all the inclusions are more (less) conducting than the matrix phase, Eq. (263) coincides with the lower (upper) bound (119), the Hashin–Shtrikman bound lower (upper) bound for a multiphase material.

The MG approximation can be extended to macroscopically-anisotropic materials that consist of $n-1$ different types of unidirectionally aligned isotropic ellipsoidal inclusions of the same shape. The result is given by

$$\sum_{i=1}^{n-1} \phi_i (\mathbf{g}_e - \mathbf{g}_i) \cdot \mathbf{M}_i^{(1)} = 0, \tag{264}$$

where $\mathbf{g}_i = g_i \mathbf{U}$, and

$$\mathbf{M}_i^{(1)} = \left(\mathbf{U} + \frac{g_i - g_1}{g_1} \mathbf{A} \right)^{-1}, \tag{265}$$

with $\mathbf{A}$ being the depolarization tensor defined in Section 4.6.2.4. If the ellipsoids are randomly distributed, then one has $M_i^{(1)} = \mathrm{Tr}[\mathbf{M}_i^{(1)}]/d$, and Tr denotes the trace of the tensor. In this case, Eq. (264) reduces to a scalar equation.

4.10 The Random Walk Method

It has been known for decades (see, for example, Chandrasekhar, 1943) that from the time dependence of the mean square displacement $\langle R^2(t) \rangle$ of a random walker that diffuses in a heterogeneous medium one can compute the effective diffusivity

D_e and conductivity g_e. In particular, the effective conductivity of a d-dimensional medium is given by

$$g_e = \lim_{t \to \infty} \frac{\langle R^2(t) \rangle}{2dt}, \tag{266}$$

where t is the time and $\langle \cdot \rangle$ denotes an ensemble average. The first application of this idea to determining the effective conductivity of composite materials appeared in a paper of Haji–Sheikh and Sparrow (1967), who studied heat conduction in a composite solid. Over the past 25 years, random walk methods have been used extensively for studying transport in disordered media. Hughes (1995) provides an extensive review of fundamental properties of random walks. In this section, we describe the application of this technique to determining the effective conductivity of composite solids based on the continuum models described in Chapter 3, while in Chapter 5 the implementation of this method in discrete models of disordered materials will be described.

The essential idea is simple: One computes the mean square displacement of some random walkers in the composite material and utilizes Eq. (266) to estimate g_e. The simulations can be sped up by using the concept of *first-passage time* (FPT), which is the statement of the fact that if a random walker moves in a homogeneous region of a material (for example, within one phase of a two-phase material where the local conductivity is essentially the same everywhere within that phase), there is no need to spend unnecessary time to simulate its motion in detail. Instead, the walker should be allowed to take long steps to quickly pass through the homogeneous regions and arrive at the interface between the two phases. The necessary time for taking the long steps can often be calculated *analytically*, hence resulting in further improvement in the efficiency of the method. The only requirement for using the FPT method is that the walker should not take such long steps that would take it outside of a homogeneous region. After taking each of such long steps, the time is increased by an amount appropriate to that step. To our knowledge, the FPT technique was first used by Sahimi *et al.* (1982; see Sahimi *et al.*, 1986 for details) for simulating hydrodynamic dispersion (that is, unsteady state mixing of two miscible fluids by diffusion and convection) in flow through a porous medium. In the context of calculating the effective conductivity and diffusivity of a disordered material, Siegel and Langer (1986) and Kim and Torquato (1990) appear to be the first to have used this method. Let us now describe the details of the FPT method for calculating the effective conductivity (and hence diffusivity) of disordered materials.

We first consider a homogeneous medium and follow Torquato *et al.* (1999) to describe the essentials of the FPT method. Suppose that a random walker is diffusing in a d-dimensional homogeneous material of conductivity g. We surround the particle with a *first-passage region* Ω which has a bounding surface $\partial\Omega$, and denote by $\mathbf{r}$ and $\mathbf{r}_B$ the position of the walker inside Ω and on its boundary, respectively. Let $P(\mathbf{r}, \mathbf{r}_B, t)$ be the probability associated with the walker hitting $\partial\Omega$ in the vicinity of $\mathbf{r}_B$ for the *first time* at time t, if it starts its motion at $\mathbf{r}$. $P(\mathbf{r}, \mathbf{r}_B, t)$ is a probability density function in the variable $\mathbf{r}_B$ and a cumulative probability distribution function in the time variable t which satisfies the diffusion equation:

$$g\nabla^2 P(\mathbf{r}, \mathbf{r}_B, t) = \frac{\partial P(\mathbf{r}, \mathbf{r}_B, t)}{\partial t}, \tag{267}$$

subject to the boundary and initial conditions that

$$P(\mathbf{r}, \mathbf{r}_B, t = 0) = 0, \quad \mathbf{r} \in \Omega \tag{268}$$

$$P(\mathbf{r}, \mathbf{r}_B, t) = \delta(\mathbf{r} - \mathbf{r}_B), \quad \mathbf{r} \text{ on } \partial\Omega, \quad t > 0. \tag{269}$$

Now, suppose that $C(\mathbf{r}, t)$ is the cumulative probability distribution function associated with the walker, starting at $\mathbf{r}$, to first hit any point on the surface $\partial\Omega$ at time t. Clearly, $C(\mathbf{r}, t)$ is the integral of $P(\mathbf{r}, \mathbf{r}_B, t)d\mathbf{r}_B$, integrated over $\partial\Omega$. Hence, it is not difficult to show that $C(\mathbf{r}, t)$ also satisfies the same unsteady-state diffusion equation as $P(\mathbf{r}, \mathbf{r}_B, t)$ with the same initial condition, but with the boundary condition that, $C(\mathbf{r}, t) = 1$ for $\mathbf{r}$ in $\partial\Omega$

An important physical quantity is the *average hitting time*, or the average FPT, $t_h(\mathbf{r})$, the average time that a random walker takes, starting from $\mathbf{r}$, to hit the surface $\partial\Omega$ for the *first time*. It is not difficult to see that

$$t_h(\mathbf{r}) = \int_0^\infty \frac{\partial C}{\partial t} dt. \tag{270}$$

In particular, we are interested in the average hitting time t_0 when the walker begins its walk from the origin, $t_0 = t_h(\mathbf{0})$. For example, for a d-dimensional sphere of radius R centered at the origin, one has $t_0 = R^2/(2dg)$, where g is the sphere's conductivity. In principle, one can solve the unsteady-state diffusion equation for $C(\mathbf{r}, t)$ from the solution of which the average hitting time can be computed. However, Torquato *et al.* (1999) showed that t_h satisfies the following *steady-state* diffusion equation

$$g\nabla^2 t_h(\mathbf{r}) = -1, \quad \mathbf{r} \in \Omega, \tag{271}$$

$$t_h(\mathbf{r}) = 0, \quad \mathbf{r} \text{ on } \partial\Omega. \tag{272}$$

Another important property is the probability density $w(\mathbf{r}, \mathbf{r}_B)$ associated with hitting the vicinity of a particular position $\mathbf{r}_B$ on the surface $\partial\Omega$ for the first time, if the walker starts at $\mathbf{r}$. It is not difficult to see that, $w(\mathbf{r}, \mathbf{r}_B) = \int_0^\infty \partial P/\partial t \, dt = P(\mathbf{r}, \mathbf{r}_B, t = \infty)$, implying that $w(\mathbf{r}, \mathbf{r}_B)$ satisfies the following boundary-value problem,

$$\nabla^2 w(\mathbf{r}, \mathbf{r}_B) = 0, \quad \mathbf{r} \in \Omega, \tag{273}$$

$$w(\mathbf{r}, \mathbf{r}_B) = \delta(\mathbf{r} - \mathbf{r}_B). \tag{274}$$

Of particular interest is the probability density $w(\mathbf{r}, \mathbf{r}_B)$ when the walker begins its walk at the origin, $w(\mathbf{r}_B) = w(\mathbf{r} = \mathbf{0}, \mathbf{r}_B)$.

A final quantity of interest is the jumping probability $p(\mathbf{r})$, the probability that the random walker, starting at $\mathbf{r}$, arrives on a certain portion of the first-passage surface $\partial\Omega_0$ for the first time. Clearly,

$$p(\mathbf{r}) = \int_{\partial\Omega_0} w(\mathbf{r}, \mathbf{r}_B) d\mathbf{r}_B, \tag{275}$$

which implies that $p(\mathbf{r})$ satisfies a steady-state diffusion equation with the boundary condition that

$$p(\mathbf{r}) = \begin{cases} 1, & \mathbf{r} \text{ on } \partial\Omega_0, \\ 0, & \text{otherwise.} \end{cases} \tag{276}$$

We now consider a two-phase, d-dimensional heterogeneous material of conductivities g_1 and g_2. Suppose that the random walker is in the vicinity of the interface between the two phases. We surround the walker with a first-passage region Ω having a bounding surface $\partial\Omega$ that encompasses both phases. Let Ω_i be that portion of Ω that contains phase i (with $i = 1, 2$) with a corresponding surface $\partial\Omega_i$, and Γ be the interface surface. Equation (271) is then modified to

$$g_i \nabla^2 t_h(\mathbf{r}) = -1, \quad \mathbf{r} \in \Omega_i, \tag{277}$$

subject to the boundary condition (272) and the interface conditions

$$t_h|_1 = t_h|_2, \quad \mathbf{r} \text{ on } \Gamma, \tag{278}$$

$$\left.\frac{\partial t_h}{\partial n_1}\right|_1 = \frac{g_2}{g_1} \left.\frac{\partial t_h}{\partial n_1}\right|_2, \quad \mathbf{r} \text{ on } \Gamma, \tag{279}$$

where n_1 is the unit outward normal from Ω_i. The jump probability $w(\mathbf{r}, \mathbf{r}_B)$ satisfies the same equation and boundary condition as before, but with the additional interface conditions similar to (278) and (279). The probability $p_i(\mathbf{r})$ that the random walker, initially at $\mathbf{r}$, hits the first-passage surface $\partial\Omega_i$ for the first time is governed by an equation similar to (275) but integrated over $\partial\Omega_i$. The boundary conditions for, e.g., $p_1(\mathbf{r})$, are given by

$$p_1(\mathbf{r}) = \begin{cases} 1, & \mathbf{r} \text{ on } \partial\Omega_1, \\ 0, & \mathbf{r} \text{ on } \partial\Omega_2. \end{cases} \tag{280}$$

Moreover, the interface conditions,

$$p|_1 = p|_2, \quad \mathbf{r} \text{ on } \Gamma, \tag{281}$$

$$\left.\frac{\partial p}{\partial n_1}\right|_1 = \frac{g_2}{g_1} \left.\frac{\partial p}{\partial n_1}\right|_2, \quad \mathbf{r} \text{ on } \Gamma, \tag{282}$$

must be satisfied. From the obvious relation, $\sum_i p_i(\mathbf{r}) = 1$, one can determine $p_2(\mathbf{r})$ given $p_1(\mathbf{r})$.

Let us demonstrate the FPT method with two examples (Torquato *et al.*, 1999). Consider a first-passage square with a side length of $2L$ in a homogeneous medium of conductivity g, with the origin of the coordinate system being at the square's center. The average hitting time, the solution of Eqs. (271) and (272), can be derived easily by the method of separation of variables (see, for example, Carslaw and Jaeger, 1959). Then, the average hitting time for a walk starting at the origin is found by setting $\mathbf{r} = \mathbf{0}$ in the solution, yielding

$$t_0 = \frac{L^2}{2g} - \frac{16L^2}{g\pi^3} \sum_{n=0}^{\infty} \frac{(-1)^n}{(2n+1)^3 \cosh[(2n+1)\pi/2]}. \tag{283}$$

Moreover, the series in Eq. (283) can be summed numerically yielding, $t_0 \simeq (0.295L^2)/g$. The probability density $w(\mathbf{r}, \mathbf{r}_B)$ can also be obtained by solving Eq. (273) by the method of separation of variables, subject to the boundary conditions that, $w(x, -L) = w(x, L) = w(-L, y) = 0$, and $w(L, y) = \delta(y - y_B)$. The solution is given by

$$w(x, y, y_B) = \frac{1}{L} \sum_{n=1}^{\infty} \frac{\sinh[n\pi(x+L)/(2L)] \sin[n\pi(y+L)/(2L)] \sin[n\pi(y_B+L)/(2L)]}{\sinh(n\pi)}, \tag{284}$$

from which the probability that the random walker lands for the first time at any point along the side $x = L$ is obtained, using Eq. (275), to be $p = 1/4$.

If the first-passage square contains two phases, then, depending on how the two phases are distributed in the square, the above results must be somewhat modified. For example, suppose that the upper and lower halves of the square are occupied by phases 1 and 2, respectively. Then, Eqs. (278) and (279) must be solved by the method of separation of variables, which is straightforward. Setting **r=0** then yields the following exact result

$$t_0 = \frac{2}{g_1 + g_2} t_{h0}, \tag{285}$$

where t_{h0} is the solution for the homogeneous square given by Eq. (283) for $g = 1$. It is easy to show that Eq. (285) is also applicable to the case in which each phase occupies two diagonal quadrants of the square. The jumping probability density function $w(x, y)$ is obtained by solving Eq. (273) with the method of separation of variables, subject to the interface and boundary conditions (274) and the analogues of (278) and (279) for $w(x, y)$. If we set $x = y = 0$ in the resulting solution and consider the boundary points y_B to be on the $x = L$ side, we obtain

$$w(y_B) = \begin{cases} (2g_2)(g_1+g_2)^{-1} w_h(y_B), & -L \le y_B < 0, \\ (2g_1)(g_1+g_2)^{-1} w_h(y_B), & 0 < y \le L, \end{cases} \tag{286}$$

where $w_h(y_B)$ is the solution for the case of homogeneous square, Eq. (284). Observe that in this case $w(y_B)$ is discontinuous at $y_B = 0$. The same solution, Eq. (286), is obtained if we consider the boundary points y_B to be on the side $x = -L$. The probability p that the random walker lands for the first time along the side $x = L$ is $p = 1/4$. If we consider the boundary point x_B on the $y = L$ side, we obtain

$$w(x_B) = \frac{2g_1}{g_1 + g_2} w_h(x_B), \quad -L \le y \le L. \tag{287}$$

The probability p that the random walker lands for the first time at any point along the side $y = L$ is given by

$$p = \int_{-L}^{L} w(x_B)dx_B = \frac{g_1}{2(g_1 + g_2)}. \tag{288}$$

A similar approach can be used if the two phases are distributed differently than what we just considered. If, for example, phase 1 occupies three quadrants of the square and phase 2 one quadrant, we obtain

$$t_0 = \frac{4}{3g_1 + g_2} t_{h0}, \tag{289}$$

and the corresponding expressions for $w(x_B)$, $w(y_B)$, and p can also be obtained.

More generally, suppose that $\Omega = \Omega_1 \bigcup \Omega_2$ is a small spherical first-passage region of radius R centered at the interface at position $\mathbf{x}_0$, and $\partial\Omega_i$ is the surface of Ω_i excluding the two-phase interface. One can show (Kim and Torquato, 1992) that, for a curved but smooth interface, the probability p_1 (p_2) that the random walker, initially at $\mathbf{x}$ near $\mathbf{x}_0$, eventually arrives at the surface $\partial\Omega_1$ ($\partial\Omega_2$) is given by

$$p_1 = \begin{cases} S_1(S_1 + \sigma S_2)^{-1}\left(1 + \sigma \sum_{n=0}^{\infty} B_{2n+1} r^{2n+1}\right), & \mathbf{x} \subset \Omega_1 \\ S_1(S_1 + \sigma S_2)^{-1}\left(1 - \sum_{n=0}^{\infty} B_{2n+1} r^{2n+1}\right), & \mathbf{x} \subset \Omega_2, \end{cases} \tag{290}$$

where

$$B_{2n+1} = \frac{(-1)^n (2n)!}{2^{2n+1}(n!)^2} \frac{4n+3}{n+1}. \tag{291}$$

Here S_i is the area of the surface $\partial\Omega_i$ in phase i, $r = |\mathbf{x} - \mathbf{x}_0|/R$, and $\sigma = g_2/g_1$. The average hitting time t_h for the random walker, initially at $\mathbf{x}$, to hit $\partial\Omega = \partial\Omega_1 \bigcup \partial\Omega_2$ for the first time is given by

$$t_h = \begin{cases} t_{01}(V_1 + V_2)(V_1 + \sigma V_2)^{-1}\left[1 - r^2(3\sigma - 1)/2 + (\sigma - 1)\sum_{n=0}^{\infty} C_{2n+1} r^{2n+1}\right], & \mathbf{x} \subset \Omega_1, \\ t_{01}(V_1 + V_2)(V_1 + \sigma V_2)^{-1}\left[1 + r^2(1 - 3\sigma^{-1})/2 - (1 - \sigma^{-1})\sum_{n=0}^{\infty} C_{2n+1} r^{2n+1}\right], & \mathbf{x} \subset \Omega_2, \end{cases} \tag{292}$$

where V_i is the volume of region Ω_i, $t_{01} = R^2/(6g_1)$, and

$$C_{2n+1} = \frac{(-1)^{n+1}(2n)!}{2^{2n+2}(n!)^2} \frac{3(4n+3)}{(n+1)(n+2)(2n-1)}. \tag{293}$$

Even if the interface is not smooth, Eqs. (290) and (292) can still be used by making the sphere radius R smaller.

In practice, in a FPT simulation one constructs the largest (imaginary) concentric sphere of radius R around a randomly chosen point in phase i, that just touches the multiphase interface. Suppose that the random walker is initially at the center of the imaginary sphere. The average time $t_{0i} = R^2/(2dg_i)$ for the particle to reach a randomly-selected point on the surface of the sphere is recorded, and the process of constructing the sphere and calculating the time a point on its surface is reached is repeated, until the random walker comes within a very small distance of the interface between the two phases. One then computes the average hitting time t_h

associated with imaginary concentric sphere of radius R in the small neighborhood of the interface, and the probability of crossing the interface, both of which were given above. If the random walker crosses the interface and enters a new phase, it finds itself in a new homogeneous phase, and therefore the process of constructing the imaginary spheres is repeated. The effective conductivity of the heterogeneous material, in the limit $t \to \infty$, is then given by

$$\frac{g_e}{g_1} = \frac{\langle \sum_i t_{01}(R_i) + \sum_j t_{01}(R_j) + \sum_k t_{01}(R_k) \rangle}{\langle \sum_i t_{01}(R_i) + \sum_j t_{01}(R_j)/\sigma + \sum_k t_h(R_k) \rangle}, \tag{294}$$

where the summations over i and j are for the random walker's paths in phase 1 and 2, respectively, the summation over k is for the paths that cross the interface boundary, and $\langle \cdot \rangle$ denotes an average over all realizations of the disordered medium.

The same principles can be used for estimating the effective conductivity tensor $\mathbf{g}_e$ of an anisotropic material. In this case, the generalization of Eq. (266) for the ij component of the effective conductivity tensor, in the limit of long times (or, equivalently, large random walker displacements), is given by

$$(g_e)_{ij} = \frac{x_i x_j}{2t}, \tag{295}$$

where x_i is the displacement of the random walker in the ith direction, and t is the *total* time that the random walker spends in order to make a total mean square displacement $\langle R^2(t) \rangle = \sum_i x_i^2$. The FPT technique can, of course, be used for anisotropic materials as well, and the analogue of Eq. (294) for $(g_e)_{ij}$ is given by

$$(g_e)_{ij} = \frac{\langle x_i x_j \rangle}{2\langle \sum_i t_{01}(R_i) + \sum_j t_{01}(R_j) + \sum_k t_h(R_k) \rangle}, \tag{296}$$

where the notations are exactly the same as in Eq. (294). As already mentioned, Figures 4.11 and 4.12 compare the effective conductivities of an anisotropic material, obtained by the FPT simulations, with the upper and lower bounds derived in Section 4.6.2.3.

Another limiting case that is of interest is one in which one phase of a two-phase heterogeneous material is superconducting. In this limit the equations that were derived above, and in particular Eqs. (290) and (292), are not useful. In this case, Kim and Torquato (1990) proposed the following recipe.

(1) If the random walker is moving in the non-superconducting region, then the procedure is still the same as what was described above.
(2) When the random walker is within a prescribed small distance from the interface between the two phases, it is absorbed into the superconducting phase and leaves it after spending a time t_s in this phase.
(3) If the random walker is already in the superconducting phase (for example, if its motion is commenced within this phase), then step (2) is repeated.

As pointed out by Kim and Torquato (1993), calculation of t_s is not straightforward. Consider, for example, the case of a dispersion of overlapping, equisized superconducting spheres of radius R dispersed in a uniform matrix of conductivity g_1. If the overlapping spheres can form large clusters, then one needs to compute t_s for each cluster size separately. This requires that one first identify the cluster of the overlapping spheres, a task which can be accomplished using the Hoshen–Kopelman algorithm discussed in Chapter 2 (see Section 2.5.4). As an example, consider a superconducting dimer of two overlapping spheres, and suppose that the random walker is within a small prescribed distance $\delta_1 R$ (typically δ_1 is about 10^{-4}) from the interface. Then, instead of computing V_1 and V_2 to be used in Eqs. (290) and (292), one must compute two other volumes, V_i and V_o. Here, V_i is the union volume of the two overlapping spheres *plus* the imaginary concentric inner shell of thickness $\delta_1 R$. If we calculate another quantity V_s, the volume of another imaginary concentric outer shell of thickness $\delta_2 R$ (typically δ_2 is about 10^{-2}), then $V_o = V_i + V_s$. The time t_s is then given by

$$t_s = t_h \left[1 - \left(\frac{V_i}{V_o} \right)^{2/3} \right], \tag{297}$$

where t_h is the average hitting time for the random walker, initially at the center of mass of the homogeneous region of volume V_o and conductivity g_1, to first hit the surface of this volume. The numerical results shown in Figure 4.12 were obtained by this method.

A random walk method is particularly useful for estimating the electrical conductivity of composite materials made of an insulating matrix and a sample-spanning conducting phase. Well-known examples of such systems are porous media made of an insulating matrix and a pore space saturated with a conducting fluid (see, for example, Sahimi, 1995b for a detailed review). A traditional method, such as the finite-element technique, is notoriously time consuming for such materials as a very fine finite-element mesh with roughly about 1 billion nodes would be required to accurately solve the Laplace equation, a prospect which is currently impractical. For this reason alone, a random walk method is the preferred technique for estimating diffusivity and conductivity of heterogeneous materials.

4.11 The Effective Dielectric Constant

Most of the results and techniques for estimating the effective conductivity of heterogeneous materials that have been described so far have their exact analogues for the effective dielectric constant ϵ_e. Therefore, we summarize such results for ϵ_e and discuss their implications. However, before doing so, we describe and discuss an elegant formulation of the problem of determining ϵ_e that was developed by Bergman (1978). He showed that one can separate out the dependence of ϵ_e on the material's morphology by defining a set of geometrical characteristic functions that have very general analytical properties. In addition, the poles and residues of these functions contain experimentally-measurable information on low-frequency excitations in the material. We discuss Bergman's formulation for a two-phase material with dielectric constants ϵ_1 and ϵ_2, although his results are valid for a general multiphase system.

4.11.1 *Spectral Representation*

To begin with, let us define two variables $s \equiv \epsilon_2/(\epsilon_2 - \epsilon_1)$ and $t = 1 - s \equiv \epsilon_1/(\epsilon_1 - \epsilon_2)$. Bergman (1978) showed that one can have four spectral representations of ϵ_e given by

$$F_1(t) \equiv 1 - \frac{\epsilon_e}{\epsilon_1} = \sum_i \frac{a_i}{t - t_i}, \tag{298}$$

$$F_2(s) \equiv 1 - \frac{\epsilon_e}{\epsilon_2} = \sum_i \frac{b_i}{s - s_i}, \tag{299}$$

$$F_3(s) \equiv 1 - \frac{\epsilon_1}{\epsilon_e} = \sum_i \frac{c_i}{s - \hat{s}_i}, \tag{300}$$

$$F_4(t) \equiv 1 - \frac{\epsilon_2}{\epsilon_e} = \sum_i \frac{d_i}{t - \hat{t}_i}. \tag{301}$$

The poles s_i, t_i, $\hat{s}_i$, and $\hat{t}_i$, and the residues a_i, b_i, c_i, and d_i are all real and in the interval [0, 1], are determined by the statistical properties of the material's microstructure, and are completely independent of ϵ_1 and ϵ_2. In addition, the residues also satisfy the following rules that relate them to the volume fractions of the phases:

$$\sum_i b_i = \sum_i d_i = \phi_1, \tag{302}$$

$$\sum_i a_i = \sum_i c_i = \phi_2. \tag{303}$$

Moreover, any additional information about the material will result in some relations between the residues and the poles. For example, for a d-dimensional material that has cubic or isotropic rotational symmetry, one has

$$\sum_i b_i s_i = \frac{\phi_1 \phi_2}{d}. \tag{304}$$

The absorption spectrum of a composite material is directly related to the spectral representation of its effective dielectric constant.

To make further progress, we assume, for the sake of convenience, that our system forms the dielectric of a parallel plate capacitor. The electric field $\mathbf{E}(\mathbf{r})$ in the system, which is curl-free (i.e., $\nabla \times \mathbf{E} = \mathbf{0}$), must satisfy

$$\nabla \cdot [\epsilon(\mathbf{r})\mathbf{E}] = 0. \tag{305}$$

We take the applied potential to be zero on one plate and L on the other. The effective dielectric constant (permittivity) is defined by (see also Chapter 1)

$$\epsilon_e \langle E^2 \rangle \equiv \frac{1}{\Omega} \int \epsilon(\mathbf{r}) \mathbf{E}^2(\mathbf{r}) d\mathbf{r}, \tag{306}$$

where Ω is the volume of the system. By definition, $\mathbf{E}(\mathbf{r})$ is a homogeneous function of order zero, while ϵ_e, similar to the conductivity of a two-component system described in Chapter 2 [see Eqs. (2.42) and (2.43)], is a homogeneous function of order one in ϵ_i ($i = 1, 2$). Bergman (1978) used this fact to substitute for ϵ_i and ϵ_e the following variables:

$$h = \frac{\epsilon_1}{\epsilon_2}, \quad m(h) \equiv \frac{\epsilon_e}{\epsilon_2} = \frac{1}{\Omega} \int \theta_h (\nabla \psi)^2 d\mathbf{r}, \tag{307}$$

with

$$\theta_h \equiv \frac{\epsilon(\mathbf{r})}{\epsilon_2} = m_1 h + m_2, \quad \mathbf{E} = -|\langle \mathbf{E} \rangle|, \tag{308}$$

where m_i is defined by Eq. (59). Here ψ is the solution of the following boundary value problem:

$$\nabla \cdot (\theta_h \nabla \psi) = 0, \tag{309}$$

$$\psi(0) = 0, \tag{310}$$

$$\psi(L) = L, \tag{311}$$

$$\frac{\partial \psi}{\partial \mathbf{n}} = 0 \quad \text{on the walls}, \tag{312}$$

where $\mathbf{n}$ is normal to the walls. One can also define the inverse variable, $\hat{h} = h^{-1}$, and reformulate the problem through the displacement field $\mathbf{D}(\mathbf{r})$ and its average.

A few examples may help the reader to better understand the spectral representation of ϵ_e. For random platelets perpendicular to two capacitor plates, which is equivalent to a set of parallel capacitors, we have

$$\epsilon_e = \phi_1 \epsilon_1 + \phi_2 \epsilon_2, \tag{313}$$

which implies that

$$F_2(s) = \frac{\phi_1}{s}, \tag{314}$$

so that there is a pole at zero with the residue being ϕ_1. The other spectral functions, $F_1(t)$, $F_3(s)$, and $F_4(t)$, are also easily determined. For the complementary case of a set of capacitors in series,

$$\frac{1}{\epsilon_e} = \frac{\phi_1}{\epsilon_1} + \frac{\phi_2}{\epsilon_2}, \tag{315}$$

so that

$$F_2(s) = \frac{\phi_1}{s - \phi_2}. \tag{316}$$

As the third example, we consider the low-volume fraction distribution of d-dimensional spherical inclusions of dielectric constant ϵ_2, distributed in a matrix of dielectric constant ϵ_1. In this limit the Maxwell–Garnett formula (see Section 4.9.4) provides reasonably accurate predictions for ϵ_e:

$$\left(\frac{1}{d\epsilon_1}+\frac{1}{\epsilon_e-\epsilon_1}\right)^{-1}=\phi_2\left(\frac{1}{d\epsilon_1}+\frac{1}{\epsilon_2-\epsilon_1}\right)^{-1}, \tag{317}$$

which implies that the approximate spectral representation of ϵ_e is given by

$$F_2(s)=\frac{\phi_2}{s-\phi_1/d}. \tag{318}$$

Equation (318) is an approximate representation of $F_2(s)$, since the MG formula provides only approximate predictions for ϵ_e. The final example is a simple-cubic array of spheres inserted in a uniform matrix, described in 4.3.1.1, for which the MG formula, Eq. (317), is in fact an approximate solution. One can now show that [see Eq. (30)]

$$F_2(s)=\frac{\phi_2-b_1\phi_2^{13/3}}{s-(1-\phi_2)/3+b_2\phi_2^{10/3}}+\frac{b_1\phi_2^{13/3}}{s-3/7+b_3\phi_2^{7/3}-b_2\phi_2^{10/3}}+\cdots, \tag{319}$$

where b_1, b_2 and b_3 are constants. On examining Eqs. (318) and (319), one can see that there is a drastic difference between the exact and approximate spectral representations of ϵ_e, and in fact, as Eqs. (298)–(301) imply, the exact spectral representation of ϵ_e for a material with an arbitrary microstructure is an infinite series. Hinsen and Felderhof (1992) computed by computer simulations the spectral functions of a system of uniform polarizable spheres distributed randomly in a uniform background.

4.11.2 *Perturbation Expansion*

A perturbation expansion, similar to that developed for the effective conductivity g_e, can also be developed for the dielectric constant ϵ_e. Our discussion here follows that of Felderhof (1982) whose formulation of this problem follows closely those of Dederichs and Zeller (1973) and Kröner (1977), who developed similar perturbation expansions for the elastic moduli of heterogeneous materials (see Chapter 7). According to this formulation, the effective dielectric constant tensor of a heterogeneous material is given by

$$\boldsymbol{\epsilon}_e=\langle\epsilon(\mathbf{I}-\mathcal{L}G_0\delta\epsilon)^{-1}\rangle, \tag{320}$$

where $\mathbf{I}$ is the identity operator, G_0 is the Green function for a reference material with uniform dielectric constant ϵ_0, $\delta\epsilon=\epsilon(\mathbf{r})-\epsilon_0$, and $\mathcal{L}$ is an operator defined by, $\mathcal{L}f=f-\langle f\rangle$. Note that, $\langle\mathcal{L}f\rangle=0$, a property which will be used later in our analysis. Because of its homogeneity and isotropy, $\boldsymbol{\epsilon}_e$ is proportional to $\mathbf{U}$, the unit tensor. We now expand Eq. (320) in powers of $(\epsilon_1-\epsilon_2)/\epsilon_1$ by taking the reference medium to be one with dielectric constant ϵ_1, in which case, $\delta\epsilon = (\epsilon_2 - \epsilon_1)m_2(\mathbf{r})$, where $m_2(\mathbf{r})$ is defined by Eq. (59). The Green function is then given by

$$G_0(\mathbf{r}-\mathbf{r}') = -\frac{1}{3\epsilon_1}\mathbf{U}\delta(\mathbf{r}-\mathbf{r}') + \frac{1}{4\pi\epsilon_1}\nabla\nabla\frac{1}{|\mathbf{r}-\mathbf{r}'|}. \tag{321}$$

The first three terms of the expansion are then given by

$$\epsilon_e = \langle\epsilon\rangle\mathbf{U} + \langle\delta\epsilon\mathcal{L}G_0\delta\epsilon\rangle + \langle\delta\epsilon\mathcal{L}G_0\delta\epsilon\mathcal{L}G_0\delta\epsilon\rangle + \cdots, \tag{322}$$

where $\langle\mathcal{L}f\rangle = 0$ has been used. Equation (322) represents a formal perturbation expansion for ϵ_e. Further progress can be made if the microstructure of the material is specified (see, for example, Torquato, 1984), and a procedure similar to the one for the effective conductivity is followed. However, we now turn our attention to deriving rigorous upper and lower bounds for the effective dielectric constant, since these bounds are usually close to each other and thus provide accurate estimates of ϵ_e, unless percolation effects are important, in which case Chapters 5 and 6 provide detailed formulation and discussion of the problem of determining ϵ_e.

4.11.3 Rigorous Bounds

The spectral representation of the dielectric constant can be used for developing rigorous bounds for ϵ_e. The formulation and treatment of the problem that we discuss here are due to Felderhof (1982). The series expansion of, e.g., $F_1(t)$, in the inverse powers of t yields

$$F_1(t) = \sum_{j=1}^{\infty} f_j t^{-j}, \tag{323}$$

which, when compared with Eq. (298), implies that

$$\sum_i a_i t_i^{j-1} = f_j, \quad j = 1, 2, \cdots \tag{324}$$

The upper bound on ϵ_e, for $0 < \epsilon_2 < \epsilon_1$, is obtained by minimizing $F_1(t)$ for $t > 1$. To do this, one varies the $\{a_i\}$ and the $\{t_i\}$ subject to any known constraints imposed on the system. If we ignore the constraints (324), then the linear variations of (298) is given by

$$\delta F_1 = \sum_i \left[\frac{\delta a_i}{t - t_i} + \frac{a_i \delta t_i}{(t - t_i)^2}\right]. \tag{325}$$

We now suppose that f_1 is known. Then, for $j = 1$ Eq. (324) yields, $\sum_i a_i = f_1$, which in differential form means that, $\sum_i \delta a_i = 0$. We choose t_0 to be the smallest pole and eliminate δa_0 from Eq. (325) by use of constraints (324) (for $j = 1$). The result is

$$\delta F_1 = \sum_{i>0}\left[\delta a_i \frac{t_i - t_0}{(t - t_i)(t - t_0)} + \frac{a_i \delta a_i}{(t - t_i)^2}\right]. \tag{326}$$

The analysis is continued further by treating δa_i and δt_i as independent. Since $t > 1$ and t_0 is the smallest pole, the implication is that $(t_i - t_0)/[(t - t_i)(t - t_0)] > 0$, so that F_1 decreases by lowering a_i (for $i > 0$) to their minimum value, $a_i = 0$. The term $a_0 \delta t_0/(t - t_0)^2$ also decreases by decreasing t_0 to zero. Therefore, when f_1 is known, the constraints (324) (for $j = 1$) yield $a_0 = f_1$, and the minimum value of F_1, denoted by $F_1^{(1)}$, is given by

$$F_1^{(1)}(t) = \frac{f_1}{t}. \tag{327}$$

If f_2 is also known, then the second constraint, Eq. (324) with $j = 2$, yields, in the differential form,

$$\sum_i (a_i \delta t_i + t_i \delta a_i) = 0, \tag{328}$$

which is used to eliminate δt_0 from (326). The result is

$$\delta F_1 = \sum_{i>0} \left[\delta a_i \frac{(t_i - t_0)^2}{(t - t_0)^2 (t - t_i)} + a_i \delta a_i \frac{(t_0 - t_i)(t_0 + t_i - 2t)}{(t - t_i)^2 (t - t_0)^2} \right]. \tag{329}$$

Using arguments similar to those utilized for deriving Eq. (327), we find that values of a_0 and t_0 are determined by the two constraints $a_0 = f_1$ and $a_0 t_0 = f_2$, so that the minimum value of F_2 is given by

$$F_1^{(2)}(t) = \frac{f_1}{t - f_2/f_1}. \tag{330}$$

Finally, if f_3 is also known, the third constraint, Eq. (324) with $j = 3$, yields, in differential form, $\sum_i (2t_i a_i \delta t_i + t_i^2 \delta a_i) = 0$. Proceeding along the same lines, we can show that in this case the minimum value of F_2 is given by

$$F_1^{(3)}(t) = \frac{f_1}{t} + \frac{f_2}{t(t - f_3/f_2)}. \tag{331}$$

The coefficients f_1, f_2 and f_3 are now obtained by comparing the perturbation expansion of ϵ_e with its spectral representation. We derive the second-order bounds which, as expected, will be identical with the Hashin–Shtrikman bounds derived in Section 4.6.1.1. Comparison of Eq. (322) with (298) and (323) yields the following results

$$f_1 = \phi_2, \tag{332}$$

$$f_2 \mathbf{U} = -\epsilon_1 \langle m_2' G_0 m_2' \rangle = \frac{1}{3} \phi_1 \phi_2, \tag{333}$$

$$f_3 \mathbf{U} = \epsilon_1^2 \langle m_2' G_0 m_2 G_0 m_2' \rangle, \tag{334}$$

where $\langle \mathcal{L} f \rangle = 0$ has been used again. Equation (333) indicates that $f_2 = -\mathbf{a}_2^{(i)}$ (in the isotropic limit), where $\mathbf{a}_2^{(i)}$ is defined by Eq. (98). Using Eq. (333) in (330), we find the second-order upper bound

$$\epsilon_e^U = \epsilon_1 \left[1 + 3\phi_2 \frac{\epsilon_2 - \epsilon_1}{\phi_1 \epsilon_2 + (2 + \phi_2)\epsilon_1} \right], \tag{335}$$

which is valid for $\epsilon_1 > \epsilon_2 > 0$. Equation (335) is identical with the Hashin–Shtrikman upper bound, if we take note of the fact that the bounds (112) were derived assuming that $\epsilon_2 > \epsilon_1$, and thus the bound (335) corresponds to the lower bound of (112). Thus, the spectral representation of the dielectric constant provides us with another method of deriving rigorous upper and lower bounds to the effective transport properties of heterogeneous materials. Felderhof (1982) showed that, under certain conditions, the above second-order bounds coincide with the third-order bounds for the effective dielectric constant, and therefore the second-order bounds can in fact be highly accurate.

4.12 Optical Properties

We now consider electromagnetic properties of composite materials at finite frequencies. To be concrete, we consider a two-phase material that consists of phases 1 and 2 with volume fractions $\phi_1 = \phi$ and $\phi_2 = 1 - \phi$, respectively. The real dielectric function of the composite material is then $\epsilon_i(\mathbf{r}, \omega) = \epsilon_1(\omega)$ or $\epsilon_2(\omega)$, while the real conductivity is $g_i(\mathbf{r}, \omega) = g_1(\omega)$ or $g_2(\omega)$, where ω is the frequency. For simplicity, we assume that the magnetic permeabilities are equal to unity.

The transport of electromagnetic waves through the two-phase composite is described by the macroscopic Maxwell equations. Assuming monochromatic fields of the form $\mathbf{E}(\mathbf{r}, t) = \mathbf{E}(\mathbf{r}) \exp(-i\omega t)$, and adopting the usual convention that the physical fields are the real parts of complex quantities, these equations take the following forms (in Gaussian units)

$$\nabla \cdot (\epsilon_i \mathbf{E}) = 4\pi\rho, \tag{336}$$

$$\nabla \cdot \mathbf{B} = 0, \tag{337}$$

$$\nabla \times \mathbf{E} = \frac{i\omega}{c}\mathbf{B}, \tag{338}$$

$$\nabla \times \mathbf{B} = \frac{4\pi}{c} g_i \mathbf{E} - \frac{i\omega}{c}\epsilon_i \mathbf{E}, \tag{339}$$

which are supplemented by the continuity equation,

$$\nabla \cdot (g_i \mathbf{E}) - i\omega\rho = 0. \tag{340}$$

Substituting Eq. (340) into (336) yields

$$\nabla \cdot \mathbf{D} = 0, \tag{341}$$

where the free current density and polarization current have been combined into a single effective displacement field,

$$\mathbf{D} = \left(\epsilon_i + \frac{4i\pi g_i}{\omega} \right) \mathbf{E}. \tag{342}$$

Thus, if we introduce a complex dielectric function

$$\epsilon(\mathbf{r},\omega)=\epsilon_i(\mathbf{r},\omega)+\frac{4i\pi}{\omega}g_i(\mathbf{r},\omega), \tag{343}$$

Maxwell's equations take the form

$$\nabla\cdot(\epsilon\mathbf{E})=0, \tag{344}$$

which has the same form as Eq. (305), and

$$\nabla\cdot\mathbf{B}=0, \tag{345}$$

$$\nabla\times\mathbf{E}=i\frac{\omega}{c}\mathbf{B}, \tag{346}$$

$$\nabla\times\mathbf{B}=-i\frac{\omega\epsilon}{c}\mathbf{E}. \tag{347}$$

If the frequency is sufficiently low, the inductive term $(i\omega/c)\mathbf{B}$ in Faraday's law can be neglected, in which case the electric and displacement fields satisfy the equations, $\nabla\cdot\mathbf{D}=0$, and $\nabla\times\mathbf{E}=\mathbf{0}$, with $\mathbf{D}(\mathbf{r},\omega)=\epsilon(\mathbf{r},\omega)\mathbf{E}(\mathbf{r},\omega)$, which are formally identical to the static limit discussed earlier in this chapter, and is known as the quasi-static approximation (QSA). In general, this approximation is reasonable if the typical linear dimension of the inclusions is small compared to the wavelength or the penetration depth of the radiation in either phase of the material. For inclusions with sizes that are of the order of a few hundred angstroms, the QSA is reasonably accurate even at visible or near-ultraviolet frequencies. Given the validity of the QSA, we can use all the static-limit results to study electromagnetic wave propagation in composite materials, in the long-wavelength limit. The propagation of waves is properly described in terms of an effective dielectric constant ϵ_e, which is a complex function of the two complex variables ϵ_1 and ϵ_2, as well as the volume fraction ϕ and the material's morphology. Similarly, the MG approximation and the EMA can be extended to treat isotropic 3D composites, provided that the conductivities are replaced by dielectric functions.

4.12.1 Conductor–Insulator Composites

As the first example, consider a composite material with a volume fraction ϕ of metal with a Drude dielectric function

$$\epsilon_1(\omega)=1-\frac{(\omega_p/\omega)^2}{1+i\omega_\tau/\omega} \tag{348}$$

and a fraction $(1-\phi)$ of the insulator with dielectric constant $\epsilon_2(\omega)=1$, where ω_p is the plasma frequency and $\tau=1/\omega_\tau$ a characteristic relaxation time. In a bulk free-electron metal, such as Al, $\omega_p\sim 10^{15}\text{sec}^{-1}$ and $\omega_p\tau\sim 100$. In a small particle of such bulk metal, $\omega_p\tau$ may be reduced from this value by surface scattering. When $\phi\ll 1$, ϵ_e is accurately predicted by the MG approximation, provided that the metal particles can be assumed to be spherical. In this limit,

$$\epsilon_e = 1 + 3\phi\frac{\epsilon_1 - 1}{\epsilon_1 + 2} + O(\phi^2). \tag{349}$$

Equation (349) has two interesting frequency regimes. Frequencies such that $\epsilon_1 + 2 \simeq 0$ correspond to the so-called surface plasmon resonance. In this limit the denominator in Eq. (349) approaches zero, as the result of which the real part of ϵ_e becomes very large. For a metal-insulator composite, this occurs near $\omega = \omega_p/\sqrt{3}$. For non-spherical inclusions, this resonance splits into several peaks that occur at other frequencies. The surface plasmon resonance characteristically appears as a strong absorption line. This absorption is responsible for the beautiful ruby colors seen in dilute suspensions of small gold particles in a transparent host, such as glass. The absorption coefficient α is the fraction of energy absorbed per unit length of material, and is given by

$$\alpha = 2\frac{\omega}{c}\,\mathrm{Im}\sqrt{\epsilon_e}, \tag{350}$$

or, when $\phi \ll 1$,

$$\alpha \simeq 3\phi\frac{\omega}{c}\,\mathrm{Im}\left(\frac{\epsilon_1 - 1}{\epsilon_1 + 2}\right), \tag{351}$$

on using Eq. (349). Note that in the QSA, the radius of the spherical inclusion cancels out (although the particle's shape is still important).

A dilute suspension of spheres also exhibits interesting behavior at $\omega\tau \ll 1$. For most metals, this corresponds to the far infrared. By substituting Eq. (348) into (351), one finds

$$\alpha = \frac{9\omega^2\phi}{4\pi g_e c} \equiv C_e\omega^2\phi, \tag{352}$$

where $g_e = \omega_p^2\tau/(4\pi)$ is the static conductivity. Experimental data do indicate an absorption coefficient that varies roughly linearly with the volume fraction ϕ, as predicted by Eq. (352), but the measured magnitude of α is much larger than that predicted by the QSA, with the discrepancy being typically a factor of 10^5 or larger.

At higher metal's volume fraction, the qualitative optical behavior of the composite material indicates a number of striking features as a function of ϕ. These features are illustrated by approximate calculations using the EMA which serves as a useful interpolation scheme. The phenomena predicted are not, however, artifacts of the EMA, but are quite general and connected with the material's percolation properties.

Figure 4.13 presents $\mathrm{Re}[g_e(\omega)] = (\omega/4\pi)\mathrm{Im}[\epsilon_e(\omega)]$ versus frequency for several values of ϕ, as predicted by the EMA. For $\phi < \phi_c$ ($\phi_c = 1/3$ in the EMA; see Section 4.9.1), $\mathrm{Re}(g_e)$ exhibits a single broad peak, confined within the frequency range $0 < \omega < \omega_p$. This is a band of surface plasmon resonances, broadened from the low-volume fraction limit by electromagnetic interactions among individual

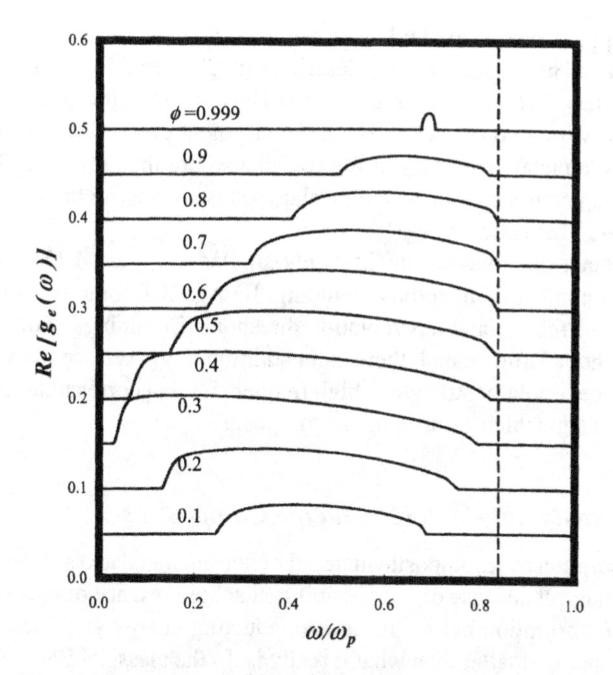

FIGURE 4.13. The real part of the conductivity $g_e(\omega)$ of a composite material that consists of a Drude metal of volume fraction ϕ and insulators of volume fraction $1 - \phi$, as predicted by the effective medium approximation. The Drude peaks are on the vertical axis at $\omega = 0$ which are, however, too narrow to be shown on the scale of the figure. The dashed line denotes the plasma frequency ω_p. The individual curves have been displaced vertically by constant amounts relative to one another (after Stroud, 1979).

grains. For $\phi > \phi_c$, a Drude peak, centered at $\omega = 0$, develops in addition to the surface plasmon band, signaling the fact that the composite is conducting at zero frequency. The peak appears at the percolation threshold of a metal-insulator composite above which the material has non-zero DC conductivity. As ϕ increases further, the integrated strength of the Drude peak grows, which is consistent with an increasing DC conductivity. The surface plasmon peak eventually shrinks and narrows to a band centered at $\omega = \omega_p\sqrt{2/3}$, corresponding to a void resonance, representing oscillating charge in the vicinity of a spherical void in an otherwise homogeneous metal.

Similar behavior is exhibited by the energy loss function, $-\text{Im}[1/\epsilon_e(\omega)]$, which shows characteristic structure related to the connectedness of various phases of the composite in a homogeneous metal, and has peaks at the plasmon resonances where $\text{Re}(\epsilon_e) \simeq 0$. For a metal described by the Drude dielectric function, this peak occurs at $\omega = \omega_p$, which persists in the composite for $\phi > \phi_c^*$, where ϕ_c^* is the volume fraction of metal above which the insulator no longer forms a sample-spanning cluster (for example, $\phi_c^* = 2/3$ according to the EMA). Besides the sharp plasmon peak, there is also a broad band in $-\text{Im}(1/\epsilon_e)$ from localized surface plasmons.

The disappearance of the bulk peak for $\phi < \phi_c^*$ may be understood by using the concepts of percolation theory described in Chapter 2. Near the plasma frequency one has, $\mathrm{Re}(\epsilon_1) \simeq 0$, while $\epsilon_2 = 1$. Thus, the metallic component acts as an insulator, with zero conductivity or displacement current, while the insulator behaves like a metal: The roles of the two phases are thus reversed. The zero in ϵ_e at $\omega = \omega_p$ persists so long as the insulator phase is present only in the form of finite clusters; that is, for $\phi > \phi_c^*$.

Experimental data that are qualitatively similar to Figure 4.13 have been reported for a number of materials including 3D Ag/KCI composites which have a bicontinuous regime and a percolation threshold (Cummings *et al.*, 1984). Because Ag is not a Drude metal, there is, in addition to the surface plasmon peak, a region of large oscillator strength at high frequencies that corresponds to interband transitions in Ag, which persists in the composite.

4.12.2 Conductor–Superconductor Composites

Optical absorption by a composite material of normal metal and a superconducting phase is of interest because of the possibility that the presence of the normal metal can produce absorption below the superconducting energy gap, thereby causing that gap to appear smaller than what it really is. In the classical Bardeen–Cooper–Schrieffer (BCS) theory of superconductivity, $\mathrm{Re}[g_s(\omega)]$ is characterized by two parts: (1) A delta function at $\omega = 0$, corresponding to the pure inductive response typical of infinite conductivity, and (2) a gap of width $2\Delta/\hbar$, where Δ is the BCS energy gap below which $\mathrm{Re}[g_s(\omega)] = 0$ and there is no absorption. Mattis and Bardeen (1958) showed that $g_1(\omega) \equiv \mathrm{Re}[g(\omega)]$ takes the following form in the superconducting state at temperature $T = 0$:

$$\left(\frac{g_{1s}}{g_n}\right)_{T=0} = \left(1 + \frac{2\Delta}{\hbar\omega}\right) E(k) - \frac{4\Delta}{\hbar\omega} K(k), \quad \hbar\omega \geq 2\Delta, \tag{353}$$

where

$$k = \left|\frac{\hbar\omega - 2\Delta}{\hbar\omega + 2\Delta}\right|, \tag{354}$$

and E and K are the standard elliptic integrals, with g_n being the conductivity of the superconductor in its normal state. For $\hbar\omega < 2\Delta$, g_{1s} vanishes. If we include the inductive delta function, the total conductivity in the superconductive state would be

$$g_t(\omega) = \frac{i}{\omega} G_0 + g_s(\omega), \tag{355}$$

where $g_s = g_{1s} + i g_{2s}$, and g_{2s} is related to g_{1s} by the Kramers–Kronig relation.

In a composite that consists of superconducting and insulating materials, a novel type of surface plasmon resonance is possible that has so far been reported only in some of the high-temperature superconductors (see, for example, Noh *et al.*, 1989). To understand the origin of this effect, we write $G_0 = \pi\Delta g_n/\hbar$, the correct

BCS form for the inductive part at low temperature. Then, the dielectric function of the superconductor is

$$\epsilon_s(\omega) = \epsilon_{\text{ph}}(\omega) + \frac{4i\pi}{\omega} g_t(\omega), \tag{356}$$

where ϵ_{ph} is that part of the dielectric function which is due to non-superconducting processes, such as phonon excitations. If we use Eq. (355), neglect the absorptive part, and approximate ϵ_{ph} by unity, we obtain

$$\epsilon_s(\omega) \simeq 1 - \frac{4\pi^2 g_n \Delta}{\omega^2 \hbar}, \tag{357}$$

which is identical in form to a Drude dielectric function with an infinite relaxation time. Thus, if a spherical grain of such a superconductor is inserted in an insulator with dielectric constant ϵ_i, we expect a surface plasmon resonance in the form of a sharp absorption line at $\epsilon_s + 2\epsilon_i \simeq 0$, or

$$\omega_{sp} = \left(\frac{4\pi^2 g_n \Delta}{3\hbar\epsilon_i}\right)^{1/2}. \tag{358}$$

This line can be detected only if it occurs below the superconducting gap; that is, for $\omega_{sp} < 2\Delta/\hbar$, or

$$\frac{\Delta}{\hbar} > \frac{\pi^2 g_n}{3\epsilon_i}, \tag{359}$$

implying a large Δ or a small g_n. In conventional low-temperature superconductors in which this inequality is not satisfied, the resonance is lost in the single-particle excitations above the gap. However, such an absorption spike has indeed been reported by Noh *et al.* (1989) in small spherical particles of $YBa_2Cu_3O_{7-\delta}$.

4.12.3 Anisotropic Materials

Several optical materials are composed of optically anisotropic constituents. Examples include intercalated graphites, many high-temperature superconductors, and quasi-linear organic conductors. In these materials, the dielectric function is a second-rank tensor with three distinct non-zero principal values. A polycrystal of such an anisotropic dielectric is essentially a composite material which, as discussed above, can be treated by the anisotropic EMA (with all the conductivities replaced by dielectric functions).

Calculations based on this approach and various generalizations to non-spherical grains have been used for interpreting the optical properties of the quasi-planar high-temperature superconductor $YBa_2Cu_3O_{7-\delta}$, which is a planar material with a highly anisotropic conductivity in both its normal and superconducting states. These calculation postulate ellipsoidal grains oriented so that the principal axes of the conductivity tensor are parallel to the principal axes of the ellipsoid within each grain. An important feature is the finite absorption found below $2\Delta/\hbar$, manifested in a finite $g_1(\omega)$ below the gap (see, for example, Walker and Scharnberg,

1990). An analysis that involves oriented anisotropic grains of high-temperature superconductors was developed by Diaz–Guilera and Tremblay (1991).

Optical anisotropy can also be produced by the application of a magnetic field. One such change is the Faraday effect, which is the rotation of light (either on transmission or reflection) on passing through a medium in the presence of a magnetic field. Such a field is treated formally by adding to the zero-field dielectric tensor an additional antisymmetric contribution. If the material is isotropic in the absence of the field, the problem is essentially an AC generalization of the Hall effect that will be described and discussed in Chapter 5. In a non-absorbing material, the components of the antisymmetric tensor are usually purely imaginary, so that left and right circularly polarized waves travel with different velocities. Faraday rotation in composite materials has been studied theoretically in the dilute limit using the MG approximation (Hui and Stroud, 1987), and for both magnetic and non-magnetic particles at higher volume fractions using the EMA (Xia *et al.*, 1990). Experimental studies are of practical interest because of the possibility of obtaining optical materials with large Faraday rotation per unit thickness and low absorption.

4.12.4 The Cole–Cole Representation

Cole and Cole (1941) suggested a method for analyzing complex dielectric constant of composite materials which appears to be extremely accurate. Their method is based on plotting the imaginary part of the effective dielectric constant versus its real part as the frequency ω is varied. Often, if one plots separately the real or imaginary part versus frequency, one may obtain excellent agreement with experimental data or an exact result. However, the true departures from the data or the exact theory are picked up only when one utilizes the Cole–Cole plot.

As an example, consider the MG approximation to the effective dielectric constant. Writing $\epsilon_e = \epsilon' + i\epsilon''$, it is straightforward to show that the MG formula for ϵ_e, Eq. (262), can be expressed in the following form:

$$\epsilon' + i\epsilon'' = 1 - \frac{\phi_2}{t - u_0}, \tag{360}$$

where $u_0 = (1 - \phi_2)/3$ is the MG pole in the spectral representation of ϵ_e (see Section 4.11.1), and $t = \epsilon_1/(\epsilon_1 - \epsilon_2)$. Writing $t(\omega) = t'(\omega) + it''(\omega)$, it follows that the MG trajectory is a circle given by

$$\left[\frac{\epsilon'(\omega)}{\epsilon_1} - 1\right]^2 + \left[\frac{\epsilon''(\omega)}{\epsilon_1} - \frac{\phi_2}{2t''}\right]^2 = \left(\frac{\phi_2}{2t''}\right)^2. \tag{361}$$

The Cole–Cole plot of the MG approximation also affords us an opportunity for comparing it with an exact result due to Felderhof and Jones (1989), so as to study its deviations from the exact values. Felderhof and Jones derived their exact result for low values of ϕ_2 by taking into account the effect of two-body interactions (which are of course absent in both the MG approximation and the EMA), which can be thought of as the coupling of the surface plasmon modes of two spheres.

Their result can be written as

$$\epsilon' + i\epsilon'' = 1 - \frac{\phi_2}{t - u_0 - \frac{1}{3}C(t)}, \tag{362}$$

where $C(t)$ contains the effect of two-body interactions and is frequency dependent. If we now compare Eq. (362) to (360), we see that the shift of the pole from u_0 to $u_0 - \frac{1}{3}C(t)$ shrinks the radius of the circle in the Cole–Cole plot, if the frequency-dependence of $C(t)$ is not too strong. Of course, as ϕ_2 increases, deviations from a circular Cole–Cole plot emerge, and in fact for large ϕ_2 the plot will not be a circle.

4.13 Beyond the Quasi-static Approximation: Mie Scattering

One may examine the validity of the QSA by analyzing a dilute suspension of inclusions in an otherwise homogeneous material, which is subjected to an incoming linearly polarized monochromatic plane electromagnetic wave with electric field $\mathbf{E} = \mathbf{E_0} \exp(ikz - i\omega t)$. We consider spherical inclusions with complex dielectric function $\epsilon_2(\omega)$ and a host of dielectric function ϵ_1. This is one rare case for which an exact analytical solution was obtained by G. Mie in 1908. Because of the symmetry of the system, we can write the electromagnetic fields in terms of two scalar functions that define the electric and magnetic Hertz vectors:

$$\mathbf{\Pi}_{\text{elec}} = \Pi_1 \hat{\mathbf{r}}, \tag{363}$$

$$\mathbf{\Pi}_{\text{mag}} = \Pi_2 \hat{\mathbf{r}}, \tag{364}$$

where all the transverse electric and magnetic modes are given in terms of the scalar functions Π_1 and Π_2, respectively. These functions are the solution of the Helmholtz equation:

$$(\nabla^2 + k^2)\Pi = 0, \tag{365}$$

outside the sphere, and

$$(\nabla^2 + \eta^2 k^2)\Pi = 0, \tag{366}$$

inside the sphere, where η is the refractive index of the sphere. The incident field $\mathbf{E}_0$ is readily decomposed into incident Hertz fields, as [r and φ are the polar coordinates in the (x, y) plane]

$$r\Pi_1^0 = \frac{1}{k^2} \sum_{n=1}^{\infty} \frac{i^{n-1}(2n+1)}{n(n+1)} \psi_n(kr) P_n^{(1)}(\cos\theta)\cos\varphi, \tag{367}$$

$$r\Pi_2^0 = \frac{1}{Z_1 k^2} \sum_{n=1}^{\infty} \frac{i^{n-1}(2n+1)}{n(n+1)} \psi_n(kr) P_n^{(1)}(\cos\theta)\sin\varphi. \tag{368}$$

Here, $\psi_n(x) = xj_n(x) = \sqrt{\pi x/2}J_{n+1/2}(x)$ are the Riccati–Bessel functions, $Z_1 = \sqrt{\mu_1/\epsilon_1}$ is the impedance of the matrix (outside the sphere), with ϵ_1 and μ_1 being its permittivity and magnetic permeability, respectively, and $P_i^{(j)}$ are the Legendre functions. Similarly, the scattered field outside the sphere is given by the same form but with Mie scattering coefficients a_n and b_n that contain information on the scattering:

$$r\Pi_1^{\rm scatt} = -\frac{1}{k^2}\sum_{n=1}^{\infty}\frac{i^{n-1}(2n+1)}{n(n+1)}a_n\zeta_n(kr)P_n^{(1)}(\cos\theta)\cos\varphi, \tag{369}$$

$$r\Pi_2^{\rm scatt} = -\frac{1}{Z_1k^2}\sum_{n=1}^{\infty}\frac{i^{n-1}(2n+1)}{n(n+1)}b_n\zeta_n(kr)P_n^{(1)}(\cos\theta)\sin\varphi, \tag{370}$$

where $\zeta_n = xh_n^{(1)}(x) = \sqrt{\pi x/2}H_{n+1/2}^{(1)}(x)$, with $H_i^{(j)}(x)$ being the Hankel functions. Inside the sphere we have the following fields,

$$r\Pi_1^{\rm inside} = \frac{1}{k^2\eta^2}\sum_{n=1}^{\infty}\frac{i^{n-1}(2n+1)}{n(n+1)}c_n\psi_n(kr)P_n^{(1)}(\cos\theta)\cos\varphi, \tag{371}$$

$$r\Pi_2^{\rm inside} = \frac{1}{Z_1k^2\eta^2}\sum_{n=1}^{\infty}\frac{i^{n-1}(2n+1)}{n(n+1)}d_n\psi_n(kr)P_n^{(1)}(\cos\theta)\sin\varphi, \tag{372}$$

with the Mie coefficients c_n and d_n containing all the information on the wave propagation inside the sphere. These coefficients, together with a_n and b_n, are determined by the boundary conditions that **E** and **H** must be continuous at the interface between the sphere and the matrix (i.e., at $r = R$). Of main interest to us are a_n and b_n which are given by

$$a_n = \frac{\psi_n'(y)\psi_n(x) - \eta\psi_n(y)\psi_n'(x)}{\psi_n'(y)\zeta_n(x) - \eta\psi_n(y)\zeta_n'(x)}, \tag{373}$$

$$b_n = \frac{\eta\psi_n'(y)\psi_n(x) - \psi_n(y)\psi_n'(x)}{\mu\psi_n'(y)\zeta_n(x) - \psi_n(y)\zeta_n'(x)}, \tag{374}$$

where $x = kR$, $y = k\eta R$, and R is the sphere's radius. The coefficients a_n and b_n are proportional to the n-pole electric and magnetic portions of the wave scattered from the sphere. Having determined these coefficients, the fields **E** and **H** are then given by

$$\mathbf{E} = \nabla\times\nabla\times(r\mathbf{\Pi}_{\rm elec}) + i\omega\mu\nabla\times(r\mathbf{\Pi}_{\rm mag}), \tag{375}$$

$$\mathbf{H} = \nabla\times\nabla\times(r\mathbf{\Pi}_{\rm mag}) - i\omega\mu\nabla\times(r\mathbf{\Pi}_{\rm elec}), \tag{376}$$

where $(\epsilon,\mu) = (\epsilon_1,\mu_1)$ inside the matrix, and $(\epsilon,\mu) = (\epsilon_2,\mu_2)$ inside the sphere.

The *extinction coefficient* $\alpha_{\rm tot}$, defined as the ratio of power lost out of the incident beam per unit volume and the incident power per unit area, is given by the standard expression

$$\alpha_{\rm tot} = \frac{4\pi n_0}{k^2}\mathrm{Re}[\mathcal{A}(0)], \tag{377}$$

$$\mathcal{A}(0) = \frac{1}{2}\sum_{n=1}^{\infty}(2n+1)(a_n + b_n), \tag{378}$$

where $\mathcal{A}(0)$ is the forward scattering amplitude, a_n and b_n are the electric and magnetic multipole coefficients given by Eqs. (373) and (374), and n_0 is the number density of spheres.

Although $\alpha_{\rm tot}$ represents the sum of losses due to scattering and absorption, scattering losses are negligible at long wavelengths (i.e., when $kR \ll 1$). The two dominant contributions to the absorption are made by the electric and magnetic dipole terms, a_1 and b_1. When $x \ll 1$ and $y \ll 1$, an expansion of the spherical Bessel and Hankel functions yields

$$a_1 \simeq -\frac{2i}{3}x^3\frac{\epsilon_1 - 1}{\epsilon_1 + 2}, \tag{379}$$

$$b_1 \simeq -\frac{i}{45}x^5(\epsilon_1 - 1). \tag{380}$$

All other coefficients (for example, the electric quadrupole coefficient a_2) vary as x^5 or a higher power of x. Thus, the dominant coefficient at low frequency is a_1, the electric dipole term (except in one special case discussed below). We then obtain

$$\alpha_{\rm tot} = 3\phi\frac{\omega}{c}\mathrm{Im}\left(\frac{\epsilon_1 - 1}{\epsilon_1 + 2}\right), \tag{381}$$

which is identical to the absorption coefficient (351) predicted by the quasi-static MG approximation. Thus, the QSA is generally valid in the limit $kR \ll 1$, the limit in which $\alpha_{\rm tot}$ is independent of R. This scale independence is expected because in the quasi-static limit a composite material can be described by an effective dielectric function that is unchanged if all the dimensions of the composite are uniformly multiplied by a scale factor.

On the other hand, in a metal-insulator composite at low frequencies, the contribution α_m to the extinction coefficient made by b_1 can be comparable to the electric dipole contribution, which is the same as the QSA or even much larger. If we substitute Eqs. (379) and (380) into (377), we obtain, $\alpha_e = C_e\omega^2\phi$, and $\alpha_m = C_m\omega^2\phi$, where C_e is defined by Eq. (352) and

$$C_m = \frac{2\pi g_e R^2}{5c^3}. \tag{382}$$

The magnetic dipole contribution also varies as $\omega^2\phi$, but, unlike α_e, it increases with particle radius because the induced eddy currents dissipate more energy in larger particles. For particles of conductivity comparable to that of Al at room temperature, C_m may exceed C_e at a particle radius as small as 30Å. For 100 Å particles, inclusion of the magnetic dipole absorption can give an enhancement of far-infrared absorption of the order of 10^2 over the QSA. This extra magnetic dipole (or eddy current) absorption undoubtedly contributes to the well-known discrepancy between the observed and the quasi-static far-infrared absorption in small metal particles. However, it is still inadequate to explain a factor of $10^4 - 10^6$ seen in many experiments.

4.14 Dynamical Effective-Medium Approximation

Several attempts have been made to extend the EMA in order to obtain more accurate predictions for $\epsilon_e(\epsilon_1, \epsilon_2, \mathbf{q}, \omega)$, where $\mathbf{q}$ represents all the modes. To be more precise, suppose that a plane wave of frequency ω propagates through a composite material. This wave is then attenuated and retarded due to multiple scattering and absorption in the material. The scattered fields can propagate over a wide range of wave vectors, and therefore the important question to ask is: How does one define an average that can take into account all the scattered fields in order to obtain the correct effective dielectric constant? For example, one can define the average on a mode-by-mode basis:

$$\langle \mathbf{D}(\epsilon_i, \mathbf{q}, \omega) \rangle = \epsilon_e(\epsilon_1, \mathbf{q}, \omega) \langle \mathbf{E}(\epsilon_i, \mathbf{q}, \omega) \rangle, \tag{383}$$

and then average over all the modes $\mathbf{q}$. However, the distribution of the modes is non-trivial, and its knowledge does in fact require the complete solution of the multiple scattering problem in the composite material, a prospect which is currently out of reach. Another important issue to address is the following. Since the wave experiences both absorption and scattering, the modes $\mathbf{q}$ are complex, and therefore the question is: Does one average over the real part or the imaginary part?

To address such issues, Stroud and Pan (1978) proposed selecting the effective medium to have a dielectric function ϵ_e such that the forward scattering from the spheres, i.e., $\mathcal{A}(0)$ [see Eq. (378)], vanishes on the average. They obtained their approximate solution by imposing the consistency condition through the Fourier components $\mathbf{D}(\mathbf{q}, \omega)$ and relating it to the electric field $\mathbf{E}(\mathbf{q}, \omega)$ via $\epsilon_e(\omega)$ at *one* magnitude of $\mathbf{q}$—that of the propagating mode itself—i.e., $k_e = \omega\sqrt{\epsilon_e}/c$. Therefore,

$$\int \mathbf{D}(\mathbf{r}, \omega) \exp(-ik_e z) d\mathbf{r} = \epsilon_e \int \mathbf{E}(\mathbf{r}, \omega) \exp(-ik_e z) d\mathbf{r}, \tag{384}$$

which implies that the average is automatically taken through the Fourier transform. Note that, in general, when the magnetic properties are also important, the effective wave vector is defined by

$$k_e = \frac{\omega}{c}\sqrt{\epsilon_e \mu_e}, \tag{385}$$

which indicates that one must also have a similar hypothesis for the field $\mathbf{B}$:

$$\int \mathbf{B}(\mathbf{r}, \omega) \exp(-ik_e z) d\mathbf{r} = \mu_e \int \mathbf{H}(\mathbf{r}, \omega) \exp(-ik_e z) d\mathbf{r}. \tag{386}$$

However, Eq. (386) also implies that, instead of attempting to directly compute ϵ_e and μ_e, one should calculate an effective value k_e for the wave vector. But, how does one do so?

To address this question, Mahan (1988) avoided including the magnetic-dipole scattering in an effective dielectric function, and instead proposed the introduction of *separate* effective dielectric function and magnetic permeability, each of which is determined by independent, EMA-like equations. In other words, he assumed that both of the averages given by Eqs. (384) and (386) are *independent*.

Even Eq. (336) cannot be used directly, because, in principle, the fields **D** and **E** must be known. These fields can, however, be computed only if we can completely solve the multiple scattering problem. In this case, a technique similar to that used for deriving the EMA in the static limit (see 4.9.1) can be used, namely, a spherical grain is embedded in the effective medium that has an effective dielectric constant $\epsilon_e(\omega)$, which is yet to be determined. This assumption, together with the superposition principle for each of the spherical grains (assumed not to be interacting with each other, so that it suffices to consider a single sphere), enable us to carry out the integrations in Eq. (384). One breaks up the integral of the field **E** into a sum over the individual grains:

$$\mathbf{I}_i = \int_{\Omega_i} \mathbf{E}_i(\mathbf{r}, \omega) \exp(-ik_e z) d\mathbf{r}, \tag{387}$$

where Ω_i is the volume of the region over which the integration is carried out, which must be at least of the order of several wavelengths. The effective dielectric constant is then given by

$$\epsilon_e(\epsilon_1, \omega) = \frac{\sum_i \epsilon_i \mathbf{I}_i}{\sum_i \mathbf{I}_i}. \tag{388}$$

Equation (388) can be thought of as the dynamical version of the MG approximation. It is, however, flawed, because it violates the important condition that the propagating mode, defined by $k_1(\epsilon_1)$ through an analogue of Eq. (385), is inconsistent with ϵ_e. In a similar way, the dynamical version of the EMA can be obtained:

$$\sum_i (\epsilon_i - \epsilon_1) \mathbf{I}_i = \mathbf{0}. \tag{389}$$

The only remaining issue is then evaluating the integral (387) which is possible once the electric field **E** is specified.

Numerically, the approximations developed by Stroud and Pan (1978) and by Mahan (1988) yield rather similar results. The conditions under which one can treat the effective medium by separate effective dielectric functions and permabilities were discussed by Lamb *et al.* (1980). Let us mention that we will derive in Chapter 6 a different effective-medium approximation for predicting frequency-dependent transport properties of disordered materials.

4.15 The Effect of Large-Scale Morphology

There can be large-scale structures in a composite material that are of the same order as the wavelength λ of electromagnetic radiation in the composite, even if the individual grains are small, so that the QSA breaks down again. Examples of such materials include large (for example, fractal) clusters of particles with a cluster linear dimension that is roughly about λ. As we should know by now, in a randomly disordered composite, the percolation correlation length ξ_p diverges as the percolation threshold ϕ_c is approached from either side, so that sufficiently near the threshold, $\xi_p > \lambda$.

The optical properties of a composite material in this regime were analyzed by Yagil *et al.* (1991). In their analysis, the incident radiation near ϕ_c samples the dielectric constant, not of the composite as a whole, but rather of only a small piece of volume $\sim L_\xi^d$, where L_ξ is the smallest of the three lengths, λ (the electromagnetic wavelength in the composite), ξ_p (the percolation correlation length), and $L(\omega)$. In turn, $L(\omega)$ is the *anomalous diffusion length*, defined as the distance a charge carrier travels within the metallic portion of the composite in one AC cycle (see Chapter 6, Section 6.1.9, for a detailed discussion of anomalous diffusion). Such a small piece of the material deviates from the bulk in two ways: Its dielectric function differs from that of an infinite sample, and it has a distribution of possible values for ϵ_e described by a probability density function. Both effects arise from large fluctuations exhibited by different finite-size samples of the composite near ϕ_c. The dielectric constant ϵ_e for a piece of size L_ξ was then assumed to be characterized by a finite-size scaling function near ϕ_c, of the form

$$\epsilon_e(\epsilon_1, \epsilon_2, \phi, L_\xi) = \left(\frac{\epsilon_1}{\epsilon_2}\right)^{-a} F_\pm\left(\frac{\epsilon_1/\epsilon_2}{\xi_p^{-a}}, \frac{\xi_p}{L_\xi}\right), \tag{390}$$

where F_+ and F_- are two scaling functions that describe the dielectric constant above and below ϕ_c, respectively. Equation (390) was tested numerically in 2D and was found to be valid.

4.16 Multiple-Scattering Approach

Over the past 35 years, numerous other approximate schemes have been proposed for estimating various properties of composite materials. Some of these methods are either simple extensions of the MG approximation or the EMA, while others are considerably more elaborate. In this section, we briefly describe one of these approaches, namely, the multiple-scattering approach which has been used widely in the condensed matter theory.

An example of such methods is one for materials that consist of a random distribution of particles of dielectric function ϵ_2 and fixed shapes (for example, spherical particles), embedded in a matrix of a different dielectric function ϵ_1. Such a composite material can be thought of as a macroscopic version of a morphologically-disordered *atomic* system, such as a liquid metal or an amorphous solid. In this analogy, the particles play the role of the atoms, while the difference $\epsilon_2 - \epsilon_1$ corresponds to the scattering potential. This analogy can be exploited for estimating the effective dielectric constant of the material by writing down an integral equation for the electric field **E**, which is obtained by combining the Maxwell equations, Eq. (346) and (347):

$$\nabla \times \nabla \times \mathbf{E} - k^2\mathbf{E} = k^2 \delta\epsilon \mathbf{E}, \tag{391}$$

where $k^2 = k_0^2\epsilon_0, k_0^2 = \omega^2/c^2, \delta\epsilon = [\epsilon(\mathbf{r}) - \epsilon_0]/\epsilon_0, \epsilon(\mathbf{r})$ is the position-dependent dielectric constant, and ϵ_0 is a reference dielectric constant to be selected in some convenient way. If we introduce a 3×3 tensor Green function **G** defined by

$$\nabla \times \nabla \times \mathbf{G} - k^2\mathbf{G} = -k^2\delta^3(\mathbf{r} - \mathbf{r}')\mathbf{U}, \tag{392}$$

where **U** is the 3×3 unit tensor, Eq. (391) is converted into an integral equation for **E**:

$$\mathbf{E}(\mathbf{r}) = \mathbf{E}_0(\mathbf{r}) + \int \mathbf{G}(k; \mathbf{r} - \mathbf{r}') \cdot \delta\epsilon(\mathbf{r}')\mathbf{E}(\mathbf{r}')d^3\mathbf{r}', \tag{393}$$

where $\mathbf{E}_0$ is the electric field in the homogeneous material. Equation (393) is a vector analogue of the integral Schrödinger equation, with **E** playing the role of the quantum-mechanical wave function ψ. The Green function **G** has been calculated for many choices of boundary and initial conditions. For example, the outgoing-wave Green function is readily shown to be given by

$$G_{ij}(k; \mathbf{R}) = -(k^2\delta_{ij} + \nabla_i\nabla_j)\frac{\exp(ikR)}{4\pi R}, \tag{394}$$

where $\mathbf{R} \equiv \mathbf{r} - \mathbf{r}'$, $R = |\mathbf{R}|$, and i and j denote Cartesian coordinates.

Given the solution of Eq. (393), one may, in the following manner, define an effective frequency- and wave number-dependent dielectric tensor $\boldsymbol{\epsilon}_e(\mathbf{q}, \omega)$. Suppose that the homogeneous solution is a monochromatic field, $\mathbf{E}_0(\mathbf{r}, t) = \mathbf{E}_0 \exp(i\mathbf{q} \cdot \mathbf{r} - i\omega t)$. Then, $\boldsymbol{\epsilon}_e(\mathbf{q}, \omega)$ is defined by an equation similar to (383) in which $\langle\cdot\rangle$ indicates an ensemble average. Given this definition, the longitudinal and transverse responses may differ. In the limit of small q, however, one may expect the q-dependence to disappear and the longitudinal and transverse dielectric functions to become equal. If we introduce a 3×3 scattering matrix **T**, defined by

$$\delta\epsilon(\mathbf{r})\mathbf{E}(\mathbf{r}) = \int \mathbf{T}(\mathbf{r}, \mathbf{r}')\mathbf{E}_0(\mathbf{r}')d\mathbf{r}', \tag{395}$$

then, $\boldsymbol{\epsilon}_e(\mathbf{q}, \omega)$ is expressed in terms of **T** through the following equation

$$\boldsymbol{\epsilon}_e(\mathbf{q}, \omega) = \epsilon_0\mathbf{U} + \langle\mathbf{T}(\mathbf{q}, \omega)\rangle[\mathbf{U} + \mathbf{G}(\mathbf{q}, \omega)\langle\mathbf{T}(\mathbf{q}, \omega)\rangle]^{-1}, \tag{396}$$

where $\mathbf{G}(\mathbf{q}, \omega)$ is the Fourier-transformed solution of Eq. (391).

Given Eq. (396), we may convert the problem of determining the effective dielectric constant to one of searching for an effective wave vector k_e. The natural way to define k_e is to look for a k such that the average scattering matrix $\langle\mathbf{T}\rangle$

vanishes. The corresponding Green function **G** then describes the average propagation characteristics of a wave in the disordered material and the corresponding $k_0 = k_e$. Likewise, the effective dielectric function ϵ_e, is just ϵ_0, the reference dielectric constant that causes the average scattering matrix to vanish. For sufficiently long wavelengths, the resulting dieletric constant will presumably be the same for both longitudinal and transverse waves.

For a suspension of identical spheres, the scattering matrix **T** can be expanded in a multiple-scattering series involving the scattering matrices $\mathbf{t}_i$ of the individual spheres. Because this series has the same form as in liquid metals, one can adapt liquid-metal approximations to compute and analyze dielectric suspensions. Davis and Schwartz (1986) used this approach to calculate ϵ_e in the long-wavelength regime, $kR \ll 1$ (where R is the sphere's radius), using Eq. (396). They used several approximate schemes for $\langle \mathbf{T} \rangle$ developed in the theory of liquid metals, including those known as the *quasi-crystalline approximation* and the EMA of Roth (1974) which, however, is an approximation entirely distinct from the usual EMA described above. Another multiple-scattering approach was developed by Lamb *et al.* (1980) who studied both ordered and disordered suspensions of spheres in the long-wave length limit, including both two- and three-body correlation functions. Their method yields predictions that differ significantly from those provided by the simpler MG approximations, and agree better with experimental data.

Summary

Many theoretical approaches for estimating transport and optical properties of disordered materials have been developed over the past several decades. Some of these methods, that utilize the continuum models of heterogeneous materials that were described in Chapter 3, were developed, analyzed and discussed in this chapter. They range from exact results for certain types of morphologies, to rigorous upper and lower bounds to the effective properties, and several mean-field and effective-medium approximations. The exact results are useful when a material's morphology resembles the models for which the exact solutions have been derived. The bounds provide in many cases useful estimates for the effective transport properties, and are also useful for checking the merit of a new theory. The mean-field and effective medium approximations are easy to use, can be used as interpolation formulae and, under certain conditions, provide relatively accurate estimates of the effective transport properties. These methods have provided us with a wealth of information on the effective transport and optical properties of disordered materials, as a result of which we now have a much deeper understanding of how a material responds to a given external excitation. Chapter 5 will describe the complementary approaches that are based on discrete models of disordered materials.

5
Effective Conductivity and Dielectric Constant: The Discrete Approach

5.0 Introduction

In Chapter 4 we described and discussed various methods of estimating the effective conductivity, dielectric constant, and optical properties of heterogeneous materials, based on the continuum models of such materials that were described in Chapter 3. The purpose of the present chapter is to essentially carry out the same program of analyses as in Chapter 4, except that we describe and discuss the solution of these problems using the discrete or lattice models of disordered materials. One main advantage of the discrete approach is that it allows one to make direct contact with the effect of percolation and clustering on the effective properties of disordered materials. Recall from Chapter 2 that percolation and clustering effects are relevant and important when there is a large difference between the properties of the phases of a disordered multiphase material. Under this condition, many of the methods that were described in Chapter 4 cannot provide accurate estimates of the effective transport and optical properties. Another advantage of the discrete models is their relative simplicity and the ease with which they can be utilized, which also explains their popularity over the past three decades. Since we will describe and discuss estimation and computation of the effective linear vector transport properties, such as the elastic moduli, in Chapters 7–9, we restrict our attention in this chapter to scalar steady-state transport, and in particular conduction, and describe and discuss various methods of estimating the effective conductivity and dielectric constant of heterogeneous materials. Since methods of estimating the effective conductivity and dielectric constant of disordered solids are completely similar, we only describe the methods for the effective conductivity. In addition, we consider not only the electrical and thermal conductivities, but also the Hall conductivity, as well as the classical (non-quantum mechanical) aspects of superconductivity. Chapter 6 will describe and discuss the analysis of the effective transport, optical, and other dynamical properties of heterogeneous material under the condition that they are frequency- and/or time-dependent.

5.1 Experimental Data for Conduction in Heterogeneous Materials

Before we begin our theoretical discussions, let us review some of the experimental studies of conduction in composite solids in order to better demonstrate the relevance of percolation and clustering effects. Our review of this problem is by no means exhaustive, as there is a very extensive literature on the subject. Therefore, we confine our attention to reviewing only some of the key results that have been reported over the past 30 years. Toward this end, we consider three types of heterogeneous materials, namely, powders, polymeric materials, and metal-insulator composites, and restrict most of our discussions to the power-law behavior of their effective conductivity in the critical region near the percolation threshold p_c, where percolation effects are dominant and universal properties, independent of the morphology of the materials, are observed. Non-universal features, such as the location of the percolation threshold and the *numerical value* of the effective transport properties away from p_c are predicted well by the effective-medium approximation and many of the methods described in Chapter 4, the discrete versions of which are described in detail later in this chapter.

Over 30 years ago, Ziman (1968) proposed that the mobility of an electron in a disordered semiconductor can be predicted by percolation theory. Eggarter and Cohen (1970) then suggested that the mobility, and hence the conductivity, should be proportional to the percolation probability $P(p)$, introduced in Section 2.6.1. However, in a classical paper, Last and Thouless (1971) reported measurements of the effective conductivity of a conducting colloidal graphite paper in which holes had been punched randomly, and showed that the conductivity decreases, as the percolation threshold is approached, much slower than $P(p)$, due to the fact that near p_c the conducting paths are very tortuous. Their simple and beautiful experiment helped the beginning of an era in which percolation gradually became an indispensable tool for explaining and modeling of transport properties of two-phase composite materials in which there is a large contrast between the properties of the two phases.

5.1.1 Powders

Malliaris and Turner (1971) were probably the first to link percolation concepts with experimental data for electrical conductivity of a compacted powder of spherical particles. They prepared powders of high-density polyethylene particles with radius R_p and nickel particles with radius R_n, and reported that the electrical conductivity of the powder was essentially zero unless, "the composition of the metal reached a critical value." This critical value, which is clearly the percolation threshold of the metallic phase, was found to depend on R_p/R_n. The electrical resistivity of the powder dropped 20 orders of magnitude at the percolation threshold of the samples with nickel segregated on the outside of the polyethylene spheres, and at a different percolation threshold for samples prepared in such a way that the nickel penetrated the spheres. Another early study of the conductivity of a percolation composite was reported by Fitzpatrick *et al.* (1974), who utilized a random close packing of equal-size spheres, as part of a freshman physics course at Harvard University. The conducting spheres were made of aluminum, the insulating ones of acrylic plastic. The effective conductivity of the packing was measured as a function of the volume fraction ϕ of the conducting spheres, and was found to vanish at a finite ϕ, indicating a percolation transition at that point.

The first systematic and extensive study of percolation and conduction properties of random packings of particles appears to have been carried out by Ottavi *et al.* (1978), who used equal-size molded plastic spheres with a fraction of them electroplated with a copper coating to make them conducting. Their data indicated that the effective conductivity g_e of the sample vanishes at $p_c \simeq 0.29 \pm 0.02$. Since the filling factor f_l (see Section 2.10) of such packings is about 0.6, one obtains a critical volume fraction $\phi_c = f_l p_c \simeq 0.17$, in agreement with the prediction of Scher and Zallen (1970) for the percolation threshold of random 3D percolating continua.

Another set of experiments was carried out by Deptuck *et al.* (1985). Sintered, submicron silver powder with a volume fraction $\phi \sim 0.4$ is commonly used for millikelvin and submillikelvin cryostats. The sinter remains percolating for ϕ as low as 0.1. Submicron copper-oxide-silver powder is routinely used in heat exchangers for optimizing heat transfer to dilute liquid ^3He-^4He mixtures. In these composites, the silver component remained sample-spanning with a critical volume fraction $\phi_c < 0.1$. Deptuck *et al.* carried out a systematic study of the electrical conductivity (and the Young's modulus; see Chapter 9) of such powders, and reported that near ϕ_c the conductivity followed the power law (2.36) with $\mu \simeq 2.15 \pm 0.25$, completely consistent with $\mu \simeq 2.0$ for 3D percolation (see Table 2.3). Similar data were reported by Lee *et al.* (1986) who used silver-coated-glass Teflon composites. The Teflon powder used was composed of particles with diameters of about $1 \mu m$. The powder was then mixed with glass spheres coated with a 600 Å silver layer which provided high conductivity. The conductivity of the mixture was then measured as a function of the volume fraction of the conducting particles. Lee *et al.* reported a percolation threshold $\phi_c \simeq 0.17$, in good agreement with the percolation threshold of 3D continuum percolation predicted by Scher and Zallen (1970). Moreover, from their data they estimated that, $\mu \simeq 2.0 \pm 0.2$, in excellent agreement with the critical exponent of the conductivity of 3D percolation (see Table 2.3). Lee *et al.*, as well as Deprez *et al.* (1989), who measured the electrical conductivity of sintered nickel samples, also reported that the power law (2.36) was satisfied over the *entire range* of the volume fraction, thus demonstrating the practical usefulness of such universal power laws.

5.1.2 Polymer Composites

Hsu and Berzins (1985) studied percolation effects in a polymer composite, namely, perfluorinated ionomers with the general formula $[(CF_2)_nCF]_m$-O-R-SO_3X, where n is typically 6-13, R is a perfluoro alkylene group which may contain ether oxygen, and X is any monovalent cation. These composites are comprised of carbon-fluorine backbone chains with perfluoro side chains containing sulfonate, carboxylate, or sulfonamide groups, possess exceptional transport, chemical, and mechanical properties, and have been used as membrane separators, acid catalysts, and polymer electrodes. Percolation effects are important in these composites because of a spontaneous phase separation occurring in the wet polymer, which is the conductive aqueous phase that is distributed randomly in the insulative fluorocarbon phase. Hsu and Berzins (1985) measured the electrical conductivity (and elastic moduli) of these polymer composites and found their measurements to be consistent with the predictions of percolation theory. For example, they found that for $10^{-2} \leq \phi - \phi_c \leq 0.8$, a very broad region, the conductivity of the polymers followed the power law (2.36), indicating again the broad applicability and usefulness of such universal power laws.

Michels *et al.* (1989) reported extensive data for the electrical conductivity of a well-known composite, namely, carbon-black/polymer composites. From a master batch of well-dispersed carbon black (CB), casting of crosslinked samples were prepared, and for each CB concentration the effective conductivity was measured. At a critical volume fraction well below 2% (in weight) CB, the composite's conductivity exhibited a sharp rise to values 11 to 14 orders of magnitude larger than that of the matrix which was a thermosetting polymer, hence indicating the formation of a sample-spanning cluster of CB. Near the critical volume fraction of CB, the effective conductivity of the sample was found to follow Eq. (2.36) with, $\mu \simeq 2.1 \pm 0.2$, in excellent agreement with the numerical estimate of μ for 3D percolation (see Table 2.3).

5.1.3 *Conductor–Insulator Composites*

Thin metal films are two-phase mixtures of metal and non-metal components which have interesting structural and transport properties. Their electrical conductivity changes continuously as the volume fraction of the metallic phase is varied. If the volume fraction of the metallic phase is large enough, the metal phase forms a sample-spanning cluster in which the non-metallic phase is dispersed in the form of isolated islands. In this regime, the electrical conductivity of the system is large and its temperature coefficient of resistance is positive. If the volume fraction of the metallic phase is close to its percolation threshold, one has a complex system of isolated metallic islands, a tortuous and sample-spanning cluster of the metallic component, and islands of the non-metallic component. If the non-metallic phase forms a sample-spanning cluster, the system is in the dielectric regime, and its electrical conductivity is very small, if not zero. Transmission electron microscopy shows that thin metal films have a discontinuous structure in which the metallic phase constitutes only a fraction of the total volume of the system. Such films have a nearly two-dimensional (2D) structure and are formed in the early stages of film growth by a variety of techniques, such as evaporation or sputtering. The metallic phase first forms isolated islands that at later stages of the process join together and form a continuous film. Generally speaking, a metal film is considered thin if its thickness is less than the percolation correlation length ξ_p of the 3D film.

Thick films of metallic and non-metallic components, i.e., those with a thickness greater than ξ_p, also have interesting properties and have been studied for a long time. They can also be produced by co-sputtering or co-evaporation of two components that are insoluble in each other, one being the non-metallic component and the other one the metallic phase. They have been studied for their novel superconducting and magnetic properties. However, the discovery by Abeles *et al.* (1975) that the electrical conductivity of such films near p_c follows the power law (2.36) made them the subject of many studies. Such thick films are mainly composed of a metal with a bulk lattice structure. The non-metallic phase is amorphous and often in the form of isolated islands. They are often called *granular metal films* because inspection of their structure shows that the metal grains are often surrounded by very thin amorphous layers of the non-metallic component, so that metallic grains remain separated. Granular composites always have high percolation thresholds, which can be close to the random close packing fractions, if the grain size is constant and the insulator is relatively thick. Examples of such composites are Ni-SiO_2, Al-Si, and Al-Ge. Unlike thin metal films, one cannot easily find clusters of metallic grains in granular metals, and therefore it is more difficult to interpret their morphology in terms of the concepts of percolation theory. Let us now discuss several properties of metal films and relate them to percolation quantities.

The first application of percolation to the interpretation of various properties of very thin (2D) metal films appears to have been made by Liang *et al.* (1976). They prepared ultrathin films which were 1.5 mm wide with pre-evaporated indium electrodes spaced 25 mm apart. The substrate used was SiO films deposited on a microscope slide glass. Several types of metallic compounds were evaporated and studied. They found that the sudden drop of the resistivity of semi-metal bismuth ultrathin films was very steep, signaling a percolation transition. When they plotted the conductivity of the composite versus the area fraction ϕ of the conducting phase, they found that near the percolation threshold the conductivity follows the power law (2.36) with $\mu \simeq 1.15$, reasonably close to the corresponding exponent for 2D percolation (see Table 2.3).

Papandreou and Nédellec (1992) prepared Pd films with a typical thickness of about 100Å by depositing them by electron gun evaporation on quartz substrate for measuring their conductivity, and on NaCl substrate for TEM observations. The substrates were coated with a SiO layer of 150 Å, which ensured similar growth conditions, regardless of the substrate below SiO. The samples were then irradiated under normal incidence with 100 keV Xe ion beam, after which their conductivities were measured. It was found that, as the percolation threshold was approached, the conductivity of the samples vanished according to the power law (2.36) with $\mu \simeq 1.3$, in complete agreement with the prediction of 2D percolation (see Table 2.3).

As already mentioned above, in a seminal paper, Abeles *et al.* (1975) studied the growth of the grains of the metals W or Mo in the insulators Al_2O_3 or SiO_2 using co-sputtering. They varied the volume fraction ϕ of the metal in the mixture in the range $0.1 < \phi \leq 1$ and obtained very finely dispersed grain structures. With annealing, one can change the size of the metal grains over a wide range. The advantage of using the W or Mo is that annealing does not cause the metal to precipitate—it remains uniformly dispersed within the insulator. The conductivity of the films was measured for several volume fractions of the metallic phase. It was found that the effective conductivity of the composite follows the power law (2.36) with $\mu \simeq 1.9 \pm 0.1$, in good agreement with the critical exponent of 3D percolation (see Table 2.3).

Results and data similar to those of Abeles *et al.* were also reported by Kapitulnik and Deutscher (1983), Tessler and Deutscher (1989), and Kapitulnik *et al.* (1990). Kapitulnik and Deutscher (1983) prepared Al-Ge films by co-evaporating them onto glass substrates from two electron beam guns through a mask with slits, and measured their electrical conductivity. The conductivity of the thick samples near the metal-insulator transition followed the power law (2.36) with $\mu \simeq 2.1 \pm 0.5$, completely consistent with the prediction of 3D percolation (see Table 2.3). The crossover between 2D and 3D films was also studied by Kapitulnik and Deutscher (1983). More recent experiments and their comparison with the predictions of percolation theory and lattice models will be described and discussed throughout this chapter. Metal-insulator composites will also be further analyzed in the next chapter, and also Chapter 3 of Volume II.

Hence, there should be little, if any, doubt that percolation theory and the discrete models provide a satisfactory explanation of the behavior of the effective conductivity of a wide variety of two-phase materials with large contrast between the conductivities of the phases. We now describe theoretical methods and computer simulation techniques for predicting the effective conductivity (and hence the effective dielectric constant) of disordered materials.

5.2 Conductivity of a Random Resistor Network

For simplicity, we consider a d-dimensional cubic lattice of $(L+1)^d$ sites, in which the sites $\mathbf{s} = (s_1, s_2, \cdots, s_d)$ have integer coordinates defined by $0 \leq s_j \leq L$. We impose a potential drop ΔV between the faces $s_1 = 0$ and $s_1 = L$, so that a potential gradient $\Delta V/L$ is imposed on the lattice. If L is large enough, the boundary conditions in the other direction will not matter, but in practice one uses finite-size lattices, and therefore periodic boundary conditions, by which the faces $s_j = 0$ and $s_j = L$ are connected for $j = 2, \cdots, d$, are used in order to eliminate sample-size effects. Suppose that, in response to the applied potential gradient $\Delta V/L$, a total current I flows in the network. Then, the current per unit area (that is, per site) of the face $s_1 = 0$ is $I/(L+1)^{d-1}$. The conductivity g of the lattice is then defined by

$$\lim_{L\to\infty} \frac{I/(L+1)^{d-1}}{\Delta V/L} = \lim_{L\to\infty} \frac{L^{2-d} I}{\Delta V}. \tag{1}$$

If the lattice is disordered (i.e., if the conductance of its bounds are distributed according to some probability density function), then we are interested in the mean value of the above limit which is the effective conductivity g_e of the lattice:

$$g_e = \lim_{L\to\infty} \left\langle \frac{L^{d-2} I}{\Delta V} \right\rangle, \tag{2}$$

where the average is taken with respect to the distribution $f(g)$ of the bond conductances. Note that $G = I/\Delta V$ is the *conductance* of the lattice.

It can be shown that if g is the conductance of the individual bonds of the lattice, then the mean conductance $\langle G \rangle$ and the effective conductivity g_e of the lattice satisfy the following inequalities, which are known as the series and parallel bounds:

$$L^{-1}(L+1)^{d-1}\langle g^{-1}\rangle^{-1} \leq \langle G \rangle \leq L^{-1}(L+1)^{d-1}\langle g \rangle, \tag{3}$$

$$\langle g^{-1}\rangle^{-1} \leq g_e \leq \langle g \rangle, \tag{4}$$

where $\langle \cdot \rangle$ denotes an average over the distribution of the conductances g of the bonds. As discussed in Section 4.6, the bounds in (4) are in fact the one-point bounds first derived by Wiener (1912).

5.3 Exact Solution for Bethe Lattices

Stinchcombe (1973, 1974) derived an exact solution for the conductance and effective conductivity of a Bethe lattice of coordination number Z, an endlessly branching network without any closed loops; see Figure 2.5. His treatment of the problem is somewhat involved, and therefore we describe a simpler analysis due to Heinrichs and Kumar (1975), which will also be used in the next chapter, and also in Chapter 3 of Volume II for analyzing more complex nonlinear transport processes on Bethe lattices.

Consider a branch of a Bethe lattice of coordination number Z which starts at the origin O. The conductance of the branch can be computed by simply realizing that, it is the equivalent conductance of one bond, say OA, of the branch that starts at O with conductance g_i in series with the branch that starts at A, the conductance of which is G_i. Suppose now that the lattice is grounded at infinity and that the voltage at O is unity. Then, the total conductance G of the network between O and infinity is that of $(Z-1)$ branches which are in parallel. Hence,

$$G = \left[\sum_{i=1}^{Z-1}\left(\frac{1}{g_i}+\frac{1}{G_i}\right)\right]^{-1}. \tag{5}$$

For an infinitely large Bethe lattice, G and G_i are statistically equivalent. Thus, if $H(G)$ represents the statistical distribution of G, we must have

$$H(G) = \int \cdots \int \delta\left\{G - \left[\sum_{i=1}^{Z-1}\left(\frac{1}{g_i}+\frac{1}{G_i}\right)\right]^{-1}\right\} \prod_{i=1}^{Z-1} f(g_i)H(G_i)dg_i dG_i, \tag{6}$$

where $f(g)$ is the probability density function for the bond conductances. If we now take the Laplace transform of both sides of Eq. (6), we obtain

$$\tilde{H}(s) = \int_0^\infty \exp(-sG)H(G)dG = \left[\int\int \exp\left(-\frac{sgG}{g+G}\right) f(g)H(G)dg dG\right]^{Z-1}, \tag{7}$$

where s is the Laplace transform variable conjugate to G. Most of the properties of interest can be obtained from the solution of the integral equation (7), which is an exact equation valid for any $f(g)$. In particular, if we consider a percolation-type distribution, i.e., a conductance distribution for a conductor-insulator material,

$$f(g) = (1-p)\delta_+(0) + ph(g), \tag{8}$$

where $h(g)$ is any normalized distribution, then, Eq. (7) becomes

$$\tilde{H}(s) = \left[1 - p + p\int\int \exp\left(-\frac{sgG}{g+G}\right) h(g)H(G)dgdG\right]^{Z-1}, \tag{9}$$

which can be further simplified if $h(g) = \delta(g - g_0)$,

$$\tilde{H}(s) = \left[1 - p + p\int_0^\infty \exp\left(-\frac{sg_0G}{g_0+G}\right) H(G)dG\right]^{Z-1}. \tag{10}$$

5.3.1 The Microscopic Conductivity

In general, one can define and calculate two different conductivities for a Bethe lattice. One is g_m, the *microscopic conductivity* of the lattice, which is obtained by grounding the lattice at infinity and imposing a unit voltage at site O of the lattice. g_m is then defined as the current that flows out along one of the outgoing bonds connected to O. It is not difficult to see that $g_m = Z\langle G\rangle/(Z-1)$. Therefore,

$$g_m = -\frac{Z}{Z-1}\left[\frac{d\tilde{H}(s)}{ds}\right]_{s=0}$$

$$= Z\left[\int\int \left(\frac{gG}{g+G}\right) f(g)H(G)dgdG\right]^{Z-2}. \tag{11}$$

If $f(g)$ is a percolation type distribution, Eq. (8), then, it can be shown that, near the percolation threshold $p_c = (Z-1)^{-1}$ of the lattice,

$$g_m = \frac{2c(Z-1)^2}{h_{-1}(Z-2)}(p-p_c)^2, \tag{12}$$

where $c \simeq -0.761$ and

$$h_{-1} = \int_0^\infty \frac{h(g)}{g}\,dg. \tag{13}$$

Equations (12) and (13) imply that, if h_{-1} is finite, then the microscopic conductivity g_m follows a power law near p_c with a critical exponent of 2. Observe that this critical exponent is the same as with the exponent $\mu \simeq 2$ for the effective conductivity g_e of three-dimensional (3D) systems (see Table 2.3), and therefore the effective *macroscopic* conductivity of 3D networks and the *microscopic* conductivity of the Bethe lattices follow the same power law near the percolation threshold. On the other hand, as discussed in Chapter 2, the bond-percolation threshold of a 3D lattice is approximated well [see Eq. (2.22)] by $p_c \simeq 3/(2Z_3)$, where Z_3 is the coordination number of the 3D lattice. Since for a Bethe lattice $p_c = (Z_b - 1)^{-1}$, where Z_b is the coordination number of the Bethe lattice, one can select the coordination number Z_b of the Bethe lattice in such a way that its percolation threshold would match closely that of a corresponding 3D network. The appropriate choice of Z_b is then given by

$$Z_b = \frac{2}{3} Z_3 + 1. \tag{14}$$

For example, a five-coordinated ($Z_b = 5$) Bethe lattice ($p_c = 0.25$) and the simple-cubic lattice ($Z_3 = 6$, $p_c \simeq 0.249$) have essentially the same bond percolation threshold. The conclusion is that, the microscopic conductivity of a Bethe lattice, the coordination number of which has been selected in such a way that its percolation threshold approximates well that of a 3D lattice, provides an excellent approximation to the effective conductivity of the 3D lattice. This has indeed been found to be the case (Heiba *et al.*, 1982, 1992).

Equation (10) can further be analyzed by the following method (Heinrichs and Kumar, 1975). We view a Bethe lattice of coordination number Z as a collection of Z subnetworks, and consider its behavior near p_c. Then, the percolation probability $P(p)$ (see Section 2.6.1), i.e., the probability that, when the fraction of the conducting bonds is p, a site belongs to the sample-spanning (infinite) conducting cluster, can be written as

$$P(p) = 1 - R(p)^Z, \tag{15}$$

where $R(p)$ is the probability that a site does *not* belong to a conducting cluster on one of the Z subnetworks. With this definition, it is not difficult to see that $R(p)$ is the root of the following equation:

$$R(p) = 1 - p + pR^{Z-1}. \tag{16}$$

Heinrichs and Kumar (1975) assumed that

$$\tilde{H}(s) = [R(p) + \phi(s)]^{Z-1}, \quad \phi(s) \ll 1. \tag{17}$$

With the help of Eqs. (15)–(17), Eq. (10) is converted into a differential equation for $\phi(s)$:

$$\frac{spR^{Z-2}}{g_0 p_c}\frac{d^2\phi}{ds^2} + \left(\frac{p}{p_c}R^{Z-2} - 1\right)\phi(s) + \frac{p(1-p_c)R^{Z-3}}{2p_c^2}\phi^2(s) + O(\phi^3) = 0, \tag{18}$$

where $p_c = (Z-1)^{-1}$, and a term which is proportional to $\phi d\phi/ds$ which, compared to $\phi^2(s)$, is small near p_c has been neglected, as in the critical region near the percolation threshold $\tilde{H}(s)$ is a slowly-varying function. The boundary conditions are, $\phi(0) = 1 - R(p)$ which follows from the normalization condition, $\tilde{H}(0) = 1$, and $\phi(\infty) = 0$. Equation (18) can be solved for any $p > p_c$ and g_0. A similar, but more tedious, method can be developed for the more general percolation distribution (8) (Heiba *et al.*, 1982, 1992) and for any p (see also Essam *et al.*, 1974).

5.3.2 *Effective-Medium Approximation*

As discussed in Chapter 4 for continuum models of disordered materials, and as we show later in this chapter, a simple and useful approximation for the conductivity of a disordered network can be obtained by an effective-medium approximation (EMA). A systematic derivation of the EMA for 2D and 3D lattices will be given shortly. For now, we utilize Eq. (7) in order to derive an EMA for the Bethe lattices. In an EMA approach, the probability density $H(G)$ is expected to be sharply peaked around a mean value G^*, and thus we may approximate $H(G)$ by $H(G) \simeq \delta(G - G^*)$, so that $\tilde{H}(s) = \exp(-sG^*)$, in which case Eq. (7) becomes

$$\exp(-sG^*) = \left[\int\int \exp\left(-\frac{sgG^*}{g+G^*}\right) f(g)dg\right]^{Z-1}. \tag{19}$$

We must first determine G^*. To do so, we take the derivative of Eq. (19) with respect to the Laplace transform variable s and evaluate the result at $s = 0$ [see Eq. (11)] to obtain

$$\int_0^\infty \left(\frac{gG^*}{g+G^*} - \frac{G^*}{Z-1}\right) f(g)dg = 0. \tag{20}$$

The effective conductivity g_e of the network is obtained if we set in Eq. (20), $f(g) = \delta(g - g_e)$ (since, as discussed below, in the EMA approach all bonds of the network have the same conductance g_e), which then yields, $G^* = g_e(Z-2)$. Substituting this result into Eq. (20) then yields the EMA for the effective conductivity of a Bethe lattice of coordination number Z (Stinchcombe, 1973,1974; Heinrichs and Kumar, 1975):

$$\int_0^\infty \frac{g - g_e}{g + (Z-2)g_e} f(g)dg = 0, \tag{21}$$

which, as shown below, is completely similar to the EMA for regular (2D or 3D) networks. If the percolation distribution (8) is substituted into Eq. (21), we find that it predicts the percolation threshold of a Bethe lattice of coordination number Z is, $p_c = (Z-1)^{-1}$, an exact result [see Eq. (2.22)]. However, Eq. (21) also predicts that near p_c the effective conductivity g_e varies linearly with $(p - p_c)$, which is incorrect.

5.3.3 *Conductor–Insulator Composites*

In addition to the microscopic conductivity defined and analyzed above, one may also define and compute the usual macroscopic or effective conductivity g_e, which is also the usual conductivity of 3D (bulk) materials. For a Bethe lattice, g_e is the average current density per unit applied field. However, unlike g_m which, as discussed above, provides an excellent approximation for the effective macroscopic conductivity of 3D systems, the effective conductivity of a Bethe lattice is only of theoretical interest. In particular, it can be shown (de Gennes, 1976a; Straley, 1977a) that near p_c

$$g_e \sim (p - p_c)^3, \tag{22}$$

so that the critical exponent μ for the power-law behavior of the effective conductivity near p_c, defined by Eq. (2.36), is $\mu = 3$. Equation (22) represents the power-law behavior of the effective conductivity near p_c in the mean-field approximation. Other properties of g_e for Bethe lattices have been discussed by Straley (1977a).

5.3.4 *Conductor–Superconductor Composites*

Before moving on to the next topic, let us point out that one may also consider the problem of computing the effective conductivity of a Bethe lattice in which the bonds are either normal conductors with a fraction $(1 - p)$ or are superconducting with a fraction p. Similar to the conductor-insulator problem described above, one may compute two effective conductivities for the conductor-superconductor problem. One is the microscopic conductivity g_m, while the second one is the usual effective macroscopic conductivity g_e. All the formulations developed above are equally applicable to conductor-superconductor problem. The only change that one must make is using the conductance distribution $f(g) = p\delta(g - \infty) + (1 - p)\delta(g - g_0)$. In particular, one finds that near p_c

$$g_m \sim |p - p_c|^{-1}, \tag{23}$$

so that the microscopic conductivity diverges at p_c with a critical exponent of 1. On the other hand, the effective macroscopic conductivity follows the following power law near p_c (Straley, 1977a):

$$g_e \sim \ln |p - p_c|, \tag{24}$$

and therefore the critical exponent s, defined by Eq. (2.37) for conductor-superconductor mixtures near p_c, is given by $s = 0$, since as $p \to p_c$, $\ln |p - p_c|$ diverges slower than any power law. This value of s should be compared with $s \simeq 0.73$ for 3D conductor-superconductor percolation composites (see Table 2.3). The conclusion is that, unlike the case of conductor-insulator mixtures described above, neither g_m nor g_e provides a reasonable approximation to the effective conductivity of a 3D conductor-superconductor mixture.

5.4 Exact Results for Two-Dimensional Composites

In addition to the exact solution of the conduction problem on a Bethe lattice, there are a few other exact results that hold for 2D systems. What follows is a brief description of these results and their implications.

5.4.1 *Exact Duality Relations*

Composite materials with a 2D microgeometry possess a special symmetry called *duality*. We already discussed in Section 4.8 the implications of this special property for continuum models of heterogeneous materials. One example of such a system is a 3D composite with cylindrical symmetry in which all interfaces are parallel to a fixed direction. Another example is a thin conducting film with a thickness which is negligible compared to other microgeometric scales, such as channel widths and grain sizes. Even though the film is part of a 3D system, the only important electric current fields lie in the plane of the film. As discussed in Section 4.8, this symmetry was first discovered by Keller (1964), and independently by Dykhne (1971), for the case in which all components of the composite have scalar conductivity tensors. This result was later generalized by Mendelson (1975a,b) to the case of general anisotropic conductivity tensors, including non-symmetric tensors. A further generalization was made by Milton (1988), who also provided additional historical references. These results were already described in Section 4.8, where we also described how they lead to some exact results for the effective conductivity of 2D composites.

A duality transformation is also possible for the lattice or discrete models. However, because the discretized versions of the local electric field $\mathbf{E}(\mathbf{r})$ and the current $\mathbf{I}(\mathbf{r})$ are the voltage V_i and current I_i in a bond i, one must also rotate each bond by 90° (see the discussion in Section 4.8) when performing the duality transformation. This alters in most cases the structure of the network, whereas in the continuum case discussed in Section 4.8 the morphology is left unchanged. For example, even in the case of a square network (for which the dual network is also a square network), if the original network is random, the dual network will usually be a different random square network. Thus, one of the implicit assumptions used to derive the duality relations presented in Section 4.8 is not satisfied. Although this turns out to be unimportant for random resistor networks in which the external field is $H = 0$, it is an important consideration for the choice of an appropriate random resistor network model for $H \neq 0$. In particular, some random resistor network models for the Hall effect that are not self-dual have turned out to belong to a universality class that differs from that of continuum composites, as far as the critical behavior of the Hall conductivity near p_c is concerned. We will discuss this issue in Section 5.17.

Consider a 2D network in which the probability density function $f(g)$ for the bond conductances is given by

$$f(g) = (1-p)\delta(g-g_1) + p\delta(g-g_2). \tag{25}$$

The effective conductivity g_e of a square network with bond conductance distribution (25) satisfies the following equation,

$$g_e(p)g_e(1-p) = g_1 g_2. \tag{26}$$

In particular, setting $p = 1/2$, we obtain

$$g_e(p=1/2) = \sqrt{g_1 g_2}. \tag{27}$$

These results can be directly deduced from the duality relation, Eq. (4.224). Equation (26) is in fact more general: Since the triangular and hexagonal lattices are also dual of each other, one can also write (Bernasconi *et al.*, 1977)

$$g_e^{\text{hexag}}(p)g_e^{\text{triang}}(1-p) = g_1 g_2. \tag{28}$$

An immediate consequence of these results is that, for the square lattice, and hence (due to the universality) for all 2D systems, the critical exponents μ and s for the power-law behavior of the effective conductivity of conductor-insulator, and conductor-superconductor composites near p_c must be equal (Straley, 1977b):

$$\mu = s, \quad \text{two-dimensional composites,} \tag{29}$$

and that if, for the region near p_c, we write, $g_e = a_0(p - p_c)^\mu$ for conductor-insulator mixtures, and $g_e = a_\infty(p_c - p)^{-s}$ for conductor-superconductor composites, then $a_0 a_\infty = 1$. All the numerical simulations confirm these important results.

5.4.2 *Log-Normal Conductance Distribution*

Although the duality relations are interesting and useful, they cannot predict, except at $p = 1/2$, the effective conductivity of any 2D lattice. There is, however, an important case for which there exists an exact relation that can predict the conductivity of a square network. Suppose that the bond conductance distribution in a square network is given by

$$f(g) = \frac{1}{\sqrt{2\pi}\, g\sigma} \exp\left[-\frac{(\log g - \log g_0)^2}{2\sigma^2}\right], \tag{30}$$

then

$$g_e = g_0. \tag{31}$$

5.5 Green Function Formulation and Perturbation Expansion

The idea that in order to calculate effective transport properties of a disordered material, one may replace it with a hypothetical homogeneous material which somehow mimics the behavior of the original disordered material, and then calculate the homogeneous material's properties, has a long history, some of which was already mentioned in Chapter 4. As discussed there, there are two approaches for implementing this idea. In the approach developed by Maxwell–Garnett (1904), one considers isolated inclusions that are embedded in a continuous matrix consisting of a single phase; the effective properties of the disordered material are derived by placing a sphere (or an ellipse) of the homogeneous or the effective medium in this matrix. Such an approach has also been called the *average t-matrix approximation*, or the *non-self-consistent approximation*, and was described and discussed in Section 4.9.4. In the second approach, originally developed by Bruggeman (1935), each inhomogeneity is embedded in the effective medium itself, the unknown properties of which are determined in such a way that the volume-averaged field, when the averaging is done with respect to the inhomogeneities, is equal to uniform field in the effective medium. This technique is usually called the *effective-medium approximation* (EMA). In effect, the EMA is an ingenious way of transforming a many-body problem into a *one-body* problem. Bruggeman's EMA, which in Section 4.9 was derived for continuum models of disordered materials, was independently rederived by Landauer (1952) [see Landauer (1978) for a history of EMAs]. In this section we derive a perturbation expansion, the lowest order of which yields the EMA for discrete models of disordered materials.

The continuity equation requires that at every site of a resistor network one must have

$$\sum_j g_{ij}\left(P_j - P_i\right) = 0, \tag{32}$$

where P_i is the field (voltage, temperature, etc.) at site i, and the sum is over all the nearest-neighbor sites of i. Here g_{ij} represents the local conductance between sites i and j which varies in space according to a probability density function. The exact and *explicit* solution of Eq. (32) for an arbitrary distribution of g_{ij} is not available in any dimension $d \geq 2$. However, one can develop an exact but *implicit* solution of this equation which can then be used as the starting point for construction of a hierarchy of analytical approximations. We first develop this solution for an isotropic material, following the work of Blackman (1976), and then analyze the same problem for anisotropic materials.

To facilitate the solution of Eq. (32) we introduce a *uniform* or an *effective medium* in which all the bond conductances are equal to an effective conductance g_e which is yet to be determined. The effective conductivity of the material will in fact be proportional to, or identical with (depending on the topology of the lattice), g_e. Therefore,

$$\sum_j g_e(P_j^0 - P_i^0) = 0, \tag{33}$$

where P_i^0 is the field at site i of the effective (uniform) network. Subtracting Eq. (32) from (33) yields

$$\sum_j [\Delta_{ij}(P_j - P_i) - (P_i - P_i^0) + (P_j - P_j)] = 0, \tag{34}$$

where $\Delta_{ij} = (g_{ij} - g_e)/g_e$. Equation (34) is now rewritten as

$$Z_i(P_i - P_i^0) - \sum_j (P_j - P_j^0) = -\sum_j \Delta_{ij}(P_i - P_j), \tag{35}$$

where Z_i is the coordination number of site i. We now introduce a *site-site* Green function G_{ij} defined by

$$Z_i G_{ik} - \sum_j G_{jk} = -\delta_{ik}, \tag{36}$$

where, physically, G_{ij} is the field at j as a result of injecting a unit flux at i. With the aid of the Green function, Eq. (35) becomes

$$P_i = P_i^0 + \sum_j \sum_k G_{ij} \Delta_{jk} \left(P_j - P_k\right). \tag{37}$$

The analysis so far is exact and applies to *any* network, since we have not specified Z_i or G_{ij}. Thus, Eq. (37) represents an exact, albeit implicit, solution of Eq. (32).

We can rewrite Eq. (37) in terms of $Q_{ij} = P_i - P_j$, where Q_{ij} is proportional to the flux between i and j, by noting that $Q_{ij} = -Q_{ji}$. Hence,

$$Q_{ij} = Q_{ij}^0 + \sum_{[lk]} Q_{lk}(G_{il} + G_{jk} - G_{jl} - G_{ik}), \tag{38}$$

where $[lk]$ indicates that the bond connecting nearest-neighbor sites l and k is counted only once in the sum. We denote bonds with Greek letters and assign direction to them and let

$$\gamma_{\alpha\beta} = (G_{il} + G_{jk}) - (G_{jl} + G_{ik}), \tag{39}$$

be a *bond-bond* Green function, where i and l (j and k) are network sites with tails (heads) of arrow on bonds α and β, respectively. Equation (38) is finally rewritten as

$$Q_\alpha = Q_\alpha^0 + \sum_\beta \Delta_\beta \gamma_{\alpha\beta} Q_\beta. \tag{40}$$

According to Eq. (40), the "flux" Q_α in bond α is the sum of the flux in the same bond but in the effective (uniform) network, plus the perturbations that are caused by the distribution of the conductances.

5.5.1 Properties of the Green Functions

The derivation of the lattice Green functions G_{ik} will be described in detail in Section 6.1.4, where we derive them for the more general problem of unsteady-state diffusion and conduction in disordered lattices. In the limit of long times, $t \to \infty$, the resulting Green functions reduce to those that are needed here. More specifically, the Green functions G_{ik} that are needed here are the limit $\epsilon = 0$ of those for the unsteady-state problem that will be derived for a variety of 2D and 3D regular lattices in 6.1.4. Here we discuss some specific properties of the Green functions that are needed for evaluating the accuracy of the above perturbation expansion for the effective conductivity. In our discussion we closely follow Blackman (1976).

Figure 5.1 shows a reference bond labelled 0 and a set of neighboring bonds labelled $1, 2, \cdots$ on a square lattice. Neighboring bonds in the other three quadrants are symmetrically equivalent to those in Figure 5.1, and hence are not shown. This figure can be reflected about a vertical axis through bond 0, and also about a horizontal axis through bond 0, but keeping the direction of the bonds intact. Using the Green functions given in Chapter 6 (in the limit $\epsilon = 0$), it is straightforward

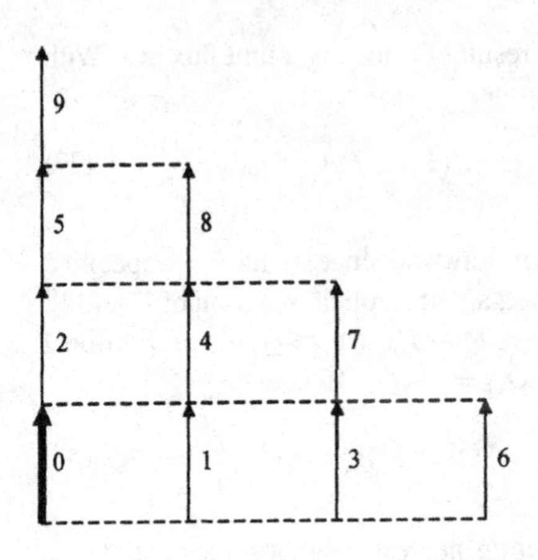

FIGURE 5.1. One quadrant of a portion of a square lattice, with a reference bond 0 and a set of its neighboring parallel bonds.

to show that

$$\gamma_{00} = \gamma_{04} = 0,$$
$$\gamma_{01} = -\gamma_{02} = \frac{1}{2} - \frac{2}{\pi}, \quad \gamma_{03} = -\gamma_{05} = \frac{5}{2} - \frac{8}{\pi}, \tag{41}$$
$$\gamma_{06} = -\gamma_{09} = \frac{25}{2} - \frac{118}{3\pi}, \quad \gamma_{07} = -\gamma_{08} = \frac{14}{3\pi} - \frac{3}{2}.$$

More generally, if the integers (m, n) label the tail of bond α with respect to the tail of bond 0, then, as shown in Section 6.1.6,

$$\gamma_{0\alpha}(m, n) = \frac{1}{2}\int_0^\infty (I_m I_{n+1} + I_m I_{n-1} - 2I_m I_n)\exp(-2t)dt. \tag{42}$$

Moreover, it is not difficult to show that

$$\gamma(m, n) = -\gamma(n, m), \quad \gamma(m, n) = \gamma(-m, -n), \quad \gamma(m, m) = 0. \tag{43}$$

Consider now the cluster of bonds shown in Figure 5.2, which includes a reference bond labelled 0 and the first two shells of perpendicular bonds. It is not difficult to show that

$$\gamma_{01} = \gamma_{03} = -\gamma_{02} = -\gamma_{04} = \frac{1}{2} - \frac{1}{\pi}$$
$$\gamma_{05} = \gamma_{07} = \gamma_{09} = \gamma_{0,11} = 1 - \frac{3}{\pi}$$
$$= -\gamma_{06} = -\gamma_{08} = -\gamma_{0,10} = -\gamma_{0,12}. \tag{44}$$

More generally,

$$\gamma_{0\alpha}(m, n) = \frac{1}{2}\int_0^\infty (I_{m+1} I_n + I_m I_{n-1} - I_m I_n - I_{m+1} I_{n-1})\exp(-2t)dt, \tag{45}$$

based on which it is not difficult to show that

$$\gamma_{m,n} = \gamma_{-(m+1),(1-n)} = \gamma_{-n,-m} = \gamma_{(n-1),(m+1)}$$
$$= -\gamma_{m,(1-n)} = -\gamma_{-(m+1),n} = -\gamma_{n,(m+1)} = -\gamma_{(n-1),-m}. \tag{46}$$

FIGURE 5.2. A reference bond 0 and a set of perpendicular bonds in the square lattice.

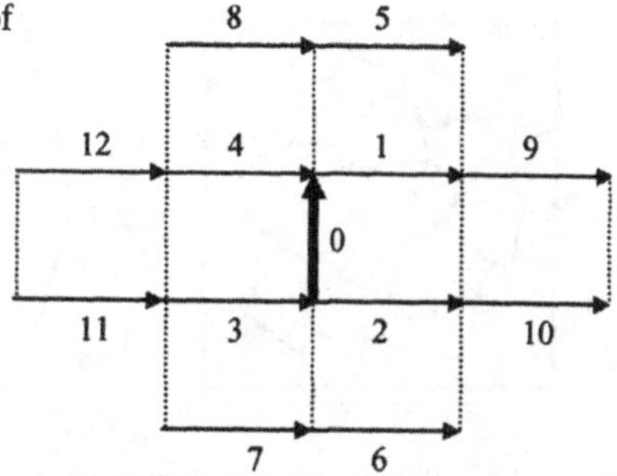

As Eq. (40) indicates, in assessing the accuracy of the perturbation expansion, one must also deal with summation over bonds. It can be shown that, for the square and simple-cubic lattices,

$$\sum_{\beta} \gamma_{\alpha\beta} = 0, \tag{47}$$

which is true for β summed over all bonds of the lattice, over bonds parallel or perpendicular to α, over a continuous chain of bonds joining the two boundaries, and if the sum is over α instead of β. One can also show that

$$\sum_{\beta,\delta,\cdots,\epsilon} \gamma_{\alpha\beta}\gamma_{\beta\delta}\cdots\gamma_{\epsilon\alpha} = (\gamma^n)_{\alpha\alpha} = (-1)^n \frac{1}{2}, \tag{48}$$

where n is the number of terms in the product so that, for example,

$$\sum_{\beta} \gamma_{\alpha\beta}\gamma_{\beta\alpha} = \frac{1}{2}, \tag{49}$$

and

$$\sum_{\beta} (\gamma_{\alpha\beta})^n = (-\frac{1}{2})^n. \tag{50}$$

Equation (48) also holds for the simple-cubic lattice, if we replace $1/2$ with $1/3$.

5.6 Effective-Medium Approximation

An EMA is now constructed by the following algorithm. Figure 5.3 shows the schematics of the method. One solves Eq. (40) for an arbitrary cluster of bonds, i.e., for *arbitrary* Δ_α, selects any bond α, and requires that

$$\langle Q_\alpha \rangle = Q_\alpha^0, \tag{51}$$

where $\langle \cdot \rangle$ denotes an average over all possible values of the bond conductance g_α in the set. Equation (51) requires that the average of the fluctuations in the "flux" Q_α across bond α from its value in the effective network vanish, as a result which one is led to a *self-consistent* determination of g_e. Therefore, one obtains a

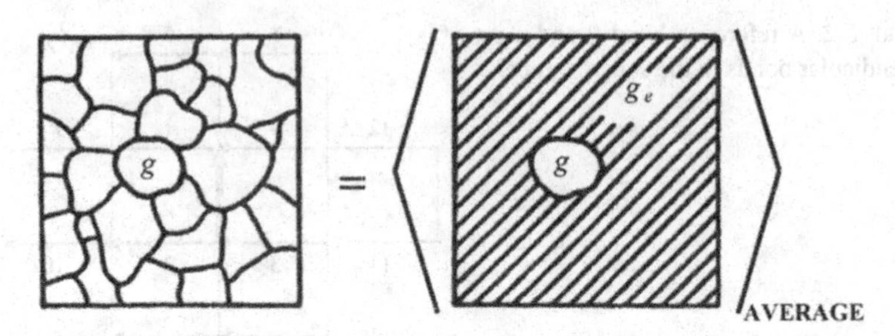

FIGURE 5.3. Schematic implementation of the effective-medium approximation, transforming a disordered material (left) to a homogeneous effective material (right).

hierarchy of approximations built on cluster of bonds, such that the larger the set of bonds that one uses in Eq. (51), the more accurate the approximate estimate of g_e would be. In fact, if we could use every bond of the network in such an averaging, the resulting g_e would be exact. In practice, this is neither possible nor necessary, because doing so would make it impossible to derive an analytical approximation, and therefore one assigns to all but a *finite* set of bonds in the network the effective conductance g_e (so that $\Delta_\alpha \neq 0$ only for a finite cluster of bonds), and proceeds as above, now averaging over the finite cluster of the bonds in order to determine g_e. The g_e so computed represents an approximation to the true value of the effective conductivity of the network. To obtain rapid convergence toward the true value of g_e with increasing size of the cluster of bonds, one must choose the cluster carefully to ensure that it preserves all symmetry properties of the network (see Section 5.6.7).

In the simplest approximation, only a single bond α has a conductance g differing from g_e, in which case equations (40) and (51) yield

$$\left\langle \frac{1}{1-\gamma_{\alpha\alpha}\Delta_\alpha} \right\rangle = 1, \tag{52}$$

or

$$\left\langle \frac{\Delta_\alpha}{1-\gamma_{\alpha\alpha}\Delta_\alpha} \right\rangle = 0. \tag{53}$$

Since $\Delta_\alpha = (g_\alpha - g_e)/g_e$, Eq. (53) is equivalent to

$$\left\langle \frac{g_\alpha - g_e}{(1+\gamma_{\alpha\alpha})g_e - \gamma_{\alpha\alpha} g_\alpha} \right\rangle = 0. \tag{54}$$

From Eq. (39) we obtain

$$\gamma_{\alpha\alpha} = 2[G(0,0) - G(1,0)]. \tag{55}$$

Equation (36) implies that, for periodic networks of coordination number Z, $Z[G(0,0) - G(1,0)] = -1$, so that

$$\gamma_{\alpha\alpha} = -\frac{2}{Z}. \tag{56}$$

If $f(g)$ represents the probability density function of the bond conductances, then Eq. (54) becomes

$$\int_0^\infty \frac{g - g_e}{g + (Z/2 - 1)g_e} f(g)dg = 0, \tag{57}$$

an equation that was first derived by Kirkpatrick (1971,1973) for random resistor networks using a different derivation. To distinguish between this EMA and those that will be described below, in which a cluster of bonds is utilized in the calculations, we sometimes refer to Eq. (57) as the *single-bond* EMA.

We may also derive Eq. (57) by a method due to Kirkpatrick (1973). We described here his method of deriving Eq. (57), and will utilize it shortly in order to derive an EMA for regular networks with *two* different coordination numbers, such as the kagomé lattice which is characterized by coordination numbers $Z_1 = 6$ and $Z_2 = 3$, and for topologically-disordered networks, such as the Voronoi network described in Chapters 2 and 3. Consider a regular network of coordination number Z, e.g., a simple-cubic network, in which the bond conductances are distributed according to a probability density function. Now construct the effective-medium equivalent of the network by replacing the conductances of all the bonds with g_e. Applying an external potential gradient to this network induces a uniform potential V_e along those bonds that are oriented along the external field. Take one such bond and replace its conductance g_e by its true value g in the original heterogeneous network (see Figure 5.4). This replacement induces an *extra* current $I_0 = V_e(g_e - g)$ in the bond, which in turn induces an extra or excess potential V_0 along the bond. As discussed above, in the EMA approach one calculates

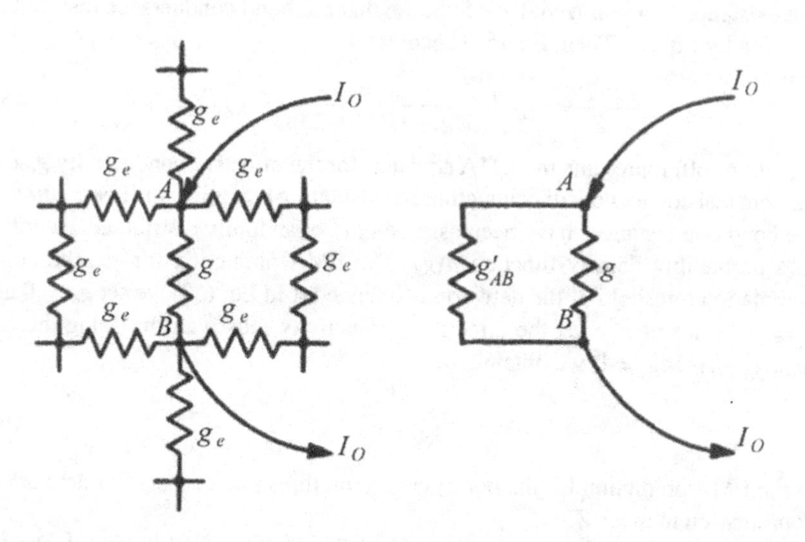

FIGURE 5.4. Embedding a bond and its true conductance g in an effective-medium network in which all bonds have the same conductance g_e (left), and its equivalent circuit (right). I_0 is the injected current and g'_{AB} is the equivalent conductance of the *rest* of the network.

the *excess* potential V_0 and insists that its average must vanish, so that

$$\int_0^\infty V_0 f(g) dg = 0 \tag{58}$$

from which g_e is determined. To calculate V_0 one only needs to calculate g'_{AB}, the conductance of the network between A and B since $V_0 = I_0/(g + g'_{AB})$. But because $g_{AB} = g'_{AB} + g_e$, all one needs to do is calculating g_{AB}, the conductance between A and B in the effective network, which can be determined by a symmetry argument: Express the current distribution in the network with $g = g_e$ as the sum of two contributions, a current I_0 injected at A and extracted at infinity, and an equal current injected at infinity and extracted at B. In both cases, the current flowing through each of the Z equivalent bonds at the point of current entrance is I_0/Z, so that a total current of $2I_0/Z$ flows through AB. Thus, $g_{AB} = Zg_e/2$, $g'_{AB} = (Z/2 - 1)g_e$, and therefore

$$V_0 = V_e \frac{g_e - g}{g + (Z/2 - 1)g_e}. \tag{59}$$

Substituting Eq. (59) into (58) yields (57).

5.6.1 Conductor–Insulator Composites

An important test for any theory of transport in a disordered medium is its ability for predicting the percolation threshold p_c of the medium. Despite its simplicity and lack of detailed information about the microstructure of the material, or the network that represents it (other than its coordination number Z), the EMA predicts the existence of a non-trivial p_c. Suppose that the bond conductance distribution is given by Eq. (8). Then, Eq. (57) becomes

$$\frac{1 - p}{1 - Z/2} + p \int_0^\infty \frac{g - g_e}{g + (Z/2 - 1)g_e} h(g) dg = 0. \tag{60}$$

Equation (60) represents the EMA equation for the effective conductivity g_e of a random resistor network of conducting and insulating bonds in *bond percolation*, as the bond conductances have been assumed to be randomly distributed according to a probability density function $h(g)$. The EMA prediction for p_c, the bond-percolation threshold of the network, is obtained if in Eq. (60) we set $g_e = 0$ and $p = p_c$, since at $p = p_c$ the effective conductivity must vanish. Using the fact that, $\int_0^\infty h(g) dg = 1$, we obtain

$$p_c = \frac{2}{Z}, \tag{61}$$

as the EMA prediction for the bond percolation threshold of a regular network of coordination number Z.

Equation (61) has a simple interpretation. Since $\langle Z \rangle = p_c Z$ represents the *average* coordination number $\langle Z \rangle$ of the network at the percolation threshold, Eq. (61)

implies that, for a network to be conducting, one must have $\langle Z\rangle \geq 2$. Recall from Chapter 2 [see Eq. (2.23)] that for d-dimensional networks $\langle Z\rangle = p_c Z \simeq d/(d-1)$ represents a very accurate approximation for the average coordination number at the bond-percolation threshold. Therefore, the EMA prediction, Eq. (61), is accurate for 2D networks, since its predictions become identical with Eq. (2.23), but not so for 3D networks, where the EMA overestimates the bond percolation thresholds [compare the predictions of Eq. (61) with those listed in Table 2.2].

Another important test of the accuracy of the EMA is its ability for predicting the power-law behavior of the effective conductivity g_e near p_c. Recall from Chapter 2 that near p_c the effective conductivity follows power-law (2.36) with $\mu(d=2) \simeq 1.3$ and $\mu(d=3) \simeq 2.0$. If in Eq. (60) we set $h(g) = \delta(g-g_0)$, we obtain

$$\frac{g_e}{g_0} = \frac{p-2/Z}{1-2/Z} = \frac{p-p_c}{1-p_c}, \tag{62}$$

and thus, according to the EMA, in any dimension d the effective conductivity g_e varies linearly with $p-p_c$, implying that the EMA predicts that the critical exponent μ that characterizes the power-law behavior of g_e near the percolation threshold is $\mu = 1$, regardless of the dimensionality d. Even if we replace $h(g) = \delta(g-g_0)$ with a more general distribution, as long as this distribution satisfies the constraint (2.44), we would still obtain $\mu = 1$. The main exception is when $h(g) = (1-\alpha)g^{-\alpha}$, where $0 \leq \alpha < 1$ is a parameter of the distribution. Such a distribution violates constraint (2.44), and in this case the EMA predicts (Kogut and Straley, 1979) that

$$g_e \sim (p-p_c)^{1+\alpha/(1-\alpha)}, \tag{63}$$

so that, $\mu = 1+\alpha/(1-\alpha)$, and thus the exponent μ is no longer universal and depends on the details of the conductance distribution [the factor $\alpha/(1-\alpha)$ represents the EMA prediction for departure from the universal value of μ]. While the EMA prediction of a non-universal behavior for μ, when one uses such power-law conductance distributions, is qualitatively correct, the predicted exponent, $\mu = 1/(1-\alpha)$, is not correct; see the discussion in Section 2.11, and in particular the inequalities (2.77). In general, as Koplik (1981) demonstrated, the EMA is very accurate if the system is not close to its percolation threshold, *regardless* of the structure of $h(g)$. Its predictions are also more accurate for 2D networks than for 3D ones. Later in this chapter, we will describe several methods for improving the accuracy of the EMA's predictions.

We point out that the EMA satisfies the exact duality relations, Eqs. (26)–(28). For example, if we substitute probability density function (25) into Eq. (57), we find that, for $p = 1/2$, the EMA prediction for g_e is exactly the same as in Eq. (27). Moreover, it is not difficult to show that Eq. (57) also satisfies the duality relation for the effective conductivities of the hexagonal and triangular networks, Eq. (28). Thus, the 2D EMA with $Z = 4$ is the simplest approximation that satisfies all the known exact results.

5.6.2 *Conductor–Superconductor Composites*

Equation (57) can also be used for predicting the effective conductivity of a network of normal conductors and superconductors. If p is the fraction of the superconducting bonds, then with $f(g) = (1-p)h(g) + p\delta(g-\infty)$ Eq. (57) becomes

$$p + (1-p)\int_0^\infty \frac{g - g_e}{g + (Z/2-1)g_e} h(g)dg = 0, \tag{64}$$

which can be used for predicting the conductivity of the network for any $h(g)$. It is straightforward to show that Eq. (64) predicts that g_e diverges at $p = p_c = 2/Z$, the EMA prediction for the bond-percolation threshold of the network. Moreover, if $h(g) = \delta(g - g_0)$, then

$$g_e = \frac{g_0}{1 - p - p(Z/2-1)} = \frac{p_c g_0}{p_c - p}. \tag{65}$$

According to Eq. (65), regardless of its dimensionality, the conductivity of a network of conducting and superconducting bonds diverges as $p \to p_c$ with an exponent $s = 1$. This value of s should be compared with those of 2D and 3D systems listed in Table 2.3, and will not change if any other distribution $h(g)$ is used, so long as it satisfies the constraint (2.44). If, however, $h(g) = (1-\alpha)g^{-\alpha}$ (with $0 \leq \alpha < 1$), which does violate the constraint (2.44), then (Kogut and Straley, 1979)

$$g_e \sim |p - p_c|^{-[1+(2-\alpha)/(\alpha-1)]}, \tag{66}$$

so that in this case, $s = 1 + (2-\alpha)/(\alpha-1)$, and therefore $(2-\alpha)/(\alpha-1)$ represents the EMa prediction for the departure from the universal value of the exponent s. This EMA-predicted non-universality is qualitatively correct, although the predicted non-universal value of s is not. Therefore, once again, the EMA predicts the correct qualitative behavior, but fails to provide accurate estimate for the associated critical exponent.

5.6.3 *Resistor Networks with Multiple Coordination Numbers*

Over the past three decades the EMA has been used heavily for predicting the effective conductivity and other transport properties of *regular* networks that are characterized by a single coordination number Z. However, despite their significance, little work has been done on regular networks that are characterized by at least two coordination numbers Z_1 and Z_2, or on *topologically-disordered* networks. An example of the former is the kagomé lattice, while the Voronoi network (see Figure 2.14) is an example of the latter type of networks. We now extend the basic EMA, Eq. (57), to these types of network.

To accomplish this task, we utilize the method that was used by Kirkpatrick (1973) for deriving Eq. (57), and was described in the paragraph after this equation. Consider a resistor network which is either topologically disordered, or has an ordered structure which is characterized by two coordination numbers Z_1 and Z_2.

We replace all the conductances by their effective-medium value g_e, and consider one bond of the network such that the coordination numbers of its two ends are Z_1 and Z_2, which are either fixed everywhere, as in the kagomé lattice, or are distributed according to a joint probability density function $\mathcal{P}(Z_1, Z_2)$. In networks in which Z_1 and Z_2 are fixed, the angle between the bonds are also fixed. If the distribution of the angles between the bonds is essentially uniform, as in the case of the Voronoi network, then injecting a current I_0 at one end of a bond with a coordination number Z_1 distributes it essentially as I_0/Z_1 in each bond connected to that site. Therefore, following Kirkpatrick's argument (see above), the total current through bond AB (see Figure 5.4) is now $I_0/Z_1 + I_0/Z_2$ and hence the analogue of Eq. (59) is given by

$$V_0 = V_e \frac{g_e - g}{g + [Z_1 Z_2/(Z_1 + Z_2) - 1]g_e}. \tag{67}$$

Therefore the single-bond EMA for a topologically-random network, obtained by substituting Eq. (67) into (58), is given by

$$\int\int\int \frac{g - g_e}{g + [xy/(x + y) - 1]g_e} f(g)\mathcal{P}(x, y)dgdxdy = 0. \tag{68}$$

Note that one must take the average of V_0 with respect to *both* $f(g)$ and $\mathcal{P}(Z_1, Z_2)$. For a lattice, such as the kagomé lattice, in which both Z_1 and Z_2 are fixed everywhere, $\mathcal{P}(x, y) = \delta(x - Z_1)\delta(y - Z_2)$, and Eq. (68) simplifies to

$$\int_0^\infty \frac{g - g_e}{g + [Z_1 Z_2/(Z_1 + Z_2)]g_e} f(g)dg = 0. \tag{69}$$

If we compare Eq. (68) with Eq. (53), we see that, since for topologically-disordered networks the coordination numbers Z_1 and Z_2 are distributed variables, the bond-bond Green function $\gamma_{\alpha\alpha}$ is a stochastic function given by

$$\gamma_{\alpha\alpha} = -\frac{Z_1 + Z_2}{Z_1 Z_2}, \tag{70}$$

which reduces to $\gamma_{\alpha\alpha} = -2/Z$ given above when $Z_1 = Z_2 = Z$. Thus, one may also define quantities such as $\langle\gamma_{\alpha\beta}\rangle$ and $\langle\gamma_{\alpha\alpha}\rangle$, where $\langle\cdot\rangle$ denotes an average over the distribution of the coordination numbers. More generally, the bond-bond green function $\gamma_{\alpha\beta}$ (and hence the site-site green function G_{ij}) must be averaged over the distribution of the coordination numbers and the angles between the bonds of the network.

Equations (68)–(70) were tested by Sahimi and Tsotsis (1997) using numerical simulations and 2D $L \times L$ Voronoi networks, where L refers to the square root of the number of Poisson points (or polygons) in the network. The bond-bond Green function $\gamma_{\alpha\alpha}$ was calculated by solving Eq. (36) for the site-site Green function G_{ik}, from which $\gamma_{\alpha\alpha}$ was computed via Eq. (55). Since the local environment and topology around each bond of the network are different, one obtains a *statistical distribution* of values of $\gamma_{\alpha\alpha}$. The resulting frequency distribution of $\gamma_{\alpha\alpha}$ is shown in Figure 5.5. It is seen that $\gamma_{\alpha\alpha}$ can take on values anywhere from -0.26 to -0.43, depending on where the reference bond α is located, although most of the values

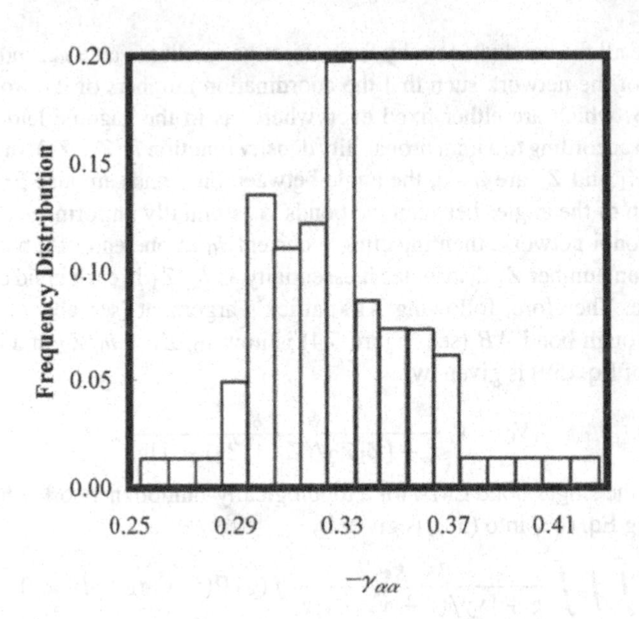

FIGURE 5.5. Frequency distribution of the bond-bond Green function $\gamma_{\alpha\alpha}$ for the two-dimensional Voronoi network.

are concentrated in the relatively narrow range between -0.3 and -0.37. If $\langle\gamma_{\alpha\alpha}\rangle$ is calculated from the frequency distribution shown in Figure 5.5, one obtains $\langle\gamma_{\alpha\alpha}\rangle \simeq -0.322$. On the other hand, if one uses Eq. (70) and averages $\gamma_{\alpha\alpha}$ over the distributions of Z_1 and Z_2, one obtains $\gamma_{\alpha\alpha} \simeq -0.33$, a difference of about 2% from its value obtained from the frequency distribution of Figure 5.6, which is presumably due to the finite size of the networks used in the simulations. One may also be tempted to use the naive approximation, $\langle\gamma_{\alpha\alpha}\rangle \simeq -2/\langle Z\rangle$, where $\langle Z\rangle$ is the average coordination number of the Voronoi network. Since $\langle Z\rangle \simeq 6.29$ (Jerauld *et al.*, 1984a), one obtains $\langle\gamma_{\alpha\alpha}\rangle \simeq -0.317$, surprisingly close to the numerical estimates. Note that, although in general the coordination numbers Z_1 and Z_2 of the two end sites of any bond in the Voronoi network are correlated with each other, as Jerauld *et al.* (1984a) demonstrated, the correlations are weak.

The numerical accuracy of Eq. (68) was also tested by considering conduction in a 2D Voronoi network. A simple conductance distribution, $f(g) = p\delta(g-1) + (1-p)\delta_+(0)$, was used, p was varied and the effective conductivity g_e was calculated by imposing on the network a unit potential gradient in one direction and periodic boundary condition in the other direction, and by performing *two* distinct averagings. One was over the spatial distribution of the conducting bonds in a given Voronoi network, while the second averaging was over different realizations of the Voronoi networks of a given linear size L used in the computations. The second averaging is necessitated by the fact that, since the initial Poisson points, that are used to construct the network, are randomly distributed, the topology of a Voronoi network varies from realization to realization.

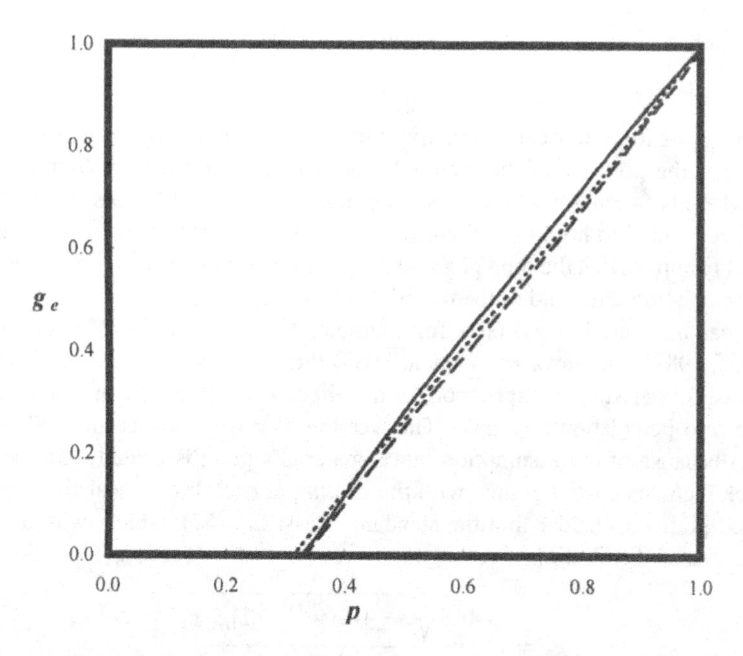

FIGURE 5.6. Comparison of the EMA predictions for the effective conductivity g_e of the two-dimensional Voronoi network, versus the fraction p of the conducting bonds. Solid curve and large broken lines show, respectively, the results for the case in which the distributions of the two coordination numbers of the reference bond are considered independently, and for the correlated case, while the middle curve represents the results for the case in which an average coordination number has been used.

Figure 5.6 presents the results. Equation (68) was also used to calculate g_e using three different approaches. In one approach the correlations between Z_1 and Z_2 were taken into account, by numerically constructing the joint distribution function $\mathcal{P}(Z_1, Z_2)$ of the two coordination numbers of any bond. In the second approach Z_1 and Z_2 were assumed to be independent of each other, so that $\mathcal{P}(Z_1, Z_2) = \zeta(Z_1)\zeta(Z_2) = \zeta(Z)^2$, where $\zeta(Z)$ is the distribution of the local coordination numbers of the network. In the third approach Eq. (57) was used with $Z = \langle Z \rangle \simeq 6.29$, the average coordination number of the Voronoi network. As Figure 5.6 indicates, there are some differences between the three sets of predictions and also with the Monte Carlo results, although the differences are certainly not too large. Therefore, use of an average coordination number in Eq. (57) provides accurate estimates of the effective conductivity of topologically-disordered networks.

5.6.4 Materials with Zero Percolation Threshold

The percolation threshold of some composite materials, especially porous ones, is zero, or nearly zero, i.e., the materials remain conducting so long as an infinitesimal volume fraction of them is conducting. For example, for many porous materials, especially rock samples, a useful empiricism is the Archie's law:

$$g_e = g_f \phi^m, \tag{71}$$

where g_f is the electrical conductivity of the fluid saturating the porous media, and ϕ is the porosity of the material. The exponent m has been found to vary anywhere between 1.3 and 4, depending upon a variety of factors. Archie's law has been found to hold even for a wide variety of porous media. It is clear that Eq. (71) implies that the fluid phase in the porous medium remains connected, i.e., the percolation threshold of the material is zero.

It has been established (see, for example, Sen *et al.*, 1981; Mendelson and Cohen, 1982; Yonezawa and Cohen, 1983) that a modification of the EMA can be used for deriving an expression for the effective conductivity of materials that have zero percolation threshold. This version of EMA is called the *self-similar* EMA because of the assumption that a material's grain is coated with the fluid which includes coated grains, with the coating at each level consisting of other coated grains. Consider first the standard EMA, Eq. (57), which, with a binary conductance distribution, $f(g) = p\delta(g - g_f) + (1 - p)\delta(g - g_s)$, predicts that

$$g_e = \frac{-A \pm \sqrt{A^2 + 4(\gamma^{-1} - 1)g_f g_s}}{2(\gamma^{-1} - 1)}, \tag{72}$$

where $\gamma^{-1} = -\gamma_{\alpha\alpha}^{-1} = Z/2$, and $A = [g_f - (\gamma^{-1} - 1)g_s - \gamma^{-1}(g_f - g_s)p]$, with g_s being the conductivity of the material's solid matrix. Equation (72) has two solutions, but only one of them is physically meaningful. In the limits $p \to 0$ and $p \to 1$, we obtain

$$g_{(1)} = g_f \left[1 + (1 - p)\gamma^{-1} \frac{g_s - g_f}{g_s + (\gamma^{-1} - 1)g_f}\right], \tag{73}$$

$$g_{(2)} = g_s \left[1 + p\gamma^{-1} \frac{g_f - g_s}{g_f + (\gamma^{-1} - 1)g_s}\right]. \tag{74}$$

The basic idea behind the self-similar EMA is as follows. One starts with a pure fluid system, replaces, step by step, small portions of it by pieces of solid matrix of the material, and applies the EMA at each step. Suppose that $g^{(i)}$ is the conductivity of the mixture at a step i, and replace a small volume Δq_i of the system by the material's grains. Equation (57) yields

$$\frac{g^{(i)} - g^{(i+1)}}{g^{(i)} + (\gamma^{-1} - 1)g^{(i+1)}}(1 - \Delta q_i) + \frac{g_s - g^{(i+1)}}{g_s + (\gamma^{-1} - 1)g^{(i+1)}}\Delta q_i = 0. \tag{75}$$

If Δq_i is small enough, we obtain from Eq. (75)

$$g^{(i+1)} = g^{(i)} \left[1 + \gamma^{-1} \frac{g_s - g^{(i)}}{g_s + (\gamma^{-1} - 1)g^{(i)}}\Delta q_i\right]. \tag{76}$$

Note that, if the volume fraction of the fluid at the ith stage is p_i, we have, $\Delta p_i q_i = p_i - p_{i+1}$. If we now use Eq. (76) repeatedly, we obtain, for the case of

$g_s = 0$,

$$g^{(n+1)} = \sum_{i=1}^{n} \left[1 + \frac{\gamma^{-1}}{\gamma^{-1} - 1} \frac{p_{i+1} - p_i}{p_i} \right] g^{(0)}, \tag{77}$$

where $g^{(0)} = g_f$ and $p_0 = 1$ ($p_0 = 1$ corresponds to pure fluid, the system we started with). If we take the limit $\Delta q_i \to 0$, then, Eq. (75) becomes a differential equation,

$$\frac{dg^{(i)}}{g^{(i)}} = -\frac{dp_i}{p_i} \frac{g_f - g^{(i)}}{g_f + (\gamma^{-1} - 1) g^{(i)}} \gamma^{-1}, \tag{78}$$

which, after integrating and using the boundary condition that, $g^{(i)} = g_f$ for $p_i = 1$, yields an equation similar to (71) with $\phi \to p$, so that if we interpret p as the porosity ϕ of the material, then the EMA produces the Archie's law with

$$m = \frac{\gamma^{-1}}{\gamma^{-1} - 1}. \tag{79}$$

Equation (79) indicates that, consistent with the experimental data, m is not universal but depends on the connectivity of the system.

What is the geometrical interpretation of this result? For a network model of a porous material, Yonezawa and Cohen (1983) presented an interesting interpretation which can be summarized as follows. At the ith stage, where every bond has a conductance $g^{(i)}$, a small fraction Δq_i of the bonds is replaced by a resistor with conductance g_s and the EMA is used for estimating the conductivity of the new system. This is equivalent to putting on each bond a resistor parallel to the original one. The conductance $g^{(ia)}$ of an added resistor should be

$$g^{(ia)} = \gamma^{-1} g^{(i)} \frac{g_2 - g^{(i)}}{g_2 + (\gamma^{-1} - 1) g^{(i)}} \Delta q_i, \tag{80}$$

if the original resistor belongs to the host medium; otherwise, it is given by Eq. (76). The implication is that even when $g_s = 0$, the application of the EMA at each stage makes the link between the nodes conducting, due to the addition of a parallel conducting resistor, and thus there is always a sample-spanning cluster of conducting bonds. Translating this for a porous material-fluid system, it implies that this procedure guarantees the continuity of the fluid phase and the granularity of the rock grains. For a more detailed discussion of this problem and the application of Eq. (71), see Sahimi (1995b).

5.6.5 *Comparison with the Experimental Data*

The EMA has been used for estimating the effective conductivity of a wide variety of composites. The list of all the papers that have reported such calculations and have compared the predictions to the experimental data is too long to be given here. The EMA has also been used for estimating the effective diffusivity of molecules diffusing in a porous material, as well as the effective permeability of such materials (see, for example, Koplik *et al.*, 1984, Doyen, 1988). In all cases, the EMA has been found to provide accurate predictions for the effective transport properties away from the percolation threshold. If the material does not behave as a percolation system (i.e., if its constituents, components, or phases, are all well-connected and conducting), then the EMA always provides accurate predictions for its transport properties, unless there are strong correlations in the material's morphology which the EMA cannot account for.

However, as discussed above, the EMA does have two main weaknesses, namely, its overestimation of the percolation threshold p_c of 3D materials, even when their morphology is nearly random, and its inability for predicting the correct value of the critical exponent μ that characterizes the power-law behavior of the effective conductivity near p_c. In a series of papers, McLachlan and co-workers (see McLachlan, 1988; Wu and McLachlan, 1997, and references therein) modified the basic EMA, Eq. (57), in order to eliminate these two weaknesses. Although the modifications that McLachlan and co-workers proposed were intended for the EMA for a continuum (see Section 4.9), they are equally applicable to resistor network models and, moreover, since the modified EMA does take into account the power-law behavior of the effective conductivity of a material near p_c, we describe it in the present chapter.

Consider a two-component material that consists of a low- and a high-conductivity component, with volume (area) fractions p and $(1-p)$, respectively. McLachlan's modified EMA is given by

$$p\frac{g_l-\sigma_e}{g_l+[p_c/(1-p_c)]\sigma_e}+(1-p)\frac{g_h-\sigma_e}{g_h+[p_c/(1-p_c)]\sigma_e}=0, \tag{81}$$

which is directly obtained from Eq. (57) if we take the conductance distribution $f(g)$ to be, $f(g)=p\delta(g-g_l)+(1-p)\delta(g-g_h)$, and replace the factor $(Z/2-1)$ by $p_c/(1-p_c)$. This replacement is plausible if we recall that within the EMA, $Z/2-1=1/p_c-1=(1-p_c)/p_c$. Here, $g_l=\sigma_l^{1/\mu}$, $g_h=\sigma_h^{1/\mu}$, and $\sigma_e=g_e^{1/\mu}$, where σ_l, σ_h, and g_e are the conductivities of the low-conductivity, high-conductivity, and the effective conductivity of the material itself, respectively, and $p_c=1-p_c'$, with p_c' being the percolation threshold or the critical volume fraction at which the high-conductivity component forms a sample-spanning cluster across the material. Note that, $p_c/(1-p_c)=p_c'/(1-p_c')$, and that if we set $\mu=1$, we recover the standard EMA derived above. Note also that, if $\sigma_l=0$ and σ_h is finite, Eq. (81) becomes

$$g_e=\sigma_h(1-p/p_c)^{\mu}=(\sigma_h/p_c^{\mu})|p_c-p|^{\mu}, \tag{82}$$

which is the correct equation and is in agreement with Eq. (2.36). Many other properties of the modified EMA are discussed by Michels (1992) and Wu and McLachlan (1997).

Figure 5.7 compares the predictions of Eq. (81) with the experimental data, in which the parameters p_c and μ were treated as adjustable parameters. The sharp decrease in the effective conductivity signals the percolation threshold at which a sample-spanning cluster of the low-conductivity material was formed. The data

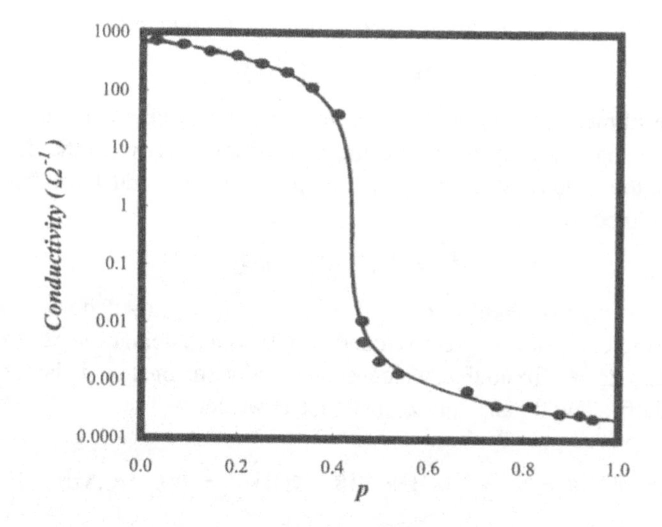

FIGURE 5.7. Comparison of the predictions of the modified EMA, Eq. (81), for the effective conductivity of a two-dimensional square tessellation of a sheet, in which a randomly-selected fraction p of the squares were made of graphite (low-conductivity phase), while the rest were made of copper (high-conductivity phase), with the experimental data (circles) (after McLachlan, 1988).

were measured on a 2D circuit board, arranged as a square tessellation, in which a randomly-selected fraction p of the squares represented the low-conductivity material. The two components were graphite (low-conductivity material) and copper (high-conductivity material). As Figure 5.7 indicates, the agreement between the predictions and the data is excellent over the entire range of the graphite volume fraction. Moreover, when μ was treated as an adjustable parameter, all the samples (the conductivity of one of which is shown in Figure 5.7) yielded, $1.19 \leq \mu \leq 1.3$, which are in agreement with what is expected for 2D percolation, $\mu \simeq 1.3$ (see Table 2.3). Many other sets of experimental data, for both 2D and 3D materials, have been shown to be predicted rather accurately by Eq. (81); see McLachlan (1988), Michels (1992), and Wu and McLachlan (1997) for more details.

5.6.6 *Accuracy of the Effective-Medium Approximation*

Due to its simplicity, it is of great interest to understand the conditions under which the EMA can provide accurate predictions for the effective conductivity of a disordered material, modeled either by a random resistor network, or by an appropriate continuum model of the types that were described in Chapter 3. Although many authors have studied this issue, in this section we briefly discuss the work of Luck (1991) who carried out what we consider a systematic investigation of this problem, and was able to quantify the conditions under which the EMA can be expected to be a very accurate approximation. What follows is a summary of his analysis.

The bond conductances are assumed to take the following form

$$g_i(\mathbf{x}) = g_0[1 + \epsilon(\mathbf{x})], \tag{83}$$

where the (dimensionless) fluctuations are assumed to be small and have zero average, so that g_0 is by definition the average of $g_i(\mathbf{x})$ over the distribution $f(g)dg$ of the bond conductances. The moments of the conductance fluctuations are now defined by

$$M_k = \langle [\epsilon_i(\mathbf{x})]^k \rangle, \quad k \geq 1, \tag{84}$$

where the average is taken over the distribution $f(g)dg$; by definition, $M_1 = 0$. The goal is to evaluate the effective conductivity as a systematic series expansion in the moments M_k. To do this, we consider a d-dimensional cubic lattice, so that the Kirchhoff's law for any site of the lattice is written as

$$\sum_{i=1}^{d} \{g_i(\mathbf{x})[V(\mathbf{x}+\mathbf{e}_i) - V(\mathbf{x})] + g_i(\mathbf{x}-\mathbf{e}_i)[V(\mathbf{x}-\mathbf{e}_i) - V(\mathbf{x})]\} = 0, \tag{85}$$

where $V(\mathbf{x})$ is the voltage at $\mathbf{x}$, and $\mathbf{e}_i$ $(1 \leq i \leq d)$ represents a basis of unit vectors. An arbitrary (but small enough) electric field $\mathbf{E}$ is imposed on the network, and a solution is sought which has the following form,

$$V(\mathbf{x}) = \mathbf{E} \cdot \mathbf{x} + W(\mathbf{x}), \quad \mathbf{E} \cdot \mathbf{x} = \sum_{i=1}^{d} E_i x_i, \tag{86}$$

where the voltage fluctuation $W(\mathbf{x})$ has the same periodicity L (L is the linear size of the sample) in each direction, and its average is zero, i.e., $\sum_{\mathbf{x}} W(\mathbf{x}) = 0$.

With the help of definitions (83) and (84), Eq. (85) is rewritten as

$$-\Delta W(\mathbf{x}) = \sum_{i=1}^{d} E_i[\epsilon_i(\mathbf{x}) - \epsilon_i(\mathbf{x}-\mathbf{e}_i)] + \sum_{i=1}^{d} \{\epsilon_i(\mathbf{x})[W(\mathbf{x}+\mathbf{e}_i) - W(\mathbf{x})] + \epsilon_i(\mathbf{x}-\mathbf{e}_i)[W(\mathbf{x}-\mathbf{e}_i) - W(\mathbf{x})\}, \tag{87}$$

where $\Delta W(\mathbf{x}) = \sum_{i=1}^{d} [W(\mathbf{x}+\mathbf{e}_i) + W(\mathbf{x}-\mathbf{e}_i) - 2W(\mathbf{x})]$ is essentially the discrete Laplace operator on a lattice. Equation (87) is now Fourier transformed over the L^d sites of the lattice. Thus, defining

$$\hat{W}(\boldsymbol{\omega}) = \sum_{\mathbf{x}} e^{-i\boldsymbol{\omega}\cdot\mathbf{x}} W(\mathbf{x}), \tag{88}$$

and noting that in the limit $L \to \infty$

$$\frac{1}{L} \sum_{\boldsymbol{\omega}} \to \int d\hat{\boldsymbol{\omega}} = \frac{1}{(2\pi)^d} \int d\omega_1 \cdots d\omega_d,$$

the Fourier transform of Eq. (87) is then given by

$$K(\boldsymbol{\omega})\hat{W}(\boldsymbol{\omega}) = \sum_{i=1}^{d} (1 - e^{-i\omega_i}) \left\{ E_i \hat{\epsilon}_i(\boldsymbol{\omega}) + \int \hat{\epsilon}_i(\mathbf{p}) \left[e^{i(\boldsymbol{\omega}-\mathbf{p})_i} - 1 \right] \hat{W}(\boldsymbol{\omega} - \mathbf{p}) d\hat{\mathbf{p}} \right\}, \tag{89}$$

where

$$K(\omega) = 2\sum_{i=1}^{d}(1 - \cos\omega_i). \tag{90}$$

In general, the conductivity tensor g_{ij} of a finite sample is defined by

$$\mathcal{I}_i = \frac{1}{L^d}\sum_{\mathbf{x}} I_i(\mathbf{x}) = \sum_{j=1}^{d} g_{ij}E_j, \tag{91}$$

where $I_i(\mathbf{x})$ is the current flowing out of $\mathbf{x}$ in the positive ith direction. In terms of the Fourier-transformed quantities, $\mathcal{I}_i$ is given by

$$\mathcal{I}_i = g_0\left[E_i + \frac{1}{L^d}\int \hat{\epsilon}_i(-\omega)(e^{i\omega_i} - 1)\hat{W}(\omega)d\hat{\omega}\right]. \tag{92}$$

To develop the series expansion for the effective conductivity, one writes, $\hat{W} = \hat{\epsilon} + \hat{\epsilon} * \hat{W}$, where all the integrals and prefactors are hidden in a convolution-like operator $*$. This implicit equation, when solved iteratively, yields, $\hat{W} = \hat{\epsilon} + \hat{\epsilon} * \hat{\epsilon} + \hat{\epsilon} * \hat{\epsilon} * \hat{\epsilon} + \cdots$ Thus, one obtains $\hat{W}(\omega)$ as an infinite power series in the function $\hat{\epsilon}(\omega)$, which involves an increasing number of complex multiple integrals. Having determined this power series, one computes the corresponding series for the effective conductivity via Eq. (92). After some tedious computations, one finally obtains the following sixth-order series expansion (higher-order terms can of course be computed, if need be)

$$g_e = g_0\left(1 + \sum_{i=2}^{6}\mathcal{M}_i + \cdots\right), \quad \mathcal{M}_2 = -\frac{M_2}{d}, \quad \mathcal{M}_3 = \frac{M_3}{d^2}, \tag{93}$$

with

$$\mathcal{M}_4 = -\frac{1}{d^3}\left[M_4 + (d - 3 + J_1)M_2^2\right], \quad \mathcal{M}_5 = \frac{1}{d^4}\left[M_5 + (3d - 10 + 4J_1 + J_3)M_2M_3\right],$$

$$\begin{aligned}\mathcal{M}_6 = -\frac{1}{d^5}\{&M_6 + (4d - 15 + 6J_1 + 3J_3)M_2M_4 + (2d - 10 + 4J_1 + J_2 + 2J_3)M_3^2 \\ &+[2d^2 - 12d + 30 + 3(d - 6)J_1 + (d - 9)J_3 + 4J_4]M_2^3\}.\end{aligned} \tag{94}$$

Here,

$$\begin{aligned} J_1 &= d^3\sum_{\mathbf{x}}[\mathcal{G}_{ii}(\mathbf{x})]^3, \quad J_2 = d^5\sum_{\mathbf{x}}[\mathcal{G}_{ii}(\mathbf{x})]^5, \\ J_3 &= d^4\sum_{\mathbf{x}}\sum_{j=1}^{d}[\mathcal{G}_{ij}(\mathbf{x})]^4, \\ J_4 &= d^5\sum_{\mathbf{x},\mathbf{y}}\sum_{j=1}^{d}\mathcal{G}_{ii}(\mathbf{x}-\mathbf{y})[\mathcal{G}_{ij}(\mathbf{x})]^2[\mathcal{G}_{ij}(\mathbf{y})]^2, \end{aligned} \tag{95}$$

where

$$\mathcal{G}_{ij}(\mathbf{x}) = \int e^{i\mathbf{p}\cdot\boldsymbol{\omega}} \frac{K_{ij}(\mathbf{p})}{K(\mathbf{p})} d\hat{\mathbf{p}}, \tag{96}$$

$$K_{ij}(\boldsymbol{\omega}) = 4e^{i(\omega_i - \omega_j)/2} \sin(\omega_i/2)\sin(\omega_j/2), \tag{97}$$

The quantity $\mathcal{G}_{ij}(\mathbf{x})$ is closely related to $G(\mathbf{x})$, the continuum analogue of the lattice Green function G_{ij} defined earlier, i.e., $G(\mathbf{x})$ is the solution of, $\nabla^2 G(\mathbf{x}) = -\delta(\mathbf{x})$, with $G(\mathbf{x}) \propto 1/\mathbf{x}$ in 3D and $G(\mathbf{x}) \propto \ln(1/\mathbf{x})$ in 2D (see also Section 5.8). In fact,

$$\mathcal{G}_{ij}(\mathbf{x}) = G(\mathbf{x}) - G(\mathbf{x}+\mathbf{e}_i) - G(\mathbf{x}-\mathbf{e}_j) + G(\mathbf{x}+\mathbf{e}_i-\mathbf{e}_j). \tag{98}$$

The same type of analysis can be carried out for deriving a series expansion for the effective conductivity within the EMA. To do this, we write down an equation for the bond conductances $g_i(\mathbf{x})$

$$g_i(\mathbf{x}) = g_e^{\text{EMA}}[1 + \eta_i(\mathbf{x})], \tag{99}$$

which is similar to Eq. (83), with $\eta_i(\mathbf{x})$ representing the bond conductance fluctuations, but, of course, *without* the assumption that $\langle g_i \rangle = g_e^{\text{EMA}}$, since, as discussed above, the EMA prediction for the effective conductivity is *not* equal to the average of the conductance distribution. The final expansion foe g_e^{EMA} is given by

$$g_e^{\text{EMA}} = g_0\left(1 + \sum_{i=2}^{6} \mathcal{M}_i^{\text{EMA}} + \cdots\right), \quad \mathcal{M}_2^{\text{EMA}} = -\frac{M_2}{d}, \quad \mathcal{M}_3^{\text{EMA}} = \frac{M_3}{d^2}, \tag{100}$$

$$\mathcal{M}_4^{\text{EMA}} = -\frac{1}{d^3}\left[M_4 + (d-2)M_2^2\right],$$

$$\mathcal{M}_5^{\text{EMA}} = \frac{1}{d^4}\left[M_5 + (3d-5)M_2M_3\right], \tag{101}$$

$$\mathcal{M}_6^{\text{EMA}} = -\frac{1}{d^5}\left[M_6 + 2(2d-3)M_2M_4 + (2d-3)M_3^2 + (2d^2-8d+7)M_2^3\right].$$

If we now compare Eqs. (100) and (101), the EMA series expansion, with the exact series expansion, Eqs. (93) and (94), we see that the two series are *identical* up to and including the third term of the expansion. The difference between the two series begins at the fourth term which, however, depends on the dimensionality d of the system. Indeed, it is easy to see, by comparing the two series expansions, that, within the EMA,

$$J_1 = J_2 = J_3 = J_4 = 1, \quad \text{EMA}. \tag{102}$$

For 1D systems, J_i $(i = 1, \cdots, 4)$ are *exactly* equal to 1, which explains why the EMA is exact for $d = 1$. For the square lattice, it is not difficult to show that, $J_1 = J_2 = 1$ and $J_3 = J_4 \simeq 1.092$, which explain the excellence of the EMA for predicting the effective conductivity of a square lattice (or a 2D isotropic continuum). For the simple-cubic lattice, $J_1 \simeq 0.9237825$, $J_2 \simeq 0.9869128$, $J_3 \simeq 1.2635747$, and $J_4 \simeq 1.0561133$, which explain the relatively large deviations of the EMA predictions from the actual numerical results.

5.6.7 Cluster Effective-Medium Approximation

As discussed earlier, the predictions of the single-bond EMA are not very accurate near the percolation threshold p_c. This is particularly true for 3D systems. The reason for the failure of the EMA near p_c is that, near the percolation threshold the currents and potentials in the conducting bonds are highly correlated, so that the state of any bond affects that of many others in its vicinity. However, such correlations are completely absent in the single-bond EMA, as one embeds only one bond with a random conductance in a uniform (effective-medium) network and estimates the effective conductivity of the system based on the requirement that the average of the potential or flux fluctuations along only one bond must vanish. Over the past two decades considerable effort has been devoted to improving the predictions of the EMA near p_c. A possible way of achieving this goal is to use higher order EMAs based on the perturbation expansion developed above. According to Eq. (40), the current Q_α is written in terms of the fluctuations in the conductance of the bond α itself and those of the other bonds β that surround α. Thus, instead of embedding in the effective-medium only a single bond α with a random conductance, one should embed a *cluster* of such bonds in the effective network, calculate Q_α for a reference bond α within the cluster, and insist that, $\langle Q_\alpha \rangle = Q_\alpha^0$, where the averaging is taken with respect to the distribution of the conductances of *all* the bonds in the cluster. In practice, one must apply Eq. (40) to every bond of the cluster and calculate Q_α numerically. Since near p_c the currents and potentials in the conducting bonds are correlated, a cluster of bonds includes to some extent the effect of such correlations, and thus with an appropriately-selected cluster one may obtain more accurate predictions for the effective conductivity of the network. This approach has been used by several authors (Blackman, 1976; Turban, 1978; Ahmed and Blackman, 1979; Sheng, 1980; Nagatani, 1981; Sahimi, 1984b). In general, such a cluster of bonds must meet two important criteria:

(1) The symmetry of the cluster, with respect to the current, must be such that the average of the voltage drop on each bond of the cluster is equal to zero, with the same value of the conductance g_e of the effective medium for all the bonds of the cluster.

(2) The cluster must preserve the symmetry of the lattice itself. For the triangular network, the smallest of such clusters is a triangles of three bonds (Turban, 1978), with which the cluster EMA predicts the bond-percolation threshold of the lattice to be, $p_c = 2\sin(\pi/18) \simeq 0.3473$, the exact value (see Table 2.1), which should also be compared with the prediction of the single-bond EMA, $p_c = 2/6 = 0.3333$. The smallest of such symmetric clusters for the hexagonal lattice is a three-bond cluster in which the bonds are all connected to the same site. Such a cluster EMA predicts (Turban, 1978) $p_c = 1 - 2\sin(\pi/18) \simeq 0.6527$ for the bond-percolation threshold of the hexagonal lattice, which is also the exact value (see Table 2.1), and should be compared with $p_c = 2/3 = 0.6666$, the prediction of the single-bond EMA.

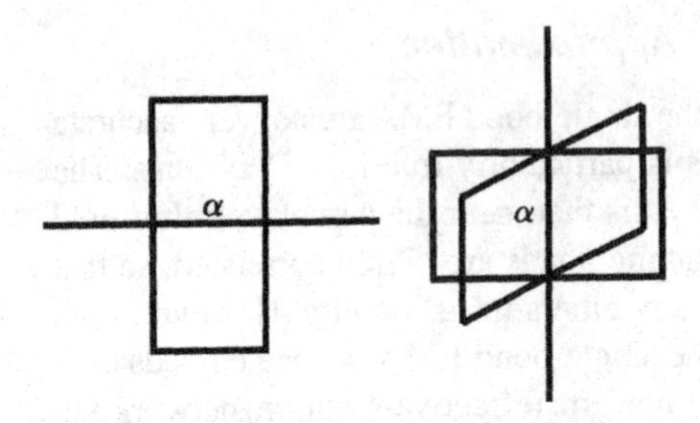

FIGURE 5.8. The smallest clusters of bonds in two and three dimensions that preserve the symmetries of the square and simple-cubic networks. The reference bond is denoted by α.

These clusters are small enough to allow one to obtain an analytical equation for the effective conductivity of the networks. However, for the square and simple-cubic lattices the problem is more difficult. Examples of such clusters are shown in Figure 5.8. The 2D cluster preserves (Blackman, 1976) the exact bond-percolation threshold of the network, $p_c = 1/2$, which is also predicted by the single-bond EMA, while the 3D cluster predicts (Ahmed and Blackman, 1979) $p_c \simeq 0.317$ for the bond-percolation threshold of the lattice, which should be compared with the prediction of the single-bond EMA, $p_c = 1/3 \simeq 0.333$, and with the currently-accepted numerical estimate (see Table 2.2), $p_c \simeq 0.2488$. Note that the 3D cluster contains 15 bonds, and hence the averaging of Q_α must be carried out over $2^{15} = 32,768$ possible configurations, and therefore the computations quickly become very complex. At the same time, the convergence of the predictions of the cluster EMA to the numerical simulation results is slow. This slow convergence can be understood by examining the bond-bond Green functions $\gamma_{\alpha\beta}$, given by Eqs. (41)–(50). It is not difficult to see that, even for a pair of bonds α and β that are at a relatively large distance from each other, $\gamma_{\alpha\beta}$ is still significant, so that a very large number of terms in the perturbation expansion (40) must be included, if the predictions of the cluster EMA are to be close to the numerical simulation results and, in particular, if the percolation threshold is to be predicted reasonably accurately. Moreover, even a cluster EMA still predicts an asymptotic value of 1 for the critical exponent μ that characterizes the power-law behavior of the effective conductivity near p_c. In this regard, the difference between the single-bond EMA and the cluster EMA is that, in the latter case the range of $(p - p_c)$ over which $\mu = 1$ is smaller than in the former case.

We may conclude that, although a cluster EMA is a systematic method for improving the predictions of the single-bond EMA, it is not the most efficient way of doing so, if accurate predictions of p_c and μ are of prime interest. There are a few other approximate methods that provide more accurate predictions for the effective conductivity of random resistor networks which will be discussed later in this chapter.

5.6.8 Coherent-Potential Approximation

Let us point out that the EMA is equivalent to the coherent-potential approximation (CPA), well-known in the studies of alloys (for reviews see, for example, Elliot *et al.*, 1974; Yonezawa, 1982). The CPA, first developed by Soven (1967) and Taylor (1967), was originally intended for calculating one-particle properties of elementary excitations, such as phonons and excitons, for substitutionally disordered alloys. It has been used for calculating, for example, optical absorption spectra of mixed ionic crystals, such as KCl-KI, magnetic properties of NiFe alloys, and the shape memory of CuAlNi alloys.

However, while for alloys difficulties are frequently encountered in attempts to improve upon the CPA, the perturbation expansion described here, and in particular EMA and its cluster extension, do not suffer from the same difficulties. The common method for improving the CPA predictions is a cluster calculation in order to improve the predictions near a band edge, to produce structure in a density of states, or to describe some properties where correlations, which are absent from both the basic (single-bond) EMA and CPA, are important. However, while the cluster CPA suffers from spurious non-analytic behavior in the associated Green functions for energies that are off the real axis (see, for example, Nickel and Butler, 1973 for a discussion of this point), the above perturbation expansion does not suffer from such difficulties, and therefore it is, in principle, straightforward (though tedious) to improve upon the predictions of the EMA by a cluster expansion. In Chapter 6, where frequency- and time-dependent properties will be computed by the generalizations of the perturbation expansion and the EMA described in this chapter, the similarities and differences with the CPA will become more transparent.

5.7 Effective-Medium Approximation for Site Percolation

Although random resistor networks with bond-percolation disorder have been studied extensively, and have been able to provide accurate description of transport processes in heterogeneous materials, study of such networks with site-percolation disorder is also of great interest, as they are in fact the exact lattice analogue of the continuum models described in Chapters 3 and 4. In particular, an EMA for random resistor networks with site-percolation disorder is a useful tool for investigating the effective transport properties of disordered materials. Development of such an EMA has been accomplished by several groups (Watson and Leath, 1974; Butcher, 1975; Bernasconi and Wiesmann, 1976; Yuge, 1977; Joy and Strieder, 1978, 1979; Sahimi *et al.*, 1984; Söderberg *et al.*, 1985).

For concreteness, we consider the simple-cubic network, but our final result will be general and applicable to any network. Development of an EMA for site percolation in this (and any other) lattice is in fact equivalent to developing a cluster EMA, since for site percolation one must consider a cluster that consists of a central site and its nearest-neighbors, embedded in the effective-medium network, as shown in Figure 5.9 for the simple-cubic lattice. The following analysis is based on the unpublished work of Sahimi (1981). We number the bonds of the cluster (that are connected to the central site) from 1 through 6, and assume that the applied external field is along, for example, the 2–4 direction shown in the figure. To develop the EMA for this cluster, we must utilize Eq. (40) to solve for

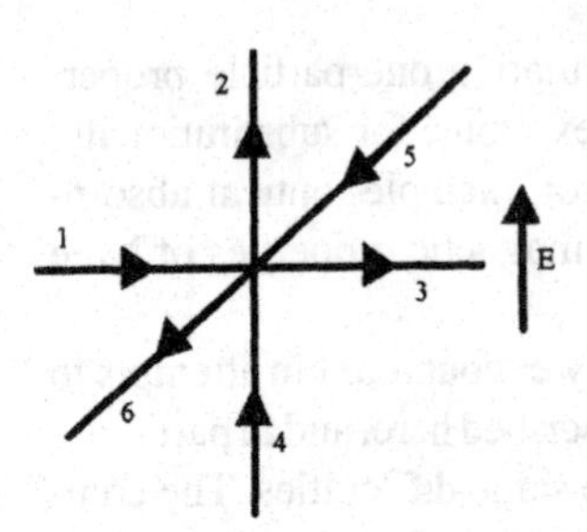

FIGURE 5.9. A cluster of six bonds used in the derivation of the EMA for site-disordered systems. E is the applied field.

Q_α. Given the configuration of the cluster shown in Figure 5.9, it is clear that $Q_1 = -Q_3$, $Q_5 = -Q_6$, and $Q_1 = Q_5$, and therefore we only need to solve for Q_1, Q_2 and Q_4. Thus, writing Eq. (40) for bond 1 yields

$$Q_1 = Q_1^0 + \Delta(\gamma_{11}Q_1 + \gamma_{12}Q_2 + \gamma_{13}Q_3 + \gamma_{14}Q_4 + \gamma_{15}Q_5 + \gamma_{16}Q_6). \quad (103)$$

Given the direction of the applied field shown in Figure 5.9, $Q_1^0 = 0$. We now take advantage of the properties of the bond-bond Green function discussed in 5.3.1 by noting that, $\gamma_{14} = \gamma_{15} = -\gamma_{16} = -\gamma_{12}$. Denoting $\gamma_{11} = \gamma_1$, $\gamma_{12} = \gamma_2$ and $\gamma_{13} = \gamma_3$, Eq. (103) is simplified to

$$[1 + \Delta(2\gamma_2 + \gamma_3 - \gamma_1)]Q_1 - \Delta\gamma_2 Q_2 + \Delta\gamma_2 Q_4 = 0. \quad (104)$$

Similarly, taking into account the fact that, $\gamma_{23} = \gamma_{26} = -\gamma_{25} = -\gamma_{12} = -\gamma_2$, and $\gamma_{24} = \gamma_{13} = \gamma_3$, the equation for Q_2 is given by

$$-4\Delta\gamma_2 Q_1 + (1 - \gamma_1\Delta)Q_2 - \Delta\gamma_3 Q_4 = Q^0. \quad (105)$$

Finally, since, $\gamma_{41} = \gamma_{45} = -\gamma_{43} = -\gamma_{46} = -\gamma_{12} = -\gamma_2$, and $\gamma_{42} = \gamma_{13} = \gamma_3$, the equation for Q_4 is simplified to

$$4\Delta\gamma_2 Q_1 - \Delta\gamma_3 Q_2 + (1 - \gamma_1\Delta)Q_4 = Q^0 \quad (106)$$

Solving Eqs. (104)–(106), we obtain an expression for $\Delta Q = (Q_2 + Q_4) - 2Q^0$. The effective conductivity g_e is then estimated by requiring that, $\langle \Delta Q \rangle = 0$, where the averaging is taken over the conductance distribution. This procedure then yields

$$\int \frac{g - g_e}{g + (\Gamma - 1)g_e} f(g)dg = 0, \quad (107)$$

which is identical with Eq. (57), except that $Z/2$ of Eq. (57) has been replaced by Γ, a quantity which is given by

$$\Gamma = -\frac{1}{\gamma_1 + \gamma_3} = -\frac{1}{\gamma_3 - 2/Z}, \quad (108)$$

where we used $\gamma_1 = \gamma_{11} = -2/Z$, as derived earlier. Moreover, it is straightforward to show (see Sections 5.3.1 and 6.1.4) that for a d-dimensional simple-cubic lattice,

$$\gamma_3 = \frac{1}{2}\left\{\int_0^\infty \left[I_0^d(t) - 2I_0^{d-1}(t)I_1(t) + I_0^{d-2}(t)I_2^2(t)\right]\exp(-dt)dt\right\}, \quad (109)$$

where I_m is the modified bessel function of order m.

It remains to only specify the conductance distribution in order to investigate the accuracy of the EMA for site percolation. In a site-disordered network, a bond is conducting if and only if its two end sites are present or active. Since a site is active with probability p, a bond would be conducting with probability p^2, and therefore

$$f(g) = (1 - p^2)\delta_+(g) + p^2 h(g), \tag{110}$$

so that Eq. (107) becomes

$$\frac{1-p^2}{1-\Gamma} + p^2 \int_0^\infty \frac{g - g_e}{g + (\Gamma - 1)g_e} h(g)dg = 0, \tag{111}$$

where $h(g)$ is any normalized probability density function. Equation (111) is the analogue of Eq. (60) for bond percolation. In particular, if $h(g) = \delta(g - g_0)$, Eq. (111) becomes

$$\frac{g_e}{g_0} = \frac{\Gamma}{\Gamma - 1}(p^2 - \Gamma^{-1}), \tag{112}$$

which should be compared with Eq. (62) for bond percolation. According to Eq. (112), the EMA prediction for the site-percolation threshold is given by

$$p_c = \Gamma^{-1/2}, \tag{113}$$

which should be compared with Eq. (61), the EMA prediction for bond percolation. In particular, one obtains $p_c = 1/\sqrt{2} \simeq 0.707$ for the hexagonal lattice, $p_c = 1 - 2/\pi \simeq 0.602$ for the square lattice, $p_c = [(5\pi + 6\sqrt{3})/(6\pi)]^{1/2} \simeq 0.531$ for the triangular lattice, and $p_c \simeq 0.459$ for the simple-cubic lattice. The 2D results are all within a few percent of the known exact results or numerical estimates listed in Table 2.1, while the estimate for the simple-cubic lattice is too high, if we compare it with the numerical estimate, $p_c \simeq 0.3116$. Therefore, similar to the EMA for bond percolation, the EMA for site percolation is very accurate in 2D, but not as accurate in 3D.

Note, however, that, unlike Eqs. (60) and (62) for bond percolation, Eqs. (111) and (112) do produce most of the curvature of the conductivity curve in the critical region near p_c. However, they still predict a value of 1 for the critical exponent μ, since the predicted slope dg_e/dp at $p = p_c$ is $2\Gamma^{1/2}/(\Gamma - 1)$, instead of zero. However, the initial slope at $p = 1$ is predicted exactly; Eq. (112) predicts that $dg_e/dp = 2\Gamma/(\Gamma - 1)$ at $p = 1$, which is the exact value (Harris and Kirkpatrick, 1977). Figure 5.10 presents the comparison between the EMA predictions and the results of computer simulations for the effective conductivity of site-disordered simple-cubic lattices.

Let us point out that Eq. (107) can also be used for estimating the effective conductivity of conductor-superconductor composites. Instead of the conductance distribution (110), the appropriate bond conductance distribution in this case is, $f(g) = p^2\delta(g - \infty) + (1 - p^2)h(g)$, so that Eq. (107) becomes

$$p^2 + (1 - p^2) \int_0^\infty \frac{g - g_e}{g + (\Gamma - 1)g_e} h(g)dg = 0, \tag{114}$$

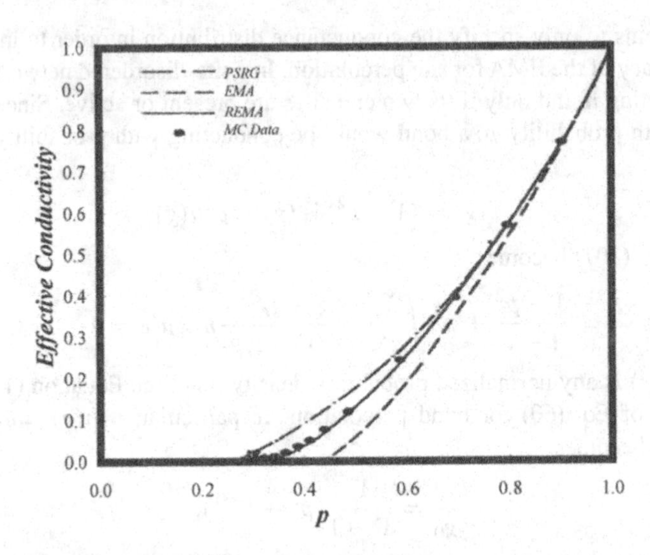

FIGURE 5.10. Effective conductivity of a site-disordered simple-cubic lattice, as predicted by the position-space renormalization group (PSRG), effective-medium approximation (EMA), and renormalized EMA, and the comparison with the Monte Carlo (MC) data (after Sahimi *et al.*, 1984).

which should be compared with Eq. (111). It is straightforward to show that Eq. (114) predicts that g_e diverges at a percolation threshold p_c given by Eq. (113).

5.8 Effective-Medium Approximation for Correlated Composites

Resistor networks with bond-percolation disorder do not contain any correlations, since the conductance of any bond is selected completely at random. Resistor networks with site-percolation disorder contain short-range correlations, because a bond is conducting only if its two end sites are active. However, both models predict percolation thresholds that are typically large compared with what is actually seen in experiments on many composite solids. These experiments indicate that many conducting composites have very low percolation thresholds that cannot be modeled by random site- or bond-percolation. On the other hand, as discussed in Chapter 2, correlations do change the location of the percolation threshold, and usually lower it. Therefore, a more realistic model, which may be more relevant to real materials, is one in which there are correlations between various portions of the material. One such model, which is directly relevant to hopping conductivity of disordered materials described in Chapter 6 (see Section 6.2), was proposed by Kirkpatrick (1973), which was already described in 2.12.1. Recall that in his model, each site of a network is assigned a random number s_i uniformly distributed in $(-1, +1)$. Random numbers b_{ij} for the bonds connecting sites i and j are then

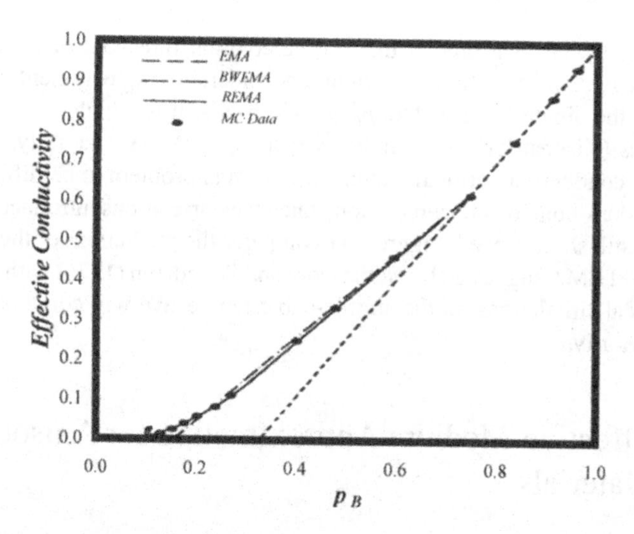

FIGURE 5.11. The same as in Figure 5.10, but for correlated-bond percolation problem. p_B is the fraction of the conducting bonds. BWEMA denotes the effective-medium approximation developed by Bernasconi and Wiesmann (1976) (after Sahimi *et al.*, 1984).

calculated from the numbers s_i and s_j via

$$b_{ij} = \frac{1}{2}(|s_i| + |s_j| + |s_i - s_j|). \tag{115}$$

All bonds with $b_{ij} > \Delta$ are then removed, where Δ is a selected limit. It is not difficult to see that the fraction $p(\Delta)$ of bonds is given by (Kirkpatrick, 1973; Bernasconi and Wiesmann, 1976)

$$p(\Delta) = \begin{cases} \frac{3}{4}\Delta^2, & 0 \le \Delta \le 1, \\ \Delta - \frac{1}{4}\Delta^2, & 1 \le \Delta \le 2. \end{cases} \tag{116}$$

The single-bond EMA, Eqs. (57), (61), or (62), can then be used for predicting the effective conductivity of this model. In particular, Eq. (62) would predict that, $g_e/g_0 = \frac{1}{2}[3p(\Delta) - 1]$ for the simple-cubic lattice which, for $p(\Delta) \ge 3/4$ ($\Delta \ge 1$), provides a more than satisfactory approximation to $g_e(p)$ (see Figure 5.11). However, for $p(\Delta) < 3/4$ ($\Delta < 1$), this EMA breaks down completely, as there are many isolated sites with $|s_i| < \Delta$, and therefore the fraction of the conducting bonds is overestimated. Bernasconi and Wiesmann (1976) noticed that, in this case, a bond that connects sites i and j would be conducting if and only if $s_i < \Delta$ *and* $s_j < \Delta$. Thus, they proposed that for $p(\Delta) < 3/4$ the system should be treated as a site-disordered network (whereas for $p(\Delta) \ge 3/4$ the system is similar to a bond-disordered network) with p^2 in Eq. (110) replaced by $\frac{4}{3}p(\Delta)$, since $p(\Delta) = \frac{3}{4}\Delta^2$ and $\Delta < 1$ plays the role of p. Thus, the conductance distribution should read

$$f(g) = [1 - \frac{4}{3}p(\Delta)]\delta_+(g) + \frac{4}{3}p(\Delta)h(g). \tag{117}$$

This model would then predict that the percolation threshold of the correlated network is $p_c \simeq 0.158$ which, although a considerable improvement on the prediction of the single-bond EMA, $p_c = 1/3$, it is still too high since numerical simulations (Kirkpatrick, 1973) indicate that, $p_c \simeq 0.103$. Naturally, we do not expect the conductivity critical exponent μ for this problem to be different from that of random-bond or site percolation, since the correlations introduced into this model are still short-ranged. Figure 5.11 compares the predictions of the EMA and the modified EMA suggested by Bernasconi and Wiesmann (1976) with the results of numerical simulations on the simple-cubic lattice. We will come back to this problem shortly.

5.9 Effective-Medium Approximation for Anisotropic Materials

All the theories of conduction described so far are appropriate for isotropic materials. Many composite materials are anisotropic, and the effect of their anisotropy on their transport properties is significant and must be modeled appropriately. Examples are abundant. In many catalytic materials, transport of molecules can be confined to a certain direction of plane, so that the material is effectively anisotropic. Many solid-state materials, such as $(Sn)_x$ and TCNQ salts, can be obtained only as small crystals, and their effective transport properties must be measured in compact powders which are highly anisotropic. Carbon-black polyvinylchloride and other polymer composites are highly anisotropic. Therefore, an EMA for predicting the conductivity tensor of such materials can be very useful. Although one may consider at least five types of anisotropies (Straley, 1980b), we consider here one particular type which is important to real materials—one in which a material is characterized by direction-dependent conductance distributions.

We extend the perturbation expansion developed in Section 5.3 for isotropic materials to anisotropic systems (Mukhopadhyay and Sahimi, 2000). To match the disordered network to an effective anisotropic one, we introduce d effective conductivities $g_{e_1}, \cdots, g_{e_d}$, so that Eq. (33) is rewritten as

$$\sum_{\ell=1}^{d} \sum_{j} g_{e_\ell} (P_j^0 - P_i^0)_\ell = 0, \tag{118}$$

where $(P_j^0 - P_i^0)_\ell$ indicates that the quantity must be evaluated for those nearest neighbor sites j of i that are in the ℓth direction. Therefore, the analogue of Eq. (34) for an anisotropic network is given by

$$\sum_{\ell=1}^{d} \sum_{j} [\Delta_{ij\ell}(P_j - P_i)_\ell + (P_j - P_i)_\ell - (P_j^0 - P_i^0)_\ell] = 0, \tag{119}$$

where $\Delta_{ij\ell} = (g_{ij} - g_{e\ell})/g_{e\ell}$. The rest of the analysis is somewhat involved but parallels that of the isotropic case. After some algebra we obtain

$$P_i = P_i^0 + \sum_{\ell=1}^{d} \sum_j \sum_l G_{ij} \Delta_{jl\ell} (P_j - P_i)_\ell, \tag{120}$$

where the Green function G_{ij} is now for the conduction problem in an anisotropic material. The analysis in terms of the "fluxes" $(Q_{ij})_\ell$ is more complex, since we now have to introduce direction dependent fluxes $Q_{ij\ell}$. Since for the construction of an EMA one can directly work with the fields P_i^0 and P_i, and the results would be identical with those obtained with the fluxes, we do not pursue the formulation of the anisotropic networks in terms of the fluxes.

The same idea that was used for isotropic networks can also be utilized for anisotropic lattices in order to derive an EMA, which we refer to it as the anisotropic EMA (AEMA). The only difference is that, since we must compute d effective conductivities $g_{e\ell}$, one for each principal direction of the lattice, we need d self-consistent equations. Thus, for each principal direction, Eq. (120) is written for two adjacent sites, from which the flux between them is calculated. This is then repeated for all the bonds in the cluster. Equation (51) is then applied to a reference bond α in each principal direction, i.e.,

$$\langle Q_\alpha \rangle_\ell = (Q_\alpha^0)_\ell \quad \ell = 1, \cdots, d. \tag{121}$$

It is now straightforward to show that the working equation for a single-bond AEMA is the analogue of Eq. (54) for anisotropic networks:

$$\left\langle \frac{(\gamma_{\alpha\alpha})_\ell (g_\alpha - g_{e\ell})}{(\gamma_{\alpha\alpha})_\ell g_\alpha - [1 + (\gamma_{\alpha\alpha})_\ell] g_{e\ell}} \right\rangle = 0, \quad \ell = 1, \cdots, d. \tag{122}$$

If a material is characterized by direction-dependent conductance distributions $f_\ell(g_\alpha)$, then, Eq. (122) becomes

$$\int_0^\infty \frac{g_\alpha - g_{e\ell}}{g_\alpha + S_\ell(g_{e_1}, \cdots, g_{e_d})} f_\ell(g_\alpha) dg_\alpha = 0, \quad \ell = 1, \cdots, d, \tag{123}$$

where

$$S_\ell = -\frac{1 + (\gamma_{\alpha\alpha})_\ell}{(\gamma_{\alpha\alpha})_\ell} g_{e\ell}. \tag{124}$$

Since $\gamma_{\alpha\alpha} = -2/Z$, we see that for isotropic networks

$$S = (\frac{Z}{2} - 1) g_e, \tag{125}$$

so that the main difference between the EMA and AEMA is in the quantity S_ℓ.

5.9.1 The Green Functions

We confine our attention to a d-dimensional simple-cubic network, although our analysis is general and applicable to any regular lattice. The governing equation for the Green functions is the anisotropic version of Eq. (36), written for a d-dimensional simple-cubic network [see also Eq. (6.14)], i.e.,

$$(2g_{e_1} + \cdots + 2g_{e_d})G(m_1, \cdots, m_d) - g_{e_1}[G(m_1+1, \cdots, m_d) + G(m_1-1, \cdots, m_d)] + \cdots + g_{e_d}[G(m_1, \cdots, m_d+1) + G(m_1, \cdots, m_d-1)] = -\delta_{m_1 0} \cdots \delta_{m_d 0}. \tag{126}$$

If we use a discrete Fourier transformation

$$\tilde{G}(\theta_1, \cdots, \theta_d) = \sum_{m_1=-\infty}^{\infty} \cdots \sum_{m_d=-\infty}^{\infty} e^{i\mathbf{m}\cdot\Theta} G(m_1, \cdots, m_d) \tag{127}$$

where $\mathbf{m} = (m_1, \cdots, m_d)$ and $\Theta = (\theta_1, \cdots, \theta_d)$, we obtain

$$2(g_{e_1} + \cdots + k_{e_d})\tilde{G} - 2(g_{e_1}\cos\theta_1 + \cdots + g_{e_d}\cos\theta_d)\tilde{G} = -1 \tag{128}$$

and hence

$$G(m_1, \cdots, m_d) = -\frac{1}{2}\frac{1}{(2\pi)^d}\int_{-\pi}^{\pi} \cdots \int_{-\pi}^{\pi} \frac{\exp^{-i\mathbf{m}\cdot\boldsymbol{\Theta}}}{\sum_{j=1}^{d} g_{e_j}(1-\cos\theta_j)} d\boldsymbol{\Theta}, \tag{129}$$

which should be compared with Eq. (6.17) for the isotropic lattices.

We first calculate the bond-bond Green function $(\gamma_{\alpha\alpha})_1$ in a principal directions of a square network. It is easily seen that the analogue of Eq. (55) for an anisotropic network is given by

$$(\gamma_{\alpha\alpha})_1 = g_{m_1}\left[2G(0,0) - 2G(m_1, \cdots, m_d)\mathrm{a}\Big|_{\substack{m_1=1\\ m_i=0}}\right], \tag{130}$$

with $i \neq \alpha$, so that for a square network

$$\begin{aligned}(\gamma_{\alpha\alpha})_1 &= 2g_{e_1}[G(0,0) - G(1,0)] \\ &= -\frac{g_{e_1}}{\pi^2}\int_0^{\pi}\int_0^{\pi} \frac{1-\cos\theta_1}{g_{e_1}(1-\cos\theta_1) + g_{e_2}(1-\cos\theta_2)} d\theta_1 d\theta_2. \end{aligned} \tag{131}$$

After some algebra Eq. (131) leads us to

$$\begin{aligned}(\gamma_{\alpha\alpha})_1 &= -\frac{1}{\pi}\left(\pi - 2\arcsin\sqrt{\frac{g_{e_2}}{g_{e_1}+g_{e_2}}}\right) \\ &= -\frac{1}{\pi}\left(\pi - 2\arctan\sqrt{\frac{g_{e_2}}{g_{e_1}}}\right),\end{aligned} \tag{132}$$

which reduces to $\gamma_{\alpha\alpha} = -1/2$ for the isotropic case ($g_{e_1} = g_{e_2}$), in agreement with $\gamma_{\alpha\alpha} = -2/Z = -1/2$ for $Z = 4$, computed earlier. Similar calculations yield

$$(\gamma_{\alpha\alpha})_2 = -\frac{1}{\pi}\left(\pi - 2\arctan\sqrt{\frac{g_{e_1}}{g_{e_2}}}\right), \tag{133}$$

which can also be obtained from Eq. (132) by cyclic permutation of the indices 1 and 2.

Given these Green functions, we can develop the relevant expressions for S_ℓ for the square and simple-cubic networks. For example, for the square network in the direction $\ell = 1$ we have

$$S_1 = g_{e_1} \frac{\arctan (g_{e_2}/g_{e_1})^{1/2}}{\arctan (g_{e_1}/g_{e_2})^{1/2}}, \tag{134}$$

and a cyclic permutation of the indices 1 and 2 leads to a formula for S_2. Equation (134) was first given by Bernasconi (1974).

In the case of a simple-cubic network, however, the equations are more complex. Using Eqs. (128) and (129), one finds that for this network

$$(\gamma_{\alpha\alpha})_1 = -\frac{g_{e_1}}{\pi^3} \int_0^\pi \int_0^\pi \int_0^\pi \frac{1 - \cos\theta_1}{g_{e_1}(1 - \cos\theta_1) + g_{e_2}(1 - \cos\theta_2) + g_{e_3}(1 - \cos\theta_3)} d\theta_1 d\theta_2 d\theta_3. \tag{135}$$

Similar expressions can be obtained for the other two directions. The integrals in (135) can be written in terms of the elliptic integrals (see below). To obtain the corresponding S_ℓ for the simple-cubic network we introduce a parameter R_ℓ such that

$$\frac{1}{R_\ell} = S_\ell + g_{e_\ell}. \tag{136}$$

Using Eqs. (124), (135) and (136) we obtain, after some algebra, for the direction $\ell = 1$

$$R_1 = \frac{1}{\pi^2 (g_{e_2} g_{e_3})^{1/2}} \int_0^\pi (1 - \cos x) \eta^{1/2} K(\eta) dx \tag{137}$$

where $K(\eta)$ is the complete elliptic integral of the first kind,

$$K(\eta) = \int_0^{\pi/2} \frac{d\tau}{(1 - \eta^2 \sin^2 \tau)^{1/2}} \tag{138}$$

and the parameter η is given by

$$\eta = \left[\frac{4 g_{e_2} g_{e_3}}{(t - g_{e_1} \cos x)^2 - (g_{e_3} - g_{e_2})^2} \right]^{1/2}, \quad t = \sum_{\ell=1}^{3} g_{e_\ell}. \tag{139}$$

We then obtain

$$S_1 = g_{e1} \frac{\arctan \, [g_{e1}^{-1} (g_{e1} g_{e2} + g_{e1} g_{e3} + g_{e2} g_{e3})]^{1/2}}{\arctan \, [g_{e1} (g_{e1} g_{e2} + g_{e1} g_{e3} + g_{e2} g_{e3})]^{-1/2}}, \tag{140}$$

and cyclic permutations of the indices lead to the corresponding formulae for R_2, R_3, S_2 and S_3. The formulation presented here provides a general framework for analyzing conduction in anisotropic networks and materials.

As an example, consider bond conductance distributions, $f_1(g) = p\delta(g - 10) + (1 - p)\delta_+(g)$, and $f_2(g) = p\delta(g - 1) + (1 - p)\delta_+(g)$. Figure 5.12 compares the AEMA predictions with the results of computer simulations on the square lattice; the agreement between the two is very good. Note also that, unlike the isotropic case, the conductivity does not vary linearly with p.

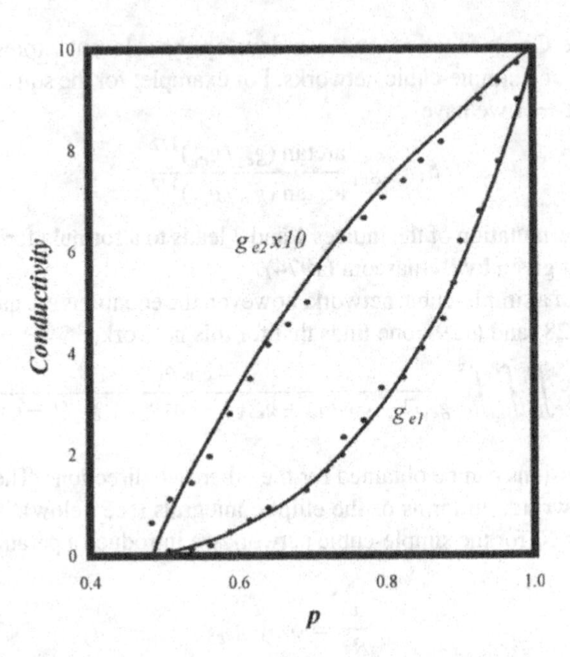

FIGURE 5.12. Comparison of the EMA predictions for the effective conductivities g_{e1} and g_{e2} for anisotropic bond-disordered square networks (solid curves) with the simulation data (symbols) (after Bernasconi, 1974).

Let us point out that the AEMA developed here has been extended by Koelman and de Kuijper (1997) to anisotropic materials that contain many components. However, the anisotropy that was considered in their work was due to shapes and orientations of the constitutive materials, e.g., ellipsoidal grains in a powder or a porous material, which is different from what we discuss here. The anisotropy of such materials persists even at the percolation threshold. We postpone the discussion of this type of anisotropy until Chapter 6, where we discuss the dielectric constant of composite materials. The EMA for the dielectric constant of such anisotropic materials is completely analogous to one for the effective conductivity tensor, and therefore need not be discussed separately.

5.9.2 *Conductivity Anisotropy Near the Percolation Threshold*

The conductivity anisotropy $\mathcal{R}$ is defined by

$$\mathcal{R} = \frac{g_{e\ell}}{g_{et}} - 1, \tag{141}$$

where $g_{e\ell}$ and g_{et} are the effective conductivities in the longitudinal direction (parallel to the applied field) and transverse directions (perpendicular to the applied field), respectively. For the type of anisotropy considered here, the anisotropy vanishes as the percolation threshold p_c is approached according to the following power law (Shklovskii, 1978)

$$\mathcal{R} \sim (p - p_c)^{\zeta}, \tag{142}$$

where ζ is presumably a universal exponent, independent of the morphology of the system. Note that, the AEMA also predicts that the conductivity anisotropy vanishes at p_c. If the material is not thick, then its finite thickness will give rise to conductivity anisotropy. It is not difficult to show that

$$\frac{g_{e1}}{g_{e2}} \sim \left(\frac{\ell_2}{\ell_1}\right)^{(\mu_3-\mu_2)/\nu_3}, \tag{143}$$

where g_{e1} (g_{e2}) is the conductivity of a sample of thickness ℓ_1 (ℓ_2), μ_2 (μ_3) is the critical exponent of the conductivity for 2D (3D) percolation, and ν_3 is the critical exponent for the percolation correlation length in 3D.

Accurate estimates of the anisotropy exponent ζ were obtained by Mukhopadhyay and Sahimi (1994), who reported that

$$\zeta(d = 2) = \zeta_2 \simeq 0.82, \quad \zeta(d = 3) = \zeta_3 \simeq 0.52. \tag{144}$$

That $\zeta_2 > \zeta_3$ is surprising, since one expects that on a 3D network which contains a much larger number of current-carrying transport paths than a 2D network, the anisotropy vanish faster, i.e., $\zeta_3 > \zeta_2$. Since vanishing of $\mathcal{R}$ is due to the increasing tortuosity of the sample-spanning conducting cluster, as p_c is approached, we must have a measure of the tortuosity of percolation clusters and their backbone in order to understand why $\zeta_2 > \zeta_3$. One such measure may be the length L_{min} of the minimum path or chemical distance between two points of the cluster, described in Section 2.8. In general, L_{min} is greater than the Pythagorean distance between the two points. According to Eq. (2.50), near the percolation threshold p_c,

$$L_{min} \sim (p - p_c)^{-\nu D_{min}}, \tag{145}$$

where D_{min} is the fractal dimension of the minimum paths. Using the numerical estimates of ν and D_{min} listed in Table 2.3, we obtain, $L_{min} \sim (p - p_c)^{-1.5}$ for 2D percolation, and $L_{min} \sim (p - p_c)^{-1.18}$ for 3D percolation. That is, as p_c is approached, L_{min} increases less strongly in 3D than in 2D, which may be interpreted as implying that near p_c the tortuosity of 3D transport paths increases less strongly than that of 2D paths. As a result, at a fixed distance from p_c, the tortuosity of 2D transport paths is *larger* than that of 3D paths, and therefore the anisotropy $\mathcal{R}$ must vanish faster in 2D than in 3D.

5.9.3 Comparison with the Experimental Data

Smith and Lobb (1979) measured the effective conductivities of 2D conductor-insulator networks generated photolithographically from laser speckle patterns. When they measured the conductivity of isotropic samples, they found that it vanishes at $p_c \simeq 0.59$, in agreement with site-percolation threshold of a square network (see Table 2.1), with a critical exponent $\mu \simeq 1.3$, in complete agreement with the 2D percolation conductivity (see Table 2.3). When they measured the conductivities of anisotropic samples, they found that the conductivity anisotropy, measured in terms of the difference between the conductivity of the system in different directions, decreased as p_c was approached, in agreement with Eq. (142).

Troadec *et al.* (1981) measured thermal and electrical conductivity of conducting (WTe_2) and semi-conducting (WSe_2) powders, characterized both by a geometrical anisotropy (grain shapes and sizes) and by anisotropy in a transport property. These mixtures belong to the family of dichalcognides of transition metals, TX_2. Their morphology is layered with a hexagonal arrangement within each plane, which gives rise to the anisotropy of the system. The main difference between the crystallographic structures of WSe_2 and WTe_2 is in the coordination number of the W atoms. In WSe_2, W is at the center of a trigonal prism of Se atoms, whereas in WTe_2 W has an octahedral environment but is not precisely at its center. Moreover, WSe_2 has a semi-conductor character, in contrast with the semi-metallic character of WTe_2. Troadec *et al.* used powder grains that were single crystal platelets having a thickness of about 1 to 2 μm and a horizontal hexagonal shape with a largest dimension of about 20 μm. The mixed powder was outgassed under secondary vacuum for two hours before being sintered in the same pressure and temperature conditions. The packing fraction was about 92%. They measured anisotropy in the electrical conductivity, as defined by Eq. (141), which exhibited a *minimum* at the percolation threshold of the WTe_2 phase. This is expected, since as p_c of a phase is approached, its conducting paths become so tortuous that the current cannot distinguish between different directions.

Experimental data similar to those of Troadec *et al.* were also reported by Balberg *et al.* (1983), who measured the resistivity of a composite composed of elongated carbon black aggregates embedded in an insulating plastic—polyvinylchloride. Because of the elongation of the aggregates, the system is anisotropic. Measurements of the anisotropy of the system produced results that indicated that the anisotropy vanishes at p_c.

5.10 Cumulant Approximation

In a series of papers, Hori and Yonezawa developed a method—the cumulant approximation (CA)—which takes into account the effect of clustering of the conducting regions of a composite, or the conducting bonds in a resistor network model, near the percolation threshold. As discussed above, such clustering effects, while highly important near p_c, are completely absent in the single-bond EMA. Although the cluster EMA described above does take such effects into account, the convergence of its predictions toward the actual values is relatively slow, and thus a faster-converging method would be desirable. A summary of the basic CA ideas is given by Hori and Yonezawa (1977), where references to their earlier papers can also be found. Although the CA method was originally developed for continuum models of composite materials, we describe the theory here for percolation composites, so that a direct comparison with the EMA and its cluster version can be made. We follow Hori and Yonezawa (1977) closely and describe the CA method.

Consider a 3D (2D) heterogeneous material with volume (surface area) Ω (S) which will eventually be assumed to be infinitely large, so that, for example, the ensemble average of the conductivity,

$$\langle g \rangle = \sum_i \phi_i g_i \tag{146}$$

can be replaced by the spatial average, where g_i and ϕ_i are the conductivity and volume fraction of the ith phase, respectively. The effective conductivity g_e is then defined through the usual relation, namely, $\langle I_i \rangle = g_e \langle E_i \rangle$, where I_i and E_i are the electric current density and field, respectively. The conductivity $g(\mathbf{r})$ at point $\mathbf{r}$ is now split into constant and fluctuating parts, $g(\mathbf{r}) = g_0 + \delta g(\mathbf{r})$. Then, the governing equation for the electrostatic potential $\Phi(\mathbf{r})$ is given by

$$g_0 \frac{\partial^2 \Phi(\mathbf{r})}{\partial x_i^2} + \frac{\partial}{\partial x_i}\left[\delta g(\mathbf{r}) \frac{\partial \Phi(\mathbf{r})}{\partial x_i}\right] = 0, \tag{147}$$

where summation over the index i is implied. A Green function $G(\mathbf{r})$ is now introduced via

$$g_0 \frac{\partial^2 G(\mathbf{r})}{\partial x_i^2} + \delta(\mathbf{r}) = 0, \tag{148}$$

[compare Eq. (148) with Eq. (36), its discrete analogue] in terms of which we can write down a formal solution for $\Phi(\mathbf{r})$:

$$\Phi(\mathbf{r}_1) = \Phi_0(\mathbf{r}_1) + \int G(\mathbf{r}_{12}) \frac{\partial}{\partial x_{2,i}}\left[\delta g(\mathbf{r}_2) \frac{\partial \Phi(\mathbf{r}_2)}{\partial x_{2,i}}\right] d\mathbf{r}_2, \tag{149}$$

where the integral is over the volume Ω (surface S), and $\mathbf{r}_{12} = \mathbf{r}_2 - \mathbf{r}_1$. The Green function $G(\mathbf{r})$ is given by the well-known equations,

$$G(\mathbf{r}) = \begin{cases} (2\pi g_0)^{-1} \log(\frac{1}{r}) & \text{2D}, \\ (4\pi g_0)^{-1} \frac{1}{r}, & \text{3D}. \end{cases} \tag{150}$$

The corresponding equation for the electric field is given by

$$E_i(\mathbf{r}_1) = \langle E_i(\mathbf{r}_1) \rangle + \int \frac{\partial G(\mathbf{r}_{12})}{\partial x_{1,i}} \frac{\partial}{\partial x_{2,j}} [\delta g(\mathbf{r}_2) E_j(\mathbf{r}_2)], \tag{151}$$

which, after some algebra, yields

$$E_i(\mathbf{r}_1) = E_i^0(\mathbf{r}_1) + \int \mathcal{G}_{ij}(\mathbf{r}_{12}) \delta g(\mathbf{r}_2) E_j(\mathbf{r}_2) d\mathbf{r}_2. \tag{152}$$

Here

$$\begin{aligned} \mathcal{G}_{ij}(\mathbf{r}_{12}) &= -\frac{\partial^2 G(\mathbf{r}_{12})}{\partial x_{1,i} \partial x_{2,j}} \\ &= -\frac{\delta_{ij}\delta(\mathbf{r}_{12})}{d g_0} + \mathcal{P}\left[\frac{1}{2^{d-1}\pi g_0 r_{12}^d}\left(d \frac{x_{12,i}}{r_{12}} \frac{x_{12,j}}{r_{12}} - \delta_{ij}\right)\right], \end{aligned} \tag{153}$$

where d is the dimensionality of the composite, and $\mathcal{P}$ denotes the principal value of the integration around the singular point $\mathbf{r}_{12} = \mathbf{0}$. In operator notations, we can rewrite Eq. (152) as

$$\mathbf{E} = \mathbf{E}^0 + \mathcal{G}\,\delta\mathbf{g}\,\mathbf{E}, \tag{154}$$

where the meaning of the operator is clear when we compare Eq. (154) with (152). Similarly,

$$\mathbf{I} = \mathbf{I}^0 + \boldsymbol{\Gamma}\,\delta\mathbf{R}\,\mathbf{I}, \tag{155}$$

where R is the resistance, $R_0\boldsymbol{\Gamma} = -\mathbf{1} - g_0\mathcal{G}$, and $R_0 = (g_0)^{-1}$.

5.10.1 *The Lorentz Field*

One can now define a Lorentz field $\mathcal{L}$ such that

$$\mathcal{L} = \frac{\delta g(\mathbf{r}) + dg_0}{dg_0}\mathbf{E} = \frac{g(\mathbf{r}) + (d-1)g_0}{dg_0}\mathbf{E}, \tag{156}$$

and

$$\kappa(\mathbf{r}) = \frac{d\delta g(\mathbf{r})}{\delta g(\mathbf{r}) + dg_0} = \frac{d[g(\mathbf{r}) - g_0]}{g(\mathbf{r}) + (d-1)g_0}. \tag{157}$$

The effective constant κ^* satisfies, $\langle\kappa\mathcal{L}\rangle = \kappa^*\langle\mathcal{L}\rangle$, and therefore

$$\kappa^* = \frac{d(g^* - g_0)}{g^* + (d-1)g_0}. \tag{158}$$

In operator notations, Eq. (154) can be rewritten as

$$\mathcal{L} = \mathbf{E}^0 + \boldsymbol{\Lambda}\kappa\mathcal{L}, \tag{159}$$

where

$$\boldsymbol{\Lambda} = \frac{\mathbf{1}}{d} + g_0\mathcal{G}. \tag{160}$$

Equations (154), (155), and (159) are then written in a general form

$$\mathbf{X} = \mathbf{X}^0 + \mathbf{A}\delta\alpha\mathbf{X}, \tag{161}$$

where $\mathbf{A}$ is the appropriate Green function.

5.10.2 *Perturbation Expansion*

Successive iteration of Eq. (161) yields

$$\mathbf{X} = \sum_{n=0}^{\infty}(\mathbf{A}\delta\alpha)^n\mathbf{X}^0, \tag{162}$$

whence

$$\langle\mathbf{X}\rangle = \sum_{n=0}^{\infty}\langle(\mathbf{A}\delta\alpha)^n\rangle\mathbf{X}^0, \tag{163}$$

$$\delta\boldsymbol{\alpha}^*\langle\mathbf{X}\rangle = \sum_{n=0}^{\infty}\langle\delta\boldsymbol{\alpha}(\mathbf{A}\delta\boldsymbol{\alpha})^n\rangle\mathbf{X}^0. \tag{164}$$

Therefore, if we eliminate $\mathbf{X}^0$ from these equations, we obtain

$$\delta\boldsymbol{\alpha}^* \equiv \sum_{n=1}^{\infty}\delta\boldsymbol{\alpha}^{(n)}, \tag{165}$$

where

$$\begin{aligned}\delta\boldsymbol{\alpha}^{(1)} &= \langle\delta\boldsymbol{\alpha}\rangle,\\ \delta\boldsymbol{\alpha}^{(2)} &= \langle\delta\boldsymbol{\alpha}\mathbf{A}\delta\boldsymbol{\alpha}\rangle - \langle\delta\boldsymbol{\alpha}\rangle\mathbf{A}\langle\delta\boldsymbol{\alpha}\rangle,\\ \delta\boldsymbol{\alpha}^{(3)} &= \langle\delta\boldsymbol{\alpha}\mathbf{A}\delta\boldsymbol{\alpha}\mathbf{A}\delta\boldsymbol{\alpha}\rangle - \langle\delta\boldsymbol{\alpha}\mathbf{A}\delta\boldsymbol{\alpha}\rangle\mathbf{A}\langle\delta\boldsymbol{\alpha}\rangle - \langle\delta\boldsymbol{\alpha}\rangle\mathbf{A}\langle\delta\boldsymbol{\alpha}\mathbf{A}\delta\boldsymbol{\alpha}\rangle + \langle\delta\boldsymbol{\alpha}\rangle\mathbf{A}\langle\delta\boldsymbol{\alpha}\rangle\mathbf{A}\langle\delta\boldsymbol{\alpha}\rangle.\end{aligned} \tag{166}$$

We now replace the moments by the corresponding cumulants, or semi-invariants, which are useful in the sense that they vanish identically for independent groups of random variables. Hence, Eqs. (166) become

$$\begin{aligned}\delta\boldsymbol{\alpha}^{(1)} &= \langle\delta\boldsymbol{\alpha}\rangle_c,\\ \delta\boldsymbol{\alpha}^{(2)} &= \langle\delta\boldsymbol{\alpha}\mathbf{A}\delta\boldsymbol{\alpha}\rangle_c,\\ \delta\boldsymbol{\alpha}^{(3)} &= \langle\delta\boldsymbol{\alpha}\mathbf{A}\delta\boldsymbol{\alpha}\mathbf{A}\delta\boldsymbol{\alpha}\rangle_c + \langle\delta\boldsymbol{\alpha}\mathbf{A}\langle\delta\boldsymbol{\alpha}\rangle_c\mathbf{A}\delta\boldsymbol{\alpha}\rangle_c.\end{aligned} \tag{167}$$

Thus, a cumulant diagram can be constructed by representing the various contributions to $\delta\boldsymbol{\alpha}_e$, using the following rules:

(1) Points $\mathbf{r}_1, \mathbf{r}_2, \cdots$ are represented by means of nodes on the horizontal base line.
(2) The Green function $\mathbf{A}(\mathbf{r}_{i,i+1})$ is assigned to a propagator that connects $\mathbf{r}_i$ and $\mathbf{r}_{i+1}$.
(3) The s-point cumulant $\langle\delta\alpha(\mathbf{r}_i)\delta\alpha(\mathbf{r}_j)\cdots\rangle$ is assigned to a cross vertex from s interaction lines start; and then
(4) the operator product of all the Green functions and cumulants described above is taken.

Figure 5.13 shows the diagrammatic equation. We now consider a few examples to demonstrate the application of the cumulant expansion.

5.10.3 Computation of the Lowest-Order Terms

We can now compute the lower-order terms of the cumulant expansion for several models of disordered materials. One main requirement for being able to do so is that the model should consist of symmetric cells. Such a model, which is statistically homogeneous, was already described in Section 3.4.6. Recall that in this model, (1) the space is completely covered by non-overlapping cells within which the material property is constant, and (2) the properties of each cell are independent of those of any other cell, and also independent of the geometrical distribution of the cells. We now describe computation of the lower-order terms of the cumulant expansion for a few models.

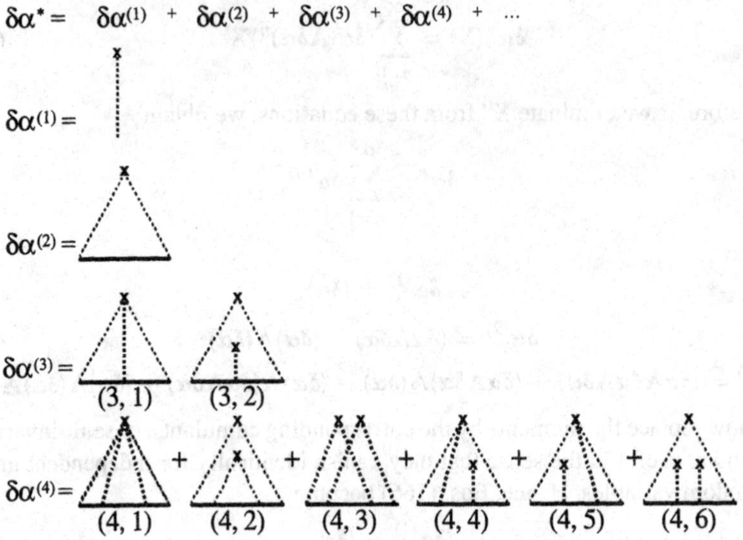

$\delta\alpha^* = \delta\alpha^{(1)} + \delta\alpha^{(2)} + \delta\alpha^{(3)} + \delta\alpha^{(4)} + \cdots$

$\delta\alpha^{(1)} =$

$\delta\alpha^{(2)} =$

$\delta\alpha^{(3)} =$ (3, 1) (3, 2)

$\delta\alpha^{(4)} =$ (4, 1) + (4, 2) + (4, 3) + (4, 4) + (4, 5) + (4, 6)

FIGURE 5.13. The diagrammatic representation of Eqs. (165)–(167) (after Hori and Yonezawa, 1977).

Statistically isotropic materials

In this case

$$\langle \delta\alpha(\mathbf{r}_1)\delta\alpha(\mathbf{r}_2)\rangle_c = \langle (\delta\alpha)^2\rangle_c C(\mathbf{r}_{12}), \tag{168}$$

where $C(\mathbf{r}_{12})$ is the normalized two-point correlation function (see Chapter 2). Therefore,

$$\delta\alpha^{(2)} = -\frac{1}{d}\frac{\langle (\delta\alpha)^2\rangle_c}{\alpha_0}. \tag{169}$$

However, third- or higher-order terms cannot be determined analytically, unless more information on the morphology of the material is supplied.

Ellipsoidal (Elliptic) cell materials with uniform orientations

Such a material is anisotropic. The second- and third-order terms are given by

$$\delta\alpha_{ij}^{(2)} = -A_{ij}\frac{\langle (\delta\alpha)^2\rangle_c}{\alpha_0}, \tag{170}$$

where A_{ij} is the depolarization or demagnetization tensor of the cells (see Section 4.6.2.4), and

$$\delta\alpha_{ij}^{(3,1)} = A_{ik}A_{kj}\frac{\langle (\delta\alpha)^3\rangle_c}{\alpha_0^2}, \tag{171}$$

$$\delta\alpha^{(3,2)} = A_{ij}\frac{\langle (\delta\alpha)^2\rangle_c\langle \delta\alpha\rangle_c}{\alpha_0^2}. \tag{172}$$

Ellipsoidal (Elliptic) cell materials with random orientations

Such materials are isotropic, and therefore all one must do is averaging the tensor A_{ij}. For the second-order term, we have

$$\langle A_{ij}\rangle = \frac{\delta_{ij}}{d}, \tag{173}$$

while for the third-order term,

$$\langle A_{ik}A_{kj}\rangle = \frac{1}{d}\sum_{k=1}^{d} A_k^2\delta_{ij}, \tag{174}$$

where A_k is the eigenvalue of the tensor A_{ij}.

Spherical cell materials

Since a spherical cell material is statistically isotropic, the second-order term is given by Eq. (169), while the third-order terms are given by

$$\delta\alpha^{(3,1)} = \frac{1}{d^2}\frac{\langle(\delta\alpha)^3\rangle_c}{\alpha_0^2}, \tag{175}$$

$$\delta\alpha^{(3,2)} = \frac{1}{d}\frac{\langle(\delta\alpha)^2\rangle_c\langle\delta\alpha\rangle_c}{\alpha_0^2}. \tag{176}$$

Given these results, it is straightforward to derive the EMA within the CA. For example, since it is not possible to sum up all terms in the perturbation series, one simple approximation is to count only the single-site moment diagrams. The summation is then given by

$$\delta g^* = \langle\delta g\rangle - \frac{\langle(\delta g)^2\rangle}{dg_0} + \frac{\langle(\delta g)^3\rangle}{(dg_0)^2} - \cdots \tag{177}$$

We now choose δg^* self-consistently, i.e., we take $g^* = g_0$, to immediately obtain

$$\left\langle \frac{g-g^*}{g+(d-1)g^*}\right\rangle = 0, \tag{178}$$

which is the continuum version of the EMA (see Section 4.9.1) and identifies g^* as the effective conductivity. To obtain the lattice version of the EMA, d must be replaced by $Z/2$. The EMA can also be derived based on the Lorentz field described above. In this case, the requirement $\langle\kappa\rangle = \langle\kappa\rangle_c = 0$ immediately reproduces the EMA.

5.10.4 Bond Percolation

Having calculated the lower-order terms of the cumulant expansion, one can now obtain an equation for the effective conductivity of a bond-percolation network. It is not difficult to see that, with a bond-conductance distribution, $f(g) = (1 -$

$p)\delta(g) + p\delta(g-1)$, the CA provides the following equation for the effective conductivity g_e,

$$g_e = \int_0^1 \left(1 + \frac{1-p}{p} x^{\gamma_0/g_e}\right)^{-1} dx, \tag{179}$$

where $\gamma_0 = -1/d = -2/Z$ is the same quantity as $\gamma_{\alpha\alpha}$ that appeared in the perturbation expansion for the effective conductivity of random resistor networks. The emergence of this quantity in the CA is not really surprising, since the Green function $G(\mathbf{r})$, defined by Eq. (148) and given by Eq. (150), is nothing but the discrete analogue of the Green function G_{ik} defined by Eq. (36). According to Eq. (179), the effective conductivity g_e vanishes at $p = p_c$, where p_c is given by

$$p_c = 1 - \exp(-2/Z). \tag{180}$$

Thus, for a simple-cubic network ($Z = 6$), one obtains, $p_c \simeq 0.283$, which is much closer to the numerical estimate (see Table 2.2), $p_c \simeq 0.249$, than is the EMA prediction, $p_c = 1/3$. On the other hand, for the square network, Eq. (180) predicts that, $p_c \simeq 0.393$, whereas the EMA predicts the exact results, $p_c = 1/2$. Therefore, while for 2D systems the CA is less accurate than the EMA, the opposite is true in 3D, which is a desirable feature of the CA.

5.11 Position-Space Renormalization Group Methods

Our discussions so far should have made it clear to the reader that the main assumption behind the EMA is that fluctuations in the potential field are small, since in deriving the EMA we require the average of the fluctuations to be zero. However, if the fluctuations are large, as in the case of, for example, a composite near its percolation threshold, or a material with a broad distribution of the conductances, the EMA breaks down and loses its accuracy. In such cases, a position-space renormalization group (PSRG) method is more appropriate because, in averaging the properties of the system, this method takes into account the properties of the *pre-averaged* material. It also predicts non-analytic power-law behavior for the transport properties near the percolation threshold, a distinct advantage over the EMA which always predicts the critical exponents of the conductivity to be unity.

Consider, for example, a square or a cubic network in which each bond is conducting with probability p. The idea in any PSRG method is that, since the network is so large that we cannot calculate its properties exactly, we partition it into $b \times b$ or $b \times b \times b$ cells, where b is the number of bonds in any direction, and calculate their effective properties, which are hopefully representative of the properties of the original network. However, the effective properties of the RG cells are calculated in a particular way which we will describe shortly. The shape of the RG cell can be selected arbitrarily, but it should be chosen in such a way that it preserves, as much as possible, the symmetry properties of the original network. For example, since the square network is self-dual, and because self-duality plays an important role in its percolation and transport properties, we use self-dual cells

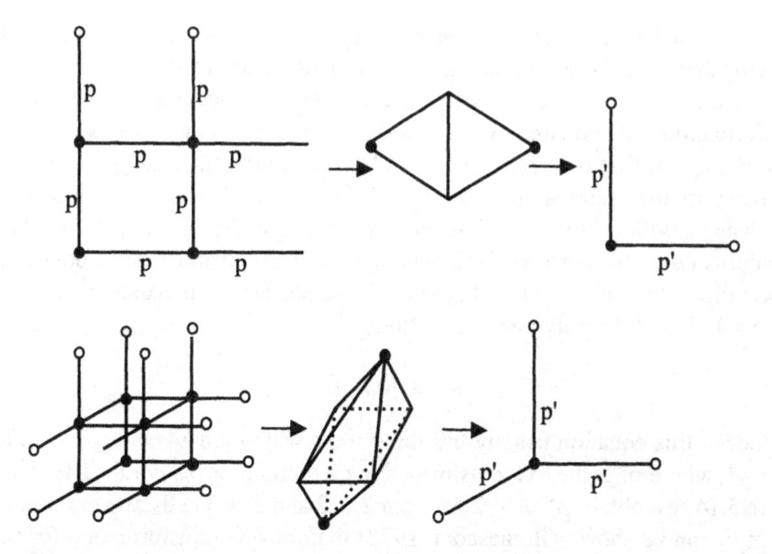

FIGURE 5.14. Two- and three-dimensional RG cells with scale factor $b = 2$, their equivalent circuits (middle) and renormalized shape (right).

to represent this network (Bernasconi, 1978). Figure 5.14 shows examples of the RG cells with $b = 2$ in the square and simple-cubic networks, where the 2D cell is self-dual.

The next step in a PSRG method is to replace each cell with one bond in each principal direction; see Figure 5.14. If in the original network each bond conducts with probability p, then the bonds that replace the cells would be conducting with probability $p' = R(p)$, where $R(p)$ is called the *renormalization group transformation*, and represents the *sum* of the probabilities of *all* the conducting configurations of the RG cell. To compute $R(p)$, we proceed as follows. Since we are interested in percolation and conduction in the network, which is supposed to be represented by the RG cell, we solve for the percolation and conduction problems in each cell by applying a fixed potential gradient across the cell in a given direction. For example, as far as percolation and conduction are concerned, the 2×2 RG cell of Figure 5.14 is equivalent to the circuit shown there, which is usually called the Wheatstone bridge. Thus, for the $b = 2$ cell, we only need to deal with a circuit of 5 bonds, and for the $2 \times 2 \times 2$ cell we construct an equivalent 12-bonds circuit (see Figure 5.14). To compute $R(p)$, we determine all the conducting configurations of such circuits, with some bonds conducting and some insulating. Thus, for the 2×2 RG cell we obtain

$$p' = R(p) = p^5 + 5p^4q + 8p^3q^2 + 2p^2q^3, \tag{181}$$

where $q = 1 - p$. It is easy to see how Eq. (181) is derived: There is only one conducting RG cell configuration with all the five bonds conducting (probability p^5), 5 conducting configurations with 4 bonds conducting and one bond insulating (probability $5p^4q$), and so on.

As discussed in Chapter 2, the sample-spanning cluster at p_c is self-similar, implying that the RG transformation should remain invariant at p_c. The invariance of $R(p)$ should also be true at $p = 1$ and $p = 0$, because under any reasonable RG transformation full and empty networks should be transformed to full and empty networks again. The points $p = 0$, 1, and p_c are thus called the *fixed points* of the transformation, and are denoted by p^*. Since the RG transformation should not change anything at these points, the implication is that at the fixed points the probability (p) of having a conducting bond in the RG cell and that of a bond in the renormalized cell $[p' = R(p)]$ should be the same. Thus, the fixed points should be the solution of the polynomial equation

$$p^* = R(p^*), \tag{182}$$

and indeed this equation usually has three roots which are $p^* = 0$, $p^* = 1$, and $p_c = p^*$, where p^* is the RG transformation prediction for p_c. For the RG cells of Figure 5.14, we obtain $p^* = 1/2$ for both 2×2 and 3×3 cells, the exact result. In fact, it can be shown (Bernasconi, 1978) that the RG transformation for such 2D self-dual cells for *any* b always predicts $p_c = 1/2$. With the $2 \times 2 \times 2$ RG cell we obtain $p^* \simeq 0.208$, which should be compared with the numerical estimate $p_c \simeq 0.249$.

So far, everything appears nice and simple. In fact, PSRG methods seem so simple that small-cell calculations can be done on the "back of the envelope," with the results being reasonably accurate. The reader may guess that one should calculate $p_c = p^*$ for several cell sizes b and then use finite-size scaling (see Section 2.9) according to which

$$p_c - p_c(b) \sim b^{-1/\nu}, \tag{183}$$

where p_c is the true percolation threshold of the network, and ν is the critical exponent of percolation correlation length ξ_p, to extrapolate the results to $b \to \infty$ and estimate p_c. Indeed, Reynolds *et al.* (1980) used this method and obtained very accurate estimates of site and bond percolation thresholds of the square network. However, as always, life is more complex than we like it to be! Ziff (1992) showed that as $b \to \infty$, the probability $R(p_c)$ approaches a *universal* value of 1/2 for *all* 2D networks. This implies that, *asymptotically*, Eq. (182) is *wrong*, because p_c is *not* universal and depends on the structure of the network, whereas $R(p_c)$ *is* universal (as $b \to \infty$). Ziff's discovery does not violate the universality of $R(p_c)$; it only fixes its value. Stauffer *et al.* (1994) showed that $R(p_c)$ is also universal in 3D.

How can one use an RG transformation for estimating p_c? Although, due to Ziff's discovery, the answer to this question during 1992-1994 was not clear, with most of the predictions that had been obtained [based on Eq. (182)] in the 1970s and early 1980s in doubt, Sahimi and Rassamdana (1995) showed that, although $R(p_c)$ is universal, for *any* $0 < \alpha < 1$ the following equation

$$R(p_c) = \alpha, \tag{184}$$

always provides an estimate of p_c in (0,1). Therefore, one can determine $R(p_c)$ for a series of RG cells of increasing linear sizes b, solve Eq. (184) for the corresponding values of p_c, and use Eq. (183) to extrapolate the results to $b \to \infty$. In other words, although Eq. (182) is wrong *theoretically*, it can still provide estimates of p_c! Moreover, it appears that, among all the possible values, $\alpha = p_c$ still provides the fastest convergence to the true value of p_c. In addition, one has a bonus in that, the critical exponent ν of the percolation correlation length can be estimated from $R(p_c)$. Suppose that the percolation correlation length in the original and renormalized lattices (RG cells) are ξ_p and ξ'_p, respectively. Since each bond of the RG cells is replaced by another bond b times its length, we must have,

$$\xi'_p = \frac{1}{b}\xi_p. \tag{185}$$

On the other hand, $\xi_p \sim (p - p_c)^{-\nu}$ and, because of the universality of the percolation exponents, $\xi'_p \sim [R(p) - R(p_c)]^{-\nu}$. If we linearize $R(p)$ and $R(p_c)$ around $p_c = p^*$, and use Eq. (185), we obtain

$$\nu = \frac{\ln b}{\ln \lambda}, \tag{186}$$

where $\lambda = dR(p)/dp$, evaluated at $p = p^*$. For example, the 2×2 and 3×3 self-dual RG cells yield, $\nu \simeq 1.43$ and 1.38, respectively, which should be compared with the exact value $\nu = 4/3$ (see Table 2.3). For the $2 \times 2 \times 2$ cell of Figure 5.14, one obtains $\nu \simeq 1.03$, which should be compared with the numerical estimate, $\nu \simeq 0.88$, listed in Table 2.3. In fact, PSRG methods can also predict the other percolation critical exponents, such as (see Section 2.7.1) β and β_B.

We now discuss the PSRG approach for calculating the conductivity of a random resistor network. In this approach, one begins with the original probability density function $f_0(g)$ for the bond conductances g of the RG cell and replaces it with a new distribution $f_1(g)$, the probability distribution for the conductance of a bond in the renormalized cell, which is calculated by determining the equivalent conductance of the RG cell. Thus, for an n-bond RG cell, one obtains a recursion relation relating $f_1(g)$ to $f_0(g)$:

$$f_1(g) = \int f_0(g_1)dg_1 f_0(g_2)dg_2 \cdots f_0(g_n)dg_n \delta(g_p - g'), \tag{187}$$

where $g_1, \ldots, g_n$ are the conductances of the n bonds of the RG cell, and g' is the equivalent conductance of the RG cell. For example, for the 5-bond cell of Figure 5.14 one has

$$g' = \frac{g_1(g_2g_3 + g_2g_4 + g_3g_4) + g_5(g_1 + g_2)(g_3 + g_4)}{(g_1 + g_4)(g_2 + g_3) + g_5(g_1 + g_2 + g_3 + g_4)}. \tag{188}$$

and if, $f_0(g) = (1 - p)\delta(g) + p\delta(g - g_0)$, then [using Eq. (187)]

$$f_1(g) = [1 - R(p)]\delta(g) + 2p^3q^2\delta\left(g - \frac{1}{3}g_0\right) + 2p^2(1 + 2p)q^2\delta\left(g - \frac{1}{2}g_0\right)$$
$$+4p^3q\delta\left(g - \frac{3}{5}g_0\right) + p^4\delta(g - g_0), \tag{189}$$

which is already more complex than $f_0(g)$. One now iterates Eq. (187) to obtain a new distribution $f_2(g)$ by substituting $f_1(g)$ into its right-hand side. The iteration process should continue until a distribution $f_\infty(g)$ is reached, the shape of which does not change under further iterations. This is called the *fixed-point distribution* and the conductance of the original network is simply an average of $f_\infty(g)$. However, it is difficult to analytically iterate Eq. (187) many times. The common practice is to replace the distribution after the ith iteration by an *optimized* distribution $f_i^o(g)$ which closely mimics the properties of $f_i(g)$. The optimized $f_i^o(g)$ is usually taken to have the following form

$$f_i^o(g) = [1 - R(p)]\delta(g) + R(p)\delta[g - g^o(p)], \tag{190}$$

where $g^o(p)$ is an *optimized* conductance. In the past, various schemes have been proposed for determining $g^o(p)$, one of the most accurate of which was proposed by Bernasconi (1978), according to which, if after i iterations of Eq. (187), $f_i(g)$ is given by

$$f_i(g) = [1 - R(p)]\delta(g) + \sum_i a_i(p)\delta(g - g_i), \tag{191}$$

then, $g^o(p)$ is approximated by

$$g^o(p) \simeq \exp\left[\frac{1}{R(p)}\sum_i a_i(p)\ln g_i\right]. \tag{192}$$

Once $g^o(p)$ is calculated, Eq. (187) is iterated again, the new distribution $f_{i+1}(g)$ and its optimized form $f_{i+1}^o(g)$ are determined, and so on. In practice, after a few iterations, even an initially broad $f_0(g)$ converges quickly to a stable and narrow distribution with a shape that does not change under further rescaling. The conductivity of the resistor network is simply a suitably-defined average of this distribution.

Renormalization methods are usually very accurate for 2D resistor networks, and are flexible enough to be used for anisotropic systems. However, they have two drawbacks for 3D systems. The first is that the predictions of the PSRG methods for percolation networks with *any* type of $b = 2$ RG cell are not accurate. Moreover, even after the first iteration of Eq. (187) the renormalized conductance distribution $f_1(g)$ is very complex; if we begin with a binary distribution, $f_1(g)$ will have *seventy-three* components of the form $\delta(g - g_i)$, with $i = 1, \cdots, 73$. Hence, analytical calculation of $f_2(g)$ is very difficult. The second drawback is that, even for a $b = 3$ RG cell, analytical determination of the RG transformation becomes very difficult, because the total number of possible configurations of the RG cell is of the order 10^{11}. Thus, one must resort to a Monte Carlo renormalization group method (Reynolds *et al.*, 1980), which is, however, not any simpler than the simple Monte Carlo method itself.

Young and Stinchcombe (1975) were the first to use the PSRG methods for computing critical exponents of 2D percolation networks. Stinchcombe and Watson (1976) were the first to use such methods to compute the conductivity exponent

μ. Their work was followed by several others who proposed various variants of the PSRG methods for calculating both percolation and conduction properties of random resistor networks and other disordered media (see for example, Payandeh, 1980; Sahimi *et al.*, 1984; Sahimi, 1988). Stanley *et al.* (1982) and Family (1984) reviewed most of the literature on this subject.

5.12 Renormalized Effective-Medium Approximation

To circumvent the difficulties that the PSRG methods encounter for 3D resistor networks, Sahimi *et al.* (1983c, 1984) proposed a new method that combined the EMA and PSRG methods, and is called the renormalized EMA (REMA). Their method took advantage of two facts.

(1) Each time a resistor network is renormalized, it becomes less critical in the sense that, its associated percolation correlation length ξ_p' is smaller than the original correlation length ξ_p by a factor of b (the linear size of the RG cell); see Eq. (185).
(2) As discussed and demonstrated above, the EMA is usually accurate away from the percolation threshold, if the morphology of the material does not contain any correlations.

Therefore, if one employs the EMA with the first iteration $f_1(g)$ of the original bond conductance distribution $f_0(g)$, instead of $f_0(g)$ itself, the performance of the EMA must improve. That is, in the REMA the conductance distribution that one uses in Eq. (57) is $f_1(g)$, instead of $f_0(g)$. Because the bonds of the renormalized resistor network are b times longer than the original ones, this necessitates a rescaling of conductivities of the renormalized network to replicate the old one: The REMA conductivity is taken to be the same as that for the original resistor network at $p = 1$.

The REMA is sharply more accurate than *both* the EMA and PSRG. For bond percolation, the REMA predicts the percolation threshold p_c to be the root of the following equation

$$R(p_c) = \frac{2}{Z}, \quad \text{bond percolation,} \tag{193}$$

which should be compared with Eq. (61), the prediction of the EMA. For example, with the 3D RG cell of Figure 5.14, the RG transformation is given by

$$R(p) = p^{12} + 12p^{11}q + 66p^{10}q^2 + 220p^9q^3 + 493p^8q^4 + 776p^7q^5$$
$$+856p^6q^6 + 616p^5q^7 + 238p^4q^8 + 48p^3q^9 + 4p^2q^{10}, \tag{194}$$

so that, using Eq. (194) in (193) with $Z = 6$, one obtains $p_c \simeq 0.267$, only 7% larger than the numerical estimate (see Table 2.2), $p_c \simeq 0.249$ for the simple-cubic network. Moreover, so long as one uses the type of 2D self-dual RG cells that are shown in Figure 5.14, the REMA predicts the exact bond-percolation threshold of the square lattice for *any* cell size b. As for site percolation, it is not difficult to show, using Eq. (107), that the REMA predicts that the site-percolation threshold of a lattice is the root of the following equation,

$$R(p_c) = \Gamma^{-1/2} \quad \text{site percolation,} \tag{195}$$

which should be compared with Eq. (113), the EMA prediction for the site-percolation threshold. Here, Γ is defined by Eq. (108), and $R(p)$ is the RG transformation for site percolation. For example, if we use a $2 \times 2 \times 2$ RG cell for site percolation in the simple-cubic lattice, we obtain

$$R(p) = p^8 + 8p^7q + 28p^6q^2 + 56p^5q^3 + 54p^4q^4 + 24p^3q^5 + 4p^2q^6, \tag{196}$$

which should be compared with Eq. (194), the RG transformation for a $2 \times 2 \times 2$ cell for bond percolation. Using Eq. (196) in (195) we obtain, $p_c \simeq 0.367$ for site-percolation threshold of a simple-cubic lattice, which is 17% larger than the numerical estimate $p_c \simeq 0.3116$ (see Table 2.2), and represents a sharp improvement over $p_c \simeq 0.46$ that is predicted by the EMA (see Section 5.5). Finally, the REMA prediction for the correlated bond-percolation threshold of a simple-cubic lattice is $p_c \simeq 0.105$, in excellent agreement with the numerical estimate, $p_c \simeq 0.103$ (see Section 5.6). Therefore, the REMA predictions for the percolation thresholds of various percolation models in both 2D and 3D lattices are accurate and close to the best numerical estimates currently available.

Using the REMA, Sahimi *et al.* (1983c, 1984) and Sahimi (1988) also obtained very accurate predictions of the effective conductivity of various 2D and 3D networks. Figure 5.15 compares the predictions of the REMA for the effective conductivity of a square network in bond percolation with those of the EMA

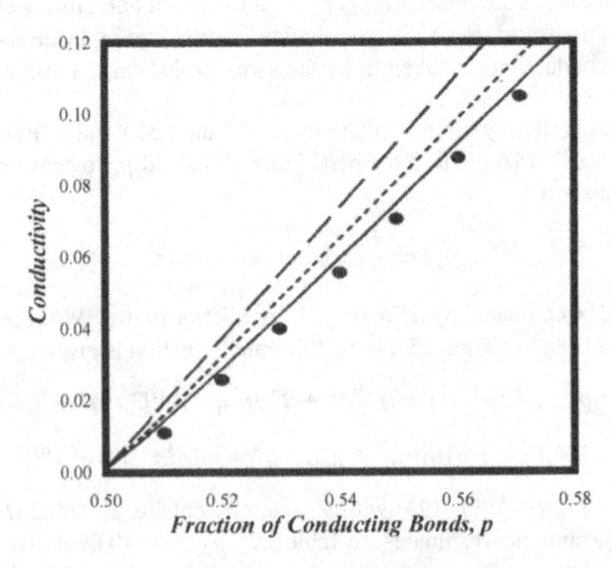

FIGURE 5.15. Comparison of the Monte Carlo data for the effective conductivity of a square network (symbols) with the predictions of the EMA (the left curve), and REMA with $b = 2$ (the middle curve) and $b = 3$ (the right solid curve) (after Sahimi *et al.*, 1983c).

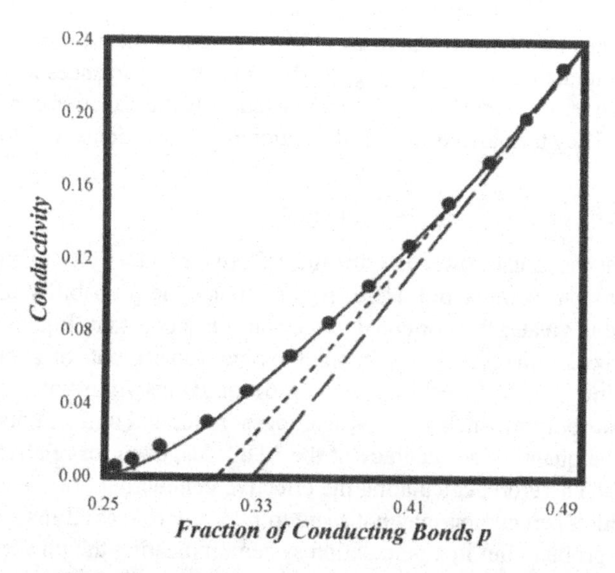

FIGURE 5.16. Comparison of the Monte Carlo data for the effective conductivity of a simple-cubic network (symbols) with the predictions of the EMA (right curve) and REMA with $b = 2$. The middle curve represents the predictions of the cumulant theory (after Sahimi *et al.*, 1983c).

and the simulation results, while Figure 5.16 does the same for the simple-cubic network. In Figure 5.10, the REMA predictions for the effective conductivity of a simple-cubic network in site percolation are compared with those of the EMA, a PSRG method, and the results of numerical simulations, while Figure 5.11 compares the REMA predictions for the effective conductivity of the same network in correlated bond percolation with those of the EMA and the modified EMA proposed by Bernasconi and Wiesmann (1976), described and discussed in Section 5.6. It is clear that in all cases the REMA provides very accurate predictions for the effective conductivities. Finally, we should mention that Zhang and Seaton (1992) extended the above REMA to an arbitrary conductance distribution, and showed the extension to be highly accurate.

5.13 The Critical Path Method

A powerful technique for analyzing conduction in heterogeneous materials is the *critical path method.* This idea was first proposed by Ambegaokar, Halperin and Langer (AHL) in 1971 who argued that conduction in a disordered solid with a *broad* distribution of conductances is dominated by those conductances with magnitudes larger than a characteristic conductance g_c, which is the smallest conductance such that the set $\{g|g > g_c\}$ forms a conducting sample-spanning cluster, which is called the critical path. Therefore, transport in a disordered medium with a broad conductance distribution can be reduced to a percolation-conduction problem with threshold g_c. This idea was later proven rigorously by Tyc and Halperin

(1989). Shante (1977) and Kirkpatrick (1979b) modified this idea by assigning g_c to all local conductances g with $g \geq g_c$, and setting all conductances with $g < g_c$ to be zero (since the contribution of such conductances to the effective conductivity is very small). They then arrived at a trial solution for the conductance of the sample given by

$$g_e = a g_c [p(g_c) - p_c]^{\mu}, \tag{197}$$

which is just the usual power law for the effective conductivity of the network near the percolation threshold. Here, $p(g_c)$ denotes the probability that a given conductance is greater than or equal to g_c, and a is a constant. Equation (197) is then maximized with respect to g_c in order to obtain an estimate of g_c and thus g_e. Computer simulations of Berman *et al.* (1986) for 2D networks with various conductance distributions, such as Gaussian, log-normal, and uniform distributions, confirmed the quantitative accuracy of the AHL idea, even for relatively narrow distributions. Therefore, calculating the effective conductivity of disordered materials in which percolation may not seem to play any role at all may be reduced to the same problem but in a percolation system, indicating the power and broad applicability of percolation theory. In Section 6.2 we will discuss in detail the application of the AHL idea for determining hopping conductivity of amorphous semiconductors.

Katz and Thompson (1986, 1987) extended the AHL method to estimating the permeability and electrical conductivity of porous materials. In such materials the local hydraulic conductance is a function of a length scale l. Therefore, the critical conductance g_c defines a corresponding characteristic length l_c. Since both flow and electrical conduction problems belong to the same class of percolation problems, the length that signals the percolation threshold in the flow problem also defines the threshold in the electrical conductivity problem. Thus, we rewrite Eq. (197) as

$$g_e = \phi g_c(l)[p(l) - p_c]^{\mu}, \tag{198}$$

where the porosity ϕ (volume fraction of the pores) ensures a proper normalization of the fluid or the electric-charge density. The function $g_c(l)$ is equal to $c_f l^3$ for the flow problem and $c_c l$ for the conduction problem. For appropriate choices of the function $p(l)$, the conductance $g_e(l)$ achieves a maximum for some $l_{max} \leq l_c$. In general, l_{max}^f for the flow problem is different from l_{max}^c for the conduction problem, because the transport paths have different weights for the two problems.

If $p(l)$ allows for a maximum in the conductance which occurs for $l_{max} \leq l_c$, then we can write

$$l_{max}^f = l_c - \Delta l_f = l_c \left[1 - \frac{\mu}{1 + \mu + l_c \mu p''(l_c)/p'(l_c)}\right], \tag{199}$$

$$l_{max}^c = l_c - \Delta l_c = l_c \left[1 - \frac{\mu}{3 + \mu + l_c \mu p''(l_c)/p'(l_c)}\right]. \tag{200}$$

If the pore size distribution (and hence the pore conductance distribution) of the material is very broad, then, $l_c \mu p''(l_c)/p'(l_c) \ll 1$, and Eqs. (199) and (200) reduce to

$$l_{max}^f = l_c\left(1 - \frac{\mu}{1+\mu}\right) \simeq \frac{1}{3}l_c, \tag{201}$$

$$l_{max}^c = l_c\left(1 - \frac{\mu}{3+\mu}\right) \simeq \frac{3}{5}l_c, \tag{202}$$

if we use $\mu \simeq 2$ for 3D percolation systems (see Table 2.3). If, for example, the morphology of the material is such that μ is non-universal (see the discussion in Section 2.11), then the non-universal value of μ, instead of $\mu \simeq 2$, must be used in Eqs. (201) and (202). Using these results, we can establish a relation between g_e and k_e, the effective permeability of the porous material. Writing

$$g_e = a_1\phi[p(l_{max}^c) - p_c]^\mu, \tag{203}$$

and

$$k_e = a_2\phi(l_{max}^f)^2[p(l_{max}^f) - p_c]^\mu, \tag{204}$$

we obtain to first order in Δl_c or in Δl_f

$$p(l_{max}^{f,c}) - p_c = -\Delta l_{f,c}\, p'(l_c). \tag{205}$$

To obtain the constants a_1 and a_2, Katz and Thompson (1986) assumed that at the local level the conductivity of the porous medium is g_f, the conductivity of the fluid (usually brine) that saturates the pore space, and that the local pore geometry is cylindrical. These assumptions imply that $a_1 = g_f$ and $a_2 = 1/32$. Therefore, one obtains

$$k_e = a_3 l_c^2 \frac{g_e}{g_f}, \tag{206}$$

where $a_3 = 1/226$. A similar argument leads to (Katz and Thompson, 1987)

$$\frac{g_e}{g_f} = \frac{l_{max}^c}{l_c}\phi S(l_{max}^c), \tag{207}$$

where $S(l_{max}^c)$ is the volume fraction (saturation) of the connected pore space involving pore widths of size l_{max}^c and larger.

Equations (206) and (207) involve no adjustable parameters. Every parameter is fixed and precisely defined. To obtain the characteristic length l_c, Katz and Thompson (1986, 1987) proposed the use of mercury porosimetry, a method that is used for characterizing the morphology of many porous materials (see, for example, Sahimi, 1993b, 1995b). Porosimetry is a percolation process in which the volume of the mercury, which is injected into the pore space of the material, is measured as a function of the pressure that is applied in order to push the mercury into the pore space. In the initial stages of this process, before a sample-spanning cluster of the pores that have been filled with the mercury, the curvature of the pressure versus volume curve is positive. There is also an inflection point in the curve beyond which the pore volume increases rapidly with the pressure. This inflection point signals the formation of the sample-spanning cluster. According to the Washburn equation (see Sahimi, 1993b, 1995b) we must have $l \geq -4\sigma \cos\theta / P_i$, where P_i is the pressure at the inflection point, σ is the surface tension between the mercury and the vacuum, and θ is the contact angle between the mercury and the pore surface. Hence

$$l_c = -\frac{4\sigma \cos\theta}{P_i} \tag{208}$$

defines the characteristic length l_c. Thus, a simple mercury porosimetry experiment yields the characteristic length l_c.

Katz and Thompson (1986, 1987) showed that, without any adjustable parameter, Eqs. (206) and (207) provide accurate predictions for a wide variety of porous materials. The critical path method has also been extended to estimating the *nonlinear* conductivity of disordered porous materials (Sahimi, 1993a).

5.14 Numerical Computation of the Effective Conductivity

Having described the analytical approaches for estimating the effective conductivity g_e of resistor networks, we now discuss the numerical methods. There are several efficient numerical methods for computing g_e, and what follows is a description of each.

5.14.1 The Conjugate-Gradient Method

One of the main numerical methods for estimating the effective conductivity of a disordered material is direct computer simulations using a resistor network model of the material. The current I_{ij} in a bond ij of the network is given by, $I_{ij} = g_{ij}(V_i - V_j)$, where g_{ij} is the bond's conductance, and V_i is the voltage at site i. The net current reaching any site of the network must be zero, and therefore

$$\sum_j I_{ij} = \sum_j g_{ij}(V_i - V_j) = 0, \tag{209}$$

where the sum is over all the sites j that are connected to i by a bond. Writing down Eq. (209) for every interior node of the network results in a set of simultaneous equations for nodal voltages, from the solution of which the voltage distribution in the network and thus its effective conductivity are calculated. The boundary conditions are usually a fixed current injected into the network at one of its sites, and extracting it at another site, or a fixed voltage difference across the network in one direction and periodic boundary conditions in the other directions. The conductance of the bonds can be selected from an arbitrary probability density function which represents the conductance distribution of the disordered material.

Equations (209) can be solved by several iterative or exact methods. Gaussian elimination has the advantage that it yields the exact numerical solution of the voltage distribution throughout the network. It requires, however, large computer memory. Moreover, its required computer time scales as $O(n^2)$, where n is the number of equations to be solved. For these reasons, Gaussian elimination is not used when n is very large. One may also solve Eqs. (209) by an over-relaxation method which has the advantages of being simple and requiring minimal computer memory. However, it suffers from critical slowing down as the percolation threshold is approached, i.e., it requires an increasingly larger number of iterations to converge to a solution with the required accuracy, and also becomes increasingly less accurate as p_c is approached. Currently, the most accurate and efficient numerical method for solving Eqs. (209) is by the conjugate-gradient method which is *guaranteed* to converge to the true solution in *at most n* iterations. A detailed description of this method and its virtues is given in Chapter 9 of Volume II, and therefore, we do not discuss it here.

If the conductance distribution is very broad and contains correlations at many length scales, then one must use a very large network with millions of sites in order to accurately model the material, and thus solving for the voltage distribution becomes a very time-consuming problem. For such cases, an efficient method due to Mehrabi and Sahimi (1997) can be used for calculating the effective conductivity of the network. In this method, instead of solving for the voltage distribution in the original large network, one first identifies the high conductance regions using a wavelet transformation and coarsens the lattice. That is, since the low conductance regions do not contribute significantly to the overall conductivity of the lattice, they do not need to be represented in the lattice by a fine lattice structure. In addition, those regions of the lattice in which the variation of the conductance is not very severe do need to be represented by a very fine grid structure. Thus, one can eliminate the lattice sites in such regions and reduce the number of sites at which the voltage is computed. In this way, the number of the equations that must be solved is reduced drastically (by 2-3 orders of magnitude), reducing the size of the computational problem to a manageable level. Then, the conjugate-gradient method is applied to the coarsened lattice. Mehrabi and Sahimi (1997) and Ebrahimi and Sahimi (2002) showed that this method reduces the computation times by orders of magnitude.

Let us point out that, although the first application of the resistor network models to estimating the effective transport properties of disordered materials and media is generally attributed to Kirkpatrick (1971), many years before him, Fatt (1956) had used 2D network models for estimating the effective permeability of disordered porous materials, a problem that is completely isomorphic to the resistor network problem and computation of the effective conductivity. Several papers published in 1960s also reported similar computations (see, for example, Nicholson, 1968; Rink and Schopper, 1968; Greenberg and Brace, 1969).

5.14.2 Transfer-Matrix Method

In this method, which was first proposed by Derrida and Vannimenus (1982), the effective conductivity of the network is calculated *exactly as the network is constructed.* We follow Derrida *et al.* (1984) and describe the method for a 2D

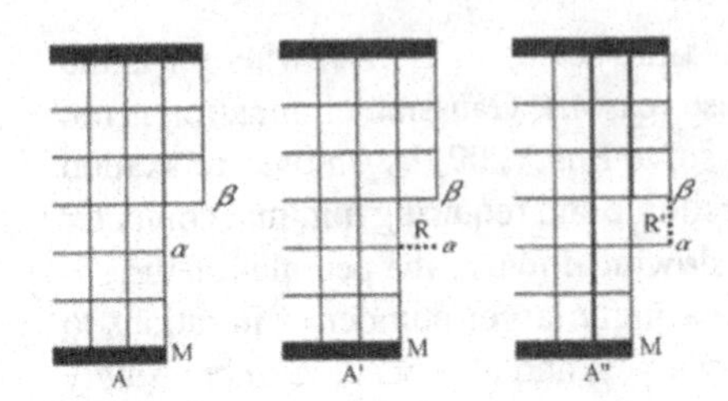

FIGURE 5.17. Construction of the random network in the transfer-matrix method (after Derrida *et al.*, 1984).

network; its generalization to 3D will then be obvious. For the sake of efficiency, the computations are usually done for a strip, although the network can have any shape or structure.

Figure 5.17 presents the strip which is built by adding one bond at a time to it, starting from the line on the left. At each stage of its construction, the left part of the strip is described by a $M \times M$ matrix **A**, where M is the number of sites in a section (column), including the sites of the first (top) and the last (bottom) rows. To construct **A** we proceed as follows. If we wish to estimate the conductivity of the left part of the strip, we attach to sites $1, 2, \cdots, M$ a wire which imposes a potential V_α on site α, as a result of which a current I_β flows in the wire attached to site β. Since the problem is linear, we must have, $I_\beta = \sum_\alpha A_{\alpha\beta} V_\alpha$. On the other hand, the currents depend only on the *difference* between the potentials, and thus we can always set $V_M = 0$. The only thing that we now need to specify is how **A** is transformed as we add more bonds to the network (strip).

There are two types of bond that one adds to a 2D strip. If we add a horizontal bond with resistance $R = 1/g$ (where g is selected according to its statistical distribution) to site α (see Figure 5.17), then **A** is transformed to **A**$'$ with entries A'_{ij} that are related to those of **A** by

$$A'_{ij} = A_{ij} - \frac{A_{i\alpha} A_{j\alpha} R}{1 + A_{\alpha\alpha} R}, \tag{210}$$

which can be proven easily by using the usual laws of connecting resistors together. Similarly, if we add a vertical bond with resistance R' between α and β, then

$$A''_{ij} = A'_{ij} + \frac{(\delta_{\alpha j} - \delta_{\beta j})(\delta_{\alpha i} - \delta_{\beta i})}{R'}, \tag{211}$$

where δ_{ij} is the usual Kronecker's delta. Note that **A** is symmetric, making the computations more efficient. The conductivity of a strip of length $L \gg 1$ is then given by

$$\sigma = \lim_{L \to \infty} \frac{A(1, 1)}{L}. \tag{212}$$

Thus, calculation of g_e becomes extremely simple. We do not store **A** for the entire $M \times L$ system, but only for the last two sections of it. For example, as soon as the construction of the strip reaches, e.g., the fourth section ($M = 4$), the first two sections are forgotten, and only the last two are kept. Moreover, the size of **A** is *independent* of L, so that one can use an extremely large L without worrying about the required computer memory. Derrida *et al.* (1984) present a computer program for transfer matrix calculation of the effective conductivity of a $M \times M \times L$ simple-cubic network. The calculations are similar to the 2D case, except that M should be replaced by $M^2 + 2$, the number of sites in a section (plane). Equation (210) is then used for all the bonds in the x-direction (the direction of macroscopic potential gradient), while Eq. (211) is used for those in the y- and z-directions.

We should point out that the transfer-matrix approach is even more efficient when it is used for computing the effective conductivity of a mixture of normal conductors and superconductors. This fact has been demonstrated by Herrmann *et al.* (1984), Normand *et al.* (1988), and Normand and Herrmann (1990). The latter two groups actually designed a special-purpose computer that computed the conductivity exponents μ and s based on the transfer-matrix method.

In its original form developed by Derrida and Vannimenus (1982), the transfer matrix method could be used to calculate only g_e, providing no information on the voltage distribution throughout the network, as **A** stores information only about the last two sections of the network. In its modified form developed by Duering and Bergman (1990), such distributions can also be calculated.

5.14.3 Network Reduction: The Lobb–Frank–Fogelholm Methods

Lobb and Frank (1984) developed an efficient method for exact computation of the effective conductivity of 2D lattices, which consisted of applying a sequence of transformations to the bonds of the lattice. The final outcome of this sequence is reduction of the lattice to a single bond that has the same conductance as the entire lattice. The details of the method are described by Frank and Lobb (1988) whose paper we follow here.

A well-known transformation in the theory of electrical circuits is the $Y - \nabla$ transformation, which is defined in both directions and is given by

$$Y \to \nabla : \begin{cases} G_A = g_2 g_3 \Delta_1 \\ G_B = g_1 g_3 \Delta_1 \\ G_C = g_1 g_2 \Delta_1, \end{cases} \tag{213}$$

$$\nabla \to Y : \begin{cases} g_1 = G_B G_C \Delta_2 \\ g_2 = G_A G_C \Delta_2 \\ g_3 = G_A G_B \Delta_2, \end{cases} \tag{214}$$

where, $\Delta_1 = (g_1 + g_2 + g_3)^{-1}$ and $\Delta_2 = G_A^{-1} + G_B^{-1} + G_C^{-1}$. We should keep in mind that, similar to the case of series and parallel arrangements of resistors, the conductor configurations are completely equivalent, as far as the external circuit is concerned.

Figure 5.18 shows the primary transformation used in the Lobb–Frank method, which can be decomposed into three separate parts: One first applies a $\nabla \to Y$ transformation, then redefines the lattice point, and finally applies a $Y \to \nabla$ transformation again. In going from Figure 5.18(a) to 5.18(b), conductances 6,7, and 8 are determined from 1, 2, and 3 by Eq. (214). In the transformation from Figure 5.18(c) to 5.18(d), conductances 9, 10, and 11 are calculated by Eq. (213).

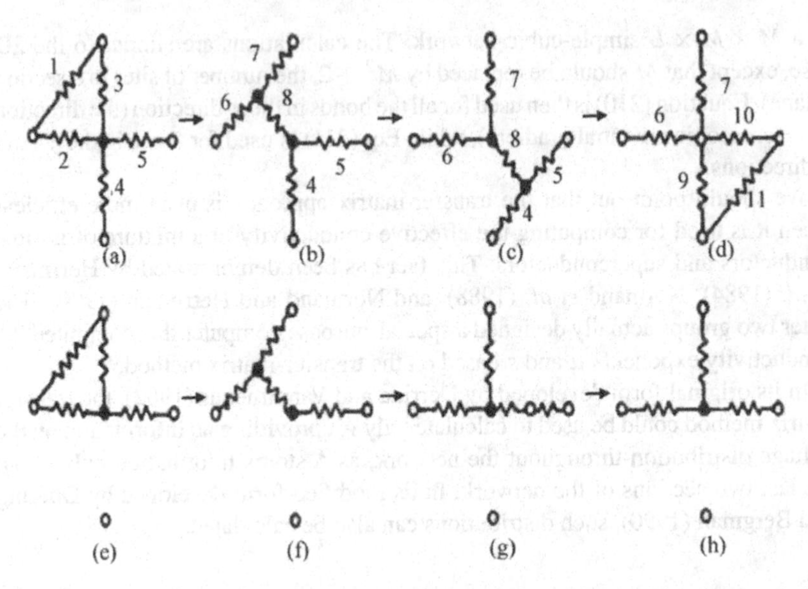

FIGURE 5.18. The propagator transformation which moves a diagonal bond through the lattice. The general case is illustrated in (a)–(d), while (e)–(h) show the case in which some bonds are insulating and hence missing (after Frank and Lobb, 1988).

If some of the bonds are not conducting, then the propagator usually becomes simpler in the sense of not resulting in a new diagonal conductor. An example of this is shown in Figures 5.18(e)–(h), where one bond is insulating. In this case, the first two parts of the propagator are the same as before, but the final $Y \to \nabla$ transformation is reduced to a simple combination of two conductors in series, with no diagonal bond.

To illustrate the method, consider a small cell on the square lattice shown in Figure 5.18. Starting with the full lattice in Figure 5.18(a), the two conductors attached to the top node of the leftmost column are combined in series to form a diagonal bond, which is then propagated through the lattice, as shown in Figure 5.18(b), until there is no diagonal bond left which, in the present example, terminates at an edge. If there are some insulating bonds, the propagator may terminate somewhere inside the lattice. The same procedure is now followed in Figures 5.18(c)–(g) for eliminating the uppermost site of the leftmost column by a series of reduction. The resulting conductor is then propagated until it terminates, at which point one obtains the configuration in Figure 5.18(h), which is easily reduced to a single conductor with a conductance equivalent to the conductance of the original lattice. Note that this algorithm is not restricted to 2D lattices, but can be used with any 2D system in which the $Y \to \nabla$ and $\nabla \to Y$ transformations are well-defined.

The algorithm proposed by Fogelholm (1980) is very similar to that of Lobb and Frank (1984), except that it is also applicable to 3D systems. It is based on the fact that an internal node in a network can be eliminated if new impedances are inserted between certain of the remaining nodes. Specifically, if node s_0 is directly connected to the nodes $s_1, s_2, \cdots, s_n$ through conductances $g_1, g_2, \cdots, g_n$, its *valence* is n, and may be eliminated by

(1) deleting s_0 and $g_1, g_2 \cdots, g_n$, and
(2) inserting the conductance $g_{ij} = g_i g_j/(g_1 + g_2 + \cdots + g_n)$ between each pair (s_i, s_j) of neighboring sites.

Thus, the conductance of the sample can be computed by successively eliminating all sites except the terminal nodes on two opposite faces of the network. The result is a single bond between the two terminal nodes, if the sample is conducting, and no bond at all if the material is insulating (i.e., if the conducting bonds do not form a sample-spanning cluster). The algorithm is highly efficient, it yields the exact conductance of a given sample, and in fact its efficiency increases as the percolation threshold p_c is approached. Fogelholm (1980) utilized this method for 2D lattices, while Gingold and Lobb (1990) used it in their 3D simulations.

5.14.4 Random Walk Method

As discussed in Section 4.10, from the time dependence of the mean square displacement of a random walker the effective diffusivity, and hence the effective conductivity g_e, can be calculated. The first application of this idea for determining the effective conductivity of a heterogeneous material appeared in a paper of Haji–Sheikh and Sparrow (1967), who studied heat conduction in a composite solid. Since then, the random walk method has been extensively used for studying transport, and in particular conduction, in disordered materials. In the context of percolation and resistor network models of disordered materials, Brandt (1975) appears to be the first to have used this method, although the method was popularized by de Gennes (1976b) who made an analogy between the motion of a random walker in a disordered medium and that of an ant in a labyrinth. Mitescu and Rousseng (1976) followed de Gennes' idea and performed relatively extensive simulations on percolation clusters. The developments in this area have been reviewed extensively by Havlin and Ben–Avraham (1987), Haus and Kehr (1987) and Hughes (1995,1996).

Bunde *et al.* (1985) considered a general two-component material with conductivities g_A and g_B, and formulated a random-walk model for calculating the effective conductivity of such a mixture. In this model, one first sets up a time scale. We know that, $g_A \sim D_A \sim \ell^2/t_A$, where D_A is the diffusivity in the region A, ℓ is the mean-free path (or the distance travelled by a random walker), and τ_A the characteristic time for travelling a distance ℓ. Therefore,

$$\frac{\tau_A}{\tau_B} = \frac{g_A}{g_B}. \tag{215}$$

Next, one must specify the transition probability W_i, the probability that the random walker takes a step to its ith neighbor. This probability is given by

$$W_i = \frac{p_i}{\sum_i p_i}, \tag{216}$$

where $p_i = g_A$ or g_B. More generally, if each bond ij of a lattice, that is connected to one of its sites i, has a conductance g_{ij}, then the transition probability W_{ij} for a step from i to j is given by

$$W_{ij} = \frac{g_{ij}}{\sum_j g_{ij}}. \tag{217}$$

Therefore, one may use a random walk method to simulate diffusion in a composite solid with a continuous (or discrete) distribution of local conductances. The effective conductivity of the system will then be exactly equal to its effective diffusivity. In practice, one must use certain tricks in order to accelerate the simulations. For example, if $g_A \ll g_B$, then the random walker may spend a long time in the A-region before making a transition to the B-region, and vice versa. Using vectorization (for use in a vector or supercomputer), one can develop a highly efficient computer algorithm for simulating random walks in a disordered medium (Sahimi and Stauffer, 1991). For the most recent application of this method to composite materials see Van Siclen (2002).

5.15 Estimation of the Critical Exponents of the Conductivity

As discussed in the Introduction, and also in Chapter 2, an important property of the effective conductivity of composite solids with percolation disorder, which are a mixture of conducting and insulating materials, is the universal power law that it follows near the percolation threshold. The power law is characterized by the critical exponent μ [see Eq. (2.36)]. If the composite consists of normal and superconducting materials, then, its effective conductivity, which diverges at the percolation threshold where the superconducting material forms a sample-spanning cluster, follows a power law near p_c and is characterized by a critical exponent s. Aside from certain continuum models that were described in Section 2.10, the critical exponents μ and s are universal. Therefore, accurate estimates of these exponents for both 2D and 3D systems have always been of great interest. In the 1970s these exponents were estimated by computing the effective conductivity for a range of $p - p_c$, and fitting the results to the corresponding power laws. However, it became gradually clear that this is an utterly inaccurate method of estimating the critical exponents. As a result, several accurate methods were developed for estimating these exponents, and what follows is a brief description of each method.

5.15.1 *Finite-Size Scaling*

We already described this method in Section 2.9, whereby one computes the effective conductivity of the networks as a function of their linear size L. The computations are carried out at the percolation threshold of the *infinite* system. For a network of conducting and insulating bonds, one has

$$g_e \sim L^{-\mu/\nu}[a_1 + a_2 g_1(L) + a_3 g_2(L) + \cdots], \tag{218}$$

where a_1, a_2, and a_3 are constants, ν is the critical exponent of percolation correlation length, and $g_1(L)$ and $g_2(L)$ represent corrections to scaling and vanish as $L \to \infty$. Often, $g_1(L)$, $g_2(L)$ and all the higher-order corrections are combined, so that Eq. (218) is rewritten as

$$g_e \sim L^{-\mu/\nu}\left(a_1 + a_2' L^{-\omega_c}\right), \tag{219}$$

where ω_c is called the leading correction-to-scaling exponent. However, as discussed in Section 2.9, Sahimi and Arbabi (1991) showed that very often $g_1(L) = L^{-1}$ and $g_2(L) = (\ln L)^{-1}$ provide very accurate fit of the data. In any event, after computing $g_e(L)$, the results are fitted to Eq. (219) [or to Eq. (218) with the aforementioned $g_1(L)$ and $g_2(L)$], leaving μ, ω_c, a_1 and a_2' as the fitting parameters to be determined.

Similarly, the effective conductivity of a network, that consists of normal conductors and superconductors, is fitted to

$$g_e \sim L^{s/\nu}\left(b_1 + b_2 L^{-\omega_s}\right), \tag{220}$$

where b_1 and b_2 are constant, and ω_s is the leading correction-to-scaling exponent for this problem, which is not necessarily equal to ω_c used in (219).

Provided that accurate data for $g_e(L)$ can be determined, these methods are highly reliable. In fact, estimates of μ and s that are listed in Table 2.3 were obtained by this method. To our knowledge, the first application of finite-size scaling to estimation of μ was made by Sarychev and Vinogradoff (1981) and Mitescu *et al.* (1982) in 2D, and by Sahimi *et al.* (1983b) in 3D. Since then, this method has been used very extensively by a large number of research groups.

5.15.2 Position-Space Renormalization Group Method

As discussed earlier, in the PSRG approach, after several iterations of Eq. (186), the fixed-point conductance distribution $f_\infty(g)$ is reached. Suppose that this distribution is obtained after m iterations, so that $f_\infty(g) = f_m(g)$. Since the shape of $f_m(g)$ does not change by further renormalization, one must have $\lambda_\mu f_{m+1}(\lambda_\mu g) \simeq f_m(g)$, where λ_μ is a constant. The critical exponent μ of the conductivity is then given by

$$\mu = \nu \frac{\ln \lambda_\mu}{\ln b}. \tag{221}$$

In practice, λ_μ is computed by the following method. If, after m iterations, the conductance distribution is $f_m(g)$, then, $\lambda_\mu^{(m)}$, the mth approximation to λ_μ, is defined by

$$\int \lambda_\mu^{(m)} f_m[\lambda_\mu^{(m)} g] h(g) dg = \int f_{m+1}(g) h(g) dg, \tag{222}$$

where $h(g)$ is a function chosen to speed up the convergence of the sequence $\lambda_\mu^{(m)}$ to λ_μ. Many such $h(g)$ have been proposed. One of the most accurate choice appears to be (Tsallis *et al.*, 1983)

$$h(g) = \frac{g}{g_0 + g}, \tag{223}$$

where g_0 is an arbitrary reference conductance. Other choices, such as $h(g) = \ln g$ and $h(g) = g^{-1}$ have also proven to be reasonably accurate.

The exponent s is computed by a similar method. In this case,

$$s = -\nu \frac{\ln \lambda_s}{\ln b}, \tag{224}$$

where λ_s is the corresponding eigenvalue for the superconductivity problem, and is estimated by an equation similar to (222), using the corresponding conductance distribution after m iterations and the same $h(g)$. For example, using this scheme with the 2×2 RG cell of Figure 5.14 yields, $\mu = s \simeq 1.32$, in excellent agreement with the numerical estimate, $\mu = s \simeq 1.3$ (see Table 2.3). The 3D RG cell of Figure 5.14 yields, $\mu \simeq 2.14$, only 7% larger than $\mu \simeq 2.0$, the most accurate numerical estimate of μ for 3D percolation network, and $s \simeq 0.76$, only 4% larger than $s \simeq 0.73$ for 3D percolation networks. Therefore, even small, suitably selected RG cells can yield reliable estimates of μ and s.

5.15.3 Series Expansion

Recall from Section 2.7.2 that, for a d-dimensional system, the conductivity exponent μ can be rewritten as

$$\mu = (d-2)\nu + \zeta, \tag{225}$$

where ζ is the exponent that describes the scaling of the resistance R_{ij} between two points i and j of a box of linear size L ($L < \xi_p$), $R \sim L^{\zeta/\nu}$. In the series expansion method, first developed by Harris and Fisch (1977; see also Fisch and Harris, 1978), one defines a percolation susceptibility χ_p by

$$\chi_p = \left\langle \sum_j s_{ij} \right\rangle, \tag{226}$$

where $s_{ij} = 1$ if the two sites i and j belong to the same percolation cluster and $s_{ij} = 0$ otherwise, and the averaging is over all configurations of the occupied sites (probability p) and unoccupied ones [probability $(1-p)$]. One also defines a resistive susceptibility $\chi_R(C)$ for a cluster C of sites via

$$\chi_R(C) = \sum_{i \in C} \sum_{j \in C} R_{ij}. \tag{227}$$

Then, the total resistive susceptibility χ_R, defined by

$$\chi_R = \left\langle \sum_j R_{ij} \right\rangle, \tag{228}$$

is obtained by summing $\chi_R(\mathcal{C})$ over all the clusters, weighting each cluster by its probability of occurrence. This is usually done in terms of cumulants, whereby one writes

$$\chi_R = \sum_{\mathcal{C}} N(\mathcal{C}; d) p^{n_b(\mathcal{C})} \chi_R^c(\mathcal{C}), \tag{229}$$

where $n_b(\mathcal{C})$ is the number of bonds in the cluster, $N(\mathcal{C}; d)$ is the number of ways *per site* a diagram, topologically equivalent to $\mathcal{C}$, can be realized on a d-dimensional simple-cubic lattice, and the sum is over all topologically *inequivalent* diagrams $\mathcal{C}$. Moreover, $\chi_R^c(\mathcal{C})$ is a cumulant defined by

$$\chi_R^c(\mathcal{C}) = \chi_R(\mathcal{C}) - \sum_{\gamma \in \mathcal{C}} \chi_R^c(\gamma), \tag{230}$$

where the sum is over all subdiagrams γ of $\mathcal{C}$. Then, the average resistance $\langle R \rangle$ is defined by

$$\langle R \rangle = \frac{\chi_R}{\chi_p} \sim |p - p_c|^{-\zeta}. \tag{231}$$

Therefore, the procedure for series analysis of resistance of random resistor networks is as follows. For each cluster $\mathcal{C}$ and fixed value of n_b (the number of bonds in the cluster), the resistance R_{ij} is computed (by solving the Kirchhoff's equations). These computations are carried out for all such clusters, from which $\chi_R(\mathcal{C})$ and hence χ_R are obtained. Writing

$$\chi_R = \sum_k \sum_l A(k, l) d^l p^k, \tag{232}$$

one obtains χ_R as a power series in p. Since, in practice, the number of possible cluster configurations increases very rapidly with n_b, the power series cannot be very long, and has been calculated (Essam and Bhatti, 1985) only up to the first 16 terms. Another series is obtained for χ_p, the computation of which is comparatively simple, since it involves only counting the number of clusters' configurations. The resulting two power series are then analyzed by a Padé approximation method, from which the average resistance $\langle R \rangle$ and hence the resistivity exponent ζ are computed.

The series expansion method has been used for estimating the exponent μ. Its predictions are comparable with those obtained by finite-size scaling.

5.15.4 Field-Theoretic Approach

Harris *et al.* (1984) and Lubensky and Wang (1986) developed a field-theoretic approach to the percolation conductivity problem. This approach is a renormalization group treatment of the problem in the Fourier space, by which the critical exponents are computed as a power series in $\epsilon = d_u - d$, where d_u is the upper critical dimension of the problem, i.e., the dimension at which the mean-field theory becomes exact. The upper critical dimension for percolation and conduction is $d_u = 6$. The method is somewhat similar to the series expansion method in that, one first derives an ϵ-expansion for the resistivity exponent ζ, and then, by combining it with the ϵ-expansion for the correlation length exponent ν, derives the expansion for μ. The result is given by (Lubensky and Wang, 1986)

$$\mu = 3 - \frac{5}{21}\epsilon + \frac{173}{9261}\epsilon^2 + O(\epsilon^3). \tag{233}$$

Note that for $\epsilon = 0$, i.e., $d = 6$, the expansion yields $\mu = 3$, in agreement with Eq. (22) for the conductivity of Bethe lattices, which represents the mean-field limit.

Such ϵ-expansions, while predicting the general trends in the dimensionality dependence of μ, are not accurate for the practical cases of $d = 2$ and 3. For example, for $d = 3$, Eq. (233) yields, $\mu \simeq 2.45$, which should be compared with the numerical estimate (see Table 2.3), $\mu \simeq 2$. However, such expansions do serve a useful purpose in the following sense. An unsolved problem is whether μ is related to the static exponents of percolation, such as ν and β. Over the years, many relations between μ and the static exponents of percolation have been proposed. However, such relations, if exact, must be consistent with Eq. (233) in that, if one substitutes the ϵ-expansions for the percolation exponents in the proposed relations, one must recover Eq. (233). To date, no such relation has been found (see Sahimi, 1984a and references therein).

5.15.5 *Comparison with the Experimental Measurements*

Let us now compare the numerical estimates of the conductivity exponents μ and s with the experimental estimates. We will show in Section 6.5 that the static dielectric constant of a heterogeneous material follows a power law near the percolation threshold p_c, and is characterized by the superconductivity exponent s. Experimental verification of this important result will also be discussed in detail in Chapter 6. Moreover, as we will discuss in Section 9.2.5.4, the exponent s may also be relevant to the divergence of the viscosity of a gelling solution at the gel point. Therefore, we do not discuss the relevance of s to, and its comparison with, the experimental data, and consider only the exponent μ.

We already compared in Section 5.1 the numerical estimate of μ, obtained with random resistor networks and various analytical and numerical computations that were described throughout this chapter, with extensive experimental data that have been reported over the past three decades. These data have established unequivocally the relevance of percolation, the EMA, and in particular the conductivity exponent μ, to conductor-insulator composites. Hence, let us make a comparison with one of the most recent sets of data, reported by Andrade *et al.* (1998) on the conductivity of pure vapor-grown carbon fibers (VGCF) under compression, and as a conducting component of a polymer composite. They measured the dependence of the effective conductivity on the apparent density ϕ of unmixed VGCF, when submitted to a uniaxial and low-speed compaction process. Near the percolation threshold, the contact resistance in the conductivity measurement device plays a significant role. Figure 5.19 presents their data. It is seen that the change

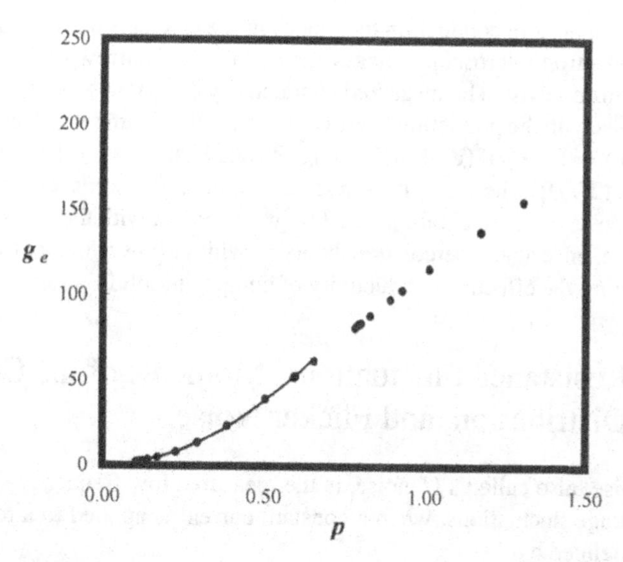

FIGURE 5.19. Effective conductivity of compressed vapor-grown carbon fibers (VGCF) (in $\Omega^{-1}\mathrm{cm}^{-1}$) as a function of the density p (in gr/cm^3) of the unmixed VGCF. The solid curve shows the percolation prediction, $g_e \sim (p - p_c)^\mu$ with $p_c \simeq 1.91 \pm 0.06$ (after Andrade *et al.*, 1998).

in the conductivity of VGCF during compression indicates the existence of two regimes, where a characteristic nonlinear dependence of the effective conductivity g_e on the density ϕ at low consolidation levels is followed by a roughly linear dependence of g_e on ϕ at higher pressures. The linear regime, far from $\phi_c \simeq 0.019$, can be accurately predicted by the EMA. If the data of Figure 5.19 near ϕ_c are fitted to

$$g_e \sim (\phi - \phi_c)^\mu, \tag{234}$$

then, one obtains, $\mu \simeq 1.91 \pm 0.06$, consistent with the conductivity exponent of 3D percolation systems (see Table 2.3), $\mu \simeq 2.0$. The same behavior was observed when VGCF was used as a conducting filler for a polymer composite, poly-propylene.

Even when the morphology of the material is such that it gives rise to non-universal power laws for the material's conductivity near the percolation threshold, experimental data and the predictions of percolation theory that were discussed in Section 2.11 have been found to be consistent with each other. An example is the work of Lee *et al.* (1986) who measured the effective conductivity of both sliver-coated-glass-Teflon composites and indium-glass composites. In one experiment with the conducting sphere-insulator composites, the silver-coated glass spheres were randomly distributed inside the Teflon host, with the conduction occurring through the contact points between the spheres. The conductivity exponent was measured to be, $\mu \simeq 2.0 \pm 0.2$, in excellent agreement with that of 3D percolation (see Table 2.3). In the second set of experiments with the insulating sphere-conductor composites, the hard glass spheres were randomly distributed inside the indium, and therefore conduction took place in the channels between the spheres. Scanning-electron microscope images showed that the narrow necks of the conducting paths do exist. The measured conductivity exponent was, $\mu \simeq 3.1 \pm 0.3$, which agrees with the prediction (see Section 2.11) that, with a conductance distribution $f(g) = (1-p)\delta(g) + p(1-\alpha)g^{-\alpha}$, one has, $\mu_\alpha \simeq \mu + \alpha/(1-\alpha)$ [see the bounds (2.77)], where μ is the universal conductivity exponent. Since for this composite, $\alpha \simeq 1/2$, we obtain $\mu_\alpha \simeq 3.0$, in agreement with the measurements.

Therefore, once again, percolation theory provides a powerful tool for modeling and predicting the effective conductivity of composite solids.

5.16 Resistance Fluctuations, Moments of the Current Distribution, and Flicker Noise

Flicker noise, also called $1/f$ noise, is the measured low-frequency spectrum of excess voltage fluctuations, when a constant current is applied to a resistor. The spectrum defined by

$$S(f) = \int_0^\infty \langle V(t)V(0)\rangle \exp(ift)dt, \tag{235}$$

almost always behaves as

$$S(f) \sim \frac{1}{f^a}, \tag{236}$$

with a usually being close to unity, hence the name $1/f$. Here $V(t)$ is the voltage at time t, and $\langle\cdot\rangle$ denotes an average over the time. The origin of this power law behavior of $S(f)$ has been very controversial (for a review see, for example, Weissman, 1988). However, two properties of $1/f$ noise are well-understood:

(1) Flicker noise is resistance noise. That is, if the current I is constant and there are voltage fluctuations δV, then δV is caused by resistance fluctuations δR, since according to Ohm's law, $\delta V = I\delta R$. This is confirmed by the facts that, (i) the noise spectrum is proportional to I^2, and that (ii) the spectrum of the resistance fluctuations, which can be obtained from $1/f$-noise experiments, can also be measured directly with no applied current from higher-order equilibrium correlation function (Voss and Clarke, 1976).

(2) Experiments (see, for example, Black *et al.*, 1981, and references therein) show that, resistance fluctuations at low frequencies are correlated only over microscopic scales.

These two facts were exploited by Rammal *et al.* (1985a,b) for studying $1/f$ noise in the framework of random resistor networks via the following model. Each bond ij of a network has a resistance r_{ij} and a small fluctuating part δr_{ij}. The fluctuations are assumed to be uncorrelated random variables with mean zero and a covariance, $\langle \delta r_{ij}\delta r_{kl}\rangle = \sigma_{ij}^2\delta_{ik}\delta_{jl}$. If all the bonds have the same resistance, then $\sigma_{ij}^2 = \sigma^2$. The mechanism for the fluctuations in the resistance is considered immaterial. If the resistance of the network and its fluctuations are, respectively, R and δR, then, we define the noise spectrum S_R,

$$S_R = \frac{1}{L^d}\frac{\langle \delta R \delta R\rangle}{R^2} = \frac{1}{L^d}\frac{\langle \delta V \delta V\rangle}{V^2} = \frac{1}{L^d}\frac{\langle \delta I \delta I\rangle}{I^2}, \tag{237}$$

as the relevant quantity for measuring the resistance fluctuations.

To compute the resistance noise in a resistor network, the following important theorems are utilized.

5.16.1 Tellegen's Theorems

B.D.H. Tellegen in 1952 proposed and proved three important theorems for resistor networks that are used for calculating resistance fluctuations (see, Spence, 1970, for a clear exposition of Tellegen's theorems). These theorems are as follows.

Theorem 1: In a given resistor network, the branch currents i_α and the voltage difference v_α across the corresponding branches satisfy the following sum rule:

$$\sum_\alpha i_\alpha v_\alpha = 0. \tag{238}$$

Theorem 2: If a set of voltages v''_α satisfies the Kirchhoff's voltage law, i.e., $\sum_\alpha v''_\alpha = 0$, where the sum is over branches of the network that form closed loops, and if a set of currents i'_α satisfies the Kirchhoff's current law, i.e., $\sum_\alpha i'_\alpha = 0$, then

$$\sum_\alpha i'_\alpha v''_\alpha = 0. \tag{239}$$

Theorem 3: Let $\mathcal{L}_1$ and $\mathcal{L}_2$ be two linear operators acting on the voltages or on the currents. Then, within the conditions of theorem 2, one has

$$\sum_\alpha (\mathcal{L}_1 i_\alpha)(\mathcal{L}_2 v_\alpha) = \sum_p (\mathcal{L}_1 i_p)(\mathcal{L}_2 v_p), \tag{240}$$

where i_p and v_p refer to the current and voltage in different ports p. In particular,

$$\sum_\alpha [(\mathcal{L}_1 i_\alpha)(\mathcal{L}_2 v_\alpha) - (\mathcal{L}_2 i_\alpha)(\mathcal{L}_1 v_\alpha)] = \sum_\alpha [(\mathcal{L}_1 i_p)(\mathcal{L}_2 v_p) - (\mathcal{L}_2 i_p)(\mathcal{L}_1 v_p)]. \tag{241}$$

These theorems have many applications, among which is the Cohn's theorem that can be used for computing S_R.

5.16.2 Cohn's Theorem

Suppose that in a one-port resistor network (one inlet and one outlet) I is the external current and V the voltage between the two contacts. According to Cohn's theorem (Cohn, 1950), the resistance fluctuation in the network is given by

$$\delta R = \sum_\alpha \delta R_\alpha \left(\frac{i_\alpha}{I}\right)^2, \tag{242}$$

where δR_α is the fluctuation of the resistance in the branch α, and i_α is the current through this branch in the original unperturbed resistor network. Equation (242) is a direct result of Eq. (241). From Cohn's theorem, one obtains, for a one-port network,

$$S_R = \frac{1}{L^d}\frac{1}{R^2}\sum_\alpha\sum_\beta \langle \delta R_\alpha \delta R_\beta \rangle \left(\frac{i_\alpha}{I}\right)^2 \left(\frac{i_\beta}{I}\right)^2 = \frac{1}{L^d}\frac{\sum_\alpha\sum_\beta \langle \delta R_\alpha \delta R_\beta \rangle i_\alpha^2 i_\beta^2}{(\sum_\alpha R_\alpha i_\alpha^2)^2}. \tag{243}$$

In particular, for uncorrelated fluctuations in the resistances and identical resistances, $R_\alpha = r$, one obtains

$$S_R = \frac{1}{L^2}\frac{\sigma^2}{r^2}\frac{\sum_\alpha i_\alpha^4}{\left(\sum_\alpha i_\alpha^2\right)^2}, \tag{244}$$

which implies that S_R is related to the second and fourth moments of the current distribution. One can also show that

$$\frac{\sigma^2}{r^2}\frac{1}{n_c} \le S_R \le \frac{\sigma^2}{r^2}\frac{r}{R}, \tag{245}$$

where n_c is the number of conducting bonds in the network.

5.16.3 *Scaling Properties*

It is easy to see that, for a d-dimensional uniform network of linear size L with uncorrelated resistance fluctuations, one has, $S_R = (\sigma^2/r^2)L^{-d}$, i.e., S_R scales as the inverse of the network's volume L^d. The question is: In a self-similar cluster of conducting bonds, such as the sample-spanning percolation clusters at p_c (or above p_c for $L < \xi_p$), how does S_R scale with L? Rammal and Toulouse (1983) showed that the resistance $R(L)$ of the sample scales with L as

$$R(L) \sim L^{-\beta_L}, \tag{246}$$

where

$$\beta_L = \frac{D_f}{D_s}(D_s - 2). \tag{247}$$

Here D_f is the fractal dimension of the cluster, $D_s = 2D_f/D_w$ is called the *spectral dimension* of the cluster, and D_w is the *random walk fractal dimension*. We will show in detail in Sections 6.1.9 and 6.1.10 that, diffusion in self-similar and fractal clusters is anomalous, i.e., the mean square displacement of a diffusant in the cluster at time t scales with t as t^{2/D_w} with $D_w > 1$, rather than t. It will also be shown there that, for the sample-spanning percolation cluster at p_c (or above p_c for $L < \xi_p$), one has $D_w = 2 + (\mu - \beta)/\nu$, where β, ν and μ are the usual percolation exponents defined by Eqs. (2.30), (2.31) and (2.36). Since for this cluster we also have [see Eq. (2.47)], $D_f = d - \beta/\nu$, Eq. (246) becomes the familiar equation

$$R(L) \sim L^{\mu/\nu - d + 2}, \tag{248}$$

so that, $\beta_L = -\mu/\nu + d - 2 = \tilde{\zeta}$, where $\tilde{\zeta}$ is the familiar resistivity exponent. The noise spectrum then satisfies the following scaling equation,

$$S_R(\lambda L) \sim \lambda^{-b} S_R(L), \tag{249}$$

where λ is a scaling factor, so that

$$S_R \sim L^{-b+d} \quad L \gg 1, \tag{250}$$

where b is a new exponent that characterizes the scale-dependence of S_R. For a percolation network, Eq. (249) can be written as

$$S_R(p, L) = L^{-b+d} h(\xi_p/L), \tag{251}$$

where $h(x)$ is a universal scaling function with the properties that, $h(x) \sim 1$ for $x \gg 1$ ($\xi_p \gg L$), so that $S_R(p, L) \sim L^{-b}$, while for $x \ll 1$, $h(x) \sim x^{d-b}$. Therefore, near p_c,

$$S_R \sim (p - p_c)^{\nu(b-d)} \sim (p - p_c)^{-\kappa}. \tag{252}$$

The exponent b (and hence κ) is independent of all other percolation exponents defined so far. In general

$$-\beta_L \leq b \leq D_{bb}, \tag{253}$$

where D_{bb} is the fractal dimension of the backbone (current-carrying part) of the sample-spanning percolation cluster. Moreover, Wright *et al.* (1986) derived the following rigorous bounds for κ,

$$(3d - 4)\nu + 1 - 2\mu \leq \kappa \leq 2(d - 1)\nu - \mu, \tag{254}$$

which are consistent with (253). Numerical simulations indicate that, $\kappa \simeq 1.12$ and 1.56 in 2D and 3D, respectively.

More generally, let us define the $2q$th moment of the current distribution by

$$M_{2q} = \sum_b i_b^{2q}, \tag{255}$$

where the sum is over all the current-carrying bonds. Then, for a self-similar cluster, we expect to have

$$M_{2q} \sim L^{-\tilde{\tau}_q}, \tag{256}$$

where, for the sample-spanning percolation cluster, $\tilde{\tau}_q = \tau_q/\nu$. The exponents τ_q are all independent of each other, and thus each moment of the current distribution describes one aspect of physics of the cluster or the transport process through it. For example, Eq. (255) implies that M_0 is simply the total number of current-carrying bonds of the cluster, i.e., the total number of the bonds in the backbone of the cluster, and therefore, $-\tilde{\tau}_0 = D_{bb}$. Moreover, $\tilde{\tau}_1 = \beta_L$, which, for the sample-spanning percolation cluster, becomes, $\tilde{\tau}_1 = -\mu/\nu + d - 2$. In general,

$$\tau_{q-1} \leq \tau_q \leq \frac{q}{q-1}\tau_{q-1} - \frac{1}{q-1}\tau_0. \tag{257}$$

These exponents have also been calculated by Park *et al.* (1987) by an ϵ-expansion, and by Blumenfeld *et al.* (1987) by a series expansion.

Wright *et al.* (1986) and Tremblay *et al.* (1992) also considered resistance fluctuations in resistor networks of normal and superconducting bonds. In particular, if for the superconducting percolation networks one has the power law, $S_R \sim (p_c - p)^{-\kappa'}$, then, because of the duality relation for 2D systems, one has $\kappa = \kappa'$.

Let us mention that de Arcangelis *et al.* (1985, 1986) studied a closely related quantity, namely, the moments of the voltage distribution in a resistor network and found, not surprisingly (in view of our discussion), that the voltage distribution is also characterized by an infinite hierarchy of exponents which are closely related to that of the current distribution.

5.16.4 Comparison with the Experimental Data

The above predictions for the noise spectrum S_R have been tested against experimental measurements by several research groups. To discuss the experimental data, let us write S_R in terms of the resistance R of the material. Since near p_c one has, $R \sim (p - p_c)^{-\mu}$, we can write

$$S_R \sim R^{\kappa/\mu} \sim R^w, \tag{258}$$

so that $w \simeq 0.86$ and 0.78 for 2D and 3D percolation networks. To our knowledge, these values have never been observed in any experiment. On the other hand, Tremblay *et al.* (1986) considered resistance fluctuations in continuum models of disordered materials near p_c and, similar to what was discussed for the effective conductivity of such models in Chapter 2 (see Section 2.11), derived the appropriate exponents w for these models. They predicted that for the Swiss-cheese (SC) model, $w \simeq 3.2$ and 2.1, while for the inverted SC model, $w \simeq 0.86$ and 2.4, in 2D and 3D, respectively. Note that, similar to the effective conductivity, the value of w for the inverted SC model in 2D is the same as that of 2D lattice models. Some of these predictions were confirmed experimentally. For example, Chen and Chou (1985) reported measurements of the resistance fluctuations in carbon powder-wax composites, and obtained $w \simeq 2.2 \pm 0.2$, in good agreement with the 3D value of w for the inverted SC model. On the other hand, Koch *et al.* (1985) measured resistance fluctuations in thin samples of gold. First, evaporated films were prepared, and then their resistance was increased by ion milling. Their measurements indicated that, $w \simeq 2.0 \pm 0.1$, which agrees neither with the prediction of the SC model nor with that of the inverted SC model. Similar discrepancies were reported by Garfunkel and Weissman (1985).

Summarizing, the resistance fluctuations can be expressed in terms of the moments of the current distribution in a resistor network. For percolation networks (and more generally, self-similar clusters), the current distribution is described by an infinite hierarchy of independent exponents, which can be estimated both by numerical simulations and direct experimental measurements.

5.17 Hall Conductivity

An effective tool for understanding electrical transport in a composite material that consists of good and poor conductors is to study the material's properties in the presence of a magnetic field H. In the presence of such a field, the conductivity of the material is described by a tensor $\mathbf{g}_e$ which has non-zero off-diagonal components, even if the material is isotropic. Some of the off-diagonal terms are symmetric and even in the magnetic field H, while others are antisymmetric and odd in H. If we choose H to be, for example, parallel to the $z-$direction, then $\mathbf{g}_e$ will have the following form

$$\mathbf{g}_e = \begin{pmatrix} g & 0 & 0 \\ 0 & g & 0 \\ 0 & 0 & g \end{pmatrix} + \begin{pmatrix} 0 & g_h & 0 \\ -g_h & 0 & 0 \\ 0 & 0 & 0 \end{pmatrix} + \begin{pmatrix} \delta g_\perp & 0 & 0 \\ 0 & \delta g_\perp & 0 \\ 0 & 0 & \delta g_\parallel \end{pmatrix} \quad (259)$$

where g is the Ohmic conductivity (at $H = 0$), g_h is the Hall conductivity, and $\delta g_\perp$ and $\delta g_\parallel$ are the transverse and longitudinal magnetoconductivities, respectively. Strictly speaking, Eq. (259) is applicable only to an isotropic material, so that even crystals with cubic symmetry are excluded. However, for cubic symmetry the form of the antisymmetric part of Eq. (259) is also correct at low magnetic fields.

Suppose now that an electric field $\mathbf{E}$ has also been imposed on the material. Then, the material has an Ohmic conductivity g_e, and a Hall conductivity $\mathbf{g}_{eh}$. For an isotropic material and small H, $\mathbf{g}_{eh}$ is proportional to H and is defined by the Kirchhoff's law relating the current density $\mathbf{I}$ and the electric field $\mathbf{E}$:

$$\mathbf{I} = g_e\mathbf{E} + \mathbf{E} \times \mathbf{g}_{eh}. \quad (260)$$

If Eq. (260) is inverted for small H, the Hall coefficient $R_H = g_{eh}/(Hg_e^2)$ appears along the Ohmic resistivity $R_e = 1/g_e$: $\mathbf{E} = R_e\mathbf{I} + R_H(\mathbf{H} \times \mathbf{I})$. Thus, the main goal is to calculate the Hall conductivity and Hall coefficient in disordered composites. Volger (1950) and Juretschke *et al.* (1956) pioneered such calculations by treating various types of disorder in 3D composites. In particular, Juretschke *et al.* (1956) derived the exact general solution for the Hall effect at low H in isotropic 2D composite films. However, their results were forgotten for a long time until they were rediscovered in the 1970s.

To make further progress, one should restrict oneself to low values of H, so that the three terms of the sum in Eq. (259) are proportional to successively higher powers of H, namely, H^0, H^1, and H^2. This then allows one to solve for the electric field by a perturbation method, leading to the following expressions for

g_{eh} and $\delta\mathbf{g}_e$:

$$g_{eh} = \frac{1}{\Omega}\int g_h(\mathbf{r})\frac{[\mathbf{E}^{(x)} \times \mathbf{E}^{(y)}]_z}{E_0^2}\,d\Omega, \tag{261}$$

$$(\delta g_e)_{ab} = \frac{\langle E^{(a)} \cdot \delta\mathbf{g} \cdot E^{(b)}\rangle}{E_0^2}$$

$$+\frac{1}{\Omega E_0^2}\int d\Omega \int [\mathbf{g}_h(\mathbf{r}) \times \mathbf{E}^{(a)}(\mathbf{r})] \cdot \nabla\nabla' G(\mathbf{r}, \mathbf{r}'|g) \cdot [\mathbf{g}(\mathbf{r}') \times \mathbf{E}^{(b)}(\mathbf{r}')]\,d\Omega'. \tag{262}$$

Here, Ω is the volume of the material, $\mathbf{E}^{(a)}(\mathbf{r})$ is the local electric field for the ohmic problem (i.e., when $H = 0$) in the composite when the average or applied field $\mathbf{E}_0$ lies along the a-axis, and G is the Green function for the same problem which is the solution of the following boundary value problem:

$$\nabla \cdot [g(\mathbf{r})\nabla G(\mathbf{r}, \mathbf{r}'|g)] = -\delta(\mathbf{r} - \mathbf{r}'),$$

$$G = 0 \quad \text{on the boundaries,} \tag{263}$$

which is a slight generalization of Eq. (148). Although one is able to write down such formal expressions as (261) and (262), in practice neither the Green function G nor the local fields $\mathbf{E}^{(a)}(\mathbf{r})$ are straightforward to derive for any system but the most trivial ones. Nevertheless, as discussed below, these expressions are useful for a variety of reasons.

5.17.1 *Effective-Medium Approximation*

Long after the work of Volger (1950) and Jerutschke *et al.* (1956), Stachowiak (1970) carried out a mean-field type computation of the effective conductivity of a composite material in the presence of a magnetic field. Soon after this work, Cohen and Jortner (1973) developed an EMA for the problem of Hall conductivity of disordered materials, which represents an extension of the EMA discussed earlier in this chapter and in Chapter 4. The goal is to compute the Hall conductivity $\mathbf{g}_{eh}$ by an EMA, since we already know how to estimate the ohmic conductivity g_e. The derivation of Cohen and Jortner is along the lines discussed for the effective conductivity of continuum models of heterogeneous materials described in Section 4.9. They considered a composite material in which any portion of it has a magnetoconductivity $\mathbf{g}_{1h}(\mathbf{H})$ or $\mathbf{g}_{2h}(\mathbf{H})$ with probabilities p and $1-p$, respectively. Similarly, the Ohmic conductivity (the limit of zero field) of any portion of the material takes on values g_1 and g_2 with probabilities p and $1-p$, respectively. Except for a spherical inclusion, the material is now replaced by a uniform medium with magnetoconductivity $g(\mathbf{H})$. Far from the sphere, the applied field retains it uniform value $\mathbf{E}^0$. The field $\mathbf{E}_i$ inside the sphere is constant and is given by

$$\mathbf{E}_i = \mathbf{E}^0 - \frac{4}{3}\pi\mathbf{P}, \tag{264}$$

where $\mathbf{P}$ is a fictitious polarization defined such that the quantity

$$P_n = \mathbf{n} \cdot \mathbf{P} \tag{265}$$

is identical to the surface-charge density on the boundary between the sphere and the effective medium, where $\mathbf{n}$ is a unit vector normal to the boundary. The field $\mathbf{E}_o$ just inside the effective medium is given by

$$\mathbf{E}_o = \mathbf{E}^0 + \frac{4}{3}\pi\mathbf{P} \cdot (3\mathbf{nn} - \mathbf{1}). \tag{266}$$

The continuity equation then requires the normal component of the current density across the boundary to remain constant, that is,

$$\mathbf{n} \cdot \mathbf{g}_i \cdot \mathbf{E}_i = \mathbf{n} \cdot \mathbf{g} \cdot \mathbf{E}_o, \tag{267}$$

where

$$\mathbf{g} = g_d\mathbf{1} + \mathbf{g}_h. \tag{268}$$

Here g_d is the diagonal part of $\mathbf{g}$, and $\mathbf{g}_h$ is antisymmetric and linear in H, so that

$$\mathbf{g}_h \cdot \mathbf{V} = A\mathbf{H} \times \mathbf{V}, \quad \mathbf{V} \cdot \mathbf{g}_h = A\mathbf{V} \times \mathbf{H}, \tag{269}$$

where $\mathbf{V}$ is any vector. Combining the above equations yields an equation for $\mathbf{P}$:

$$4\pi\mathbf{P} = \left[g_d\mathbf{1} - \frac{1}{3}(\mathbf{g} - \mathbf{g}_i) \right]^{-1} \cdot (\mathbf{g}_i - \mathbf{g}) \cdot \mathbf{E}^0, \tag{270}$$

where the conductivity tensors are treated as 3×3 matrices.

The effective medium is now defined by requiring that, $\langle \mathbf{P} \rangle = \mathbf{0}$, where the average is taken with respect to the distribution of the conductivities. This requirement then results in

$$\left\langle [g_d\mathbf{1} - \frac{1}{3}(\mathbf{g} - \mathbf{g}_i)]^{-1} \cdot (\mathbf{g}_i - \mathbf{g}) \right\rangle = 0, \tag{271}$$

which, when solved to the first order in the field for $\mathbf{g}_h$, yields

$$\mathbf{g}_h = \frac{\langle \mathbf{g}_{ih}/(g_i + 2g)^2 \rangle}{\langle (g_i + 2g)^{-2} \rangle}, \tag{272}$$

where $\mathbf{g}_{ih} = \mathbf{g}_{1h}$, $\mathbf{g}_{2h}$, and $g_i = g_1$, g_2, with probabilities p and $1 - p$, respectively.

Stroud (1980) extended this EMA to composite materials that are composed of normal and superconducting components. He ignored quantum effects (hence treated the problem "classically") and, in order to take into account the possibility of the material being anisotropic, considered ellipsoidal inclusions rather than the spherical ones considered by Cohen and Jortner. In his model the (volume) fraction of the normal metal or conductor is $1 - p$ and has a conductivity tensor $\mathbf{g}_m(B)$, where B is the magnitude of the local magnetic induction. The components of the tensor are given by

$$g_m^{xx} = g_m^{yy} = \frac{g_0}{1 + \tilde{H}^2}, \quad g_m^{zz} = 0,$$

$$g_m^{xy} = -g_m^{yx} = \frac{g_0 \tilde{H}}{1+\tilde{H}^2},$$

where $\tilde{H}$ is the dimensionless field, and g_0 is a constant conductance. The remaining components of the tensor are zero. The (volume) fraction of the superconducting component of the material is p. The composite is characterized by an effective conductivity tensor $\mathbf{g}_e$ with components that are predicted by the EMA to be given by

$$\begin{aligned} g_e^{xx} + p\Gamma_{xx}^{-1} &= g_m^{xx}, \\ g_e^{zz} + p\Gamma_{zz}^{-1} &= g_m^{zz}, \\ g_e^{xy} = -g_e^{yx} &= g_m^{xy}, \end{aligned} \tag{273}$$

with $g_e^{yy} = g_e^{xx}$, where

$$\Gamma_{xx} = \frac{1}{2g_e^{zz}\mathcal{R}}\left[1 - \frac{1}{\sqrt{\mathcal{R}(1-\mathcal{R})}}\sin^{-1}\sqrt{\mathcal{R}}\right], \tag{274}$$

$$\Gamma_{zz} = -\frac{1}{g_e^{zz}\mathcal{R}}\left[1 - \sqrt{\frac{1-\mathcal{R}}{\mathcal{R}}}\sin^{-1}\sqrt{\mathcal{R}}\right], \tag{275}$$

where $\mathcal{R} = 1 - g_e^{xx}/g_e^{zz}$ is the anisotropy. Equations (273) can be obtained in a straightforward manner from the EMA for anisotropic materials, developed by Stroud (1975) and derived in Section 4.9.2, by assuming that the conductivity of one of the two components is infinite. It can be shown that the result $g_e^{xy} = g_m^{xy}$ is *exact*, not just within the EMA.

Examination of Eqs. (273) reveals an interesting phenomenon. The EMA predicts that, in the presence of a strong magnetic field, addition of a superconducting material to a normal conductor *increases* the resistivity of the composite. This effect, which arises from the current distortions near the superconducting inclusions that cause extra dissipation in the normal conductor (hence increasing its resistivity), is in qualitative agreement with the experimental observations. Moreover, according to the EMA, at a fixed value of p, the (volume) fraction of the superconductors, the resistance of the material rises at first linearly with the field, with the slope being proportional to p, and then saturates at large fields. If we define a Hall coefficient $R_H(p, \tilde{H}) = 1/(g_e^{xx}\tilde{H})$ (which is slightly different from what was introduced earlier), then according to the EMA,

$$R_H(p, 0) = \lim_{\tilde{H}\to 0} \{\tilde{H}^{-1} g_e^{xy}(p, \tilde{H})/[g_e^{xx}(p, \tilde{H})]^2\} = \frac{R_H(0,0)}{p_c^2}(p_c - p)^2, \tag{276}$$

where p_c is the EMA prediction for the percolation threshold of the system. That R_H vanishes at p_c (of the superconducting component) is expected, since for $p > p_c$ the composite cannot support a potential difference across itself. In strong fields,

$$R_H \sim \frac{g_e^{xy}}{(g_e^{xx})^2}\tilde{H}^{-1} \sim \tilde{H}^{-2}, \tag{277}$$

since g_e^{xx} saturates and g_e^{xy} decreases as $\tilde{H}^{-1}$. Moreover, $R_H(p, 0)$ satisfies the *exact* relation,

$$R_H(p, 0)g_e(p)^2 = \lim_{\tilde{H}\to 0} \frac{g_m^{xy}}{\tilde{H}} = \text{constant}, \tag{278}$$

where $g_e(p)$ is the effective conductivity of the composite in zero field. Since, $g_e(p) \sim (p_c - p)^{-s}$, where s is the superconductivity exponent defined and discussed earlier in this chapter and in Chapter 2, we obtain

$$R_H \sim (p_c - p)^{2s}, \tag{279}$$

which is not only valid in any dimension, but is also consistent with the EMA prediction, Eq. (276), since the EMA predicts that $s = 1$.

5.17.2 Network Model

Unlike the case of the Ohmic conductivity g_e, the presence of the magnetic field makes the development of a network model for calculating g_{eh} and R_H a quite complex task. For example, the general circuit element of a bond of the network cannot be a simple resistor, but must be a conductance *matrix*. Bergman *et al.* (1990) proposed a network model for calculating g_{eh} and R_H which satisfies all the necessary requirements for being an appropriate model. In what follows we describe this model.

The early network models for computing the Hall conductivity and Hall coefficient suffered from several shortcomings. For example, they allowed a Hall current to flow in a conducting bond as a result of a non-zero field across a neighboring perpendicular bond, even if this bond was insulating. A refinement of this model was one in which each element of the network was either a pair (in 2D) or a triplet (in 3D) of identical, mutually perpendicular bonds which represent electrically unconnected Ohmic conductors when $H = 0$. When $H \neq 0$, a Hall current was allowed to flow in each of these bonds as a result of a non-zero voltage on one of the others, when H had a component that was perpendicular to both bonds. However, despite many desirable properties, this model too suffered from a major shortcoming, in that it always resulted in a network that was split up into two or more unconnected pieces that occupied the same region of space, which is a particularly severe shortcoming for composites that are near their percolation threshold. These problems were all addressed in Bergman *et al.*'s model. They considered a square or a simple-cubic lattice and defined the electric field in a direction $\mathbf{b}$ perpendicular to a given bond b to be a weighted average over the voltages on the four bonds in direction $\mathbf{b}$ that are nearest neighbor to b. The weight of each of these bonds is $1/4$ if it is of the same type as b and 0 otherwise. In this model, an electric field in one component cannot produce a Hall current in another component, thus correcting the first major shortcoming of the earlier models mentioned above. Moreover, it does not result in split networks, hence addressing the second deficiency of the earlier models. An example of $x - y$ layer of a sample of linear size $L = 4$ in a simple-cubic network is shown in Figure 5.20.

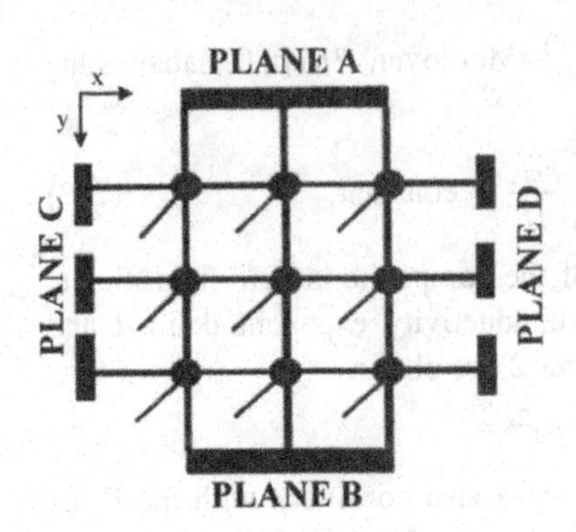

FIGURE 5.20. One layer of the appropriate network model for computing the Hall conductivity (after Bergman *et al.*, 1990).

These models are all restricted to low values of H. To our knowledge, a network model for computing the Hall conductivity and Hall coefficient, in the presence of a strong magnetic field, has not been developed yet.

5.17.3 Exact Duality Relations

As discussed earlier in Section 5.4.1, and also in Section 4.8, 2D composites possess a special symmetry called duality, which enables one to derive exact results for such materials. To discuss the implications of duality for the Hall conductivity, we consider a material with a conductivity tensor $\mathbf{g}$ that is isotropic but may have an antisymmetric off-diagonal part, due to the presence of a magnetic field perpendicular to the plane of the system, taken to be the $x-y$ plane, so that

$$\mathbf{g} = \begin{pmatrix} g & g_h \\ -g_h & g \end{pmatrix}. \tag{280}$$

Suppose that the effective conductivity of the material is $\mathbf{g}_e$. Then, the dual of the material is another composite with an effective conductivity $\mathbf{g}' = [\mathbf{g}_e(\mathbf{g})]^{-1}$. We thus write the components of the resistivity tensor as

$$\mathbf{R}_e(\mathbf{g}) = [\mathbf{g}_e(\mathbf{g})]^{-1} \equiv \begin{pmatrix} R_e^{xx} & -R_e^{xy} \\ R_e^{xy} & R_e^{xx} \end{pmatrix}. \tag{281}$$

Consider now a two-component composite in which each portion of the material has good Ohmic and Hall conductivities g_1 and g_{1h} with probability p and poor Ohmic and Hall conductivities g_2 and g_{2h} with probability $1-p$. Since

$$R_e^{xx} = g_e\left(\frac{g_1}{g_1^2+g_{1h}^2}, \frac{-g_{1h}}{g_1^2+g_{1h}^2}, \frac{g_2}{g_2^2+g_{2h}^2}, \frac{-g_{2h}}{g_2^2+g_{2h}^2}\right),$$

$$R_e^{xy} = -g_{eh}\left(\frac{g_1}{g_1^2+g_{1h}^2}, \frac{-g_{1h}}{g_1^2+g_{1h}^2}, \frac{g_2}{g_2^2+g_{2h}^2}, \frac{-g_{2h}}{g_2^2+g_{2h}^2}\right), \tag{282}$$

we take $g_{2h} \sim g_2^2$, and let both $g_2 \to 0$ and $g_{2h} \to 0$, so as to obtain

$$\frac{g_2}{g_2^2 + g_{2h}^2} \to \infty, \quad \text{and} \quad \frac{g_{2h}}{g_2^2 + g_{2h}^2} \to \frac{g_{1h}}{g_1^2 + g_{1h}^2},$$

It then follows that (for $p > p_c$)

$$R_e^{xx} = g_e\left(\frac{g_1}{g_1^2 + g_{1h}^2}, 0, \infty, 0\right) = \frac{1}{g_e}\left(\frac{g_1^2 + g_{1h}^2}{g_1}, 0\right) = R_e(R_1^{xx}, \infty) \equiv R_e(R_1^{xx}), \tag{283}$$

$$R_e^{xy} = \frac{g_{1h}}{g_1^2 + g_{1h}^2} = R_1^{xy}, \tag{284}$$

where R_1^{xx} and R_1^{xy} are the components of the resistivity tensor, and R_e is the effective resistivity at $H = 0$. Equation (284) implies that the Hall resistivity of the composite is the same as that of the conducting component. Likewise, Eq. (283) indicates that the relative magnetoresistance of the composite is just that of the good conductor:

$$\frac{R_e^{xx}(H) - R_e^{xx}(0)}{R_e^{xx}(0)} = \frac{R_1^{xx}(H) - R_1^{xx}(0)}{R_1^{xx}(0)}. \tag{285}$$

For weak magnetic fields, that is, when $g_h \ll g$, we can derive explicit results when both components have non-zero conductivities. In that case, we can write

$$\mathbf{g}' \equiv (\mathbf{g})^{-1} \simeq \begin{pmatrix} 1/g & -g_h/g^2 \\ g_h/g^2 & 1/g \end{pmatrix} \tag{286}$$

A careful examination of the duality relation, $\mathbf{g}_e' = (\mathbf{g}_e)^{-1}$, together with the fact that the ratio $(g_{eh} - g_{2h})/(g_{1h} - g_{2h})$ depends only on g_2/g_1, leads to the following exact result

$$\frac{g_{eh} - g_{2h}}{g_{1h} - g_{2h}} = \frac{g_e^2 - g_2^2}{g_1^2 - g_2^2}. \tag{287}$$

5.17.4 Scaling Properties

The study of the scaling behavior of g_{eh} and R_H near p_c was initiated by Levinshtein *et al.* (1976), Shklovskii (1977), and Straley (1980a). However, a complete scaling theory was developed by Bergman and Stroud (1985). Consider again the two-component network in which each bond has good Ohmic and Hall conductivities g_1 and g_{1h} with probability p, and poor Ohmic and Hall conductivities g_2 and g_{2h} with probability $1 - p$. Bergman and Stroud (1985) proposed that for $g_{2h}/g_{1h} \ll 1$ and for p close to p_c

$$\frac{g_{eh} - g_{2h}}{g_{1h} - g_{2h}} = |p - p_c|^{\tau} F\left[\frac{g_{2h}/g_{1h}}{|p - p_c|^{\mu+s}}\right], \tag{288}$$

where μ and s are the usual conductivity exponents, ans τ is a new critical exponent characterizing the power-law behavior of g_{eh} near p_c when $p > p_c$, $g_2 = 0$, and

$g_{2h} = 0$, i.e.,

$$g_{eh} \sim (p - p_c)^{\tau}. \tag{289}$$

Equation (288) has been written in complete analogy with Eq. (2.43). The scaling function $F(x)$ has the properties that,

$$F(x) \sim \begin{cases} \text{constant} & \text{if } x \ll 1 \text{ and } p > p_c, \quad \text{regime I}, \\ x^2 & \text{if } x \ll 1 \text{ and } p < p_c, \quad \text{regime II}, \\ x^{\tau/(\mu+s)} & \text{if } x \gg 1 \text{ and } p \simeq p_c, \quad \text{regime III}. \end{cases} \tag{290}$$

However, a better way of understanding Eq. (288) is in terms of the Hall coefficient R_H. To each bond of a network we assign a Hall coefficient $R_{1h} = g_{1h}/g_1^2$ with probability p or $R_{2h} = g_{2h}/g_2^2$ with probability $1 - p$. Then, according to the scaling theory of Bergman and Stroud (1985) one has

$$R_H \sim \begin{cases} a_1 R_{1h}|p - p_c|^{-\psi} + b_1 R_{2h}(g_2/g_1)^2 |p - p_c|^{-2\mu} & \text{regime I}, \\ a_2 R_{1h}|p - p_c|^{-\psi} + b_2 R_{2h}|p - p_c|^{2s} & \text{regime II}, \\ a_3 R_{1h}(g_2/g_1)^{-\psi/(\mu+s)} + b_3 R_{2h}(g_2/g_1)^{2s/(\mu+s)} & \text{regime III}, \end{cases} \tag{291}$$

where the a_i and b_i are constant, and ψ is a critical exponent that characterizes the power-law behavior of R_H as p_c is approached from below, and is related to μ and τ by

$$\psi = 2\mu - \tau. \tag{292}$$

It is not yet clear whether τ (or ψ) is an independent exponent, or is related to the other percolation exponents defined so far. Straley (1980a) proposed that

$$\psi = 2[\mu - (d - 1)\nu], \tag{293}$$

where ν the critical exponent of percolation correlation length. Equation (293) predicts that $\psi(d = 3) \simeq 0.48$, fully compatible with the numerical estimate, $\psi \simeq 0.49 \pm 0.06$, and that in the mean-field limit, $\psi = 1$, in agreement with the exact calculation of Straley (1980b) using a Bethe lattice. However, Eq. (293) also predicts that $\psi(d = 2) \simeq -0.066$, in disagreement with the exact result $\psi(d = 2) = 0$ obtained by Shklovskii (1977). As pointed out by Bergman and Stroud (1985), the important point to remember is that in the regime I the ratio of the second to the first term is of the order of $(g_{2h}/g_{1h})|p - p_c|^{-\tau}$, and therefore, depending on the morphology of the material, either term may dominate, so that the experimental verification of these scaling laws is not straightforward.

5.17.5 Comparison with Experimental Data

One of the first experimental studies of Hall conductivity and Hall coefficient in a material with percolation disorder was carried out by Levinshtein *et al.* (1975). They performed Hall experiments on 20 mm × 60 mm conducting graphite paper sheets with holes randomly punched in them. In their 3D experiments a compressed

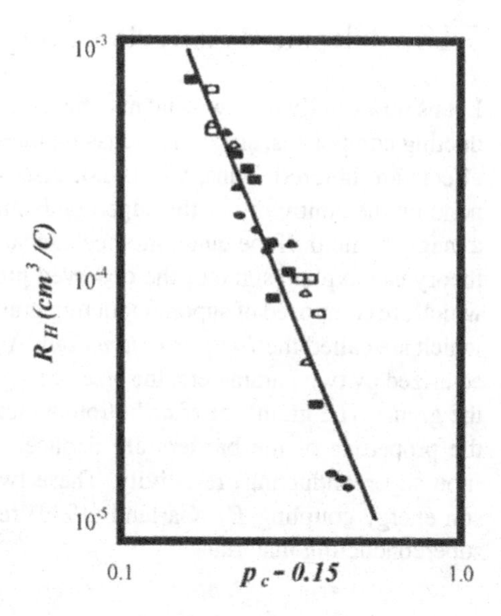

FIGURE 5.21. Logarithmic plot of the effective Hall coefficient R_H versus $p - 0.15$, where p is the volume fraction of Al. Different symbols refer to various substrates (after Dai *et al.*, 1987).

stack of 15 individually punched graphite sheets was used. The results confirmed the divergence of R_H as p_c is approached. Sichel and Gittleman (1982) measured g_{eh} and R_H in granular metals Au-SiO_2 and W-Al_2O_3 near their percolation thresholds.

Quantitative experimental verification of Eqs. (291) was provided by several groups. Palevski *et al.* (1984) showed that the Hall coefficient does indeed remain finite in a 2D (thin) metal film. Dai *et al.* (1987) studied the same problem in a 3D Al-Ge metal-insulator composite. In regime I, the poor conductor was Al-doped-Ge which was found to dominate the Hall effect. To make the first term of Eq. (291), for regime I, dominant, Dai *et al.* dissolved the metallic aluminum in KOH, which left the doped Ge as the good conductor, the poor conductor role now being played by the vacuum. In such a situation, the first term of Eqs. (291) for regime I is dominant, and therefore the critical exponents ψ and τ can be directly measured. Figure 5.21 shows the measured Hall coefficient R_H versus $p - p_c$, where p is the volume fraction of Al, and $p_c \simeq 0.15$ is the percolation threshold. The data yielded, $\tau \simeq 3.8 \pm 0.2$ and $\psi \simeq 0.38 \pm 0.05$, which should be compared with accurate simulation results based on percolation networks, $\tau \simeq 3.56 \pm 0.06$ and $\psi \simeq 0.49 \pm 0.06$. Rohde and Micklitz (1987) also measured the Hall conductivity g_{eh} on the metallic side of a granular rare-gas metal mixture, Sr:Ar, and reported that, $\tau \simeq 2.7 \pm 0.2$, which is somewhat low, and $\psi \simeq 0.49 \pm 0.07$, which agrees with the numerical simulation and the scaling theory of Straley (1980a). Thus, theory and experiment are in general agreement with each other. Since the EMA of Cohen and Jortner (1973) and of Stroud (1980) can be used for accurately estimating g_{eh} and R_H away from p_c, we now have a fairly complete theory of Hall effect in composite materials.

5.18 Classical Aspects of Superconductivity

Let us now briefly review what may be called the classical properties of superconducting composites, since in discussing these properties the quantum mechanical effects are ignored. Thus, we do not discuss the quantum properties which depend on the continuity of the superconducting wave function and its behavior in a magnetic field. If the quantum mechanical effects are ignored, then, percolation theory can explain some of the observed properties of granular superconductors, which are composed of superconducting grains separated by thin insulating grains, which are called the *Josephson junctions*. A granular superconductor can be characterized by two parameters, the size of its grains, and the energy barrier between the grains. The grain size distribution is measured by electron microscopy, while the properties of the barriers are deduced from measurements of normal-state (non-superconducting) resistivity. These two parameters determine the Josephson energy coupling E_j. Garland (1989) reviewed many granular properties of superconducting materials.

5.18.1 *Magnetoconductivity*

As usual, let us consider a disordered binary composite in d dimensions ($d = 2$ or 3) described by a conductivity field $g(\mathbf{r})$ that can take on two values, $g_1 = \infty$ with probability p and a finite value g_2 with probability $1 - p$. The effective conductivity $g_e(p)$ must therefore diverge when $p > p_c$, the percolation threshold for the superconducting component. For $p < p_c$, g_e is finite, with an expected asymptotic behavior given by Eq. (2.37) and characterized by the exponent s which depends on both the dimensionality and the microstructure of the composite; its values for 2D and 3D materials are listed in Table 2.3. In the Swiss-cheese model which consists of a random distribution of overlapping d-dimensional spheres of normal conductors in a superconducting matrix, or in the inverted Swiss-cheese model, in which the roles of the matrix and the spheres are reversed, the numerical value of s may be different from the lattice values listed in Table 2.3.

However, apart from an experimental study by Deutscher and Rappaport (1979), there appears to be little experimental evidence confirming this picture. In both 2D and 3D, an experimental test of this prediction would be complicated by the proximity effect, which causes the superconducting region to grow a distance ξ_s into the surrounding normal region, where ξ_s, is the temperature-dependent normal-conductor coherence length.

A better realization of such classical effects may be obtained upon application of a magnetic field. How does a magnetic field affect the behavior of a granular superconductor? There are two types of homogeneous superconductors that behave quite differently in an external magnetic field.

(1) *Type*-I superconductors remain superconducting and expel the magnetic field up to a critical field H_c, since below H_c the flux is excluded from the superconductor (except for a thin layer). They then abruptly cross over to the normal state.

(2) *Type*-II superconductors allow the magnetic field to penetrate into the system if H is greater than a lower critical field H_{c1}, but lose their superconductivity at an *upper critical field* H_{c2}. Between H_{c1} and H_{c2}, the flux penetrates partially and inhomogeneously, typically forming a lattice of quantized flux lines. Such a field fully penetrates the superconducting composite, and to a first approximation the magnetic induction B can be assumed to be uniform and equal to the applied H in the normal component. In a magnetic field, the conductivity g_N of the normal component becomes a tensor, as does g_e. Among the fields satisfying this condition, there are two ranges to consider.

(i) At sufficiently low fields, the off-diagonal elements of both g_e and g_N are linear in H, while the diagonal elements are unchanged to first order in H. Under such conditions, it can be shown that $g_{e,ij} = g_{N,ij}$ for $i \neq j$, leading to the interesting prediction that, near p_c, the Hall coefficient of the composite below the percolation threshold for superconductivity follows the relation [see Eq. (279)]

$$R_H(p) \equiv \frac{g_{e,12}}{Hg_{e,11}^2} \sim R_H(1)(p_c - p)^{2s}. \tag{294}$$

This prediction also apparently remains to be tested in 2D or 3D.

(ii) At higher fields, one expects both a transverse and a longitudinal magnetoresistance in the composite when a superconducting material is added, even if the pure normal conductor has no magnetoresistance because, as discussed above, the superconducting inclusions distort the nearby current lines, leading to increased dissipation. This conclusion was confirmed by measurements of Resnick *et al.* (1979) on a composite of Pb in an Al matrix and, as discussed above, was also confirmed by the EMA calculations of Stroud (1980).

5.18.2 Magnetic Properties

The starting point for studying the magnetic properties of superconductors is the London penetration depth of a composite that contains a volume fraction p of a superconductor and $1 - p$ of a normal conductor. In the London theory of superconductivity, the basic electrodynamic properties are described by two macroscopic equations:

$$\mathbf{E} = \frac{\partial}{\partial t}\left(\frac{m}{n_s q^2}\mathbf{I}_s\right), \tag{295}$$

$$\mathbf{B} = -c\nabla \times \left(\frac{m}{n_s q^2}\mathbf{I}_s\right). \tag{296}$$

Equation (295) simply describes the undamped response of n_s particles per unit volume, each of mass m and charge q, to the force generated by an electric field $\mathbf{E}$. If we combine Eq. (296) with the Maxwell equation,

$$\nabla \times \mathbf{B} = \frac{4\pi}{c}\mathbf{I} \tag{297}$$

we obtain

$$\left(\nabla^2 - \frac{1}{\Lambda^2}\right)\mathbf{B} = 0, \tag{298}$$

where

$$\Lambda = \sqrt{\frac{mc^2}{4\pi n_s q^2}} \tag{299}$$

is the *London penetration depth* of the superconductor. Physically, Eq. (298) implies that the magnetic field decays exponentially within the superconductor with a decay length Λ.

To obtain an analogous quantity in a normal conductor-superconductor composite, we write Eq. (295) in the frequency domain as

$$\mathbf{I}_S = g_S \mathbf{E}, \tag{300}$$

$$g_S = \frac{iA}{\omega}, \tag{301}$$

where $A = c^2/(4\pi\Lambda^2)$, and g_S is clearly an imaginary, frequency-dependent (inductive) conductivity of the superconducting component.

5.18.3 *Comparison with the Experimental Data*

Let us now consider some percolation properties of superconducting materials that have been tested either by experimental observations or by numerical simulations.

5.18.3.1 The London Penetration Depth

For $p > p_c$, where p_c is the percolation threshold of the superconducting component, the conductivity of the composite at sufficiently low frequencies is dominated by g_S. We therefore deduce that

$$g_e \equiv \frac{iA_e}{\omega} \sim g_S(p - p_c)^{\mu}, \tag{302}$$

which implies that $A_e \sim A(p - p_c)^{\mu}$, or

$$\Lambda_e \sim (p - p_c)^{-\mu/2}. \tag{303}$$

Experimental test of Eq. (303) would require making a series of samples with different p, and then measuring the penetration depth of each sample. Similar to the measurement of the exponent s below p_c, such a direct test would be complicated by the proximity effect. Measurements of the temperature-dependent penetration depth in a single sample of composite superconductor have also been interpreted in terms of a temperature-dependent volume fraction $p(T)$ of the superconductor. Assuming that $p(T)$ is smooth, the normal conductor-superconductor transition occurs at a critical temperature T_c such that, $p(T_c) = p_c$. If we expand $p(T)$

around T_c, we obtain, $p(T) - p(T_c) = \alpha(T - T_c)$, where $\alpha < 0$. This argument then yields

$$\Lambda(T) \sim (T_c - T)^{-\mu/2}, \tag{304}$$

near T_c, in contrast to the usual Bardeen–Cooper–Schrieffer temperature dependence, $\Lambda(T) \sim (T_c - T)^{-1/2}$. It also predicts that for the resistivity above T_c, $R(T) \propto (T - T_c)^s$, which differs from the predictions of superconducting fluctuation theory. Both predictions rely, of course, on a somewhat arbitrary (linear) translation from volume fraction p to temperature T, and it remains to be seen whether this assumption is valid.

5.18.3.2 The Specific Heat

Deutscher *et al.* (1980) proposed a percolation model for the onset of superconductivity by assuming that the grains are coupled if $E_j > k_B T$, where k_B is the Boltzmann's constant, and E_j is the Josephson energy coupling. Because the coupling energy depends on T, more and more grains become coupled as T is lowered. The coupling is obtained randomly with a probability that, similar to what was discussed above, depends on T. This assumption defines a percolation process, because when the coupling probability is equal to the percolation threshold, a sample-spanning cluster of coupled superconducting grains is formed. If the temperature is still lowered, the sample-spanning cluster of the superconducting grains grows in size. The distributions of the grain sizes and the junction resistances give rise to randomness in the coupling energy. In Deutscher *et al.*'s model, the coupling between the grains can be very strong, giving rise to a broad grain size distribution. In this case the randomness in the coupling energy is due to the dependence on the grain size of the temperature at which a grain becomes superconducting. If, on the other hand, the grain size distribution is narrow, the coupling would be weak, which was also the limit that Deutscher *et al.* considered. They used a dense random packing of hard spherical grains with a size of about 30 Å, small enough that only the sample-spanning cluster of the superconducting grains was the main contributor to the specific heat of the system, the main property that was calculated. The coupling energy was estimated by assuming that the normal-state resistances of the junctions follow a Gaussian distribution with a variance proportional to the mean. The mean of the distribution was treated as the only adjustable parameter of the model. Based on this model, the temperature dependence of the specific heat of the system was calculated, and was found to be in excellent agreement with the experimental data for granular Al-Al_2O_3 films.

5.18.3.3 The Critical Current

Although there is Josephson tunneling through the junctions, it does not generate any potential difference between the grains up to a *critical current* I_c. If the temperature of the system is lowered below the critical temperature T_c of the system, the number of superconducting links N_ℓ (per unit cross section area) increases. Percolation theory provides estimates for N_ℓ, the critical current I_c, and the critical temperature T_c at which the material becomes superconducting. In the percolation model of Deutscher *et al.* (1980) T_c is close to the temperature at which the specific heat attains its maximum. Alternatively, T_c may be defined as the temperature at which 50% of transition in the resistivity occurs. Near p_c one has the scaling law

$$I_c \sim (p - p_c)^v, \tag{305}$$

where v is a new critical exponent. Near the percolation threshold, the granular superconductor can be viewed as a network of superconducting blobs connected by links (see Section 2.6.3). The links are quasi-linear chains of Josephson junctions. The current that flows through parallel links is inversely proportional to N_ℓ, and the distance between the links is roughly proportional to the percolation correlation length ξ_p. Thus, for a d-dimensional system, N_ℓ is of the order $1/\xi_p^{d-1}$. Since $I_c \sim N_\ell$, we obtain

$$I_c \sim (p - p_c)^v \sim (p - p_c)^{\nu(d-1)}, \tag{306}$$

so that v is a purely topological exponent. Equation (306) was confirmed by Monte Carlo simulations in 2D percolation networks (Kirkpatrick 1979a, Lobb and Frank 1979), and by careful experiments of Deutscher and Rappaport (1979) who prepared thin (2D) superconductor–semiconductor films of Pb-Ge and measured their critical currents. They reported $v \simeq 1.3 \pm 0.1$, in very good agreement with the 2D prediction, $v = \nu = 4/3$.

5.18.3.4 The Critical Fields

The variation of the magnetic induction B with H in typical type-I and type-II superconductors is well known. This behavior can be interpreted in terms of a field-dependent magnetic permeability $k_e = B/H$ which, for $H < H_c$ (or $H < H_{c1}$) is zero. Above H_c in type-I superconductors, k_e rises discontinuously to unity. In type-II superconductors, on the other hand, k_e rises continuously above H_{c1}, reaching unity at H_{c2}. According to this analysis, H_c or H_{c1} may be viewed as a type of breakdown field above which the superconducting component begins to become a normal conductor. Thus, this problem is seemingly similar to the phenomenon of dielectric breakdown in a network of insulating and conducting bonds below the conductivity threshold, which will be described in Chapter 5 of Volume II. The field $\mathbf{H}$ is curl free ($\nabla \times \mathbf{H} = \mathbf{0}$) because the screening currents are viewed as magnetization currents and therefore do not constitute a source for $\mathbf{H}$. In a similar manner, the analogue of $\mathbf{I}$ is the divergenceless field $\mathbf{B}$.

If this analogy is valid, one may deduce $H_{c1}(p)$ from known results of the dielectric breakdown problem. Thus, at a fixed composite volume, we expect to have

$$H_{c1}(p) \sim H_{c1}(0) \left[p_c^{(n)} - p\right]^y, \tag{307}$$

where $y = \nu$; the derivation of this equation (or its analogue for the dielectric breakdown problem) will be described in Chapter 5 of Volume II. Here, $p_c^{(n)}$ is the volume fraction of the superconducting material at which the normal conductor first forms a sample-spanning cluster, and ν is the usual percolation correlation length exponent. In 3D, as well as in some 2D systems, $p_c^{(n)} > p_c$, and therefore there exists a finite range of volume fractions where the composite is an electrically-perfect conductor but has zero critical field, since at such volume fractions the composite is bicontinuous and hence can transport both supercurrent (through the superconducting component) and flux (through the normal component). In a real composite, the validity of this analysis presumably depends on the manner in which the magnetic field is introduced. If a bicontinuous composite is cooled in a field, the flux expelled from the superconducting component might still continue to thread the composite even below T_c, just as in a single superconducting loop. But if the field is turned on below T_c, it will be screened out by the induced supercurrents. Moreover, if the flux through a given link of normal conductor is sufficiently small, and if the superconducting grains are all large compared to the penetration depth Λ_S of the superconducting material (as assumed in the above discussion), this model must be modified by requirement of flux quantization through each link.

A characteristic length scale of granular superconductors is their superconducting coherence length ξ_s, already mentioned above, which can be defined as the diffusion length over a characteristic time t_c for the relaxation of the order parameter at a given T. Therefore, one has $\xi_s \sim \sqrt{Dt_c}$, where D is a diffusion coefficient, and $t_c \sim (T_c - T)^{-1}$. If $\xi_s > \xi_p$, one is in the homogeneous regime in which $H_{c2} \sim (T_c - T)$ near T_c. The upper critical field of a type-II superconductor is proportional to ξ_s^{-2}. In the inhomogeneous regime, where $\xi_s \ll \xi_p$, one has fractal or anomalous diffusion (see Sections 6.1.9 and 6.1.10 for a comprehensive discussion), instead of Fickian diffusion. Since ξ_s is defined as a diffusion length scale, ξ_s^2 should be proportional to $\langle R^2(t)\rangle$, the mean square displacement of a diffusing particle at time t which, for anomalous diffusion, is given by, $\langle R^2(t)\rangle \sim t^{2/(2+\theta)}$, where $\theta = (\mu - \beta)/\nu$, with μ, β and ν being the usual percolation exponents. Therefore, one obtains, $H_{c2} \sim \xi_s^{-2} \sim t_c^{-2/(2+\theta)}$, implying that

$$H_{c2} \sim (T_c - T)^{2/(2+\theta)}. \tag{308}$$

Equation (308) was confirmed experimentally by Gerber and Deutscher (1989), who measured the upper critical field of thin semicontinuous lead films on Ge substrates and found that $H_{c2} \sim (T_c - T)^{\gamma}$ with $\gamma \simeq 0.66 \pm 0.06$. Since in 2D, $\theta \simeq 2.87$, the percolation prediction, $\gamma = 2/(2+\theta) \simeq 0.7$, is in very good agreement with the measured value.

5.18.3.5 Differential Diamagnetic Susceptibility

The last property that we consider is the differential diamagnetic susceptibility χ of a superconducting composite near the percolation threshold p_c. It is known that in bulk superconductors near T_c, χ increases sharply in magnitude due to diamagnetic fluctuations, i.e., momentary fluctuations of the normal conductor into the superconducting state. A loosely analogous phenomenon occurs in composites, as was first discussed by de Gennes (1981) who noted that for $p < p_c$, the superconducting component is present only in the form of finite clusters, some of which contain closed loops. In the presence of an applied DC magnetic field, diamagnetic screening currents can flow in these loops, giving rise to a finite diamagnetic susceptibility χ. As $p \to p_c$, the loops become larger and larger, as does χ. One therefore expects χ to diverge at p_c according to a power law:

$$\chi_d \sim (p_c - p)^{-b}, \tag{309}$$

where b depends on the dimensionality of the system, and is a measure of the number and size of loops of the network. Equation (309) was first proposed by de Gennes (1981), who also suggested that $b = 2\nu - \mu$ in all dimensions. Rammal *et al.* (1983) proposed an alternative relation, $b = 2\nu - \mu - \beta$, where μ, β, and ν are the usual percolation exponents. The de Gennes formula agrees slightly better with the numerical simulations in 2D, while that of Rammal *et al.* agrees better in 3D. Lagarkov *et al.* (1987) suggested yet another relation, $b = \nu$ in 2D and $b = 0$ (i.e., a logarithmic divergence) in 3D, and also argued that the divergence is removed if one considers the inductive interaction between screening currents induced in different clusters (an effect omitted in all the other estimates which considered such clusters as independent).

Summary

Several methods for estimating transport properties of disordered materials, both analytical and numerical, are now available. These methods are based on the discrete models of heterogeneous materials, and have provided much better understanding of the transport properties of composites. Some of these methods, such as the effective-medium approximations, provide accurate estimates of the transport properties if the material is not near its percolation threshold and its morphology does not contain correlations, while others, such as the renormalized EMA, are even more accurate and capable of providing accurate predictions for the transport properties under a wide variety of conditions. However, when the heterogeneities are broadly distributed, the EMA breaks down and loses its accuracy. These methods, together with those that were described in Chapter 4, have provided a fairly complete understanding of many transport properties of disordered materials.

6
Frequency-Dependent Properties: The Discrete Approach

6.0 Introduction

In this chapter we continue the discussions that we began in Chapters 4 and 5, and describe methods for computing the effective dynamical properties of heterogeneous materials, and in particular their effective conductivity, including hopping and AC conductivity, the effective diffusivity, and the dielectric constant. The main difference between what we discuss in this chapter and what was considered in Chapters 4 and 5 is that, here we consider mostly frequency- and/or time-dependent properties. Of particular interest to us are dynamical properties of *strongly heterogeneous* materials, i.e., those with large contrast between the properties of their constituent phases. The prototype for such heterogeneities is percolation disorder, and thus we emphasize predicting and computing frequency- and/or time-dependent transport properties in materials with this type of disorder. As we emphasize throughout this book, one main reason for considering percolation-type disorder is that, we believe that if a theoretical approach (or a computational one for that matter) can provide reasonable estimates of the transport and optical properties of materials with this type of disorder, it should be at least equally accurate for midler types of disorder. As shown in Chapters 4 and 5, far from the percolation threshold p_c, the effective properties of composite materials, under steady-state conditions, are accurately predicted by the effective-medium approximation (EMA) and similar mean-field theories, rigorous bounds and cluster expansions. One of the main goals of this chapter is to describe the application of the EMA and similar approximations to time- and/or frequency-dependent phenomena. However, because as discussed in Chapters 4 and 5, the EMA and similar theories are not accurate near p_c, we also describe numerical methods of computing the effective properties and emphasize, as we did in Chapter 5, the scaling properties of the composite materials in the critical region near p_c, where percolation and clustering effects are dominant and universal properties, independent of the microscopic details of the materials, are prevalent. As we discuss later in this chapter, it is often true that what is usually referred to as the critical region is quite broad, and therefore the universal power laws, that are presumably valid only near p_c, can be very useful for predicting the effective properties of composite materials, and developing correlations for them, over a broad range of the volume fractions of the composites' phases. We begin this chapter by discussing time-dependent diffusion processes in heterogeneous materials, after which other types of transport phenomena will be considered.

6.1 Diffusion in Heterogeneous Materials

Because in a heterogeneous material the local diffusivity varies spatially, the relevant diffusion equation is, $\partial P/\partial t = \nabla \cdot (\mathcal{D}\nabla P)$, where $\mathcal{D}$ denotes a spatially-varying diffusion coefficient. Discretizing this equation by a finite-element or finite-difference method yields a master equation:

$$\frac{\partial P_i}{\partial t} = \sum_j W_{ij}[P_j(t) - P_i(t)], \tag{1}$$

where W_{ij} is related to the transport coefficients $\mathcal{D}_i$ and $\mathcal{D}_j$ at points i and j of the system. For example, for a linear chain, $W_{ij} = (\mathcal{D}_i + \mathcal{D}_j)/(2\delta^2)$, where δ is the distance between i and j. Equation (1) can also be viewed as a probabilistic description of time-dependent diffusion (and conduction) in a disordered material modeled by a disordered lattice or network, as $P_i(t)$ can be thought of as the probability that a diffusing particle would be found at site i at time t, given that it started its motion at the origin. According to this interpretation then, W_{ij} would be the transition rate between sites i and j, i.e., the probability of diffusing or hopping from i to j, taken to be non-zero only when sites i and j are nearest neighbors. More general situations in which the diffusants can jump to second nearest-neighbor sites, third nearest-neighbor sites, and so on, can also be considered, although the computations will be considerably more complex. In general, the transition rates do not have to be symmetric (i.e., in general, $W_{ij} \neq W_{ji}$), and in fact finite-difference or finite-element discretization of the continuum diffusion equation with a spatially-varying diffusivity will lead to an equation which is more complex than Eq. (1) as it will contain asymmetric transition rates. However, for now we ignore such complexities. Equation (1) is sometimes referred to as the *symmetric random hopping model* (see also Section 6.2.2). The probabilistic interpretation of Eq. (1) also allows us to study diffusion (and conduction) in heterogeneous materials by random walk (Brownian) simulation, the essentials of which were already described in Sections 4.10 and 5.14.4.

Let us point out that, in addition to modeling of ordinary diffusion in solid materials, Eq. (1) can also be used for investigating conduction in a variety of other materials. For example, large conductivity observed in superionic conductors is often interpreted in terms of the symmetric random hopping model, Eq. (1). This interpretation is appropriate so long as the average time $\langle t_h \rangle$ between the hops is considerably larger than the actual hopping time t_h. If the frequency-dependent conductivity is of interest, then, as Kimball and Adams (1978) showed, Eq. (1) can be used only if it yields a conductivity which increases monotonically with frequency up to frequencies comparable with those of phonons. We will come back to this point later in this chapter.

The exact solution of Eq. (1) is not available for any two- or three-dimensional (3D) model of heterogeneous materials. Aside from numerical simulations, a powerful method of analyzing transient diffusion (and conduction) in heterogeneous materials is by perturbation expansion. The general technique for developing perturbation expansion for such problems was already described in Section 5, where we analyzed steady-state conduction. In this chapter we extend that formulation to time-dependent diffusion (and conduction) problem which parallels closely the perturbation expansion developed in Chapter 5, with the main difference being the fact that, due to transient nature of the problem, certain subtleties arise that must carefully be treated; these subtleties will be discussed below. The formulation presented below and its predictions were first developed by Sahimi *et al.* (1983a).

6.1.1 Green Function Formulation and Perturbation Expansion

Taking the Laplace transform of Eq. (1) yields

$$\lambda \tilde{P}_i(\lambda) - \delta_{i0} = \sum_j W_{ij}[\tilde{P}_j(\lambda) - \tilde{P}_i(\lambda)], \tag{2}$$

where λ is the Laplace transform variable conjugate to t, and the initial condition $P_i(0) = \delta_{i0}$ has been used. A *uniform* lattice or network is now introduced in which all the transition rates are equal to $\tilde{W}_e(\lambda)$, with site occupation probabilities being $\tilde{P}_i^0(\lambda)$, so that

$$\lambda \tilde{P}_i^0(\lambda) - \delta_{i0} = \sum_j \tilde{W}_e(\lambda)[\tilde{P}_j^0(\lambda) - \tilde{P}_i^0(\lambda)], \tag{3}$$

where the effective transition rate $\tilde{W}_e$ is yet to be determined. If one subtracts Eq. (2) from (3), after some rearrangements one obtains

$$(Z_i + \epsilon)[\tilde{P}_i(\lambda) - \tilde{P}_i^0(\lambda)] - \sum_j [\tilde{P}_j(\lambda) - \tilde{P}_j^0(\lambda)] = -\sum_j \Delta_{ij}[\tilde{P}_i(\lambda) - \tilde{P}_j(\lambda)], \tag{4}$$

where Z_i is the coordination number of site i, $\epsilon = \lambda/\tilde{W}_e$, and $\Delta_{ij} = (W_{ij} - \tilde{W}_e)/\tilde{W}_e$. A *site-site* Green function G_{ik} is now introduced by the equation

$$(Z_i + \epsilon)G_{ik} - \sum_j G_{jk} = -\delta_{ik}. \tag{5}$$

G_{ik} is the response of the system at k if a unit current is injected into the network at i. In terms of G_{ik}, Eq. (4) is rewritten as

$$\tilde{P}_i(\lambda) = \tilde{P}_i^0(\lambda) + \sum_j \sum_k G_{ij}\Delta_{jk}[\tilde{P}_j(\lambda) - \tilde{P}_k(\lambda)]. \tag{6}$$

Similar to the steady-state transport discussed in Chapter 5, Eq. (6) is an exact but implicit solution of Eq. (1) which is applicable to *any* lattice, irrespective of its dimensionality or topological structure. However, for topologically-disordered lattice, such as the Voronoi network (see, for example, Sections 2.10 and 2.11), further progress requires a statistical treatment of the Green function appropriately coupled to the topological disorder of the network. The statistical treatment of G_{ik} will be discussed later in this chapter, while its derivation for regular lattices is described in the next section.

Equation (6) can also be rewritten in terms of the "flux," $\tilde{Q}_{ij}(\lambda) = \tilde{P}_i(\lambda) - \tilde{P}_j(\lambda)$. We denote bonds with Greek letters, assign directions to them, and let $\gamma_{\alpha\beta} = (G_{il} + G_{jk}) - (G_{jl} + G_{ik})$ be a *bond-bond* Green function, where i and l (j and k) are the network sites with tails (heads) of arrows on bonds α and β, respectively. Similar to G_{ij}, the bond-bond Green function $\gamma_{\alpha\beta}$ is the response of the system in bond β if a unit current is injected into the bond α. In terms of $\gamma_{\alpha\beta}$ one can rewrite Eq. (6) as

$$\tilde{Q}_\alpha(\lambda) = \tilde{Q}^0_\alpha(\lambda) + \sum_\alpha \Delta_\beta \gamma_{\alpha\beta} \tilde{Q}_\beta(\lambda). \tag{7}$$

Similar to Eq. (6), (7) is also exact, and expresses the flux $\tilde{Q}_\alpha$ in the bond α in the heterogeneous network as the sum of the flux $\tilde{Q}^0_\alpha$ in the same bond but in the uniform medium and the fluctuations in the flux through that bond that arises as the result of the heterogeneity of the network.

6.1.2 Self-Consistent Approach

The effective transition rate $\tilde{W}_e$ is calculated by the following *self-consistency* condition: Solve Eq. (7) for an arbitrary cluster of bonds with transition rates Δ_β. Then select any bond α of the cluster and require that

$$\langle \tilde{Q}_\alpha \rangle = \tilde{Q}^0_\alpha, \tag{8}$$

where $\langle \cdot \rangle$ denotes an average over all possible transition rates of all the bonds in the cluster. Equation (8), when rewritten as $\langle \tilde{Q}_\alpha - \tilde{Q}^0_\alpha \rangle = 0$, implies that, on average, the fluctuations in the flux in bond α that are caused by the heterogeneities must vanish. If one substitutes the effective transition rate so obtained in Eq. (3) and invert the Laplace transform, one obtains

$$\frac{\partial P_i(t)}{\partial t} = \sum_j \int_0^t W_e(t - \tau)[P_j(\tau) - P_i(\tau)]\mathrm{d}\tau. \tag{9}$$

Equation (9), which is referred to as a *generalized master equation*, indicates that matching a heterogeneous medium to a uniform one induces *memory effects*.

Before proceeding further, we must point out that, although the effective transition rate W_e that one determines by the self-consistent condition (8) is independent of spatial position, it must be time-dependent. To see this, we need only consider the most naive approximate analysis of Eq. (1) in which all the transition rates W_{ij} are replaced by a single, constant (independent of time) rate W_c. The resulting equation is the discretized equation for a homogeneous material and can therefore predict *only* macroscopic diffusive behavior. Consider now a two-phase material with conducting and insulating phases, modeled by a random network with a distribution of the transition rates which is percolation-like, i.e., only a fraction $p < 1$ of all bonds of the network are open or conducting (have $W_{ij} > 0$). Percolation theory has taught us that if $p < p_c$, no connected path of open bonds spans the lattice, and thus macroscopic diffusion is precluded, whereas the naive replacement of all the transition rates with a constant W_c still predicts macroscopic diffusion, and thus such an approximation breaks down. Only when the analysis is performed in the Laplace transform space with a time-dependent $\tilde{W}_e(\lambda)$, are these difficulties circumvented.

6.1.3 *Self-Consistent Approximation, Generalized Master Equation and Continuous-Time Random Walks*

Kenkre *et al.* (1973) showed that, for a periodic lattice, a generalized master equation of the following form

$$\frac{\partial}{\partial t}P(l,t) = \int_0^t \phi(t-\tau)\sum_{l'}[W(l-l')P(l',\tau) - W(l'-l)P(l,\tau)]d\tau, \quad (10)$$

can be put into a one-to-one correspondence with a continuous-time random walk (CTRW) on the same lattice (for an in-depth study of CTRWs see, for example, Hughes, 1995). In a CTRW the random walker waits for a time t before making a jump from one site to another. The waiting time is a statistically-distributed quantity characterized by a *waiting time density* $\psi(t)$. In Eq. (10), $W(l-l')$ is the transition probability from site l' to site l, and $\phi(t)$ is a memory kernel which, in the Laplace transform state, is related to the waiting time density by

$$\tilde{\psi}(\lambda) = \frac{\tilde{\phi}(\lambda)}{\lambda + \tilde{\phi}(\lambda)}. \quad (11)$$

If we restrict our attention to periodic lattices with lattice sites defined by position vectors $\mathbf{l}$ and diffusion commencing from site $\mathbf{l} = \mathbf{0}$, we have

$$\lambda\tilde{P}(\mathbf{l},\lambda) - \delta_{\mathbf{l},\mathbf{0}} = \tilde{W}_e(\lambda)\sum_{\mathbf{l}'}[\tilde{P}(\mathbf{l}',\lambda) - \tilde{P}(\mathbf{l},\lambda)]. \quad (12)$$

This equation can be analyzed as a generalized master equation, or its equivalent CTRW, characterized by a waiting time density $\psi(t)$, with

$$\tilde{\psi}(\lambda) = \frac{Z\tilde{W}_e(\lambda)}{\lambda + Z\tilde{W}_e(\lambda)}, \quad (13)$$

where Z is the coordination number of the lattice.

6.1.4 *The Green Functions*

We now describe how the Green function G_{ik} is derived for a variety of periodic lattices. Such functions arise in a variety of contexts, such as numerical analysis, theory of lattice dynamics within the harmonic approximation (see, for example, Maradudin *et al.*, 1971), and in the theory of lattice random walks (see, for example, Hughes, 1995). Although such Green functions have been extensively studied, it is useful to collect here most of the results of interest.

Consider first a simple-cubic lattice in d dimensions, so that each site has a coordination number $Z = 2d$. Because G_{ik} depends only on the relative positions of sites i and k, there is no loss of generality in writing $k = (0, \cdots, 0)$ and $i = (m_1, m_2, \cdots, m_d)$, with m_i being integers, and defining $G_{ik} \equiv G(m_1, \cdots, m_d)$. Equation (5) reduces to the difference equation

$$\begin{aligned}(2d+\epsilon)G(m_1, m_2, \cdots, m_d) - [G(m_1+1, m_2, \cdots, m_d) + G(m_1-1, m_2, \cdots, m_d)\\ +G(m_1, m_2+1, \cdots, m_d) + G(m_1, m_2-1, \cdots, m_d)\\ +\cdots + G(m_1, m_2, \cdots, m_d+1) + G(m_1, m_2, \cdots, m_d-1)] = -\delta_{m_1 0}\delta_{m_2 0}\cdots\delta_{m_d 0}.\end{aligned} \tag{14}$$

We now define a discrete Fourier transform by

$$\hat{G}(\theta_1, \theta_2, \cdots, \theta_d) = \sum_{m_1=-\infty}^{\infty}\sum_{m_2=-\infty}^{\infty}\cdots\sum_{m_d=-\infty}^{\infty} \exp(im_1\theta_1 + im_2\theta_2 + \cdots + im_d\theta_d)G(m_1, m_2, \cdots, m_d), \tag{15}$$

with the help of which Eq. (14), in the Fourier space, becomes

$$[2d+\epsilon - 2(\cos\theta_1 + \cos\theta_2 + \cdots + \cos\theta_d)]\hat{G}(\theta_1, \theta_2, \cdots, \theta_d) = -1. \tag{16}$$

Therefore, by inverting the Fourier transform one obtains

$$\begin{aligned}G(m_1, \cdots, m_d) &= -\frac{1}{2}\frac{1}{(2\pi)^d}\int_{-\pi}^{\pi}\int_{-\pi}^{\pi}\cdots\int_{-\pi}^{\pi}\frac{\exp(-im_1\theta_1 - im_2\theta_2 - \cdots - im_d\theta_d)}{d+\frac{1}{2}\epsilon - (\cos\theta_1 + \cos\theta_2 + \cdots + \cos\theta_d)}d\theta_1\cdots d\theta_d\\ &= -\frac{1}{2}\frac{1}{(2\pi)^d}\int_{-\pi}^{\pi}\int_{-\pi}^{\pi}\cdots\int_{-\pi}^{\pi}\exp(-im_1\theta_1 - im_2\theta_2 - \cdots - im_d\theta_d)d\theta_1 d\theta_2\cdots d\theta_d\\ &\quad\times\int_0^{\infty}\exp[-(d+\frac{1}{2}\epsilon)t + (\cos\theta_1 + \cdots + \cos\theta_d)t]dt\\ &= -\frac{1}{2}\int_0^{\infty}\exp[-(dt+\frac{1}{2}\epsilon)]I_{m_1}(t)\cdots I_{m_d}(t)dt,\end{aligned} \tag{17}$$

where we have used the integral representation of the modified Bessel function,

$$I_m(t) = \frac{1}{2\pi}\int_{-\pi}^{\pi}\cos(m\theta)\exp(t\cos\theta)d\theta. \tag{18}$$

Integral representations of the Green function for other periodic lattices are obtained by mapping the sites of these lattices onto a subset of the sites of a simple-cubic lattice. For example, any site j of the BCC lattice is coupled to eight sites, the coordinates of which differ from those of j by $(\pm 1, \pm 1, \pm 1)$, while any site j of the FCC lattice is coupled to 12 sites with coordinates that differ from those of j by $(\pm 1, \pm 1, 0)$, $(\pm 1, 0, \pm 1)$, or $(0, \pm 1, \pm 1)$. Given such considerations, it is straightforward to show that for the BCC lattice,

$$G(\mathbf{m}) = -\frac{1}{(2\pi)^3} \int_{-\pi}^{\pi} \int_{-\pi}^{\pi} \int_{-\pi}^{\pi} \frac{\exp(-im_1\theta_1 - im_2\theta_2 - im_3\theta_3)}{8 + \epsilon - 8\cos\theta_1 \cos\theta_2 \cos\theta_3} d\theta_1 d\theta_2 d\theta_3, \tag{19}$$

while for the FCC lattice,

$$G(\mathbf{m}) = -\frac{1}{(2\pi)^3} \int_{-\pi}^{\pi} \int_{-\pi}^{\pi} \int_{-\pi}^{\pi} \frac{\exp(-im_1\theta_1 - im_2\theta_2 - im_3\theta_3)}{12 + \epsilon - 4(\cos\theta_1 \cos\theta_2 + \cos\theta_2 \cos\theta_3 + \cos\theta_3 \cos\theta_1)} d\theta_1 d\theta_2 d\theta_3 \tag{20}$$

where $\mathbf{m} = (m_1, m_2, m_3)$. These Green functions vanish at those sites $\mathbf{m}$ of the underlying simple-cubic lattice that do not belong to the BCC or FCC lattice. Ishioka and Koiwa (1978) derived the Green function for the diamond lattice.

The triangular lattice may be analyzed as a sheared square lattice with one diagonal interaction or, more simply, as a square lattice with site j coupled to the six sites with coordinates that differ from those of j by $(\pm 2, 0)$ or $(\pm 1, \pm 1)$. Then,

$$G(m_1, m_2) = -\frac{1}{(2\pi)^2} \int_{-\pi}^{\pi} \int_{-\pi}^{\pi} \frac{\exp(-im_1\theta_1 - im_2\theta_2)}{6 + \epsilon - 4\cos\theta_1 \cos\theta_2 - 2\cos 2\theta_2} d\theta_1 d\theta_1, \tag{21}$$

with the integral vanishing identically for those (m_1, m_2) which do not correspond to sites of the triangular lattice. At all other sites, when $\epsilon = 0$, the integrand has non-integrable singularities at the central point $(\theta_1, \theta_2) = (0, 0)$, and also at the four corners $(\theta_1, \theta_2) = (\pm\pi, \pm\pi)$

The hexagonal lattice is more difficult to analyze, because it is *not* translationally invariant. Horiguchi (1972) derived the Green function for this lattice, and here we merely quote the results:

$$G_{\text{hex}}(2m_1, m_2) = -\frac{1}{(2\pi)^2} \int_{-\pi}^{\pi} \int_{-\pi}^{\pi} \frac{(3 + \epsilon)\exp(-im_1\theta_1 - im_2\theta_2)}{(3 + \epsilon)^2 - 1 - 4\cos\theta_1 \cos\theta_2 - 4\cos^2\theta_2} d\theta_1 d\theta_2, \tag{22}$$

$$G_{\text{hex}}(2m_1 + 1, m_2) = -\frac{1}{(2\pi)^2} \int_{-\pi}^{\pi} \int_{-\pi}^{\pi} \frac{\exp(-im_1\theta_1 - im_2\theta_2)[1 + 2\exp(-i\theta_1)\cos\theta_2]}{(3 + \epsilon)^2 - 1 - 4\cos\theta_1 \cos\theta_2 - 4\cos^2\theta_2} d\theta_1 d\theta_2. \tag{23}$$

A transparent derivation of the results for the hexagonal and triangular lattices was presented by Sahimi *et al.* (1983a), who also discussed many other properties of these lattice Green functions.

6.1.5 *Effective-Medium Approximation*

In practice, one cannot solve Eq. (8) for an *arbitrary* cluster of bonds, and thus suitable approximate schemes must be developed. As described in Section 5.6, in an EMA approach one assigns to all but a finite cluster of bonds in the network the effective transition rate $\tilde{W}_e(\lambda)$ (so that $\Delta_\beta \neq 0$ only for a finite set of bonds), and proceeds as above, now averaging over the transition rates of the bonds in the cluster in order to determine $\tilde{W}_e(\lambda)$. In the simplest case (the lowest-order approximation), only a single bond α has its transition rate W differing from $\tilde{W}_e(\lambda)$; this is called the *single-bond* EMA, but we simply refer to it as the EMA. In this case Eq. (8) yields

$$\left\langle \frac{1}{1-\gamma_{\alpha\alpha}\Delta_\alpha} \right\rangle = 1. \tag{24}$$

If $f(W)$ represents the statistical distribution of the transition rates, one obtains

$$\int_0^\infty \frac{f(W)}{1-\gamma_{\alpha\alpha}\Delta_\alpha} dW = 1. \tag{25}$$

We must point out that if, instead of the fluxes $\tilde{Q}_\alpha$, we require that $\langle \tilde{P}_i \rangle = \tilde{P}_i^0$, we would obtain the same equation as (25). Both methods imply that $\tilde{W}_e(\lambda)$ should be calculated by requiring that the average of the excess flux, $\langle \tilde{Q}_\alpha - \tilde{Q}_\alpha^0 \rangle$, or excess "potential," $\langle \tilde{P}_i - \tilde{P}_i^0 \rangle$, vanish. It is not difficult to show that, for periodic lattices,

$$\gamma_{\alpha\alpha} = -2/Z + (2\epsilon/Z)G(\epsilon) = -p_c + (2\epsilon/Z)G(\epsilon), \tag{26}$$

where $G(\epsilon) = -G_{ii}(\epsilon) > 0$, and $p_c = 2/Z$ is the EMA prediction of the bond percolation threshold of the periodic lattices already derived in Section 5.6. Equation (25) is general and can be used with an arbitrary $f(W)$. Let us note that, as pointed out in Chapter 5, the EMA is equivalent to the *coherent-potential approximation*.

6.1.6 The Mean Square Displacement

The properties of diffusion in a heterogeneous material can be characterized in terms of the mean square displacement $\langle R^2(t) \rangle$ of the diffusants. To calculate $\langle R^2(t) \rangle$, we apply a discrete Fourier transform to $\tilde{P}(\mathbf{l}, \lambda)$,

$$\hat{P}(\boldsymbol{\omega}, \lambda) = \sum_{\mathbf{l}} \exp(i\boldsymbol{\omega} \cdot \mathbf{l}) \tilde{P}(\mathbf{l}, \lambda), \tag{27}$$

so that

$$\hat{P}(\boldsymbol{\omega}, \lambda) = \{\lambda + Z\tilde{W}_e(\lambda)[1 - \Lambda(\boldsymbol{\omega})]\}^{-1}, \tag{28}$$

where

$$\Lambda(\boldsymbol{\omega}) = \frac{1}{Z} \sum_{\mathbf{l}'} \exp[i\boldsymbol{\omega} \cdot (\mathbf{l}' - \mathbf{l})], \tag{29}$$

is called the *structure factor* of the lattice. Then, the Laplace transform of the mean square displacement $\langle R^2(t) \rangle$ of the diffusing particles is given by

$$\tilde{R^2}(\lambda) = \sum_{\mathbf{l}} l^2 \tilde{P}(\mathbf{l}, \lambda) = -\nabla_{\boldsymbol{\omega}}^2 \hat{P}(\boldsymbol{\omega}, \lambda)|_{\boldsymbol{\omega}=\mathbf{0}}. \tag{30}$$

If the bonds of the lattice all have unit length, we can easily see that $-\nabla_{\omega}^2 \Lambda(\omega) = 1$ at $\omega = \mathbf{0}$, and therefore,

$$\langle R^2(t) \rangle = \mathcal{L}^{-1}\left[\frac{Z}{\lambda^2}\tilde{W}_e(\lambda)\right], \tag{31}$$

where $\mathcal{L}^{-1}$ denotes an inverse Laplace transform. Therefore, the behavior of the diffusion process can be characterized by the mean square displacement $\langle R^2(t) \rangle$ or, equivalently, the waiting time distribution $\psi(t)$, both of which can be obtained once W_e has been computed. Therefore, whether the computed $\langle R^2(t) \rangle$ or $\psi(t)$ is exact or an approximate estimate depends on whether W_e has been calculated exactly or approximately. Clearly, the self-consistent method and its EMA simplification provide a means of computing all the important properties of diffusion processes in a heterogeneous material.

Once the mean square displacement $\langle R^2(t) \rangle$ is computed through Eq. (31), the effective diffusivity D_e can also be calculated, as it is related to $\langle R^2(t) \rangle$ through the following equation

$$\langle R^2(t) \rangle = 2dD_e t. \tag{32}$$

Note that if we compare Eqs. (31) and (32), we see that, for d-dimensional simple-cubic lattices (for which $Z = 2d$), $D_e = W_e$.

6.1.7 *Difference Between Transport in Low- and High-Dimensional Materials*

Before proceeding with analyzing the predictions of the EMA, let us point out an important difference between diffusion in two- or lower-dimensional systems and that in $d = 3$. It is necessary to keep $\mathrm{Re}(\epsilon) \geq 0$, and insist that, when $d \leq 2, \epsilon \neq 0$. Therefore, in 2D the behavior of the Green functions near $\epsilon^+ = 0$ is particularly important, as this limit corresponds to the time $t \to \infty$. It can be shown that (Sahimi *et al.*, 1983a)

$$G(m_1, m_2) \sim \frac{A}{4\pi}\ln(\epsilon), \quad \text{as } \epsilon \to 0^+ \tag{33}$$

where A is a constant given by

$$A = \begin{cases} \sqrt{3}, & \text{hexagonal lattice} \\ 1, & \text{square lattice} \\ 1/\sqrt{3}, & \text{triangular lattice} \end{cases} \tag{34}$$

In 3D, $G_{ii} \to c$ as $\epsilon \to 0^+$, where c is a constant with $-c = 0.44822, 0.25273, 0.17415$, and 0.11206, for the diamond, simple cubic, BCC and FCC lattices, respectively.

The continuity of the Green function in three or more dimensions means that the steady-state injection problem

$$ZG_{ik} - \sum_j G_{jk} = -\delta_{ik} \tag{35}$$

can be treated as the small λ limit of the general time-dependent problem defined by Eq. (5). However, for $d = 1$ and 2, one should, in principle, treat the steady state problem separately. In this case, a Green function for the injection problem, Eq. (35), is constructed from that of the time-dependent problem, Eq. (5), by subtracting G_{kk} and letting $\epsilon \to 0^+$. For example, the time-dependent Green function for the square lattice

$$G(m_1, m_2) = -\frac{1}{2}\int_0^\infty \exp(-2t - \frac{1}{2}\epsilon) I_{m_1}(t) I_{m_2}(t) dt, \tag{36}$$

is replaced, for $\epsilon = 0$, by the Green function

$$G(m_1, m_2) = \frac{1}{2}\int_0^\infty \exp(-2t)[I_0^2(t) - I_{m_1}(t) I_{m_2}(t)] dt. \tag{37}$$

It can be shown (see, for example, Sahimi *et al.*, 1983a; Hughes, 1995) that the divergence of the Green function at $\epsilon = 0$ for $d \leq 2$ ensures that the diffusants will return to the origin of their motion, while its finiteness for $d = 3$ establishes that the random walk of the diffusants is *transient*, i.e., the diffusants need not return to the origin of their motion and will ultimately escape to infinity.

6.1.8 Predictions of the Effective-Medium Approximation

Equation (25) can be used for (approximate) analysis of diffusion in any heterogeneous material. However, a stringent test of the accuracy of the EMA is its ability for predicting the correct modes of transport in heterogeneous materials with percolation disorder. Therefore, we consider a composite material for which

$$f(W) = (1 - p)\delta_+(W) + ph(W), \tag{38}$$

i.e., a fraction $1 - p$ of the material (bonds) does not conduct (i.e., it does not allow diffusion or conduction to take place), while the local transition rates (or, equivalently, the local diffusivity or conductivity) of the rest of the material (bonds) are distributed according to $h(W)$, where $h(W)$ is any normalized probability density function. Then, Eq. (25) becomes

$$p\int_0^\infty \frac{h(W)}{1 - p_c + p_c\epsilon G + p_c(1 - \epsilon G)(W/\tilde{W}_e)} dW = \frac{p - p_c + \epsilon p_c G}{1 - p_c + \epsilon p_c G}, \tag{39}$$

which was first derived by Odagaki and Lax (1981), Summerfield (1981), and Webman (1981) by methods that are different from what is described here. We now summarize the predictions of the EMA for 1D, 2D, and 3D systems. More details are given by Sahimi *et al.* (1983a).

6.1.8.1 One-Dimensional Materials

Let us point out at the outset that, diffusion in a 1D disordered material is *not* merely of academic interest. Over the past twenty years, certain classes of materials have been discovered to which 1D or quasi-1D description applies. The quasi-1D

π-conjugated Polyacetylene (PA) and Polydiacetylene (PDA), which have found many applications, have many features that are the direct result of their low dimensionality. In particular, to describe transport properties of such materials, the description of their morphology by a 1D or quasi-1D system is essential. Many other materials can be described by such low-dimensional models, some of which were already mentioned in Chapter 3.

We should also point out that diffusion in disordered chains has been studied very extensively, as a result of which many interesting predictions, some of which exact, have emerged. For example, Sinai (1982) considered a diffusion process in a linear chain in which the diffusant, which is at site i at time t, moves to site $i+1$ with probability p_i or to site $i-1$ with probability $1-p_i$, where the probabilities p_i are distributed according to a probability density function $f(p_i)$. Subject to certain constraints on $f(p_i)$, Sinai found that the mean square displacement of the diffusant at time t is give by, $\langle R^2(t)\rangle \sim \ln^2 t$, i.e., the diffusion process is very slow and its effective diffusivity at long times is *zero*. Derrida and Pomeau (1982) considered the same problem, but removed the constraints imposed on $f(p_i)$. They showed that there exists a finite velocity to the right (to the left) only if $\langle(1-p_i)/p_i\rangle < 1$ $[\langle p_i/(1-p_i)\rangle < 1]$. If both of these inequalities are violated, then $\langle R^2(t)\rangle \sim t^x$, where x depends on $f(p_i)$. Denteneer and Ernst (1983) carried out further analysis of this problem and obtained certain exact results. Movaghar *et al.* (1987) analyzed the problem for the case in which there is a drift in the system (caused by, for example, an external potential), and one in which there is trapping in the system, i.e., the diffusants react and disappear if they hit certain sites of the system. Most of such results were reviewed by Alexander *et al.* (1981) and, more recently, by Hughes (1996) who presents an elegant and rigorous analysis of this problem, and therefore we do not discuss them here. On the other hand, the EMA provides a very accurate, yet simple, description of diffusion in 1D or quasi-1D disordered materials, and therefore we discuss its predictions in detail.

For a linear chain, $p_c = 1$, and therefore Eq. (39) reduces to

$$\int_0^\infty \frac{f(W)}{\epsilon G(\epsilon) + [1-\epsilon G(\epsilon)](W/W_e)} dW = 1, \tag{40}$$

where

$$G(\epsilon) = \frac{1}{\sqrt{\epsilon(4+\epsilon)}}, \tag{41}$$

which can be obtained easily from Eq. (17), the Green function for a d-dimensional simple-cubic lattice. Consider the case in which the chain is uninterrupted, i.e., $p = 1$. Then, so long as

$$f_{-1} = \int_0^\infty \frac{f(W)}{W} dW < \infty, \tag{42}$$

one has, as $\lambda \to 0^+$,

$$\tilde{W}_e(\lambda) \to \tilde{W}_e(0) = D_e = \frac{1}{f_{-1}}, \tag{43}$$

which is an *exact* result. However, if $f_{-1} = \infty$, we have, as $\lambda \to 0^+$,

$$\tilde{W}_e(\lambda) \sim 2[f(0)\ln(\lambda^{-1})]^{-1}, \tag{44}$$

so that the transport process evolves *subdiffusively* (that is, the growth of the mean square displacement with the time t is slower than linearly):

$$\langle R^2(t)\rangle \sim \frac{2}{f(0)}\frac{t}{\ln t}. \tag{45}$$

More severe subdiffusive behavior arise if we consider

$$f(W) = (1-\alpha)W^{-\alpha}, \quad 0 < \alpha < 1, \tag{46}$$

which, as discussed in Section 2.11, is of the type of distribution that arises in the Swiss-cheese model of a disordered material. We then find that

$$\tilde{W}_e(\lambda) \sim \left(\frac{\sin \pi\alpha}{2^\alpha c\pi}\right)^{2/(2-\alpha)} \lambda^{\alpha/(2-\alpha)}, \tag{47}$$

and

$$\langle R^2(t)\rangle \propto t^{2(1-\alpha)/(2-\alpha)}. \tag{48}$$

These results, which are exact, were first derived by Alexander *et al.* (1981).

If the chain is interrupted, i.e., if $p < 1$, and if h_{-1} is finite, then it can be shown that

$$\tilde{W}_e(\lambda) \sim \frac{p(2-p)}{4(1-p)^2}\lambda, \tag{49}$$

and

$$\langle R^2(t)\rangle = \frac{p(2-p)}{2(1-p)^2}, \tag{50}$$

so that the mean square displacement remains finite, which is not surprising since the chain is a collection of finite clusters or segments, and therefore $\langle R^2(t)\rangle$ must reflect the mean size of the clusters.

6.1.8.2 Two- and Three-Dimensional Materials

Before we proceed with presenting and discussing the EMA predictions, we should point out that extensive simulations of diffusion in 2D or 3D disordered lattices have been carried out by various research groups. The techniques for such simulations were already described in Sections 4.10 and 5.14.4. For composite materials with percolation disorder, the scaling properties of the effective diffusivity near the percolation threshold was discussed in Section 2.7.2. The scaling theory for the time-dependence of the mean square displacement will be described below. Many other properties of diffusion and random walks in disordered media were reviewed and discussed by Havlin and Ben–Avraham (1987) and Haus and Kehr (1987). Thus, in this section, we consider the predictions of the EMA which provides an approximate analytical solution of this problem.

For 2D and 3D systems, three distinct types of transport can occur.

(1) Consider first the case of *diffusive transport*, where $\tilde{W}_e(\lambda) \to \tilde{W}_e(0) > 0$ as $\lambda \to 0^+$. Given the limiting behavior of $G(\epsilon)$ described above, we obtain

$$\tilde{W}_e(0)(p/p_c)\int_0^\infty \frac{h(W)}{(1-p_c)\tilde{W}_e(0)/p_c + W}dW = \frac{p-p_c}{1-p_c}, \tag{51}$$

which can yield a solution for $\tilde{W}_e(0)$ only if $p > p_c$. That is, diffusive transport is possible only above the percolation threshold. If h_{-1} is finite, then

$$D_e = \tilde{W}_e(0) \sim \frac{p-p_c}{h_{-1}(1-p_c)}, \quad p \to p_c^+, \tag{52}$$

which implies that D_e vanishes linearly with $p - p_c$ as $p \to p_c$. Since D_e is proportional to the effective conductivity g_e of the system, this results implies that the critical exponent $\mu = 1$ [see Eq. (2.36)], which is incorrect. Therefore, the prediction of the EMA near p_c cannot be very accurate, a conclusion that was already reached in Chapter 5.

If, however, $h_{-1} = \infty$, then the results are completely different. It can be shown that if $h(W) \to h_0 > 0$, then

$$D_e \propto \tilde{W}_e(0) \sim \frac{1}{h_0(1-p_c)}\frac{p-p_c}{-\ln(p-p_c)}, \tag{53}$$

whereas if $h(W)$ is of the type that arises in the Swiss-cheese model of heterogeneous materials, i.e., if $h(W) = (1-\alpha)W^{-\alpha}$, then

$$D_e \propto \tilde{W}_e(0) \sim \left(\frac{\sin \pi\alpha}{c\pi p_c^\alpha}\right)^{1/(1-\alpha)} \frac{(p-p_c)^{1/(1-\alpha)}}{1-p_c}, \tag{54}$$

so that the EMA predicts that $\mu = 1/(1-\alpha)$, i.e., μ is non-universal, which is in qualitative (but not quantitative) agreement with what we discussed in Section 2.11, where we considered transport in continuum models of heterogeneous materials.

(2) The next transport regime to be considered is *localization*, i.e., when $\langle R^2(t)\rangle$ is finite as $t \to \infty$. In the language of CTRWs, localization is equivalent to

$$\int_0^\infty \psi(t)dt < 1, \text{ localization.} \tag{55}$$

Given this condition, we obtain in 2D

$$D_e \propto \tilde{W}_e(\lambda) \sim \frac{p_c A}{4\pi(p_c - p)} \ln(1-p/p_c)^{-1}\lambda, \tag{56}$$

$$\lim_{t\to\infty} \langle R^2(t)\rangle \sim \frac{p_c AZ}{4\pi(p_c-p)} \ln(1-p/p_c)^{-1}, \tag{57}$$

where the constant A is given by Eq. (34). The corresponding results for 3D are

$$D_e \propto \tilde{W}_e(\lambda) \sim \frac{p_c G(0)}{p_c - p}\lambda, \tag{58}$$

$$\lim_{t\to\infty} \langle R^2(t)\rangle \sim \frac{p_c Z G(0)}{p_c - p}. \tag{59}$$

Thus, localization is possible only if $p < p_c$. Note that Eqs. (56) and (58) both predict that $D_e \to 0$ as $\lambda \to 0^+$ (i.e., as $t \to \infty$), which is the correct expected behavior.

(3) The third type of transport occurs at $p = p_c$. Consider first the case for which h_{-1} is finite. Then, the EMA predicts that in 2D

$$D_e \propto \tilde{W}_e(\lambda) \sim \left[\frac{p_c A \ln\lambda}{8\pi h_{-1}(1-p_c)}\right]^{1/2} \lambda^{1/2}, \tag{60}$$

Then, as $t \to \infty$,

$$\langle R^2(t)\rangle \sim \frac{Z}{\pi}\left[\frac{p_c A \ln t}{2h_{-1}(1-p_c)}\right]^{1/2} t^{1/2}, \tag{61}$$

while the corresponding results for 3D systems are

$$D_e \propto \tilde{W}_e(\lambda) \sim \left[\frac{p_c G(0)}{h_{-1}(1-p_c)}\right]^{1/2} \lambda^{1/2}, \tag{62}$$

and as $t \to \infty$,

$$\langle R^2(t)\rangle \sim \frac{2Z}{\sqrt{\pi}}\left[\frac{p_c G(0)}{h_{-1}(1-p_c)}\right]^{1/2} t^{1/2}. \tag{63}$$

In terms of the waiting time distribution of the associated CTRW, the EMA predicts that

$$\psi(t) \sim \begin{cases} t^{-3/2}(\ln t)^{1/2}, & \text{2D}, \\ t^{-3/2}, & \text{3D}. \end{cases} \tag{64}$$

If $h(W) \to h_0$ as $W \to 0^+$, then in 2D

$$D_e \propto \tilde{W}_e(\lambda) \sim \left[\frac{p_c A \lambda}{4\pi h_0(1-p_c)}\right]^{1/2}, \tag{65}$$

and in 3D,

$$D_e \propto \tilde{W}_e(\lambda) \sim \left[\frac{2p_c G(0)\lambda}{h_0(1-p_c)\ln(1/\lambda)}\right]^{1/2}, \tag{66}$$

The mean square displacement, as $t \to \infty$, is then given by

$$\langle R^2(t)\rangle \sim \begin{cases} t^{1/2}, & \text{2D}, \\ t^{1/2}(\ln t)^{-1/2}, & \text{3D}. \end{cases} \tag{67}$$

The associated waiting time distributions are given by

$$\psi(t) \sim \begin{cases} t^{-3/2}, & \text{2D}, \\ t^{-3/2}(\ln t)^{-1/2}, & \text{3D}. \end{cases} \tag{68}$$

Finally, if $h(W) = (1-\alpha)W^{-\alpha}$, then

$$D_e \propto \tilde{W}_e(\lambda) \sim \left[\frac{A \sin \pi\alpha}{4\pi^2(2-\alpha)}\right]^{1/(2-\alpha)} \left(\frac{p_c}{1-p_c}\right)^{(1-\alpha)/(2-\alpha)} (\lambda \ln 1/\lambda)^{1/(2-\alpha)}, \tag{69}$$

$$\langle R^2(t)\rangle \sim t^{(1-\alpha)/(2-\alpha)} (\ln t)^{1/(2-\alpha)}, \tag{70}$$

for 2D materials, and

$$D_e \propto \tilde{W}_e(\lambda) \sim \left[\frac{G(0) \sin \pi\alpha}{\pi(1-\alpha)}\right]^{1/(2-\alpha)} \left(\frac{p_c}{1-p_c}\right)^{(1-\alpha)/(2-\alpha)} \lambda^{1/(2-\alpha)}, \tag{71}$$

$$\langle R^2(t)\rangle \propto t^{(1-\alpha)/(2-\alpha)}, \tag{72}$$

for 3D material.

6.1.9 Anomalous Diffusion

As the above discussions indicate, at $p = p_c$, the mean square displacement $\langle R^2(t)\rangle$ does not grow linearly with the time, rather it increases with t *slower than linearly*. This unusual phenomenon is called *anomalous diffusion* (Gefen *et al.*, 1983) or *fractal diffusion* (Sahimi *et al.*, 1983a). To better characterize this phenomenon, we introduce a *random walk fractal dimension* D_w defined by

$$\langle R^2(t)\rangle \sim t^{2/D_w}, \tag{73}$$

so that Fickian diffusion corresponds to $D_w = 2$. It is then clear that the EMA predicts that, so long as $h_{-1} < \infty$, one has a universal value, $D_w = 4$. However, if $h_{-1} = \infty$, then D_w will be non-universal. In particular, if $h(W) = (1-\alpha)W^{-\alpha}$, where $0 < \alpha < 1$, then the Eqs. (70) and (72) imply that the EMA predicts that

$$D_w = \frac{2-\alpha}{2(1-\alpha)}. \tag{74}$$

For $p > p_c$ diffusion will still be anomalous so long as $\langle R^2(t)\rangle^{1/2} \ll \xi_p$, where ξ_p is the correlation length of percolation. Therefore, one may define a time scale t_{co} such that for $t \ll t_{co}$ diffusion is anomalous, whereas for $t \gg t_{co}$ diffusion is Fickian. The time scale t_{co} is then the time at which a crossover between Fickian and anomalous diffusion takes place. It can be shown that, if h_{-1} is finite, then the EMA predicts that

$$[t_{co}(p-p_c)^2]^{-1} \ln[t_{co}(p-p_c)] \sim 1, \tag{75}$$

in 2D, and

$$t_{co} \sim (p-p_c)^{-2}, \tag{76}$$

in 3D, so that in both cases the crossover time diverges at $p = p_c$.

The divergence of t_{co} is caused by two factors. (1) As p_c is approached, the correlation length ξ_p diverges [see Eq. (2.32)], and thus the material is inhomogeneous on a rapidly increasing length scale. (2) As $p \to p_c$, the number of dead-end bonds or sites also increases rapidly. Such bonds or sites slow down the diffusion process. The divergence of the time scale t_{co} for the crossover between Fickian and anomalous diffusion has an important implication for interpreting experimental data for the diffusivity in that, one must ensure that Fickian diffusion regime has been reached before interpreting the experimental data as implying a constant diffusivity D_e.

6.1.10 Scaling Theory of Anomalous Diffusion

The accuracy of the predictions of the EMA for the time dependence of $\langle R^2(t)\rangle$ can be tested against the scaling theory of diffusion in disordered materials with percolation disorder. Suppose that diffusion takes place only on the sample-spanning percolation cluster. From Eq. (2.40) one has

$$D_e \sim (p - p_c)^{\mu-\beta} \sim \xi_p^{-\theta}, \tag{77}$$

where $\theta = (\mu - \beta)/\nu$, and μ, β and ν are, respectively, the critical exponents of the conductivity, strength of the percolation system, and the percolation correlation length ξ_p. If the mean square displacement $\langle R^2(t)\rangle^{1/2} \gg \xi_p$, then diffusion must be Fickian so that the mean square displacement grows linearly with the time t, and therefore

$$\langle R^2(t)\rangle \propto D_e t \propto t(p - p_c)^{\mu-\beta}. \tag{78}$$

On the other hand, if the size of the cluster on which diffusion is taking place is finite, which happens when $p < p_c$, then after a long enough time the diffusant has explored the cluster completely, and since the typical radius of such clusters is ξ_p, we must have

$$\langle R^2(t)\rangle = \xi_p^2 \propto |p - p_c|^{-2\nu}. \tag{79}$$

We may combine Eqs. (78) and (79) and write down a general scaling equation for the mean square displacement:

$$\langle R^2(t)\rangle^{1/2} = t^x f[(p - p_c)t^y]. \tag{80}$$

For $p > p_c$ and long times one must have $f(z) \sim z^{(\mu-\beta)/2}$ as $z \to \infty$, in order for Eq. (80) to be consistent with Eq. (78), and thus, $x + y(\mu - \beta)/2 = 1/2$. On the other hand, for $p < p_c$ the mean square displacement must become independent of t, which would be the case if $f(z) \to z^{-x/y}$ as $z \to -\infty$, implying that $x/y = \nu$. Thus, solving for x and y we obtain $x = \nu/(2\nu + \mu - \beta)$ and $y = x/\nu$. Since $\langle R^2(t)\rangle \sim t^{2x}$ we obtain

$$\langle R^2(t)\rangle \sim t^{2\nu/(2\nu+\mu-\beta)}, \tag{81}$$

which, when compared with Eq. (73), implies that the random walk fractal dimension is given by

$$D_w = 2 + \frac{\mu - \beta}{\nu} = 2 + \theta. \tag{82}$$

This result is usually attributed to Gefen *et al.* (1983). Equation (82) can also be derived by the following, more intuitive method. Since fractal or anomalous diffusion occurs only if $\langle R^2(t)\rangle^{1/2} \ll \xi_p$, one can replace ξ_p with $\langle R^2(t)\rangle^{1/2}$ [since for $\langle R^2(t)\rangle^{1/2} \ll \xi_p$ the only relevant length scale of the system is $\langle R^2(t)\rangle^{1/2}$] and write $D_e \sim \langle R^2(t)\rangle^{-\theta/2}$ [see Eq. (77)]. On the other hand, Eq. (32) implies that $D_e \sim d\langle R^2(t)\rangle/dt \sim \langle R^2(t)\rangle^{-\theta/2}$, which after integration yields, $\langle R^2(t)\rangle \sim t^{2/(2+\theta)}$, and thus we recover Eq. (82).

Since, as discussed in Chapter 2, the exponents μ, β and ν depend on the dimensionality d of a system, the random walk fractal dimension D_w also varies with d. Only in the mean-field approximation, valid for $d \geq 6$, which predicts that $\mu = 3$, $\beta = 1$ and $\nu = 1/2$, does $D_w = 6$ become independent of d, but this prediction also disagrees with that of EMA, $D_w = 4$. However, using the numerical values of the exponents given in Table 2.3, one finds that $D_w(d = 2) \simeq 2.87$ and $D_w(d = 3) \simeq 3.8$, so that the 3D value of D_w is only about 5% smaller than the EMA prediction $D_w = 4$, and thus this prediction of EMA is reasonably accurate.

From the scaling equation (80) one can also obtain the time t_{co} at which a crossover between fractal and Fickian diffusion takes place. It is clear that $t_{co} \sim \xi_p^{D_w}$, and therefore

$$t_{co} \sim (p - p_c)^{-(2\nu+\mu-\beta)}. \tag{83}$$

Equation (83) implies in 3D, $t_{co} \sim (p - p_c)^{-3.34}$, which should be compared with the prediction of the EMA, Eq. (76). Thus, once again the EMA prediction is qualitatively correct.

The results so far pertain to diffusion on the sample-spanning percolation cluster. One can also consider diffusion on *all* the clusters. In this case $\langle R^2(t)\rangle$ must be averaged over all the clusters. As discussed in Section 2.6.1, the probability that a site belongs to a cluster of size s is sn_s, where n_s is the number of clusters of size s. If R_s^2 is the mean square displacement of a particle on a cluster of size s, then

$$\langle R^2(t)\rangle(t \to \infty, p < p_c) = \sum_s sn_s R_s^2 \propto |p - p_c|^{\beta-2\nu}. \tag{84}$$

Since diffusion is taking place on all the clusters, then, above p_c the diffusivity is directly proportional to g_e, the effective conductivity, and therefore in this case $D_e \sim (p - p_c)^\mu$ and $\langle R^2(t)\rangle \sim D_e t \sim t(p - p_c)^\mu$, which can be combined with Eq. (84) and be written as a single scaling equation,

$$\langle R^2(t)\rangle^{1/2} \propto t^k f[(p - p_c)t^y]. \tag{85}$$

Similar to Eq. (80), the exponents k and y can be obtained by demanding that, the scaling function $f(z)$ must be such that Eq. (85) can reproduce the results for $p > p_c$ $(z \to +\infty)$ and $p < p_c$ $(z \to -\infty)$. One then obtains $k = (\nu - \beta/2)/(2\nu + \mu - \beta)$ and $y = (2\nu + \mu - \beta)^{-1}$. Since $\langle R^2(t)\rangle \sim t^{2k}$, we obtain

$$\langle R^2(t)\rangle \sim t^{(2\nu-\beta)/(2\nu+\mu-\beta)}, \tag{86}$$

which should be compared with Eq. (81). Equation (86) predicts that in 3D one must have $\langle R^2(t) \rangle \sim t^{0.33}$, if we use the 3D values of the various exponents, which should be compared with the prediction of Eq. (80), $\langle R^2(t) \rangle \sim t^{0.527}$. Clearly, diffusion on the finite clusters further slows down the transport process. Note that, Eq. (86) implies that in the mean-field approximation, when $\nu = 1/2$, $\beta = 1$ and $\mu = 3$ (see Table 2.3), one obtains $k = 0$. However, this does not imply that in this limit $\langle R^2(t) \rangle$ approaches a constant, rather it can be shown that in this limit $\langle R^2(t) \rangle \sim \ln t$ (i.e., the mean square displacement grows with t slower than any power of t). Note also that, since the EMA replaces a heterogeneous material by a uniform one, it cannot distinguish between diffusion on the sample-spanning cluster and diffusion on all the clusters.

Let us point out that a percolation system at $p = p_c$, or one above the percolation threshold but at length scales smaller than the correlation length ξ_p, is not the only system that can give rise to anomalous diffusion. In fact, diffusion in *any* material with a fractal morphology is anomalous. Many aspects of anomalous diffusion in such materials were discussed in detail by Havlin and Ben–Avraham (1987) and Haus and Kehr (1987).

6.1.11 Comparison with the Experimental Data

Equations (81) and (86) and the predicted values of D_w have been tested and confirmed by numerical simulations of diffusion in percolation clusters. However, an interesting experimental verification of these results was presented by Knackstedt *et al.* (1995) who prepared a porous material, a ternary microemulsion comprised of three components, didodecyl dimthyl ammonium bromide, water, and cyclohexane, which was a bicontinuous water-oil system, but by tuning the volume fraction of the three components could, at high water contents, undergo a structural transition to disconnected water droplets in oil, i.e., the water phase would undergo a percolation transition, thus allowing measurement of the various mechanical and transport properties, including the diffusivity. Knackstedt *et al.* (1995) measured water self-diffusion D_e in such a system by pulsed field gradient spin-echo technique at controlled temperature of $25 \pm 0.5°C$. By varying the length of the gradient pulse $\mathcal{T}$ and maintaining a constant gradient pulse interval Δ, the diffusion coefficient was measured. The decay I of the echo density is given by

$$I = I_0 \exp[-D_e G^2 \mathcal{T}^2 \gamma^2 (\mathcal{T} - \Delta/3)], \tag{87}$$

where G is the gradient strength, γ is the gyromagnetic ratio of the observed nucleus (^1H in this case), and I_0 is the signal intensity in the absence of a gradient pulse. The gradient strength was calibrated with a sample of H_2O for which D_e was known.

Diffusion in this material corresponds to the case in which transport takes place on all the clusters, and therefore the relevant equation is (86). Anomalous diffusion was obtained at the intermediate values of the water volume fractions, 13.5%–14.1%. Since $D_e \sim d\langle R^2(t) \rangle/dt \sim t^{2k-1}$, a logarithmic plot of D_e versus t should be a straight line with the slope $2k - 1$. Knackstedt *et al.*'s measurements yielded $2k \simeq 0.3 \pm 0.1$, in good agreement with the prediction of the scaling theory $2k \simeq 0.33$ discussed above.

6.1.12 The Governing Equation for Anomalous Diffusion

Since anomalous diffusion results in an effective diffusivity which depends on the time, the classical diffusion equation with a constant diffusivity can no longer describe the diffusion process. Therefore, an important question has been the form of the equation that governs anomalous diffusion, a problem that has ben studied for many years. Havlin and Ben–Avraham (1987) provided a detailed discussion of various approaches to this problem. For simplicity we consider the problem in radial geometry and consider $P(r, t)$, the probability of finding the diffusant at position r at time t. It now appears that the correct governing equation for anomalous diffusion is given by (Metzler *et al.*, 1994; Zeng and Li, 2000)

$$\frac{\partial^{2/D_w} P(r,t)}{\partial t^{2/D_w}} = \frac{1}{r^{D_s-1}} \frac{\partial}{\partial r}\left[r^{D_s-1} \frac{\partial P(r,t)}{\partial r}\right]. \tag{88}$$

Here, $D_s = 2D_f/D_w$, where D_f is the fractal dimension of the system (for example, the fractal dimension of the sample-spanning percolation cluster). The quantity D_s is called the *spectral* or *fracton* dimensionality and, as discussed in Section 6.6.4, plays an important role in the dynamical properties of heterogeneous materials. The derivative on the left-hand side of Eq. (88) is called *fractional derivative* and is defined by

$$\frac{\partial^\gamma P(r,t)}{\partial t^\gamma} = \frac{1}{\Gamma(1-\gamma)} \frac{\partial}{\partial t} \int_0^t \frac{P(r,\tau)}{(t-\tau)^\gamma} d\tau, \tag{89}$$

where $\Gamma(x)$ is the gamma function. Using the Laplace transform, the asymptotic solution of Eq. (88) can be obtained with the result being

$$P(r,t) \sim t^{-D_s/2} \exp\left[-c\left(\frac{r}{t^{1/D_w}}\right)^{D_w/(D_w-1)}\right], \tag{90}$$

where c is a constant. It can now be easily shown that

$$\langle R^2(t)\rangle = \int_0^t r^2 P(r,t) r^{D_f-1} dr \sim t^{2/D_w}, \tag{91}$$

in agreement with Eq. (73).

6.2 Hopping Conductivity

Hopping refers to sudden displacement of a charge carrier from one position to another. In the simplest model, non-interacting carriers are placed on a simple-cubic lattice and are allowed to jump to only their nearest neighbors. The jump rates, i.e., the transition probabilities per unit time, are assumed to be symmetric, i.e., they are equal for forward and backward jumps, but, more generally, they can be unequal if, for example, an external field is imposed on the system. The major shortcomings of this model are (Dyre and Schrøder, 2000) that, (1) it ignores the fact that the charge carriers repel each other; (2) it allows an arbitrary number of charge carriers at each site, but—regardless of whether the carriers are ions or localized electrons—there is room for just one at each site, and (3) it assumes constant site energies. Thus, a more realistic model has the site energies to vary spatially, and allows only one carrier at each site. Such a model is, in general, nonlinear. However, it can be linearized with respect to an external electric field, in which case the model becomes equivalent to a symmetric hopping model. This linearization was first carried out in the context of development of a model for hopping conductivity of amorphous semiconductors, which we now discuss in detail.

Hopping conduction in amorphous semiconductors was first associated (Hung and Gliessman, 1950) with the observation that the activation energy of the conductivity in doped Ge exhibits a break at low temperatures T, and was attributed to a distinct mechanism of conduction. Mott (1956) and Conwell (1956) proposed a model of conduction in which electrons conduct by thermally-activated tunneling from a filled site to a vacant one, a process that is usually called *phonon-assisted hopping*. Their model was modified and extended by Miller and Abrahams (1960; see also the discussion by Butcher, 1976) who developed a model consisting of two parts, the quantum mechanical theory of the wave functions and of the transition rates W_{ij} from a localized state i to another localized state j, and a statistical mechanical theory of transport that employs such transition rates. They showed that their model is equivalent to a random resistor network, so that it can be used for computing the hopping conductivity of disordered solids, although, as we discuss below, it became clear later that the Miller–Abrahams resistor network model suffers from certain deficiencies.

In the early 1970s, it was realized independently by Ambegaokar, Halperin, and Langer (1971), Shklovskii and Efros (1971), and Pollak (1972) that hopping conduction in amorphous semiconductors can be modeled successfully by percolation theory. Since their seminal papers, several electronic properties of such materials have been successfully predicted by percolation theory. This section summarizes the important elements of this successful application of percolation to a technologically important problem. Our discussion is by no means exhaustive as the number of published papers on the subject is too large. Thorough discussions of the subject are given by Shklovskii and Efros (1984) and Böttger and Bryksin (1985), while an older account of some of such percolation predictions is given by Pollak (1978).

In what follows we first describe the derivation of the Miller–Abrahams resistor network, and establish contact with what we discussed in the previous section to show that an EMA can be used for estimating the hopping conductivity. We then describe the percolation model of hopping conductivity of amorphous semiconductors.

6.2.1 The Miller–Abrahams Network Model

The starting point is the Boltzmann's equation

$$\frac{\partial P_i}{\partial t} = \sum_j [W_{ji} P_j (1 - P_i) - W_{ij} P_i (1 - P_j)], \tag{92}$$

where the notation is the same as in Section 6.1. It is implicitly assumed that the occupation probabilities are uncorrelated. If repulsion can cause strong correlations, then the exclusion factor $(1 - P_i)$ should be omitted. In the linear (ohmic) regime, the current is proportional to the applied field, and therefore we can linearize Eq. (92) by writing

$$P_i = P_i^0 + \Delta P_i, \tag{93}$$

$$W_{ij} = W_{ij}^0 + \Delta W_{ij}, \tag{94}$$

where 0 denotes the equilibrium value, and Δ an increment proportional to the applied field. It is clear that this linearization implies that $\Delta W_{ij} = -\Delta W_{ji}$. We thus obtain the linearized version of Eq. (92):

$$\frac{\partial \Delta P_i}{\partial t} + \sum_j A_{ij} \Delta P_i - \sum_j A_{ji} \Delta P_j = \sum_j B_{ji} \Delta W_{ji}, \tag{95}$$

where

$$A_{ij} = W_{ij}^0 (1 - P_j^0) + W_{ji} P_j^0, \tag{96}$$

$$B_{ij} = P_i^0 (1 - P_j^0) + P_j^0 (1 - P_i^0). \tag{97}$$

Equation (95) is a set of linear equations for the unknowns ΔP_i. The equilibrium values P_i^0 are given by the Fermi distribution,

$$P_i^0 = \frac{1}{\exp(E_i / k_B T) + 1}, \tag{98}$$

where E_i is the energy of a carrier on site i, as measured from the Fermi level, and k_B is the Boltzmann's constant. The equilibrium values W_{ij}^0 are given by

$$W_{ij}^0 = \frac{u_{ij}}{|\exp[(E_j - E_i)/k_B T] - 1|}, \tag{99}$$

with

$$u_{ij} = u_{ji} = \frac{1}{\tau_0} \exp(-2 r_{ij} / a). \tag{100}$$

In Eq. (100) $1/\tau_0$ is of the order of a phonon frequency, r_{ij} is the distance between i and j, and a is a Bohr radius. It is assumed that τ_0 depends only weakly on r_{ij} and T. We point out that the above linearization is based on the non-trivial assumption that the site occupation number, which is either 0 or 1, can be replaced by a continuous variable.

Suppose that $\mathbf{F}$ is the intensity of an applied field, and $\mathbf{r}_i$ is the position of site i. The applied field changes the energy difference Δ_{ij} between the energies of sites i and j. Thus, in a linearized theory we should have

$$\Delta W_{ij} = \frac{dW_{ij}}{d\Delta_{ij}} e\mathbf{F} \cdot (\mathbf{r}_i - \mathbf{r}_j) = \frac{e\mathbf{F} \cdot (\mathbf{r}_i - \mathbf{r}_j)}{\sinh^2(\Delta_{ij}/k_BT)} u_{ij}, \tag{101}$$

where e is the charge of an electron. Miller and Abrahams defined a new variable V_i such that

$$P_i = P_i^0 + \Delta P_i \equiv \frac{1}{|\exp[(E_i - eV_i)/k_BT] + 1|}, \tag{102}$$

so that to first order

$$\Delta P_i = \frac{dP_i^0}{dE_i} eV_i = \frac{eV_i}{4k_BT \cosh^2(E_i/2k_BT)}. \tag{103}$$

The variable V_i is, in the linear regime, proportional to $\mathbf{F}$. Thus, one can transform the set of linear equations for P_i to another set for V_i, which is then given by

$$D_i \frac{\partial V_i}{\partial t} = \sum_j D_{ji} V_j - \sum_j D_{ij} V_i + \sum_j G_{ij} \mathbf{F} \cdot \mathbf{r}_{ij}, \tag{104}$$

where $D_i = P_i^0(1 - P_i^0)$, $D_{ij} = D_i A_{ij}$, and $G_{ij} = B_{ij} W_{ij}^0 W_{ji}^0 / u_{ij}$.

We now construct a network model for calculating the hopping conductivity. Consider first the steady-state case. We define a temperature-dependent conductance g_{ij} by

$$\frac{k_BT g_{ij}}{e^2} = P_i^0(1 - P_j^0) W_{ij}^0 = P_j^0(1 - P_i^0) W_{ji}^0. \tag{105}$$

If we substitute Eq. (105) into the steady-state limit of Eq. (104), we obtain

$$\sum_j \left\{ \left[V_i - \frac{\mathbf{F} \cdot \mathbf{r}_i (W_{ij}^0 + W_{ji}^0)}{u_{ij}} \right] - \left[V_j - \frac{\mathbf{F} \cdot \mathbf{r}_j (W_{ij}^0 + W_{ji}^0)}{u_{ij}} \right] \right\} g_{ij} = 0, \tag{106}$$

where $(W_{ij}^0 + W_{ji}^0)/u_{ij} = \coth(|\Delta_{ij}|/2k_BT)$. We are mainly interested in the regime for which $\coth(|\Delta_{ij}|/2k_BT) \sim 1$, in which case Eq. (106) becomes

$$\sum_j [(V_i - \mathbf{F} \cdot \mathbf{r}_i) - (V_j - \mathbf{F} \cdot \mathbf{r}_j)] g_{ij} = 0. \tag{107}$$

Equation (107) represents a network of resistors if we think of $(V_i - \mathbf{F} \cdot \mathbf{r}_i)$ as the potential at site i. Then, $Z_{ij} = 1/g_{ij}$ is the resistance between sites i and j, and Eq. (107) is simply Kirchhoff's equation for site i. Miller and Abrahams treated Z_{ij} more generally and considered it as an impedance, which is what we also do here.

For unsteady-state conduction, the time dependent term of Eq. (104) does not vanish, and Eq. (107) must be rewritten as

$$\frac{P_i^0 e^2(1-P_i^0)}{k_B T}\frac{\partial V_i}{\partial t} = \sum_j [(V_i - \mathbf{F}\cdot\mathbf{r}_i) - (V_j - \mathbf{F}\cdot\mathbf{r}_j)]g_{ij}. \tag{108}$$

To construct a network model for this case, we define a capacitance $C = P_i^0 e^2(1 - P_i^0)/k_B T$ with a potential V_i across it, and refer all the potentials to "ground" potential which is zero. Because $\mathbf{F}\cdot\mathbf{r}_i$ is the applied potential at i, it is represented as an output from a generator which is connected in series with C between the ground and site i. There is an impedance Z_{ij} connected between any two junctions i and j. There is also a capacitor C_i in series with a generator connected to the ground. Using the expressions for P_i^0 and W_{ij}^0, and restricting our attention to the case in which various site energies are of the order of, or larger than, $k_B T$, we obtain

$$Z_{ij} = k_B T\frac{\exp[(|E_i| + |E_j| + |E_i - E_j|)/2k_B T]}{e^2 u_{ij}}, \tag{109}$$

$$C_i = \frac{e^2}{k_B T}\exp(-E_i/k_B T). \tag{110}$$

Using Eq. (100), Eq. (109) is also rewritten as

$$Z_{ij} = \frac{k_B T}{e^2}\exp(E_{ij}/k_B T + 2r_{ij}/a)\tau_0, \tag{111}$$

where E_{ij} is either the energy of the site farther from the Fermi energy, or $E_{ij} = \frac{1}{2}(|E_i| + |E_j| + |E_i - E_j|)$. Equations (109) and (110) have an important implication: Even if the site energies E_i are moderately distributed, the exponential dependence of the impedance and capacitance on E_i implies that the distributions of Z_{ij} and C_i can be enormously broad. This fact is used to reduce the computations of the effective properties of the network, since the broadness of the distribution of Z_{ij} implies that there are many small conductances that can be removed from the network without affecting its overall conductivity. The network that is obtained by removing such conductances is called the *reduced network*.

The procedure for constructing the reduced network is as follows (Pollak, 1978). As the first step, one selects only the largest capacitances (for example, those within a given factor) in the network. All such capacitances are then considered as equal with a common value C, while the rest of the capacitances and their associated sites are deleted from the network. One then discards all impedances (resistances) Z_{ij} that are larger than $Z_\ell = 2/(\omega C)$, where ω is the frequency at which the effective conductivity of the network is to be computed, and replaces all impedances that are smaller than Z_ℓ by shorts. At high frequencies Z_ℓ is very small, and therefore there are very few resistances smaller than or equal to Z_ℓ. But as the frequency is lowered, clusters of connected resistors begin to appear which gradually merge together and form a sample-spanning cluster when a critical frequency ω_c and the corresponding critical impedance $Z_c = 2/(\omega_c C)$ are reached. The reduction procedure becomes ineffective for frequencies $\omega < \omega_c$ because replacing the resistances by shorts would produce a macroscopic short throughout the network.

6.2.2 The Symmetric Hopping Model

It is clear that, if in Eq. (108) we take $\mathbf{F} = \mathbf{0}$, we obtain a symmetric hopping model in which [aside from the constant factor $P_i^0 e^2(1 - P_i^0)/(k_B T)$] V_i plays the same role as that of P_i in Eq. (1), and therefore the formulation and analysis that were developed in 6.1 can be immediately utilized for computing the hopping conductivity of a disordered solid by the symmetric hopping model.

However, Eq. (1) is applicable when there is no external electric field in the system. If the field is non-zero, then the jumps in the field direction are obviously more likely, and as a result one has a net current in the system, which may seem to make the problem very complex. However, the *fluctuation-dissipation theorem* relates the frequency-dependent conductivity $\hat{g}_e(\omega)$ in terms of the equilibrium (i.e., zero-field) current autocorrelation function (Kubo, 1957). According to this theorem, if n is the density of the charge carriers, then for a system of non-interacting carriers one has

$$\hat{g}_e(\omega) = \frac{ne^2}{k_B T}\hat{D}_e(\omega), \tag{112}$$

where the frequency-dependent diffusivity $\hat{D}_e(\omega)$ is defined by

$$\hat{D}_e(\omega) = \int_0^\infty \langle v(0)v(t)\rangle e^{-i\omega t}\,dt. \tag{113}$$

Here, $v(t)$ is the velocity projected onto a fixed direction in space. At zero frequency Eq. (113) reduces to the well-known identity for the static diffusivity, in which case Eq. (112) reduces to the Nernst–Einstein relation.

6.2.2.1 Exact Solution for One-Dimensional Materials

One can derive an exact solution for the symmetric hopping model in a linear chain (Odagaki and Lax, 1980). Consider a linear chain in which each bond is present (active) with probability p and disconnected with probability $(1 - p)$. The transition rate along the active bonds is W_0. The frequency-dependent effective diffusivity of the chain can be written as

$$\hat{D}_e(\omega) = \sum_{n=1}^{\infty} np_n \hat{D}_n(\omega), \tag{114}$$

where n is the size of a cluster of active bonds, i.e., $n - 1$ active bonds terminated by an inactive bond at each end of the cluster, p_n is the probability of finding such a cluster, and $\hat{D}_n(\omega)$ is the frequency-dependent diffusivity for a cluster of size n which is given by

$$\hat{D}_n(\omega) = -\frac{\omega^2}{2n}\sum_{j,j_0}^{n}(j - j_0)^2 \tilde{P}(j, i\omega|j_0), \tag{115}$$

which is derived easily by combining Eqs. (30) and (31), and taking $\lambda = i\omega$ and $\hat{D}_n(\omega) = \tilde{W}_e(i\omega)$. Here, j_0 is the point at which the hopper begins its motion, and $\tilde{P}(j, i\omega|, j_0)$ is the Laplace transform of the probability that the hopper is at point j, given that it began its motion at j_0. The probability of finding a cluster of $n-1$ active bonds, terminated at its two ends by two inactive bonds, is given by

$$p_n = (1-p)^2\, p^{n-1}. \tag{116}$$

It remains to calculate $\tilde{P}(j, \lambda|j_0)$. Equation (2) can be written in a matrix form, $\mathbf{A}\tilde{\mathbf{P}}(\lambda, j_0) = \boldsymbol{\delta}(j_0)$, where $\mathbf{A}$, for a 1D system, is a $n \times n$ tridiagonal matrix. Since the transition rates for all the active bonds is assumed to be W_0, calculating the eigenvalues of $\mathbf{A}$ is not difficult. Suppose that $\boldsymbol{\Gamma} = (\Gamma_m, \Gamma_{3m}, \cdots, \Gamma_{(2n-1)m})$ is the eigenvector of $\mathbf{A}$. Then, it is not difficult to show (see, for example, Noble and Daniel, 1977) that, $\Gamma_m = \cos(m\pi/2n)$ (with $m = 0, 1, 2, \cdots, n-1$), and that the eigenvalue corresponding to $\boldsymbol{\Gamma}$ is simply $\lambda + 2W_0(1 - \Gamma_{2m})$. Therefore,

$$\tilde{P}(j, \lambda|j_0) = \frac{1}{n}\left\{\frac{1}{\lambda} + 2\sum_{m=1}^{n-1}\frac{\Gamma_{(2j-1)m}\Gamma_{(2j_0-1)m}}{\lambda + 2W_0(1-\Gamma_{2m})}\right\}. \tag{117}$$

Replacing λ with $i\omega$ and using Eqs. (116) and (117) in (115), we obtain the following equation for the frequency-dependent diffusivity of a cluster (finite chain) of size n (Odagaki and Lax, 1980):

$$\hat{D}_n(\omega) = 1 + \frac{1}{n}\sqrt{1 + 4/i\hat{\omega}}\left(\frac{1}{z_+^{2n}+1} - \frac{1}{z_-^{2n}+1}\right), \tag{118}$$

where $z_\pm = [(i\hat{\omega})^{1/2} \pm (4 + i\hat{\omega})^{1/2}]/2$, and $\hat{\omega} = \omega/W_0$. If we then substitute Eq. (118) into (114), we obtain the desired expression for the frequency-dependent effective diffusivity. In particular, we find that in the low-frequency limit,

$$\hat{D}_e(\omega) \sim \frac{p(1+p)^2}{4(1-p)^4}\hat{\omega}^2 + i\frac{p}{2(1-p)^2}\hat{\omega}, \tag{119}$$

while at high frequencies

$$\hat{D}_e(\omega) \sim p - \frac{2p(1-p)}{\hat{\omega}^2} + i\frac{2p(1-p)}{\hat{\omega}}. \tag{120}$$

For comparison, we also list the EMA predictions. In the low-frequency limit,

$$\hat{D}_e(\omega) = \frac{p(2-p)}{8(1-p)^4}\hat{\omega}^2 + i\frac{p(2-p)}{4(1-p)^2}\hat{\omega}, \tag{121}$$

while in the high-frequency limit,

$$\hat{D}_e(\omega) = p - \frac{2p(1-p)(2-p)}{\hat{\omega}^2} + i\frac{2p(1-p)}{\hat{\omega}}. \tag{122}$$

These results can be obtained from Eq. (38). Therefore, once again, the EMA predictions are qualitatively, but not quantitatively, correct.

6.2.2.2 Exact Solution for Bethe Lattices

An exact solution of the symmetric hopping model can be derived if one assumes that the disordered solid can be represented by a Bethe lattice of coordination number Z. We already described in Section 5.3 how the solution of the steady-state conduction problem is derived for the Bethe lattice (see also Chapter 3 of Volume II for the solution of the steady-state *nonlinear* conduction on the Bethe lattice). In this section, we generalize the method of Chapter 5 in order to derive an exact solution of the symmetric hopping model in Bethe lattices.

To begin with, note that Eq. (2) is equivalent to an electrical network, with $\tilde{P}_j(\lambda)$ corresponding to the voltage at site j and W_{ij} the conductivity of the bond joining sites i and j. The terms $-\delta_{i0}$ and $\lambda\tilde{P}_i(\lambda)$ represent, respectively, a current source and an electrical connection to the ground, with λ being a conductivity if it is real, or more generally an admittance if it is complex. Consider an infinite Bethe lattice of random transition rates (conductances) with a statistical distribution $f(W)$ and a grounded conductance (admittance) connected to each site of the lattice. Then, consider a site O and let G_i be the total conductance looking outward into the network from any site A which is a nearest neighbor of O (with conductance W_i). The conductance of the branch that starts at O is in parallel with the grounded conductance λ. Therefore, the conductance G, as seen from site O, is given by

$$G = \sum_{i=1}^{Z-1}\left(\frac{1}{G_i} + \frac{1}{W_i}\right)^{-1} + \lambda = \sum_{i=1}^{Z-1}\left(\frac{G_i W_i}{G_i + W_i} + \frac{\lambda}{Z-1}\right). \tag{123}$$

However, G and G_i are statistically equivalent and therefore the problem is simply to determine the probability distribution of G_i and G self-consistently. That is, the distribution $H(G_i, \lambda)$ of G_i, when substituted into the right-hand side of Eq. (123), must yield the same distribution for G on the left-hand side of (123), implying that

$$H(G,\lambda) = \int \cdots \int \delta\left\{G - \left[\sum_{i=1}^{Z-1}\left(\frac{G_i W_i}{G_i + W_i} + \frac{\lambda}{Z-1}\right)\right]\right\}\prod_{i=1}^{Z-1} f(W_i)H(G_i,\lambda)dW_i dG_i. \tag{124}$$

If we now take the Laplace transform of Eq. (124), exploit the facts that G_i and W_i are statistically independent, that all the W_i and G_i are identically distributed, and replace λ by $i\omega$, we obtain

$$\tilde{H}(s, i\omega) = \int_0^{\infty} \exp(-sG)H(G, i\omega)dG$$
$$= \left\{\int\int \exp\left[-s\left(\frac{GW}{G+W} + \frac{i\omega}{Z-1}\right)\right] f(W)H(G,i\omega)dG dW\right\}^{Z-1}. \tag{125}$$

Following our discussions in Section 53, the frequency-dependent effective conductivity $\hat{g}_e(\omega)$ of a 3D lattice can be accurately approximated by the microscopic conductivity $\hat{g}_m(\omega)$ of a Bethe lattice of coordination number Z, provided that Z is selected in such a way that the bond percolation threshold of the lattice is essentially the same as that of the Bethe lattice. How to ensure the proximity of the percolation threshold of a 2D or 3D lattice to that of the Bethe lattice was also discussed in Section 5.3. As discussed there, $\hat{g}_m$ is essentially the average $\langle G\rangle$ of the distribution $H(G, i\omega)$ and is given by, $\hat{g}_m = Z\langle G\rangle/(Z-1)$. Using the properties of the Laplace transform, it is straightforward to show that (Sahimi, 1980, unpublished notes) the frequency-dependent effective hopping conductivity, in the Bethe lattice approximation, is given by

$$\hat{g}_e(\omega) = Z\left[\int\int f(W)H(G, i\omega)\left(\frac{GW}{G+W} + \frac{i\omega}{Z-1}\right)dWdG\right]^{Z-2}, \quad (126)$$

which provides accurate estimates of $\hat{g}_e(\omega)$. Movaghar *et al.* (1980a,b) derived the same results as presented here using an entirely different approach.

We may use the same line of analysis that was discussed in Section 5.3 in order to derive an EMA for hopping conduction on a Bethe lattice. The idea is that, for a fixed frequency, in an effective medium the probability distribution $H(G, i\omega)$ is expected to achieve its maximum around a mean value G^*, and thus it may be approximated as a simple delta function. Then, the resulting equation will be similar to what was presented in Section 5.3, except that it also depends on the Laplace transform variable λ. A slightly different EMA, derived by Movaghar *et al.* (1980a,b), is given by

$$\hat{g}_e(\omega) = \int\left[\frac{1}{W} + \frac{1}{i\omega + (Z-1)\hat{g}_e(\omega)}\right]^{-1} f(W)dW. \quad (127)$$

We will compare the predictions of the model of hopping conduction on a Bethe lattice with the experimental data in Section 6.2.3.

6.2.2.3 Perturbation Expansion and Effective-Medium Approximation

Given Eqs. (112) and (113), it should be clear how to estimate the frequency-dependent hopping conductivity by a perturbation expansion. All one must do is replacing in Eqs. (7) and (31) the Laplace transform variable λ by $i\omega$. Certain subtleties that arise from such a replacement must be carefully addressed. In particular, Eq. (25) can be used for estimating the frequency-dependent diffusivity $\hat{D}_e(\omega)$, and hence the conductivity [via Eq. (112)], by an EMA. Indeed, this technique was used by many in the early 1980s for calculating the frequency-dependent hopping conductivity. All the strengths and shortcomings of the EMA that were described in Chapter 5 and in Section 6.1 above are also true for the $\hat{g}_e(\omega)$ predicted by the EMA.

We consider here d-dimensional lattices of coordination number Z with percolation-type disorder, i.e., with a distribution of transition rates given by, $f(W) = (1-p)\delta_+(W) + p\delta(W - W_0)$, and summarize the main predictions of the EMA for frequency-dependent hopping conductivity $\hat{g}_e(\omega)$ of such models which were derived by, among others, Sahimi *et al.* (1983a) and Odagaki *et al.* (1983). In general, in the low-frequency regime, the leading terms of the frequency-dependent hopping conductivity $\hat{g}_e(\omega)$ is written as

$$\hat{g}_e(\omega) = \hat{g}_e(0) + \mathcal{R}(Z, d, p)h_r(\omega) + i\mathcal{I}(Z, d, p)h_i(\omega), \quad (128)$$

TABLE 6.1. The EMA predictions for asymptotic form of $\mathcal{R}(Z, d, p)$ and $\mathcal{I}(Z, d, p)$ [see Eq. (128)] for $p < p_c$, where d and Z are the lattice dimensionality and coordination number, respectively.

d	$\mathcal{R}(Z, d, p)$	$\mathcal{I}(Z, d, p)$
$1 < d < 2$	$\frac{2}{dZ^2}\left[\frac{c\Gamma(1-d/2)}{Z}\right]^{4/d}\frac{1-p}{(p_c-p)^{1+4/d}}$	$\frac{1}{Z}\left[\frac{2c\Gamma(d/2)\Gamma(1-d/2)}{Z(p_c-p)}\right]^{2/d}$
$d = 2$	$\frac{4c^2[\ln(p_c-p)]^2}{4(p_c-p)^3}$	$\frac{-2c\ln(p_c-p)}{Z^2(p_c-p)}$
$d > 2$	$\frac{4I_w^2(1-p)}{Z^4(p_c-p)^3}$	$\frac{2I_w}{Z^2(p_c-p)}$

where $\mathcal{R}, \mathcal{I}, h_r$ and h_i are functions that depend on d and p. Obviously, for $p < p_c$ one has $\hat{g}_e(0) = 0$. The EMA predicts that, for $p < p_c$, and regardless of d, one has

$$h_r(\omega) = \omega^2, \quad h_i(\omega) = \omega. \tag{129}$$

Table 6.1 summarizes the EMA predictions for $p < p_c$. In this table, $p_c = 2/Z$ is the EMA prediction of the bond percolation threshold [see Eq. (5.61)], c is a constant of order 1, and I_w is called the Watson integral (Hughes, 1995) which represents the value of the Green functions G, derived in Section 6.1.4, at the origin and in the limit $\lambda = 0$. For the main three 3D lattices, one has

$$\begin{aligned} I_{\mathrm{SC}} &= \frac{1}{\pi^3}\int\int_0^{\pi}\int \frac{du\,dv\,dw}{3-\cos u-\cos v-\cos w} \\ &= \frac{\sqrt{6}}{96\pi^3}\Gamma\left(\frac{1}{24}\right)\Gamma\left(\frac{5}{24}\right)\Gamma\left(\frac{7}{24}\right)\Gamma\left(\frac{11}{24}\right) \simeq 0.50546, \end{aligned} \tag{130}$$

$$I_{\mathrm{BCC}} = \frac{1}{\pi^3}\int\int_0^{\pi}\int \frac{du\,dv\,dw}{1-\cos u\cos v\cos w} = \frac{1}{4\pi^3}\Gamma\left(\frac{1}{4}\right)^4 \simeq 1.3932, \tag{131}$$

$$\begin{aligned} I_{\mathrm{FCC}} &= \frac{1}{\pi^3}\int\int_0^{\pi}\int \frac{du\,dv\,dw}{3-\cos u\cos v-\cos v\cos w-\cos w\cos u} \\ &= \frac{3}{2^{14/3}\pi^4}\Gamma\left(\frac{1}{3}\right)^6 \simeq 0.4482, \end{aligned} \tag{132}$$

where $\Gamma(x)$ is the gamma function.

Table 6.2 summarizes the EMA predictions for frequency-dependence of the hopping conductivity at $p = p_c$, while Table 6.3 does the same but for $p > p_c$. Unlike the percolation state of the system for which there are three distinct regimes (below, at, and above p_c), the EMA predicts that above p_c, and depending on the dimensionality of the material, there are five distinct functions that yield the frequency-dependence of the hopping conductivity. In Table 6.3, m_1 is a constant given by

$$m_1 = \int_{-\infty}^{1} \frac{\mathcal{N}(x)}{(1-x)^2}dx, \tag{133}$$

TABLE 6.2. The EMA predictions for the frequency-dependent hopping conductivity at the percolation threshold p_c of a d-dimensional lattice of coordination number Z.

d	$\hat{g}_e(\omega) - \hat{g}_e(\omega = 0)$
$1 < d < 2$	$\frac{1}{Z}\left[\frac{2Z}{Z-2}c\Gamma(d/2)\Gamma(1-d/2)\right]^{2/(d+2)}[\cos(d\pi/d+2) + i\sin(d\pi/d+2)]\,\omega^{d/(d+2)}$
$d = 2$	$\left[\frac{c}{2Z(Z-2)}\right]^{1/2}(1+i)(-\omega\ln\omega)^{1/2}$
$d > 2$	$\left[\frac{c}{Z(Z-2)}\right]^{1/2}(1+i)\omega^{1/2}$

TABLE 6.3. The EMA predictions for the frequency-dependent hopping conductivity of a d-dimensional lattice of coordination number Z above the percolation threshold p_c, in the low-frequency limit.

d	$\hat{g}_e(\omega) - \hat{g}_e(\omega = 0)$
$1 < d < 2$	$\frac{2c\Gamma(d/2)\Gamma(1-d/2)(Z-2)^{2-d/2}(1-p)}{Z^{d-1}(p-p_c)^{d/2}}[\cos(d\pi/4) + i\sin(d\pi/4)]\,\omega^{d/2}$
$d = 2$	$\frac{2c(1-p)}{Z(Z-2)(p-p_c)}(-\omega\ln\omega)i + \frac{\pi c(1-p)}{(Z-2)(p-p_c)}\omega$
$2 < d < 4$	$\frac{2I_w(1-p)}{Z(Z-2)(p-p_c)}\omega i + \frac{2c\Gamma(d/2-1)\Gamma(2-d/2)(1-p)}{Z^{d-1}(Z-2)^{2-d/2}(p-p_c)^{d/2}}\cos(d\pi/4)\omega^{d/2}$
$d = 4$	$\frac{2I_w(1-p)}{Z(Z-2)(p-p_c)}\omega i - \frac{2c(1-p)}{Z^3(p-p_c)^2}\omega^2\ln\omega$
$d > 4$	$\frac{2I_w(1-p)}{Z(Z-2)(p-p_c)}\omega i + \frac{2(1-p)}{Z^3(p-p_c)^2}\left[\frac{2I_w^2}{Z(p-p_c)} + m_1\right]\omega^2$

where $\mathcal{N}(x)$ is the density of states of the homogeneous (fully-connected) lattice at $p = 1$. For example, for the simple-cubic lattice in d dimensions, one has

$$\mathcal{N}(x) = \frac{\Gamma(d+1)}{\sqrt{\pi}\Gamma(d/2)}(1-x^2)^{d/2-1}. \tag{134}$$

The density of states for other regular 3D homogeneous lattices was calculated by Lax (1955). According to Table 6.3, the EMA predicts that, above the percolation threshold p_c and a certain dimensionality of the system, the real and imaginary part of $\hat{g}_e(\omega)$ depend on the frequency as ω^2 and ω, respectively. Table 6.3 indicates that, according to the EMA, this "upper dimensionality" is $d = 4$. However, this upper dimensionality should not be confused with the upper critical dimensionality of percolation which, as mentioned in Chapter 2, is $d = 6$.

6.2.3 *The Asymmetric Hopping Model: Perturbation Expansion*

In the presence of an external electric field, the symmetric hopping model is no longer valid, as hops in the direction of the applied field are more likely than those in the opposite direction. For this case, the theory of hopping conductivity developed by Movaghar and Schirmacher (1981) turns out to be very accurate.

The appropriate master equation is (92) which is, however, nonlinear and very difficult to treat theoretically. However, it can be linearized in a manner similar to what we described above. Then, the frequency-dependent hopping conductivity $\hat{g}_e(\omega)$ is given by (Butcher, 1976)

$$\hat{g}_e(\omega) = \frac{e^2(i\omega)^2}{6\Omega k_B T}\left\langle \sum_{ij} F_i \mathbf{R}_{ij}^2 \hat{G}_{ij}(\omega) \right\rangle, \tag{135}$$

where Ω is the volume of the system, the angular bracket denotes a configurational average, $\mathbf{R}_{ij}$ is the distance between i and j, and the rest of the notation is as before. The quantity F_i is given by, $F_i = h(E_i)[1 - h(E_i)]$, where, $h(E_i) = \{\frac{1}{2}\exp[\beta(E_i - E_F)] + 1\}^{-1}$, with E_F being the energy at the Fermi level. The quantity $\hat{G}_{ij}(\omega)$ is similar to the Green function defined earlier (in which the Laplace transform variable λ has been replaced by $i\omega$). It can be interpreted as a Green function of the operator $(i\omega\mathbf{I} - \mathcal{H})^{-1}$, where $\mathcal{H}$ is a Hamiltonian with diagonal elements, $\mathcal{H}_i = -\sum_n \mathcal{H}_{in}$ and off-diagonal elements, $\mathcal{H}_{ij} = h(E_i)[1 - h(E_j)]W_{ij}/F_i$. Alternatively, $\hat{G}_{ij}$ may be thought of as the Fourier–Laplace transform of the time-dependent Green function $G_{ij}(t)$ of the linear master equation with asymmetric transition rates $\mathcal{H}_{ij}$ and $\mathcal{H}_{ji}$. For convenience, we delete ˆ and use $G_{ij}(\omega)$ in place of $\hat{G}_{ij}(\omega)$.

To analyze the problem, Movaghar and Schirmacher (1981) developed a perturbation expansion for $G_{ij}(\omega)$ given by

$$G_{ij}(\omega) = G_{ii}\delta_{ij} + G_{ii}\mathcal{H}_{ij}G_{jj}^{(i)} + \sum_{l\neq i\neq j} G_{ii}\mathcal{H}_{il}G_{ll}^{(i)}\mathcal{H}_{lj}G_{jj}^{(i,l)} + \cdots \tag{136}$$

where $G_{kk}^{(n,k,\cdots,s)}$ represents, for example, the exact local Green function with transitions forbidden to the sites $n, l, \cdots$, and s. Therefore, for example,

$$G_{kk}^{(n)} = \frac{1}{i\omega + \sum_s \mathcal{H}_{ks} - \Delta_k^{(n)}}, \tag{137}$$

with

$$\Delta_k^{(n)} = \sum_{l\neq n} \mathcal{H}_{kl}G_{ll}^{(n,k)}\mathcal{H}_{lk} + \sum_{l\neq s\neq n} \mathcal{H}_{kl}G_{ll}^{(n,k)}\mathcal{H}_{ls}G_{ss}^{(n,k,l)}\mathcal{H}_{sk} + \cdots \tag{138}$$

All repeated indices are eliminated in the summation. It is also useful to rewrite the Green function expansion in the following form,

$$G_{ij}(\omega) = G_{ii}\delta_{ij} + G_{ii}g_{ij}G_{jj}(ij) + \sum_{l\neq i\neq j} G_{ii}g_{il}G_{ll}(il)g_{lj}^{(i)}G_{jj}(il; lj) + \cdots \tag{139}$$

where, for example, $G_{jj}(il; lj)$ refers to the exact local Green function with transition to sites i and l excluded, and the bonds $(i - l)$ and $(l - j)$ removed from the system. Here,

$$g_{il} = \frac{\mathcal{H}_{il}/\mathcal{H}_{li}}{1/\mathcal{H}_{li} + G_{ll}(il)}, \tag{140}$$

$$g_{ij}^{(i)} = \frac{\mathcal{H}_{lj}/\mathcal{H}_{jl}}{1/\mathcal{H}_{jl} + G_{jj}(il; lj)}. \tag{141}$$

One may also rewrite these quantities in the following forms,

$$G_{ii} = \left(i\omega + \sum_k g_{ik} - C_i \right)^{-1}, \tag{142}$$

$$G_{jj}(ij) = \left(i\omega + \sum_{k \neq i} g_{jk}^{(i)} - C_j^{(i)} \right)^{-1}, \tag{143}$$

with

$$C_i = \Delta_i - \sum_l \mathcal{H}_{il} G_{ll}^{(i)} \mathcal{H}_{li}. \tag{144}$$

Although these equations are all exact, regardless of the structure of the lattice used, it is very difficult in practice to exactly compute all of their terms, and hence one must resort to approximate schemes, either in summing the terms in the perturbation expansions or in the structure of the lattice used for the computations. For example, if the topology of the material can be approximated by a Bethe lattice or Cayley tree of coordination number Z, then, one may obtain exact results. We already discussed such an exact solution for the symmetric hopping model. We now briefly discuss such a solution for the asymmetric case.

6.2.3.1 Exact Solution for Bethe Lattices

Absence of closed loops of bonds on a Bethe lattice implies that all the Cs in Eqs. (142) and (143) are identically zero, simplifying greatly the analysis of the problem. Equation (138) is rewritten as

$$G_{ij}(\omega) = \frac{\delta_{ij}}{i\omega + \sum_l g_{il}} + \frac{1}{i\omega + \sum_l g_{il}} g_{ij} \frac{1}{i\omega + \sum_{s \neq i} g_{js}}$$
$$+ \sum_{k \neq i \neq j} \frac{1}{i\omega + \sum_l g_{il}} g_{ik} \frac{1}{i\omega + \sum_{n \neq i} g_{kn}} g_{kj} \frac{1}{i\omega + \sum_{s \neq k} g_{js}} + \cdots \tag{145}$$

where

$$g_{il} = \left(\frac{\mathcal{H}_{il}}{\mathcal{H}_{li}} \right) \left(\frac{1}{\mathcal{H}_{li}} + \frac{1}{i\omega + \sum_{s \neq i} g_{ls}} \right)^{-1}. \tag{146}$$

If we now write, $\mathbf{R}_{ij}^2 \simeq n\langle \mathbf{R}^2 \rangle$, where n is the number of steps from i to j and $\langle \mathbf{R}^2 \rangle$ is a mean square nearest-neighbor distance, and use the facts that, $F_0 G_{0n} = F_n G_{n0}$, and $\sum_k G_{sk} = 1/i\omega$, Eq. (135) simplifies to

$$\hat{g}_e(\omega) = \frac{\langle \mathbf{R}^2 \rangle}{6\Omega} \frac{e^2}{k_B T} \left\langle \sum_n \frac{1 - i\omega G_{nn}}{G_{nn}^{(1)}} F_n \right\rangle, \tag{147}$$

which can also be written as

$$\hat{g}_e(\omega) = \frac{\langle \mathbf{R}^2 \rangle}{6\Omega} \frac{e^2}{k_B T} \left\langle \sum_n \left[\sum_\ell{}' F_n g_{nl} + i\omega \left(1 - \frac{G_{nn}}{G_{nn}^{(1)}} F_n \right) \right] \right\rangle, \tag{148}$$

where the prime on the summation means that one neighbor is excluded from the sum. The quantity $F_n g_{nl}$ is given by

$$F_n g_{nl} = \left(\frac{1}{\tau_{nl}} + \frac{1}{i\omega F_l + \sum_{s \neq n} F_l g_{ls}} \right)^{-1}, \tag{149}$$

where, $\tau_{ij} = h(E_i)[1 - h(E_j)]W_{ij}$. In the static limit, $\omega = 0$, Eq. (148) is equivalent to the microscopic conductivity of a Bethe lattice of coordination number Z, derived by Stinchcombe (1973,1974) and described in Section 5.3.

6.2.3.2 Two-Site Self-Consistent Approximation

As discussed by Movaghar and Schirmacher (1981), the exact solution for the Bethe lattices provides a clue for developing approximate solutions for real 2D or 3D lattices, since it provides a means of eliminating the effect of the closed loops of bonds on such lattices. The simplest of such approximations is a two-site self-consistent approximation. Since the quantity C in Eqs. (142) and (143) represent the effect of the closed loops, the first step in constructing such approximate schemes is to eliminate them from these equations. Consider first the case of symmetric hopping rate. Setting the Cs to be zero, the total self-energy Σ_i now becomes

$$\Sigma_i = \sum_l W_{il} - \Delta_i = \sum_l g_{il}, \tag{150}$$

where

$$g_{il} = \left[\frac{1}{W_{il}} + G_{ll}(il) \right]^{-1} = \frac{W_{il}[i\omega/\Sigma(il) + 1]\Sigma(il)}{W_{il} + G_{ll}^{-1}(il)}, \tag{151}$$

where $\Sigma(il)$ is defined, in analogy to Eq. (150), as $[G_{ll}^{-1}(il) - i\omega]$. The factor $\Sigma(il)$ in the numerator of (151) is now iterated to infinite order, and terms that correspond to closed loops are eliminated. This results in a quantity $g_{il}^{(R)}$, the reduced g_{il}, given by

$$g_{il}^{(R)} = \sum_{i \neq l \neq k \neq s \cdots} \frac{W_{il}[1 + i\omega/\Sigma_l(il)]W_{lk}[1 + i\omega/\Sigma_k(il; lk)]W_{ks} \cdots}{[W_{il} + G_{ll}^{-1}(il)][W_{lk} + G_{kk}^{-1}(il; lk)] \cdots}, \tag{152}$$

where no repeated indices are allowed.

The next step of constructing the approximation is to insert

$$G_{nn}(il; lk; \cdots; sn) = G_{nn}(sn) \tag{153}$$

in Eq. (152) and reduce these quantities in the same way. Physically, Eq. (153) implies that hops to sites $i, l, k, \cdots$, that were forbidden before, are now allowed.

It also implies that all the gs in (141) are equal. Therefore,

$$G_{ij}(\omega) = G_{ii}\delta_{ij} + G_{ii}g_{ij}^{(R)}G_{jj}(ij) + \sum_{l\neq i\neq j} G_{ii}g_{il}^{(R)}G_{ll}(il)g_{lj}^{(R)}G_{jj}(lj) + \cdots \tag{154}$$

where

$$G_{jj}(ij) = \left(i\omega + \sum_{s\neq i} g_{js}^{(R)}\right)^{-1}, \tag{155}$$

which is the same as Eq. (142), except that the effect of the closed loops (i.e., the C_i) has been eliminated and g_{ik} has been replaced by its reduced value. Equation (155) is now identical with that of a Bethe lattice. Therefore, defining $g_i = \sum_s g_{is}^{(R)}$ and using Eq. (152), we can write

$$g_i = \sum_{i\neq s\neq k\neq \cdots} \frac{W_{is}(1+i\omega/g_s)W_{sk}(1+i\omega/g_k)\cdots}{(W_{is}+i\omega+g_s)(W_{sk}+i\omega+g_k)\cdots}. \tag{156}$$

The effective hopping conductivity would then be $\hat{g}_e(\omega) = \langle g_1\rangle$, where $\langle\cdot\rangle$ denotes an average over the distribution of the transition rates W_{ij}. To obtain the corresponding expression for the asymmetric case, we make the replacements,

$$g_i \to \sum_s F_i g_{is}^{(R)}, \quad W_{ij} \to \tau_{ij}. \tag{157}$$

Physically, this approximation corresponds to breaking up the closed loops into products of two-site contributions, as is evident in Eq. (156).

6.2.3.3 Two-Site Effective-Medium Approximation

We now apply the EMA directly to the Green function expansion, Eq. (154) (Movaghar and Schirmacher, 1981). The EMA that is obtained, and also the one that will be derived in the next section, are *not* identical with the usual EMA that was developed for the DC conductivity, because the EMA technique is applied directly to the Green function expansion, rather than the flux, as is usually done. This leads us to

$$\langle G_{ij}\rangle_{ij} = \left(i\omega + \sum_n \langle g_{in}\rangle\right)^{-1} \langle g_{ij}\rangle_{ij} \left(i\omega + \sum_{l\neq i}\langle g_{jl}\rangle\right)^{-1} + \cdots \tag{158}$$

The indices below the angular brackets indicate that the sites i and j are not to be averaged out. Here, the configurational average of a quantity Q is defined as

$$\langle Q(\mathbf{R}_1, \mathbf{R}_2, \cdots; E_1, E_2, \cdots)\rangle = \int QC(\mathbf{R}_1, \mathbf{R}_2, \cdots, \mathbf{R}_N)h(E_1)h(E_2)\cdots h(E_N)\prod_{i=1}^{N} d\mathbf{R}_i dE_i, \tag{159}$$

where $h(E)$ is a normalized energy distribution function, and $C(\mathbf{R}_1, \mathbf{R}_2, \cdots)$ denotes the N-site correlation function, which is typically approximated as a product of pair-distribution functions C_{ij}:

$$C(\mathbf{R}_1, \mathbf{R}_2, \cdots, \mathbf{R}_N) = C_{12}(\mathbf{R}_1 - \mathbf{R}_2)C_{23}(\mathbf{R}_2 - \mathbf{R}_3) \cdots C_{N-1,N}(\mathbf{R}_{N-1} - \mathbf{R}_N). \tag{160}$$

We first assume symmetric transition rates with a structure given by

$$W_{ij} = W(\mathbf{R}_{ij})\nu(|E|/k_B T), \tag{161}$$

and take the Fourier transform of (158), using (159) and (160); we find that

$$G_{\mathbf{k}=0}(\omega) = [i\omega + ng_{\mathbf{k}=0}(\omega) - ng_{\mathbf{k}}(\omega)]^{-1} - [i\omega + ng_{\mathbf{k}=0}(\omega)]^{-1}, \tag{162}$$

where

$$G_{\mathbf{k}}(\omega) = n \int \exp(-i\mathbf{k} \cdot \mathbf{R}_{ij}) C_{ij}(\mathbf{R}_i - \mathbf{R}_j) \langle G_{ij} \rangle_{ij} \, d\mathbf{R}_{ij}, \tag{163}$$

$$g_{\mathbf{k}}(\omega) = \int \exp(-i\mathbf{k} \cdot \mathbf{R}_{ij}) C_{ij}(\mathbf{R}_i - \mathbf{R}_j) \langle g_{ij} \rangle_{ij} \, d\mathbf{R}_{ij}. \tag{164}$$

Therefore, the effective hopping conductivity is given by

$$\hat{g}_e(\omega) = -\frac{e^2}{k_B T} \frac{(i\omega)^2}{6\Omega} \frac{\partial^2 G_{\mathbf{k}}}{\partial \mathbf{k}^2} |_{\mathbf{k}=0} = \frac{e^2}{k_B T} \frac{N_c(1-n_c)}{6\Omega} \left\langle \sum_j \mathbf{R}_{ij}^2 g_{ij}^{(R)} \right\rangle, \tag{165}$$

where $n_c = N_c/N$ is number of hopping particles per sites, N_c is the total number of particles, while N is the total number of sites. The corresponding expression for the asymmetric case is given by

$$\hat{g}_e(\omega) = \frac{e^2}{k_B T} \frac{1}{6\Omega} \left\langle \sum_{i,j} F_i \mathbf{R}_{ij}^2 g_{ij}^{(R)} \right\rangle. \tag{166}$$

6.2.3.4 Energy-Dependent Effective-Medium Approximation

In the symmetric case, when all the F_ns are equal and, $F_n = n_c(1 - n_c)$, Eq. (149) simplifies to

$$g_{nl} = \left(\frac{1}{W_{nl}} + \frac{1}{i\omega + \sum_{k \neq n} g_{lk}} \right)^{-1}, \tag{167}$$

from which we immediately obtain

$$\langle g_{ij} \rangle_{ij} = \left(\frac{1}{W_{ij}} + \frac{1}{i\omega + \langle \sum_{l \neq i} g_{jl} \rangle_j} \right)^{-1}. \tag{168}$$

If we substitute (168) in (165), we obtain

$$\hat{g}_e(\omega) = \frac{e^2}{k_B T} \frac{N_c(1-n_c)}{\Omega} \hat{D}_e(\omega), \tag{169}$$

where

$$\hat{D}_e(\omega) = \frac{\langle \hat{\mathbf{R}}^2(\omega) \rangle}{6a_p} \hat{g}_1(\omega). \tag{170}$$

Here $a_p = e$, where e is the natural logarithm base. The quantity $\hat{g}_1(\omega)$, which is defined by Eq. (156), satisfies the following self-consistency equation

$$\hat{g}_1(\omega) = na_p[\hat{g}_1(\omega) + i\omega] \int\int \frac{h(E)C_{ij}(\mathbf{R}_{ij})W_{ij}}{i\omega + \hat{g}_1(\omega) + W_{ij}} d\mathbf{R}_{ij} dE. \tag{171}$$

while

$$\langle \mathbf{R}^2(\omega) \rangle = \left\langle \mathbf{R}_{ij}^2 \frac{W_{ij}}{i\omega + W_{ij} + \hat{g}_1(\omega)} \right\rangle \left\langle \frac{W_{ij}}{i\omega + W_{ij} + \hat{g}_1(\omega)} \right\rangle^{-1}. \tag{172}$$

The case of asymmetric hopping is more complex. In this case, the starting point, in analogy with Eq. (168), is

$$\langle F_i g_{ij} \rangle_{ij} = \left(\frac{1}{\tau_{ij}} + \frac{1}{i\omega F_j + \langle \sum_{l \neq i} F_j g_{jl} \rangle_j} \right)^{-1}, \tag{173}$$

which, when substituted into (166), yields

$$\hat{g}_e(\omega) = \frac{e^2}{k_B T} \frac{n^2}{6} \int\int\int \frac{h(E)h(E')C_{ij}(\mathbf{R}_{ij})\mathbf{R}_{ij}^2}{[\tau(E, E'; \mathbf{R})]^{-1} + [i\omega F(E') + \hat{g}_1(E', \omega)]^{-1}} d\mathbf{R}_{ij} dE dE'. \tag{174}$$

Here, $\hat{g}_1(E', \omega)$ satisfies the following integral equation

$$\hat{g}_1(E', \omega) = na_p \int\int \frac{h(E'')C_{ij}(\mathbf{R}_{ij})}{[\tau(E', E''; \mathbf{R})]^{-1} + [i\omega F(E'') + \hat{g}_1(E'', \omega)]^{-1}} d\mathbf{R}_{ij} dE''. \tag{175}$$

We must emphasize again that these EMAs are *not* the same as those that were derived previously, because of the way that the averaging procedure is carried out.

The predictions of these equations will be compared with numerical simulations and experimental data later in this chapter.

6.2.4 *Variable-Range Hopping: The Critical Path Method*

Mott (1968) pointed out that the exponential dependence of the resistances on the site energies cannot be ignored in most cases, because if the activation energy of a nearest-neighbor site is large, then a hop to a distant site with a lower energy may be easier than one to a nearest-neighbor site. How far the hopper can go depends on the ease of activation to higher energies, and therefore the hopping distance and the resistance both depend on the temperature. This mechanism of hopping conduction, which is usually referred to as the *variable-range hopping*, is in contrast with the original work of Miller and Abrahams that was restricted to nearest-neighbor hopping. It is now generally believed that variable range hopping is the appropriate mechanism at low temperatures, whereas nearest-neighbor hopping may be appropriate at high temperatures. Moreover, Mott showed that at low temperatures

$$g_e = g_0 \exp[-(T_0/T)^\alpha], \tag{176}$$

which is an important characteristic of variable-range hopping conductivity, and is now one of most famous characteristics of hopping conductivity of amorphous semiconductors. In general, α depends on the density of states near the Fermi level. In Mott's theory the density of states was assumed to be constant, which results in $\alpha = 1/(d+1)$ for a d-dimensional system. Here, g_0 is a constant, and

$$T_0 = \frac{\Lambda a^3}{k_B \mathcal{N}}, \tag{177}$$

where $\mathcal{N}$ is the density of states at the Fermi level E_F, and Λ is a dimensionless parameter. Equation (176) is particularly accurate for amorphous Ge in the range $60\text{K} \leq T \leq 300\text{K}$, with $T_0 \simeq 7 \times 10^7$K. A similar temperature dependence of g_e has also been found for amorphous silicon and carbon, and vanadium oxide. However, as Pollak (1978) pointed out, unless g_e is measured over several orders of magnitudes, a $T^{-1/4}$ behavior should be considered with caution, and should not automatically be interpreted as evidence for variable-range hopping conductivity. Hill (1976) analyzed most of the published experimental data and showed that most of them do follow Eq. (176) with $\alpha = 1/4$. The conditions under which the behavior of g_e might deviate from $\alpha = 1/4$ will be discussed shortly.

Over a decade after the original work of Miller and Abrahams (1960), Ambegaokar *et al.* (1971), Shklovskii and Efros (1971), Pollak (1972), and Brenig *et al.* (1971) realized that the transport paths that Miller and Abrahams had thought to be carrying most of the current in the network carry little, if any, current in most situations. The reason that the Miller–Abrahams paths carry very little or no current is that, if one always proceeds through the nearest-neighbor sites, as in the Miller–Abrahams model, one is certain to arrive at a site where a nearest-neighbor site is a large distance away (in terms of the required energy), and therefore it may be more efficient to go through further neighbor sites, which is in agreement with Mott's arguments.

Let us first discuss how Eq. (176) can be derived using the energy-dependent EMA derived in the last section. If we assume that

$$W_{ij} = \nu_0 \exp(-2\alpha|R_{ij}|) \exp(-|E|/k_B T), \tag{178}$$

where α is a constant, with

$$h(E) = \begin{cases} W_0^{-1}, & 0 \leq W \leq W_0, \\ 0, & \text{otherwise} \end{cases} \tag{179}$$

assume further that the heterogeneity of the material is random, so that, $C_{ij}(\mathbf{R}_{ij}) = 1$, and substitute these in Eq. (171) in the limit $\omega = 0$, we obtain

$$\frac{4\pi k_B n a_p T}{8\alpha^3 W_0} \int_0^{W_0/k_B T} \int_0^\infty \frac{x^2}{(g_1/\nu_0)\exp(x+y)+1} dx dy = 1. \tag{180}$$

In the low-temperature limit—the regime of validity of variable-range hopping—$W_0/k_BT \to \infty$, we can solve for g_1 to find

$$g_1 = \nu_0 \exp[-(T/T_0)^{1/4}], \tag{181}$$

$$T_0 = \frac{24\alpha^3 a_p W_0}{\pi k_B n}. \tag{182}$$

Equation (181) is of course in complete agreement with (176). In fact, if we set $a_p = 1$, then the predicted T_0, Eq. (182), will also be in complete agreement with Mott's prediction, Eq. (177). This simple example demonstrates the predictive power of the approximations that were derived and discussed above. We shall come back to this point later in this section.

Let us now describe and discuss the percolation model for calculating the hopping conductivity of amorphous semiconductors. We consider the zero-frequency limit, $\omega = 0$, and follow Ambegaokar *et al.* (1971) (hereafter referred to as AHL) whose work is very elegant and conceptually simple. AHL argued that an accurate estimate of g_e is the critical percolation conductance g_c, which is the smallest value of the conductance such that a subset of the network with bond conductances $g_{ij} > g_c$ still contains a conducting sample-spanning cluster. Thus, they divided the network into three parts:

(1) A set of isolated clusters of high conductivity, each cluster consisting of a group of sites connected together by conductances $g_{ij} \gg g_c$.
(2) A small number of resistors with g_{ij} of the order of g_c, which connect together a subset of high conductance clusters to form the sample-spanning cluster, called the *critical subnetwork*, which is essentially the same as the static limit of reduced network discussed above.
(3) The remaining resistors with $g_{ij} \ll g_c$.

It is clear that the resistors in group (2) dominate the overall conductance of the network. Thus, the condition that $g_{ij} > g_c$, together with Eq. (111), can be expressed as

$$\frac{r_{ij}}{r_m} + \frac{|E_i| + |E_j| + |E_i - E_j|}{2E_m} \leq 1, \tag{183}$$

where $r_m = a\ln(g_0/g_c)/2$ is the maximum distance between any two sites between which a hop can occur, and $E_m = k_BT\ln(g_0/g_c)$ is the maximum energy that any initial or final state can have.

To construct the critical subnetwork, AHL began with an empty network and, starting with the smallest ones, inserted resistors in the network one by one. As more resistors are inserted, clusters of connected resistors begin to form until the critical resistance $Z_c = 1/g_c$ is reached at which a sample-spanning cluster is formed. To compute Z_c, AHL used Eq. (111), assuming that τ_0 is constant. Moreover, they also assumed that the density of states $\mathcal{N}$ is constant near the Fermi level, $\mathcal{N}(E) = \mathcal{N}(E_F)$. Thus, Z_{ij} is a monotonic function of the random variable $\zeta = E_{ij}/k_BT + 2r_{ij}/a$, with its critical value Z_c defining a corresponding critical value ζ_c. Then, around each site i such that $E_i < \zeta k_B T/2$, a sphere of radius

$$r_i = \frac{a}{2}\left(\frac{\zeta}{2} - \frac{E_i}{k_B T}\right), \tag{184}$$

is drawn which increases with increasing ζ. When two spheres overlap, a bond is inserted between the two sites at their centers. This happens only if inequality (183) is satisfied. Percolation occurs at ζ_c, corresponding to a critical radius r_c. It should be clear that this problem, as formulated by AHL, is a site percolation process. Pollak (1972) treated it as a bond percolation problem which allowed him to include the effect of short-range correlations, but the essence of his basic results was the same as that of AHL. Equation (184) indicates that there is a maximum radius, $r_m = a\zeta_c/4$, and a maximum energy, $E_m = \zeta_c k_B T/2$, with $\zeta_c = 2\ln(g_0/g_c)$. Thus, if two sites are separated by a distance larger than r_m, or farther from energy level E_m, they will not contribute significantly to the effective hopping conductivity g_e. The volume of the sphere defined by Eq. (184) is $(\pi/6)a^3(\zeta_c/2 - E_i/k_B T)^3$, and its average value $\langle\Omega\rangle$, where the averaging is taken over all the sites with a sphere with a non-zero radius (i.e., those for which $E_i < \zeta k_B T/2$), is given by

$$\langle\Omega\rangle = \frac{\pi}{384}a^3\zeta_c^3 = \frac{\pi}{48}r_m^3. \tag{185}$$

On the other hand, ϕ_c, the volume fraction of the spheres at the percolation threshold, is given by, $\phi_c \simeq n\langle\Omega\rangle$, where, assuming that the density of states is constant, $n = \mathcal{N}\zeta_c k_B T$ is the fraction of the sites with a sphere, i.e., those with an energy in the interval $(-E_m, E_m)$. AHL estimated that $\phi_c \simeq 1/4$, somewhat larger than $\phi_c \simeq 0.15 - 0.17$ mentioned in Section 2.10.1 and estimated by Scher and Zallen (1970) for 3D percolating continua. One can also calculate the number of bonds per sites B_c of the network. We already learned in Section 2.5.2 that for bond percolation $B_c \simeq d/(d-1)$. For 3D amorphous materials (or continuum percolation), $B_c = 4\pi n r_c^3/3$, and computer simulations of Pike and Seager (1974) indicated that $B_c \simeq 2.8$ for 3D systems. On the other hand, B_c is related to the density of states $\mathcal{N}(E)$ by

$$B_c = \frac{4\pi n}{3}\frac{\int_{-E_m}^{+E_m}\mathcal{N}(E_i)dE_i\int_{-E_m}^{+E_m}(r_m^3 + 3r_m^2 D_m + 3r_m D_m^2)\mathcal{N}(E_j)dE_j}{\int_{-E_m}^{+E_m}\mathcal{N}(E_i)dE_i}, \tag{186}$$

where D_m is the mean size of the (spheres around the) sites. Combining $\phi_c = n\langle\Omega\rangle \simeq 1/4$ with Eq. (185), or using Eq. (186) together with the lattice or continuum value of B_c and the appropriate expression for r_m, one finally obtains

$$g_e \simeq g_c = g_0 \exp[-(T_0/T)^{1/4}], \tag{187}$$

$$T_0 = \frac{16a^3}{k_B \mathcal{N}}. \tag{188}$$

Comparison of Eq. (188) with (177) indicates that the percolation model of AHL predicts that $\Lambda \simeq 16$. Thus, not only does their model predict the $T^{-1/4}$ behavior proposed by Mott, but also provides an estimate of the temperature T_0 defined by Eq. (176).

The pre-exponential factor g_0 has also been estimated by several research groups, since quantitative prediction of g_c requires an accurate estimate of g_0. For example, using some of the ideas of Kurkijärvi (1974), Shklovskii and Efros (1975) proposed that $Z_0 = 1/g_0 = r_c(2r_c/a)^\nu R_c$, where ν is the correlation length exponent of 3D percolation [see Eq. (2.32)], and R_c is the resistance for $\zeta = \zeta_c$. Kirkpatrick (1974) suggested the same expression, except that in his equation ν was replaced by $(\mu - 1)$, where μ is the critical exponent of the conductivity of 3D percolation [see Eq. (2.36)]. Pollak (1972), Butcher and McInnes (1978), and Butcher (1980) also calculated the pre-exponential factor g_0, although their results did not involve any critical exponent of percolation. The predictions of Butcher, and of Movaghar *et al.*, Eq. (182), are particularly accurate.

6.2.4.1 Effect of a Variable Density of States

In the percolation model described above, it was assumed that the density of states is constant near the Fermi level. Although the basic $T^{-1/4}$ law has been observed in many systems (see, for example, Knotek *et al.* 1973, Viščor and Yoffe 1982, among others), criticism of this formulation was raised by many (see, e.g., Szpilka and Viščor 1982), mainly because the predicted and measured values of g_0 differed by several orders of magnitude. It was suggested that the assumption of a constant density of states in the above percolation model may not be justified, and one must use a variable density of states, in which case one would have to use Eq. (186).

Ortuno and Pollak (1983) investigated this problem and proposed that, if the density of states is concave, then an appropriately modified percolation model can explain the experimental data and remove the disagreement between the predictions and the data. They used Eq. (186) with $\mathcal{N}(E) = \mathcal{N}(E_F)\exp(-E/E_0)$, where E_0 is the exponential decay rate, treated the problem in detail, compared the predictions with the experimental data for amorphous Si and amorphous Ge, and found good agreement between the predictions and the data. Moreover, the predicted value of g_0 was of the same order of magnitude as that of the experimental data.

Sheng and Klafter (1983) studied hopping conductivity of disordered granular materials. In such disordered materials, conduction results from tunneling of electrons and holes from charged grains to neutral ones. Thus, electrons must be transferred from one neutral grain to another. This requires each grain to be characterized by a charging energy $E_c = e^2/(\varepsilon D)$, where ε is the dielectric constant, and D is the grain size. A disordered granular material is thus characterized by a distribution $h(E_c)$ (which is of course related to the grain size distribution), and a density of states related to this distribution given by

$$\mathcal{N}(E) = \frac{1}{\langle \Delta E \rangle} \int_0^E h(E_c) dE_c, \tag{189}$$

where $\langle \Delta E \rangle$ is the average electronic level separation inside the conducting grains. Thus, any distribution of energies $h(E_c)$ can be immediately translated into one for the density of states. Normally, one expects that $\mathcal{N}(0) = 0$. However, in any composite material one can have energy states other than those in the conducting grains. That is, one can have impurities that could contribute a non-zero $\mathcal{N}(E)$ at $E = 0$, for which there is in fact some experimental evidence. Thus, Sheng and Klafter (1983) assumed that $\mathcal{N}(E) = \mathcal{N}_0 + 1/\langle \Delta E \rangle$ and a log-normal distribution for $h(E_c)$, where $\mathcal{N}_0 = \mathcal{N}(E_F)$, and calculated the hopping conductivity of a granular material, using Eq. (186). They used a two parameter fit of the results, with the fitting parameters being the width of the distribution $h(E)$, and $x = \pi n D_m/6$, and showed that varying the two parameters enables one to obtain a variety of temperature-dependence of the conductivity, ranging from $T^{-1/4}$ to $T^{-1/2}$.

6.2.4.2 Effect of Coulomb Interactions

The $T^{-1/2}$ behavior of hopping conductivity has been observed in several materials. Although, as discussed above, a variable density of states, such as $\mathcal{N}(E) \sim E^\beta$ or $\mathcal{N}(E) = \mathcal{N}_0 + cE^\beta$, can explain such data, the origin of such power-laws was not clear for some time until it was explained by Efros and Shklovskii (1975). They suggested that Coulomb interactions (which are long-range interactions) between localized electrons can generate a soft gap, called the Coulomb gap, in the density of states near the Fermi level. This means that in a narrow gap centered around the Fermi level, the density of states cannot be constant and must vary with the energy, whereas outside the gap the density of states vanishes. If δ is the width of the gap, Efros and Shklovskii (1975) showed that $\mathcal{N}(\delta) \sim \delta^{d-1}$ in d dimensions. As a result, they suggested that in 3D there must be a crossover from the $T^{-1/4}$ behavior at relatively high temperatures, where the Coulomb gap is not effective, to $T^{-1/2}$ at lower temperatures. Although a $T^{-1/2}$ behavior had been reported by several groups, the temperature below which the gap could be detected by conductivity measurements is usually too low in amorphous semiconductors. Generally speaking, the Coulomb gap does not affect the hopping conductivity of amorphous semiconductors and, moreover, it cannot be found in good metals. It can only affect those materials that have localized electronic states. For example, the Coulomb gap affects the hopping conductivity of doped crystallic semiconductors which provide experimental evidence for the crossover from $T^{-1/4}$ to $T^{-1/2}$ behavior.

6.2.4.3 Comparison with the Experimental Data

We should keep in mind that Eq. (187) yields a *lower bound* to the true hopping conductivity of a solid material with microscopic conductances that vary over a broad range. That this equation is a lower bound to the true g_e is because the critical subnetwork corresponds to replacing all $g_{ij} < g_c$ by 0, and all $g_{ij} \geq g_c$ by g_c in the original network. Equation (187) is exact only in the limit $T \to 0$. If $T > 0$, hops with conductance less than g_c also contribute to the effective conductivity, which means that the optimal cutoff should be somewhat larger than g_c. Moreover, the percolation approach of AHL cannot be used for 1D or quasi-1D conductors, since percolation disorder divides a linear chain into finite segments and the problem becomes meaningless, unless of course the frequency ω is non-zero. The physical systems to which this situation may be relevant are a class of compounds that consist of weakly coupled parallel chains of strongly coupled atoms or molecules. The conductivity of such materials is highly anisotropic, and therefore they may be treated as essentially 1D conductors. Well-known examples of such materials are salts of the organic ion-radical tetracyanoquinodimethane (TCNQ) and the square planar complexes of transition metals, such as platinum and iridium. Shante (1977) proposed a modification of the percolation model of AHL that could take into account the effect of low dimensionality of the material. The model is a bundle of chains in which hopping occurs along the chains. The interchain hopping was also allowed, although it was assumed to be much more difficult than intrachain hopping. Shante's model allowed the possibility of such intrachain hoppings in either 1D or 2D. At low temperatures Shante's model corresponded to 2D and 3D percolation model, and consequently $T^{-1/3}$ and $T^{-1/4}$ behavior was obtained. At high temperatures the percolation model is no longer applicable, and a T^{-1} behavior was obtained. Figure 6.1 compares the predictions of his model with the experimental data for N-methylphenazinium-TCNQ compounds; the agreement is excellent.

The results of AHL can also be predicted by the perturbation expansion of Movaghar and co-workers described above. One writes W_{ij} as

$$W_{ij} = \nu_0 \exp[-(|E_i| + |E_j| + |E_i - E_j|)/2k_BT - 2\alpha|\mathbf{R}_{ij}|], \tag{190}$$

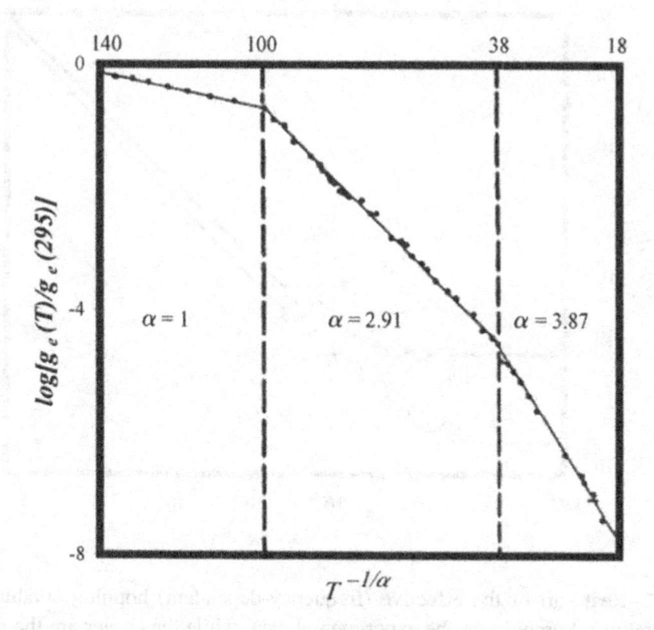

FIGURE 6.1. Temperature dependence of the hopping conductivity of NMP-TCNQ. Temperature is in Kelvin, and the straight lines are the best fits of the data (after Shante, 1977).

and takes the same approach that was used for deriving Eqs. (180)–(182). Movaghar *et al.* (1980a) considered low-temperature hopping near E_F, the energy at the Fermi level, and assumed that the density of states is constant. Combining Eqs. (169) and (170), it is straightforward to obtain the following result:

$$\hat{g}_e(\omega) = \frac{\langle \mathbf{R}^2 \rangle}{6} \frac{e^2}{k_B T} [nN(E_F) k_B T \nu_0 \hat{g}_1(\omega)], \tag{191}$$

where

$$\hat{g}_1(\omega) = \frac{4Ta_p}{T_0} [i\omega + \hat{g}_1(\omega)] \int_0^\infty \frac{x^3 \exp(-x)}{i\omega + \hat{g}_1(\omega) + \exp(-x)} dx. \tag{192}$$

The experimental value, $T_0 = 3.6 \times 10^8$K for germanium was used in place of T_0/a_p. Movaghar *et al.* (1980a) also utilized $\nu_0 = 10^{21}$ sec^{-1}, and employed Eq. (172) for computing $\mathbf{R}^2$. Figure 6.2 presents the predicted real part of the frequency-dependent effective conductivity as a function of ω, at two different temperatures, and compares them with the experimental data for Ge. The agreement between the predictions and the data is excellent. In general, the real part of $\hat{g}_e(\omega)$ behaves as

$$\text{Re}[\hat{g}_e(\omega)] \sim \omega^x, \tag{193}$$

where x is a temperature-dependent exponent, and $10^2 \le \omega \le 10^5$ Hz. Figure 6.3

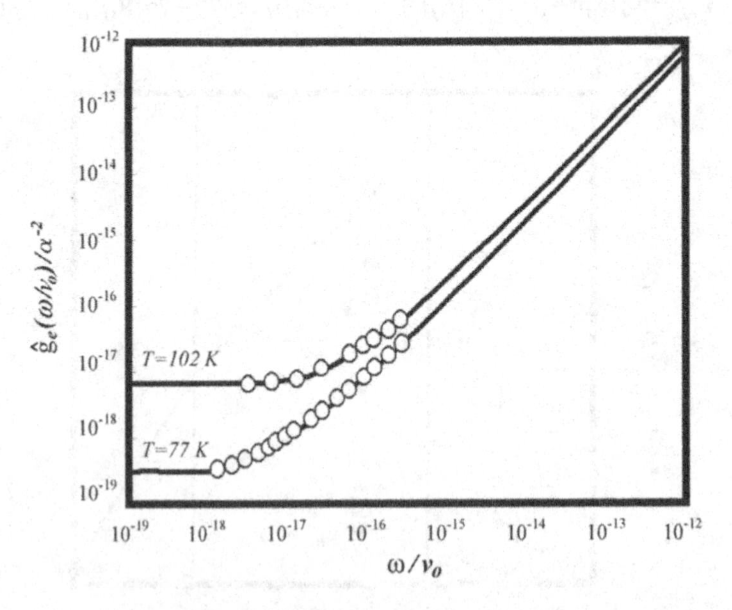

FIGURE 6.2. Real part of the effective (frequency-dependent) hopping conductivity for two temperatures. Symbols are the experimental data, while the curves are the theoretical predictions. The absolute magnitude of the DC conductivity has been fitted to one curve (after Movaghar *et al.*, 1980a).

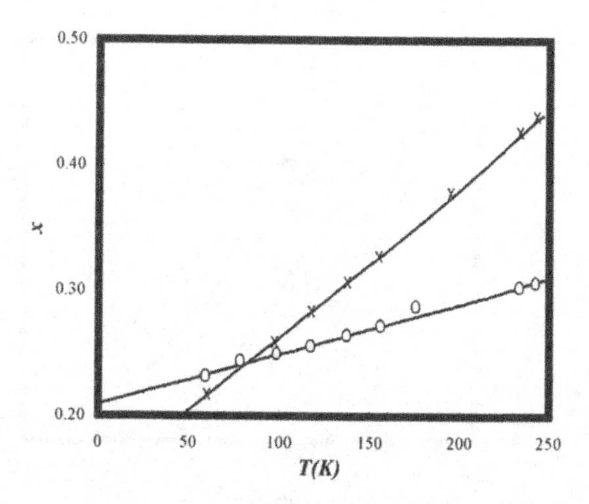

FIGURE 6.3. Temperature dependence of the exponent x [see Eqs. (193) and (194)] for ($\times$) $\omega/\nu_0 = 10^{-8}$ and ($\bigcirc$) $\omega/\nu_0 = 10^{-6}$, with $T_0 = 5 \times 10^7$K (after Movaghar *et al.*, 1980a).

shows the predictions of Movaghar *et al.*'s model for the temperature-dependence of x which indicate that,

$$x = a_1 - a_2 T, \tag{194}$$

where $a_1 \simeq 0.8$ for both cases, consistent with many experimental data. For both cases $T_0 = 5 \times 10^7$ K was used in the computations, and $\omega/\nu_0 = 10^{-8}$ and 10^{-6} were assumed. The linear dependence of x on T has also been observed experimentally, and therefore in this sense also the theory is very accurate.

Maloufi *et al.* (1988) used the theory of Ortuno and Pollak with a variable density of states to fit their conductivity data for amorphous Si_ySn_{1-y} with $y = 0.47 - 1$. Figure 6.4 presents the fits of their data by this theory, and the agreement appears to be excellent over much of the temperature range. As another example of a variable density of states, consider the case for which $\mathcal{N}(E) \sim |E|^\beta$, where β is a positive constant. This case was analyzed by Hamilton (1972) and Pollak (1972). Using Eq. (186) one obtains an equation similar to (176) but with $\alpha = (\beta + 1)/(\beta + 4)$ [which does reduce to Eq. (176) when $\beta = 0$]. The limit $\beta \to \infty$ is also interesting because it corresponds to a system that has a sudden onset of states away from the Fermi level, hence yielding a T^{-1} behavior for such materials. An experimental realization of a power-law density of states was provided by Redfield (1973), who carried out a careful study of hopping conductivity of heavily doped and strongly compensated GaAs. His results indicated a $T^{-1/2}$ behavior rather than $T^{-1/4}$. Redfield showed that, although a $T^{-1/4}$ might look plausible, his data could be fitted extremely accurately by a $T^{-1/2}$ law; see Figure 6.5. His data can be easily explained with a variable density of states, $\mathcal{N} \sim |E|^2$.

If, instead of the density of states that Sheng and Klafter used, we utilize $\mathcal{N}(E) = \mathcal{N}_0 + cE^\beta$, where c is a constant, we obtain Eq. (176) with $\alpha = (\beta + 1)/(\beta + 2)$. Such a density of states may correspond to a granular material with a broad particle

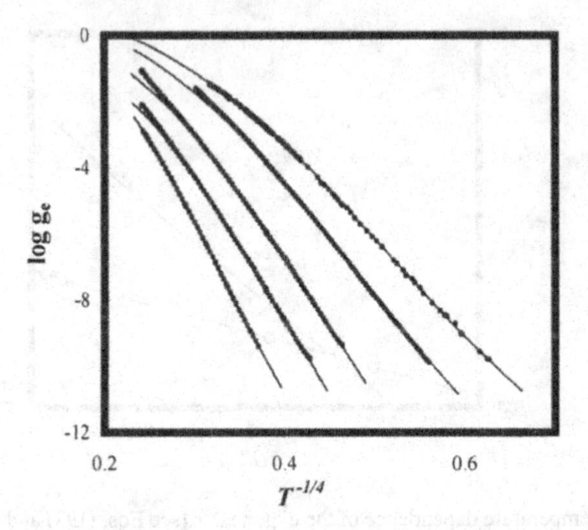

FIGURE 6.4. Comparison of the predicted temperature dependence of hopping conductivity g_e of $a\text{−}Si_ySi_{1-y}$ (the lines), obtained with an exponential density of states, with the experimental data (symbols). The data are, from left, $y = 0.47$, 0.62, 0.77, 0.9, and 1. T is in Kelvin (after Maloufi *et al.*, 1988).

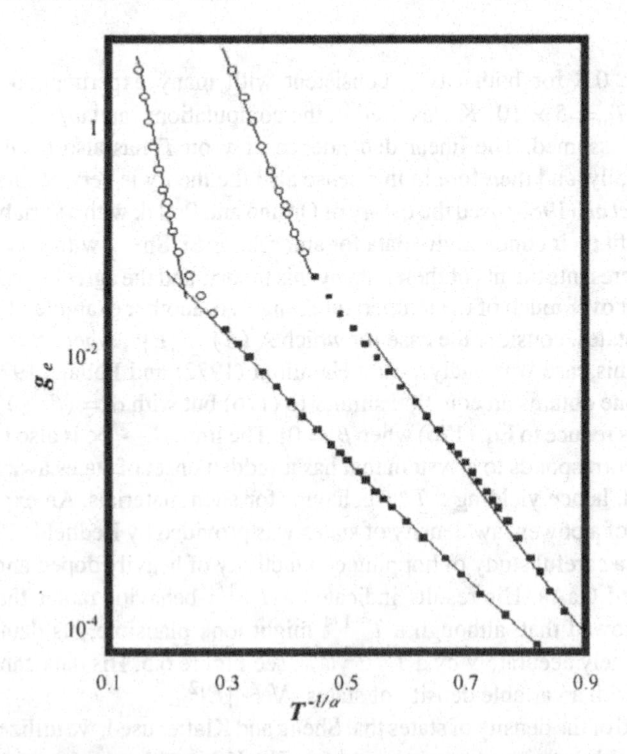

FIGURE 6.5. Temperature dependence of hopping conductivity g_e of doped and compensated GaAs. Symbols represent the experimental data, the line on the left is their $T^{-1/2}$ fit, while that on the right is their $T^{-1/4}$ fit (after Redfield, 1973).

size distribution. Mehbod *et al.* (1987) showed that this type of density of states and the resulting exponent α can fit very well their experimental data for hopping conductivity of polymer-conducting-carbon-black composites. Such materials contain randomly dispersed particles in the matrix with a broad size distribution. The polymeric matrices that were used by Mehbod *et al.* were polystyrene, polyethylene, ethylene-propylene copolymer, and styrene-butadiene copolymer.

The first convincing experimental evidence for the crossover, predicted by Efros and Shklovskii (1975) as the result of Coulomb interactions, was probably provided by Zhang *et al.* (1990). They measured the resistivity of five insulating compensated n-type CdSe samples, which are doped semiconductors. Their data clearly indicated a crossover from $T^{-1/4}$ behavior to $T^{-1/2}$ as the temperature was decreased by about three orders of magnitude from 15 to 0.04 K. Over this temperature range the hopping energy becomes comparable with, and then smaller than, the energy gap discussed by Efros and Shklovskii (1975). The crossover temperature T_{co} was found to decrease as the concentration of the donor was increased, which is expected. Aharony *et al.* (1992) and Meir (1996) proposed a scaling theory for this crossover by suggesting that the effective resistivity $Z_e = 1/g_e$ of the sample follows the following scaling law

$$\ln(Z_e/Z_0) = Ah(T/T_{co}), \tag{195}$$

where the scale factor A and the crossover temperature T_{co} both depend on the sample properties, but the scaling function $h(x)$ is *universal* and has the limiting behaviors, $h(x) \sim x^{-1/4}$ for $x \gg 1$ and $h(x) \sim x^{-1/2}$ for $x \ll 1$. Moreover, using the percolation model, Aharony *et al.* proposed that

$$h(x) = \frac{x + [\sqrt{x+1} - 1]}{x[\sqrt{1+x} - 1]^{1/2}}, \tag{196}$$

and provided expressions for A and the crossover temperature T_{co}. The implication of Eqs. (195) and (196) is that, if $\ln(Z_e/Z_0)$ for various samples are plotted as a function of T/T_{co}, then all the data should collapse onto a single curve. Figure 6.6 presents such a collapse for the data of Zhang *et al.*; it is clear that the data collapse is essentially complete.

6.2.4.4 Fractal Morphology and Superlocalization

As discussed above, a $T^{-1/2}$ behavior for the hopping conductivity can be explained in terms of the Coulomb interactions and gap. However, Coulomb interactions do not play any role in many semiconductors which do exhibit a $T^{-1/2}$ behavior. There has been some experimental evidence that, if the material has a fractal morphology, then its hopping conductivity may exhibit the $T^{-1/2}$ behavior. Lévy and Souillard (1987) suggested that in a material with fractal morphology impurity quantum states are *superlocalized*, i.e., their wave functions decay with the distance r as $\exp(-r^\gamma)$, with $\gamma > 1$ (recall that in the classical Anderson localization, $\gamma = 1$). Ever since the publication of their paper, the possibility of such superlocalization of electrons has been extensively studied. For example, Harris and Aharony (1987) and Aharony and Harris (1990) argued that one must

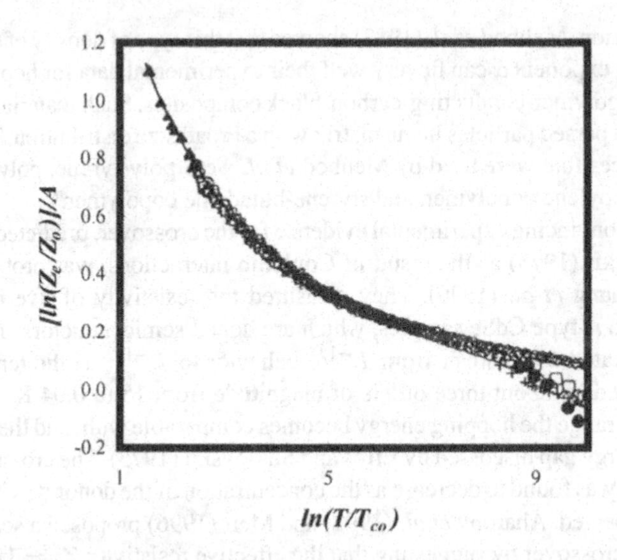

FIGURE 6.6. Temperature dependence of hopping conductivity g_e of carbon-black polymer composite. Temperature is in Kelvin (after van de Putten *et al.*, 1992).

distinguish between the behavior of a *typical* geometry, and that of a system with the *average* geometry, where the averaging is taken with respect to all the possible geometries of the system. Moreover, they argued that the latter yields $\gamma = 1$, i.e., there is no superlocalization of electrons. Early numerical studies of this problem were carried out by Lambert and Hughes (1991), but the most definitive study was probably carried out by Kantelhardt and Bunde (1997), who investigated localization of electronic eigenfunctions and vibrational excitations (usually called *fractons*; see Section 6.6) on percolation clusters, both in the topological (r_t) and Euclidean (r_e) space, and found three different regimes. In the first regime, which corresponds to small r_t and r_e, electrons and fractons are superlocalized in the r_t space, but the localization is not exponential in the r_e space. In the intermediate regime, stretched exponential *sublocalization* was found in both spaces with $\gamma \simeq 0.6$. In the third regime, corresponding to large r_t and r_e, the average behavior was found to depend strongly on the number of percolation configurations used in the averaging, but the asymptotic regime was a simple exponential decay with $\gamma = 1$, in agreement with the argument of Aharony and Harris. We will come back to this issue in Section 6.6.7.2.

However, none of these results supports the view that a material with a fractal morphology can by itself explain the observed $T^{-1/2}$ behavior of hopping conductivity of semiconductors in which Coulomb interactions are unimportant. van der Putten *et al.* (1992) measured the conductivity of carbon-black-polymer composites as a function of carbon black concentration from a point in the vicinity of the percolation threshold p_c up to $33p_c$, and in the temperature range from 4 K to 300 K. Their data indicated that α, the exponent defined by Eq. (176), is about

2/3. To interpret this value, they assumed that, $\mathcal{N}(E) \sim E^{\beta}$, and proposed that

$$\alpha = \frac{\gamma(\beta+1)}{D_f + \gamma(\beta+1)}, \tag{197}$$

where D_f is the fractal dimension of 3D sample-spanning percolation cluster at p_c. Therefore, with $\alpha = 2/3$ and $D_f \simeq 2.53$ one obtains $\beta \simeq 2.75$. Thus, their data can be best explained by a *combination* of a percolating fractal morphology (i.e., one at p_c) *and* a power-law density of states. Moreover, if we take, $\beta = 1$, then Eq. (197) predicts that $\alpha \simeq 1/2$, indicating that a linear density of states *and* a percolating fractal morphology may be responsible for the observed $T^{-1/2}$ behavior of hopping conductivity of semiconductors in which Coulomb interactions are unimportant. However, Aharony *et al.* (1993) argued that there are more than one way of interpreting the data of van der Putten *et al.* (1992), and therefore their interpretation may not provide the true solution to this controversy. Hence, the problem, at the time of writing this book, had remained, to our knowledge, unsolved. For the most recent experimental data for the conductivity of carbon-black polymer composites see Mandal *et al.* (1997) who, however, claimed that their data can be understood on the basis of superlocalization.

6.2.5 *Continuous-Time Random Walk Model*

The last model we discuss for computing the frequency-dependent hopping conductivity is the continuous-time random walk (CTRW) model, originally developed by Montroll and Weiss (1965) and extended by Scher and Lax (1973a). We already discussed in Section 6.1.3 the connection between the CTRW model, the generalized master equation that arises because of averaging the disorder in the material and replacing it by an effective homogeneous medium, and the self-consistent approximation or the EMA, with the key equation being Eq. (13) that establishes this connection. Note that, to the extent that $\tilde{W}_e$ can be estimated accurately, the waiting time density $\psi(t)$ is also computed accurately. If we solve for the effective transition rate $\tilde{W}_e(\lambda)$ and replace λ by $i\omega$, assuming a simple-cubic lattice model ($Z = 2d$), we obtain

$$\hat{W}_e(i\omega) = \frac{i\omega\hat{\psi}(i\omega)}{2d[1-\hat{\psi}(i\omega)]}. \tag{198}$$

The equation that was derived by Scher and Lax (1973a) for the frequency-dependent effective diffusivity is completely similar to Eq. (198) (recall that, as discussed in Section 6.1.6, for a simple-cubic lattice, $D_e = W_e$) and is given by

$$\hat{D}_e(\omega) = \frac{i\omega\hat{\psi}(i\omega)}{2d[1-\hat{\psi}(i\omega)]}\langle \hat{r}^2(\omega)\rangle, \tag{199}$$

which is essentially the same as Eq. (198), but corrected by a factor $\langle \hat{r}^2(\omega)\rangle$ which provides the effective diffusivity with the proper dimensions. Here

$$\langle \hat{r}^2(\omega)\rangle = \sum_{\mathbf{s}} \frac{s^2\hat{\psi}(\mathbf{s}, i\omega)}{\hat{\psi}(i\omega)}, \tag{200}$$

where $\hat{\psi}(\mathbf{s}, i\omega)$ is the causal Fourier transform of $\psi(\mathbf{s}, t)$, such that $\psi(\mathbf{s}, t)\Delta t$ is the probability that the time between hops occurs in the interval $(t, t + \Delta t)$, resulting in a vector displacement $\mathbf{s}$, and

$$\psi(t) = \sum_{\mathbf{s}} \psi(\mathbf{s}, t), \tag{201}$$

is the usual waiting time density. The frequency dependence of $\langle \hat{r}^2(\omega) \rangle$ is weak. It typically varies by at most a factor of 3 over several orders of magnitude variations in the frequency ω, and hence it is assumed that $\langle \hat{r}^2 \rangle$ is a constant, $\langle \hat{r}^2(\omega) \rangle = \langle \hat{r}^2(\omega = 0) \rangle = \langle r^2 \rangle$.

As an example, consider a waiting time density $\psi(t)$ given by

$$\psi(t) = \frac{W_0(W_0 t)^\alpha}{\Gamma(\alpha + 1)} \exp(-W_0 t), \tag{202}$$

where W_0 can be thought of as the minimum transition rate in the system, α is an arbitrary parameter, and $\Gamma(x)$ is the gamma function. Since, $\hat{\psi}(i\omega) = (1 + i\omega/W_0)^{-\alpha-1}$, we immediately obtain

$$\hat{D}_e(\omega) = \frac{\langle r^2 \rangle}{6} \frac{i\omega}{(1 + i\omega/W_0)^{\langle t \rangle W_0 - 1}}, \tag{203}$$

with

$$\langle t \rangle = \int_0^\infty \psi(t) dt = \frac{d\hat{\psi}(i\omega)}{d(i\omega)}|_{\omega=0} = \frac{\alpha + 1}{W_0} \tag{204}$$

being the mean waiting time between the hops. As discussed by Scher and Lax (1973a), one may consider three distinct limits:

(1) Consider first the limit $\langle t \rangle W_0 \to \infty$, which corresponds to a highly peaked distribution. Then, for a fixed $\langle t \rangle$, we let $W_0 \to \infty$ which results in $\hat{\psi}(i\omega) \to \exp(-i\omega\langle t \rangle)$. Therefore, $\psi(t) \to \delta(t - \langle t \rangle)$, implying that

$$\hat{D}_e(\omega) = \frac{1}{6}\langle r^2 \rangle \left[\frac{1}{2}\omega \cot\left(\frac{1}{2}\omega\langle t \rangle\right) - \frac{1}{2}i\omega \right]. \tag{205}$$

Thus, in this limit the hops occur at fixed time intervals, $t = n\langle t \rangle$.

(2) We now consider the limit $\langle t \rangle W_0 = 1$, which means that, $\psi(t) = W_0 \exp(-W_0 t)$, and therefore

$$\hat{D}_e(\omega) = \frac{1}{6}\langle r^2 \rangle W_0. \tag{206}$$

Thus, with an exponential waiting time density, the effective diffusivity is *independent* of the frequency. This is due to the fact that the parameter W_0^{-1} is simply a time scale, and there is only one transition rate in the system, whereas the frequency response of the material is connected with the possibility of more than one, or a range of, transition rate.

(3) Consider now the limit $\langle t \rangle W_0 \sim 0$. It can be shown that, within the CTRW formulation, a waiting time distribution with such a characteristic is necessary to

describe impurity hopping conduction. To see this, consider the high-frequency limit of Eq. (203), $|i\omega/W_0| \gg 1$. Then

$$\hat{D}_e(\omega) \sim \langle r^2\rangle W_0(\omega/W_0)^{1-\langle t\rangle W_0} \exp[i\pi/2(1-\langle t\rangle W_0)]. \tag{207}$$

If we compare Eq. (207) to (193), we may identify the exponent x with $1-\langle t\rangle W_0$, and therefore $\mathrm{Im}[\hat{D}_e(\omega)]/\mathrm{Re}[\hat{D}_e(\omega)] = \tan\frac{1}{2}\pi x$. Therefore, $1-\langle t\rangle W_0$ must be small in order for the CTRW model, with this type of waiting time density, to be able to describe hopping conduction. As mentioned in Section 6.2.4, a typical value of x is $0.7-0.9$, and therefore one must have, $1-\langle t\rangle W_0 \sim 0.1-0.3$.

In their second paper, Scher and Lax (1973b) considered modeling of hopping conduction in amorphous semiconductors, such as Si and Ge. They derived a waiting time density given by

$$\frac{\psi(t)}{W_0} = 2\eta - \left(\frac{1}{4}\eta + 4\eta^2\right) W_0 t + \cdots, \tag{208}$$

where W_0 was taken to be temperature dependent, $W_0 \propto \exp(-E_a/k_B T)$ (E_a is the activation energy), and $\eta = \frac{1}{2}\pi n a^3$, with n and a being the number of donors (per unit volume) and the Bohr radius, respectively, with typical values, $\eta/a \sim 10^{-5}-10^{-1}$. Using the appropriate experimental values for E_a and N_d, they obtained accurate estimates of the hopping conductivity over wide ranges of the frequency and temperature which were in good agreement with the experimental data of Pollak and Geballe (1961).

6.3 AC Conductivity

In the literature, sometimes no distinction is made between the AC conductivity of a disordered solid and its frequency-dependent hopping conductivity. The reality is that there is not much that distinguishes the two quantities from each other, since hopping is just a mechanism of conduction in materials. However, we consider the AC conductivity separately because frequency-dependent conductivity of disordered materials appears to possess certain universal properties that have always been discussed under the title "universal properties of AC conduction" (see, for example, Dyre and Schr⊘der, 2000, for an informative review), rather than "universal properties of hopping conduction."

If $\mathbf{I}$ is the current density, to which there are contributions from both free and bound charges, and $\mathbf{E}$ is the electric field, then the effective conductivity g_e is defined by the usual relation, $\mathbf{I} = g_e\mathbf{E}$. In general, g_e depends on the frequency ω, in which case one writes, $\mathbf{I}_0 = \hat{g}_e(\omega)\mathbf{E}_0$, where $\mathbf{I} = \mathrm{Re}(\mathbf{I}_0 e^{i\omega t})$, and similarly for the electric field. With the aid of $\hat{g}_e(\omega)$, we can define the complex frequency-dependent dielectric constant $\varepsilon(\omega)$:

$$\hat{g}_e(\omega) - \hat{g}_e(0) = i\omega[\varepsilon(\omega) - 1]\varepsilon_0, \tag{209}$$

where ε_0 is the vacuum permittivity. If there are no free charges, then, $\hat{g}_e(0) = 0$, and $\mathbf{I} = \partial \mathbf{P}/\partial t$, where $\mathbf{P}$ is the dipole density, in which case Eq. (209) reduces to the standard equation, $\mathbf{D}_0 = \varepsilon(\omega)\varepsilon_0\mathbf{E}_0$, where $\mathbf{D}_0$ is the complex amplitude of the displacement vector, $\mathbf{D} = \varepsilon_0\mathbf{E} + \mathbf{P}$.

If we write the complex dielectric constant in the standard notation, $\varepsilon(\omega) = \varepsilon'(\omega) + i\varepsilon''(\omega)$, then $-\varepsilon''(\omega)$ is usually referred to as the *dielectric loss*, since it determines the dissipation in excess of the DC dissipation. At frequencies $\omega \ll 10^{13}$ Hz, much below the phonon frequencies, the bound charge dielectric constant ε_∞ is independent of frequency. To distinguish it from the overall effective conductivity $\hat{g}_e(\omega)$, we denote the AC conductivity by $\sigma(\omega)$ which, by definition, is the free charge carrier contribution to the effective conductivity $\hat{g}_e(\omega)$. Since $\hat{g}_e(\omega) = \sigma(\omega) + i\omega(\varepsilon_\infty - 1)\varepsilon_0$, Eq. (209) implies that

$$\sigma(\omega) - \sigma(0) = i\omega[\varepsilon(\omega) - \varepsilon_\infty]\varepsilon_0. \tag{210}$$

If we now write $\sigma(\omega)$ in the standard notation, $\sigma(\omega) = \sigma'(\omega) + i\sigma''(\omega)$, then, $\sigma'(\omega)$ must be positive because thermodynamics constraints require positive dissipation. Below phonon frequencies, the charge carrier displacements always lag behind the electric field, but the lag is at most one-quarter of a period, and therefore the current attains its maximum *before* the field, implying that $\sigma''(\omega) > 0$, and reflecting a capacitance response rather than an inductive one.

6.3.1 Universality of AC Conductivity

Many disordered solid materials appear to possess very similar AC conductivities. The examples, and the references where the details of the experimental evidence can be found, are too numerous to be given here, and therefore we only mention a few of the well-known examples of such materials and the most recent references that we are aware of. Such disordered materials include ion conducting glasses (Roling, 1998), amorphous semiconductors (Long, 1991), polycrystalline semiconductors (Kunar and Srivastava, 1994), electron conducting polymers (Jastrzebska *et al.*, 1998), ion conducting polymers (Rozanski *et al.*, 1995), transition metal oxides (Suzuki, 1980), metal cluster compounds (van Staveren *et al.*, 1991), organic-inorganic composites (Bianchi *et al.*, 1999), and doped single-crystal semiconductors at helium temperature (Pollak and Geballe, 1961).

Figure 6.7 presents the AC conductivity of an ion conducting melt, $0.4Ca(NO_3)_2 - 0.6KNO_3$, a highly viscous melt (Howell *et al.*, 1974), while Figure 6.8 shows the same for a polycrystalline diamond film, an electron conducting material (Fiegl *et al.*, 1994). There are many other such examples of disordered materials, all exhibiting very similar AC conductivities. The similarities between the two figures are striking, particularly in the light of the fact that ion conduction is a classical barrier-crossing process, whereas electron conduction in heterogeneous solids is typically by quantum-mechanical tunneling between localized states, a completely different mechanism. Therefore, the question is: What is the origin of this striking similarity? Dyre (1988) pointed out that the common features among all the materials with similar AC conductivity are very broad distributions of transition rates,

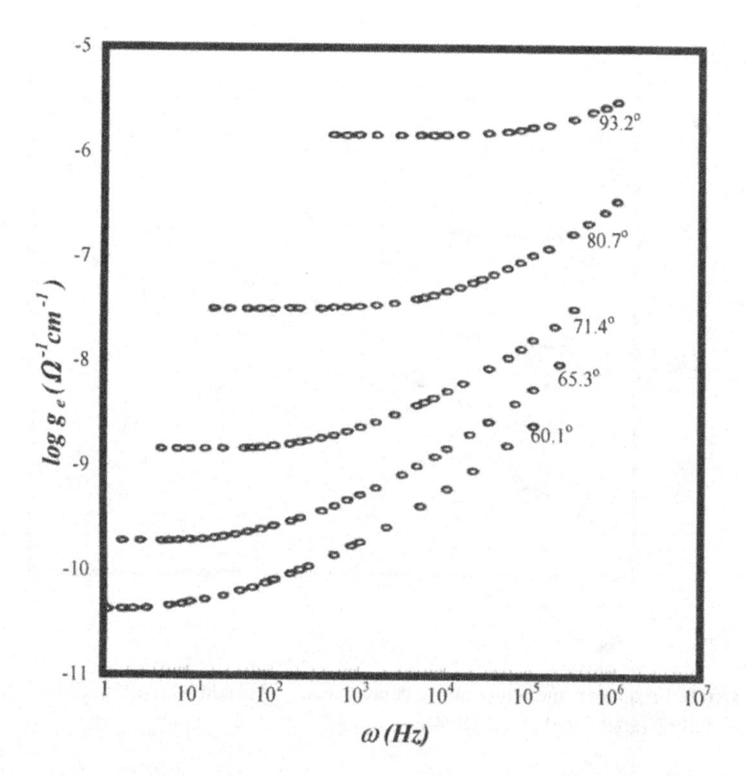

FIGURE 6.7. Frequency and temperature dependence of AC conductivity of $0.4Ca(NO_3)_2$-$0.6KNO_3$ highly porous melt (after Howell *et al.*, 1974).

tunneling rates, and/or local mobilities, and therefore they may be responsible for this apparent universality of the AC conductivity.

This universality was apparently first discovered by Taylor (1956) for ion conducting glasses, who showed that the dielectric loss for various glasses fall on a single curve when plotted versus scaled frequency. Taylor also pointed out that the activation energy of the DC conductivity was the same as that of the frequency that marks the onset of AC conduction. Subsequently, Isard (1961) noted that if one plots the normalized AC conductivity versus $C\omega/\sigma(0)$, it exhibits universal behavior. That is, one can write

$$\hat{\sigma} = \frac{\sigma(\omega)}{\sigma(0)} = F\left[C\frac{\omega}{\sigma(0)}\right], \tag{211}$$

where $F(x)$ is the apparent universal function representing the master curve. C is called the Taylor-Isard scaling constant, and has been taken by many to be proportional to the inverse of the temperature T. Figure 6.9 presents an example of the Taylor–Isard master curve for eight different ion conducting glasses (Roling, 1998), while Figure 6.10 depicts the same for Poly(methylthiophene) (Rehwald *et al.*, 1987). To explain the origin of the apparent universality of the AC conductivity,

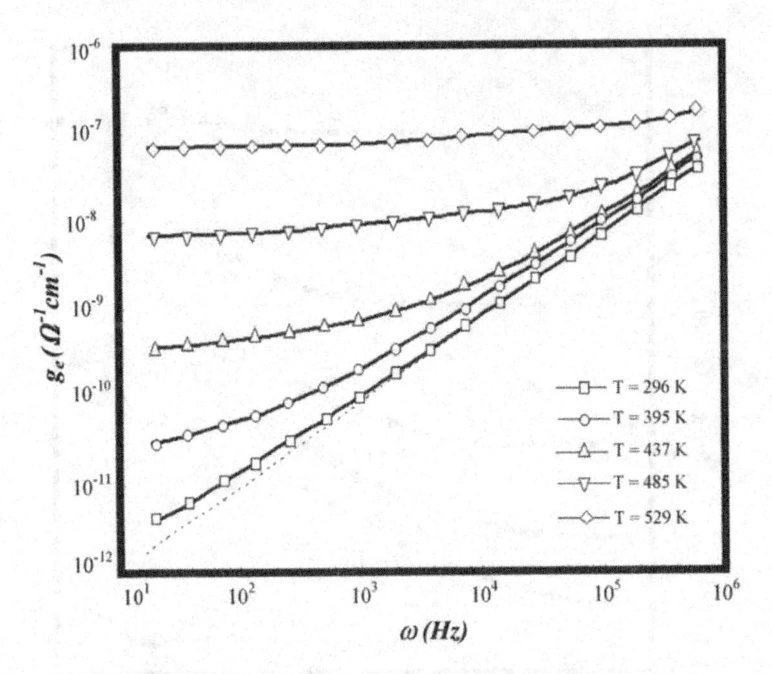

FIGURE 6.8. Frequency and temperature dependence of AC conductivity of polycrystalline diamond films (after Fiegl *et al.*, 1994).

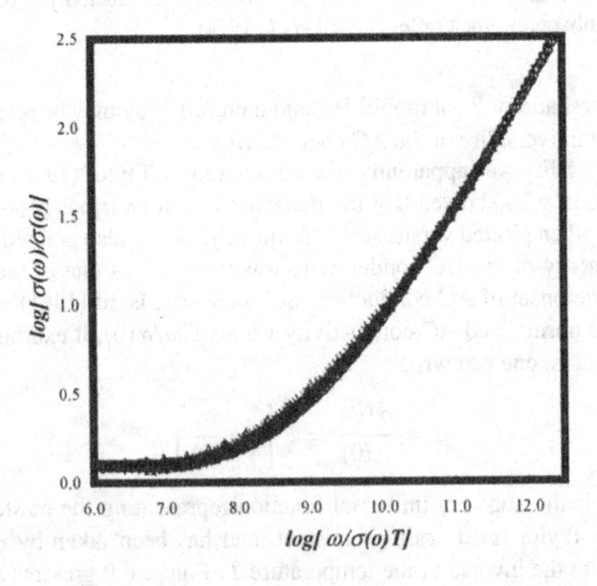

FIGURE 6.9. AC conductivity master curve for eight different ion conducting oxide glasses (after Roling, 1998).

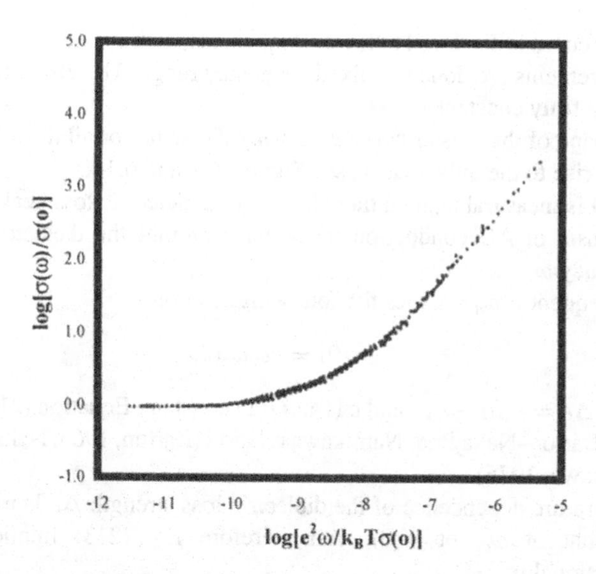

FIGURE 6.10. AC conductivity master curve for conducting polymer Poly(methylthiophene). The data were collected for temperatures 210K $\leq T \leq$ 295K (after Rehwald *et al.*, 1987).

Dyre and SchrØder (2000) listed *sixteen* common AC characteristics of the vast majority of disordered solids, which are as follows.

(1) $\sigma'(\omega)$, the real part of the AC conductivity increases with ω, while $\sigma''(\omega) \geq 0$.

(2) As mentioned earlier, the real part of the hopping conductivity depends on the frequency ω by a power law; see Eq. (193). Hence, we may write

$$\sigma'(\omega) \sim \omega^x. \tag{212}$$

(3) The exponent x is typically between 0.6 and 1.0.

(4) Deviations from Eq. (212) occur when the exponent x is a weakly increasing function of ω.

(5) As discussed earlier, x is a decreasing function of the temperature; see Eq. (194).

(6) $x \simeq 1$ when the DC conductivity is essentially zero.

(7) When $x \simeq 1$, $\sigma'(\omega)$ is almost independent of the temperature.

(8) As the frequency decreases, the conductivity begins to become independent of ω.

(9) In a log-log plot, the power-law regime of $\sigma(\omega)$ has a much weaker temperature dependence than does the DC conductivity.

(10) In a log-log plot, the shape of $\sigma'(\omega)$ is independent of T, making it possible to construct a master curve for it. A similar principle also holds for $\sigma''(\omega)$. Although this may seem an apparent contradiction with the experimental observations that the exponent x depends on the temperature, it is in fact not so because the power-law (212) is only an approximation. If T is lowered the

master curve is displaced to lower frequencies; at the same time $x \to 1$ if the measurements are done in a fixed frequency range. Therefore, the exponent x is not truly constant.

(11) The shape of the master curve is *roughly* the same for all disordered solids, giving rise to the universality; see Figures 6.9 and 6.10.

(12) If $\sigma(0)$ is measurable, then there is always a dielectric loss peak.

(13) The onset of AC conduction takes place around the dielectric loss peak frequency ω_m.

(14) The frequency ω_m satisfies the following relation

$$\sigma(0) = c\varepsilon_0\omega_m\Delta\varepsilon, \tag{213}$$

where $\Delta\varepsilon = \varepsilon(0) - \varepsilon_\infty$, and c is an $O(1)$ constant. Equation (213) is known as the Barton–Nakajima–Namikawa relation (Barton, 1966; Nakajima, 1972; Namikawa, 1975).

(15) Temperature dependence of the dielectric loss strength $\Delta\varepsilon$ is much weaker than that of ω_m or $\sigma(0)$, and therefore Eq. (213) implies a rough proportionality,

$$\sigma(0) \sim \omega_m. \tag{214}$$

(16) Both $\sigma(0)$ and ω_m have an Arrhenius-type temperature dependence with the same activation energy.

These sixteen common AC characteristics of disordered solid materials provide strong empirical support for the universality of AC conductivity. There is also strong theoretical evidence for this universality that will be discussed shortly. However, as noted by Dyre and Schrøder (2000), there are also some "dissenting" views, emphasizing the *differences* between various disordered solids and their AC conductivity (see, for example, Elliot, 1994; Macdonald, 1997; Ngai and Moynihan, 1998), since there are disordered solids for which the exponent x is slightly *larger* than 1.0 (see, for example, Cramer and Buscher, 1998), and those for which the DC conductivity is *not* of Arrhenius type (for example, group-IV amorphous semiconductors, or fast ion conducting glasses). But, the vast majority of disordered solids do seem to exhibit universal AC conductivity behavior.

Sidebottom (1999) proposed the following universal scaling representation of the AC conductivity:

$$\hat{\sigma} = \frac{\sigma(\omega)}{\sigma(0)} = F\left[\varepsilon_0\Delta\varepsilon\frac{\omega}{\sigma(0)}\right], \tag{215}$$

that is, the Taylor–Isard constant C is given by, $C = \varepsilon_0\Delta\varepsilon$. The proof of this equation was provided by Schrøder and Dyre (2000). Suppose that $\hat{\omega}$ is the scaled frequency. Then, one can expand the master curve to first order, $\hat{\sigma} = 1 + i\hat{\omega}A + \cdots$ Since, $\sigma(\omega) = \hat{\sigma}\sigma(0)$, we have, $\sigma(\omega) = \sigma(0) + i\hat{\omega}A\sigma(0)$. On the other hand, Eq. (210) implies that to first order in ω, $\sigma(\omega) = \sigma(0) + i\omega\varepsilon_0\Delta\varepsilon$, which, when compared with the last equation, implies that $\hat{\omega} = A^{-1}[\varepsilon_0\Delta\varepsilon/\sigma(0)]\omega$, and therefore $C = \varepsilon_0\Delta\varepsilon$.

We now discuss theoretical evidence and computer simulations that support the universality of AC conductivity.

6.3.2 Resistor–Capacitor Model

Consider a disordered solid with a spatially varying free charge conductivity $g(\mathbf{r})$, which is independent of frequency, and a uniform bound charge dielectric constant ε_∞, so that $\mathbf{I}(\mathbf{r}, t) = g(\mathbf{r})\mathbf{E}(\mathbf{r}, t)$ and $\mathbf{D}(\mathbf{r}, t) = \varepsilon_\infty\varepsilon_0\mathbf{E}(\mathbf{r}, t)$. If we combine these equations with $\mathbf{E} = \nabla\phi$ (where ϕ is a potential), $\nabla \cdot \mathbf{D} = \rho$ (where ρ is the free charge density), and the continuity equation, $\partial\rho/\partial t + \nabla \cdot \mathbf{I} = 0$, we obtain

$$\nabla \cdot \{[i\omega\varepsilon_0\varepsilon_\infty + g(\mathbf{r})]\nabla\phi\} = 0. \tag{216}$$

If we discretize this equation by, for example, a finite-difference method, we obtain a network, the 2D version of which is shown in Figure 6.11 (Fishchuk, 1986; Dyre, 1993). Each bond of the network consists of a capacitor proportional to $\varepsilon_0\varepsilon_\infty$ and a resistor proportional to the local resistivity $1/g(\mathbf{r})$. The currents through the resistors are the free charge currents. In a periodic field the current through the capacitors is proportional to $i\omega\varepsilon_0\varepsilon_\infty$ times the potential drop. However, since $\mathbf{D} = \varepsilon_0\varepsilon_\infty\mathbf{E}$, and $\mathbf{E}$ is proportional to the potential drop, the current through the capacitors should be proportional to $\partial\mathbf{D}/\partial t$.

Consider now a $L \times L$ or $L \times L \times L$ network of this type. We impose a periodic potential across two opposing faces of the network that act as electrodes, the

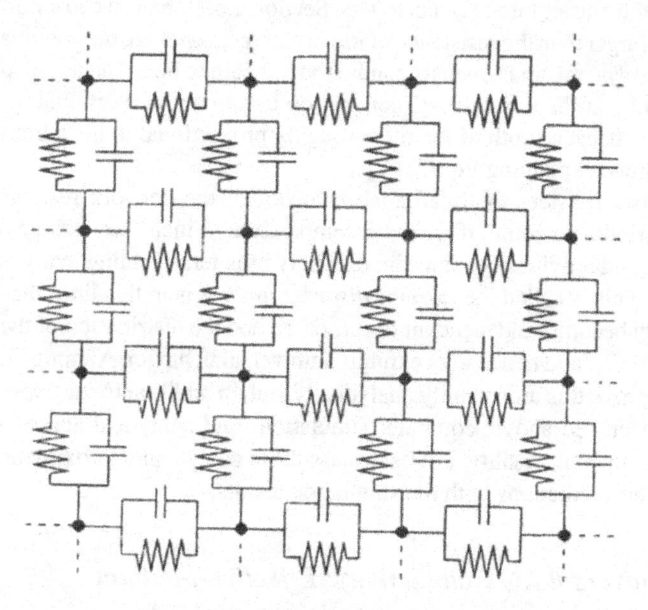

FIGURE 6.11. Two-dimensional network model that arises from discretizing Eq. (216). All capacitors are equal, proportional to the bound charge dielectric constant, while each resistor is proportional to the inverse free charge conductivity at the corresponding position in the material.

admittance between which is $Y(\omega)$. Then,

$$\sigma(\omega) = \frac{Y(\omega)}{L^{d-2}} - i\omega\varepsilon_0\varepsilon_\infty, \tag{217}$$

In a realistic simulation one should take, $g(\mathbf{r}) = g_0 \exp(-\beta E_a)$, where g_0 varies locally (which can be selected from a given distribution), $\beta = (k_B T)^{-1}$ and E_a is the activation energy. Alternatively, one may fix g_0 and take E_a to be a spatially varying quantity, selected from a given distribution; this is what we assume in most of our discussions. It is clear that, within this model, the mechanism of conduction (classical or quantum mechanical) is immaterial. Except in 1D, one must use large-scale simulations in order to compute $\sigma(\omega)$. However, in the limit of high frequencies, the capacitor admittances are so large that they dominate, resulting in a spatially homogeneous electric field or potential. Therefore, the average resistor current is determined by the average free charge conductivity, i.e., $\sigma(\infty) = \langle g \rangle$.

The resistor-capacitor model also provides the answer to a crucial question. As discussed above, at low frequencies the AC conductivity is independent of ω. The question then is: What determines the onset of AC conduction? At very low frequencies each capacitor has an admittance with a numerical value which is smaller than that of its partner resistor. As ω is increased, an increasing number of the capacitors will have admittances that are larger than the resistance of their partner resistors. As long as such capacitors are isolated, or have not formed a sample-spanning cluster, the overall conductivity of the system remains independent of frequency. However, when the links of the sample-spanning cluster, i.e., those that, if cut, split the cluster into two pieces (see Section 2.6.3), have a capacitor with an admittance larger than the resistance of its partner resistor, then, the potential distribution in the system, and hence its conductivity, begin to be frequency-dependent. The frequency at the onset of AC conduction is roughly proportional to the DC conductivity, because both of them are roughly proportional to the admittance of the links, hence explaining Eq. (214).

Let us now consider the qualitative behavior of the network response to an imposed periodic potential. If we lower temperature, β increases and the distribution of the conductivities becomes increasingly broader, spanning many orders of magnitude. This is called the *extreme disorder limit.* It is in this limit that the AC conductivity becomes independent of β (i.e., T) and the distribution of the activation energies E_a, and hence $g(\mathbf{r})$ exhibits a universal behavior. Although it is very difficult to prove this universality analytically, but, in addition to the experimental evidence discussed above, computer simulations and analytical approximations also support the universality. Let us discuss these analytical approximations, and compare their predictions with the simulation results.

6.3.3 *Universal AC Conductivity: Effective-Medium Approximation*

As discussed in Chapter 5 and earlier in the present chapter, in the EMA for randomly varying admittances y on a d-dimensional simple cubic lattice, the effective admittance y_m is computed from [see Eqs. (4.239) and (4.240)]

$$\left\langle \frac{y - y_m}{y + (d-1)y_m} \right\rangle = 0, \tag{218}$$

where $\langle \cdot \rangle$ indicates an average over the distribution of the admittances. We now substitute the admittance $y \propto g(\mathbf{r}) + i\omega\varepsilon_0\varepsilon_\infty$ into Eq. (218). Then, $y_m(\omega) \propto \sigma(\omega) + i\omega\varepsilon_0\varepsilon_\infty$. It is then straightforward to see that Eq. (218) becomes

$$\left\langle \frac{g - \sigma}{g + (d-1)\sigma + id\omega\varepsilon_0\varepsilon_\infty} \right\rangle = 0. \tag{219}$$

Equation (219) is exact for $d = 1$ and also becomes exact for any d in the limit $\omega \to \infty$, i.e., it reproduces the exact results, $\sigma(\infty) = \langle g \rangle$, mentioned above (Dyre, 1993). Note that the total admittance Y of the network is $Y = L^{d-2} y_m$.

We now show that the EMA predicts a universal AC conductivity at low temperatures. The discussion that follows is a summary of the analysis presented by Dyre (1993). Let $s = i\omega\varepsilon_0\varepsilon_\infty$. Because, $g - \sigma = g + (d-1)\sigma + ds - d(s + \sigma)$, Eq. (219) is rewritten as

$$\frac{1}{d(s+\sigma)} = \left\langle \frac{1}{g(E_a) + (d-1)\sigma + ds} \right\rangle_{E_a}, \tag{220}$$

where the average is taken over the distribution $h(E_a)$ of the activation energies. In the limit $\beta = 1/k_BT \to \infty$ the local conductivity $g(E_a) = g_0 \exp(-\beta E_a)$ varies rapidly and, depending on E_a, for given σ and s there are two extreme possibilities, either $g(E_a) \ll (d-1)\sigma + ds$ or $g(E_a) \gg (d-1)\sigma + ds$. In the former case $g(E_a)$ can be ignored, while in the latter case the denominator of Eq. (220) becomes very large and there is little contribution to the right-hand side. The energy $E_g(s)$ that separates the two cases is obtained from $g = (d-1)\sigma + ds$, so that

$$E_g(s) = -\frac{1}{\beta} \ln \left[\frac{(d-1)\sigma + ds}{g_0} \right]. \tag{221}$$

On the other hand, Eq. (220), for large values of β, becomes

$$\frac{d-1}{d} + \frac{s}{d(s+\sigma)} = \int_{E_g(s)}^{\infty} h(E)dE. \tag{222}$$

If we set $s = 0$ in Eq. (222) and subtract the result from Eq. (222), we obtain

$$\frac{s}{d(s+\sigma)} = \int_{E_g(s)}^{E_g(0)} h(E)dE. \tag{223}$$

For large values of β, $E_g(s) \simeq E_g(0)$, and the integral can be replaced by $h[E_g(0)][E_g(0) - E_g(s)]$ [assuming that $h(E_a)$ is smooth around $E_a = 0$], and therefore

$$\begin{aligned} \frac{s}{d(s+\sigma)} &= -\frac{h[E_g(0)]}{\beta} \left\{ \ln \left[\frac{\sigma(0)(d-1)}{g_0} \right] - \ln \left[\frac{\sigma(d-1) + ds}{g_0} \right] \right\} \\ &= \frac{h[E_g(0)]}{\beta} \ln \left[\frac{\sigma}{\sigma(0)} + \frac{ds}{\sigma(0)(d-1)} \right]. \end{aligned} \tag{224}$$

Therefore, if one introduces the dimensionless variables,

$$\hat{\sigma} = \frac{\sigma}{\sigma(0)}, \quad \hat{s} = \frac{\beta s}{d\sigma(0)h[E_g(0)]}, \tag{225}$$

then, Eq. (224), in the limit $\beta \to \infty$ (low temperatures) becomes

$$\hat{\sigma} \ln(\hat{\sigma}) = \hat{s}. \tag{226}$$

Equation (226) states that, for $d > 1$, the AC conductivity of disordered solids at low temperatures is universal, independent of the distribution $h(E_a)$ of the activation energies. Equation (226) was derived by Bryksin (1980), Fishchuk (1986), and Dyre (1988) in various contexts. More recently, Bleibaum *et al.* (1996) derived an equation similar to (226) by a completely different method.

Equation (226) can also provide information on the scaling of the AC conductivity with frequency. Recall [see Eq. (212)] that the real part σ' of the AC conductivity depends on the frequency by a power law, and similarly for its imaginary part. Therefore, if we write $\hat{\sigma} \sim \hat{s}^u$, then,

$$\hat{\sigma} = \frac{\hat{s}}{\ln(\hat{\sigma})} \simeq \frac{\hat{s}}{\ln(\hat{s})}, \tag{227}$$

so that

$$u = \frac{d\ln(\hat{\sigma})}{d\ln(\hat{s})} \simeq 1 - \frac{1}{\ln(\hat{s})}, \quad \hat{s} \gg 1. \tag{228}$$

$\hat{s}$ is imaginary at real frequencies. Therefore, if we write $\hat{s} = i\hat{\omega}$, we obtain

$$\hat{\sigma} \simeq \frac{i\hat{\omega}}{\ln(i\hat{\omega})} = \frac{i\hat{\omega}}{\ln(\hat{\omega}) + i\pi/2}, \quad \hat{\omega} \gg 1, \tag{229}$$

and therefore

$$\hat{\sigma}' \simeq \frac{\pi}{2}\frac{\hat{\omega}}{\ln^2(\hat{\omega})}, \quad \hat{\sigma}'' \simeq \frac{\hat{\omega}}{\ln(\hat{\omega})}, \quad \hat{\omega} \gg 1. \tag{230}$$

Equations (230) imply that, if one defines the exponents x and y by

$$\hat{\sigma}' \sim \hat{\omega}^x, \quad \hat{\sigma}'' \sim \hat{\omega}^y, \tag{231}$$

then, for $\hat{\omega} \gg 1$

$$x = 1 - \frac{2}{\ln(\hat{\omega})}, \quad y = 1 - \frac{1}{\ln(\hat{\omega})}, \quad \hat{\omega} \gg 1, \quad \text{EMA}. \tag{232}$$

Figure 6.12 compares the results of computer simulations, using 200×200 square networks, with the predictions of the EMA, Eq. (219) (Dyre, 1993). In these simulations, $\beta = 5$, 10, 20, 40, 80, and 160, four different distributions of the activation energy were utilized which were, (i) asymmetric Gaussian; (ii) symmetric Gaussian; (iii) power-law, $h(E_a) \sim (1 + E_a)^{-4}$, and (iv) triangle, and the results were averaged over 10 different realizations of the network. In each case, the distribution was centered around a value $\langle E_a \rangle$. The agreement between the predictions and the simulation results is excellent, particularly in the light of the fact that the EMA has no adjustable parameter. Note that the AC conductivity becomes universal at low temperatures (high values of β).

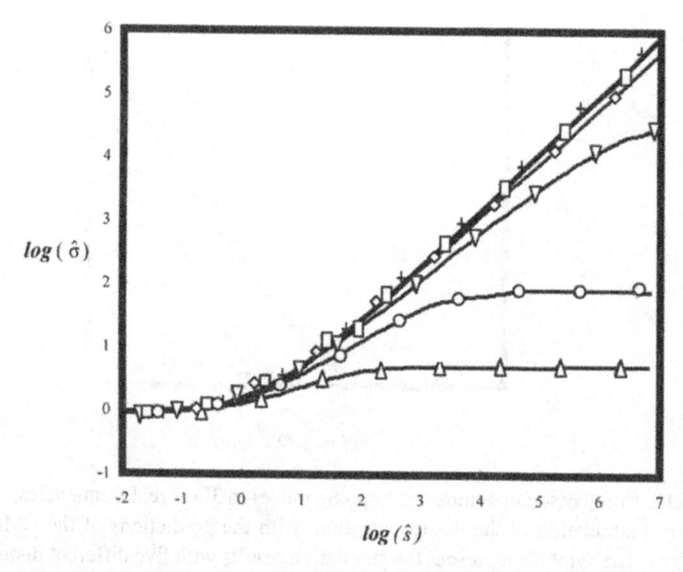

FIGURE 6.12. Logarithmic plot of the dimensionless conductivity $\hat{\sigma}$ (symbols) versus real dimensionless Laplace frequencies $\hat{s}$, and its comparison with the EMA predictions (curves). The distribution of the activation energies is asymmetric Gaussian, and symbols represent the results for $\beta = 5$ (△), 10 (◯), 40 (◇), 80 (□), and 160 (+). The results with other types of the activation energy distributions (see the text) are similar. (after Dyre, 1993).

6.3.4 Universal AC Conductivity: Symmetric Hopping Model

The symmetric hopping model described in Sections 6.1 and 6.2 can also be used for modeling the AC conductivity. All the results discussed previously for this model are also applicable to the AC conduction. In particular, one has two exact results:

$$\sigma(0) = \langle W^{-1} \rangle^{-1}, \quad d = 1. \tag{233}$$

$$\sigma(\infty) = \langle W \rangle, \quad d \geq 1, \tag{234}$$

where W is the transition rate between the lattice sites, and the averages are taken with respect to the statistical distribution $f(W)$. Moreover, the various EMAs developed for the symmetric hopping model can also be used for estimating the AC conductivity. As shown by Dyre (1994), in the low-temperature limit, they all lead to Eq. (226), and hence the symmetric hopping model also predicts universal behavior for AC conductivity. Figure 6.13 presents the results of simulation of the 3D symmetric hopping model for various distributions of the activation energies [that is, if one assumes that $W = W_0 \exp(-\beta E_a)$], reported by Schr⊘der and Dyre (2000), and compares them with the EMA predictions. Clearly, the EMA provides reasonably accurate predictions for the simulation results and, moreover, the symmetric hopping model also predicts universal behavior for the AC conductivity, independent of the statistical distribution of the activation energies. Summerfield

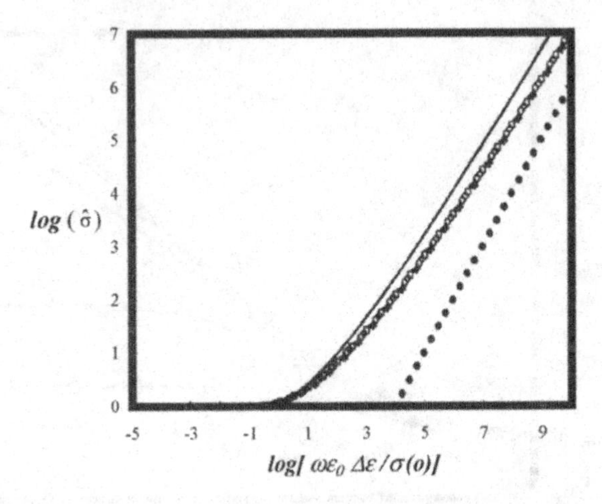

FIGURE 6.13. Comparison of symmetric hopping model in 3D at real frequencies, obtained by numerical simulation of the master equation, with the predictions of the EMA (solid upper curve). The symbols represent the simulation results with five different distributions for the activation energy. Solid circles to the right indicate a straight line with slope one (after Schrøder and Dyre, 2000).

and Butcher (1982) and Summerfield (1985) were the first to point out the possible universality of AC conductivity based on the symmetric hopping model.

6.3.5 Role of Percolation in Universality of AC Conductivity

By now it should be clear to the reader how the activation energy $E_{\rm DC}$ of the DC conductivity, defined by, $\sigma(0) \propto \exp(-\beta E_{\rm DC})$ for large values of β, can be determined by percolation. If $h(E_a)$ is the statistical distribution of the activation energies, then according to the critical path concept described in Section 6.2.4, $E_{\rm DC}$ is the solution of the following equation:

$$\int_0^{E_{\rm DC}} h(E)dE = p_c, \tag{235}$$

where p_c is the bond percolation threshold of the system. The argument leading to Eq. (235) can be extended to the symmetric hopping model as well (Dyre and Schrøder, 2000): The main contributions to the (zero-field) mean square displacement are made by charge carriers that use the links of the network that have the largest transition rates. Thus, optimal carriers hop preferably on the percolation cluster which is defined by marking links in order of decreasing transition rates until percolation (which is how the cluster is constructed in the critical path method; see Section 6.2.4). Such optimal carriers must occasionally overcome the largest barriers on the percolation cluster, which act as bottlenecks, and therefore $E_{\rm DC}$, as given by Eq. (235), determines the rate of the mean square displacement, hence the diffusivity and thus the conductivity.

Computer simulations of Brown and Esser (1995) indicate that, at extreme disorder, the DC current follows almost 1D paths, and thus a simplistic approach is to take these paths as strictly 1D. If so, one obtains what Dyre and Schrøder (2000) call the *percolation path approximation* (PPA). To quote them, "The universal AC conductivity is equal to that of extreme disorder limit of a 1D model with a sharp upper cutoff in the activation energy probability distribution." Then, it is straightforward to show that (Dyre, 1988, 1994)

$$\frac{1}{\sigma + s} = c \int_{-\infty}^{E_{\mathrm{DC}}} \frac{h(E)}{g(E) + s} dE, \tag{236}$$

where c is a constant the value of which is unimportant. In a fixed range of frequencies around the transition frequency, the dominant contribution to Eq. (236) at low temperatures is due to energies that are close to E_{DC}. Thus, one may replace $h(E)$ by $h(E_{\mathrm{DC}})$. It is then easy to show that

$$\hat{\sigma} = \frac{\hat{s}}{\ln(1 + \hat{s})}, \tag{237}$$

or, if we set $\hat{s} = i\hat{\omega}$, we obtain (Dyre, 1988)

$$\hat{\sigma} \ln(1 + i\hat{\omega}) = i\hat{\omega}. \tag{238}$$

Thus, the PPA also predicts a universal AC conductivity at low temperatures. Equation (238) provides a slightly more pronounced frequency dependence of the conductivity than Eq. (226), the prediction of the EMA.

One may apply the PPA to hopping. To do so, the AC conductivity in 1D with a sharp activation energy cutoff must be computed in the limit of extreme disorder. Dyre and Schrøder (1996) showed that the EMA in *one dimension* represents the simulation results very accurately. It is not difficult to show that, under these conditions, the EMA for AC conductivity of 1D materials predicts the following universal AC conductivity:

$$\sqrt{\hat{\sigma}} \ln(1 + \sqrt{i\hat{\omega}\hat{\sigma}}) = \sqrt{i\hat{\omega}}, \tag{239}$$

which can be considered as the PPA-EMA for the hopping model. Compared to Eq. (226), the prediction of the EMA, Eq. (239) provides a somewhat less pronounced frequency dependence of the conductivity.

We may conclude, based on the available experimental and computer simulation results and the predictions of various analytical approximations, that the AC conductivity of the vast majority of disordered solids, in the limit of low temperatures, exhibits universal behavior.

An important question is: How robust is the universality of AC conductivity? Dyre and Schrøder (2000) provided the answer to this question by making an interesting analogy between the universality of AC conductivity and that of second-order phase transitions which is highly robust to very extensive modification of the system. The universality in the latter case is usually associated with the existence of a diverging length scale as a critical point is approached. For example, as we learned in Chapter 2, as the percolation threshold is approached, the percolation correlation length diverges [see Eq. (2.32)]. Is there such a diverging length scale in the AC conductivity problem? Consider, for example, the hopping model and define ℓ_m^2 to be the mean square displacement of the hopper at time $t = 1/\omega_m$, where ω_m is defined through Eq. (213). Recall that ω_m is the frequency that defines the onset of AC conductivity. Both the EMA and computer simulations indicate that ℓ_m^2 diverges as $\beta \to \infty$, and therefore the universality of AC conductivity can also be associated with a diverging length scale, which in turn implies that this universality is very robust.

For the most recent discussion of interpretation of experimental data for the AC conductivity see Léon *et al.* (2001).

6.4 Dielectric Constant and Optical Properties

Although as pointed out in Chapters 4 and 5, the analogy between various linear transport processes in heterogeneous materials implies that most of the theoretical methods that have been developed so far for predicting the effective conductivity have their analogues for the effective dielectric constant, we wish, in this section, to describe and discuss the discrete models for computing frequency-dependent effective dielectric constant (or, more generally, dielectric *function*) of composite materials. We first describe numerical calculation of the dielectric constant, and then discuss various analytical approximations that can be used for estimating this important property.

6.4.1 Resistor–Capacitor Model

Let us first describe and discuss how the dielectric constant of a composite material is computed via a discrete model. Consider a disordered material, such as a mixture of a conducting component and a dielectric one. A typical conducting grain size a in the material is much smaller than the wavelength of the light in the visible and infrared spectral ranges. If so, one can introduce a potential $\phi(\mathbf{r})$ for the local electric field and write the local current density $\mathbf{I}$ as, $\mathbf{I}(\mathbf{r}) = g(\mathbf{r})[-\nabla\phi(\mathbf{r}) + \mathbf{E}_0]$, where $\mathbf{E}_0$ is the external field, and $g(\mathbf{r})$ is the local conductivity at $\mathbf{r}$. In the quasi-static limit, computation of the field distribution reduces to finding the solution of the Poisson equation since, due to current conservation, $\nabla \cdot \mathbf{I} = 0$, one has

$$\nabla \cdot \{g(\mathbf{r})[-\nabla\phi(\mathbf{r}) + \mathbf{E}_0]\} = 0. \tag{240}$$

The local conductivity $g(\mathbf{r})$ is equal to either g_c or g_d, representing the conducting and dielectric components, respectively. We rewrite Eq. (240) in terms of the local dielectric constant,

$$\varepsilon(\mathbf{r}) = \frac{4\pi i}{\omega}\, g(\mathbf{r}), \tag{241}$$

so that

$$\nabla \cdot [\varepsilon(\mathbf{r})\nabla\phi(\mathbf{r})] = \mathcal{E}, \tag{242}$$

where $\mathcal{E} = \nabla \cdot [\varepsilon(\mathbf{r})\mathbf{E}_0]$. While the external field $\mathbf{E}_0$ is real, $\phi(\mathbf{r})$ is, in general, a complex function since ε_c is complex in the optical and infrared spectral ranges. Since Eq. (242) is very difficult to solve analytically, one discretizes it in order to solve it by numerical simulations. If, for example, a standard 5-point (in 2D) or 7-point (in 3D) finite-difference discretization is used, then, a discrete model on a simple-cubic lattice is obtained in which the conducting and dielectric particles are represented by conducting and dielectric bonds of the lattice. In this way, Eq. (242), in discretized form, takes on the form of Kirchhoff's equations defined on a lattice. Assuming that the external electric field $\mathbf{E}_0$ is directed along a specific principal axis of the lattice, say the z-axis, one obtains

$$\sum_j \varepsilon_{ij}(\phi_j - \phi_i) = \sum_j \varepsilon_{ij} E_{ij} \tag{243}$$

where ϕ_i is the electric potential at site i of the lattice, and the sum is over the nearest neighbors j of the site i. For the bonds ij in the $\pm z$-direction, the electromotive force E_{ij} is given by, $E_{ij} = \pm E_0 a_0$ (where a_0 is the spatial period of the lattice), while $E_{ij} = 0$ for the other bonds that are connected to site i. Thus, the composite material is modeled by a resistor-capacitor network in which the bond permittivities ε_{ij} are statistically independent and a_0 is equal to the conducting grain size, $a_0 = a$. In the case of a two-phase conductor-dielectric random composite, the permittivities ε_{ij} are equal to either ε_c or ε_d with probabilities p and $1-p$, respectively. We also write the permittivities in the standard form, $\varepsilon_j = \varepsilon'_j + i\varepsilon''_j$, where $j = c$ or d.

To make further progress, we take the lattice to be a simple-cubic lattice with a very large but finite number of sites N, and rewrite Eq. (243) in a matrix form:

$$\mathcal{H}\boldsymbol{\phi} = \mathcal{E}, \tag{244}$$

where $\boldsymbol{\phi} = \{\phi_1, \phi_2, \ldots, \phi_N\}$, and the elements of the vector $\mathcal{E}$ are, $\mathcal{E}_i = \sum_j \varepsilon_{ij} E_{ij}$. Here $\mathcal{H}$ is a $N \times N$ matrix such that for $i \neq j$, $\mathcal{H}_{ij} = -\varepsilon_{ij} = \varepsilon_d > 0$ and $\varepsilon_c = (-1 + i\kappa)|\epsilon'_c|$ with probabilities p and $1-p$, respectively, and $\mathcal{H}_{ii} = \sum_j \varepsilon_{ij}$, where j refers to nearest neighbors of site i, and κ is the usual loss factor, $\kappa = \varepsilon''_c / |\varepsilon'_c| \ll 1$. The diagonal elements $\mathcal{H}_{ii}$ are distributed between $2d\varepsilon_c$ and $2d\varepsilon_d$, where d is the dimensionality of the space.

Similar to the dielectric constant, we write $\mathcal{H} = \mathcal{H}' + i\kappa\mathcal{H}''$, where $i\kappa\mathcal{H}''$ represents losses in the system. The Hamiltonian $\mathcal{H}'$ formally coincides with the Hamiltonian of the problem of metal-insulator transition (Anderson transition) in quantum systems, i.e., it maps the quantum-mechanical Hamiltonian for the Anderson transition problem with both on- and off-diagonal correlated disorder onto the present problem. This mapping will be discussed in more detail in Chapter 3 of Volume II.

Once a lattice model is set up, one may calculate the effective dielectric constant of the composite by several methods. Direct numerical simulations, similar to those for computing the effective conductivity of a composite that were described in Chapter 5, are, in principle, not difficult, except that one must deal with complex numbers, and therefore the size of the lattice that can be used in the simulations may be somewhat limited. One may also tackle the problem by several analytical approximations. Two of such approximations will described shortly.

6.4.2 *Resistor–Capacitor–Inductor Model*

A more general network model, which can be used for studying the optical properties of heterogeneous materials, is the resistor-inductor-capacitor network (Koss and Stroud, 1987; Zeng, Hui and Stroud, 1989). In particular, one can use such networks to model a metal-insulator composite and its absorption properties. To see this, recall, as pointed out in Section 4.13, that such composites possess surface plasmon modes or Mie resonances, which are easily understood by considering a (spherical) metallic grain described by a Drude dielectric function

$$\varepsilon_m = 1 - \frac{(\omega_p/\omega)^2}{1 + i\omega_\tau/\omega}, \tag{245}$$

embedded in an insulating matrix of dielectric constant of unity. Here ω_p is the plasma frequency and $\omega_\tau = 1/\tau$, with τ being a characteristic relaxation time. If, in the long-wavelength limit, the composite is subjected to an external electric field $\mathbf{E}_0 \exp(i\omega t)$, then the field inside the metallic particle will be uniform, given by

$$\mathbf{E}_i = \frac{3\mathbf{E}_0}{2 + \varepsilon_m} \exp(i\omega t). \tag{246}$$

In the limit $\omega_p \gg \omega_\tau$, the real part of $3/(2 + \varepsilon_m)$ becomes very large at frequencies $\omega \sim \omega_p/\sqrt{3}$, the surface plasmon frequency of a small spherical metallic particle. As a result, the absorption coefficient $\alpha = (\omega/c)\text{Im}(\sqrt{\varepsilon})$ will exhibit a strong peak near $\omega = \omega_p/\sqrt{3}$.

Such a composite can be modeled by a network of resistors, inductors, and insulators (Koss and Stroud, 1987; Zeng, Hui and Stroud, 1989). The bonds of the network present either a metallic component or an insulating material. An insulating bond is represented simply as a capacitor with an admittance

$$y_i = i\omega C_i, \tag{247}$$

where C_i is the capacitance. A metallic bond is composed of a resistance R and an inductance L in series, the two in parallel with a capacitance C_m that admits displacement current, so that the total admittance y_m of a metallic bond is given by

$$y_m = \frac{1 + i\omega R C_m - \omega^2 C_m L}{R + i\omega L}, \tag{248}$$

and thus $\tau = L/R$ is the relaxation time of the system. It is easy to see that such a network can model a composite of Drude metal and insulating material, if the dielectric function of the insulating material is selected to be $\varepsilon_i = 1$, and the frequency units are such that $\omega_p = 1$. In principle, the capacitances C_i and C_m, the resistance R, and the inductance L can vary from bond to bond.

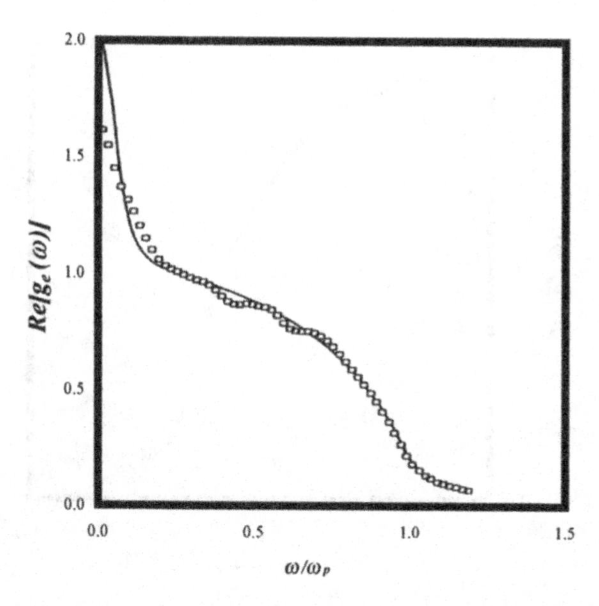

FIGURE 6.14. Real part of the frequency-dependent conductivity for a Drude metal-insulator composite, obtained by numerical simulations on a square lattice, for metal volume fraction $p = 0.6$, and its comparison with the EMA predictions (curve). The relaxation time is 10 (after Zeng *et al.*, 1989).

Once such a network is constructed, a variety of numerical methods can be used to compute its effective properties. Direct numerical simulation, while certainly possible, is hugely time consuming. More efficient methods include the Frank–Lobb method, and the transfer-matrix method, both of which were described in Chapter 5. One can also use the EMA (see below) to compute the effective properties of the network, and it turns out that EMA can provide highly accurate predictions for these properties. Figure 6.14 presents the real part of the effective conductivity $\hat{g}_e(\omega)$ of the network (which, of course, corresponds to absorption in the system), as a function of frequency (Zeng, Hui and Stroud, 1989). The simulations were carried out for a metallic fraction $p = 0.6$ on a 100×100 square network with $\tau = 10$. Also shown are the predictions of the EMA which are in excellent agreement with the simulation results. The results shown in this figure are typical for $p > p_c$, exhibiting a Drude peak which is due to a non-zero DC conductivity. For $p \leq p_c$, the real part of the conductivity exhibits a broad surface plasmon resonance peak, an example of which is shown in Figure 6.15 for $p = 0.5$. The peak becomes narrower as p decreases to small values.

Let us now concentrate on a conductor-dielectric composite, represented by a resistor-capacitor network, and describe how its effective properties can be computed. However, the methods that we describe are equally applicable to resistor-inductor-capacitor networks.

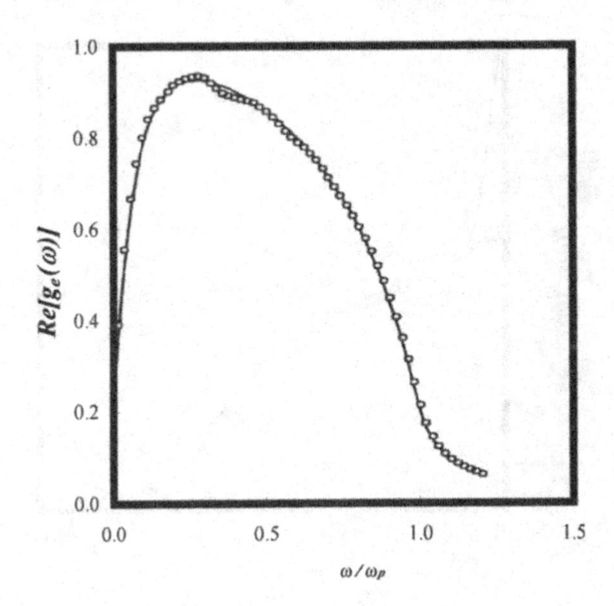

FIGURE 6.15. Same as Figure 6.14, but for metal volume fraction $p = 0.5$ (after Zeng *et al.*, 1989).

6.4.3 Position-Space Renormalization Group Approach

Computing the effective properties of a conductor-dielectric composite by a position-space renormalization group (PSRG) method is similar to what we described in Section 5.11 for calculating the effective conductivity of disordered solids. Here, we describe one such PSRG approach due to Wilkinson *et al.* (1983).

Consider a square lattice consisting of two types of bonds with conductances g_1 and g_2. The bonds with $g_1 = g_c$ are purely resistive and occur with probability p. The second type of the bonds are purely capacitive with $g_2 = i\omega C$, where ω is the frequency and C is the capacitance, and occur with probability $q = 1 - p$. An example is show in Figure 6.16. Suppose that a 2×2 RG cell, of the type shown in Figure 6.17, is used, which is similar to the self-dual RG cell used in Section 5.11 for computing the effective conductivity. Each cell is renormalized into conductive or capacitive bonds in the horizontal and vertical directions, also shown in Figure 6.17. The equivalent conductances g_1' and g_2' of the renormalized bonds are, in general, complex variables that occur with probabilities p' and $q' = 1 - p'$, respectively. For convenience, we shall refer to g_1' and g_2' as the conductor and capacitor, respectively, even though they both are complex variables. Since the horizontal and vertical directions are independent of each other and are equivalent, one can consider just one of them, say the vertical direction. Thus, as discussed in Section 5.11, we impose a voltage drop between the top and bottom of the RG cell, so that the effective conductance of the cell is equivalent to that of a Wheatstone bridge which contains 5 bonds, and therefore there are $2^5 = 32$ possible configurations of the bridge. These configurations are divided into two

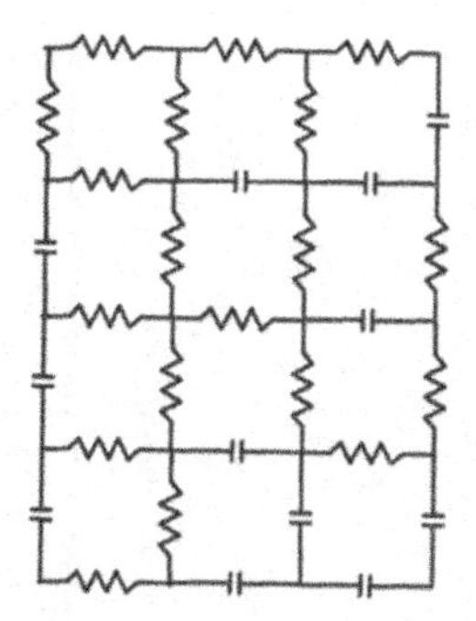

FIGURE 6.16. Two-dimensional resistor-capacitor model in which each bond is occupied by a conductor with probability p or a capacitor with probability $1 - p$ (after Wilkinson *et al.*, 1983).

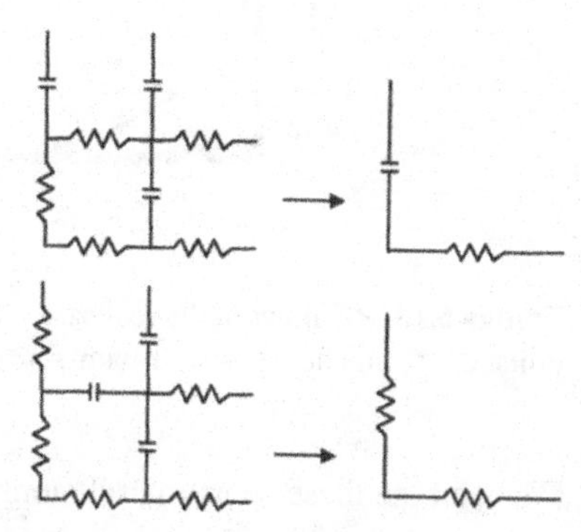

FIGURE 6.17. Two examples of a RG transformation in the bimodal approximation (after Wilkinson *et al.*, 1983).

groups: The conducting ones in which the resistive component g_1 forms a conducting path, and the insulating configurations when it does not. This grouping of the configurations is especially accurate when $|g_2| < g_1$, and therefore the results discussed below are accurate when, $\omega < g_c/C$. The regime $\omega > g_c/C$ can also be studied by simply interchanging the roles of g_1 and g_2.

If $\alpha = g_2/g_1$, then three recursion relations are obtained,

$$p' = f(p), \quad p'g_1' = g_1 f_1(p, \alpha), \quad q'g_2' = g_2 f_2(p, \alpha), \tag{249}$$

where

$$f(p) = p^5 + 5p^4q + 8p^3q^2 + 2p^2q^3, \tag{250}$$

which is obviously the same as the one that we obtained in Chapter 5 [see Eq. (5.181)], as it should be, and

$$f_1(p, \alpha) = p^5 + p^4q + 4p^4q\left(\frac{5\alpha+3}{3\alpha+5}\right) + 2p^3q^2\left(\frac{\alpha+1}{2}\right)$$

$$+4p^3q^2\left(\frac{2\alpha^2+5\alpha+1}{\alpha^2+5\alpha+2}\right) + 2p^3q^2\left(\frac{3\alpha+1}{\alpha+3}\right) + 2p^2q^3\left(\frac{\alpha+1}{2}\right), \tag{251}$$

$$f_2(p, \alpha) = q^5 + pq^4 + 4pq^4\left(\frac{3\alpha+5}{5\alpha+3}\right) + 2p^2q^3\left(\frac{2}{\alpha+1}\right)$$

$$+4p^2q^3\left(\frac{\alpha^2+5\alpha+2}{2\alpha^2+5\alpha+1}\right) + 2p^2q^3\left(\frac{\alpha+3}{3\alpha+1}\right) + 2p^3q^2\left(\frac{2}{\alpha+1}\right). \tag{252}$$

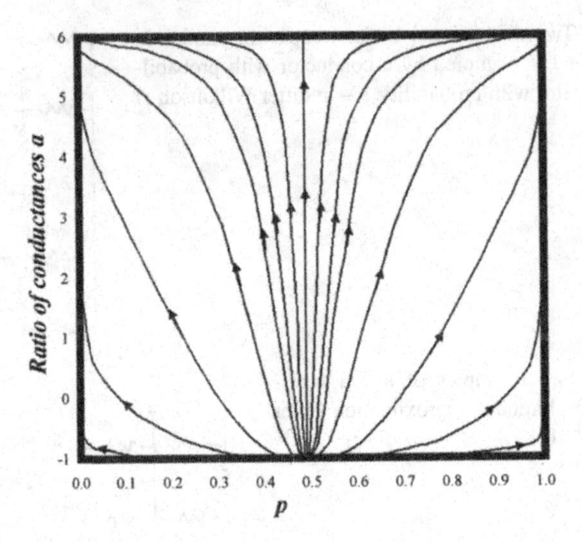

FIGURE 6.18. RG flows in the real case. The point $p = 1/2$ and $\alpha = 0$ is the unstable fixed point. All points flow to $\alpha = 1$ with $p = 0$ or 1 (after Wilkinson *et al.*, 1983).

We combine these recursion relations into a single relation,

$$\frac{\alpha'}{\alpha} = \frac{p' f_2(p, \alpha)}{q' f_1(p, \alpha)}, \tag{253}$$

so that Eqs. (250) and (253) define the RG flows in the space of three parameters which are p, $\mathrm{Re}(\alpha)$, and $\mathrm{Im}(\alpha)$.

Next, the fixed points of these recursion relations are determined. We consider first the flows for the case in which g_1 and g_2 are both purely resistive. Then, the flow diagram is shown in Figure 6.18. Note that there are six fixed points which are, $p = 0, 1/2$, and 1 with $\alpha = 0$ or 1. The unstable fixed point at $p = 1/2$ and $\alpha = 0$ corresponds to the exact percolation threshold of the square lattice. The fixed points at $\alpha = 1$ and $p = 0$ or 1 are stable. In the complex α plane, the flows are more complicated, but are qualitatively the same as in Figure 6.18. In particular, the system always flows to $\alpha = 1$ with $p = 0$ or 1.

The recursion relations are now linearized in the manner that was discussed in Chapter 5. The results are,

$$p' - p_c \simeq r(p - p_c), \quad r = 13/8 \tag{254}$$

$$\frac{g_1'}{g_1} = 2f_1(1/2, \alpha) \simeq a_1 + b_1\alpha + \cdots, \quad a_1 = 17/30 \tag{255}$$

$$\frac{g_2'}{g_2} = 2f_2(1/2, \alpha) \simeq a_2 + b_2\alpha + \cdots, \quad a_2 = 23/12 \tag{256}$$

$$\frac{\alpha'}{\alpha} = \frac{f_2(1/2, \alpha)}{f_1(1/2, \alpha)} = A + B\alpha + \cdots, \quad A = 115/34. \tag{257}$$

Suppose that p is slightly larger than p_c and that $\alpha = i\omega C/g_c$ is small (so that ω is very small). Then, Eq. (254) predicts that after n iterations, where $n \sim -\ln(2\Delta p)/\ln(r)$ (with $\Delta p = p - p_c$), the flow reaches the vicinity of $p = 1$ and the transformed $g_1(n)$ yields the effective conductance $g_e \sim g_1(n) \sim a_1^n g_c \sim g_c \Delta p^\mu$ with $\mu = -\ln(a_1)/\ln r \sim 1.17$, where μ is the usual critical exponent that characterizes the power-law behavior of the effective conductivity near p_c; see Eq. (2.36). This estimate of μ should be compared with the accepted value, $\mu \simeq 1.3$, for 2D percolation (see Table 2.3). We can also check to see whether α actually remains small after the n iterations, since

$$\alpha(n) = \frac{i\omega C}{g_c} \prod_{m=0}^{n-1} [A + B\alpha(m)] \sim \frac{i\omega C}{g_c} A^n \left[1 + \frac{i\omega C}{g_c} \frac{B}{A} \frac{A^n - 1}{A - 1}\right], \tag{258}$$

where $\alpha(m) \sim (i\omega C/g_c)A^m$ was used. Then, the criterion $|\alpha(n)| \ll 1$ becomes $(i\omega C/g_c)A^n \ll 1$, or

$$\omega \ll \omega_c \equiv \frac{g_c}{C} (\Delta p)^{\mu+s}. \tag{259}$$

Here, $s = \ln(a_2)/\ln(r) \simeq 1.34$, is the critical exponent that is associated with the power-law behavior of the conductivity of a percolation network of conductors-superconductors near p_c, defined by Eq. (2.37). The relation between this exponent and the dielectric constant of a composite will be discussed in more details in Section 6.5. As discussed in Sections 2.7.2 and 5.4, because of a duality relation for 2D percolation systems, one must have $\mu = s$, and therefore this PSRG treatment of the problem violates the duality relation. However, a slightly different treatment of the problem, along the lines that were described in Section 5.11, preserves the duality. Since in the present analysis we are interested mostly in the qualitative aspects of the phenomenon, the slight violation of the duality relation is not a serious setback to our analysis. Equation (259) indicates that as p_c is approached, the range of ω for which our criterion is valid shrinks.

The effective capacitance C_e can also be computed. In order to do this, we retain the second term of Eq. (255) and write

$$g_e \sim g_c \prod_{m=0}^{n-1} [a_1 + b_1\alpha(m)] \sim g_c a_1^n \left[1 + \frac{i\omega C}{g_c} \frac{b_1}{a_1} \frac{A^n - 1}{A - 1}\right], \tag{260}$$

so that

$$C_e \equiv \frac{1}{\omega} \operatorname{Im}(g_e) \sim C(\Delta p)^{-s}, \tag{261}$$

which implies that the effective capacitance of the composite diverges at p_c. Equation (261) is a general result which will be discussed in more detail in Section 6.5.

A similar analysis can be carried out in the insulating regime, $p < p_c$. In this case the probability flows to $p = 0$, and therefore g_e must be calculated from the iterated value of g_2. The effective capacitance C_e is still given by Eq. (261), while

$$g_e \sim \frac{\omega^2 C^2}{g_c} |\Delta p|^{-(\mu+2s)}. \tag{262}$$

The PSRG approach can also be used for computing the complex conductance in the frequency region above ω_c where, $|\Delta p^{\mu+s}| \ll \omega C/g_c < 1$. Within the linearized approximation, the two conductances never become equal since g_1 remains real while g_2 remains imaginary. However, one may estimate the number of iterations for the completely nonlinear map [Eq. (261)] to reach $\alpha = 1$, which is the number of iterations n for which $|\alpha(n)| \sim 1$. Equation (254) predicts this value of n to be, $n \sim -\ln(\omega C/g_c)/\ln(A)$, and therefore,

$$g_e \sim a_1^n g_c \sim g_c \left(\frac{\omega C}{g_c}\right)^{\mu/(\mu+s)}, \tag{263}$$

which, when used in Eq. (261), yields

$$C_e \sim C \left(\frac{\omega C}{g_c}\right)^{-s/(\mu+s)}. \tag{264}$$

We should keep in mind that the values of the exponents μ and s that are to be used in these equations are those that are predicted by the PSRG method, namely, $\mu \simeq 1.17$ and $s \simeq 1.34$. However, as shown in Section 6.5, the general forms of Eqs. (261)–(264) agree completely with the scaling properties of g_e and C_e. Note also that g_e is, in general, a complex quantity, so that the effective conductance of the network is actually the real part of g_e.

As discussed in Section 5.11, a distinct advantage of such PSRG methods is that, not only can they be used for estimating the effective properties of a system near p_c, but also away from p_c. To do this, one starts with specific values of $g_1 = g_c$, $g_2 = i\omega C$, and p, and iterates the nonlinear recursion relations (249). The effective conductivity g_e is the limiting value of the iteration process. Figure 6.19 presents typical results obtained by this method.

6.4.4 *Effective-Medium Approximation*

The EMA can also be used for estimating the effective dielectric constant, or the effective complex conductance g_e. With a binary distribution, $f(g) = p\delta(g - g_1) + (1-p)\delta(g - g_2)$, and $g_1 = g_c$, Eq. (5.57) [or Eq. (25) in the limits $\lambda = 0$ and $Z = 4$] predicts that

$$g_e = g_c[x + (x^2 + \alpha)^{1/2}], \tag{265}$$

where $x = (p - 1/2)(1 - \alpha)$. The EMA predicts equations similar to (261) and (262), except that, as discussed in Chapter 5, it also predicts that, $\mu = s = 1$. Moreover, the EMA is exact in the limits $p = 0$ and 1. As $p \to 0$, the EMA predicts that

$$g_e = i\omega C\left(1 + 2p\frac{1-\alpha}{1+\alpha}\right) + O(p^2), \tag{266}$$

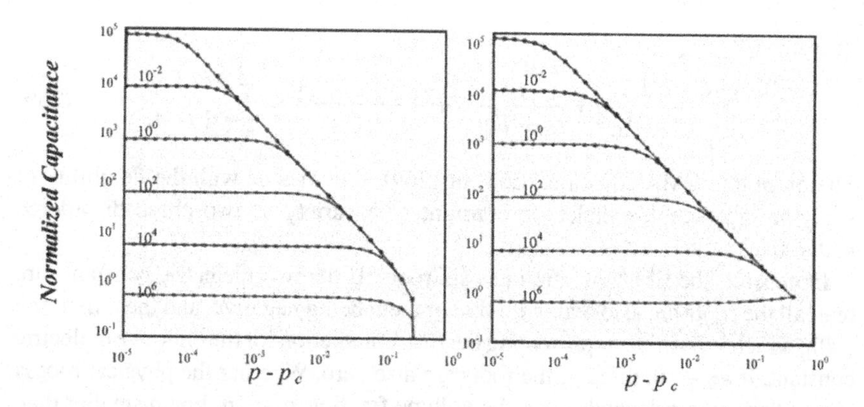

FIGURE 6.19. Frequency dependence of the normalized capacitance C_e. Numbers on top of the curves indicate the frequency ω. The results on the left and right figures are, respectively, for $p > p_c$ and $p < p_c$ (after Wilkinson *et al.*, 1983).

while in the limit $p \to 1$,

$$g_e = g_c\left(1 - 2q\frac{1-\alpha}{1+\alpha}\right) + O(q^2). \tag{267}$$

The PSRG approach discussed above produces the same types of results for $p = 0$ and 1, except that, unlike the EMA, it is not exact in these limits. An advantage of the EMA, especially when one deals with complex conductances, is that, it can be used with ease with any conductance distribution, whereas if the conductance distribution is anything but a simple binary distribution, the computations become very complicated, if any method other than the EMA is used.

A more general EMA for a two-phase material was suggested by Norris *et al.* (1985), and is as follows. Suppose that ε_1 and ε_2 are the complex dielectric constants of the two components of the material, and let p_1 and $p_2 = 1 - p_1$ be their respective volume fractions. As usual, we must specify the shape of the inclusion that is inserted in the effective medium, in order to compute the material's effective dielectric constant self-consistently. We take the shapes to be ellipsoids that are randomly distributed in space. We also assume that each phase has only one particular shape, and that the shape is the same for both phases. The ellipsoids are characterized by depolarization coefficients $A_i > 0$ (see Section, 4.6.2.4), with $i = 1, 2$, and 3 for the three major axes of the ellipsoid with $\sum_i A_i = 1$. For example, for d-dimensional spheres, $A_i = 1/d$. If $D_i = A_i/(1 - A_i)$, then the EMA equation is given by

$$\sum_{i=1}^{3}\left[\frac{p_1(\varepsilon_e - \varepsilon_1)}{(1 - A_i)\varepsilon_e + A_i\varepsilon_1} + \frac{p_2(\varepsilon_e - \varepsilon_2)}{(1 - A_i)\varepsilon_e + A_i\varepsilon_2}\right] = 0, \tag{268}$$

or

$$\sum_{i=1}^{3} \frac{1}{(1-A_i)^2}\left(\frac{p_1}{\varepsilon_e + D_i\varepsilon_1} + \frac{p_2}{\varepsilon_3 + D_i\varepsilon_2}\right) = \sum_{i=1}^{3} \frac{1}{1-A_i}. \tag{269}$$

Therefore, the EMA equation, (268) or (269), provides us with the flexibility of studying the complex dielectric constant of a variety of two-phase disordered materials.

In general, the EMA equation has six roots. If the two dielectric constants are real, all the roots are also real, with five of them being negative, and the remaining sixth, which is positive, representing the EMA prediction for the effective dielectric constant. If $\varepsilon_2 \to 0$, three of the roots are also zero. Whether the physical root is one of three zeros depends upon the volume fraction $p_2 \equiv p$. For p smaller than the percolation threshold p_c, the physical root is larger than zero, but for $p \geq p_c$ the physical root is zero.

6.4.5 *Random Walk Model*

If one applies an external electric field to a polymeric or glassy solid, the induced polarization decays slowly with time when the external field is removed. Williams and Watts (1970) and Williams *et al.* (1971) found empirically that the probability $\phi(t)$ that a polar molecule, or a polar group on a polymer molecule, has not relaxed by the time t after the external field has been removed is given by

$$\phi(t) = \exp[-(t/t_0)^\beta], \tag{270}$$

where t_0 is a characteristic time, and β an exponent, both of which depend on the temperature and the nature of the material. For simple materials one has $\beta = 1$, which is often referred to as the *Debye model,* while for glasses and polymers, $0 < \beta < 1$.

The function $\phi(t)$ can be directly related to the frequency-dependent dielectric constant $\varepsilon(\omega)$ of the material based on the linear response theory. Let us write down a most general linear equation that relates the electric displacement $\mathbf{D}(t)$ to the electric field $\mathbf{E}(t)$:

$$\mathbf{D}(t) = \mathbf{E}(t) + \int_0^\infty f(t')\mathbf{E}(t-t')dt'. \tag{271}$$

Gaussian units have been utilized to ensure that $\mathbf{D}$ and $\mathbf{E}$ have the same units. For an oscillating field one has, $\mathbf{D}(t) = e^{i\omega t}\mathbf{D}$ and $\mathbf{E}(t) = e^{i\omega t}\mathbf{E}$, so that, $\mathbf{D} = \varepsilon(\omega)\mathbf{E}$, where

$$\varepsilon(\omega) = 1 + \int_0^\infty e^{-i\omega t} f(t)dt. \tag{272}$$

In the static limit we have

$$\varepsilon_s = 1 + \int_0^\infty f(t)dt, \tag{273}$$

so that

$$\frac{\varepsilon(\omega)-1}{\varepsilon_s-1}=\int_0^\infty e^{-i\omega t}f(t)\left\{\int_0^\infty f(t')dt'\right\}^{-1}dt. \tag{274}$$

One now interprets

$$f(t)\left\{\int_0^\infty f(t')dt'\right\}^{-1}=-\frac{d\phi(t)}{dt} \tag{275}$$

as a probability density function for the relaxation time, so that $\phi(t)$ is the probability that relaxation has not occurred up to time t. Since in an actual experiment the electric field is supplied by a capacitor of finite size, the measured real part of the dielectric constant is increased by an additive contribution, so that we can write

$$\frac{\varepsilon(\omega)-\varepsilon_\infty}{\varepsilon_s-\varepsilon_\infty}=-\int_0^\infty e^{-i\omega t}\frac{d\phi(t)}{dt}dt. \tag{276}$$

It remains to determine the form of $\phi(t)$. To do this, models have been developed in which the solid has been assumed to contain defects that move in the material (as a result of, for example, thermal disturbances). The dipoles are assumed to be frozen in place. The defects keep moving in the solid until they reach the vicinity of a dipole where they stop their motion. Among the many theoretical attempts for understanding this phenomenon, the work of Shlesinger and Montroll (1984) and Shlesinger (1984) is particularly appealing. They modeled the defects' motion in terms of N non-interacting continuous-time random walkers moving on a lattice. A walker stops when it encounters the dipole, at which time relaxation occurs at once. Suppose that the lattice contains $\Omega+1$ sites, one of which (s_0, say) is occupied by the dipole. The concentration of the defects is $c=N/\Omega$, and we are interested in the joint limit $N\to\infty$ and $\Omega\to\infty$.

Suppose that the walkers (defects) are at time $t=0$ at sites $s_1, s_2, \cdots, s_N$, randomly and independently of each other. Then, the probability that the walker (defect), initially at site s_j, reaches site s_0 (the dipole) in the time interval $(0, t)$ is given by

$$\frac{1}{\Omega}\sum_{s_j\neq s_0}\int_0^t f(s_0|s_j,t')dt',$$

and therefore, the probability $\phi(t)$ of the dipole surviving for at least a time t before a walker (defect) reaches it is given by

$$\begin{aligned}\phi(t)&=\prod_{j=1}^{N}\left[1-\frac{1}{\Omega}\sum_{s_j\neq s_0}\int_0^t f(s_0|s_j,t')dt'\right]\\&=\left[1-\frac{1}{\Omega}\sum_{s\neq s_0}\int_0^t f(s_0|s,t')dt'\right]^N.\end{aligned} \tag{277}$$

Taking the limit $N \to \infty$ and $\Omega \to \infty$, and recalling that, $(1 + x/N)^N \to e^x$, we find that

$$f(t) = \exp[-ch(t)], \tag{278}$$

with

$$h(t) = \sum_{s \neq s_0} \int_0^t f(s_0|s, t')dt'. \tag{279}$$

The basic idea behind this method of deriving an expression for $f(t)$ [and hence $\phi(t)$] has been utilized by many authors; see Scher *et al.* (1991) for a review.

Suppose now that $P_n(s|s_0)$ is the probability that the walker is at site s after n steps, given that the walk started at s_0. A generating function for $P_n(s|s_0)$ is defined by

$$P_n(s|s_0; \xi) = \sum_{n=0}^{\infty} P_n(s|s_0)\xi^n. \tag{280}$$

It can then be shown that, if $\psi(t)$ is the waiting time density for the continuous-time random walkers, then, $\tilde{h}(\lambda)$, the Laplace transform of $h(t)$, is given by

$$\tilde{h}(\lambda) = \frac{1}{\lambda P[s_0|s_0; \tilde{\psi}(\lambda)]} \left\{ \frac{1}{1 - \tilde{\psi}(\lambda)} - P[s_0|s_0; \tilde{\psi}(\lambda)] \right\}. \tag{281}$$

Therefore, the long-time behavior of $h(t)$, and hence $\phi(t)$, can now be extracted by the usual Tauberian Theorem arguments.

Let us examine the transient random walks, i.e., those for which the probability $\mathcal{R}$ of return to the origin of the walk is less than 1 (Hughes, 1995) which, for unbiased nearest-neighbor walks, happen on periodic lattices of dimensionality $d \geq 3$. For such walks, $\mathcal{R} = 1 - (ZG_0)^{-1}$, where G_0 is the value of the Green function G_{ik}, defined by Eq. (5), at the origin and in the limit $\lambda = 0$. For the three 3D cubic lattices, G_{ik} is given by Eqs. (17), (19) and (20). For the simple-cubic lattice, $G_0 = \frac{1}{2} I_{\rm SC} = 0.25273$, for the BCC lattice, $G_0 = \frac{1}{8} I_{\rm BCC} = 0.17415$, and for the FCC lattice, $G_0 = \frac{1}{4} I_{\rm FCC} = 0.11206$, where $I_{\rm SC}$, $I_{\rm BCC}$ and $I_{\rm FCC}$ are the Watson integrals given by Eqs. (130), (131) and (132). In such cases (Hughes, 1995), $P[s_0|s_0; \tilde{\psi}(\lambda)] \to (1 - \mathcal{R})^{-1}$ as $\lambda \to 0$ (i.e., in the limit of long times), so that

$$\tilde{h}(\lambda) \sim \frac{1 - \mathcal{R}}{\lambda[1 - \tilde{\psi}(\lambda)]}. \tag{282}$$

Let us now consider two limiting cases.

(1) Suppose that the mean waiting time $\langle t \rangle$ between the steps of the CTRWs is finite. Then,

$$1 - \tilde{\psi}(\lambda) \sim \lambda \langle t \rangle, \tag{283}$$

so that, $\tilde{h}(\lambda) \sim (1-\mathcal{R})/(\langle t\rangle\lambda^2)$, and therefore

$$h(t) \sim (1-\mathcal{R})\frac{t}{\langle t\rangle}. \tag{284}$$

This is the classical Debye relaxation with an exponential decay of $\phi(t)$, i.e., $\beta = 1$.

(2) If the mean waiting time between the steps of the CTRWs is infinite, then

$$1-\tilde{\psi}(\lambda) \sim A\lambda^{\beta}, \quad \text{as } \lambda \to 0, \tag{285}$$

where A is a constant. In this case one obtains the Williams–Watts law,

$$h(t) \sim \frac{(1-\mathcal{R})t^{\beta}}{A\Gamma(1+\beta)}. \tag{286}$$

Therefore, an elegant CTRW model yields the classical William–Watts law for dielectric relaxation of glasses and polymers, but also affords us the flexibility of predicting a variety of other interesting possible behaviors. Bendler and Shlesinger (1985) extended these ideas to obtain the temperature-dependence of the dielectric relaxation. Other random walk models for this phenomenon were discussed by Niklasson (1989), to whom the interested reader is referred.

6.5 Scaling Properties of AC Conductivity and Dielectric Constant

We now focus on the scaling properties of the AC conductivity and dielectric constant near the percolation threshold p_c. Some of such properties were already discussed in the previous sections and also in Section 2.7.2, but in the present section we derive and discuss them in a more systematic way. Consider first a regular or random network in which each bond has a conductance a with probability p or a conductance b with probability $q = 1 - p$. Using dimensional analysis, it is not difficult to show that the effective conductivity g_e of the network is a homogeneous function and takes on the following form

$$g_e(p, a, b) = aF(p, h), \tag{287}$$

where $h = b/a$ can assume any complex or (positive) real value. By definition, the effective conductivity g_e is invariant under the interchange of a and b, and therefore we must have

$$g_e(p, a, b) = g_e(q, b, a) \tag{288}$$

$$F(p, h) = hF(q, 1/h). \tag{289}$$

We already described and discussed in Chapters 2 and 5 two limiting cases of the system, namely, the case in which $b = 0$ and a is finite (conductor-insulator mixtures), and one in which $a = \infty$ and b is finite (conductor-superconductor mixtures). Both cases correspond to $h = 0$, and therefore the point $h = 0$ at $p = p_c$

is particularly important. Let us therefore focus our attention on this singular point. In the critical region near this point, where both $|p - p_c|$ and h are small, the effective conductivity g_e follows the following scaling equation,

$$g_e \sim a|p - p_c|^{\mu}\Phi_{\pm}(h|p - p_c|^{-\mu-s}), \tag{290}$$

where the critical exponents μ and s were already defined and described in Chapter 2 [Eqs. (2.36) and (2.37)], and Φ_+ and Φ_- are two homogeneous functions corresponding, respectively, to the regions above and below p_c. Equation (290) was first proposed by Efros and Shklovskii (1976) and Straley (1976). Similar to the exponents μ and s, $\Phi_{\pm}$ are universal and do not depend on the network type once h and $p - p_c$ are fixed.

For any fixed and non-zero value of h, the effective conductivity g_e has a smooth dependence on $p - p_c$. This becomes clearer if we rewrite Eq. (290) in the following form

$$g_e \sim ah^{\mu/(\mu+s)}\Psi\left[|p - p_c|h^{-1/(\mu+s)}\right], \tag{291}$$

where $\Psi(x) = x^{\mu}\Phi_+(x^{-\mu-s}) = (-x)^{\mu}\Phi_-[(-x)^{-\mu-s}]$. The scaling function $\Psi(x)$ possesses a Taylor expansion around $x = 0$, $\Psi(x) = \Psi(0) + \Psi_1 x + \Psi_2 x^2 + \cdots$, implying that at $p = p_c$ and for $|h| \ll 1$ we must have

$$g_e \sim \Psi(0)(a^s b^{\mu})^{1/(\mu+s)} \equiv \Psi(0)ah^u, \tag{292}$$

where

$$u = \frac{\mu}{\mu + s}. \tag{293}$$

Equation (292) implies that $\Phi_-(x) \sim \Phi_+(x) \sim \Psi(0)x^u$, which demonstrates clearly the homogeneous nature of these scaling functions. Many other properties of these scaling functions were discussed by Clerc *et al.* (1990), to whom the interested reader is referred.

These results can now be used for understanding the AC conductivity and dielectric constant of a disordered material near p_c. Among the earliest studies of this problem we should mention those of Efros and Shklovskii (1976), and Bergman and Imry (1977). A comprehensive review of this subject is provided by Clerc *et al.* (1990). Here, we only summarize the main theoretical results, and discuss their experimental verification. To understand the AC conductivity and dielectric constant near p_c, we view a and b as complex conductances. Consider a percolation network in which a fraction p of the bonds are purely resistive, while the remaining fraction $q = 1 - p$ of the bonds behave as perfect capacitors. Thus we set

$$a = \frac{1}{R}, \qquad b = iC\omega, \tag{294}$$

where R is the resistance of the bond, and C is the capacitance. The conductance ratio h is then given by

$$h = \frac{i\omega}{\omega_0}, \tag{295}$$

where $\omega_0 = 1/(RC)$. In the static limit ($\omega = 0$) the capacitors become insulators, and the model reduces to the usual conductor-insulator composite already described and discussed in Chapters 2 and 5. It is clear that this model is the resistor-capacitor ($R-C$) model already described above. One key result is that if p is close to p_c, the $R-C$ model possesses scaling properties, and the effective conductivity $g_e(p, \omega)$ of the material satisfies the following scaling equation

$$g_e(p, \omega) \sim \frac{1}{R}|p - p_c|^{\mu}\Phi_{\pm}\left(\frac{i\omega}{\omega_0}|p - p_c|^{-\mu-s}\right), \tag{296}$$

which follows directly from Eq. (290). An immediate consequence of Eq. (296) is the existence of a time scale t_s that diverges as p is approached from *both sides*

$$t_s \sim \omega_0^{-1}|p - p_c|^{-(\mu+s)}. \tag{297}$$

The significance of t_s is discussed below.

We now define the frequency-dependent complex dielectric constant $\varepsilon_e(p, \omega)$ of the system by the following equation

$$\varepsilon_e(p, \omega) = \frac{g_e(p, \omega)}{i\omega}, \tag{298}$$

which is a generalization of the usual static dielectric constant ε_0. It is not difficult to show that for an insulating dielectric material $g_e \simeq i\omega\varepsilon_0$ as $\omega \to 0$. By using general analytic properties of the effective complex dielectric constant of a disordered composite, Bergman and Imry (1977) derived the following scaling relations which can be obtained from Eqs. (296) and (298):

$$g_e(p_c, \omega) \sim \omega^x, \tag{299}$$

which was already mentioned above [see, for example, Eqs. (193) and (212)], and

$$\varepsilon_e(p_c, \omega) \sim \omega^{-y}, \tag{300}$$

where the exponents x and y are supposed to satisfy the following relation

$$x + y = 1. \tag{301}$$

Equation (301) is a direct consequence of the fact that a complex conductivity is an analytic function of $i\omega$. Bergman and Imry (1977) argued that the main contribution to the AC properties is due to polarization effects between the percolation clusters in the disordered material, based on which they proposed that

$$x = \frac{\mu}{\mu + s}, \tag{302}$$

$$y = \frac{s}{\mu + s}, \tag{303}$$

so that in 2D where $\mu = s$, one has $x = y = 1/2$. Gefen *et al.* (1983), on the other hand, who studied anomalous diffusion on percolation clusters (see Section 6.1.9), argued that the anomalous nature of diffusion and the fractal morphology of percolation clusters at length scales up to ξ_p, the correlation length of percolation, dominate the contributions to the AC properties, and proposed instead that

$$x = \frac{\mu}{\nu(2+\theta)}, \tag{304}$$

$$y = \frac{2\nu - \beta}{\nu(2+\theta)}, \tag{305}$$

which also satisfy Eq. (301), where $\theta = (\mu - \beta)/\nu$, with β, μ and ν being usual percolation exponents. We discuss the experimental verification of these equations shortly.

An important consequence of Eq. (296) is the power-law behavior of ε_0 in the critical region near p_c. Using the Taylor expansion of $\Phi_\pm$ discussed above, it is not difficult to show that

$$\varepsilon_0 \sim A_\pm C |p - p_c|^{-s}, \tag{306}$$

that is, the static dielectric constant *diverges* as p_c is approached from *both sides*, and the critical exponent that characterizes this divergence is s, the critical exponent of a percolation network of conductors-superconductors described and discussed in Chapters 2 and 5. This spectacular result was first derived by Efros and Shklovskii (1976). In Eq. (306), A_+ and A_- represent the amplitudes of ε_s above and below p_c, respectively. Although Eq. (306) is supposed to be valid in the static limit, $\omega = 0$, its validity actually extends to higher frequencies as long as $\omega \ll 1/t_s$, where t_s is given by Eq. (297). An important property of the amplitudes $A_\pm$ is that their ratio A_+/A_- is a universal quantity, independent of the microscopic details of the system. If we now write $g_e(p,\omega)$ in the usual way, namely, $g_e = g' + ig'' = i\omega\varepsilon_e(p,\omega) = i\omega(\varepsilon' - i\varepsilon'')$, then a *loss angle* δ is defined by

$$\tan\delta = \frac{g'}{g''} = \frac{\varepsilon''}{\varepsilon'}, \tag{307}$$

and it is clear that $0 \leq \delta \leq \pi$.

Consider now the effective conductivity of the $R - C$ model at p_c. According to Eq. (292),

$$g_e(p,\omega) \sim \frac{\Psi(0)}{R}\left(\frac{i\omega}{\omega_0}\right)^u, \tag{308}$$

so that at p_c the loss angle δ_c is *universal* and is given by

$$\delta_c = \frac{\pi}{2}(1-u) = \frac{\pi}{2}\frac{s}{\mu + s}. \tag{309}$$

Although Eq. (309) is supposed to be valid exactly at $p = p_c$, it is important to remember that the universal loss angle δ_c can also be measured in a broad frequency range if $|p - p_c|$ is small enough, which implies that $1/t_s \ll \omega \ll \omega_0$. Note that, since for *any* 2D system, $s = \mu$, we must have $\delta_c = \pi/4$, another remarkable result.

These scaling properties have been verified by extensive numerical simulations of 2D networks. In particular, Bug *et al.* (1986) and Laugier *et al.* (1986b) used a transfer-matrix method described in Section 5.14.2 and bond percolation networks to verify the existence of the universal scaling functions $\Phi_{\pm}$, and Koss and Stroud (1987) did the same for site percolation networks. We now discuss experimental verification of these predictions.

6.5.1 Comparison with the Experimental Data

One of the earliest experimental studies of dielectric properties was reported by Castner *et al.* (1975), who measured the static dielectric constant of n-type silicon. Approximately 1200 Å of Au was evaporated on thin-disk samples which consisted of two imperfect Schottky barriers with thin (about 5-10 Å thick) oxide barriers bounding from 0.2 to 1.0 mm of bulk semiconductor. They varied the concentration c_d of the donor, and showed that the static dielectric constant ε_s diverges as c_d approaches a critical concentration from the insulating side. Although percolation was not mentioned in this work, the divergence of ε_s was a clear indication of the percolation transition indicated by Eq. (306). To explain these data, Dubrov *et al.* (1976) developed a model in which each bond of a percolation network represented a 300 Ω resistor and a 0.5 μF capacitor in parallel. Starting with a square network with only the capacitors, they added resistors to the network at randomly-selected bonds, and made measurements of the conductivity of the system at very low frequencies. As the fraction of the resistors approached the percolation threshold of the network, the dielectric constant of the network appeared to diverge. Dubrov *et al.* did write down a scaling law for this divergence that was similar to Eq. (306), but did not attempt to estimate the associated critical exponent, since the network that was used was too small.

A definitive experimental study of dielectric constant of composite materials near p_c was undertaken by Grannan *et al.* (1981) who used small spherical Ag particles randomly distributed in a non-conducting KCl host. The metal particles were prepared by evaporating Ag in the presence of argon gas and a small amount of oxygen. The particles were polydisperse, their sizes varied between 60 Å and 600 Å, and the overall size distribution was log-normal. The composite was prepared by mixing a given amount of Ag particles and KCl powder and by compressing the mixture into a solid pellet under high pressure. The dielectric constant of the sample was measured by a capacitance bridge operated at 1 kHz. Figure 6.20 presents their data as a function of the volume fraction of Ag. The dielectric constant appears to diverge at $p_c \simeq 0.2$, somewhat larger than $\phi_c \simeq 0.15 - 0.17$ for 3D percolating continua. Figure 6.21 presents a logarithmic plot of the same data, all of which appear to lie on a straight line indicating that ε_0 diverges as p_c is approached with an exponent $s \simeq 0.73 \pm 0.01$, in perfect agreement with the 3D percolation prediction, $s \simeq 0.735$ (see Table 2.3). Similar results were obtained by Niklasson and Grangvist (1984).

Laibowitz and Gefen (1984) prepared a series of samples of Au films on Si_3N_4 with varying thicknesses which were selected to span the entire metal–insulator

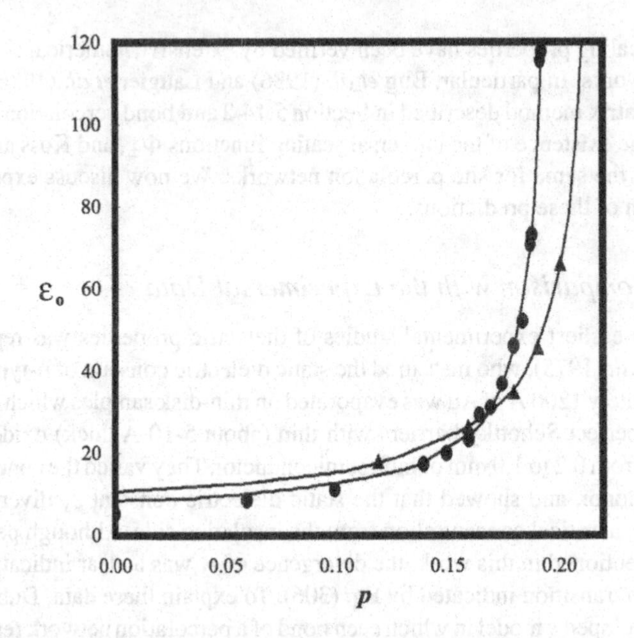

FIGURE 6.20. Static dielectric constant ε_0 of two series of Ag- KCl composites versus the volume fraction p of Ag. The curves represent the best fits of the data to Eq. (306) (after Grannan *et al.*, 1981).

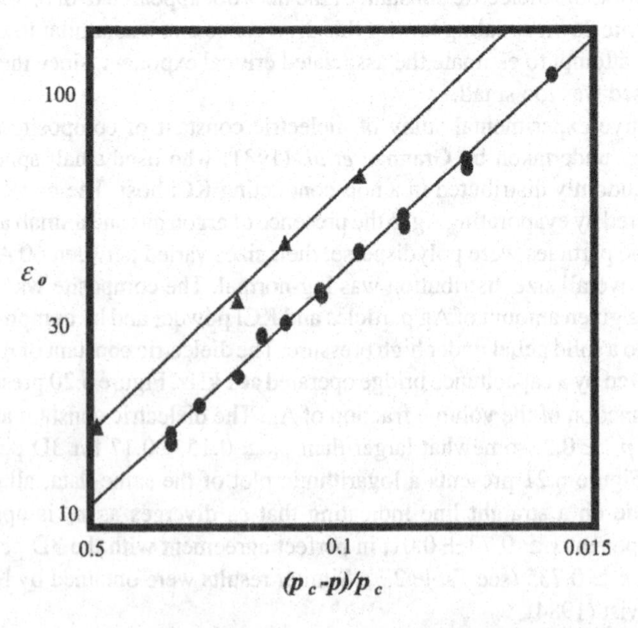

FIGURE 6.21. Logarithmic plot of the data shown in Figure 6.20 (after Grannan *et al.*, 1981).

transition. Insulating samples below p_c were easily achievable indicating that the contribution of tunneling and hopping to the conductivity can be ignored in the more metallic samples. The AC conductivity and capacitance (which is proportional to the dielectric constant) of the samples were then measured. When the data were fitted to Eqs. (299) and (300), they yielded $x \simeq 0.95 \pm 0.05$ and $y \simeq 0.13 \pm 0.05$, roughly satisfying Eq. (301). On the other hand, Eqs. (302) and (303) predict that for 2D percolation systems (where $\mu = s$) $x = y = 1/2$, while Eqs. (304) and (305) predict that $x \simeq 0.34$ and $y \simeq 0.66$, neither of which agree with the experimental data.

Song *et al.* (1986) measured the AC electrical properties of a powder mixture of amorphous carbon and Teflon in the frequency range 10 Hz to 13 MHz. Because of its stability, Teflon powder was used as the insulating component. Moreover, the low conductivity of amorphous carbon powder made it possible to easily observe the change of the conductivity as a function of p. The samples were prepared by mixing the carbon and Teflon powder to the desired volume fraction and were then compressed. The electrical conductivity g_e was then measured near p_c and was found to follow $g_e \sim (p - p_c)^\mu$ with $\mu \simeq 1.85 \pm 0.25$, in good agreement with 3D percolation conductivity (see Table 2.3). The static dielectric constant was found to diverge according to Eq. (306) with $s \simeq 0.68 \pm 0.05$, in good agreement with value of s for 3D percolation (see Table 2.3). The AC conductivity and the frequency-dependent dielectric constant were also measured, from which it was estimated that $x \simeq 0.86 \pm 0.06$ and $y \simeq 0.12 \pm 0.04$, which neither agree with Eqs. (302) and (303) which predict that $x \simeq 0.73$ and $y \simeq 0.27$, nor with the predictions of Eqs. (304) and (305), $x \simeq 0.60$ and $y \simeq 0.4$.

Partial resolution of the disagreement between theory and experimental data was provided by Hundley and Zettl (1988). They measured the AC conductivity and dielectric constant of thin Au films, similar to those of Laibowitz and Gefen (1984), but extended the frequency range to between 100 Hz and 1 GHz, and also considered both high (> 100 K) and low (< 50 K) temperatures. Their measurements, shown in Figure 6.22 for 300 K, indicated that in the intermediate frequency regime, corresponding to that of Laibowitz and Gefen and Song *et al.*, $x \simeq 1.0$ and $y \simeq 0$, in agreement with their data, while at higher frequencies $x \simeq 0.32$, in excellent agreement with the prediction of Eq. (304) for 2D systems, $x \simeq 0.34$, and $y \simeq 0.8$, in rough agreement with the prediction of Eq. (305) for 2D systems, $y \simeq 0.64$. The reasonable agreement between the results of Hundley and Zettl (1988) at high temperatures and the predictions of Eqs. (304) and (305) indicates that, at such temperatures the fractal morphology of percolation clusters plays an important role in the AC conductivity and dielectric properties of disordered materials at high frequencies. However, the data of Hundley and Zettl at low temperatures were in complete disagreement with the predictions of Eqs. (302) and (303), and also with those of (304) and (305).

On the other hand, measurements of Chen and Johnson (1990) for the AC conductivity and dielectric constant of composites of nickel particles in a matrix of polypropylene, and silver particles in potassium chloride yielded results which seem to favor Eqs. (302) and (303). For example, for the Ag-KCl composite,

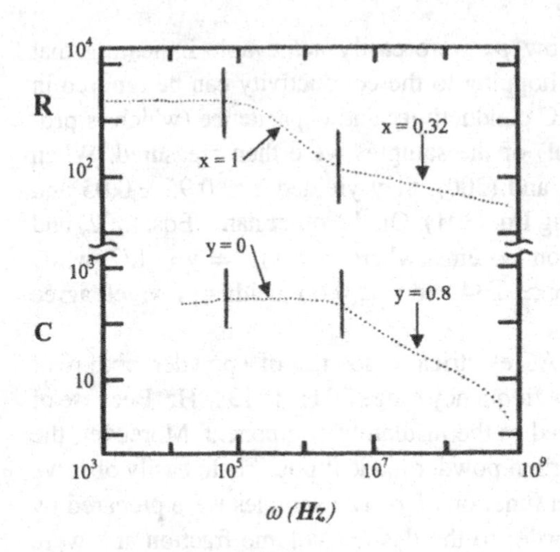

FIGURE 6.22. Frequency dependence of the resistance R and capacitance C of an Au fractal film at $T = 300$K. The critical exponents x and y are defined by Eqs. (299) and (300) (after Hundley and Zettl, 1988).

$x \simeq 0.77$ and $y \simeq 0.22$ were obtained which satisfy Eq. (301), and are in good agreement with the predictions of Eqs. (302) and (303) for 3D systems, $x \simeq 0.73$ and $y \simeq 0.27$. Yoon and Lee (1990) fabricated 2D lattices on an aluminum film by using a computer-controlled $x - y$ plotter, and measured the AC conductivity and dielectric constant of the composite for frequencies between 10 Hz and 1 MHz. Their measurements near the percolation threshold p_c yielded $x \simeq 1$ and $y \simeq 0$, in agreement with the data of Laibowitz and Gefen (1986) and satisfying Eq. (301), but in complete disagreement with the predictions of Eqs. (302) and (303), and also those of (304) and (305). Measurements of McLachlan *et al.* (1994) in a composite of NbC-KCl were also inconclusive.

To rectify this unsatisfactory situation, Sarychev and Brouers (1994) proposed that below p_c tunneling effects are important and must be taken into account. To do so, they proposed a modified scaling equation given by

$$g_e(p) \sim a|p - p_c|^{\mu}\Phi_{\pm}\left[\frac{\mathcal{T}_0^{\nu(d-2)}|p - p_c|^{\mu+m}}{\exp(-p_c\mathcal{T})}\right], \tag{310}$$

which should be compared with Eq. (290). Here $\mathcal{T}$ is called the renormalized tunneling constant which follows a power-law near p_c given by

$$\mathcal{T} \sim \mathcal{T}_0|p - p_c|^n, \tag{311}$$

where $n = (d-2)\nu + s$. The exponent m is given by, $m = \nu(d-2)[\nu(d-2) + s - 1]$. For $\mathcal{T} \gg 1$, the scaling function Φ behaves as $\Phi \propto 1$ for $p > p_c$, and $\Phi \propto g_t|p - p_c|^{\mu}$ for $p < p_c$, where g_t is the tunneling conductivity. Thus, above p_c, where tunneling is unimportant, one recovers Eq. (290). For the frequency-dependent conductivity $g_e(p, \omega)$, Sarychev and Brouers (1994) proposed that

$$g_e(p, \omega) \sim g_e(p)\Psi\left[\frac{-i\omega\varepsilon_s(p)}{g_e(p)}\right], \tag{312}$$

Although computer simulations of Sarychev and Brouers seem to support these new scaling representation of the conductivity, we are not aware of their experimental verification, since it would in fact be difficult to experimentally test the tunneling hypothesis.

Laugier *et al.* (1986a) measured the AC conductivity of random mixtures of glass microbeads, a varying fraction of them having their surface coated with silver which, however, did not change appreciably the density of the powder. The average diameter of the beads was about 30 μm, and the frequencies used were as large as 50 MHz. The loss angle δ was also measured. Equation (309) predicts that at p_c one must have $\tan \delta_c \simeq 0.45$, while the measured value was $\tan \delta_c \simeq 0.5$, in good agreement with the prediction.

Another class of disordered materials for which the AC conductivity and dielectric properties have been measured is microemulsions, which are thermodynamically stable, isotropic, and transparent dispersions of two immiscible fluids, such as water and oil, with one or more surfactants that are surface active. A water in oil (W/O) microemulsion usually consists of small spherical water droplets surrounded by a monomolecular layer of surfactant and dispersed in a continuous oil phase. The W/O microemulsions usually have a small macroscopic conductivity, because the water droplets are separated by the surfactant layers and the oil phase. Ionic surfactants can donate an ion to the water phase and increase its conductivity. If the volume fraction ϕ_w of the water phase exceeds a critical value ϕ_{wc}, the conductivity increases sharply, usually by several orders of magnitudes, indicating that the charge carriers are able to move along connected paths in the microemulsion, and hence the conductivity transition in microemulsions is a percolation phenomenon.

van Dijk (1985) was probably the first to measure the dielectric constant of a microemulsion at ϕ_{wc}. His material was a microemulsion of AOT, sodium di-2-ethylhexylsulfosuccinate (an anionic surfactant with a SO_3^- head group and two hydrocarbon tails), water and iso-octane. The volume fraction of water was varied by changing the amount of oil and keeping the molar ratio water/AOT constant. Figure 6.23 presents the data which exhibit a sharp peak for the dielectric constant and a dramatic increase for the electrical conductivity, both at ϕ_{wc}, in agreement with the predictions of percolation. Moreover, according to Eq. (309), at ϕ_{wc} the loss angle δ_c is independent of frequency ω, and van Dijk's data indicated that, over more than one order of magnitude variations in ω, this is indeed the case. From his data one obtains $u \simeq 0.62 \pm 0.02$, reasonably close to the percolation prediction $u = \mu/(\mu + s) \simeq 0.73$. More extensive measurements of the properties of the same microemulsions were reported by van Dijk *et al.* (1986).

Moha–Ouchane *et al.* (1987), Clarkson and Smedley (1988), and Peyrelasse *et al.* (1988) all measured the AC conductivity and dielectric constant of the same microemulsion that van Dijk *et al.* used. However, they found that the static dielectric constant diverges at ϕ_{wc} with an exponent $s' \simeq 1.6$, significantly larger than $s \simeq 0.73$, predicted by Eq. (306). Grest *et al.* (1986) argued that in microemulsions one must take into account the effect of *cluster diffusion* that rearranges the system

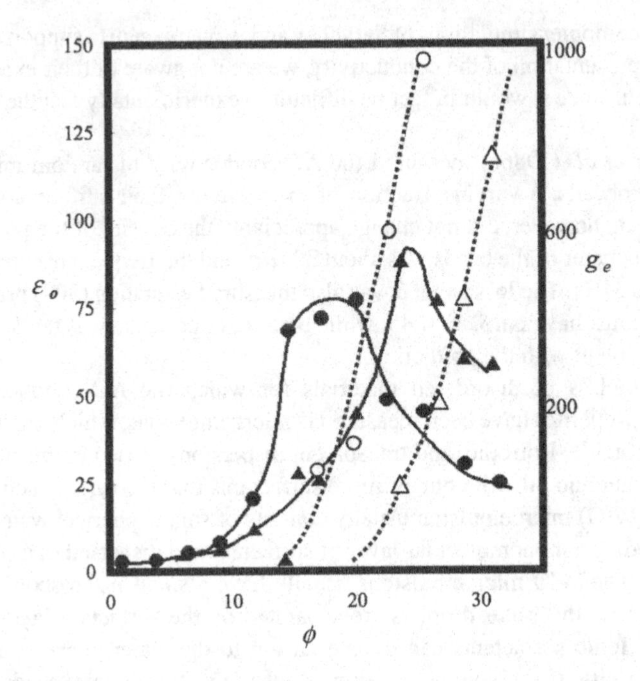

FIGURE 6.23. Static dielectric constant ε_0 and DC conductivity g_e of two different microemulsions as a function of the volume fraction ϕ of water. The data are for $W_0 = [H_2O]/[AOT] = 25$ and $T = 318$K (circles), and for $W_0 = 35$ and $T = 313$K (triangles) (after van Dijk, 1985).

and dynamically changes its structure. Thus, they proposed a dynamic percolation model in which the percolation clusters diffuse randomly in the network. Based on this argument and their model, Grest *et al.* (1986) proposed that the exponent s, characterizing the divergence of the static dielectric constant, must be replaced by $s' = 2\nu - \beta$, where β and ν are the usual percolation exponents. However, even $s' \simeq 1.35$ (using the 3D values of β and ν; see Table 2.3) seems to be in disagreement with the measurements of Moha–Ouchane *et al.*, Clarkson and Smedley, and Peyrelasse *et al.* To our knowledge, the disagreement between these measurement and the predictions based on the exponent s or s' remains unexplained.

On the other hand, Bhattacharya *et al.* (1985), who measured the static dielectric constant ε_s of a W/O microemulsion *above* p_c, reported that ε_s diverges as p_c is approached from above [hence providing the first experimental verification of Eq. (306) for $p > p_c$] with an exponent $s' \simeq 1.3$, which is in excellent agreement with the theoretical prediction of Grest *et al.* (1986). Bhattacharya *et al.* (1985) also reported that the value of the exponent y defined by Eq. (300) to be, $y \simeq 0.29 \pm 0.04$, which is in good agreement with the prediction of Eq. (303). For an extensive discussion of various properties of microemulsion systems, and the comparison between experimental data and the predictions of percolation see Clarkson (1988).

6.6 Vibrational Density of States: The Scalar Approximation

In Chapters 7 and 8 we will describe useful models for investigating mechanical properties of heterogeneous materials. In particular, Chapter 8 will describe the discrete models that are based on a lattice or network representation of a material in which each bond is an elastic element, such as a spring. How such a model comes about (that is, how it is derived from the classical equations of linear elasticity) will be described in Chapter 8, but for now we take this model for granted in order to describe an important property of solids, namely, their vibrational density of states.

Consider a $L \times L \times L$ network (or a $L \times L$ in 2D) in which each site contains a particle of mass m, with the nearest-neighbor particles connected to each other by springs. Suppose that $\mathbf{u}_i = (u_{ix}, u_{iy}, u_{iz})$ is the vector displacement of the particle at i. Then, the equation of motion for this particle at time t is given by Newton's law

$$m\frac{\partial^2 \mathbf{u}_i}{\partial t^2} = \sum \mathbf{F}, \tag{313}$$

where $\mathbf{F}$ represents any force that acts on the particle. If the springs that connect the particles are harmonic and can only tolerate stretching forces, then $\sum \mathbf{F}$ is given by the Hook's law (force = spring constant × displacement), and the equation of motion becomes

$$m\frac{\partial^2 \mathbf{u}_i}{\partial t^2} = \sum_{<ij>} k_{ij}\left[(\mathbf{u}_j - \mathbf{u}_i) \cdot \mathbf{R}_{ij}\right], \tag{314}$$

where $\mathbf{R}_{ij}$ is a unit vector from i to j, k_{ij} is the spring (elastic) constant of the bond between i and j, and the sum is over all the bonds ij that are connected to i. For simplicity we take $m = 1$. We can of course include other types of forces in Eq. (314), e.g., the bond-bending or angle changing forces that will be described in Chapters 8 and 9.

The standard method of analyzing Eq. (313) for vibrational properties is to assume, $\mathbf{u}_i = \mathbf{A}_i \exp(-i\omega t)$, where ω is the frequency of the vibrations, and $\mathbf{A}_i$ is an unknown vector to be determined. Substituting this expression into Eq. (313) yields a set of $n = L^3$ simultaneous linear equations for the $\mathbf{A}_i$'s which has n positive eigenvalues $\omega_1^2, \omega_2^2, \cdots$, and n eigenvectors $\mathbf{A}_{e1}, \mathbf{A}_{e2}, \cdots$ Then, the solution to $\mathbf{u}_i$ is given by

$$\mathbf{u}_i = \mathrm{Re}\left[\sum_j c_j \mathbf{A}_{ej} \exp(-i\omega_j t)\right], \tag{315}$$

where c_j are complex numbers that must be determined from the initial conditions. Having determined $\mathbf{u}_i$'s, we obtain all the vibrational properties of the system. One of the most important such properties is $\mathcal{N}(\omega)$, the *vibrational density of states* (DOS). $\mathcal{N}(\omega)d\omega$ is the number of vibrational modes with a frequency between

ω and $\omega + d\omega$. The DOS is an important quantity for obtaining the specific heat and thermal conductivity of a solid material (see below), as well as the study of proteins and other biomaterials (see, for example, Elber and Karplus, 1986), and can itself be computed from the distribution of the eigenmodes ω_i. However, under certain conditions there is a much simpler way of computing the DOS which is as follows.

Consider, for example, Eq. (314) and take its Fourier transform

$$-\omega^2 \hat{\mathbf{u}}_i = \sum_{<ij>} k_{ij}[(\hat{\mathbf{u}}_j - \hat{\mathbf{u}}_i) \cdot \mathbf{R}_{ij}]. \tag{316}$$

If we write Eq. (316) for one principal direction of the network, say x, we obtain

$$-\omega^2 \hat{u}_{ix} = \sum_{<ij>} k_{ij}(\hat{u}_{jx} - \hat{u}_{ix}). \tag{317}$$

Consider now the diffusion equation in discretized form, Eq. (1), the Laplace transform of which, ignoring the term that arises from the initial condition, is given by

$$\lambda \tilde{P}_i(\lambda) = \sum_{<ij>} W_{ij}(\tilde{P}_j - \tilde{P}_i), \tag{318}$$

which is completely similar to Eq. (317), if we note that the role of $\hat{u}_{ix}$ is played by $\tilde{P}_i$, and that of $-\omega^2$ by λ. We may then interpret λ as the frequency for diffusion, just as ω is the frequency for vibrations. This analogy was first exploited by Alexander *et al.* (1978) for computing the DOS. We refer to this method of computing the DOS as the *scalar approximation*. Alexander *et al.* (1978) showed that the numerical difference between the DOS computed based on Eq. (313) and one based on the scalar approximation is of the order of the time derivatives.

To compute the DOS, we consider $\langle P_0(t) \rangle$, the average probability that a random walker (diffusant) that is moving in the network is at the origin (of its motion) at time t, where the averaging is taken over all the initial positions of the walker. For classical diffusion in macroscopically-homogeneous materials, i.e., when the random walk fractal dimension is $D_w = 2$ and the fractal dimension D_f of the system is the same as the Euclidean dimensionality d, Eq. (90) predicts that

$$\langle P_0(t) \rangle \sim t^{-d/2}. \tag{319}$$

This equation can also be derived by observing that

$$\langle P_0(t) \rangle = [\langle S(t) \rangle]^{-1}, \tag{320}$$

where $\langle S(t) \rangle$ is the mean number of distinct sites visited by the diffusant (random walker) at time t. It can be shown (Hughes, 1995) that for diffusion in macroscopically-homogeneous media, $\langle S(t) \rangle \sim t^{d/2}$, which, together with Eq. (320), immediately yields Eq. (319). As discussed above, $\mathcal{N}(\omega)$ should in principle be computed from the solution of the vector model, Eq. (313) or (314). However, it is not difficult to show that, if the scalar approximation can be used for computing the DOS, then $\mathcal{N}(\omega)$ and $\langle P_0(t)\rangle$ are related to each other through the following equation (Alexander *et al.*, 1978)

$$\mathcal{N}(\omega) = -\frac{2\omega}{\pi}\mathrm{Im}\langle \tilde{P}_0(-\omega^2)\rangle. \tag{321}$$

If we take the Laplace transform of Eq. (319) and substitute the result into Eq. (321), we obtain

$$\mathcal{N}(\omega) \sim \omega^{d-1}, \tag{322}$$

which is the well-known result for the DOS in the Debye regime. Equation (322) is valid at low frequencies (i.e., long wavelengths or large length scales over which the material is homogeneous), such that $\omega < \omega_{co}$, where ω_{co} is a cutoff or crossover frequency to be described below. It can be shown that, in the Debye regime, even if we compute $\mathcal{N}(\omega)$ based on the solution of the true vectorial model of the system [that is, Eq. (313)], it would still follow Eq. (322), provided that the material is macroscopically homogeneous. The general form of Eq. (322) is also consistent with this claim: There is nothing in this equation that indicates whether $\mathcal{N}(\omega)$ was computed from the solution of Eq. (313) or (318); only the dimensionality d of the material has entered this equation. Vibrational states in homogeneous materials that are expressed by Eq. (322), are usually called *phonons*. Later in this chapter, we will discuss the precise conditions under which the scalar approximation can be used for computing the DOS.

6.6.1 Numerical Computation

Before embarking on describing analytical theories of the DOS, let us first discuss how this quantity is computed numerically. This is an important issue, since in order to compute the DOS of a highly heterogeneous material one must use a large network, and therefore direct use of an equation such as (315), the solution of which requires computing the eigenvalues and eigenvectors of a very large matrix, becomes problematic. What makes the problem difficult is that, for systems without any particular symmetry properties (such as disordered materials considered here), it is necessary to determine *all* the eigenvalues and eigenvectors of the dynamical matrix that arises in such problems, if one is to take the direct route of computing the DOS via Eq. (315).

There are many methods for numerical computation of the DOS (see, for example, Williams and Maris, 1985; Lam and Bao, 1985; Li *et al.*, 1988, 1990; Beniot *et al.*, 1992; Royer *et al.*, 1992). Let us substitute $\lambda = -\omega^2$ into Eq. (318) and rewrite it as

$$\omega^2\tilde{P}_i = -z_i\tilde{P}_i + \sum_j W_{ij}\tilde{P}_j, \tag{323}$$

where $z_i = \sum_j W_{ij}$. Equation (323) is completely analogous to the tight-binding Hamiltonian of the electronic problem (see also Chapter 9 of Volume II) for which a considerable amount of analytical as well as numerical work has been carried out which can be used for the present problem. The most straightforward method for computing the eigenvalues and eigenvectors is to simply diagonalize the dynamical matrix that arises in Eq. (323). However, only relatively small matrices can be diagonalized directly, implying that only small networks can be used with this method.

Here we describe an efficient method which is called the *Strum sequence method* (Li *et al.*, 1988, 1990). It consists of an application of Gaussian elimination on the consecutive columns or rows of the matrix $(\omega^2\mathbf{U} - \mathcal{H})$, where $\mathbf{U}$ is the identity matrix, and the matrix $\mathcal{H}$ is such that, $\mathcal{H}_{ij} = W_{ij} - z_i\delta_{ij}$. The Gaussian elimination of the matrix's entries is continued until the matrix is reduced to a triangular form. The DOS is then computed by counting the fractional number of positive elements in the main diagonal of the reduced triangular matrix. The main advantages of this method are the economy of storage and its speed. In this method the computing time for an $M \times N$ network scales as $M^{3(d-1)} \times N$, hence allowing one to use very long strips with length N of a considerable width M. Moreover, this method does not require matrix inversion. Later in this chapter we will briefly discuss large-scale numerical computations of the DOS of disordered materials, including those modeled by percolation networks.

6.6.2 Effective-Medium Approximation

The DOS of a disordered network can be computed by an EMA (Derrida *et al.*, 1984; Sahimi, 1984b). Consider the Green function G_{ij} defined by Eq. (5). To calculate the DOS we only need to know the value of the Green function at the origin, since the Laplace transform $\langle\tilde{P}_0(\lambda)\rangle$ of $\langle P_0(t)\rangle$ is given by

$$\tilde{P}_0(\lambda) = -\frac{G_{00}(\lambda)}{\tilde{W}_e(\lambda)}, \tag{324}$$

where $\tilde{W}_e(\lambda)$ is the (Laplace-transformed) effective transition rate, the computation of which was described in detail in Section 6.1. Since $\tilde{W}_e(\lambda)$ can be computed by an EMA via Eq. (25), and because the Green functions G_{ij}, and in particular G_{00}, are known for many lattices, we can calculate $\tilde{P}_0(\lambda)$ and hence the DOS via Eq. (321). For simplicity, let us denote $G_{00}(\lambda)$ by $G(\lambda)$. We consider a network with percolation-type disorder in which a fraction p of the bonds have a non-zero transition rates, while the remaining bonds are inactive.

As discussed in detail in Section 6.1, as $p \to p_c^+$, $\tilde{W}_e(\lambda)$ remains finite and, furthermore, to leading order varies linearly with $p - p_c$, so that $\tilde{W}_e(\lambda)/(p - p_c)$ also remains finite as the percolation threshold p_c is approached. Thus, if $p - p_c \ll 1$, we must have $\tilde{W}_e(\lambda) \gg 1$. Therefore, we expand $G(\lambda)$ in powers of $\lambda/\tilde{W}_e(\lambda)$ to obtain

$$G(\lambda) = -\frac{1}{2}I_w(d) + \frac{1}{\lambda}\left[\frac{\lambda}{\tilde{W}_e(\lambda)}\right]^{d/2} I(d) + \cdots \tag{325}$$

which is valid for $2 < d < 4$, where d, the dimensionality of the system, has been treated as a continuous variable. Here I_w is the Watson integral which, for the 3D cubic lattices, is given by Eqs. (130)–(132), and

$$I(d) = \frac{\operatorname{cosec} \pi[(d-2)/2]}{2^d \Gamma(d/2)\pi^{d/2-1}}. \tag{326}$$

If we restrict our attention to a percolation-type distribution of the transition rates, $f(W) = (1-p)\delta(0) + p\delta(W-1)$, then the EMA, Eq. (25), yields

$$\tilde{W}_e(\lambda) = \frac{p + \gamma_{\alpha\alpha}}{1 + \gamma_{\alpha\alpha}}, \tag{327}$$

where $\gamma_{\alpha\alpha} = -1/d - \lambda G(\lambda)/[d\tilde{W}_e(\lambda)]$, as defined earlier by Eq. (26), if we restrict our attention to d-dimensional simple-cubic lattices. We now substitute the expansion for $G(\lambda)$ into Eq. (25) to compute $\gamma_{\alpha\alpha}$ and then insert the result into Eq. (327). Let $\tilde{W}_e(\lambda) = A_1 + A_2 + \cdots$ Then

$$A_1 = \frac{d}{2(d-1)}\left\{p - p_c + \left[(p-p_c)^2 + \frac{2\lambda(d-1)I_w(d)}{d^2}\right]^{1/2}\right\}, \tag{328}$$

and

$$A_2 = -\frac{A_1 I(d)}{d}\left(\frac{\lambda}{A_1}\right)^{d/2}\left[1 - \frac{1}{d} + \frac{\lambda I_w(d)}{2dA_1^2}\right]^{-1}. \tag{329}$$

Therefore, the following equation is obtained for $\langle \tilde{P}_0(\lambda)\rangle$

$$\langle \tilde{P}_0(\lambda)\rangle = \frac{I_w(d)}{2A_1}\left\{1 + \frac{I(d)}{d}\left(\frac{\lambda}{A_1}\right)^{d/2}\left[\left(\frac{d-1}{d}\right)A_1 + \frac{\lambda I_w(d)}{2dA_1}\right]^{-1/2}\right\}, \tag{330}$$

which is the EMA prediction for $\langle \tilde{P}_0(\lambda)\rangle$. According to Eq. (330), $\langle \tilde{P}_0(\lambda)\rangle$ becomes singular at a crossover value λ_{co} of the Laplace transform variable λ given by

$$\lambda_{co} = -\frac{d^2}{2(d-1)I_w(d)}(p-p_c)^2. \tag{331}$$

Therefore, if we set $\lambda = -\omega^2$ and use Eq. (321), we obtain our final expression for the DOS, valid for $2 < d < 4$:

$$\mathcal{N}(\omega) = \frac{2}{\pi\omega}\left[\frac{1}{2}I_w(d)(d-1)\right]^{1/2}(\omega^2 - \omega_{co}^2)^{1/2}, \quad \omega > \omega_{co}, \quad 2 < d < 4, \tag{332}$$

where $\omega_{co}^2 = -\lambda_{co}$. According to Eq. (332), in the vicinity of (or at) p_c, the DOS approaches a constant for $\omega \gg \omega_{co}$ (when $\omega_{co}/\omega \ll 1$). As discussed in the next section, Alexander and Orbach (1982) showed that, *within the scalar approximation*, the vibrational DOS of a material at its percolation threshold p_c, or at $p > p_c$ but over length scales that are smaller than ξ_p, the correlation length of percolation, is given by

$$\mathcal{N}(\omega) \sim \omega^{D_s - 1}, \quad \omega \gg \omega_{co}, \tag{333}$$

where

$$D_s = 2\frac{D_f}{D_w} \tag{334}$$

is the spectral or fracton dimension of the network, with D_f and D_w being, respectively, the fractal dimension of the sample-spanning cluster and its random walk fractal dimension. For the sample-spanning percolation cluster, the latter quantity is given by Eq. (82). Equations (333) and (334) are actually supposed to be more general and valid for any material that has a fractal morphology with a fractal dimension D_f and a random walk fractal dimension D_w. Since according to the EMA, near or at p_c the DOS $\mathcal{N}(\omega)$ approaches a constant for $\omega \gg \omega_{co}$, the implication is that the EMA predicts the spectral dimension to be

$$D_s = 1, \quad 2 < d < 4. \tag{335}$$

Since the EMA also predicts that, for the sample-spanning cluster at p_c, one has, $D_w = 4$ [see the discussion after Eq. (73)], we also obtain an EMA estimate for the fractal dimension D_f of the sample-spanning percolation cluster, $D_f = 2$ for $2 < d < 4$. This estimate should be compared with the numerical estimate of D_f in 3D, namely, $D_f \simeq 2.53$ (see Table 2.3).

As discussed in Sections 6.1.9 and 6.1.10, for the times $t \gg t_{co}$ transport in percolation networks is diffusive, while it is anomalous or fractal for $t \ll t_{co}$, where t_{co} is the crossover time. It should be clear to the reader that the time scale t_{co} is nothing but $|\lambda_{co}^{-1}|$, and therefore according to the EMA, Eq. (331),

$$t_{co} = 2d^{-2}(d-1)I_w(d)(p-p_c)^{-2}, \tag{336}$$

which should be compared with the (correct) prediction of the scaling theory, Eq. (83). For the DOS the role of t_{co} is played by the crossover frequency ω_{co} defined above. Hence, for $\omega < \omega_{co}$ and $p > p_c$, the mechanical vibrations of disordered materials, modeled by a percolation network, are of ordinary, Debye type, and are given by Eq. (322), whereas for $\omega \gg \omega_{co}$ they are described by a new regime characterized by Eq. (332). All the results presented so far were derived independently and simultaneously by Derrida *et al.* (1984) and Sahimi (1984b).

It is not difficult to calculate within an EMA the DOS for the Debye or phonon regime. In this case, the leading-order term for the expansion of $\tilde{W}_e(\lambda)$ [which is of the order of $(p-p_c)^{-1}$] does not contribute much, and therefore one must calculate the higher-order terms which, when done, results in

$$\mathcal{N}(\omega) = \frac{2}{\pi}J_d\omega^{d-1}\sin[(d/2-1)\pi]\left\{1 - \frac{\omega^2}{2d}\left[1 - \frac{1}{d} - \frac{\omega^2 I_w(d)}{2d}\right]^{-1}\right\}, \tag{337}$$

where

$$J_d = \frac{S_d}{(2\pi)^d}\int_0^\infty \frac{x^{d-1}}{x^2(x^2+1)}dx, \tag{338}$$

with $S_d = 2(\pi)^{d/2}/\Gamma(d/2)$ being the surface of the unit sphere in d dimensions.

One can also develop an expansion similar to one that led to Eq. (332) for $1 < d < 2$. One then finds (Derrida *et al.*, 1984) that at $p = p_c$, or for $p > p_c$ and $\omega > \omega_{co}$,

$$\mathcal{N}(\omega) \sim \omega^{(d-2)/(d+2)}, \quad 1 < d < 2, \tag{339}$$

which, when compared with Eq. (333), implies that

$$D_s = \frac{2d}{d+2}, \quad 1 < d < 2. \tag{340}$$

The corresponding equation for the Debye or phonon regime is

$$\mathcal{N}(\omega) \sim \frac{\omega^{d-1}}{(p-p_c)^{d/2}}, \quad \omega \ll \omega_{co}, \tag{341}$$

which, in terms of frequency-dependence of the DOS, is similar to Eq. (322), confirming once again that the EMA can produce exact results, or provide very accurate approximations, for low-dimensional systems.

Figure 6.24 presents the vibrational DOS for the simple-cubic lattice with percolation disorder, as a function of frequency ω, for several values of p, the fraction of the active bonds. There is a qualitative change in the DOS at the crossover frequency ω_{co}. For $\omega > \omega_{co}$ the DOS quickly approaches a constant value, consistent with our discussion above. For $\omega \ll \omega_{co}$ the DOS varies linearly with the frequency, whereas in the vicinity of ω_{co} (but below it) it appears to depend on the

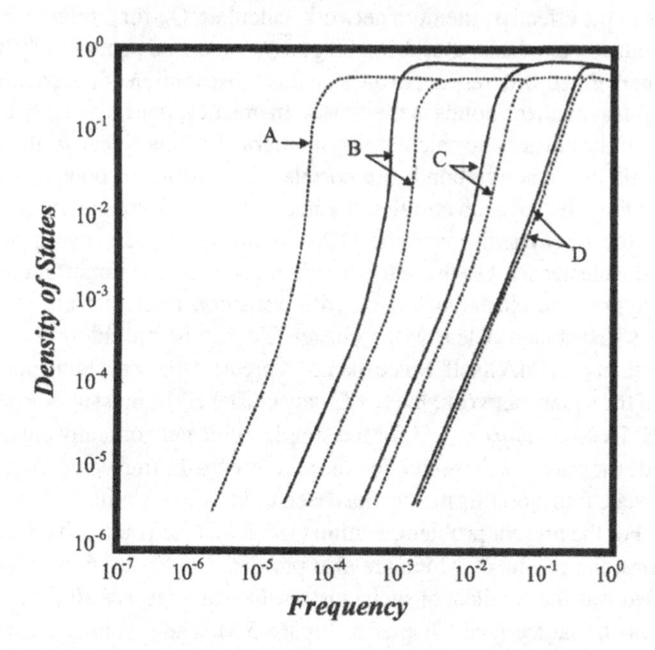

FIGURE 6.24. Density of states $\mathcal{N}(\omega)$ versus frequency, as predicted by the EMA (full curves) and cluster EMA (dotted curves) on a simple- cubic lattice. The fraction p of the conducting bonds is $p = 0.316$ (A), 0.334 (B), 0.35 (C), and 0.50 (D) (after Sahimi, 1984b).

frequency quadratically, consistent again with Eq. (322). Moreover, ω_{co} appears to vary linearly with $p - p_c$, which is again consistent with Eq. (331) (with $\lambda_{co} = -\omega_{co}^2$). Similar results are obtained for 2D systems (Derrida *et al.*, 1994; Sahimi, 1984b).

6.6.3 Cluster Effective-Medium Approximation

All of the EMA predictions for the DOS are in qualitative agreement with the predictions of the scaling theory discussed in the next section, and also with direct numerical simulations (see, for example, Lam and Bao, 1985; Li *et al.*, 1990; see also below) which again indicates the usefulness of the EMA. However, these predictions are not very accurate near p_c, which is the region where the EMA begins to be inaccurate. An obvious method of improving the performance of the EMA for predicting the DOS near p_c is to use higher order or cluster EMAs, since according to Eq. (324) the more accurate $\tilde{W}_e(\lambda)$ is computed, the more accurate the EMA predictions for the DOS will be. The development of such cluster EMAs was already discussed in detail in Section 5.6.7 for the steady-state transport. The same approach is utilized here for the present problem. As Eq. (7) indicates, the "current" $\tilde{Q}_\alpha$ is written in terms of the fluctuations in the normalized conductivities Δ_α of the bond α itself and those of the other bonds β that surround α. Thus, instead of embedding in the effective medium only a single bond α with a random conductance, which would result in Eq. (25), one can embed a cluster of such bonds in the effective-medium network, calculate $\tilde{Q}_\alpha$ for a reference bond α within the cluster, and insist that the average $\langle\tilde{Q}_\alpha\rangle$ must be equal to $\tilde{Q}_\alpha^0$, with the averaging performed with respect to the statistical distribution of the conductances (transition rates) of all the bonds in the cluster. In practice, one must apply Eq. (7) to every bond of the cluster and calculate $\tilde{Q}_\alpha$ numerically. Since near p_c the currents and potentials in the active bonds are correlated, a cluster of bonds includes to some extent the effect of such correlations, and thus one may obtain more accurate predictions for $\tilde{W}_e(\lambda)$ and hence the DOS. To obtain rapid convergence (with increasing the cluster size) to those for an infinite system, it is important to choose a suitably symmetric cluster of bonds with transition rates that are allowed to be randomly distributed. Clearly, the cluster EMA must provide more accurate predictions than the EMA itself. Since the EMA predicts the exact bond percolation threshold of the square network, $p_c = 1/2$, any cluster EMA must also do so. Since in 3D the EMA predicts $p_c = 1/3$ for the simple-cubic network, any cluster EMA must provide a more accurate prediction for p_c, closer to the true value, $p_c \simeq 0.249$. For steady-state transport this method has been used by many with various degrees of success. For the present problem, Sahimi (1984b) carried out such cluster EMA computations, the results of which are also presented in Figure 6.24. The cluster that was used was the smallest of such clusters that can preserve all the symmetry properties of the lattice (see Chapter 5, Figure 5.8). The resulting cluster EMA predicts a bond-percolation threshold, $p_c \simeq 0.316$ for the simple-cubic lattice, an improvement of about 10% over the EMA prediction. As Figure 6.24 indicates, it is clear that the cluster EMA does improve the predictions of the EMA. In particular, while the EMA cannot provide any physical predictions for $p < 1/3$, the present cluster EMA can do so for $p \geq 0.316$. However, the convergence of the predictions of the cluster EMA to the true values is relatively slow.

6.6.4 *Scaling Theory: Phonons Versus Fractons*

The spectral dimension D_s (also called the fracton dimension) is a key quantity for describing the dynamical properties of materials with a fractal morphology, in addition to the fractal dimension D_f which describes how the mass of the material depends on its length scale. As discussed in Section 6.1.12, the spectral dimension D_s characterizes anomalous diffusion in a fractal material. We already mentioned that Alexander and Orbach (1982) mapped the problem of anomalous diffusion onto the problem of vibrational DOS using the scalar approximation, and showed that the basic properties of vibrations of fractal materials, such as their DOS, the dispersion relation, and localization, are all characterized by the spectral dimension D_s. Rammal and Toulouse (1983) derived D_s via a scaling analysis, and showed that various random walk properties, such as the probability $P_0(t)$ of return to the origin, and the mean number $\langle S(t)\rangle$ of the visited sites, are governed by the spectral dimension. In this section we describe and discuss the works of Alexander and Orbach (1982) and Rammal and Toulouse (1983), and then consider their implications. Our discussions follow rather closely the comprehensive review of Nakayama *et al.* (1994).

The linear size of a region of sites visited by a diffusant or random walker is, $R = \langle R^2(t)\rangle^{1/2} \sim t^{1/D_w} \sim t^{1/(2+\theta)}$, where, $\theta = (\mu - \beta)/\nu$, as usual. The mean number $\langle S(t)\rangle$ of sites visited is therefore proportional to the volume R^{D_f} of the region, $\langle S(t)\rangle \sim R^{D_f}$, and thus (Rammal and Toulouse, 1983)

$$\langle S(t)\rangle \sim t^{D_s/2}. \tag{342}$$

As long as $D_s \leq 2$, Eq. (330) is valid. Therefore,

$$\langle P_0(t)\rangle \sim t^{-D_s/2}, \tag{343}$$

with $D_s < 2$ given by Eq. (334). In view of Eq. (321), we immediately obtain Eq. (333), the main prediction of Alexander and Orbach (1982). In analogy with the usual Debye DOS, Eq. (322), Alexander and Orbach (1982) called the related excitations "fractons" and D_s the "fracton dimension." D_s was called the "spectral dimension" by Rammal and Toulouse (1983) because it represents the DOS for the vibrational excitation spectrum. We use the terminology of Rammal and Toulouse throughout this book.

Rammal and Toulouse (1983) also derived the vibrational DOS of fractal materials using a finite-size scaling approach (see Section 2.9 for a detailed discussion of this approach). The essence of their analysis is as follows. Consider a fractal material of linear size L with fractal dimension D_f. Its DOS per particle at the lowest frequency $\Delta\omega$ is defined by

$$\mathcal{N}(\Delta\omega, L) = (L^{D_f}\Delta\omega)^{-1}. \tag{344}$$

If we assume that the dispersion relation for $\Delta\omega$ is given by

$$\Delta\omega \sim L^{-a}, \tag{345}$$

where a is an exponent to be determined, we can eliminate L from Eq. (344) to obtain

$$\mathcal{N}(\Delta\omega) \sim \Delta\omega^{D_f/a-1}. \tag{346}$$

The explicit expression for the exponent a is obtained from Eqs. (73) and (82) with the mappings, $t \to 1/\Delta\omega^2$ (see above) and $\langle R^2(t)\rangle^{1/2} \to L$. We thus obtain

$$L \sim \Delta\omega^{-2/D_w}. \tag{347}$$

Therefore

$$a = \frac{2\nu + \mu - \beta}{2\nu} = \frac{D_w}{2} = \frac{D_f}{D_s}. \tag{348}$$

Since the material is fractal and self-similar, $\Delta\omega$ can be replaced by an arbitrary frequency ω, and therefore we recover Eq. (333). Equation (346) also defines a "dispersion relation,"

$$\omega \sim [L(\omega)]^{-D_f/D_s}. \tag{349}$$

Immediately after the invention of D_s and the discussion of its importance to dynamics of fractal materials, it was observed that, if we use the most accurate estimates of the percolation exponents μ, β and ν (see Table 2.3), then $D_s = 2D_f/D_w = 2(\nu d - \beta)/(2\nu + \mu - \beta)$ takes on a numerical value which is almost independent of the dimensionality d of the system, and is given by

$$D_s \simeq 4/3, \quad \text{all } d.$$

In fact, $D_s = 4/3$ was conjectured by some to be an exact result, which would have been a significant result, because it would have led us to an exact scaling relation between the dynamical exponent μ and the static exponents ν and β. However, it is now established that $D_s = 4/3$ is *not* an exact result, albeit 4/3 is a very accurate approximate estimate of the true value of the spectral dimension D_s.

6.6.5 *Characteristics of Fractons*

Let us now discuss important properties of the fracton DOS and compare them with the corresponding Debye DOS.

6.6.5.1 Localization

Rammal and Toulouse (1983) showed that, for $D_s < 2$, fractons are spatially localized. To prove this, they used the β_L function, defined by Abrahams *et al.* (1979) in their scaling theory of localization:

$$g(L) \sim L^{\beta_L}, \tag{350}$$

where $g(L)$ is the dimensionless conductance of a sample of linear size L, i.e., a quantity of the order of $g_e L^{d-2}$, where g_e is the conductivity of the sample. It is clear that the wave functions should be localized when $\beta_L \leq 0$. In the case of percolating networks, $g_e \sim L^{-\mu/\nu}$ [see Eq. (2.54)]. Therefore, the conductance is given by

$$g(L) \sim L^{-\mu/\nu+d-2} \sim L^{D_f(D_s-2)/D_s}, \tag{351}$$

and therefore, $\beta_L = D_f(D_s - 2)/D_s$. Since as discussed above, for materials with percolation disorder, $D_s \simeq 4/3$, $\beta_L < 0$ for all Euclidean dimensionalities d. We should note that for length scales, $L \gg \xi_p$, $D_s \to d$, implying that for $d = 3$, for example, β_L may well be positive, indicating delocalized vibrations (phonons). At length scales $L \ll \xi_p$ one has, $D_s \simeq 4/3$, leading to negative values of β_L and localization. Therefore, the crossover can be thought of as a dimensionality change in so far as localization is concerned.

6.6.5.2 Dispersion Relation

The dispersion relation is given by Eq. (349). To derive this equation (Alexander, 1989), consider the vibrations of an isolated fractal cluster of linear size L. Although high-frequency modes of the vibrations are not affected by a change in the boundary conditions, the low-frequency modes will disappear from the spectrum. The crossover between the two regimes occurs at a frequency ω_L, such that

$$L \sim \Lambda(\omega_L), \tag{352}$$

where $\Lambda(\omega)$ is the wavelength. The integrated spectral weight of the missing low-frequency modes is lumped together in the center-of-mass degrees of freedom of the disconnected cluster (for example, the rotational and translational modes). Since this quantity cannot depend on L and ω_L, one must have

$$[\Lambda(\omega_L)]^{D_f} \int_0^{\omega_L} \omega^{D_s-1} d\omega \sim \Lambda^{D_f} \omega_L^{D_s} = \text{constant}, \tag{353}$$

and thus one has a dispersion relation for an arbitrary frequency ω:

$$\omega \sim \Lambda(\omega)^{-D_f/D_s}. \tag{354}$$

If we apply the argument of a frequency-dependent length scale $\Lambda(\omega)$ to waves in a finite homogeneous material (for which $D_f = D_s = d$), we obtain a length scale $\Lambda = 2\pi v_s/\omega = \lambda$, where v_s is the sound velocity, because the lowest frequency of a cluster of linear size Λ is $\omega(\Lambda) = 2\pi v_s/\Lambda$.

6.6.5.3 Crossover from Phonons to Fractons

If the wavelengths λ of excited modes in a disordered material with percolation disorder and $p > p_c$ are larger than the percolation correlation length $\xi_p(p)$, the material is homogeneous on this scale and its vibrational excitations are weakly-localized phonons. This is because the scattering is determined by the square of the mass-density fluctuations, averaged over regions of volume λ^d. Hence, even if the short-range disorder is strong, the effective strength of the disorder for phonons with $\lambda \gg \xi_p$ is very weak. If the characteristic length λ of waves becomes of the order of, or shorter than, ξ_p, the fractality of the material's morphology becomes relevant. Thus, there must be a crossover in the nature of the vibrational excitations when $\lambda(\omega_{co}) \sim \xi_p$. From Eq. (349) or (354) the crossover frequency ω_{co} is given by

$$\omega_{co} \sim (p - p_c)^{\nu D_f / D_s}. \tag{355}$$

By substituting $\omega_{co} = v_s(p)k$ (at $k \sim 1/\xi_p$) in Eq. (355), the p-dependence of the phonon velocity becomes

$$v_s(p) \sim (p - p_c)^{\nu D_f / D_s - \nu} \sim (p - p_c)^{(\mu - \beta)/2}. \tag{356}$$

Because $\mu - \beta > 0$ for all d, $v_s(p) \to 0$ as $p \to p_c$. Thus, the results for the DOS can be summarized as

$$\mathcal{N}(\omega) \sim \begin{cases} \omega^{d-1}/[v_s(p)]^d, & \omega \ll \omega_{co}, \\ \omega^{D_s - 1}, & \omega \gg \omega_{co}, \end{cases} \tag{357}$$

while the dispersion relations become

$$\omega \sim \begin{cases} v_s(p)k, & \omega \ll \omega_{co}, \\ k^{D_f/D_s}, & \omega \gg \omega_{co}, \end{cases} \tag{358}$$

where, for $\omega \gg \omega_{co}$, k does not represent the wave number due to the lack of the translational symmetry of the system, but rather it describes the inverse of the characteristic length Λ^{-1}.

6.6.6 *Large-Scale Computer Simulations*

In this section, we briefly discuss the results of computer simulations for the vibrational DOS of large percolation networks using the scalar approximation. These data provide rich insight into the fracton dynamics. We also discuss the characteristic properties of fracton wave functions. The numerical results based on a vector model, Eq. (313), will be discussed in Chapter 9.

Grest and Webman (1984) were the first to calculate the DOS of 3D percolation networks, using the scalar approximation and the standard diagonalization method (see above) for a cubic lattice of linear size $L = 18$, which was of sufficient accuracy, except at low frequencies. For larger systems and lower frequencies, they used a recursive technique (Lam and Bao, 1985) to calculate the eigenfunctions and eigenvalues. Although, as discussed above, there are considerable difficulties in simulating large-scale systems, the situation since the early work of Grest and Webman has been changing as array-processing supercomputers have become available. Several numerical methods have been reported to overcome such difficulties, and have been used for investigating fracton dynamics. The list of these papers is too long to be given here; the interested reader should consult the comprehensive review of Nakayama *et al.* (1994). Here, we only consider a few of more notable of such efforts and discuss their results.

Li *et al.* (1990) used the Sturm sequence method described above to calculate the integrated DOS. They treated a square lattice of size 160×640 with site-percolation disorder. Royer *et al.* (1992) used the spectral moment method, which allowed them to work with a very large square network of linear size $L = 1415$. Yakubo and Nakayama (1989) and Yakubo *et al.* (1990a,b) succeeded in treating systems with $N = 10^6$ sites, using the forced oscillator method of Williams and Maris (1985). This algorithm is based on the principle that a complex mechanical system, when driven by a periodic external force of frequency ω, will respond with a large amplitude in those eigenmodes that are close to this frequency. Yakubo *et al.* (1991) formulated a method for judging the accuracy of the calculated eigenmodes and eigenfrequencies. Their algorithm can be readily vectorized for use on an array-processing supercomputer, and has the great advantage that it becomes more accurate with increasing number of the sites.

Figure 6.25 presents the DOS at the site percolation threshold $p_c \simeq 0.593$ of the square network (Yakubo and Nakayama, 1989). The line through the solid circles has a slope of about 1/3, which holds even in the low-frequency region, because the lower cutoff frequency ω_L (see above) is determined from the finite size of the clusters. The DOS of a square network with site-percolation disorder and a fraction $p = 0.67$ of the active sites is shown in Figure 6.26 (Yakubo and Nakayama, 1989). These results indicate that the frequency dependence of the DOS is characterized by two regimes. In the frequency region $\omega_{co} \ll \omega \ll 1$, the DOS is closely proportional to $\omega^{1/3}$, in agreement with Eq. (333) with $D_s \simeq 4/3$. The

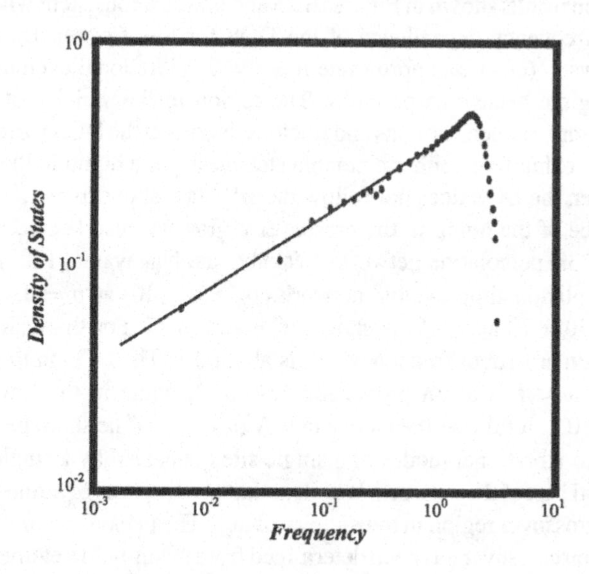

FIGURE 6.25. Density of states at the site percolation threshold of the square lattice, $p \simeq 0.593$. The straight line represents $\mathcal{N}(\omega) \propto \omega^{1/3}$ (after Yakubo and Nakayama, 1989).

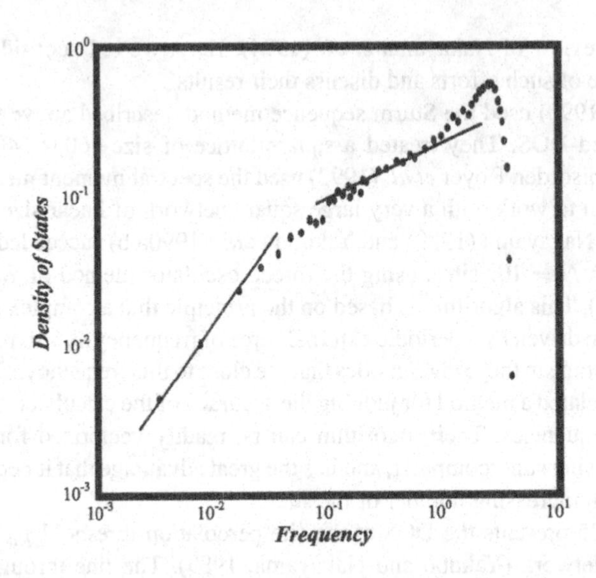

FIGURE 6.26. Same as Figure 6.25, but at $p = 0.67$. The straight line on the left corresponds to $\mathcal{N}(\omega) \propto \omega$, while the one on the right represents $\mathcal{N}(\omega) \propto \omega^{1/3}$ (after Yakubo and Nakayama, 1989).

crossover frequency ω_{co} corresponds to the mode of wavelength λ which is equal to the percolation correlation length ξ_p. Therefore, for $\omega \ll \omega_{co}$ the DOS should be given by the conventional Debye law, Eq. (322), and $\mathcal{N}(\omega) \sim \omega^{D_s-1}$ for $\omega \gg \omega_{co}$ The simulation results shown in Figures 6.25 and 6.26 are consistent with this view, because the frequency dependence of the DOS for low frequencies ($\omega \ll \omega_{co}$) clearly follows $\mathcal{N}(\omega) \sim \omega$, appropriate foe $d = 2$. Vibrational excitations in this frequency regime behave as phonons. The region in the vicinity of ω_{co} is the crossover region between phonons and fractons. Note that the DOS varies smoothly in this region, exhibiting neither a notable steepness nor a hump in the vicinity of ω_{co}. Moreover, the DOS does not follow the $\omega^{1/3}$ law above $\omega \sim 1$.

The absence of the hump in the crossover region has also been demonstrated in the case of 3D percolation networks (Yakubo and Nakayama, 1989). The DOS of a site-percolation simple-cubic network ($p_c \simeq 0.3116$) at $p = 0.4$, computed with a $70 \times 70 \times 70$ network, is shown in Figure 6.27. For this value of p the phonon-fracton crossover frequency ω_{co} is about 0.1. The DOS in the frequency region $0.1 < \omega < 1$ is again proportional to $\omega^{1/3}$, while in the low-frequency regime ($\omega \ll 0.1$) it follows the Debye law, $\mathcal{N}(\omega) \sim \omega^2$. The sharp peak at $\omega = 1$ is attributed to vibrational modes of a single site connected by a single bond to a relatively rigid part of the network. It is clear that no steepness or hump of the DOS exists in the crossover region in the vicinity of ω_{co}. The behavior of the DOS at the phonon-fracton crossover has been determined from mean-field treatments (Loring and Mukamel, 1986; Korzhenevskii and Luzhkov, 1991). In particular, Loring and Mukamel (1986) suggested a smooth transition of the DOS at the phonon-fracton

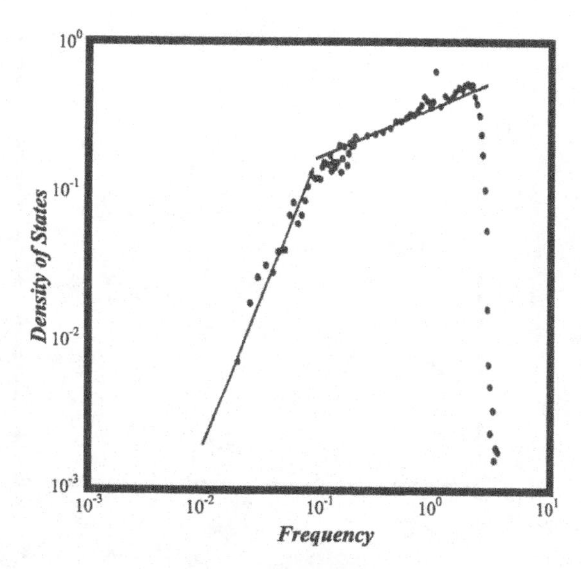

FIGURE 6.27. Density of states for site percolation in the simple-cubic network at $p = 0.4$. The straight line on the left corresponds to $\mathcal{N}(\omega) \propto \omega^2$, while the one on the right represents $\mathcal{N}(\omega) \propto \omega^{1/3}$ (after Yakubo and Nakayama, 1989).

crossover, in contrast to the prediction of the EMA discussed above, or that of the scaling theory (Aharony *et al.*, 1985). We will come back to this point later.

The DOS calculated with bond-percolation networks exhibits some interesting differences with site-percolation networks discussed above. The difference is due to the fact that site percolation generates short-range correlations that are absent in bond percolation, since in the former case a bond is active only if its end sites are also active. In Figure 6.28 the DOS for the square and simple-cubic lattices at their bond-percolation thresholds (see Table 2.2) are shown. For the 2D case, a spectral dimension $D_s \simeq 1.33 \pm 0.01$ is obtained, whereas for the 3D case, $D_s \simeq 1.31 \pm 0.02$ is estimated.

6.6.7 Missing Modes

The above simulation results for the DOS, using the scalar approximation, confirm that the crossover between the phonon and fracton regimes is smooth, with no visible accumulation of modes or a hump in the DOS around the crossover frequency ω_{co}, which is in agreement with neither the early predictions based on scaling considerations (Alexander *et al.*, 1983; Aharony *et al.*, 1985), nor with arguments based on the EMA computations described above, nor with the numerical results obtained by a recursion technique (Lam and Bao, 1985). The scaling considerations attributed the origin of this hump to the fact that the crossover from fractons to phonons is accompanied by missing modes in the normalized DOS. We now discuss the possible whereabouts of these missing modes, following the work of Yakubo *et al.* (1990a,b).

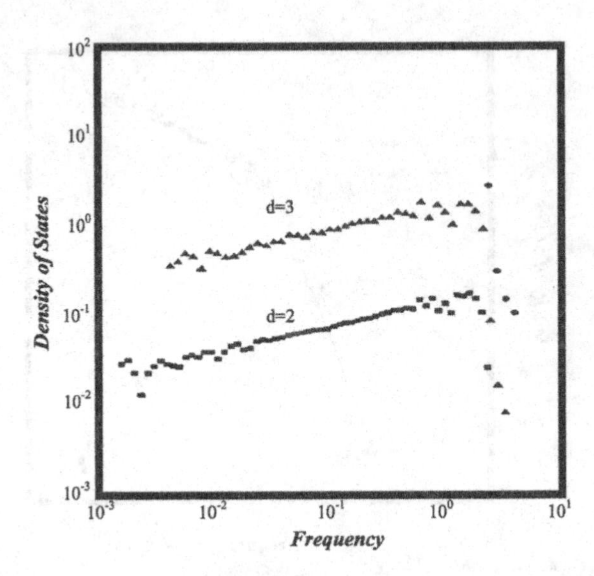

FIGURE 6.28. Density of states at the bond percolation threshold of d-dimensional cubic lattices (after Nakayama, 1992).

We first discuss the early scaling arguments about missing modes. In this analysis, the DOS of a percolation network above p_c was characterized by two regimes: The fracton DOS, $\mathcal{N}_{\rm fr}(\omega, p) \sim \omega^{D_s-1}$, for high frequencies ($\omega > \omega_{co}$), and the phonon DOS, $\mathcal{N}_{\rm ph}(\omega, p) \sim \omega^{d-1}$, for low frequencies ($\omega < \omega_{co}$). Assuming strict similarity in the fractal regime, we may expect that, $\mathcal{N}_{\rm fr}(\omega, p) = \mathcal{N}_{\rm fr}(\omega, p_c)$, where the DOS per particle is normalized, $\int_0^\infty \mathcal{N}(\omega)d\omega = 1$. Because $d > D_s$, we have $\mathcal{N}_{\rm ph} < \mathcal{N}_{\rm fr}$, when $\mathcal{N}_{\rm fr}$ is extrapolated to phonon frequencies. As the integration of $\mathcal{N}_{\rm fr}(\omega, p_c)$ is normalized to unity, some modes must be missing for $\mathcal{N}(\omega, p > p_c)$, in view of the existence of the phonon regime, and therefore their spectral weight must be recovered somewhere. It was argued that the most reasonable place for their accumulation is near ω_{co}, leading to a hump in the DOS and to a corresponding hump in the low-temperature specific heat. This is shown in Figure 6.29. However, a hump is seen neither in the simulations of the phonon-fracton crossover nor in actual experiments on silica aerogels (see below).

To explain the absence of a hump in the crossover region, we write a scaling equation for the DOS of a percolation network for $p > p_c$:

$$\mathcal{N}(\omega, p) = A(p)\omega^{D_s-1}F(\omega/\omega_{co}). \tag{359}$$

The crossover frequency ω_{co} is defined as the intercept of the asymptotic phonon and fracton lines in a double-logarithmic plot of the DOS versus ω (Alexander *et al.*, 1983). As discussed above, for an infinite percolation network, the phonon-fracton crossover frequency follows a power law given by, $\omega_{co} = c(p - p_c)^{\nu D_f/D_s}$ [see Eq. (349) if we replace $L(\omega)$ with the correlation length $\xi_p \sim (p - p_c)^{-\nu}$], where c is a constant. The scaling function $F(x)$ has the properties that, $F(x) \sim 1$ for

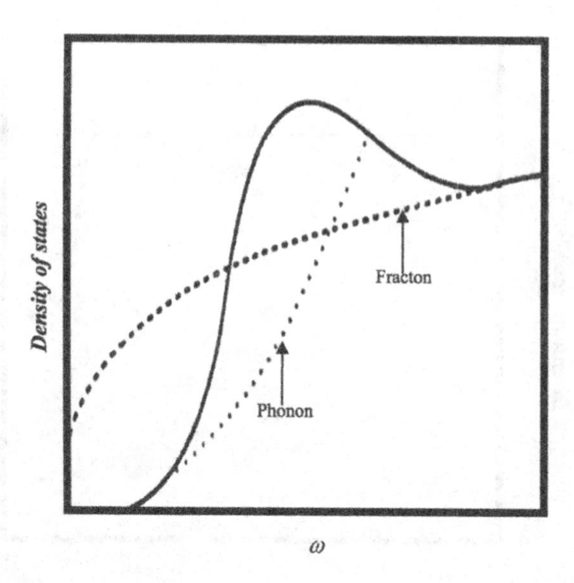

FIGURE 6.29. Schematics of density of states (solid curve) and its crossover. The dotted (dashed) curve represents the continuation of phonon (fracton) asymptotic behavior into the crossover regime (after Nakayama *et al.*, 1994).

$x \gg 1$ and $F(x) \sim x^{d-D_s}$ for $x \ll 1$. Equation (359) yields a prediction for the p-dependence of $\mathcal{N}(\omega, p)$ in the phonon regime:

$$\mathcal{N}(\omega) \sim A(p)(p-p_c)^{\nu D_f (D_s-d)/D_s} \omega^{d-1}, \tag{360}$$

which should be compared with Eq. (341), the EMA prediction. In fact, if we plot the simulation results as $\mathcal{N}(\omega)/\omega^{1/3}$ versus ω, the validity of Eq. (360) in the phonon-fracton crossover region is confirmed. This is shown in Figure 6.30 where the data for the DOS of a square network at its site percolation threshold, $p_c \simeq 0.593$, and also for $p = 0.67$, are shown. Because the ordinate in this figure is $\mathcal{N}(\omega)/\omega^{1/3}$, the fracton regime corresponds to a horizontal line of height $A(p)$. One can clearly discern the two regimes in Figure 6.30, with a crossover frequency $\omega_{co} \sim 0.1$. It is clear that these results do not exhibit any noticeable hump near ω_{co}. Furthermore, the magnitude of the DOS in the fracton regime is different for the two values of p, hence invalidating $\mathcal{N}_{\rm fr}(\omega, p) = \mathcal{N}_{\rm fr}(\omega, p_c)$ assumed by Aharony *et al.* (1985), and also justifying the non-trivial dependence of $A(p)$ on p. We also note that the two curves of Figure 6.30 could not possibly be made to scale towards the upper end of the fracton range, where modes missing from the low-frequency regime could have possibly accumulated.

The above considerations can now be put on a firm theoretical foundation. The discussion is facilitated by adopting the nodes-links-blobs model of the sample-spanning percolation clusters described in Section 2.6.3. The typical separation distance between the nodes that form the macroscopically-homogeneous network is about the correlation length ξ_p. With $\mathcal{N}_{\rm ph} = A(p)\omega/\omega_{co}^{2/3}$, and $\mathcal{N}_{\rm fr} = A(p)\omega^{1/3}$,

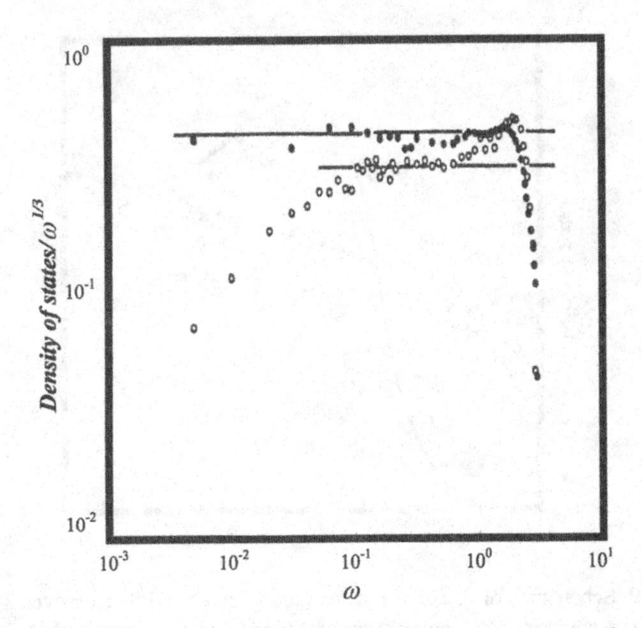

FIGURE 6.30. Density of states, normalized to one particle and divided by $\omega^{1/3}$, as a function of the frequency ω in site percolation in the square lattice. Solid and open circles represent, respectively, the results for $p = p_c = 0.593$ and $p = 0.670$. The lines are guide to the eye (after Yakubo *et al.*, 1990a).

the number of missing modes $M_{\rm ph}$ associated with the phonon regime is given by

$$M_{\rm ph} \sim \int_0^{\omega_{co}} (\mathcal{N}_{\rm fr} - \mathcal{N}_{\rm ph}) d\omega = \frac{1}{4} A(p) \omega_{co}^{D_s} = \frac{1}{4} A(p) c^{D_s} (p - p_c)^{\nu D_f}. \quad (361)$$

From Figure 6.30 one can estimate that, $A(p) \sim A(p_c) \simeq 0.4$ and $c \simeq 13$, and therefore, $M_{\rm ph} \sim 3(p - p_c)^{\nu D_f}$. The fraction of occupied sites on the sample-spanning cluster is simply $X^A(p)$, the accessible fraction defined in Section 2.6.1. Near p_c one has [see Eq. (2.30)], $X^A(p) = X_0(p - p_c)^\beta$, with $\beta = 5/36$ (see Table 2.3) and $X_0 \simeq 1.53$ for the square network. Hence, the actual number of missing modes within the correlation area is $\xi_p^2 X^A(p) M_{\rm ph} \sim 3\xi_0^2 X_0$, where ξ_0 is defined by $\xi_p = \xi_0 |p - p_c|^{-\nu}$, independent of $p - p_c$. Using $\xi_0 \simeq 0.95$ for the square network, the number of missing modes is of the order of unity. Thus, there is only one missing mode per area. One should also note that, for any non-negligible ξ_p, the number $M_{\rm ph}$ relative to the total number of modes with $\omega < \omega_{co}$ is very small compared to one.

However, the more important quantity is the number of missing modes $M_{\rm fr}$, produced by the depression of the fracton density from $A(p_c)$ to $A(p)$. If we ignore the hump near the high-frequency cutoff, $M_{\rm fr}$ will be given by

$$M_{\rm fr} \simeq \int_0^{\omega_D} [\mathcal{N}_{\rm fr}(\omega, p_c) - \mathcal{N}_{\rm fr}(\omega, p)] d\omega = 1 - \frac{A(p)}{A(p_c)}, \quad (362)$$

where for the second equality we used Eq. (360) and the fact that, $\int_0^\infty \mathcal{N}_{\rm fr}(\omega, p_c)d\omega = 1$. Here ω_D is the Debye frequency associated with the phonon velocity of sound v_s. The numerical results for $M_{\rm fr}$ on a square network and for several values of p were obtained by Yakubo *et al.* (1990a) which exhibited critical behavior near p_c, $M_{\rm fr} = M_0(p - p_c)^m$, with $m \simeq 4/3$ and $M_0 \simeq 4.1$.

These results can be explained as follows. Within an area ξ_p^2, a number of sites have higher coordination than those in the network at p_c. The number of such sites is much larger than the small number of sites that eventually form the homogeneous system, with their relative density being $1/\xi_p^2 X^A(p) \sim M_{\rm ph}$. One may naively expect the number of these sites to be proportional to the length of the perimeter, i.e., be proportional to ξ_p^{d-1}. In the case of a percolation cluster, however, the perimeter at ξ_p is a fractal of dimension $D_f - 1$ with its length being $\xi_p^{D_f-1}$. With the total number of occupied sites within this perimeter being $\xi_p^{D_f}$, the relative number of modes rejected to high frequencies is $M_{\rm fr} \sim \xi_p^{D_f-1}/\xi_p^{D_f} \sim \xi_p^{-1} \sim (p - p_c)^\nu$, in agreement with the results of Yakubo *et al.* (1991). Both the exponent and the amplitude of $M_{\rm fr}$ agree well with that of ξ_p. Although there is, strictly speaking, no well-defined perimeter at ξ_p for which the connectivity of all the sites increases, the concept remains well-defined from an average point of view. The effect of the higher connectivity is to depress the number of modes throughout the fracton regime.

6.6.8 *Localization Properties of Fractons*

Let us now describe a few important localization properties of fractons.

6.6.8.1 Mode Patterns of Fractons

Yakubo and Nakayama (1989) were the first to compute mode patterns of large-scale fractons. They used a square network at its site percolation threshold, $p_c \simeq 0.593$. If we examine a cross section of amplitudes of their calculated mode pattern for a single fractor, we see that the fracton core possesses very clear boundaries for the edges of the excitation, almost of steplike character, with a long tail in the direction of the weak segments. This is in contrast with the case of homogeneously extended modes, i.e., phonons, in which the change of their amplitudes is correlated smoothly over a long distance. Moreover, displacements of the sites in the dead-end bonds move in phase, while the vibrational amplitudes fall off sharply at their edges. Finally, the tail of the pattern extends over a very large distance with many phases changes.

6.6.8.2 Ensemble-Averaged Fractons

Since the pioneering work of Anderson (1958), localization phenomena in disordered materials, including localization of phonons, photons, and spin waves, have received much attention. In particular, it was predicted by John *et al.* (1983) that the vibrational excitations in disordered materials with dimensionality $d \leq 2$ should always be localized, with the localization length ξ_l behaving as $\xi_l \sim \exp(l/\omega^2)$ for $d = 2$ and as $\xi_l \sim \omega^{-2}$ for $d = 1$.

As already discussed above, Rammal and Toulouse (1983) applied the scaling theory of localization (Abrahams *et al.*, 1979) to fracton excitations on percolation networks. The key parameter in their scaling theory is the exponent β_L, defined by Eq. (35), which was predicted to be given by

$$\beta_L = \frac{D_f}{D_s}(D_s - 2). \tag{363}$$

Because $D_s \simeq 4/3$ for percolation in any dimension d, it is clear that fractons are always localized. In this context, Entin–Wohlman *et al.* (1985) proposed that the ensemble average of the fracton wave function Ψ on percolation networks is localized with

$$\langle \Psi_{\mathrm{fr}} \rangle \sim \exp\left\{ -\left[\frac{r}{\Lambda(\omega)} \right]^{\gamma} \right\}, \tag{364}$$

where $\Lambda(\omega)$ is the frequency-dependent fracton length scale (dispersion or localization), r a radial distance from the center, and the exponent γ denotes the strength of localization. Many studies have been carried out for estimating the numerical value of γ, some of which were discussed in Section 6.2.4.4 where we considered the effect of fractal morphology of a material on its hopping conductivity.

To study the localized nature of fractons, Nakayama *et al.* (1989) focused on the value of the exponent γ for the core region of fractons, and carried out computer simulations using a square network in site percolation in order to determine γ for the cores of the fractons. The core has a large amplitude around the center of a localized fraction ($r \simeq \Lambda$). Smoothly varying ensemble-averaged mode patterns were obtained. The resulting shape of the fracton core was calculated by averaging over 129 fractons at $\omega = 0.01$. Their computations yielded $\gamma \simeq 2.3$, larger than all other numerical estimates. In addition, γ and $\Lambda(\omega)$ were calculated for four different eigenfrequencies, excited on five independent percolation networks. Their results indicated that the localization length $\Lambda(\omega)$ depends on frequency as, $\Lambda(\omega) \sim \omega^{-0.71}$, in good agreement with the theoretical dispersion law, Eq. (349), $\Lambda(\omega) \sim \omega^{-D_s/D_f}$, with $D_s/D_f \simeq 0.705$. However, it is now understood that this large value of γ applies only to the core region of fractons, and that two exponents are required to characterize the localized nature of fractons, namely, one for the core and another one for the tail (Roman *et al.*, 1991). In fact, as discussed in Section 6.2.4.4, the asymptotic value of γ, when the vibrational equations for large percolation clusters are solved and averaged over many individual fracton modes for each frequency, is $\gamma = 1$. Note that a large crossover regime exists when $r \simeq \Lambda$, with an *effective* exponent $\gamma > 1$ (which coincides with the result of Nakayama *et al.*, 1989). However, asymptotically, for $r \gg \Lambda$, one has $\gamma = 1$.

Bunde and Roman (1992) presented an analytical explanation for the asymptotic behavior of fractons. Within the scalar approximation, the envelope function $|\Psi(r, \omega)|$ of fractons is related to $P(r, t)$, the probability of finding a random walker at position r at time t (see Section 6.1.12):

$$P(r,t) = \int_0^\infty \mathcal{N}(\omega)|\Psi(r,\omega)| \exp(-\omega^2 t) d\omega, \tag{365}$$

where $\mathcal{N}(\omega)$ is the DOS normalized to unity. If we now use Eq. (90) for $P(r,t)$, we obtain, upon taking the inverse Laplace transform of Eq. (365) using the method of steepest descent, $\gamma = 1$ with $\Lambda(\omega)^{-1} \sim \cos(\pi/D_w)\omega^{2/D_w}$ (Roman *et al.*, 1991).

6.6.9 Comparison with the Experimental Data

In this section we discuss experimental observations of fracton dynamics and the corresponding DOS. Most of such studies involve aerogels, and therefore we first briefly discuss the preparations and properties these materials.

Aerogels are highly porous solid materials that have a very tenuous structure. Their porosity can be as high as 99%, and for this reason they often have very unusual and unique properties. For example, they can be prepared in transparent form, they have very small thermal conductivity, solid-like elasticity and, because of their large porosity, they possess large internal surface area. Because of such unusual properties, they also have a wide-range of applications, from catalyst supports to thermal insulators and radiators, and detectors of Cerenkov radiation (see, for example, Fricke, 1988; Brinker and Scherer, 1990). They are prepared by a variety of methods using different materials, but silica aerogels have received the widest attention. They are produced by hydrolysis of $Si(OR)_4$, where R represents either CH_3 or C_2H_5. A catalyst, which is either an acid or a base, is also used which strongly influences the reaction. The degree of hydrolysis is controlled by the concentration ratio $Si(OR)_4/H_2O$, while the final density of the aerogel is controlled by $Si(OR)_4/ROH$. Because of the acid or base catalyst, the pH of the solution also has a strong effect on the morphology of the gel. Hydrolysis produces -SiOH groups which then polymerize into -Si-O-Si-. Then, particles start to grow in the liquid solution and after some time form a gel network. The solvent is then removed to obtain the solid porous structure. Aerogels are obtained if the solvent is removed at a temperature *above* its critical point. Comprehensive discussion of preparation of such gels is given by Courtens *et al.* (1989) to whom we refer the interested reader.

Small-angle scattering techniques using neutrons and X-rays are very well-suited for systematically investigating the morphology of silica aerogels using such techniques. It has been reported (see, for example, Schaefer and Keefer, 1986; Courtens and Vacher, 1987; Vacher *et al.*, 1988; Woignier *et al.*, 1990; Posselt *et al.*, 1992) that aerogels may have a fractal morphology. Bourret (1988) and Duval *et al.* (1992) reported high-resolution electron microscopy observations that were compatible with a fractal morphology for these materials. Beck *et al.* (1989) and Ferri *et al.* (1991) used light-scattering techniques for characterizing the geometrical features of these materials.

To characterize the structure of these materials, the structure factor $S(q)$, which describes the correlation between particles in a cluster, has been measured. This quantity is obtained from the Fourier transform of the particle density-density correlation function $C(\mathbf{r})$:

$$S(\mathbf{q}) = 1 + \frac{N}{\Omega}\int_{\Omega} |C(\mathbf{r}) - 1|e^{i\mathbf{q}\cdot\mathbf{r}}d\mathbf{r}, \tag{366}$$

where $q = |\mathbf{q}| = (4\pi/\lambda)\sin(\theta/2)$, with λ being the wavelength of the radiation scattered by the material through an angle θ. If, as discussed in Section 2.3.2, the self-similarity of a material extends up to a correlation length ξ_s, then the correlation function $C(r)$ will be given by Eq. (2.15), which is the most convenient, but not unique, choice. For small values of q with $q\xi_s \ll l$, $S(q)$ is almost independent of q. When $q\xi_s \gg 1$, one obtains, by substituting Eq. (2.15) into Eq. (366)

$$S(q) \sim q^{-D_f}, \tag{367}$$

which is similar to Eq. (2.14) [often $S(q)$ and $I(q)$, the intensity of the scattering, are used interchangeably]. According to Eq. (367), the value of D_f can be deduced from the slope of the logarithm of the observed corrected intensity versus the logarithm of q.

An example of such scattering measurements is shown in Figure 6.31 for silica aerogels (Vacher *et al.*, 1988). The various curves are labeled by the macroscopic density ρ of the sample. The solid curves represent the best fits to the data, and have been extrapolated into the particle regime ($q > 0.15\text{Å}^{-1}$) to emphasize that the fits do not apply in that region, particularly for the denser samples. Remarkably, the fractal dimension $D_f \simeq 2.4 \pm 0.03$ is independent of sample density to within experimental accuracy. Furthermore, ξ_s scales with ρ as $\xi_s \sim \rho^{-1.67}$. The departure

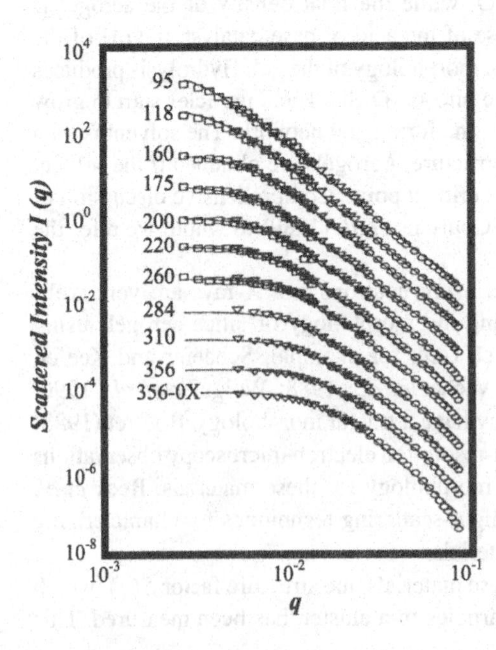

FIGURE 6.31. Scattered intensities, for eleven silica aerogel samples, versus the magnitude q of the scattering vector in Å^{-1}. The top ten curves are for untreated and neutrally reacted samples of increasing density, while the bottom one is for an oxidized sample. The curves are labeled with the density in kg/m^3 (after Vacher *et al.*, 1988).

of $S(q)$ from the q^{-D_f} dependence at large q indicates the presence of particles with radii of gyration of about a few angströms. Thus, silica aerogels exhibit three different length-scale regions. (1) At short distances, elementary particles aggregate together to form clusters with linear size ξ_s at intermediate sizes, forming a gel by connecting the clusters at larger distances. (2) At intermediate length scales, the clusters possess fractal morphology. (3) At large length scales the gel is a homogeneous porous glass; see Courtens and Vacher (1987) and Kjems (1993) for more details.

Incoherent inelastic neutron scattering experiments measure the amplitude-weighted DOS. The scattered intensity is given by

$$I(q,\omega) \sim q^2 \frac{k_s}{k_i} \frac{n(\omega)}{\omega} \sum_i e^{-2W_i} \mathcal{N}_i(\omega), \tag{368}$$

where $n(\omega)$ is the Bose–Einstein distribution function, the wave vectors k_i and k_s correspond to the incident and scattered neutrons, respectively, and $\mathbf{q} = \mathbf{k}_s - \mathbf{k}_i$. Here $\mathcal{N}_i(\omega)$ and W_i are the DOS and the Debye–Waller factor characteristic of the ith site, respectively. The sum is over the different sites (the atoms), each of which contributes to the incoherent-scattering with an intensity proportional to the amplitude of vibration at frequency ω. Incoherent neutron scattering from protons that are chemically bonded to the particle surfaces can be used to determine the DOS in porous materials (Richter and Passell, 1980). The incoherent inelastic scattering by aerogels have been measured by several groups, a list of which is too long to be given here. Buchenau *et al.* (1992) and Kjems (1993) reviewed these experiments and their implications.

The phonon-fracton crossover in silica aerogels has been studied using such inelastic neutron scattering experiments on back scattering spectrometers (see, for example, Pelous *et al.*, 1989; Conrad *et al.*, 1990), and also the spin-echo technique (Courtens *et al.*, 1990; Schaefer *et al.*, 1990). The former technique has the advantage that it enables ones to observe the low-frequency Debye regime as a constant-intensity level, extending from the elastic line to the crossover frequency, and thus any excess modes at the phonon-fracton crossover should show up as a peak in the scattering intensity at that frequency, although no such peak has yet been observed in the backscattering experiments. Rather, a gradual decrease is observed as one passes through the crossover regime. The neutron-scattering spin-echo technique has the advantage that its larger spectral range makes it a suitable tool for determining the spectral dimension D_s.

The crossover frequencies determined by both backscattering and spin-echo measurements are generally in good agreement with those determined by Brillouin scattering (see, for example, Courtens *et al.*, 1987). Some investigations at higher frequencies (Reichenauer *et al.*, 1989; Vacher *et al.*, 1989) exhibited a change of slope in the log-log plot of the DOS versus ω at $\omega = 200$ GHz, giving a stronger increase with frequency at higher frequencies, which was interpreted as a crossover from fractons to vibrational modes within the particles. The effective slope of the DOS (i.e., $D_s - 1$) was about 1.5, which appears to have originated from the contributions by both surface (which is proportional to ω) and the bulk (which is proportional to ω^2) particle modes. However, the energy resolution was insufficient to observe the crossover to the long-wavelength phonon regime. Measurements of Vacher *et al.* (1989) at higher resolutions confirmed the existence of the extended fracton region. To further discuss this phenomenon, one must in fact take into account the effect of the vectorial nature of the vibrational modes, and therefore we will come back to this issue in Chapter 9.

Neutron-scattering experiments on other disordered materials have also been analyzed in terms of the fracton theory. Freltoft *et al.* (1987) measured the low-frequency DOS for fractal silica aggregates by inelastic neutron scattering and estimated their spectral dimensions. Page *et al.* (1989) performed inelastic neutron scattering experiments on fumed silica and compared the results with analogous results for amorphous quartz. They observed no evidence for a hump in the DOS near the phonon-fracton crossover, nor did they find that the temperature and wave-vector dependence of the intensity was in agreement with simple phonon models. Fontana *et al.* (1990) reported on a study of low-frequency vibrational dynamics and electron-vibration coupling in AgI-doped glasses. By using both time-of-flight neutron-scattering and Raman-scattering spectroscopies, they were able to determine the vibrational DOS and the frequency dependence of the electron-vibration coupling. The spectral dimension of this material was estimated to be about $D_s \simeq 1.4$. Zemlyanov *et al.* (1992) employed inelastic neutron scattering measurements to study low-frequency vibrational excitations in polymethyl metacrylate. The measured DOS followed a power law in the energy, with a spectrum corresponding to a spectral dimension $D_s \simeq 1.8 \pm 0.05$. Other experimental efforts in this area were reviewed by Nakayama *et al.* (1994).

There have also been experimental studies of the dynamical properties of fractal materials without involving any scattering measurements. For example, Kopelman *et al.* (1986) measured the exciton recombination characteristics of naphthalene-doped microporous materials. This technique yields the spectral dimension of the embedded naphthalene structure, or, equivalently, the effective random-walk fractal dimension D_w of the porous network. Kopelman *et al.* obtained $1 \leq D_s \leq 2$. Fischer *et al.* (1990) studied the trap-depth distribution of dibenzofuran singlet excitons and the temperature-dependent energy migration by time-resolved spectroscopy via synchrotron radiation and two photon laser excitation, and obtained $D_s \simeq 1.14$.

6.7 The Dynamical Structure Factor

The dynamical structure factor (DSF) $S(q, \omega)$ provides deep insight into the properties of fractal materials. We thus describe and discuss the theoretical and experimental developments for the DSF. However, before doing so, we should clarify the meaning of energy width of fractons. Clearly, one can find exact eigenstates of a random medium which would have no energy width: They would be precisely defined in energy. It is only when one projects them onto plane-wave states that a lifetime is generated equally in the ω or $\mathbf{q}$ space, because of the linear phonon-dispersion relation. Therefore, when we calculate an energy width for the fractons, it is meant to be that width that a plane wave would experience.

6.7.1 *Theoretical Analysis*

In general, as described in Section 2.3.2, the intensity $I(\mathbf{q}, \omega)$ of inelastic neutron or light scattering with a frequency shift ω is proportional to the Fourier transform of the density-density correlation function, defined by $C(\mathbf{r} - \mathbf{r}', t) = \langle \rho(\mathbf{r}, t)\rho(\mathbf{r}', 0)\rangle$, where $\rho(\mathbf{r}, t)$ is the density and $\langle \cdot \rangle$ denotes an equilibrium ensemble average. We now introduce the density fluctuation $\delta\rho(\mathbf{r}, t)$ defined by, $\delta\rho(\mathbf{r}, t) = \rho(\mathbf{r}, t) - \rho(\mathbf{r})$, where $\rho(\mathbf{r})$ is the static density given by, $\rho(\mathbf{r}) = \sum_i \delta(\mathbf{R}_i - \mathbf{r})$, and $\mathbf{R}_i$ is the equilibrium position of the ith atom. The term that contributes only to elastic scattering ($\omega = 0$) is usually neglected, in which case a closely-related quantity $\mathcal{S}(q, \omega)$ (see below) is expressed in terms of the Fourier transform of the density fluctuation,

$$\mathcal{S}(\mathbf{q}, \omega) = \frac{1}{2\pi} \int \langle \delta\rho_{-q}(0)\delta\rho_q(t)\rangle \exp(-i\omega t)dt. \tag{369}$$

Because the density fluctuations, $\delta\rho(\mathbf{r}, t)$, induced by lattice vibrations with displacements $\mathbf{u}_i(t)$, is written as, $\delta\rho(\mathbf{r}, t) = \sum_i \{\delta[\mathbf{R}_i + \mathbf{u}_i(t) - \mathbf{r}) - \delta(\mathbf{R}_i - \mathbf{r}]\}$, the Fourier transform of $\delta\rho_q(t)$ becomes

$$\delta\rho_q(t) = \sum_i \left\{ e^{i\mathbf{q}\cdot[\mathbf{R}_i + \mathbf{u}_i(t)]} - e^{-\mathbf{q}\cdot\mathbf{R}_i} \right\}. \tag{370}$$

Decomposing $\mathbf{u}_i(t)$ into normal modes, $\mathbf{u}_i = \sum_\lambda \mathbf{u}_i^\lambda e^{-i\omega_\lambda t}$, one obtains

$$\delta\rho_q(t) = \sum_\lambda \rho_\lambda(\mathbf{q}, t) + O(u^2), \tag{371}$$

where, $\delta\rho_\lambda(\mathbf{q}, t) = e^{-i\omega_\lambda t}\delta\rho_\lambda(\mathbf{q})$, and

$$\delta\rho_\lambda(\mathbf{q}) = -i \sum_i (\mathbf{q} \cdot \mathbf{u}_i^\lambda) e^{-i\mathbf{q}\cdot\mathbf{R}_i}. \tag{372}$$

If we substitute Eq. (371) into (369), we obtain

$$\mathcal{S}(\mathbf{q}, \omega) = \sum_\lambda \delta(\omega - \omega_\lambda)\langle \delta\rho_\lambda(\mathbf{q})\delta\rho_\lambda(-\mathbf{q})\rangle. \tag{373}$$

Hereafter, for convenience, the reduced dynamical structure factor $S(\mathbf{q}, \omega)$ is used by factoring out from $\mathcal{S}(\mathbf{q}, \omega)$ the usual mode quantization and the thermal factor $[n(\omega) + 1]/\omega$, where $n(\omega)$ is the Bose–Einstein distribution function:

$$S(\mathbf{q}, \omega) = \frac{\omega}{n(\omega) + 1}\mathcal{S}(\mathbf{q}, \omega). \tag{374}$$

Changing the sum in Eq. (373) into a frequency integral, one has

$$S(\mathbf{q}, \omega) = \mathcal{N}(\omega)\langle \delta\rho_\lambda(\mathbf{q})\delta\rho_\lambda(-\mathbf{q})\rangle_{\lambda \simeq \omega}, \tag{375}$$

where $\langle\cdot\rangle_{\lambda\simeq\omega}$ denotes an average of $\delta\rho_\lambda(\mathbf{q})$ over all modes λ with frequencies that are close to ω. The final expression for the intensity of inelastic scattering is then given by,

$$I(\mathbf{q},\omega) \sim \frac{n(\omega)+1}{\omega}\mathcal{N}(\omega)\langle\delta\rho_\lambda(\mathbf{q})\delta\rho_\lambda(-\mathbf{q})\rangle_\omega. \tag{376}$$

In principle, $S(\mathbf{q},\omega)$ can be calculated analytically from Eq. (373) or (375) if, for a specific realization, $\delta\rho_\lambda(\mathbf{r})$, or the fracton wave function $\Psi_\lambda(\mathbf{r})$, is known. However, because of the complex characteristics of fractons, this is not straightforward. Although certain scaling arguments have been made (Aharony *et al.*, 1988), and EMA-based computations have been carried out (Polatsek and Entin–Wohlman, 1988; Entin–Wohlman *et al.*, 1989a; Polatsek *et al.*, 1989), the explicit form of the DSF is not known, except for some deterministic fractals (Entin–Wohlman *et al.*, 1989b).

6.7.2 *Scaling Analysis*

Alexander (1989) and Alexander *et al.* (1993) presented scaling analysis of the asymptotic behavior of $S(\mathbf{q},\omega)$, based on the assumption that the only relevant length scale is $\Lambda(\omega)$, defined be Eq. (352). The essence of their analysis can be summarized as follows.

If the assumption of Alexander *et al.* (1993) is valid, then $S(\mathbf{q},\omega)$ should have the following scaling form,

$$S(q,\omega) = q^y H[q\Lambda(\omega)], \tag{377}$$

where the DSF is a function of $q = |\mathbf{q}|$, due to the spherical symmetry of the averaged structure of the random networks. The asymptotic behavior of the scaling function $H(x)$ is given by

$$H(x) \sim \begin{cases} x^a & x \ll 1, \\ x^{-a'} & x \gg 1, \end{cases} \tag{378}$$

where a and a' are new scaling exponents to be given shortly. Therefore,

$$S(q,\omega) \sim \begin{cases} q^{y+a}\omega^{-aD_s/D_f}, & q\Lambda(\omega) \ll 1, \\ q^{y-a'}\omega^{-a'D_s/D_f}, & q\Lambda(\omega) \gg 1. \end{cases} \tag{379}$$

Consider first the limit $q\Lambda \ll 1$. In this case, Eq. (372) can be expanded as

$$\delta\rho_\lambda(\mathbf{q}) \simeq -e^{-i\mathbf{q}\cdot\mathbf{R}_\lambda}\sum_i(\mathbf{q}\cdot\mathbf{R}_{i\lambda})(\mathbf{q}\cdot\mathbf{u}_i^\lambda), \tag{380}$$

where $\mathbf{R}_{i\lambda} = \mathbf{R}_i - \mathbf{R}_\lambda$ and $\mathbf{R}_\lambda$ is the center of the λ-mode fracton. The summand in Eq. (380) can be written as $\mathbf{q}\cdot\{\mathbf{R}_{i\lambda}\otimes\mathbf{u}_i^\lambda\}\cdot\mathbf{q}$, where $\otimes$ denotes the dyadic product. If we choose the center of the fracton at the origin, i.e., if $\mathbf{R}_\lambda = \mathbf{0}$ and $\sum_i \mathbf{R}_{i\lambda} = \mathbf{0}$

from the condition, $\sum_i \mathbf{u}_i^\lambda = \mathbf{0}$, then

$$\delta\rho_\lambda(\mathbf{q}) \simeq -\sum_i^{\nu_\lambda} \mathbf{q}\cdot\{\mathbf{R}_{i\lambda}\otimes[\mathbf{u}_i^\lambda - \mathbf{u}_\lambda]\}\cdot\mathbf{q} \tag{381}$$

where u_λ is the amplitude at the center of the λ-mode fracton, and the sum is restricted to a vibrating region ν_λ which is the smallest zone for which the boundary condition plays no significant role for the vibration λ. If an average strain tensor $\hat{\mathbf{e}}_\lambda$ is defined by

$$\mathbf{u}_i^\lambda - \mathbf{u}_\lambda = \hat{\mathbf{e}}_\lambda\cdot\mathbf{R}_{i\lambda}, \tag{382}$$

one obtains

$$\delta\rho_\lambda(\mathbf{q}) \simeq -\mathbf{q}\cdot\left\{\sum_i^{\nu_\lambda}(\mathbf{R}_{i\lambda}\otimes\mathbf{R}_{i\lambda})\hat{\mathbf{e}}_\lambda\right\}\cdot\mathbf{q}. \tag{383}$$

Since in ν_λ, $R_{i\lambda}\sim O(\Lambda)$, the magnitude of $\delta\rho_\lambda(\mathbf{q})$ can be estimated as

$$\delta\rho_\lambda \sim q^2\Lambda^{D_f+2}\hat{e}_\lambda, \tag{384}$$

and therefore,

$$S(q,\omega)\sim\mathcal{N}(\omega)q^4[\Lambda(\omega)]^{2D_f+4}\langle(\hat{e}_\lambda)^2\rangle_\omega. \tag{385}$$

Alexander *et al.* (1993) assumed that, to leading order

$$[\langle(e_\lambda)^2\rangle_\omega]^{1/2}\sim\frac{u(\omega)}{[\Lambda(\omega)]^\sigma}, \tag{386}$$

where

$$u(\omega) = \left[\left\langle\frac{1}{N_\lambda}\sum_i^{\nu_\lambda}|\mathbf{u}_i^\lambda|^2\right\rangle_\omega\right]^{1/2}, \tag{387}$$

and N_λ is the number of sites contained in the region ν_λ of the vibrations [i.e., where $\langle N_\lambda\rangle_\omega = \Lambda^{D_f}$]. The new exponent σ characterizes an effective length relevant to an average strain. Due to the normalization condition, $\sum_i|\mathbf{u}_i^\lambda|^2 = 1$, the magnitude of $u(\omega)$ is proportional to $[\Lambda(\omega)]^{-D_f/2}$, and from the dispersion relation, Eq. (354), $\Lambda\sim\omega^{-D_s/D_f}$. Therefore,

$$S(q,\omega)\sim q^4\omega^{(D_s/D_f)(2\sigma-4)-1},\quad q\Lambda(\omega)\ll 1. \tag{388}$$

Thus, the exponents y and a are given by

$$y = 2\sigma - \frac{D_f}{D_s} = 2\sigma - \frac{1}{2}D_w, \tag{389}$$

$$a = 4 + \frac{D_f}{D_s} - 2\sigma = 4 + \frac{1}{2}D_w - 2\sigma. \tag{390}$$

A similar analysis can be carried out for the limit $q\Lambda\gg 1$. For this case, Alexander *et al.* (1993) obtained the following results,

$$S(q,\omega)\sim\omega^{D_s-1}q^{2-D_f+x}\sim\mathcal{N}(\omega)q^{2-D_f+x},\quad q\Lambda(\omega)\gg 1. \tag{391}$$

The exponents a' and x are now given by

$$a' = \frac{D_f}{D_s}(D_s - 1) = \frac{1}{2}D_w(D_s - 1). \tag{392}$$

and

$$x = 2(\sigma - 1). \tag{393}$$

To summarize, the scaling analysis predicts, by introducing the averaged strain exponent σ, that the DSF behaves as

$$S(q, \omega) \sim \begin{cases} q^4\omega^{(D_s/D_f)(2\sigma-4)-1}, & q\Lambda(\omega) \ll 1, \\ q^{2\sigma-D_f}\omega^{D_s-1}, & q\Lambda(\omega) \gg 1. \end{cases} \tag{394}$$

6.7.3 *Numerical Computation*

The dynamical structure factor $S(q, \omega)$ has been computed by computer simulations (see, for example, Nakayama *et al.*, 1994, for a review). We discuss only those computations in which large networks were utilized, since the early simulations (Montagna *et al.*, 1990; Pilla *et al.*, 1992; Mazzacurati *et al.*, 1992) used small 2D or 3D percolation networks.

Nakayama and Yakubo (1992) carried out numerical simulations to calculate $S(q, \omega)$ using 500×500 square networks at their site percolation threshold, $p_c \simeq 0.593$, with periodic boundary conditions. The DSF was calculated by the same numerical technique employed in their work on calculating the DOS described earlier. They excited several modes simultaneously with frequencies close to a fixed frequency ω, thereby decreasing slightly the monochromaticity of the excited modes. Their algorithm then automatically performed the average $\langle\cdot\rangle_{\lambda\simeq\omega}$ in Eq. (375). Using the mode-mixed displacement patterns $\{v_i^\omega\}$, which were normalized by the condition, $\sum_i (v_i^\omega)^2 = 1$, it is not difficult to see that $S(q, \omega)$ is given by

$$S(q, \omega) = \mathcal{N}(\omega)\left\langle \sum_{ij} (\mathbf{q}\cdot\mathbf{v}_i^\omega)(\mathbf{q}\cdot\mathbf{v}_j^\omega)e^{-i\mathbf{q}\cdot(\mathbf{R}_i-\mathbf{R}_j)} \right\rangle, \tag{395}$$

where $\langle\cdot\rangle$ indicates an average over realizations of the network. The scalar approximation was used, whence, $\mathbf{q}\cdot\mathbf{v}_i^\omega = qv_i^\omega$. The Fourier transform of the correlation function $\langle v_i^\omega v_j^\omega\rangle$ was calculated by assuming a spherically symmetric correlation function, $C^\omega(\mathbf{r}_i - \mathbf{r}_j) \equiv \langle v_i^\omega v_j^\omega\rangle$, which leads to

$$\sum_{ij} \langle v_i^\omega v_j^\omega\rangle \exp[-\mathbf{q}\cdot(\mathbf{R}_i - \mathbf{R}_j)] \sim \sum_{R_{ij}} R_{ij}C^\omega(R_{ij})J_0(qR_{ij}), \tag{396}$$

where $R_{ij} = |\mathbf{r}_i - \mathbf{r}_j|$, and $J_0(x)$ is the zeroth-order Bessel function of the first kind.

The results are shown in Figures 6.32 and 6.33, where the former presents the results for the q-dependence of $S(q, \omega)$ for 50 frequencies in the range,

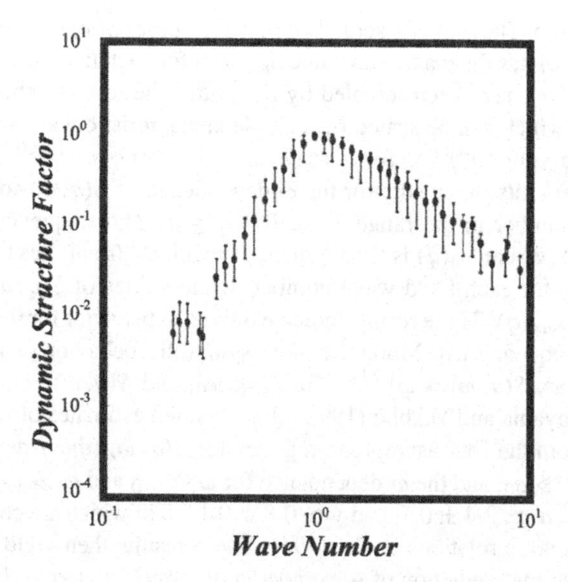

FIGURE 6.32. The dynamic structure factor $S(q, \omega)$, for 50 different frequencies ω, versus the wave number q (normalized by q_0). The results are for site percolation in a square network. The data have been averaged over a narrow range of q close to the reduced q/q_0 (after Nakayama and Yakubo, 1992).

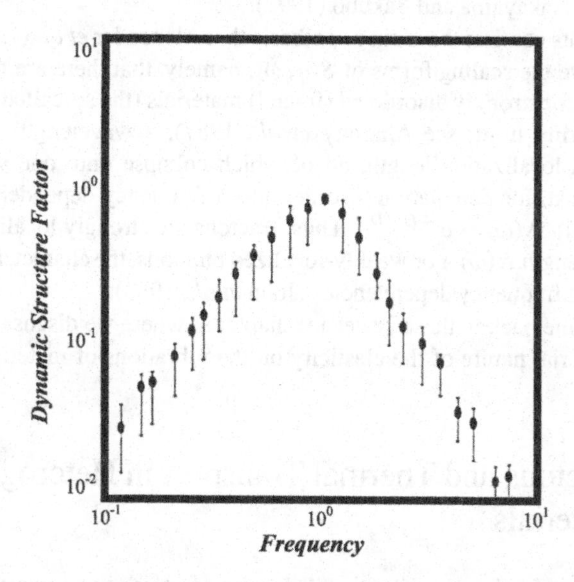

FIGURE 6.33. The dynamic structure factor $S(q, \omega)$ versus the reduced frequency ω/ω_0 for 125 different wave numbers in the range $2\pi/250 \leq q \leq \pi$. The results are for site percolation in the square network. The data have been averaged over a narrow frequency close to ω/ω_0 (after Nakayama and Yakubo, 1992).

$0.005 < \omega < 0.5$. The q-axis is actually $q/q_0(\omega)$, where $q_0(\omega)$ is the wave number at which $S(q,\omega)$ has the maximum value $S_{max}(\omega)$ for each fixed frequency, while values of $S(q,\omega)$ have been rescaled by $S_{max}(\omega)$. The results exhibit a $S(q,\omega)$ for various ω which can be scaled by a single characteristic wave number q_0. In particular, $S(q,\omega) \sim q^{4.0\pm0.1}$ for $q \ll q_0$, and, $S(q,\omega) \sim q^{-1.6\pm0.1}$ for $q \gg q_0$. Figure 6.33 presents the results for the ω-dependence of $S(q,\omega)$ for 125 values of the wave number in the range $2\pi/250 \leq q \leq \pi$. The frequency axis represents $\omega/\omega_0(q)$, where $\omega_0(q)$ is the frequency at which $S(q,\omega)$ has the maximum value $S_{max}(q)$ for each fixed wave number, while values of $S(q,\omega)$ have been re-scaled by $S_{max}(q)$. These results indicate universal behavior for the DSF scaled by a single frequency ω_0. Moreover, the asymptotic behavior of the DSF can be expressed as, $S(q,\omega) \sim \omega^{1.7\pm0.1}$ for $\omega \ll \omega_0$, and $S(q,\omega) \sim \omega^{-2.25\pm0.1}$ for $\omega \gg \omega_0$. Nakayama and Yakubo (1992) also obtained estimates of the exponents a, a' and y from the four asymptotic regimes for $S(q,\omega)$ (the q-dependence for $q \ll q_0$ and $q \gg q_0$, and the ω-dependence for $\omega \ll \omega_0$ and $\omega \gg \omega_0$) and found, $a \simeq 3.2 \pm 0.1$, $a' \simeq 2.4 \pm 0.1$, and $y \simeq 0.8 \pm 0.1$, all of which are consistent with the four asymptotic relations in Eq. (394). These results then yield, $\sigma \simeq 1.1$, in agreement with the prediction of Alexander *et al.* (1993) that, $\sigma > 1$.

Stoll *et al.* (1992) computed $S(q,\sigma)$ for d-dimensional simple-cubic networks of size 68×68 and $21 \times 21 \times 21$ in bond percolation, and employed the standard diagonalization technique. Their estimates were, $a(d=2) \simeq 3.32$, $a(d=3) \simeq 3.65$, $\sigma(d=2) \simeq 1.05$, and $\sigma(d=3) \simeq 1.11$. The 2D results are all consistent with those of Nakayama and Yakubo (1992).

These results confirm the main hypothesis that Alexander *et al.* (1993) made in order to derive the scaling forms of $S(q,\omega)$, namely, that there are three distinct length scales for strongly disordered (fractal) materials (the so-called Ioffe–Regel strong scattering limit; see Aharony *et al.*, 1987), a wavelength, a scattering length, and a localization length, all of which collapse onto one single length scale $\Lambda(\omega)$ that, for percolation disorder, has a frequency dependence given by [see Eq. (354)], $\Lambda(\omega) \sim \omega^{-D_s/D_f}$. Thus, fractons are strongly localized with the localization length $\Lambda(\omega)$. For weakly-localized phonons, the characteristic lengths have different frequency dependencies (John *et al.*, 1983).

We will come back to these issues in Chapter 9, where we discuss the effect of the true vectorial nature of the elasticity on the vibrations of materials and their DOS.

6.8 Fractons and Thermal Transport in Heterogeneous Materials

The existence of a fracton vibrational density of states has important implications for thermal transport in disordered materials which exhibit such a DOS. We now consider phonon-assisted fracton hopping and the effect of vibrational anharmonicity, and describe the characteristic hopping distance and the contribution to the thermal conductivity g.

6.8.1 Anharmonicity

The thermal conductivity g_e at temperature T is given by

$$g_e = \frac{1}{\Omega} C_{\alpha'}(T) D_{\alpha'}(T), \tag{397}$$

where, as usual, Ω is the volume of the system, $C_{\alpha'}$ is the specific heat and $D_{\alpha'}$ the diffusivity associated with the mode α'. However, in the absence of diffusion, g_e vanishes, implying that whenever the condition for localization in the sense of Anderson localization occurs (Anderson, 1958), thermal transport ceases, which means that thermal conductivity approaches zero where the mean-free path is of the order of, or less than, the wavelength for vibrational excitations. Transport in materials with a morphology that permits a crossover from phonon to fracton vibrational excitations is an example of this phenomenon since, as discussed above, fractons are strongly localized, and therefore thermal transport can occur only by the (extended) phonon normal modes.

Therefore, it is imperative to calculate g_e under these conditions. Clearly, g_{ph}, the contribution of phonons to the thermal conductivity, will increase with increasing temperature. The reason for this increase is that both the occupied mode density and the Bose factor increase as the temperature T increases. The increase continues until one exhausts all of the extended phonon states, at which time g_{ph} saturates in the Dulong–Petit regime ($k_B T \gg \hbar\omega_{\text{ph}}$). However, experiments (see below) have shown that this is not the case for aerogels, as its thermal conductivity continues to increase above the value where g_{ph} saturates. This apparent contradiction can be explained by introducing anharmonicity into the phonon-fracton excitation spectrum, which makes it possible for the fractons to contribute to thermal transport. The introduction of anharmonicity is essential for thermal transport in the fracton regime (Alexander *et al.*, 1986a,b). It allows for fracton hopping in much the same sense as the phonon-assisted electronic hopping of Mott (1969) for localized electronic states, which was described earlier in this chapter. While in ordered materials anharmonicity *reduces* thermal transport, in disordered materials thermal transport is a result of anharmonicity.

6.8.2 Phonon-Assisted Fracton Hopping

The schematics of the introduction of vibrational anharmonicity and its result are shown in Figure 6.34 (Alexander *et al.*, 1986a,b; Jagannathan *et al.*, 1989). We first introduce the corresponding Hamiltonian,

$$\mathcal{H} = C_e \sum_{\alpha,\alpha',\alpha''} (A_{\alpha,\alpha',\alpha''} b^\dagger_{\alpha'} b_{\alpha''} b_\alpha + \cdots), \tag{398}$$

where the b_α ($b^\dagger_\alpha$) operators annihilate (create) phonons or fractons, depending on whether the index α refers to modes with frequencies less than or greater than the crossover frequency ω_{co}. Because fractons are strongly localized, the two fractons in Figure 6.34 are, in general, located at different spatial positions. To compute

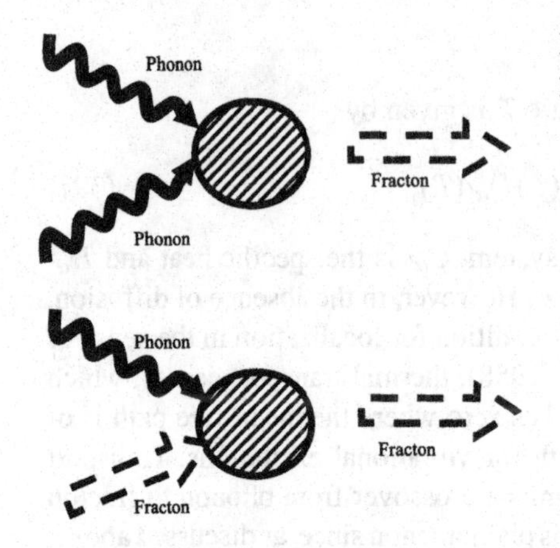

FIGURE 6.34. Schematics of anharmonic process (after Nakayama *et al.*, 1994).

their distance from each other, the concept of the most probable hopping distance, due to Mott (1969), is invoked. A region of volume ξ^{D_f} contains $\mathcal{N}_{\rm fr}(\omega'_\alpha)\Delta\omega'_\alpha$ fractons with an energy in the interval $[\omega'_\alpha, \omega_{\alpha'} + \Delta\omega_{\alpha'}]$. The differential probability of finding such fractons in a volume element $r^{D_f-1}dr$ (assuming a uniform distribution of the fractons in the volume) is then

$$dP(r,\omega_{\alpha'}) = \frac{1}{\xi^{D_f}}\mathcal{N}_{\rm fr}(\omega_{\alpha'})\Delta\omega_{\alpha'} r^{D_f-1}dr. \tag{399}$$

Consider the lower left fracton in Figure 6.34 (index α'') and assume that it lies at the origin. Then, to calculate the most probable hopping distance, we integrate Eq. (399) up to a distance $R(\alpha')$ that would give us a second fracton within the same volume with probability 1. Integrating up to the maximum distance ξ would yield $\mathcal{N}_{\rm fr}(\omega_{\alpha'})\Delta\omega_{\alpha'}$, i.e., the total number of fractons inside the volume ξ^{D_f}. Therefore,

$$R(\alpha') = \Lambda(\omega_{\alpha'})\left(\frac{\omega_{\alpha'}}{\omega_{co}}\right)^{1/D_f}, \tag{400}$$

where the uncertainty $\Delta\omega_{\alpha'}$ has been assumed to be ω_{co} (Alexander *et al.*, 1983). Note that $R_{\alpha'} > \Lambda(\omega_{\alpha'})$, so that in fact the fracton hops a significant distance relative to its localization length scale. The diffusion constant associated with the hopping of the α' fracton is then given by

$$D_{\alpha'} = \frac{R^2(\omega_{\alpha'})}{t_{\rm fr}(\omega_{\alpha'}, T)}, \tag{401}$$

where $t_{\rm fr}(\omega_{\alpha'}, T)$ is the lifetime of the fracton of energy $\omega_{\alpha'}$ at temperature T associated with its hopping a distance $R_{\alpha'}$.

These results can now be used for calculating the thermal transport associated with fracton hopping. At temperatures greater than the crossover energy, substitution of Eq. (401) into Eq. (397) yields $g_{\rm hop}$ (Jagannathan *et al.*, 1989):

$$g_{\rm hop}(T) = \frac{2^4 c\pi^3 C_e^2 D_s^2 \omega_D}{8\rho^3 v_s^2 \xi^8 \omega_{co}^3} k_B^2 T, \tag{402}$$

where all the notations are the same as before. Here, c is an adjustable parameter, ω_D is the Debye frequency associated with the phonon velocity of sound v_s, and ρ

is the mass density. An important feature of Eq. (402) is the absence of any dependence upon the fracton DOS for $\omega > \omega_{co}$, which is due to the fact that the dispersion relation for fractons [Eq. (358)] leads to a rapid spatial diminution of the fracton size with increasing fracton energy. Therefore, the fracton overlap associated with the lower vertex in Figure 6.34 deceases so rapidly with increasing fracton energy that the principal contribution to $g_{\rm hop}$ arises from fractons with $\omega \simeq \omega_{co}$. Hence, only the fractons with the lowest energies contribute to the thermal conductivity via phonon-assisted fracton hopping. This may explain the universal form found for thermal transport in amorphous materials (see also Section 6.3.1) above the plateau temperature, in which the thermal conductivity appears to increase linearly with temperature *independent* of the precise nature of the DOS.

The lower vertex in Figure 6.34 not only determines the fracton hopping rate, but also the phonon lifetime associated with fracton hopping. Because the same vertex is involved for both processes, one can express $g_{\rm hop}$ in terms of the inelastic lifetime $t_{\rm ph}(\omega_{\rm ph}, T)$ for a phonon of frequency $\omega_{\rm ph}$ at temperature T. In this case, there are essentially no adjustable parameters, and one finds (Jagannathan *et al.*, 1989) that

$$g_{\rm hop} = \frac{2^{D_f} a\gamma I\omega_D}{8D_f\Gamma(D_f/\gamma)} \frac{k_B\omega_{co}^2}{\xi} \frac{1}{\omega_{\rm ph}^2 t_{\rm ph}(\omega_{\rm ph}, T)}, \tag{403}$$

where $a = 5 - D_s - 4D_s/D_f$, I is an integral of the order of unity, and γ is the exponent for the wave function $\Psi_{\rm fr}$, Eq. (364).

6.8.3 *Dependence of Sound Velocity on Temperature*

The above results allow one to calculate the change δv_s in the velocity of sound caused by the lower vertex of Figure 6.34, in terms of the fracton hopping contribution to the thermal conductivity. Jagannathan and Orbach (1990) found that

$$\frac{\delta v_s}{v_s} = -0.1\frac{\xi^2}{2\pi^2 v_s}\frac{g_{\rm hop}(T)}{T}\frac{T}{k_B}. \tag{404}$$

The term $g_{\rm hop}(T)/T$ is independent of temperature for phonon-assisted fracton hopping, so that Eq. (404) is linear in the temperature but independent of the frequency of the sound wave. Equation (404) differs significantly from that derived by the so-called two-level models in amorphous materials (see, for example, Tielbürger *et al.*, 1992):

$$\frac{\delta v_s}{v_s} = \frac{ck_BT}{E_0}\ln(\omega t_0), \tag{405}$$

where c is a constant, E_0 is the ground-state energy of the tunneling particle, ω is the sound wave frequency, and t_0 is a few times the vibrational frequency of the tunneling particle in a single well.

The specific heat and thermal conductivity of aerogels were first measured by Calemczuk *et al.* (1987) and de Goer *et al.* (1989). These measurements were carried out over a restricted temperature range that did not cover the phonon-fracton crossover energy. The specific heat and thermal conductivity measurements by Sleator *et al.* (1991), Posselt *et al.* (1991), and Bernasconi *et al.* (1992) were

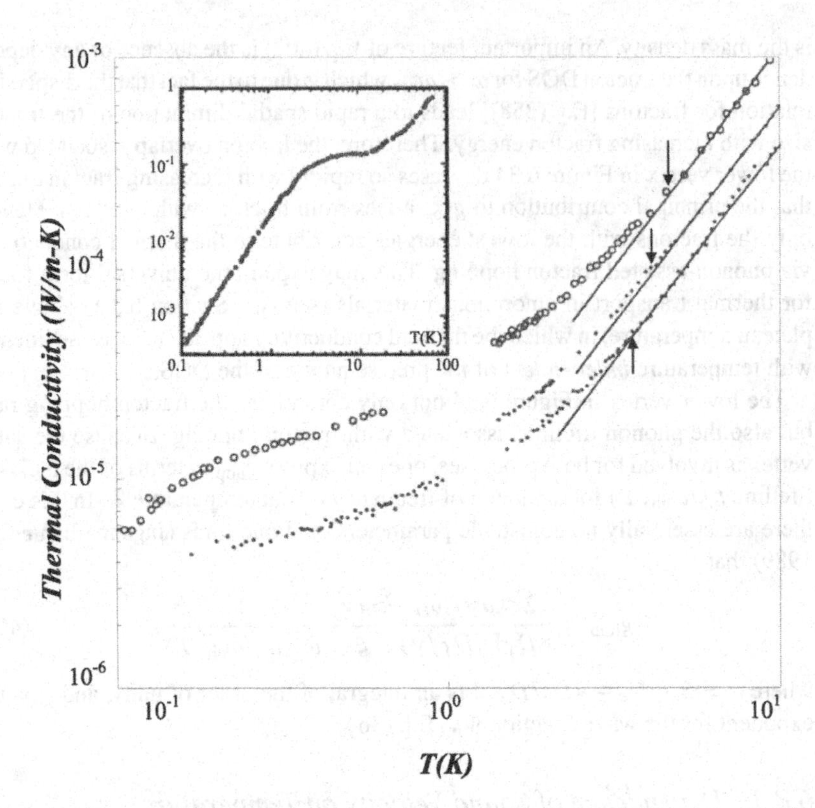

FIGURE 6.35. Temperature dependence of the thermal conductivity of silica aerogels. The symbols from the top to bottom indicate the data for samples with density 0.275, 0.190, and 0.145, all in gr/cm^3 (after Bernasconi *et al.*, 1992). The inset shows the data for vitreous silica.

carried out on the same materials, but over a temperature range that included the phonon-fracton crossover energy.

Figure 6.35 presents the experimental data of Bernasconi *et al.* (1992) for the thermal conductivity of two aerogels of density $\rho = 0.145$ g/cm^3 (low density, LD) and $\rho = 0.275$ g/cm^3 (high density, HD). These are the first measurements ever to explore the thermal conductivity of aerogels over such a temperature range. For the HD sample, $T_c = \hbar\omega_{co}/k_B = 0.37$ K, and a strong break in slope occurs around $T = 0.13$ K, which is shown in Figure 6.36. This can be interpreted as evidence of a crossover from phonon-dominated to fracton-dominated thermal transport. The sharp increases in g_e above about 3 K is attributed to particle modes, caused by the approximately 20 Å spheres that make up the fractal network in the aerogels. For the LD sample, $T_c = 0.1$ K, and the crossover from phonon- to fracton-dominated transport was not possible to measure.

The phonon-fracton model predicts the following form for the thermal conductivity:

$$g_e = g_{\text{ph}} + g_{\text{hop}}, \tag{406}$$

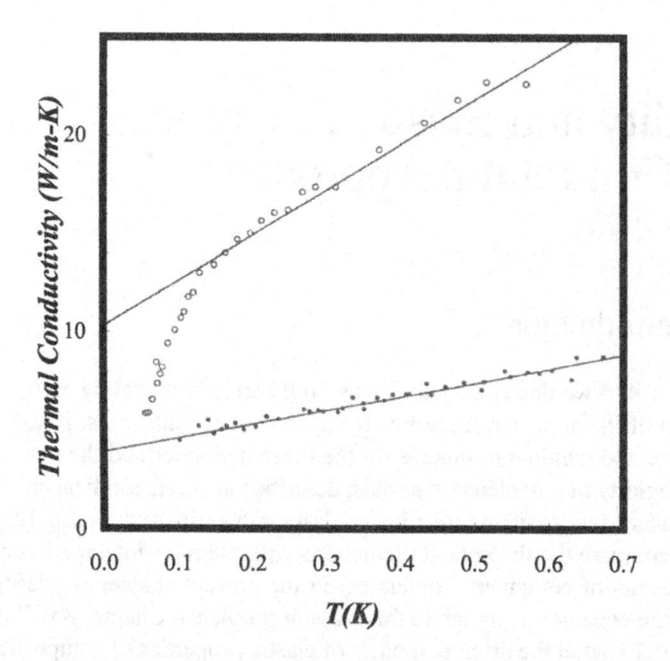

FIGURE 6.36. Thermal conductivity of silica aerogels in a narrow temperature range. The top and bottom data are for samples with density 0.275 and 0.145, both in gr/cm^3. The straight lines represent the fits of the data to Eq. (407) (after Bernasconi *et al.*, 1992).

which can be written, for $T > T_c$, as

$$g_e = A + BT. \tag{407}$$

Posselt *et al.* (1991) found that A scales with ξ in a consistent manner for both the HD and LD samples, the limiting phonon mean-free path being of the order of ξ. Thus, the predictions for thermal transport based on phonon-assisted fracton hopping appear to be quantitatively verified in the aerogels.

Summary

This chapter demonstrates that, effective-medium approximation, percolation and fractal concepts, and scaling analysis are powerful tools for accurately predicting many dynamical, frequency-dependent, properties of a wide variety of disordered materials. These predictions have been confirmed by large-scale computer simulations, as well as careful experimental measurements. Thus, we now have a much better understanding of many frequency-dependent properties of heterogeneous materials. In particular, the apparent universality of AC conductivity of many disordered materials, and their vibrational density of states are now well-understood.

7
Rigidity and Elastic Properties: The Continuum Approach

7.0 Introduction

In Chapters 4–6 we described and discussed theoretical modeling and numerical estimation of linear scalar properties of heterogeneous materials, based on both the discrete and continuum models. In the present chapter and the next two, we consider rigidity of disordered materials, describe various theoretical and numerical approaches for predicting their linear elastic properties and, as usual and when possible, compare the theoretical predictions with the relevant experimental data and the results of computer simulations. In the present chapter we describe the continuum mechanics approach to this class of problems. Chapter 8 will describe and discuss in detail the discrete models of elastic properties of composite solids, and various theoretical and computational approaches for analyzing such problems. Chapter 9 will demonstrate the validity of such models by comparing their predictions with the experimental data for the elastic properties of several important classes of materials. In Volume II, we will describe and discuss the same types of models and approaches for *nonlinear* transport and mechanical properties, and will, in particular, consider both the constitutive and threshold nonlinearities that were described in Chapter 1. The latter type of nonlinearities describe mechanical fracture, and electrical and dielectric breakdown of materials, an important class of phenomena of great industrial interest. In addition, in Volume II we will describe atomistic modeling of materials' properties.

It is of course always desirable to develop predictive theories for the mechanical properties of materials by approaching the problem in its most general form. However, although such an approach will be completely rigorous, it will also have little predictive power, since the theoretical framework worked out for the general problem will be too complex and contain too many parameters (some of which may be too difficult, or even impossible, to measure) to be useful. Therefore, one must set a less ambitious goal, one in which one does make certain simplifying assumptions and hypotheses which, of course, must be realistic from a physical view point. In this vein, linear elasticity theory has provided important contributions to our understanding of mechanical properties of a wide variety of materials by providing a realistic theoretical framework for analyzing their mechanical behavior. Moreover, when applied to many practical problems, the predictions of the theory take on relatively simple forms that can be tested and used readily. Linear elasticity theory does not of course solve all the important problems on mechanical properties of materials, as many materials do in fact behave inelastically, the description of which is well beyond the realm of the linear theory.

We begin this chapter by summarizing some elementary but important elements of continuum mechanics, which will be useful to our discussions in this chapter, after which various continuum models of linear elastic properties of heterogeneous materials will be described and discussed. For an in-depth discussion of elements of continuum mechanics of the mechanical properties the reader can consult, for example, Timoshenko and Goodier (1970).

Let us fix our notations for this chapter and in Chapters 8 and 9. Second-order tensors are denoted by either bold face capital letters or bold face Greek letters, while in most cases fourth-order tensors are denoted by bold face letters. We consider mainly two-phase materials that consist of a matrix (phase 1, volume fraction ϕ_1) and inclusions (phase 2, volume fraction ϕ_2) that are distributed in the matrix. The properties of the phases are given subscripts 1 and 2. We use ν_p to denote the Poisson's ratio. When we refer to the Poisson's ratio of a phase, we use a double subscript, e.g., ν_{p1} for phase 1, and ν_{pe} for the effective Poisson's ratio of the material.

7.1 The Stress and Strain Tensors

In our discussions, we utilize Cartesian coordinates with the usual tensor notation that involves summation over repeated indices. In its most general form, the linear elastic stress-strain relationship is expressed by

$$\boldsymbol{\sigma}(\mathbf{x}) = \mathbf{C}(\mathbf{x}) : \boldsymbol{\epsilon}(\mathbf{x}), \tag{1}$$

where : denotes the contraction with respect to two indices, so that in indexed notations, $\sigma_{ij} = C_{ijkl}\epsilon_{kl}$, with $j = 1, 2$ and 3, where $\boldsymbol{\sigma}$ and $\boldsymbol{\epsilon}$ are, respectively, the linear stress and strain tensors and $\mathbf{C}$ is the fourth-order tensor of elastic constants of the material. Equation (1) is the generalization of Hooke's law. The stress-strain relation can also be expressed in inverted form, $\boldsymbol{\epsilon}(\mathbf{x}) = \mathbf{S}(\mathbf{x}) : \boldsymbol{\sigma}(\mathbf{x})$, where $\mathbf{S}$ is the *compliance* tensor. The stiffness and compliance tensors are related by, $\mathbf{S} : \mathbf{C} = \mathbf{I}$, where $\mathbf{I}$ is the fourth-order identity tensor.

The fact that $\mathbf{C}$ and $\mathbf{S}$ are fourth-order tensors can be proven by showing that their d^4 components in d dimensions, relative to a coordinate system, transform to d^4 components in another "primed" coordinate system according to the transformation rule for fourth-order tensors, e.g., $C'_{ijkl} = l_{im}l_{jn}l_{kp}l_{lq}C_{mnpq}$, where l_{ij} are the direction cosines. Thus, in general, the tensor $\mathbf{C}$ has 81 independent components. Since the stress and strain tensors are symmetric (that is, $\sigma_{ij} = \sigma_{ji}$ and $\epsilon_{ij} = \epsilon_{ji}$), we must have $C_{ijkl} = C_{jikl}$ and $C_{ijkl} = C_{ijlk}$. These two conditions reduce the 81 independent components of C_{ijkl} to just 36 for three-dimensional (3D) elasticity. More generally, in d dimensions, the d^4 components are reduced to $[d(d+1)/2]^2$ independent components. Moreover, the stress can be expressed as a derivative of the *strain energy density function* $\mathcal{H}_s$: $\sigma_{ij} = \partial\mathcal{H}_s/\partial\epsilon_{ij}$. Therefore, it immediately follows from the Hooke's law that $\mathcal{H}_s$, up to an additive constant, is given by $\mathcal{H}_s = \frac{1}{2}\sigma_{ij}\epsilon_{ij} = \frac{1}{2}\epsilon_{kl}C_{ijkl}\epsilon_{ij} \geq 0$, and hence we have the additional symmetry that $C_{ijkl} = C_{klij}$, which, in d dimensions, further reduces the number of independent elastic constants to $\frac{1}{8}d(d+1)(d^2+d+2)$. In particular, for 3D systems, the number of independent elastic constants is reduced to 21. Note that the compliance tensor **S** possesses the same symmetries as the stiffness tensor **C**, and that $\mathcal{H}_s \geq 0$ implies that **C** and **S** are both positive definite.

If **u** is the displacement vector, then

$$\epsilon = \frac{1}{2}\left[\nabla \mathbf{u} + (\nabla \mathbf{u})^{\mathrm{T}}\right], \tag{2}$$

where T denotes the transpose operation, so that, for example,

$$\epsilon_{ij} = \frac{1}{2}\left(\frac{\partial u_i}{\partial x_j} + \frac{\partial u_j}{\partial x_i}\right),$$

with $\mathbf{x} = (x_1, x_2, x_3)$ representing the coordinates. Since the six independent components of ϵ are derived from three independent displacements, some of the strain components cannot be independent. Instead, the strain components are related through 81 so-called *compatibility equations*, according to which

$$\frac{\partial^2 \epsilon_{ij}}{\partial x_k \partial x_l} + \frac{\partial^2 \epsilon_{kl}}{\partial x_i \partial x_j} = \frac{\partial^2 \epsilon_{ik}}{\partial x_j \partial x_l} + \frac{\partial^2 \epsilon_{jl}}{\partial x_i \partial x_k}. \tag{3}$$

These 81 equations are not entirely independent of each other. One may also write the compatibility equations in terms of the stresses, using the stress-strain relations. Finally, the governing equation expressing the balance of momenta in a (linear) material is given by

$$\nabla \cdot \sigma + \mathbf{F}(\mathbf{x}, t) = \rho \frac{\partial^2 \mathbf{u}}{\partial t^2}, \tag{4}$$

where ρ is the mass density and $\mathbf{F}(\mathbf{x}, t)$ is the body force.

7.1.1 *Symmetry Properties of the Stiffness Tensor*

As described above, for 3D materials the number of independent components of **C** is 21. Any additional reduction in the number of independent components of **C** must come through the constraints imposed by the material's symmetry properties. To see how further reduction in the number of independent components of **C** can be achieved, we restrict ourselves to 3D. The generalized Hooke's law, Eq. (1), really represents only six independent equations with 21 elastic constants, which can be conveniently represented as a matrix equation that expresses a six-column vector of stresses in terms of a six-element column vector of strain. Thus, we write

$$\begin{bmatrix} \sigma_{11} \\ \sigma_{22} \\ \sigma_{33} \\ \sigma_{23} \\ \sigma_{13} \\ \sigma_{12} \end{bmatrix} = \begin{bmatrix} C_{11} & C_{12} & C_{13} & C_{14} & C_{15} & C_{16} \\ & C_{22} & C_{23} & C_{24} & C_{25} & C_{26} \\ & & C_{33} & C_{34} & C_{35} & C_{36} \\ & & & C_{44} & C_{45} & C_{46} \\ & & & & C_{55} & C_{56} \\ & & & & & C_{66} \end{bmatrix} \begin{bmatrix} \epsilon_{11} \\ \epsilon_{22} \\ \epsilon_{33} \\ \epsilon_{23} \\ \epsilon_{13} \\ \epsilon_{12} \end{bmatrix} \tag{5}$$

where the matrix **C** is symmetrical. For symmetry with respect to a plane, such that the coordinate x_3 is normal to the plane of symmetry, **C** has 13 independent components with the following form

$$\mathbf{C} = \begin{bmatrix} C_{11} & C_{12} & C_{13} & 0 & 0 & C_{16} \\ & C_{22} & C_{23} & 0 & 0 & C_{26} \\ & & C_{33} & 0 & 0 & C_{36} \\ & & & C_{44} & C_{45} & 0 \\ & & & & C_{55} & 0 \\ & & & & & C_{66} \end{bmatrix}. \tag{6}$$

This is referred to as *monoclinic symmetry*. As in the general anisotropic case, here a pure shear strain can give rise to a normal stress.

We may also consider another class of symmetry properties called *orthotropic symmetry*, which consists of symmetry with respect to three mutually orthogonal planes. Orthotropy reduces the number of independent components of $\mathbf{C}$ to 9, such that

$$\mathbf{C} = \begin{bmatrix} C_{11} & C_{12} & C_{13} & 0 & 0 & 0 \\ & C_{22} & C_{23} & 0 & 0 & 0 \\ & & C_{33} & 0 & 0 & 0 \\ & & & C_{44} & 0 & 0 \\ & & & & C_{55} & 0 \\ & & & & & C_{66} \end{bmatrix} \tag{7}$$

For symmetry with respect to a 90° rotation about one axis, say the x_1-axis, it can be shown that $\mathbf{C}$ has six independent components:

$$\mathbf{C} = \begin{bmatrix} C_{11} & C_{12} & C_{12} & 0 & 0 & 0 \\ & C_{22} & C_{23} & 0 & 0 & 0 \\ & & C_{33} & 0 & 0 & 0 \\ & & & C_{44} & 0 & 0 \\ & & & & C_{66} & 0 \\ & & & & & C_{66} \end{bmatrix}. \tag{8}$$

This is referred to as *transverse square symmetry*. For example, a tetragonal crystal has such symmetry.

Further reduction in the number of independent components of $\mathbf{C}$ can be obtained by considering transversely isotropic materials, for which one of the planes for the orthotropic case is a plane of isotropy. If, for example, x_1 is normal to the plane of isotropy, then $\mathbf{C}$ will have only 5 independent components, such that

$$\mathbf{C} = \begin{bmatrix} C_{11} & C_{12} & C_{12} & 0 & 0 & 0 \\ & C_{22} & C_{23} & 0 & 0 & 0 \\ & & C_{22} & 0 & 0 & 0 \\ & & & \frac{1}{2}(C_{22} - C_{23}) & 0 & 0 \\ & & & & C_{66} & 0 \\ & & & & & C_{66} \end{bmatrix}. \tag{9}$$

This is referred to as *transversely isotropic symmetry*. For example, a crystal with a six-fold rotational symmetry axis (hexagonal) is transversely isotropic with respect to the stiffness. Although a crystal with a three-fold rotational symmetry axis (trigonal) is *not* transversely isotropic, there are only two independent elastic moduli in the transverse plane, and therefore 2D crystals with three-fold rotational symmetry axis are elastically isotropic. For materials with transversely isotropic symmetry, there are many interrelation between the various elastic moduli and the Poisson's ratio (see, for example, Christensen, 1979, and Torquato, 2002). For example, one has, $\mu^{\mathrm{T}} = \frac{1}{2}Y^{\mathrm{T}}(1+\nu_p^{\mathrm{T}})^{-1}$, and $4/Y^{\mathrm{T}} = 1/K^{\mathrm{T}} + 1/\mu^{\mathrm{T}} + 4(\nu_p^{\mathrm{L}})^2/Y^{\mathrm{L}}$, where μ, K and Y are, respectively, the shear, bulk and Young's moduli, and superscripts L and T denote the longitudinal and transverse directions. Thus, one obtains, $\nu_p^{\mathrm{T}} = [K^{\mathrm{T}} - \mu^{\mathrm{T}} - 4K^{\mathrm{T}}\mu^{\mathrm{T}}(\nu_p^{\mathrm{L}})^2/Y^{\mathrm{L}}][K^{\mathrm{T}} + \mu^{\mathrm{T}} + 4K^{\mathrm{T}}\mu^{\mathrm{T}}(\nu_p^{\mathrm{L}})^2/Y^{\mathrm{L}}]^{-1}$. Materials with this type of symmetry can, therefore, have a transverse Poisson's ratio, $-1 \le \nu_p^{\mathrm{T}} \le 1$. The lower limit of -1 corresponds to the case when $\mu^{\mathrm{T}}/K^{\mathrm{T}} \to \infty$ and $Y^{\mathrm{L}}/\mu^{\mathrm{T}} \to \infty$, while the upper limit of $+1$ corresponds to the case in which $K^{\mathrm{T}}/\mu^{\mathrm{T}} \to \infty$ and $Y^{\mathrm{L}}/K^{\mathrm{L}} \to \infty$.

For symmetry with respect to 90° rotations about two perpendicular axes, say the x_1-axis and x_2-axis, it can be shown that $\mathbf{C}$ has three independent components:

$$\mathbf{C} = \begin{bmatrix} C_{11} & C_{12} & C_{12} & 0 & 0 & 0 \\ & C_{11} & C_{12} & 0 & 0 & 0 \\ & & C_{11} & 0 & 0 & 0 \\ & & & C_{44} & 0 & 0 \\ & & & & C_{44} & 0 \\ & & & & & C_{44} \end{bmatrix}. \tag{10}$$

This is called *cubic symmetry*.

Finally, for a completely isotropic material, $\mathbf{C}$ has only two independent components, such that

$$\mathbf{C} = \begin{bmatrix} C_{11} & C_{12} & C_{12} & 0 & 0 & 0 \\ & C_{11} & C_{12} & 0 & 0 & 0 \\ & & C_{11} & 0 & 0 & 0 \\ & & & \frac{1}{2}(C_{11}-C_{12}) & 0 & 0 \\ & & & & \frac{1}{2}(C_{11}-C_{12}) & 0 \\ & & & & & \frac{1}{2}(C_{11}-C_{12}) \end{bmatrix} \tag{11}$$

We emphasize that since $\mathbf{C}$ is symmetric, we do not specify its lower-diagonal part in Eqs. (8)–(11).

For completely isotropic materials, the stress-strain relation can be written as

$$\sigma_{ij} = \lambda\epsilon_{kk}\delta_{ij} + 2\mu\epsilon_{ij}, \quad i, j = 1, \cdots, d. \tag{12}$$

where λ is the Lamé elastic constant [in Eq. (12), the shear modulus μ is also often referred to as a Lamé constant), and δ_{ij} is the usual Kronecker delta. The strain-stress relation can also be written as

$$\epsilon_{ij} = -\frac{\nu_p}{Y}\sigma_{kk}\delta_{ij} + \frac{1+\nu_p}{Y}\sigma_{ij}, \quad i, j = 1, \cdots, d. \tag{13}$$

Note that the Poisson's ratio is defined (here) as the negative of the strain in the direction of a uniaxial stress divided by the associated transverse strain.

The stress-strain relations can also be represented in terms of the *deviatoric* and *dilatational* (hydrostatic) components of $\boldsymbol{\sigma}$ and $\boldsymbol{\epsilon}$. Thus, we write

$$s_{ij} = \sigma_{ij} - \frac{1}{d}\sigma_{kk}\delta_{ij}, \tag{14}$$

$$e_{ij} = \epsilon_{ij} - \frac{1}{d}\epsilon_{kk}\delta_{ij}, \tag{15}$$

where s_{ij} and e_{ij} are the deviatoric components of the stress and strain. Therefore,

$$s_{ij} = 2\mu e_{ij}, \quad \text{deviatoric relation,} \tag{16}$$

$$\sigma_{kk} = 3K\epsilon_{kk}, \quad \text{dilatational relation.} \tag{17}$$

We can now see the elastic moduli μ and K as representing volumetric changes in a material as a result of deforming it. Due to isotropy of the material, only two of the three parameters λ, μ and K are independent. In fact,

$$C_{11} = \lambda + 2\mu, \quad C_{12} = \lambda, \tag{18}$$

and more importantly

$$\mu = \frac{Y}{2(1+\nu_p)}, \tag{19}$$

$$K = \lambda + \frac{2}{d}\mu = \frac{Y}{d[1+(1-d)\nu_p]}. \tag{20}$$

The Poisson's ratio ν_p for an isotropic material is given by

$$-1 \le \nu_p = \frac{dK - 2\mu}{d(d-1)K + 2\mu} \le \frac{1}{d-1}. \tag{21}$$

The definition of ν_p for an anisotropic material is, of course, not unique.

The upper bound in (21) is reached when $K/\mu \to \infty$, i.e., the incompressible limit. For example, for 3D materials such as rubbery solids and liquids, $K \gg \mu$, and $\nu_p \simeq 1/2$. The lower bound in (21) is reached when $\mu/K \to \infty$. Elastically isotropic materials with $\nu_p < 0$ are called *auxetic materials*, and are rare in nature. However, as discussed in Section 9.7, such materials, which have unusual properties, have been fabricated.

7.1.2 Theorems of Minimum Energy

In analyzing the mechanical behavior of heterogeneous materials, two minimum energy theorems have proven very useful, especially for deriving rigorous upper and lower bounds to the effective elastic moduli of materials (see Section 7.5). The analogous theorems for deriving rigorous bounds for the effective conductivity were stated in Section 4.6.1.2. Here, we state these theorems without proof. The proofs are found in many books on mechanical properties of materials.

Theorem of minimum potential energy: Consider a static elasticity problem with body forces F_i and the boundary conditions

$$\sigma_i = \sigma_{ij} n_j = f_i \quad \text{on } S_\sigma \tag{22}$$

$$u_i = u \quad \text{on } S_u \tag{23}$$

where S_σ and S_u are complementary portions of the surface of the system with volume Ω, and n_j are the components of the unit outward vector normal to the surface. We define a potential energy functional

$$\mathcal{H}_p = \int_\Omega [\mathcal{H}(\epsilon_{ij}) - F_i u_i] d\Omega - \int_{S_\sigma} f_i u_i dS, \tag{24}$$

where $\mathcal{H}(\epsilon_{ij}) = \frac{1}{2} C_{ijkl} \epsilon_{ij} \epsilon_{kl}$. An admissible displacement field $\mathbf{u}^{(a)}$ is defined as any continuous field that satisfies the displacement boundary condition (23), but is otherwise arbitrary. Then, the theorem of minimum potential energy states that:

Of all the admissible displacement fields, the one that satisfies the equations of equilibrium makes the potential energy functional (24) *an absolute minimum.*

Theorem of minimum complementary energy: Likewise, we define the complementary energy functional by

$$\mathcal{H}_c = \int_\Omega \mathcal{H}_s(\sigma_{ij}) d\Omega - \int_{S_u} \sigma_i u dS, \tag{25}$$

where the strain energy $\mathcal{H}_s$ is expressed in terms of the stresses as

$$\mathcal{H}_s = \frac{1}{2} S_{ijkl} \sigma_{ij} \sigma_{kl}, \tag{26}$$

with S_{ijkl} being the tensor of elastic compliances (the inverse of the elastic constants), defined above. An admissible stress $\sigma_{ij}^{(a)}$ is one that satisfies the equilibrium equations and the stress boundary condition (22), but is otherwise arbitrary. Then, the theorem of minimum complementary energy states that:

Of all the admissible stress fields, the one that satisfies the compatibility equations makes the complementary energy functional (25) *an absolute minimum.*

7.1.3 The Strain Energy of a Composite Material

A very useful result for the analysis of heterogeneous materials is the formula that was derived by Eshelby (1956) for computing the strain energy in materials that contain inhomogeneities. Consider a material that is subjected to surface traction and contains a single inclusion of another material. The elastic strain energy of the heterogeneous material is given by

$$\mathcal{H}_s = \frac{1}{2} \int_\Omega \sigma_{ij} \epsilon_{ij} d\Omega, \tag{27}$$

where Ω is the volume of the material. Eshelby derived a formula that reduces the volume integral in Eq. (27) to a particular type of surface integral. To do this, he defined the strain energy in a material identical with the original one, but with the inclusion replaced by the homogeneous matrix material. The resulting homogeneous material is subjected to the same set of surface tractions as in the heterogeneous material, and its elastic strain energy is given by

$$\mathcal{H}_0 = \frac{1}{2}\int_{\Omega} \sigma_{ij}^0 \epsilon_{ij}^0 d\Omega. \tag{28}$$

In the first step, Eshelby (1956) showed that

$$\mathcal{H}_s = \mathcal{H}_0 + \frac{1}{2}\int_S \sigma_i^0 (u_i - u_i^0) dS, \tag{29}$$

where S represents the surface of the material, $\sigma_i = \sigma_{ii}$, and on the external surface S, $\sigma_i = \sigma_i^0$. He then showed that

$$\mathcal{H}_s = \mathcal{H}_0 + \frac{1}{2}\int_{S_i} (\sigma_i^0 u_i - \sigma_i u_i^0) dS, \tag{30}$$

where S_i is the surface of the inclusion. Eshelby also showed that if, instead of surface traction, one considers a problem that involves specified displacements on the outer boundary of the material, then

$$\mathcal{H}_s = \mathcal{H}_0 + \frac{1}{2}\int_{S_i} (\sigma_i u_i^0 - \sigma_i^0 u_i) dS. \tag{31}$$

The generalization of Eshelby's formulae to a composite with many inclusions is straightforward.

7.1.4 Volume Averaging

In order to estimate the effective mechanical properties of heterogeneous materials, one must assume the existence of a characteristic length scale for the inhomogeneities. For example, in a fibrous composite, this characteristic length is the mean distance between the fibers. For the class of heterogeneous materials that we consider in this chapter, there is a length scale, much larger than the characteristic scale for the inhomogeneities, over which the mechanical properties of a material can be averaged in an unambiguous manner. If this length scale is much smaller than the linear size of the material, then the material is macroscopically homogeneous.

Suppose that Ω is the volume of an element of a heterogeneous material over which its properties are to be averaged. If we impose a macroscopically-homogeneous stress or deformation field on this volume element, we can then define the volume-averaged stress by

$$\langle \sigma_{ij} \rangle = \int_{\Omega} \sigma_{ij} d\Omega, \tag{32}$$

and the volume-averaged strain by

$$\langle \epsilon_{ij} \rangle = \int_{\Omega} \epsilon_{ij} d\Omega, \tag{33}$$

where ϵ is an infinitesimal strain tensor. These average properties are then related to each other by

$$\langle \sigma_{ij} \rangle = (C_{ijkl})_e \langle \epsilon_{kl} \rangle, \tag{34}$$

which define the effective elastic stiffness or moduli tensor $\mathbf{C}_e$ of the material. Although the averaging procedure is conceptually simple, its implementation for anything but the simplest morphology of composite materials is a difficult task. Therefore, models of composite materials cannot have an arbitrary morphology, because otherwise predicting their effective mechanical properties, even with the current powerful computers, will be an extremely difficult task.

Because of such difficulties, one must classify the possible morphologies into several groups, each with distinct properties. According to Christensen (1979), there are at least five such groups. In the first group is the crystal grain structure of common metals, where each grain is anisotropic and the orientation of its axes or planes of symmetry is different from that of the other grains. The materials of this type contain only one phase. In the second class of materials are those that have two or more distinct phases. Each phase is continuous, and there are no distinguishing geometrical features of the interface between the phases that can help one to decide which phase is on which side of the interface. In the remaining three classes of materials' morphology are those that are composed of a continuous matrix, populated with inclusions of another phase that are spherical, cylindrical or lamellar (see Chapter 3). These three types of inclusions are in fact the limiting cases of a general ellipsoidal inclusion, with the cylindrical and lamellar inclusions being associated with prolate and oblate ellipsoids, respectively. An ellipsoidal inclusion also offers great flexibility for gradual gradation in between these limiting cases. Moreover, although a composite that consists of a matrix and such inclusions may seem rather abstract, it is in fact encountered in practical situations. For example, in some polymeric materials where kinetic driving forces are present and thermodynamical equilibrium is achieved, one always obtains either a spherical or cylindrical or lamellar structure.

Consider now a two-phase material with one phase being continuous (the matrix) and the other material being in the form of discrete inclusions. We assume that both phases are isotropic. The stress-strain relation for the inclusion phase is given by

$$\sigma_{ij} = \lambda_2 \epsilon_{kk} \delta_{ij} + 2\mu_2 \epsilon_{ij}, \tag{35}$$

while for the matrix phase one has

$$\sigma_{ij} = \lambda_1 \epsilon_{kk} \delta_{ij} + 2\mu_1 \epsilon_{ij}, \tag{36}$$

where λ and μ are the usual Lamé constants. The average stress in the material can be written as

$$\langle\sigma_{ij}\rangle = \frac{1}{\Omega}\int_V \sigma_{ij}dV + \frac{1}{\Omega}\sum_{m=1}^{N}\int_{\Omega_m}\sigma_{ij}d\Omega_m, \tag{37}$$

where $V = \Omega - \sum_{m=1}^{N}\Omega_m$, Ω_m is the volume of the mth inclusion, and N is the total number of the inclusions. If we use the stress-strain relation in Eq. (37), we obtain

$$\langle\sigma_{ij}\rangle = \frac{1}{\Omega}\int_V (\lambda_1\epsilon_{kk}\delta_{ij} + 2\mu_1\epsilon_{ij})dV + \frac{1}{\Omega}\sum_{m=1}^{N}\int_{\Omega_m}\sigma_{ij}d\Omega_m, \tag{38}$$

which can be rewritten as

$$\langle\sigma_{ij}\rangle = \frac{1}{\Omega}\int_\Omega (\lambda_1\epsilon_{kk}\delta_{ij} + 2\mu_1\epsilon_{ij})d\Omega$$
$$-\frac{1}{\Omega}\sum_{m=1}^{N}\int_{\Omega_m}(\lambda_1\epsilon_{kk}\delta_{ij} + 2\mu_1\epsilon_{ij})d\Omega_m + \frac{1}{\Omega}\sum_{m=1}^{N}\int_{\Omega_m}\sigma_{ij}d\Omega_m. \tag{39}$$

If we now use the definition of the volume-averaged properties, we obtain (Russel and Acrivos, 1972)

$$C_{ijkl}\langle\epsilon_{kl}\rangle = \lambda_1\langle\epsilon_{kk}\rangle\delta_{ij} + 2\mu_1\langle\epsilon_{ij}\rangle + \frac{1}{\Omega}\sum_{m=1}^{N}\int_{\Omega_m}(\sigma_{ij} - \lambda_1\epsilon_{kk}\delta_{ij} - 2\mu_1\epsilon_{ij})d\Omega_m, \tag{40}$$

which implies that, in order to evaluate the effective tensor $\mathbf{C}_e$, only the conditions within the inclusions are needed.

We are now in a position to begin describing and discussing various continuum models for estimating linear mechanical properties of heterogeneous materials.

7.2 Exact Results

There are several exact, and very useful, results for the effective elastic properties of both 2D and 3D composites. In this section we briefly describe these results and discuss their implications. We do not provide the proofs of these results; they can be found in the original references. To begin with, let us write down the constitutive equations for a linear elastic and isotropic material in 2D and 3D, and summarize the known results between the two that will be useful to our discussions. These equations, in explicit forms, are given by

$$\epsilon_{x_1x_1} = \frac{1}{Y^{(3)}}\left[\sigma_{x_1x_1} - \nu_p^{(3)}(\sigma_{x_2x_2} + \sigma_{x_3x_3})\right], \quad \epsilon_{x_1x_2} = \frac{1+\nu_p^{(3)}}{Y^{(3)}}\sigma_{x_1x_2}, \tag{41}$$

where Y is the Young's modulus of the material, and superscript 3 denotes the dimensionality. The rest of the governing equations are obtained by cyclic permutations, $x_1 \to x_2 \to x_3$. Similar equations hold in 2D, namely,

$$\epsilon_{x_1x_1} = \frac{1}{Y^{(2)}}\left[\sigma_{x_1x_1} - \nu_p^{(2)}\sigma_{x_2x_2}\right], \quad \epsilon_{x_1x_2} = \frac{1+\nu_p^{(2)}}{Y^{(2)}}\sigma_{x_1x_2}, \tag{42}$$

with cyclic permutations, $x_1 \to x_2$. As Eqs. (19) and (20) indicate, the bulk modulus K and the shear modulus μ can be written in terms of the Young's modulus Y and the Poisson's ratio ν_p.

7.2.1 *Interrelations Between Two- and Three-Dimensional Moduli*

When we wish to connect the elastic moduli of isotropic 2D (planar) elasticity to those of isotropic 3D systems, we must specify either *plane-strain* or *plane-stress* elasticity. The former one is relevant in studying the elastic behavior of a fiber-reinforced material, while the latter is relevant in considering two-phase composites in the form of thin sheets.

For plane strain we have $\epsilon_{x_1x_3} = \epsilon_{x_2x_3} = \epsilon_{x_3x_3} = 0$, so that Eq. (41) can be written as

$$\epsilon_{x_1x_1} = \frac{1-[\nu_p^{(3)}]^2}{Y^{(3)}}\left(\sigma_{x_1x_1} - \frac{\nu_p^{(3)}}{1-\nu_p^{(3)}}\sigma_{x_2x_2}\right), \quad \epsilon_{x_1x_2} = \frac{1+\nu_p^{(3)}}{Y^{(3)}}\sigma_{x_1x_2}, \tag{43}$$

with cyclic permutations, $x_1 \to x_2$. The longitudinal stress component is given by, $\sigma_{x_3x_3} = \nu_p^{(3)}(\sigma_{x_1x_1} + \sigma_{x_2x_2})$. If we compare Eqs. (42) and (43), we see that for plane strain

$$Y^{(2)} = \frac{Y^{(3)}}{1-[\nu_p^{(3)}]^2}, \quad \nu_p^{(2)} = \frac{\nu_p^{(3)}}{1-\nu_p^{(3)}}. \tag{44}$$

For plane stress we have, $\sigma_{x_1x_3} = \sigma_{x_2x_3} = \sigma_{x_3x_3} = 0$, so that Eq. (41) can be rewritten as

$$\epsilon_{x_1x_1} = \frac{1}{Y^{(3)}}\left[\sigma_{x_1x_1} - \nu_p^{(3)}\sigma_{x_2x_2}\right], \quad \epsilon_{x_1x_2} = \frac{1+\nu_p^{(3)}}{Y^{(3)}}\sigma_{x_1x_2}, \tag{45}$$

with cyclic permutations, $x_1 \to x_2$; moreover, $\epsilon_{x_3x_3} = -[\nu_p^{(3)}/(1-\nu_p^{(3)})](\epsilon_{x_1x_1} + \epsilon_{x_2x_2})$. If we compare Eq. (45) with Eq. (42), we find that for plane stress

$$Y^{(2)} = Y^{(3)}, \quad \nu_p^{(2)} = \nu_p^{(3)}. \tag{46}$$

The governing equations for plane stress and plane strain are sometimes combined by using a quantity κ called the Kosolov constant (Muskhelishvili, 1953), such that Eqs. (42) have the same form, but the coefficients involve κ rather than $\nu_p^{(2)}$, and the shear modulus $\mu^{(3)}$ rather than $Y^{(2)}$:

$$\epsilon_{x_1x_1} = \frac{1}{2\mu^{(3)}}\left[\sigma_{x_1x_1} - \frac{1}{4}(3-\kappa)(\sigma_{x_1x_1} + \sigma_{x_2x_2})\right], \quad \epsilon_{x_1x_2} = \frac{1}{2\mu^{(3)}}\sigma_{x_1x_2}, \tag{47}$$

together with cyclic permutations, $x_1 \to x_2$. The mapping to 2D now takes the forms

$$\mu^{(2)} = \mu^{(3)}, \quad \kappa = 3 - 4\nu_p^{(3)} = \frac{3 - \nu_p^{(2)}}{1 + \nu_p^{(2)}} \qquad \text{plane strain} \tag{48}$$

$$\mu^{(2)} = \mu^{(3)}, \quad \kappa = \frac{3 - \nu_p^{(3)}}{1 + \nu_p^{(3)}} = \frac{3 - \nu_p^{(2)}}{1 + \nu_p^{(2)}} \qquad \text{plane stress.} \tag{49}$$

7.2.2 *Exact Results for Regular Arrays of Spheres*

In Sections 4.3.1 and 4.3.2 we described in detail how the exact solution for the problem of conduction in a two-phase material, that consists of a regular array of d-dimensional spheres dispersed in a uniform background matrix, is derived. One can carry out a similar analysis for the analogous problem of computing the effective elastic moduli of a two-phase material that consists of a regular array of d-dimensional sphere (phase 2) in a uniform matrix (phase 1). We do not provide the details of such an analysis, as they are the elastic analogue of the conductivity problem, but only summarize the results.

Nunan and Keller (1984) analyzed the problem of the effective elastic moduli of the three 3D cubic arrays of *rigid* spheres in a matrix. As pointed out earlier, such composite materials are characterized by three independent elastic moduli. Therefore, the effective stiffness tensor $\mathbf{C}_e$, in component form, is given by

$$(C_e)_{ijkl} = (\lambda_1 + \mu_1\gamma)\delta_{ij}\delta_{kl} + \mu_1(1+\beta)(\delta_{ik}\delta_{jl} + \delta_{il}\delta_{jk}) + 2\mu_1(\alpha - \beta)\delta_{ijkl}. \tag{50}$$

Here, $\delta_{ijkl} = 1$, if $i = j = k = l$, and $\delta_{ijkl} = 0$, otherwise. All the notations are as before. [There is a misprint in the original paper of Nunan and Keller (1984).] The main problem is then developing expressions for the three coefficients, α, β, and γ, for which low-density asymptotic expansions were given by Nunan and Keller (1984):

$$\alpha = \frac{15\phi_2(1-\nu_{p1})}{2(4-5\nu_{p1})}\left[1 - \left(1 - \frac{3a_1}{4-5\nu_{p1}}\right)\phi_2 + \frac{3a_2}{4-\nu_{p1}}\phi_2^{5/3} + O(\phi_2^{7/3})\right]^{-1}, \tag{51}$$

$$\beta = \frac{15\phi_2(1-\nu_{p1})}{2(4-5\nu_{p1})}\left[1 - \left(1 + \frac{2a_1}{4-5\nu_{p1}}\right)\phi_2 - \frac{2a_2}{4-5\nu_{p1}}\phi_2^{5/3} + O(\phi_2^{7/3})\right]^{-1}, \tag{52}$$

$$\gamma = \frac{3\phi_2(1-\nu_{p1})}{(4-5\nu_{p1})(1-2\nu_{p1})}\left[1 - \frac{\nu_{p1}(3-2a_3)+a_3}{4-5\nu_{p1}}\phi_2 + \frac{a_4(1-2\nu_{p1})}{4-5\nu_{p1}}\phi_2^{5/3} + O(\phi_2^{7/3})\right]^{-1}. \tag{53}$$

The coefficients (a_1, a_2, a_3, a_4) for the three cubic lattices are, (−1.396, 1.714, 2.889, −3.077), (0.430, −0.520, −6.163, 3.373), and (0.388, −0.411, −5.928, 2.750) for the simple-cubic, BCC and FCC lattices, respectively.

The problem in the opposite limit, namely, the case of nearly close-packed spheres, is more complex. Flaherty and Keller (1973) and Nunan and Keller (1984) observed that the elastic interaction between spheres near the maximum close-packing sphere density is concentrated in the regions near the points of contact. Thus, a local analysis of the region between two nearly touching spheres, together with knowledge of nearest-neighbor locations, produce the dominant contribution of the interaction for the entire sphere.

For 2D systems, asymptotic expressions for the effective elastic moduli of periodic lattices of circular holes, and of superrigid (infinite elastic constants), circular inclusions that are applicable near the percolation threshold, have also ben derived. For circular voids, arranged on the sites of a hexagonal lattice near its percolation threshold $\phi_{2c} = \pi/(3^{3/2})$, the elastic moduli are given by (Day *et al.*, 1992)

$$\frac{\mu_e}{\mu_1} = \frac{\sqrt{3}(1+\nu_{p1})}{2\pi}\left(\frac{\phi_{2c}-\phi_2}{\phi_{2c}}\right)^{1/2}, \quad \frac{K_e}{K_1} = \frac{\sqrt{3}(1-\nu_{p1})}{\pi}\left(\frac{\phi_{2c}-\phi_2}{\phi_{2c}}\right)^{1/2}. \tag{54}$$

For a composite in which superrigid circles are arranged on the sites of the hexagonal lattice, one has (Chen *et al.*, 1995)

$$\frac{\mu_e}{\mu_1} = \frac{2\pi}{\sqrt{3}(3-\nu_{p1})}\left(\frac{\phi_{2c}-\phi_2}{\phi_{2c}}\right)^{-1/2}, \quad \frac{K_e}{K_1} = \frac{\pi}{\sqrt{3}(1+\nu_{p1})}\left(\frac{\phi_{2c}-\phi_2}{\phi_{2c}}\right)^{-1/2}. \tag{55}$$

The analogous results for the triangular lattice are given by (Day *et al.*, 1992)

$$\frac{\mu_e}{\mu_1} = \frac{4(1+\nu_{p1})}{3\sqrt{3}\pi}\left(\frac{\phi_{2c}-\phi_2}{\phi_{2c}}\right)^{3/2}, \quad \frac{K_e}{K_1} = \frac{1-\nu_{p1}}{\sqrt{3}\pi}\left(\frac{\phi_{2c}-\phi_2}{\phi_{2c}}\right)^{3/2}, \tag{56}$$

for circular holes distributed on the sites of the lattice near its percolation threshold $\phi_{2c} = \pi/(2\sqrt{3})$, and (Chen *et al.*, 1995)

$$\frac{\mu_e}{\mu_1} = \frac{\sqrt{3}\pi(3-\nu_{p1})}{4(1-\nu_{p1})}\left(\frac{\phi_{2c}-\phi_2}{\phi_{2c}}\right)^{-1/2}, \quad \frac{K_e}{K_1} = \frac{\sqrt{3}\pi}{1+\nu_{p1}}\left(\frac{\phi_{2c}-\phi_2}{\phi_{2c}}\right)^{-1/2}, \tag{57}$$

for superrigid circles on the sites of the triangular lattice near its percolation threshold.

7.2.3 *Exact Results for Coated Spheres and Laminates*

Another model for which exact computations can be carried out is the coated-spheres model. Recall from Section 4.4 that this model consists of composite spheres that are composed of a spherical core of a given transport property (for example, conductivity or elastic moduli) and radius a, surrounded by a concentric shell that has a different transport property and an outer radius b. The ratio a/b is fixed and in d-dimensions, the spheres volume fraction is $\phi_2 = (a/b)^d$. In Section 4.4., the exact solution of conduction in this system was derived. Similar exact computations can be carried out for some, but not all, of the effective elastic moduli of the system. Hashin and Rosen (1964) showed that one cannot exactly compute the shear modulus of this model, because the shear field outside a composite sphere is not uniform, whereas as discussed in Section 4.4, and also below, the uniformity of such fields outside the composite sphere is essential to deriving the exact solution. Christensen and Lo (1979) did show that a variant of this model, the so-called *three-phase model*, allows exact computation of the effective shear modulus μ_e, but we do not consider their model here.

Here we derive an exact expression for the effective bulk modulus K_e of the d-dimensional coated-spheres model. Consider a uniform material with (as yet) unknown bulk modulus K_e, on which a uniform hydrostatic strain field ϵ_0 has been imposed. The analysis parallels that presented in Section 4.4 for the effective conductivity. The displacement field $\mathbf{u}$ satisfies the following equations

$$(\lambda_2 + \mu_2)\nabla(\nabla \cdot \mathbf{u}) + \mu_2 \nabla^2 \mathbf{u} = \mathbf{0}, \quad r \leq a, \tag{58}$$

$$(\lambda_e + \mu_e)\nabla(\nabla \cdot \mathbf{u}) + \mu_e \nabla^2 \mathbf{u} = \mathbf{0}, \quad r \geq b, \tag{59}$$

$$(\lambda_1 + \mu_1)\nabla(\nabla \cdot \mathbf{u}) + \mu_1 \nabla^2 \mathbf{u} = \mathbf{0}, \quad a \leq r \leq b, \tag{60}$$

subject to the usual continuity of the displacement field $\mathbf{u}$ at $r = a$ and $r = b$, and that of the radial component of the stress tensor at $r = a$. Thus, analogous to Eq. (4.52) for the voltage field, the general solution for the displacement field $\mathbf{u}$ is given by

$$\mathbf{u} = \begin{cases} c_1 \epsilon_0 \cdot \mathbf{r}, & r \leq a, \\ c_2 \epsilon_0 \cdot \mathbf{r} + c_3 \dfrac{\epsilon_0 \cdot \mathbf{n}}{r^{d-1}}, & a \leq r \leq b, \\ \epsilon_0 \cdot \mathbf{r}, & r \geq b, \end{cases} \tag{61}$$

where $r = |\mathbf{r}|$, $\mathbf{n} = \mathbf{r}/r$, and the three constants c_1, c_2, and c_3 must be determined from continuity of $\mathbf{u}$ at $r = a$ and $r = b$, and continuity of the radial component of the stress at $r = a$. These constants are given by

$$c_1 = \frac{dK_1 + 2(d-1)\mu_1}{dK_2 + 2(d-1)\mu_1 - d\phi_2(K_2 - K_1)}, \tag{62}$$

$$c_2 = \frac{dK_2 + 2(d-1)\mu_1}{dK_2 + 2(d-1)\mu_1 - d\phi_2(K_2 - K_1)}, \tag{63}$$

$$c_3 = \frac{da^d(K_1 - K_2)}{dK_2 + 2(d-1)\mu_1 - d\phi_2(K_2 - K_1)}. \tag{64}$$

Having determined $\mathbf{u}$, the strain field ϵ is obtained from Eq. (2). The stress field is then given by

$$\sigma = \left(K_e - \frac{2}{d}\mu_e\right) \mathrm{Tr}(\epsilon)\mathbf{U} + 2\mu_e \epsilon, \tag{65}$$

where Tr denotes the trace of the tensor. Continuity of the radial component of the stress tensor at $r = a$ then yields

$$K_e = c_2 K_1 - \frac{2c_3\mu_1(d-1)}{db^d}, \tag{66}$$

and therefore

$$K_e = \phi_1 K_1 + \phi_2 K_2 - \frac{d\phi_1\phi_2(K_2 - K_1)^2}{d(\phi_2 K_1 + \phi_1 K_2) + 2\mu_1(d-1)}, \tag{67}$$

which is very similar to Eq. (4.55) for the effective conductivity.

Similarly, one may obtain exact results for the effective elastic moduli of two-phase laminates of isotropic phases. The calculations are similar to, but more complex than, the analogous calculations for the effective conductivity. Consider, for example, the calculation of the effective stiffness tensor $\mathbf{C}_e$ for a 2D laminate of rank 1, which is composed of alternating layers of phases 1 and 2 according to some random process such that, for example, the thickness of the layers vary in space. In plane elasticity, one has

$$\mathbf{C}_e = \begin{bmatrix} (C_e)_{11} & (C_e)_{12} & 0 \\ (C_e)_{12} & (C_e)_{22} & 0 \\ 0 & 0 & (C_e)_{66} \end{bmatrix}, \tag{68}$$

with

$$(C_e)_{11} = \left\langle \frac{1-\nu_p^2}{Y} \right\rangle^{-1}, \quad (C_e)_{12} = \langle \nu_p \rangle \left\langle \frac{1-\nu_p^2}{Y} \right\rangle^{-1}, \tag{69}$$

$$(C_e)_{22} = \langle \nu_p^2 \rangle \left\langle \frac{1-\nu_p^2}{Y} \right\rangle^{-1} + \langle Y \rangle, \quad (C_e)_{66} = \langle \mu^{-1} \rangle^{-1}, \tag{70}$$

where for any property p the average $\langle p \rangle$ is defined by, $\langle p \rangle = \phi_1 p_1 + \phi_2 p_2$. As expected, the Young's modulus of the laminate is anisotropic and depends on the direction. Thus,

$$(Y_e)_{11} = (C_e)_{11} - \frac{(C_e)_{12}^2}{(C_e)_{22}}, \quad (Y_e)_{22} = (C_e)_{22} - \frac{(C_e)_{12}^2}{(C_e)_{11}}, \tag{71}$$

whereas the effective shear modulus is given by the harmonic average of the shear moduli of the two phases, $(\mu_e)_{12} = \langle \mu^{-1} \rangle^{-1}$.

Similar results can be derived for the 3D version of the same laminate (Postma, 1955). The effective stiffness tensor $\mathbf{C}_e$ is symmetric and is given by

$$\mathbf{C}_e = \begin{bmatrix} (C_e)_{11} & (C_e)_{12} & (C_e)_{12} & 0 & 0 & 0 \\ & (C_e)_{22} & (C_e)_{23} & 0 & 0 & 0 \\ & & (C_e)_{22} & 0 & 0 & 0 \\ & & & \frac{1}{2}[(C_e)_{22} - (C_e)_{23}] & 0 & 0 \\ & & & & (C_e)_{66} & 0 \\ & & & & & (C_e)_{66} \end{bmatrix}, \tag{72}$$

with

$$(C_e)_{11} = \langle (K + \frac{4}{3}\mu)^{-1} \rangle^{-1}, \quad (C_e)_{12} = \left\langle \frac{3K - 2\mu}{3K + 4\mu} \right\rangle (C_e)_{11}, \tag{73}$$

$$(C_e)_{22} = \left\langle \frac{3K - 2\mu}{3K + 4\mu} \right\rangle^2 (C_e)_{11} + \langle K + \frac{4}{3}\mu \rangle - \left\langle \frac{(3K - 2\mu)^2}{3(3K + 4\mu)} \right\rangle, \tag{74}$$

$$(C_e)_{23} = \left\langle \frac{3K - 2\mu}{3K + 4\mu} \right\rangle^2 (C_e)_{11} + \langle K - \frac{2}{3}\mu \rangle - \left\langle \frac{(3K - 2\mu)^2}{3(3K + 4\mu)} \right\rangle, \quad (C_e)_{66} = \langle \mu^{-1} \rangle^{-1}. \tag{75}$$

The effective Young's moduli along the coordinate axes are then give by

$$(Y_e)_{11} = (C_e)_{11} - \frac{2(C_e)_{12}^2}{(C_2)_{22} + (C_e)_{23}}, \tag{76}$$

$$(Y_e)_{22} = (Y_e)_{33} =$$
$$(C_e)_{22} - \frac{(C_e)_{12}^2[(C_e)_{23} - (C_e)_{22}] + (C_e)_{23}[(C_e)_{12}^2 - (C_e)_{11}(C_e)_{23}]}{(C_e)_{11}(C_e)_{22} - (C_e)_{12}^2}. \tag{77}$$

Note that under a uniform applied field, the local fields in the laminate are piecewise constant, a property that can be taken advantage of in the construction of disordered materials with a desired effective stiffness tensor. Moreover, although when the Poisson's ratios of the two phases are equal, the effective Young's moduli of the laminate are given by arithmetic and harmonic averages of the Young's moduli of the two phases, such averages provide poor estimates of the moduli for the more general case of isotropic materials in which the Poisson's ratios of the two phases are not the same. In general (and similar to the effective conductivity), the arithmetic average $\langle Y \rangle$ *overestimates* the true Young's modulus of an isotropic two-phase material, whereas the harmonic average $\langle Y^{-1} \rangle^{-1}$ *underestimates* it.

7.2.4 Connection to Two-Dimensional Conductivity

Consider transversely isotropic fiber-reinforced composite materials with phase boundaries that are cylindrical surfaces of arbitrary shape and with generators that are parallel to one axis. Hashin (1970) observed that the governing equation for conduction in the transverse plane, i.e., the Laplace equation, is identical to that for out-of-plane shear. Therefore, determination of the effective longitudinal (that is, axial) shear modulus of the composite is mathematically equivalent to the determination of its effective *transverse* conductivity.

7.2.5 Exact Duality Relations

Certain extensions of the exact duality relations for the effective conductivity of 2D two-phase materials, that were explained in Section 4.8, have been made to their mechanical properties. Suppose that $K_e(K_1, \mu_1, K_2, \mu_2)$ and $\mu_e(K_1, \mu_1, K_2, \mu_2)$ are the effective bulk and shear moduli of a 2D, two-phase isotropic material, and that $K_e(K_2, \mu_2, K_1, \mu_1)$ and $\mu_e(K_2, \mu_2, K_1, \mu_1)$ are the corresponding effective moduli for the "dual" material, obtained from the original material by interchanging the bulk and shear moduli of its two phases. Berdichevsky (1983) proved that, if the material is incompressible, then

$$\mu_e(K_1, \mu_1, K_2, \mu_2)\mu_e(K_2, \mu_2, K_1, \mu_1) = \mu_1\mu_2. \tag{78}$$

Moreover, if the bulk moduli of the two phases are equal, then Helsing *et al.* (1997) showed that

$$Y_e(\mu_1, \mu_2)Y_e(\mu_2, \mu_1) = Y_1Y_2. \tag{79}$$

Equations (78) and (79) are completely similar to Eq. (4.224) for the effective conductivity of the same material. As in the case of the effective conductivity, there are no known duality relations for 3D materials, although certain inequalities have been derived.

7.2.6 The Cherkaev–Lurie–Milton Theorem and Transformation

Cherkaev, Lurie and Milton (1992) proved an interesting and useful theorem for 2D composites, which is applicable to linearly elastic materials with general anisotropy. The theorem requires no restrictions on the number of components or on the geometry of the material. It does, however, require that the displacements and the tractions to be continuous functions across all the interfaces, i.e., bonding at all the interfaces must be perfect. We refer to this work as the CLM theorem, which is as follows.

Suppose that a 2D composite material has spatially-varying bulk and shear moduli $K(\mathbf{r})$ and $\mu(\mathbf{r})$, and that the effective bulk and shear moduli of the material are K_e and μ_e. If

$$\frac{1}{K^{\mathrm{t}}(\mathbf{r})} = \frac{1}{K(\mathbf{r})} - C \text{ and } \frac{1}{\mu^{\mathrm{t}}(\mathbf{r})} = \frac{1}{\mu(\mathbf{r})} + C, \tag{80}$$

where C is a constant, then

$$\frac{1}{K_e^{\mathrm{t}}} = \frac{1}{K_e} - C \text{ and } \frac{1}{\mu_e^{\mathrm{t}}} = \frac{1}{\mu_e} + C, \tag{81}$$

where the superscript t denotes a transformed composite. We refer to Eq. (80) as the CLM transformation, and Eq. (81) as the CLM theorem. The CLM transformation connects composites materials that Cherkaev *et al.* (1992) call *equivalent*, which are materials that have the same stress field, for given external tractions, even though their material parameters differ. To prove the CLM theorem, one must show that the local stress tensor is the same in both materials when they are both subjected to the same external loading. Although the CLM theorem has been proven only for perfect bonding, but, as discussed by Thorpe and Jasiuk (1992), it presumably holds more generally. The constant C is constrained by

$$-\min\,[1/\mu(\mathbf{r})] < C < \min\,[1/\kappa(\mathbf{r})],$$

to ensure that the elastic moduli are positive everywhere in the transformed material. This constraint implies that the CLM theorem is most powerful when the inclusions in the composite are cavities (so that $-\infty < C < \infty$), thus being very useful to describing composite materials with percolation-type disorder, but is useless when the inclusions are completely rigid, since in this case the bounds collapse, resulting in $C = 0$.

It may be more useful to rewrite the CLM transformation in terms of the Young's modulus which then yields, $Y^{\mathrm{t}}(\mathbf{r}) = Y(\mathbf{r})$, indicating that the Young's moduli of the materials are invariant under this transformation, and therefore the CLM theorem can be written as

$$Y_e^{\mathrm{t}} = Y_e. \tag{82}$$

7.2.7 *Universality of Poisson's Ratio in Percolation Composites*

As discussed by Thorpe and Jasiuk (1992), an important application of the CLM theorem is to 2D composite materials that are prepared by removing material to form holes of any size, shape, area fraction, etc. Numerical simulations of Day *et al.* (1992) showed that, for a prescribed geometry, the relative Young's modulus Y_e/Y_1 of a 2D sheet containing such holes, overlapping or not (where Y_1 is the Young's modulus of the sheet without any holes or inclusions), is *the same* for *all* materials, *independent* of the host Poisson's ratio ν_p. The CLM theorem provides a simple proof for these numerical results as the CLM transformation leaves holes as holes. Therefore, any matrix material can be obtained by using the CLM transformation together with an overall scaling factor.

Figure 7.1 demonstrates this universality for the Young's modulus. These results, computed by Day *et al.* (1992), were obtained with the 2D Swiss-cheese model, i.e., one in which one randomly punches the system with circular holes, regardless of whether they overlap or not, and are plotted as the Young's modulus of the system versus the volume fraction ϕ_1 of the remaining material. It is clear that the results are independent of the Poisson's ratio of the matrix, so that

$$\frac{Y_e^{\mathrm{t}}}{Y_1} = \frac{Y_e}{Y_1}. \tag{83}$$

Figure 7.1 also presents the effective value ν_{pe} of the Poisson's ratio as a function of the volume fraction of the matrix. It is seen that, as the percolation threshold ϕ_c of the system is approached, where $\phi_c \simeq 0.34$, regardless of the initial state of the system (i.e., regardless of value of the Poisson's ratio ν_{p1}), the effective Poisson's ratio of the material approaches a constant value of about 1/3. To see this, note that one can write

$$\nu_{pe}^{\mathrm{t}} - \nu_{pe} = (\nu_{p1}^{\mathrm{t}} - \nu_{p1})\,\frac{Y_e}{Y_1}. \tag{84}$$

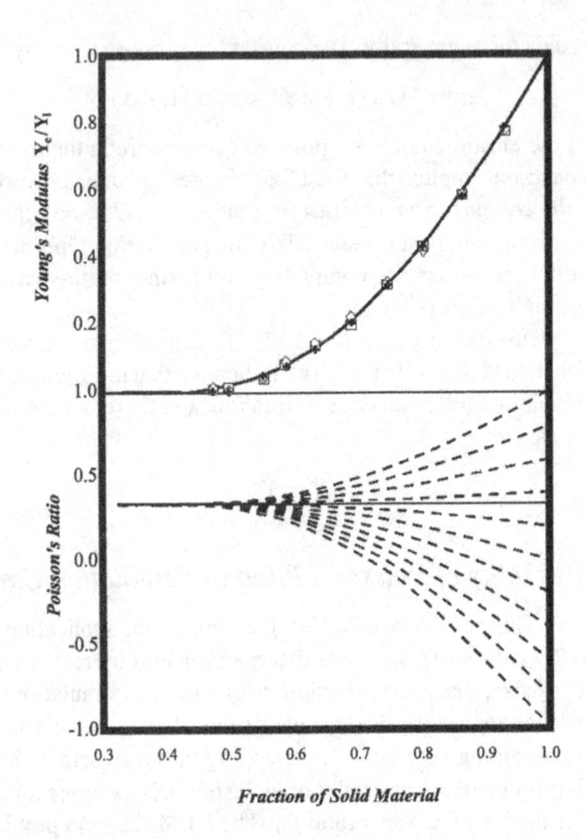

FIGURE 7.1. Young's modulus and Poisson's ratio of a sheet that contains randomly-centered circular holes. Symbols in the Young's modulus plot indicate materials with different Poisson's ratios. The plot for the Poisson's ratio was constructed from the solid reference line at $\nu_{p1} = 1/3$ (after Day *et al.*, 1992).

Since the ratio Y_e/Y_1 is universal, so must also be the value of the effective Poisson's ratio. Now, since as the percolation threshold is approached, one has, $Y_e/Y_1 \to 0$, and therefore one must also have, $\nu_{pe}^t \to \nu_{pe}$, and Figure 7.1 indicates that this is indeed the case.

A similar universal behavior was found in the lattice models of mechanical properties of disordered materials, which will be described in Chapter 8 (see Section 8.8.8). We should emphasize that the universal value of the Poisson's ratio shown in Figure 7.1 is only with respect to the elastic constants of the material. If, instead of circular holes, one punches other types of inclusions into the material, other values of the Poisson's ratio will emerge which will be universal so long as the shape of the inclusion is held fixed. Finally, one may expect the same type of universality if the inclusions are completely rigid, rather than being holes, although in this case the CLM theorem is useless for proving the universality of the Poisson's ratio.

7.2.8 Composite Materials with Equal Shear Moduli

Hill (1963,1964) and Hashin (1965) suggested that if all the components of a composite material have the same shear modulus μ, then the elastic constants of the composite can be calculated exactly, in both 2D and 3D, because under this condition the upper and lower bounds for the elastic constants collapse onto a single value. In particular, for 2D composites of this type, one has the following exact results,

$$\mu_e = \mu, \quad K_e = K_1 + \frac{1-\phi_1}{\dfrac{1}{K_2 - K_1} + \dfrac{\phi_1}{K_1 + \mu}}. \tag{85}$$

Here ϕ_1 is the volume fraction of phase 1, and K_1 and K_2 are the bulk moduli of the individual phases. Equation (85) is independent of the material's geometry. Although not obvious, the equation for K_e is symmetric under interchange of 1 and 2. To see this symmetry, Thorpe and Jasuik (1992) expressed Eq. (85) in terms of the Young's modulus and the Poisson's ratio, with the results given by

$$Y_e = \sum_{i=1}^{2} \phi_i Y_i, \quad \nu_{pe} = \sum_{i=1}^{2} \phi_i \nu_{pi}, \tag{86}$$

hence demonstrating the symmetry of the exact results in Eq. (85). The proof of Eqs. (85) and (86) via the CLM theorem is straightforward (Thorpe and Jasuik, 1992). We take the transformed composite to have $\mu_1^{\mathrm{t}} = \mu_2^{\mathrm{t}} = \infty$, and therefore

$$\mu_e^{\mathrm{t}} = \infty, \quad K_e^{\mathrm{t}} = \phi_1 K_1^{\mathrm{t}} + \phi_2 K_2^{\mathrm{t}}. \tag{87}$$

The result for μ_e^{t} is obvious, if the bonding at the interface is perfect, while the result for K_e^{t} follows because the material can only sustain a homogeneous hydrostatic strain if it is to hold together. One now applies the CLM theorem with the constant $C = 1/\mu$, and Eqs. (85) and (86) follow immediately. Note that this proof and Eqs. (86) can easily be generalized to more than two components. Equations (85) can also be generalized, although not as easily as (86).

7.2.9 Dundurs Constants

Dundurs (1967) proved that if a composite material that consists of two linearly elastic and isotropic phases is subjected to specified tractions and undergoes plane deformation, then the stress in the material exhibits a reduced dependence on the material's elastic constants. In particular, for materials of this type, the stress depends only on two dimensionless parameters, as opposed to the usual three combinations of elastic constants for the 3D case. Specifically, Dundurs (1967) showed that the stress field at point $\mathbf{r}$ in the material can be written as

$$\boldsymbol{\sigma}(\mathbf{r}) = \sigma_0 \mathbf{h}(\mathbf{r}; \alpha_{12}, \beta_{12}), \tag{88}$$

where

$$\alpha_{12} = \frac{Y_1^{-1} - Y_2^{-1}}{Y_1^{-1} + Y_2^{-1}}, \tag{89}$$

$$4\beta_{12} = \frac{K_1^{-1} - K_2^{-1}}{Y_1^{-1} + Y_2^{-1}}. \tag{90}$$

As pointed out by Thorpe and Jasuik (1992), these results follow from the CLM transformation, since Eq. (80) states that the components of the stress tensor are unchanged by this transformation, which reduces the number of parameters needed to describe the material by one. Note that α_{12} and β_{12} are invariant under the CLM transformation. Moreover, they are not unique. For example, using the relations between the elastic moduli, one can obtain another constant γ_{12} given by

$$\gamma_{12} = 4(\alpha_{12} - \beta_{12}) = \frac{\mu_1^{-1} - \mu_2^{-1}}{Y_1^{-1} + Y_2^{-1}}, \tag{91}$$

and thus any two of the three parameters α_{12}, β_{12}, and γ_{12} can be used.

An important limiting case is when one of the composites contains percolation-type disorder, i.e., when one of the two components, say 2, represents cavities (of any size or shape), so that, $K_2 = Y_2 = 0$, and hence, $\alpha_{12} = -1$ and $4\beta_{12} = -Y_2/K_2 = -2(1 - \nu_{p2})$. Thus, in this case there is only one parameter, β_{12}, that depends on the Poisson's ratio, and the stress becomes independent of the elastic constants, a famous result going back to Mitchel (1899). This result is the basis for the experimental technique of mapping out the stress field by optical birefringence, since the stress field depends only on the geometry of the holes and not on the value of the elastic constants.

We will return to discussing further properties of the CLM transformation and theorem when we describe the effective-medium approximation.

7.2.10 *Relations Between Elastic Moduli and Thermoelastic Properties*

The thermal expansion coefficient α of a material, which is a measure of its volumetric strain due to a change in temperature under traction-free conditions, is an important property. Levin (1967) showed that the effective thermal expansion coefficient α_e of a two-phase isotropic material, in which the two phases have thermal expansion coefficients α_1 and α_2, is given by

$$\alpha_e = \phi_1\alpha_1 + \phi_2\alpha_2 + \frac{\alpha_2 - \alpha_1}{K_2^{-1} - K_1^{-1}}\left(\frac{1}{K_e} - \left\langle\frac{1}{K}\right\rangle\right). \tag{92}$$

Levin's equation has been generalized to macroscopically anisotropic two-phase materials, polycrystals (Benveniste, 1996), and elastic-plastic two-phase composites (Dvorak, 1986). Rosen and Hashin (1970) showed that the effective specific heat $(C_p)_e$ of a two-phase isotropic material at constant pressure is given by

$$(C_p)_e = \phi_1 C_{p1} + \phi_2 C_{p2} + 9T \left(\frac{\alpha_2 - \alpha_1}{K_2^{-1} - K_1^{-1}} \right)^2 \left(\frac{1}{K_e} - \left\langle \frac{1}{K} \right\rangle \right), \tag{93}$$

from which, through use of thermodynamics, one can also compute the effective specific heat of the material at constant volume. The extension of these results to materials with more than two phases appears not to be possible.

Let us end this section by noting that Grabovsky *et al.* (2000) developed a systematic procedure for finding exact relations for the effective properties of materials, that are *independent of the microstructure*, such as the Levin equation, Eq. (92).

7.3 Dispersion of Spherical Inclusions

The first class of models that we consider consists of a continuous matrix of one material and spherical inclusions of another material dispersed in the matrix. A great amount of work has been done for estimating the linear elastic moduli of such models. We already summarized in Section 7.2.2 the exact results for the elastic properties of regular arrays of spheres distributed in a uniform background matrix. In this section we describe and analyze the problem of determining the elastic properties of composite materials that consist of one or two spherical inclusions inserted in a uniform background matrix, which also serve as an approximation to the true problem of determining the elastic properties of a composite material composed of a random dispersion of inclusions in a uniform matrix.

7.3.1 The Dilute Limit: A Single Sphere

The simplest model to consider is an infinite matrix of a material in which a single spherical inclusion has been inserted. Suppose that a strain field ϵ_0 is applied to the material at infinity. If $\mathbf{r}$ is the position vector emanating from the center of the sphere, then, to determine the displacement field $\mathbf{u}$ one must solve Eq. (60) for the region inside the sphere, where the Lamé constants are λ_1 and μ_1, and Eq. (58) for the region outside the sphere (within the matrix), where the Lamé constants are λ_2 and μ_2. The boundary conditions are the continuity of the displacement field $\mathbf{u}$, and of the radial component of the stress field, $\boldsymbol{\sigma} \cdot \mathbf{n}$ (where $\mathbf{n} = \mathbf{r}/|\mathbf{r}|$), at the interface between the inclusion and the matrix. In addition, as $r \to \infty$, one must have $\mathbf{u} = \boldsymbol{\epsilon}_0 \cdot \mathbf{r}$.

The solution of this problem in 3D for a hydrostatic applied field, i.e., one with $\boldsymbol{\epsilon}_0 = (\epsilon_0/d)\mathbf{U}$ (where $\mathbf{U}$ is the second-order identity tensor), and also for an applied deviatoric strain field, i.e., one with $\mathrm{Tr}(\boldsymbol{\epsilon}_0) = 0$ (where Tr denotes the trace operation), was derived a long time ago by Goodier (1933). More generally, such solutions can be derived for any $d \geq 2$. For a hydrostatic applied field, one has

$$\mathbf{u} = \begin{cases} (1 - K_{21})\epsilon_0 \cdot \mathbf{r}, & r \leq a, \\ \epsilon_0 \cdot \mathbf{r} + K_{21}\epsilon_0 \cdot \mathbf{r}\left(\frac{a}{r}\right)^d, & r \geq a, \end{cases} \tag{94}$$

where a is the radius of the sphere, and K_{21} is the d-dimensional *bulk modulus polarizability*, given by

$$K_{21} = \frac{K_2 - K_1}{K_2 + 2d^{-1}(d-1)\mu_1}. \tag{95}$$

The local strain field ϵ within the inclusion is uniform and is related to the applied strain by $\epsilon = (1 - K_{21})\epsilon_0$ for $r < a$. The strain field for the region outside the sphere is obtained by using Eq. (2) and the solution for $\mathbf{u}$, Eq. (94).

The solution for the case in which a deviatoric strain field is applied to the material is more complex. The most general form of the solution for the displacement field is now given by

$$\mathbf{u} = (1 + c_1)\epsilon_0 \cdot \mathbf{r}, \quad r \leq a, \tag{96}$$

$$\mathbf{u} = \epsilon_0 \cdot \mathbf{r} + c_2\epsilon_0 : \nabla\left(\frac{1}{r^{d-2}}\right) + c_3\epsilon_0 : \nabla\nabla\nabla\left(\frac{1}{r^{d-2}}\right) + c_4\epsilon_0 : \nabla\nabla\nabla\left(r^{d-2}\right), \quad r \geq a, \tag{97}$$

where c_1, c_2, c_3, and c_4 are four constants that must be determined from the boundary conditions. If

$$h_1 = \frac{\mu_1[d^2K_1 + 2(d+1)(d-2)\mu_1]}{2d(K_1 + 2\mu_1)}, \tag{98}$$

then in terms of the d-dimensional shear modulus polarizability μ_{21},

$$\mu_{21} = \frac{\mu_2 - \mu_1}{\mu_2 + h_1}, \tag{99}$$

the displacement field within the sphere is given by

$$\mathbf{u} = (1 - \mu_{21})\epsilon_0 \cdot \mathbf{r}, \quad r \leq a, \tag{100}$$

i.e., $c_1 = -\mu_{21}$. If

$$A = \frac{a^d(\mu_1 - \mu_2)}{\mu_1[d^3K_1 + 2d(d+1)(d-2)\mu_1] + \mu_2[2d^2(K_1 + 2\mu_1)]},$$

then, $c_2 = 2A(d+2)[dK_1 + 2(d-1)\mu_1]$, $c_3 = a^2A[dK_1 + (d-2)\mu_1]$, and $c_3 = A(d+2)[dK_1 + (d-2)\mu_1]$. Similar to the case of an applied hydrostatic field, the local strain field within the inclusion is uniform and is given by, $\epsilon = (1 - \mu_{21})\epsilon_0$, for $r < a$. The strain field for the region outside the sphere is obtained by using Eq. (2) and the solution for $\mathbf{u}$.

These results can be extended to the case in which an ellipsoidal inclusion is embedded within the matrix. The derivation of the results is, however, considerably more complex than for the spherical inclusion; see Mura (1987).

One may also think of the single inclusion model as a two-phase material with many spherical inclusions which, however, do not interact with each other. Based on this assumption, one can obtain estimates for the effective bulk and shear moduli of the material which, however, would be accurate only when the inclusion volume fraction ϕ_2 is small. As such, the approximation is a type of mean-field approach to the problem. Hence, suppose that a simple shear deformation at a large distance from the inclusion is imposed on the composite material. Then, according to the theoretical developments of Section 7.1.4, we may write

$$2\mu_e\langle e\rangle = 2\mu_1\langle e\rangle + \frac{1}{\Omega}\int_{\Omega_i}(\lambda_2\epsilon_{kk}\delta_{ij} + 2\mu_2\epsilon_{ij} - \lambda_1\epsilon_{kk}\delta_{ij} - 2\mu_1\epsilon_{ij})d\Omega_i, \tag{101}$$

where μ_e is the effective shear modulus of the composite, Ω_i is the volume of the spherical inclusion, and $\langle e\rangle$ is the mean shear strain. Eshelby (1957) showed that a single ellipsoidal inclusion in an infinite matrix is in a state of homogeneous deformation, corresponding to what is imposed on the matrix at large distances from the inclusion. Therefore, using Eshelby's result, Eq. (101) can be rewritten as

$$\mu_e\langle e\rangle = \mu_1\langle e\rangle + \phi_2(\mu_2 - \mu_1)\langle e_2\rangle, \tag{102}$$

where $\phi_2 = \Omega_i/\Omega$ is the sphere's volume fraction, and $\langle e_2\rangle$ is the uniform shear in the inclusion. Solving Eq. (102) for the effective shear modulus μ_e, we obtain

$$\frac{\mu_e - \mu_1}{\mu_2 - \mu_1} = \frac{\langle e_2\rangle}{\langle e\rangle}\phi_2, \tag{103}$$

which predicts a linear dependence of μ_e on the sphere's volume fraction. In a similar way, we obtain an expression for the effective bulk modulus K_e of the composite:

$$\frac{K_e - K_1}{K_2 - K_1} = \frac{\langle \epsilon_2\rangle}{\langle \epsilon\rangle}\phi_2, \tag{104}$$

which is completely similar to Eq. (103); here $\langle\epsilon\rangle$ is the average dilatational strain, and $\langle\epsilon_2\rangle$ is the corresponding uniform strain in the inclusion. Equations (103) and (104) resemble somewhat the Maxwell–Garnett approximation for the effective conductivity and dielectric constant of composite solids described in Section 4.9.4.

7.3.2 Nondilute Dispersions

To make further progress and understand the elastic properties of composite solids in the non-dilute limit, one must go beyond the single-sphere limit analyzed in the last section. In order to be able to do this, we first derive and discuss a general expression for the elastic moduli of a composite containing a distribution of spherical inclusions. The analysis that we present here is due to Chen and Acrivos (1978b). One goal of discussing their analysis here is to demonstrate the complexities that are involved in analyzing the elastic properties of composite materials, even when they are modeled as a seemingly simple dispersion of spheres embedded in a uniform background matrix.

With the use of the divergence theorem, Eq. (40) can be rewritten as

$$\langle\sigma_{ij}\rangle = \lambda_1\langle\epsilon_{kk}\rangle\delta_{ij} + 2\mu_1\langle\epsilon_{ij}\rangle + \frac{1}{\Omega}\sum_{m=1}^{N}\int_{A_m}[x_j\sigma_{ik}n_k - \lambda_1 n_k u_k\delta_{ij} - \mu_1(u_i n_j + u_j n_i)]dA_m, \tag{105}$$

where A_m is the surface of the mth inclusion, $\mathbf{x}$ (with components x_i) is the position vector with respect to a fixed origin, and $\mathbf{n}$ (with components n_i) is the unit outer vector normal to A_m. The quantity

$$S_{ij} = \int_{A_r}[x_j\sigma_{ik}n_k - \lambda_1 u_k n_k\delta_{ij} - \mu_1(u_i n_j + u_j n_i)]dA_r, \tag{106}$$

is called the *stresslet* and, as shown below, is the key to calculating the average stress field in the material and hence its elastic properties. Equation (105) is now rewritten as

$$\langle\sigma_{ij}\rangle = \lambda_1\langle\epsilon_{kk}\rangle\delta_{ij} + 2\mu_1\langle\epsilon_{ij}\rangle + n\langle S_{ij}\rangle, \tag{107}$$

where $n = \phi_2/\Omega_p$ is the number density of the inclusions in the material, ϕ_2 is their volume fraction, and $\Omega_p = 4\pi a^3/3$ is the volume of a single particle of radius a. The quantity $\langle S_{ij}\rangle$ is the volume average of S_{ij} over all the particles, and is given by

$$\langle S_{ij}\rangle = \int S_{ij}(C|0)P(C|0)dC.$$

Here, $P(C|0)$ is the probability density function of the configuration C of the material with the reference sphere at the origin. If the spheres volume fraction $\phi_2 \ll 1$, so that the particle-particle interactions can be neglected, $\langle S_{ij}\rangle$ can be approximated by $S_{ij}^{(0)}$, the value of S_{ij} for a single inclusion in an infinite matrix (see above). In this limit, Eq. (107) becomes

$$\langle\sigma_{ij}\rangle = \lambda_1\langle\epsilon_{kk}\rangle\delta_{ij} + 2\mu_1\langle\epsilon_{ij}\rangle + \frac{\phi_2}{v_p}S_{ij}^{(0)} + O(\phi_2^2). \tag{108}$$

If the applied strain at infinity is set equal to the bulk strain $\langle\epsilon_{ij}\rangle$, then, the solution for $S_{ij}^{(0)}$ is given by

$$S_{ij}^{(0)} = v_p\gamma_1\left(K_1 + \frac{4}{3}\mu_1\right)\langle\epsilon_{kk}\rangle\delta_{ij} + 40\pi a^3\mu_1\gamma_2(1-\nu_{p1})\left(\langle\epsilon_{ij}\rangle - \frac{1}{3}\langle\epsilon_{kk}\rangle\delta_{ij}\right), \tag{109}$$

where $\gamma_1 = 3(K_2 - K_1)/(3K_2 + 4\mu_1)$, and $\gamma_2 = (\beta - 1)/[2\beta(4 - 5\nu_{p1}) + (7 - 5\nu_{p1})]$, with $\beta = \nu_{p2}/\nu_{p1}$ being the ratio of the Poisson's ratios of the inclusions (phase 2) and the matrix (phase 1).

Therefore, if the effective moduli are defined by

$$\langle\sigma_{ij}\rangle \equiv \lambda_e\langle\epsilon_{kk}\rangle\delta_{ij} + 2\mu_e\langle\epsilon_{ij}\rangle, \tag{110}$$

we obtain

$$\frac{\mu_e}{\mu_1} = 1 + 15\gamma_2(1 - \nu_{p1})\phi_2 + O(\phi_2^2), \tag{111}$$

$$\frac{\lambda_e}{\lambda_1} = 1 + \frac{1}{\lambda_1}\left[\gamma_1\left(K_1 + \frac{4}{3}\mu_1\right) - 10\mu_1\gamma_2(1 - \nu_{p1})\right]\phi_2 + O(\phi_2^2), \tag{112}$$

which are similar to what we already obtained in the last section with a single spherical inclusion, except that they now contain an $O(\phi_2^2)$ term.

To derive the $O(\phi_2^2)$ terms of the above expressions, one must compute the dipole, $\mathbf{S}^{(1)} = \mathbf{S} - \mathbf{S}^{(0)}$ [or, in component form, $S_{ij}^{(1)} = S_{ij} - S_{ij}^{(0)}$], on the reference sphere, due to the presence of the second sphere, which, due to linearity of the problem, must be linear in $\langle \boldsymbol{\epsilon} \rangle$ (or, in component form, must be linear in $\langle \epsilon_{ij} \rangle$). Therefore, its general form is given by

$$S_{ij}^{(1)} = F_1\langle\epsilon_{kk}\rangle\delta_{ij} + F_2\langle\epsilon_{kl}\rangle\frac{r_k r_l}{R^2}\delta_{ij} + F_3\langle\epsilon_{kk}\rangle\frac{r_i r_j}{R^2} + F_4\left(\langle\epsilon_{ij}\rangle - \frac{1}{3}\langle\epsilon_{kk}\rangle\delta_{ij}\right)$$
$$+F_5\langle\epsilon_{kl}\rangle\left(\frac{r_i r_k\delta_{jl} + r_j r_k\delta_{il}}{R^2} - \frac{2}{3}\frac{r_k r_l}{R^2}\delta_{ij}\right) + F_6\langle\epsilon_{kl}\rangle\frac{r_k r_l}{R^2}\left(\frac{r_i r_j}{R^2} - \frac{1}{3}\delta_{ij}\right), \tag{113}$$

where the scalar functions F_i depend on the properties of the two spheres and $\rho = a/R$, with $R = |\mathbf{r}|$ being the distance between the centers of two spheres of radius a, assuming that the reference sphere is at the origin and the second sphere is at $\mathbf{r} = \{r_i\}$. The functions F_i are given by

$$\begin{aligned}
F_1 &= -\frac{80}{3}\pi a^3\mu_1(1 - \nu_{p1})\gamma_1\gamma_2\rho^3 + \frac{100}{3}\pi a^3\mu_1(1 - \nu_{p1})(5 - 4\nu_{p1})\gamma_2^2\rho^3 + F_1', \\
F_2 &= 40\pi a^3\mu_1(1 - \nu_{p1})\gamma_1\gamma_2\rho^3 + F_2', \\
F_3 &= 40\pi a^3\mu_1(1 - \nu_{p1})\gamma_1\gamma_2\rho^3 - 100\pi a^3\mu_1(1 - \nu_{p1})(5 - 4\nu_{p1})\gamma_2^2\rho^3 + F_3', \\
F_4 &= -200\pi a^3\mu_1(1 - \nu_{p1})(1 - 2\nu_{p1})\gamma_2^2\rho^3 + F_4', \\
F_5 &= -600\pi a^3\mu_1(1 - \nu_{p1})\nu_{p1}\gamma_2^2\rho^3 + F_5', \\
F_6 &= 1500\pi a^3\mu_1(1 - \nu_{p1})\gamma_2^2\rho^3 + F_6'.
\end{aligned} \tag{114}$$

We now write Eq. (107) as

$$\langle\sigma_{ij}\rangle = \lambda_1\langle\epsilon_{kk}\rangle\delta_{ij} + 2\mu_1\langle\epsilon_{ij}\rangle$$
$$+ nS_{ij}^{(0)} + n\int_{\mathbf{r}}\left\{S_{ij}^{(1)}(\mathbf{r}|0)P(\mathbf{r}|0) - S_{ij}^{(0)}[\boldsymbol{\epsilon}^{(1)}]P(\mathbf{r})\right\}d\mathbf{r} + O(\phi_2^3). \tag{115}$$

Here, $\boldsymbol{\epsilon}^{(1)} = \boldsymbol{\epsilon} - \langle\boldsymbol{\epsilon}\rangle$ is the excess strain at the origin due to a single sphere being at $\mathbf{r}$ undergoing the applied strain $\langle\boldsymbol{\epsilon}\rangle$. As pointed out by Jeffrey (1974), the second term of the integral in Eq. (115) is necessary in order to make the integral absolutely convergent.

We now need to compute the probabilities $P(\mathbf{r}|0)$ and $P(\mathbf{r})$ which are, respectively, the probability of having a sphere at $\mathbf{r}$ with and without a sphere at the origin. For a random distribution of spheres in a matrix (the Swiss-cheese model) one has (see Chapter 3), $P(\mathbf{r}) = n$, where n is the number density of the spheres. To specify $P(\mathbf{r}|0)$ one assumes that the second sphere can be anywhere in the matrix, but cannot overlap with the test sphere. Therefore,

$$P(\mathbf{r}|0) = \begin{cases} 0 & R < 2a \\ n & R \geq 2a, \end{cases}$$

where R is the distance between the centers of the two spheres of radius a.

If (the center of) a sphere is at $\mathbf{r} = \{r_i\}$, and the applied strain at infinity is $\langle\epsilon\rangle$, then Eshelby (1961) calculated the excess strain $\boldsymbol{\epsilon}^{(1)}$ at the origin:

$$\epsilon_{ij}^{(1)} = \frac{1}{8\pi(1-\nu_{p1})} e_{kl} \frac{\partial^4 h}{\partial r_i \partial r_j \partial r_k \partial r_l}$$
$$-\frac{1}{4\pi}\left(e_{ik}\frac{\partial^2 \tilde{h}}{\partial r_k \partial r_j} + e_{jk}\frac{\partial^2 \tilde{h}}{\partial r_k \partial r_i}\right) - \frac{\nu_{p1}}{4\pi(1-\nu_{p1})} e_{kk}\frac{\partial^2 \tilde{h}}{\partial r_i \partial r_j}, \tag{116}$$

with

$$e_{ij} = -\frac{3(1-\nu_{p1})}{1+\nu_{p1}}\gamma_1\langle\epsilon_{kk}\rangle\delta_{ij} - 15\gamma_2(1-\nu_{p1})\left(\langle\epsilon_{ij}\rangle - \frac{1}{3}\langle\epsilon_{kk}\rangle\delta_{ij}\right),$$

$$h = \begin{cases} \pi a^4 + \frac{2}{3}\pi a^2 R^2 - \frac{\pi}{15}R^4 & R \leq a \\ \frac{4\pi}{3}a^3 R + \frac{4\pi}{15}\frac{a^5}{R} & R \geq a, \end{cases}$$

$$\tilde{h} = \begin{cases} 2\pi a^2 - \frac{2\pi}{3}R^2 & R \leq a \\ \frac{4\pi a^2}{3}\frac{1}{R} & R \geq a. \end{cases}$$

Therefore, one obtains

$$\langle\sigma_{ij}\rangle = \left(A - \frac{2}{3}B\right)\langle\epsilon_{kk}\rangle\delta_{ij} + 2B\langle\epsilon_{ij}\rangle, \tag{117}$$

so that

$$K_e = A, \quad \mu_e = B, \tag{118}$$

where

$$A = K_1\left\{1 + \gamma_1\left(1 + \frac{4\mu_1}{3K_1}\right)\phi_2 + \gamma_1^2\left(1 + \frac{4\mu_1}{3K_1}\right)\right.$$
$$\left.\times\left[1 + \frac{27\int_{2a}^{\infty}\left(F_1' + \frac{1}{3}F_2' + \frac{1}{3}F_3'\right)R^2 dR}{4\pi a^6\gamma_1^2(3K_1 + 4\mu_1)}\right]\phi_2^2\right\}, \tag{119}$$

$$B = \mu_1 \left\{ 1 + 15\gamma_2(1-\nu_{p1})\phi + 30\gamma_2^2(1-\nu_{p1})(4-5\nu_{p1}) \right.$$
$$\left. \times \left[1 + \frac{3\int_{2a}^{\infty}\left(F_4' + \frac{2}{3}F_5' + \frac{2}{15}F_6'\right)R^2 dR}{80\pi a^6 \mu_1 \gamma_2^2 (1-\nu_{p1})(4-5\nu_{p1})} \right] \phi_2^2 \right\}. \tag{120}$$

Thus, the procedure to calculate the effective elastic properties of the composite material is as follows. One first solves the problem for a composite that consists of a matrix and two spherical inclusions. The solution of this problem is described and discussed in the following section. From this solution, $\mathbf{S} = \{S_{ij}\}$, $\mathbf{S}^{(0)} = \{S_{ij}^{(0)}\}$, and hence $\mathbf{S}^{(1)} = \mathbf{S} - \mathbf{S}^{(0)}$ are computed. Using Eq. (114), the functions F_i and hence, via Eqs. (115), F_i' are computed, from which the elastic moduli $K_e = A$ and $\mu_e = B$ are calculated. We now describe and discuss the technique for deriving the solution of the two-sphere problem.

7.3.3 Two Spherical Inclusions

This problem was analyzed by Chen and Acrivos (1978a). Due to linearity of the problem, its solution can be obtained as a superposition of the solutions for the following four independent applied strains:

$$(1)\ \epsilon_{ij} = \delta_{ij}, \quad (2)\ \epsilon_{ij} = \delta_{i1}\delta_{j1} + \delta_{i2}\delta_{j2} - 2\delta_{i3}\delta_{j3},$$
$$(3)\ \epsilon_{ij} = \delta_{i1}\delta_{j2} + \delta_{i2}\delta_{j1}, \quad (4)\ \epsilon_{ij} = \delta_{i2}\delta_{j3} + \delta_{i3}\delta_{j2}, \tag{121}$$

where the x_3-axis is along the line of centers of the spheres. The first two strains give rise to axisymmetric problems that can be solved for the general case of two elastic spheres in an infinite matrix. Cases (3) and (4) are, however, more complex, and therefore Chen and Acrivos (1978a) considered the limiting cases in which the two inclusions were either totally rigid or were cavities. In what follows, we summarize and discuss their solutions.

The centers of the two spheres are located at $O(x, y, z)$ and $O_1(x_1, y_1, z_1)$, where zz_1 is along the line that connects the centers, with R being the distance between them. The two sets of coordinates, one for each sphere, are related by, $x = x_1$, $y = y_1$, and $z = z_1 + R$. Therefore, in spherical coordinates, we have

$$\begin{aligned} x = r\sin\theta\cos\alpha x_1 &= r_1 \sin\theta_1 \cos\alpha \\ y = r\sin\theta\sin\alpha y_1 &= r_1\sin\theta_1\sin\alpha \\ z = r\cos\theta z_1 &= r_1\cos\theta_1 \end{aligned} \tag{122}$$

where $r \geq 0$, $r_1 \leq \infty$, $\theta \geq 0$, $\theta_1 \leq \pi$, and $0 \leq \alpha \leq 2\pi$.

In the absence of body forces, the general solution for the displacement equations,

$$\frac{\partial^2 u_j}{\partial x_i \partial x_j} + (1-2\nu_p)\frac{\partial^2 u_i}{\partial x_j^2} = 0, \tag{123}$$

is represented in terms of the Boussinesq-Papkovich stress functions (Eubanks and Sternberg, 1956) as

$$2\mu u_i = \frac{\partial \psi}{\partial x_i} + x_j \frac{\partial \tau_j}{\partial x_i} - (3 - 4\nu_p)\tau_i = \frac{\partial \psi}{\partial x_i} + \frac{\partial (x_j \tau_j)}{\partial x_i} - 4(1 - \nu_p)\tau_i, \quad (124)$$

with $\nabla^2 \psi = \nabla^2 \tau_i = 0$. It is well known (Naghdi and Hsu, 1961) that, of the four harmonic functions given in Eq. (124), only three are independent. However, by retaining all of them one can arbitrarily either eliminate one or combine two of them in order to analyze particular cases. When the problem is axisymmetric, only two functions are needed, namely ψ and τ_3, with x_3 being the axis of symmetry.

The boundary conditions that must be satisfied are that, far from the inclusions the displacement field must approach that of the corresponding loading, and that the displacement and the traction be continuous on the surface of each sphere. If the spheres are rigid, the total force and couple acting on each of them must be zero.

Axisymmetric solution for $\epsilon_{ij}^\infty = \delta_{ij}$

With this boundary condition, the corresponding displacement and stress fields are

$$u_i^\infty = x_i^* \quad (125)$$

$$\sigma_{ij}^\infty = 3K_1 \delta_{ij} \quad (126)$$

where x_i^* is at the midpoint between the two spheres' centers. In spherical coordinates with their origin at O, Eqs. (125) and (126) become

$$u_r^\infty = r - \frac{1}{2} R \cos\theta, \quad u_\theta^\infty = \frac{1}{2} R \sin\theta, \quad u_\alpha^\infty = 0 \quad (127)$$

$$\sigma_{rr}^\infty = \sigma_{\theta\theta}^\infty = \sigma_{\alpha\alpha}^\infty = 3K_1, \quad \sigma_{r\theta}^\infty = \sigma_{r\alpha}^\infty = \sigma_{\theta\alpha}^\infty = 0. \quad (128)$$

Since the solution to this case involves only two stress functions ψ and τ_3, which depend only on r and θ, the displacement field is represented as:

(a) *Outside the spheres:*

$$2\mu_1 \mathbf{u} = 2\mu_1 \mathbf{u}^\infty + \nabla\zeta + \nabla(z\psi + z_1\psi_1) - 4(1 - \nu_{p1})(\psi + \psi_1)\mathbf{e}_z \quad (129)$$

$$\zeta = \sum_{n=0}^{\infty} \left[A_n \frac{a^{n+3}}{r^{n+1}} P_n(\cos\theta) + A_n^1 \frac{a^{n+3}}{r_1^{n+1}} P_n(\cos\theta_1) \right] \quad (130)$$

$$\psi = \sum_{n=0}^{\infty} C_n \frac{a^{n+2}}{r^{n+1}} P_n \cos\theta, \quad \psi_1 = \sum_{n=0}^{\infty} C_n \frac{a^{n+2}}{r_1^{n+1}} P_n(\cos\theta_1), \quad (131)$$

where $P_n(x)$ is the Legendre polynomial of order n.

(b) *Inside sphere centered at O:*

$$2\mu_2 \mathbf{u} = \nabla\eta + \nabla(z\xi) - 4(1-\nu_{p2})\xi \mathbf{e}_z. \tag{132}$$

with

$$\eta = \sum_{n=0}^{\infty} B_n \frac{r^n}{a^{n-2}} P_n(\cos\theta), \quad \xi = \sum_{n=0}^{\infty} D_n \frac{r^n}{a^{n-1}} P_n(\cos\theta). \tag{133}$$

A similar expression can be written down for the region inside sphere centered at O_1. The relations between the harmonic functions referred to each of the two spheres are given by Hobson (1931):

$$\frac{P_n^m(\cos\theta)}{r^{n+1}} = \sum_{s=m}^{\infty} \frac{(s+n)!}{(s+m)!(n-m)!} \frac{r_1^s}{R^{s+n+1}} (-1)^{s+m} P_s^m(\cos\theta_1). \tag{134}$$

$$\frac{P_n^m(\cos\theta_1)}{r_1^{n+1}} = \sum_{s=m}^{\infty} \frac{(s+n)!}{(s+m)!(n-m)!} \frac{r^s}{R^{s+n+1}} (-1)^{n+m} P_s^m(\cos\theta). \tag{135}$$

where $r_1 > 0$ and $r < R$.

In order to satisfy the boundary conditions at the surface of the sphere centered at O, Eq. (129) should be written in terms of the (r, θ, α) coordinates. Using $z_1 = z + R$ and Eqs. (134) and (135), we obtain

$$2\mu_1 \mathbf{u} = 2\mu_1 \mathbf{u}^\infty + \nabla(\zeta - R\psi_1) + \nabla[z(\psi + \psi_1)] - 4(1-\nu_{p1})(\psi - \psi_1)\mathbf{e}_z, \tag{136}$$

$$\zeta - R\psi_1 = \sum_{n=0}^{\infty} \left[A_n \frac{a^{n+3}}{r^{n+1}} P_n(\cos\theta) + (A_n^1 a - RC_n^1) a^{n+2} \sum_{s=0}^{\infty} g_{ns}(-1)^n r^s P_s(\cos\theta) \right], \tag{137}$$

$$\psi + \psi_1 = \sum_{n=0}^{\infty} \left[C_n \frac{a^{n+2}}{r^{n+1}} P_n(\cos\theta) + C_n^1 a^{n+2} \sum_{s=0}^{\infty} g_{ns}(-1)^n r^s P_s(\cos\theta) \right], \tag{138}$$

where

$$g_{ns} = \frac{(s+n)!}{s!\, n!} R^{-(s+n+1)}.$$

The corresponding displacement and stress fields are then given by

$$u_r = u_r^\infty + \frac{1}{2\mu_1} \sum_{n=0}^{\infty} \Bigg\{ -A_n \frac{(n+1)a^{n+3}}{r^{n+2}} P_n - C_n \frac{a^{n+2}}{r^{n+1}} \frac{n+4-4\nu_{p1}}{2n+1}$$
$$\times[(n+1)P_{n+1} + nP_{n-1}] + (A_n^1 a - RC_n^1) a^{n+2} \sum_{s=0}^{\infty} g_{ns}(-1)^n r^{s-1} s P_s$$
$$+ C_n^1 a^{n+2} \sum_{s=0}^{\infty} g_{ns}(-1)^n r^s \frac{s-3+4\nu_{p1}}{2s+1} [(s+1)P_{s+1} + sP_{s-1}] \Bigg\}, \tag{139}$$

$$u_\theta = u_\theta^\infty - \frac{\sin\theta}{2\mu_1}\sum_{n=0}^{\infty}\left\{A_n\frac{a^{n+3}}{r^{n+2}}P_n' + C_n\frac{a^{n+2}}{r^{n+1}}\frac{1}{2n+1}\right.$$

$$\times[(n-3+4\nu_{p1})P_{n+1}' + (n+4-4\nu_{p1})P_{n-1}']$$

$$+(A_n^1 a - RC_n^1)a^{n+2}\sum_{s=0}^{\infty}g_{ns}(-1)^n r^{s-1}P_s' + C_n^1 a^{n+2}$$

$$\left.\times\sum_{s=0}^{\infty}g_{ns}(-1)^n r^s\frac{1}{2s+1}[(s-3+4\nu_{p1})P_{s-1}' + (s+4-4\nu_{p1})P_{s+1}')\right\} \quad (140)$$

$$u_\alpha = 0 \quad (141)$$

and

$$\sigma_{rr} = \sigma_{rr}^\infty + \sum_{n=0}^{\infty}\left\{A_n(n+1)(n+2)\frac{a^{n+3}}{r^{n+3}}P_n + +C_n\frac{a^{n+2}}{r^{n+2}}\frac{n+1}{2n+1}\right.$$

$$\times[(n+4-4\nu_{p1})nP_{n-1} + (n^2+5n+4-2\nu_{p1})P_{n+1}] + (A_n^1 a - RC_n^1)a^{n+2}$$

$$\times\sum_{s=0}^{\infty}g_{ns}(-1)^n s(s-1)r^{s-2}P_s + C_n^1 a^{n+2}\sum_{s=0}^{\infty}g_{ns}(-1)^n r^{s-1}\frac{s}{2s-1}$$

$$\left.\times[(s^2-3s-2\nu_{p1})P_{s-1} + (s+1)(s-3+4\nu_{p1})P_{s+1}]\right\} \quad (142)$$

$$\sigma_{r\theta} = \sigma_{r\theta}^\infty + \sin\theta\sum_{n=0}^{\infty}\left\{A_n(n+2)\frac{a^{n+3}}{r^{n+3}}P_n' + C_n\frac{a^{n+2}}{r^{n+2}}\frac{1}{2n+1}[(n^2+2n-1+2\nu_{p1})P_{n+1}'\right.$$

$$+(n+1)(n+4-4\nu_{p1})P_{n-1}'] - (A_n^1 a - RC_n^1)a^{n+2}\sum_{s=0}^{\infty}g_{ns}r^{s-2}(-1)^n(s-1)P_s'$$

$$\left.-C_n^1 a^{n+2}\sum_{s=0}^{\infty}g_{ns}\frac{r^{s-1}}{2s+1}[(s-3+4\nu_{p1})sP_{s+1}' + (s^2-2+2\nu_{p1})P_{s-1}']\right\}. \quad (143)$$

$$\sigma_{r\alpha} = 0. \quad (144)$$

In a similar way, the expressions for the displacement and the stress fields inside the sphere centered at O are obtained:

$$u_r^{(I)} = \frac{1}{2\mu_2}\sum_{n=0}^{\infty}\left\{D_n\frac{r^n}{a^{n-1}}\frac{n-3+4\nu_{p2}}{2n+1}[(n+1)P_{n+1} + nP_{n-1}] + B_n\frac{nr^{n-1}}{a^{n-2}}P_n\right\}, \quad (145)$$

$$u_\theta^{(I)} = \frac{-\sin\theta}{2\mu_2}\sum_{n=0}^{\infty}\left\{D_n\frac{r^n}{a^{n-1}}\frac{1}{2n+1}[(n-3+4\nu_{p2})P_{n+1}' + (n+4-4\nu_{p2})P_{n-1}']\right.$$

$$\left.+B_n\frac{r^{n-1}}{a^{n-2}}P_n'\right\}, \quad (146)$$

$$u_\alpha^{(I)} = 0 \quad (147)$$

and

$$\sigma_{rr}^{(I)} = \sum_{n=0}^{\infty} \left\{ D_n \frac{r^{n-1}}{a^{n-1}} \left[\frac{n(n+1)(n-3+4\nu_{p2})}{2n+1} P_{n+1} + \frac{n(n^2-3n-2\nu_{p2})}{2n+1} P_{n+1} \right] + B_n \frac{n(n-1)r^{n-2}}{a^{n-2}} P_n \right\}, \tag{148}$$

$$\sigma_{r\theta}^{(I)} = -\sin\theta \sum_{n=0}^{\infty} \left\{ D_n \frac{r^{n-1}}{a^{n-1}} \frac{1}{2n+1} \left[(n-3+4\nu_{p2}) n P'_{n+1} + (n^2-2+2\nu_{p2}) P'_{n-1} \right] + B_n \frac{r^{n-2}}{a^{n-2}} (n-1) P'_n \right\}, \tag{149}$$

$$\sigma_{r\alpha}^{(I)} = 0. \tag{150}$$

From the definition of the stresslet S_{ij}, Eq. (106), and the exterior solution, we can now easily determine that, for a sphere of radius a,

$$S_{ij} = \frac{4\pi a^3(1-\nu_{p1})}{1-2\nu_{p1}} \{A_0\delta_{ij} + C_1[\delta_{i1}\delta_{j1} + \delta_{i2}\delta_{j2} + (3-4\nu_{p1})\delta_{i3}\delta_{j3}]\}, \tag{151}$$

and

$$S_{ii} = \frac{4\pi a^3(1-\nu_{p1})}{1-2\nu_{p1}} [3A_0 + (5-4\nu_{p1})C_1]. \tag{152}$$

Since the loading given by Eqs. (125) and (125) is symmetric both with respect to the axis of symmetry of the two spheres and, for equal-sized spheres, to the plane of symmetry perpendicular to this axis, we have, $A_m = (-1)^m A_m^1$ and $C_m = (-1)^{m+1} C_m^1$, implying that the two sphere problem simplifies to that of determining the unknown coefficients A_m and C_m by satisfying the boundary conditions on the surface of sphere centered at O.

Substituting the exterior and the interior solutions, Eqs. (139)–(150), in the boundary conditions that require the continuity of displacement and traction at the interface between the matrix and the spheres, and using the orthogonality of the Legendre polynomials, one obtains a set of relations for the coefficients A_m, B_m, C_m and D_m. However, since the calculation of the effective bulk modulus K_e involves only S_{ii}, which depends only on A_0 and C_1, the interior coefficients B_m and D_m are eliminated and one obtains

$$A_0 + \frac{5-4\nu_{p1}}{3} C_1 = 2\mu_1\gamma_1 + \frac{2}{3}\gamma_1(1-2\nu_{p1}) \sum_{s=0}^{\infty} C_s(s+1)\rho^{s+2}, \tag{153}$$

and for $m \geq 0$

$$A_m = 2\mu_1\gamma_1\delta_{m0} + \sum_{s=0}^{\infty} a^{s+m} \{(A_s a + RC_s)[M_1 g_{ms} + M_2 a^2 g_{s(m+2)}] - C_s[M_3 g_{s(m-1)} + M_4 a^2 g_{s(m+1)} + M_5 a^4 g_{s(m+3)}]\}, \tag{154}$$

$$C_m = -\sum_{s=0}^{\infty} a^{s+m+1}\{M_6(A_s a + RC_s)g_{s(m+1)} - C_s(M_7 g_{sm} + M_8 a^2 g_{s(m+2)})\}, \tag{155}$$

where, as before, $\rho = a/R$, $\gamma_1 = 3(K_2 - K_1)/(3K_2 + 3\mu_1)$, $g_{ms} = [(s+m)!/s!m!]R^{-(s+m+1)}$, and the first few M_is are given by

$$M_1 = \frac{m(1-\beta)(m-1)(2m-1)}{2[\beta(m-1)(3m-4m\nu_{p1}+2-2\nu_{p1}) + (m^2+m+1-2m\nu_{p1}-\nu_{p1})]},$$

$$M_2 = \frac{(1-\beta)(2m+5)(m+1)(m+5-4\nu_{p1})}{2[\beta(m+1)(3m-4m\nu_{p1}+8-10\nu_{p1}) + (m^2+5m+7-2m\nu_{p1}-5\nu_{p1})]},$$

$$M_3 = \frac{m-4+4\nu_{p1}}{2m-1} M_1.$$

Equations (153)–(155) can be solved by expanding A_m and C_m as power series in ρ and obtaining a recurrence formula for the appropriate coefficients. Since to compute the stresslet S_{ij} only A_0 is needed, we quote the result:

$$\frac{A_0}{\gamma_1\mu_1} = 2 + \frac{5(1-\beta)(5-4\nu_{p1})}{2\beta(4-5\nu_{p1}) + (7-5\nu_{p1})}\rho^3 - \left[\frac{50(1-\beta)^2(5-4\nu_{p1})(2-\nu_{p1})}{[2\beta(4-5\nu_{p1}) + (7-5\nu_{p1})]^2} + \frac{20(1-\nu_{pM})(1-2\nu_{pM})}{2\beta(4-5\nu_{pM}) + (7-5\nu_{pM})}\gamma_1\right]\rho^6 + O(\rho^8). \tag{156}$$

Axisymmetric solution for $\epsilon_{ij}^{\infty} = \delta_{i1}\delta_{j1} + \delta_{i2}\delta_{j2} - 2\delta_{i3}\delta_{j3}$

In this case, the displacement and the stress fields at infinity are (with x_j^* taken from the midpoint of the centers of the two spheres)

$$u_i^{\infty} = x_1^*\delta_{i1} + x_2^*\delta_{i2} - 2x_3^*\delta_{i3}, \tag{157}$$

$$\sigma_{ij}^{\infty} = 2\mu_1(\delta_{i1}\delta_{j1} + \delta_{i1}\delta_{j2} - 2\delta_{i3}\delta_{j3}). \tag{158}$$

Since the system is also axisymmetric, the method is exactly the same as that for the first case, the only difference being in the field applied at infinity. The solutions for the unknown coefficients A_m and C_m are therefore similar to Eqs. (153)–(155), except that the terms before the summation signs are $5\mu_1(5-4\nu_{p1})\gamma_2\delta_{m0} + 6\mu_1\gamma_2\delta_{m2}$ in the equation for A_m, and $-15\mu_1\gamma_2\delta_{m1}$ in the equation for C_m, where, as before, $\gamma_2 = (\beta - 1)/[2\beta(4-5\nu_{p1}) + (7-5\nu_{p1})]$. Hence,

$$\frac{A_0}{\gamma_2\mu_1} = 5(5-4\nu_{p1}) + \left[-\frac{50(5-4\nu_{p1})(2-\nu_{p1})(1-\beta)}{2\beta(4-5\nu_{p1}) + (7-5\nu_{p1})} - 20\gamma_1(1-2\nu_{p1})\right]\rho^3 + O(\rho^5). \tag{159}$$

Asymmetric solution for $\epsilon_{ij}^{\infty} = \delta_{i1}\delta_{j2} + \delta_{i2}\delta_{j1}$

The corresponding displacement and stress at infinity are

$$u_i^{\infty} = x_2^* \delta_{i1} + x_1^* \delta_{i2}, \tag{160}$$

$$\sigma_{ij}^{\infty} = 2\mu_1(\delta_{i1}\delta_{j2} + \delta_{i2}\delta_{j1}). \tag{161}$$

Transforming to the center of the sphere centered at O in spherical coordinates yields

$$\begin{aligned} u_r^{\infty} &= \frac{r}{3} P_2^2(\cos\theta)\sin 2\alpha, \\ u_\theta^{\infty} &= -\frac{r\sin\theta}{6} P_2^{2'} \sin 2\alpha, \\ u_\alpha^{\infty} &= \frac{r}{3\sin\theta} P_2^2 \cos 2\alpha. \end{aligned} \tag{162}$$

$$\begin{aligned} \sigma_{rr}^{\infty} &= \frac{2\mu_M}{3} P_2^2 \sin 2\alpha, \\ \sigma_{r\theta}^{\infty} &= -\frac{\mu_M \sin\theta}{3} P_2^{2'} \sin 2\alpha, \\ \sigma_{r\alpha}^{\infty} &= \frac{2\mu_M}{3\sin\theta} P_2^2 \cos 2\alpha, \end{aligned} \tag{163}$$

where the P_n^m's are the associated Legendre polynomials and, $P_n^{m'}(x) = dP_n^m(x)/dx$. To satisfy the conditions at infinity, we take into account that the displacement and stresses are all proportional to $\cos 2\alpha$ or $\sin 2\alpha$, and choose the four stress functions as:

$$\psi = \psi^* \sin\alpha, \tag{164}$$

$$\tau_1 = \tau_1^* \sin\alpha, \quad \tau_2 = \tau_1^* \cos\alpha, \quad \tau_3 = \tau_3^* \sin 2\alpha, \tag{165}$$

where ψ^*, τ_1^* and τ_3^* depend only on r and θ. In Eqs. (164) and (165), the four stress functions are combined to yield three independent functions. The solution of the displacement field outside the two spheres is then represented by

$$\begin{aligned} 2\mu_1 \mathbf{u} = 2\mu_1 \mathbf{u}^{\infty} &+ \nabla(\psi^* \sin 2\alpha) + \nabla(x\tau \sin\alpha + y\tau\cos\alpha + z\xi \sin 2\alpha \\ &+ z_1\xi_1 \sin 2\alpha) - 4(1 - \nu_{p1})[\tau\sin\alpha, \tau\cos\alpha, (\xi + \xi_1)\sin 2\alpha], \end{aligned} \tag{166}$$

where

$$\begin{aligned} \psi^* &= \sum_{n=2}^{\infty} \left[A_n \frac{a^{n+3}}{r^{n+1}} P_n^2(\cos\theta) + A_n^1 \frac{a^{n+3}}{r_1^{n+1}} P_n^2(\cos\theta_1) \right], \\ \tau &= \sum_{n=1}^{\infty} \left[B_n \frac{a^{n+2}}{r^{n+1}} P_n^1(\cos\theta) + B_n \frac{a^{n+2}}{r_1^{n+1}} (\cos\theta_1) \right], \\ \xi &= \sum_{n=2}^{\infty} C_n \frac{a^{n+2}}{r^{n+1}} P_n^2(\cos\theta), \quad \xi_1 = \sum_{n=2}^{\infty} C_n^1 \frac{a^{n+2}}{r^{n+1}} P_n^2(\cos\theta_1). \end{aligned} \tag{167}$$

Since the boundary conditions on the surfaces of both spheres must be satisfied, Eq. (166) is first expressed in terms of the coordinates (r, θ, α) with respect to spheres centered at O and O_1 using $z = z_1 + R$, and the relations between the harmonic functions referred to each of the two spheres, Eqs. (134) and (135). Because the spheres are identical, the system is symmetric with respect to the plane passing through the midpoint of the distance between the spheres' centers and perpendicular to the common axis of the system. Therefore, one obtain, $A_n = (-1)^n A_n^1$, $B_n = (-1)^{n+1} B_n^1$, and $C_n = (-1)^{n+1} C_n^1$. Hence, only the boundary conditions on one of the spheres must be satisfied. For this applied strain, the only non-zero component of the stresslet of the reference sphere of radius a is

$$S_{21} = S_{12} = 8\pi a^3 (1 - \nu_{p1}) B_1. \tag{168}$$

Chen and Acrivos (1978a) derived the solutions for the limiting cases of either rigid particles or cavities.

In the case of the rigid inclusions, the displacements on the surfaces vanish, i.e., because of the symmetry of the problem, $u_i = 0$. Upon satisfying this condition on the surface of the sphere centered at O, one obtains a set of equations relating A_n, B_n and C_n which, when solved, yields the solutions in terms of power series in ρ. In particular,

$$B_1^I = 5 - \frac{25(1-2\nu_{p1})}{2(4-5\nu_{p1})}\rho^3 - \frac{15}{4-5\nu_{p1}}\rho^5 + \frac{125(1-2\nu_{p1})^2}{4(4-5\nu_{p1})^2}\rho^6 + O(\rho^8). \tag{169}$$

where the superscript I denotes multiplication of the corresponding quantity by $2(4 - 5\nu_{p1})/\mu_1$.

When the inclusions are cavities, the proper boundary condition becomes $\sigma_{ij} n_j = 0$, where $\mathbf{n}$ is the unit normal vector of the surfaces. Thus,

$$\sigma_{rr} = \sigma_{r\theta} = \sigma_{r\alpha} = 0, \quad \text{at } r = a. \tag{170}$$

After satisfying the boundary conditions, one again obtains a set of equations for the coefficients A_n, B_n and C_n, the solutions of which are power series in ρ. In particular,

$$B_1^{II} = -5 - \frac{25(1-2\nu_{p1})}{7-5\nu_{p1}}\rho^3 - \frac{30}{7-5\nu_{p1}}\rho^5 - \frac{125(1-2\nu_{p1})^2}{(7-5\nu_{p1})^2}\rho^6 - \left[\frac{300(1-2\nu_{p1})}{7-5\nu_{p1}} + \frac{75(9-44\nu_{p1}+77\nu_{p1}^2)}{2(13-7\nu_{p1})(7-5\nu_{p1})}\right]\rho^8 + O(\rho^9), \tag{171}$$

where superscript II indicates multiplication of the corresponding quantity by $(7 - \nu_{p1})/\mu_1$.

Asymmetric solution for $\epsilon_{ij}^\infty = \delta_{i2}\delta_{j3} + \delta_{j3}\delta_{j2}$

The corresponding displacement and stress at infinity are

$$u_i^\infty = x_2^*\delta_{i3} + x_3^*\delta_{i2}, \tag{172}$$

$$\sigma_{ij}^{\infty} = 2\mu_1(\delta_{i2}\delta_{j3} + \delta_{i3}\delta_{j2}). \tag{173}$$

In the new coordinates with the origin being at O, Eqs. (172) and (173) become

$$\begin{aligned} u_r^{\infty} &= \left(\frac{2}{3} r P_2^1 - \frac{1}{2} R P_1^1\right) \sin\alpha, \\ u_\theta^{\infty} &= \left(-\frac{1}{3} r P_2^{1'} + \frac{1}{2} R P_1^{1'}\right) \sin\theta \sin\alpha \\ u_\alpha^{\infty} &= \left(\frac{1}{3} r P_2^1 - \frac{1}{2} R P_1^1\right) \frac{\cos\alpha}{\sin\theta}, \end{aligned} \tag{174}$$

$$\begin{aligned} \sigma_{rr}^{\infty} &= \frac{4}{3}\mu_1 P_2^1 \sin\alpha, \\ \sigma_{r\theta}^{\infty} &= -\frac{2}{3}\mu_1 P_2^{1'} \sin\theta \sin\alpha \\ \sigma_{r\alpha}^{\infty} &= \frac{2}{3\sin\theta}\mu_1 P_2^1 \cos\alpha. \end{aligned} \tag{175}$$

Since the displacements and stresses are all proportional to either $\sin\alpha$ or $\cos\alpha$, the four stress functions are chosen as

$$\begin{aligned} \psi &= \psi^*(r,\theta)\sin\alpha, & \tau_1 &= 0, \\ \tau_2 &= \tau_2^*(r,\theta), & \tau_3 &= \tau_3^*(r,\theta)\sin\alpha. \end{aligned} \tag{176}$$

Using the multipole expansion method (see Chapter 4), the displacement field outside the two spheres is expressed as

$$\begin{aligned} 2\mu_1\mathbf{u} = 2\mu_1\mathbf{u}^{\infty} &+ \nabla(\psi^*\sin\alpha) + \nabla(y\tau + z\xi\sin\alpha + z_1\xi_1\sin\alpha) \\ &- 4(1-\nu_{p1})[0, \tau, (\xi + \xi_1)\sin\alpha], \end{aligned} \tag{177}$$

where

$$\begin{aligned} \psi^* &= \sum_{n=1}^{\infty}\left[A_n \frac{a^{n+3}}{r^{n+1}} P_n^1(\cos\theta) + A_n^1 \frac{a^{n+3}}{r_1^{n+1}} P_n^1(\cos\theta_1)\right], \\ \tau &= \sum_{n=0}^{\infty}\sum\left[B_n \frac{a^{n+2}}{r^{n+1}} P_n(\cos\theta) + B_n^1 \frac{a^{n+2}}{r_1^{n+1}} P_n(\cos\theta_1)\right], \\ \xi &= \sum_{n=1}^{\infty} C_n \frac{a^{n+2}}{r^{n+1}} P_n^1(\cos\theta), \quad \xi_1 = \sum_{n=1}^{\infty} C_n^1 \frac{a^{n+2}}{r_1^{n+1}} P_n^1(\cos\theta_1). \end{aligned} \tag{178}$$

One again follows the procedures described above by transforming the origin to O and satisfying the boundary conditions. However, unlike the previous case, there is no plane of symmetry in this system. Nevertheless, one finds that the solution of this case should be symmetric with respect to the midpoint of the spheres' centers.

Thus, one has

$$u_r(r,\theta,\alpha) = u'_r(r,\pi-\theta,\pi+\alpha),$$
$$u_\theta(r,\theta,\alpha) = u'_\theta(r,\pi-\theta,\pi+\alpha), \tag{179}$$
$$u_\alpha(r,\theta,\alpha) = u'_\alpha(r,\pi-\theta,\pi+\alpha),$$

where the primes denote the quantities referred to the origin being at O_1. Therefore, one obtains, $A_n = (-1)^n A_n^1$, $B_n = (-1)^{n+1} B_n^1$ and $C_n = (-1)^{n+1} C_n^1$, which satisfy the two-sphere problem. In order to solve for the unknown coefficients A_n, B_n and C_n, the boundary conditions on the surface of the sphere centered at O must be satisfied. The stresslet for this applied strain is then given by

$$S_{ij} = 8\pi a^3 B_1(1-\nu_{p1})(\delta_{i2}\delta_{j3} + \delta_{i3}\delta_{j2}). \tag{180}$$

For the case of the rigid inclusions, the displacement must be continuous and the net force and torque must vanish there. Since on the sphere centered at O,

$$\text{total force} = \int \sigma_{ij} n_j dS = 8\pi a^2 B_0(1-\nu_{p1})\delta_{i2},$$
$$\text{total torque} = \int \epsilon_{ijk} n_j \sigma_{km} n_m dS = 8\pi a^2 (C_1 - B_1)\delta_{i1},$$

we conclude that $B_0 = 0$ and $C_1 = B_1$, conditions that are necessary for obtaining a unique solution for rigid spheres, but not for cavities. The rigid particle displacement on sphere centered at O under the applied strain $\epsilon_{ij}^\infty = \delta_{i2}\delta_{j3} + \delta_{i3}\delta_{j2}$ is expressed in the form, $u_i^{(p)} = V^{(p)}\delta_{i2} + a\Omega^{(p)}\epsilon_{i1k} n_k$, where the function $V^{(p)}$ and $\Omega^{(p)}$ depend upon $\rho = a/R$ and the elastic moduli of the matrix. After satisfying the boundary conditions on the surface of the sphere centered at O, a set of relations for A_n, B_n and C_n is obtained which are solved using the method described above. In particular,

$$B_1^I = 5 - \frac{25(1+\nu_{p1})}{2(4-5\nu_{p1})}\rho^3 + \frac{60}{4-5\nu_{p1}}\rho^5 + \frac{125(1+\nu_{p1})^2}{4(4-5\nu_{p1})^2}\rho^6$$
$$+ \left[\frac{150(18-23\nu_{p1}+14\nu_{p1}^2)}{(4-5\nu_{p1})(11-14\nu_{p1})} - \frac{225(4+\nu_{p1})}{2(4-5\nu_{p1})^2}\right]\rho^8 + O(\rho^9), \tag{181}$$

where, as before, the superscript I denotes multiplication of the corresponding quantity by $2(4-5\nu_{p1})/\mu_1$. Similarly, in case of cavities, one obtains

$$B_1^{II} = -5 - \frac{25(1+\nu_{p1})}{7-5\nu_{p1}}\rho^3 + \frac{120}{7-5\nu_{p1}}\rho^5 - \frac{125(1+\nu_{p1})^2}{(7-5\nu_{p1})^2}\rho^6$$
$$+ \left[\frac{15(1+\nu_{p1})}{7-5\nu_{p1}} + \frac{1200(1+\nu_{p1})}{(7-5\nu_{p1})^2} - \frac{30(139+38\nu_{p1}+14\nu_{p1}^2)}{(7-5\nu_{p1})(13-7\nu_{p1})}\right]\rho^8 + O(\rho^9). \tag{182}$$

where, as before, B_1^{II} is defined as B_1 multiplied by the quantity $(7-5\nu_{p1})/\mu_1$.

7.4 Exact Strong-Contrast Expansions

Torquato (1997, 1998) considered the general problem of exact determination of the effective stiffness tensor $\mathbf{C}_e$ of a macroscopically anisotropic two-phase material with an arbitrary, but statistically homogeneous, morphology. For the case in which the variations in the phase stiffness tensors are small, formal perturbation series were derived by Dederichs and Zeller (1973), Gubernatis and Krumhansl (1975), and Willis (1981). Although such perturbation expansions, which are valid at all volume fractions, have fundamental importance, as they lead to rigorous bounds on the effective properties of heterogeneous materials, they provide accurate predictions only when the contrast between the properties of the two phases is small.

In the problems of determining the effective conductivity and elastic moduli of heterogeneous materials, one must usually deal with conditionally-convergent integrals, i.e., integrals that depend on the materials' macroscopic shape. However, this is physically unacceptable as the effective properties of a material cannot depend on its macroscopic shape. Therefore, proper techniques must be developed in order to circumvent this difficulty, and develop formulations that involve absolutely-convergent integrals. We already described in Sections 4.3 and 4.5 how this difficulty is circumvented in the problem of determining the effective conductivity of disordered materials. To remove the conditionally-convergent integrals from the solutions of these problems, a type of renormalization is carried out by identifying the contributions of such integrals and replacing them by convergent terms that make the same contributions. Torquato's solution for the elasticity problem are free of such restrictions. In what follows we describe his method and discuss the results.

7.4.1 Integral Equation for the Cavity Strain Field

Consider a large but finite-sized, ellipsoidal macroscopically anisotropic composite material in arbitrary space dimension d, comprised of two isotropic phases. The microstructure possesses a characteristic microscopic length scale which is much smaller than the smallest semi-axes of the ellipsoid, and thus the material is statistically homogeneous. For now it is assumed that the volume is finite, but the passage to the infinite-volume limit will be taken at the end of the analysis. The local stiffness tensor $\mathbf{C}(\mathbf{x})$ is given in terms of the local bulk modulus $K(\mathbf{x})$ and the local shear modulus $\mu(\mathbf{x})$ by the relation

$$\mathbf{C}(\mathbf{x}) = dK(\mathbf{x})\mathbf{\Lambda}_h + 2\mu(\mathbf{x})\mathbf{\Lambda}_s, \tag{183}$$

or in component form

$$(\Lambda_h)_{ijkl} = \frac{1}{d}\delta_{ij}\delta_{kl}, \qquad i, j, k, l = 1, 2, ..., d, \tag{184}$$

$$(\Lambda_s)_{ijkl} = \frac{1}{2}(\delta_{ik}\delta_{jl} + \delta_{il}\delta_{jk}) - \frac{1}{d}\delta_{ij}\delta_{kl}, \qquad i, j, k, l = 1, 2, ..., d. \tag{185}$$

The tensor $\mathbf{\Lambda}_h$ projects onto fields that are everywhere isotropic, i.e., hydrostatic fields, whereas the tensor $\mathbf{\Lambda}_s$ projects onto fields that are everywhere trace-free, i.e., shear fields. Thus, we refer to $\mathbf{\Lambda}_h$ and $\mathbf{\Lambda}_s$ as the *hydrostatic projection tensor* and the *shear projection tensor*, respectively. The following identities will be useful in subsequent discussions,

$$(\Lambda_h)_{ijkl} + (\Lambda_s)_{ijkl} = U_{ijkl} \tag{186}$$

$$(\Lambda_h)_{ijkl}(\Lambda_s)_{ijkl} = (\Lambda_h)_{ijmn}(\Lambda_s)_{mnkl} = 0 \tag{187}$$

$$(\Lambda_h)_{ijkl} I_{ijkl} = (\Lambda_h)_{ijkl}(\Lambda_h)_{ijkl} = 1, \tag{188}$$

$$(\Lambda_s)_{ijkl} I_{ijkl} = (\Lambda_s)_{ijkl}(\Lambda_s)_{ijkl} = \frac{1}{2}(d-1)(d+2), \tag{189}$$

$$(\Lambda_h)_{ijmn}(\Lambda_h)_{mnkl} = (\Lambda_h)_{ijkl}, \tag{190}$$

$$(\Lambda_s)_{ijmn}(\Lambda_s)_{mnkl} = (\Lambda_s)_{ijkl}, \tag{191}$$

where $U_{ijkl} = \frac{1}{2}(\delta_{ik}\delta_{jl} + \delta_{il}\delta_{jk})$ is the fourth-order unit tensor. The local stiffness tensor can be written in terms of the phase stiffness

$$\mathbf{C}^{(i)} = dK_i \mathbf{\Lambda}_h + 2\mu_i \mathbf{\Lambda}_s, \quad i = 1, 2 \tag{192}$$

by the relation

$$\mathbf{C}(\mathbf{x}) = \mathbf{C}^{(1)} m_1(\mathbf{x}) + \mathbf{C}^{(2)} m_2(\mathbf{x}), \tag{193}$$

where

$$m_p(\mathbf{x}) = \begin{cases} 1 & \mathbf{x} \text{ in phase } p \\ 0 & \text{otherwise} \end{cases}$$

is the characteristic function of phase p, with $p = 1, 2$.

The d-dimensional ellipsoidal composite is now embedded in an infinite *reference* phase q which is subjected to an applied strain field $\boldsymbol{\epsilon}^0(\mathbf{x})$ at infinity. The reference phase is taken to be either phase 1 or phase 2, i.e., $q = 1$ and 2. The local stress $\boldsymbol{\sigma}(\mathbf{x})$ is related to the local strain $\boldsymbol{\epsilon}(\mathbf{x})$ via Hooke's law, Eq. (1), and, at steady state and in the absence of body forces $\mathbf{F}$, satisfies the simplified version of Eq. (4), i.e., $\nabla \cdot \boldsymbol{\sigma} = \mathbf{0}$. If we introduce a polarization field defined by

$$\mathbf{p}(\mathbf{x}) = [\mathbf{C}(\mathbf{x}) - \mathbf{C}^{(q)}] : \boldsymbol{\epsilon}(\mathbf{x}), \tag{194}$$

we can re-express the stress as

$$\boldsymbol{\sigma}(\mathbf{x}) = \mathbf{C}^{(q)} \boldsymbol{\epsilon}(\mathbf{x}) + \mathbf{p}(\mathbf{x}). \tag{195}$$

The symmetric, second-order tensor $\mathbf{p}(\mathbf{x})$ is the *induced polarization field* relative to the medium in the absence of phase p, and hence vanishes in the reference phase q, but is non-zero in the "polarized" phase p with $p \neq q$. Throughout our discussion, the indices p and q will be used only for the polarized and reference phases, respectively. The choice of which is the reference or polarized phase is arbitrary; all the results are valid for any $p \neq q$, i.e., $p = 1$ and $q = 2$ or $p = 2$ and $q = 1$.

With the aid of Eq. (195), we can write

$$C_{ijkl}^{(q)} \frac{\partial^2 \hat{u}_k(\mathbf{x})}{\partial x_j \partial x_l} = -\frac{\partial p_{ij}(\mathbf{x})}{\partial x_j}, \tag{196}$$

with $\hat{u}_k(\mathbf{x}) \to 0$ as $\mathbf{x} \to \infty$, where $\hat{\mathbf{u}}(\mathbf{x})$ is the displacement field in excess of the displacement field $\mathbf{u}^0(\mathbf{x})$ at infinity, i.e., $\hat{\mathbf{u}}(\mathbf{x}) = \mathbf{u}(\mathbf{x}) - \mathbf{u}^0(\mathbf{x})$. In component form, the Green function $G_{ij}^{(q)}$ satisfies,

$$C_{ijkl}^{(q)} \frac{\partial^2 G_{im}^{(q)}(\mathbf{x}, \mathbf{x}')}{\partial x_j \partial x_l} = -\delta_{km}\delta(\mathbf{x} - \mathbf{x}'), \tag{197}$$

with $G_{km}^{(q)}(\mathbf{x} - \mathbf{x}') \to 0$ as $\mathbf{x} \to \infty$. The physical interpretation of the Green function is similar to that for the conduction problem discussed in Chapters 4 and 5: $\mathbf{G}(\mathbf{x}, \mathbf{x}')$ is the strain at $\mathbf{x}$ as a result of an applied stress at $\mathbf{x}'$. Therefore,

$$u_i(\mathbf{x}) = u_i^0(\mathbf{x}) + \int \frac{\partial}{\partial x_l} G_{ik}^{(q)}(\mathbf{x}, \mathbf{x}') p_{kl}(\mathbf{x}') d\mathbf{x}'. \tag{198}$$

Note that, because of the polarization $\mathbf{p}$ in Eq. (198), the integration volume extends only over the region of space occupied by the ellipsoidal composite. It is not difficult to show that in d dimensions,

$$G_{ij}^{(q)}(\mathbf{r}) = \begin{cases} \dfrac{1}{2\Omega_s \mu_q} \ln\left(\dfrac{1}{r}\right)\delta_{ij} + b_q n_i n_j & d = 2 \\ a_q \dfrac{\delta_{ij}}{r^{d-2}} + b_q \dfrac{n_i n_j}{r^{d-2}} & d \geq 3 \end{cases} \tag{199}$$

with

$$a_q = \frac{1}{2(d-2)\Omega_s \mu_q} \frac{dK_q + (3d-2)\mu_q}{dK_q + 2(d-1)\mu_q}, \tag{200}$$

$$b_q = \frac{1}{2\Omega_s \mu_q} \frac{dK_q + (d-2)\mu_q}{dK_q + 2(d-1)\mu_q}, \tag{201}$$

where $\Omega_s(d) = 2\pi^{d/2}/\Gamma(d/2)$ is the total solid angle contained in a d-dimensional sphere, $\Gamma(x)$ is the gamma function, $\mathbf{r} = \mathbf{x} - \mathbf{x}'$, and $\mathbf{n} = \mathbf{r}/|\mathbf{r}|$. Since the Green function has a singularity at $\mathbf{x}' = \mathbf{x}$, one must exclude a small region containing this point. Roughly speaking, the integral is convergent if it exists in the limit that the excluded region vanishes, independent of its shape. According to this criterion then, the integral in Eq. (198) is convergent.

To obtain the strain field, one must differentiate Eq. (162); however, because of the singular nature of the integral, one cannot simply differentiate the integrand. If we exclude a spherical region or "cavity" from the origin in Eq. (162), we find that the components of the strain field are given by

$$\epsilon_{ij}(\mathbf{x}) = \epsilon_{ij}^0(\mathbf{x}) + \int G_{ijkl}^{(q)}(\mathbf{r}) p_{kl}(\mathbf{x}') d\mathbf{x}', \tag{202}$$

where

$$G^{(q)}_{ijkl}(\mathbf{r}) = -A^{(q)}_{ijkl}\delta(\mathbf{r}) + H^{(q)}_{ijkl}(\mathbf{r}). \tag{203}$$

In Eq. (203), the components of the constant fourth-order tensor $\mathbf{A}^{(q)}$ (that arises because of the exclusion of the spherical cavity) are given by

$$A^{(q)}_{ijkl} = \frac{1}{dK_q + 2(d-1)\mu_q}(\Lambda_h)_{ijkl} + \frac{d}{(d+2)\mu_q}\left[\frac{K_q + 2\mu_q}{dK_q + 2(d-1)\mu_q}\right](\Lambda_s)_{ijkl}, \tag{204}$$

and $\mathbf{H}^{(q)}(\mathbf{r})$ is a position-dependent fourth-order tensor given by

$$\begin{aligned} H^{(q)}_{ijkl} &= \frac{1}{2}\left[\tilde{H}^{(q)}_{ijkl}(\mathbf{r}) + \tilde{H}^{(q)}_{ijlk}(\mathbf{r})\right] \\ &= \frac{1}{2\Omega_s[dK_q + 2(d-1)\mu_q]}\frac{1}{r^d}\big[\alpha_q\delta_{ij}\delta_{kl} - d(\delta_{ik}\delta_{jl} + \delta_{il}\delta_{jk}) \\ &\quad -d\alpha_q(n_kn_l\delta_{ij} + n_in_j\delta_{kl}) + \frac{1}{2}d(d-\alpha_q)(n_jn_l\delta_{ik} + n_jn_k\delta_{il} + n_in_l\delta_{jk} + n_in_k\delta_{jl}) \\ &\quad + d\alpha_q(d+2)n_in_jn_kn_l\big], \end{aligned} \tag{205}$$

where $\alpha_q = dK_q/\mu_q + d - 2$. It is understood that integrals involving the tensor $\mathbf{H}^{(q)}$ are to be carried out with the exclusion of an infinitesimal sphere in the limit that the sphere approaches zero. The components of the tensor $\tilde{\mathbf{H}}^{(q)}$ are given by

$$\tilde{H}^{(q)}_{ijkl} = \frac{1}{2}\left(\frac{\partial^2 G^{(q)}_{ik}}{\partial x_j \partial x_l} + \frac{\partial^2 G^{(q)}_{jk}}{\partial x_i \partial x_l}\right). \tag{206}$$

The symmetry of the polarization tensor $\mathbf{p}$ enables one to define the symmetric tensor $\mathbf{H}^{(q)}$ which is symmetric with respect to the first two indices and the second two indices, as well as with respect to interchange of ij and kl, i.e.,

$$H^{(q)}_{ijkl} = H^{(q)}_{jikl} = H^{(q)}_{ijlk} = H^{(q)}_{klij}.$$

Moreover, the integral of $\mathbf{H}^{(q)}(\mathbf{r})$ over the surface of a sphere of radius $a > 0$ is identically zero, i.e.,

$$\int_{r=a} \mathbf{H}^{(q)}(\mathbf{r})d\Omega_s = \mathbf{0}. \tag{207}$$

Some properties of the tensor $\mathbf{H}^{(q)}$ that will be of use in the subsequent analysis are as follows,

$$H^{(q)}_{ijkk}(\mathbf{r}) = \frac{d}{\Omega_s[dK_q + 2(d-1)\mu_q]}\frac{1}{r^d}(dn_in_j - \delta_{ij}), \tag{208}$$

$$H^{(q)}_{iikl}(\mathbf{r}) = \frac{d}{\Omega_s[dK_q + 2(d-1)\mu_q]}\frac{1}{r^d}(dn_kn_l - \delta_{kl}), \tag{209}$$

$$H^{(q)}_{iikk}(\mathbf{r}) = H^{(q)}_{ikik}(\mathbf{r}) = 0. \tag{210}$$

Moreover,

$$H_{iikl}^{(q)}(\mathbf{r})H_{kljj}^{(q)}(\mathbf{s}) = \frac{d^3}{\Omega_s^2[dK_q + 2(d-1)\mu_q]^2}\frac{1}{r^d}\frac{1}{s^d}[d(\mathbf{n}\cdot\mathbf{m})^2 - 1], \quad (211)$$

$$H_{ijkl}^{(q)}(\mathbf{r})H_{klij}^{(q)}(\mathbf{r}) = \frac{1}{4\Omega_s^2[dK_q + 2(d-1)\mu_q]^2}\frac{1}{r^d}\frac{1}{s^d}\{d(d+2)\alpha_q^2[d(\mathbf{n}\cdot\mathbf{m})^4 - 3]$$
$$- d(5d+6)\alpha_q^2[d(\mathbf{n}\cdot\mathbf{m})^2 - 1] + 2d^2(d-2)\alpha_q[d(\mathbf{n}\cdot\mathbf{m})^2 - 1]$$
$$+ d^3(d+2)[d(\mathbf{n}\cdot\mathbf{m})^2 - 1]\}, \quad (212)$$

where $\mathbf{m} = \mathbf{s}/|\mathbf{s}|$ is a unit vector in the direction of $\mathbf{s}$.

An integral equation for the "cavity" strain field $\mathbf{f}$ is now introduced. If Eq. (203) is substituted into Eq. (202), the following integral equation is obtained

$$\mathbf{f}(\mathbf{x}) = \boldsymbol{\epsilon}^0(\mathbf{x}) + \int_\epsilon \mathbf{H}^{(q)}(\mathbf{x}-\mathbf{x}') : \mathbf{p}(\mathbf{x})d\mathbf{x}', \quad (213)$$

which is related to the usual strain by

$$\mathbf{f}(\mathbf{x}) = \{\mathbf{U} + \mathbf{A}^{(q)} : [\mathbf{C}(\mathbf{x}) - \mathbf{C}^{(q)}]\} : \boldsymbol{\epsilon}(\mathbf{x}), \quad (214)$$

with

$$\int_\epsilon d\mathbf{x}' \equiv \lim_{\epsilon\to 0}\int_{|\mathbf{x}-\mathbf{x}'|>\epsilon} d\mathbf{x}', \quad (215)$$

that is, integration over the sample volume is carried out with the exclusion of an infinitesimally small sphere centered at $\mathbf{x}$ of radius ϵ, with the limit $\epsilon \to 0$ ultimately taken. We refer to $\mathbf{f}(\mathbf{x})$ as the *cavity strain field* because, as can be seen from Eq. (214), it is a modified strain field equal to the usual strain plus a contribution involving the constant tensor $\mathbf{A}^{(q)}$ which arises as a result of excluding a spherical cavity from the origin. The cavity strain field is the elasticity analog of the Lorentz electric field used in dielectric theory (see Chapters 4-6).

If we now combine Eq. (194) and (214), we obtain a relation between the stress polarization and cavity strain field, i.e.,

$$\mathbf{p}(\mathbf{x}) = \mathcal{L}^{(q)}(\mathbf{x}) : \mathbf{f}(\mathbf{x}), \quad (216)$$

where the fourth-order tensor $\mathcal{L}^{(q)}(\mathbf{x})$ is given by

$$\mathcal{L}^{(q)}(\mathbf{x}) = \{\mathbf{C}(\mathbf{x}) - \mathbf{C}^{(q)}\}\{\mathbf{I} + \mathbf{A}^{(q)} : [\mathbf{C}(\mathbf{x}) - \mathbf{C}^{(q)}]\}^{-1}, \quad (217)$$

and has the same symmetry properties as the stiffness tensor $\mathbf{C}(\mathbf{x})$. If the stiffness tensor has an isotropic form, then $\mathcal{L}^{(q)}(\mathbf{x})$ can be written as a constant tensor $\mathbf{L}^{(q)}$ multiplied by the characteristic indicator function $m_p(\mathbf{x})$, i.e.,

$$\mathcal{L}^{(q)}(\mathbf{x}) = \mathbf{L}^{(q)} m_p(\mathbf{x}), \quad (218)$$

with

$$\mathbf{L}^{(q)} = [dK_q + 2(d-1)\mu_q]\left[K_{pq}\boldsymbol{\Lambda}_h + \frac{\mu_q(d+2)}{d(K_q+2\mu_q)}\mu_{pq}\boldsymbol{\Lambda}_s\right], \quad (219)$$

where

$$K_{pq} = \frac{K_p - K_q}{K_p + \dfrac{2(d-1)}{d}\mu_q}, \tag{220}$$

$$\mu_{pq} = \frac{\mu_p - \mu_q}{\mu_p + \dfrac{\mu_q[dK_q/2 + (d+1)(d-2)\mu_q/d]}{K_q + 2\mu_q}}. \tag{221}$$

Note that K_{pq} and μ_{pq} are *not* tensors. As pointed out earlier in Section 7.3.1, and in analogy with the dielectric theory (see Chapters 4-6), one may refer to K_{pq} and μ_{pq} as the bulk modulus polarizability and the shear modulus polarizability, respectively.

7.4.2 Exact Series Expansions

The effective tensor $\mathbf{L}_e^{(q)}$ is defined via the relation linking the average polarization to the average cavity strain field, that is,

$$\langle \mathbf{p}(\mathbf{x})\rangle = \mathbf{L}_e^{(q)} : \langle \mathbf{f}(\mathbf{x})\rangle, \tag{222}$$

where

$$\mathbf{L}_e^{(q)} = \left[\mathbf{C}_e - \mathbf{C}^{(q)}\right]\left\{\mathbf{U} + \mathbf{A}^{(q)} : \left[\mathbf{C}_e - \mathbf{C}^{(q)}\right]\right\}^{-1}. \tag{223}$$

The constitutive relation (222) is independent of the shape of the ellipsoidal composite in the limit of infinite volume. In light of Eq. (223), we see that the effective tensor $\mathbf{L}_e^{(q)}$ has the same symmetry properties as the $\mathbf{C}_e$. Keeping in mind that the tensors $\mathcal{L}^{(q)}(\mathbf{x})$, $\mathbf{L}_e^{(q)}$ and $\mathbf{H}^{(q)}$ are all associated with the reference phase q, we temporarily drop, in the subsequent discussions, the superscript q when referring to these tensors.

We now wish to derive an explicit expression for the effective moduli tensor $\mathbf{L}_e$ using the solution of the integral equation (213). It is not difficult to show that

$$\mathbf{p} = \mathcal{L}\boldsymbol{\epsilon}^0 + \mathcal{L}\mathbf{H}\mathbf{p}. \tag{224}$$

A solution for the polarization $\mathbf{p}$ in terms of an operator acting on the applied strain field $\boldsymbol{\epsilon}^0$ is obtained by successive substitution using Eq. (224) with the result being

$$\mathbf{p} = \mathcal{L}\boldsymbol{\epsilon}^0 + \mathcal{L}\mathbf{H}\mathcal{L}\boldsymbol{\epsilon}^0 + \mathcal{L}\mathbf{H}\mathcal{L}\mathbf{H}\mathcal{L}\boldsymbol{\epsilon}^0 + \cdots = \mathbf{T}\boldsymbol{\epsilon}^0, \tag{225}$$

where the fourth-order tensor operator $\mathbf{T}$ is given by

$$\mathbf{T} = \mathcal{L}(\mathbf{U} - \mathcal{L}\mathbf{H})^{-1}. \tag{226}$$

More explicitly,

$$\mathbf{p}(1) = \mathcal{L}(1) : \boldsymbol{\epsilon}^0(1) + \int \mathcal{L} : \mathbf{H}(1,2) : \mathcal{L}(2) : \boldsymbol{\epsilon}^0(2)d2 + \int \mathcal{L}(1) : \mathbf{H}(1,2) : \mathcal{L}(2) : \mathbf{H}(2,3) : \mathcal{L}(3) : \boldsymbol{\epsilon}^0(3)d2d3 + \cdots = \int \mathbf{T}(1,2) : \boldsymbol{\epsilon}^0(2)d2. \tag{227}$$

where the notation has been shortened by representing $\mathbf{x}$ and $\mathbf{x}'$ by 1 and 2, respectively. Ensemble averaging Eq. (225) yields

$$\langle \mathbf{p} \rangle = \langle \mathbf{T} \rangle \boldsymbol{\epsilon}^0. \tag{228}$$

Since the tensor $\mathbf{H}$ decays as r^{-d}, $\langle \mathbf{T} \rangle$ at best involves conditionally convergent integrals and, hence, must be dependent upon the shape of the ellipsoidal composite. Thus, the relation between the polarization and the applied strain field is *non-local*. Given this non-locality, the strategy for developing a local relation between the average polarization $\langle \mathbf{p} \rangle$ and $\langle \mathbf{f} \rangle$ as prescribed by Eq. (222) is to eliminate the applied field $\boldsymbol{\epsilon}^0$ in favor of the appropriate average field. Thus, inverting Eq. (228) yields, $\boldsymbol{\epsilon}^0 = \langle \mathbf{T} \rangle^{-1} \langle \mathbf{p} \rangle$, and from the relation, $\langle \mathbf{f} \rangle = \boldsymbol{\epsilon}^0 + \mathbf{H} \langle \mathbf{p} \rangle$, we obtain

$$\langle \mathbf{f} \rangle = \mathbf{X} \langle \mathbf{p} \rangle, \tag{229}$$

where

$$\mathbf{X} = \langle \mathbf{T} \rangle^{-1} + \mathbf{H}. \tag{230}$$

More explicitly, Eq. (193) reads

$$\begin{aligned} \langle \mathbf{f}(\mathbf{x}) \rangle &= \int \mathbf{X}(1,2) \cdot \langle \mathbf{p}(2) \rangle d2 = \langle \mathcal{L}(1) \rangle^{-1} : \langle \mathbf{p}(1) \rangle \\ &- \int \Big[\langle \mathcal{L}(1) \rangle^{-1} : \langle \mathcal{L}(1) : \mathbf{H}(1,2) : \mathcal{L}(2) \rangle : \langle \mathcal{L}(2) \rangle^{-1} \\ &- \langle \mathcal{L}(1) \rangle^{-1} : \langle \mathcal{L}(1) \rangle : \mathbf{H}(1,2) : \langle \mathcal{L}(2) \rangle : \langle \mathcal{L}(2) \rangle^{-1} \Big] : \langle \mathbf{p}(2) \rangle d2 - \cdots, \end{aligned} \tag{231}$$

which means that

$$\left[\mathbf{L}_e^{(q)} \right]^{-1} = \mathbf{X}. \tag{232}$$

It is convenient to multiply this equation by the constant fourth-order tensor $\mathbf{L}^{(q)}$ from the left to yield

$$\mathbf{L}^{(q)} \left[\mathbf{L}_e^{(q)} \right]^{-1} = \mathbf{L}\mathbf{X}, \tag{233}$$

or, more explicitly,

$$\begin{aligned} \mathbf{L}^{(q)} : \left[\mathbf{L}_e^{(q)} \right]^{-1} &= \frac{1}{S_1^{(p)}(1)} \mathbf{U} - \int \left[\frac{S_2^{(p)}(1,2) - S_1^{(p)}(1) S_1^{(p)}(2)}{S_1^{(p)}(1) S_1^{(p)}(2)} \right] \mathbf{U}^{(q)}(1,2) d2 \\ &- \int \left[\frac{S_3^{(p)}(1,2,3)}{S_1^{(p)}(1) S_1^{(p)}(2)} - \frac{S_2^{(p)}(1,2) S_2^{(p)}(2,3)}{S_1^{(p)}(1) S_1^{(p)}(2) S_1^{(p)}(3)} \right] \mathbf{U}^{(q)}(1,2) : \mathbf{U}^{(q)}(2,3) - \cdots, \end{aligned} \tag{234}$$

where

$$\begin{aligned} U_{ijkl}^{(q)}(\mathbf{r}) &= L_{ijmn}^{(q)} H_{mnlk}^{(q)}(\mathbf{r}) \\ &= [d K_q + 2(d-1)\mu_q] \left\{ \left[K_{pq} - \frac{(d+2)\mu_q}{d(K_q + 2\mu_q)} \mu_{pq} \right] \frac{\delta_{ij}}{d} H_{mmkl}^{(q)}(\mathbf{r}) \right. \\ &\left. + \frac{(d+2)\mu_q}{d(K_q + 2\mu_q)} \mu_{pq} H_{ijkl}^{(q)}(\mathbf{r}) \right\}. \end{aligned} \tag{235}$$

Here the n-point correlation function $S_n^{(p)}$ for the polarized phase p is defined by

$$S_n^{(p)}(\mathbf{x}_1, \cdots, \mathbf{x}_n) = \langle m_p(\mathbf{x}_1) \cdots m_p(\mathbf{x}_n) \rangle. \tag{236}$$

As discussed in Section 3.4.1.3, $S_n^{(p)}$ is the probability of simultaneously finding n points with positions $\mathbf{x}_1, \cdots, \mathbf{x}_n$ in phase p, and is sometimes called the *n-point probability function*. For statistically-anisotropic but homogeneous materials, $S_n^{(p)}$ depends on the relative displacements $\mathbf{x}_{ij} = \mathbf{x}_j - \mathbf{x}_i$, with $1 \leq i < j \leq n$; in particular, $S_1^{(p)}$ is simply the volume fraction ϕ_p of phase p. The reason why $S_n^{(p)}$ arises in the expansion (234) is because the operator $\mathbf{X}$ contains averages over products of the position-dependent tensor $\mathbf{L}^{(q)}(\mathbf{x})$ which, in turn, depends on the characteristic function $m_p(\mathbf{x})$. Note that we still have not passed to the statistically homogeneous, infinite-volume limit.

The general term of the expansion (234) can be easily written down as

$$\phi_p^2 \mathbf{L}^{(q)} : \left[\mathbf{L}_e^{(q)}\right]^{-1} = \phi_p \mathbf{U} - \sum_{n=2}^{\infty} \mathbf{B}_n^{(p)}, \tag{237}$$

where the tensors $\mathbf{B}_n^{(p)}$ are given by

$$\mathbf{B}_2^{(p)} = \int_\epsilon \mathbf{U}^{(q)}(1,2)[S_2^{(p)}(1,2) - \phi_p^2]d2, \tag{238}$$

$$\mathbf{B}_n^{(p)} = (-1)^n \left(\frac{1}{\phi_p}\right)^{n-2} \int d2 \int dn \mathbf{U}^{(q)}(1,2) : \mathbf{U}^{(q)}(2,3) ::: $$

$$\mathbf{U}^{(q)}(n-1, m)\Delta_n^{(p)}(1, \cdots, n), \quad n \geq 3, \tag{239}$$

and $\Delta_n^{(p)}$ is a position-dependent determinant associated with phase p given by

$$\Delta_n^{(p)} = \begin{vmatrix} S_2^{(p)}(1,2) & S_1^{(p)}(2) & \cdots & 0 & 0 \\ S_2^{(p)}(1,2,3) & S_1(p)(2,3) & \cdots & 0 & 0 \\ \vdots & \vdots & \cdots & \vdots & \vdots \\ \vdots & \vdots & \cdots & \vdots & \vdots \\ \vdots & \vdots & \cdots & \vdots & \vdots \\ S_{n-1}^{(p)}(1,\cdots,n-1) & S_{n-2}^{(p)}(2,\cdots,n-1) & \cdots & S_2^{(p)}(n-2,n-1) & S_1^{(p)}(n-1) \\ S_n^{(p)}(1,\cdots,n) & S_{n-1}^{(p)}(2,\cdots,n) & \cdots & S_3^{(p)}(n-2,n-1,n) & S_2^{(p)}(n-1,n) \end{vmatrix} \tag{240}$$

A few points are worth mentioning here.

(1) Equation (237) actually represents two different series expansion, one for $p = 1$ and $q = 2$ and a second one for $p = 2$ and $q = 1$.

(2) Since the quantity within the brackets of Eq. (238) and the determinant $\Delta_n^{(p)}$ identically vanish at the boundary of the sample (due to the asymptotic properties of the $S_n^{(p)}$ described in Chapters 3 and 4), the integrals in (238) and (239)

are independent of the shape of the macroscopic ellipsoid, i.e., they are absolutely convergent and, hence, any convenient shape, such as a d-dimensional sphere, may be employed in the limit of infinite volume. Moreover, when $n \geq 3$, the limiting process of excluding an infinitesimally small cavity about $r_{ij} = 0$ in the integrals (239) is no longer necessary since $\Delta_n^{(p)}$ is identically zero for such values.

(3) As shown below, for macroscopically isotropic materials, the series (237) may be regarded as expansions that perturb about the optimal structures that realize the Hashin–Shtrikman bounds (Hashin and Shtrikman, 1963; Hashin, 1965). For macroscopically anisotropic materials, the series (237) may be regarded as expansions that perturb about the optimal structures that realize the bounds due to Willis (1977). Later in this chapter, we will derive and discuss rigorous bounds to the effective linear elastic properties of disordered materials.

(4) The n-point tensors $\mathbf{B}_n^{(p)}$ are functionals of the correlation functions $S_1^{(p)}, \cdots, S_n^{(p)}$ but, unlike their conductivity counterpart $\mathbf{A}_n^{(p)}$ (see Section 4.5.1), they also depend on the phase properties through the polarizabilities K_{pq} and μ_{pq}, since, in Eq. (239), $\mathbf{U}^{(q)}$ can be written in terms of products of the tensor $\mathbf{H}^{(q)}$ which involves terms such as $K_{pq}^m \mu_{pq}^m$, where $m = 0, 1, 2, \cdots, n$. Some of these terms will vanish identically, depending on the value of n. For example, for *any* n, all terms that involve K_{pq}^n vanish.

(5) The expansion parameters K_{pq} and μ_{pq} arise because of excluding a *spherical* cavity from the point $\mathbf{x}' = \mathbf{x}$ in the integrals in Eqs. (198) and (213). Had another cavity shape been selected, the expansion parameters would have been different.

(6) The tensors $\mathbf{B}_n^{(p)}$ for all n do not possess common principal axes, implying that, in general, the principal axes of the effective stiffness tensor $\mathbf{C}_e$ will rotate as the phase moduli ratio varies. An example is provided by materials with chirality. Nonetheless, there exists a significant class of microstructures that have the necessary symmetry for the tensors $\mathbf{B}_n^{(p)}$ to possess common principal axes. An example is provided by unidirectionally-oriented cylinders in a uniform matrix.

7.4.3 *Exact Series Expansions for Isotropic Materials*

For macroscopically isotropic materials, it is seen from Eq. (223) that

$$\mathbf{L}_e^{(q)} = [dK_q + 2(d-1)\mu_q]\left[K_{eq}\mathbf{\Lambda}_h + \frac{(d+2)\mu_q}{d(K_q + 2\mu_q)}\mu_{eq}\mathbf{\Lambda}_s\right], \tag{241}$$

where the *effective polarizabilities* K_{eq} and μ_{eq} are defined by the scalar relations (220) and (221). Therefore, series (237) becomes

$$\phi_p^2\left(\frac{K_{pq}}{K_{eq}}\mathbf{\Lambda}_h + \frac{\mu_{pq}}{\mu_{eq}}\mathbf{\Lambda}_s\right) = \phi_p\mathbf{U} - \sum_{n=2}^{\infty}\mathbf{B}_n^{(p)}. \tag{242}$$

To derive an explicit expression for the effective bulk modulus K_e, we take the quadruple dot product of the hydrostatic projection tensor $\mathbf{\Lambda}_h$ with Eq. (242) and use identities (187) and (188) to obtain

$$\phi_p^2 \frac{K_{pq}}{K_{eq}} = \phi_p - \sum_{n=3}^{\infty} C_n^{(p)}, \tag{243}$$

where the scalar microstructural coefficients $C_n^{(p)}$ are given by

$$C_n^{(p)} = \mathbf{\Lambda}_h \vdots \mathbf{B}_n^{(p)}, \tag{244}$$

and $\vdots$ denotes a quadruple dot product. Note that the series (243) begins with the third-order term, i.e., $C_2^{(p)}$ is zero because the invariant $H_{iikk}^{(q)}$ vanishes. Note also that when the shear moduli of the phases are equal ($\mu_1 = \mu_2$), the well-known exact result

$$K_{eq} = \phi_p K_{pq} \tag{245}$$

due to Hill (1963,1964), which was already discussed in Section 7.2.8, immediately follows because

$$U_{ijkl}^{(q)}(\mathbf{r}) = [dK_q + 2(d-1)\mu_q] K_{pq} \frac{\delta_{ij}}{d} H_{mmkl}^{(q)}(\mathbf{r}), \tag{246}$$

and therefore each coefficient $C_n^{(p)}$ possesses the invariant $\mathbf{\Lambda}_h \vdots \mathbf{H}^{(q)}$ which vanishes.

Classical perturbation expansions involve small parameters that are simple differences in the phase moduli, e.g., $(K_1 - K_2)$ and $(\mu_1 - \mu_2)$ (see, for example, Beran and Molyneux, 1966; Silnutzer, 1972; Milton and Phan–Thien, 1982; Milton, 1984), and have a small radius of convergence, and hence require many terms in the series if the phase moduli are appreciably different from one another. By contrast, the expansions (243) are *non-classical* in the sense that their parameters are the polarizabilities K_{eq} and μ_{eq} and, for certain morphologies (described below) can converge rapidly for *any* values of the phase moduli. It may be helpful to consider the morphologies for which the microstructural parameters $C_n^{(p)}$ vanish for any values of the phase moduli, i.e., for the class of composites for which Eq. (209) or, equivalently,

$$\frac{K_e - K_q}{K_e + \dfrac{2(d-1)}{d}\mu_q} = \frac{K_p - K_q}{K_p + \dfrac{2(d-1)}{d}\mu_q}\phi_p \tag{247}$$

exact. For $d = 2$ and $d = 3$, Eq. (247) coincides with the Hashin–Shtrikman bounds (see the discussion on such bounds later in this chapter) on the effective bulk modulus for any isotropic two-phase composite (Hashin and Shtrikman, 1963; Hashin, 1965), and hence is exact for the assemblages of coated circles ($d = 2$) and coated spheres ($d = 3$) (see Sections 4.4 and 7.2.3) that realize the bounds. The Hashin–Shtrikman bounds are also realized for certain finite-rank laminates (i.e., layered materials) in both 2D and 3D (Francfort and Murat, 1986). It is important to emphasize that for either the coated-inclusion assemblages or finite-rank laminates, one of the phases is always a disconnected, dispersed phase in a connected matrix phase. Equation (247) is the d-dimensional generalization of the Hashin–Shtrikman bounds on K_e for any $d \geq 2$; it yields, for $K_2 \geq K_1$ and $\mu_2 \geq \mu_1$, a lower bound for $q = 1$ and $p = 2$ and an upper bound for $q = 2$ and $p = 1$. Therefore, series (243) may be viewed as an expansion that perturbs around the optimal Hashin–Shtrikman structures, and thus it is expected to converge rapidly for any value of the phase moduli for dispersions in which the inclusions, taken to be the polarized phase, are prevented from forming large clusters. This will be further discussed below.

We now write down Eq. (243) through third-order terms and simplify it to find that

$$\phi_p \frac{K_{pq}}{K_{eq}} = 1 - \frac{C_3^{(p)}}{\phi_p} = 1 - \frac{(d+2)\mu_q K_{pq}\mu_{pq}}{d(K_q + 2\mu_q)} \frac{M_p}{\phi_p}, \tag{248}$$

where M_p is a three-point microstructural parameter independent of the phase moduli, given by

$$M_p = \frac{d^2}{\Omega_s^2} \int\int \frac{1}{r^d} \frac{1}{s^d} [d(\mathbf{n}\cdot\mathbf{m})^2 - 1] \left[S_3^{(p)}(\mathbf{r},\mathbf{s}) - \frac{S_2^{(p)}(\mathbf{r}) S_2^{(p)}(\mathbf{s})}{\phi_p} \right] d\mathbf{r} d\mathbf{s}, \tag{249}$$

and $\mathbf{n} = \mathbf{r}/|\mathbf{r}|$ and $\mathbf{m} = \mathbf{s}/|\mathbf{s}|$ are unit vectors. Note that M_p is the same as the parameter $A_n^{(p)}$, described in Section 4.5.1, where we considered perturbation expansions for the effective conductivity of disordered anisotropic materials. Note also that, M_1 and M_2 are not independent of one another; specifically, one has

$$M_1 + M_2 = (d-1)\phi_1\phi_2, \tag{250}$$

which is easily proven by using the fact that Eq. (236) yields the exact results for the effective bulk modulus K_e through third-order in the difference in the moduli, i.e.,

$$\begin{aligned} K_e = K_q + a_1^{(p)}(K_p - K_q) + a_2^{(p)}(K_p - K_q)^2 + a_3^{(p)}(K_p - K_q)^3 \\ + b_3^{(p)}(K_p - K_q)^2(\mu_p - \mu_q), \end{aligned} \tag{251}$$

where, $a_1^{(p)} = \phi_p$, and

$$\begin{aligned} a_2^{(p)} &= -\frac{d\phi_p\phi_q}{dK_q + 2(d-1)\mu_q}, \\ a_3^{(p)} &= \frac{d^2\phi_p\phi_q^2}{[dK_q + 2(d-1)\mu_q]^2}, \\ b_3^{(p)} &= \frac{2dM_p}{[dK_q + 2(d-1)\mu_q]^2}. \end{aligned} \tag{252}$$

Since K_e remains invariant under different labels of the reference phase, Eq. (250) follows immediately. The parameter M_p is closely related to the mictrostructural parameter ζ_p (and in particular ζ_2) that was described in detail in Section 4.5.3. In general,

$$\zeta_p = \frac{M_p}{(d-1)\phi_1\phi_2}. \tag{253}$$

Derivation of an explicit expression for the effective shear modulus μ_e proceeds along the same lines as that for K_e, namely, we take a quadruple dot product of the shear tensor $\mathbf{\Lambda}_s$ with (241) and use identities (187) and (189) to obtain

$$\phi_p^2 \frac{\mu_{pq}}{\mu_{eq}} = \phi_p - \sum_{n=3}^{\infty} D_n^{(p)}, \tag{254}$$

with the scalar coefficients

$$D_n^{(p)} = \frac{2}{(d+2)(d-1)} \mathbf{\Lambda}_s : \mathbf{B}_n^{(p)}. \tag{255}$$

Note that this series also begins with the third-order term, i.e., $D_2^{(p)}$ is zero because the invariant $H_{ikik}^{(q)}$ vanishes. Truncating the series (254) after the first term yields

$$\mu_{eq} = \phi_p \mu_{pq} \tag{256}$$

or, equivalently,

$$\frac{\mu_e - \mu_q}{\mu_e + \dfrac{\mu_q[dK_q/2 + (d+1)(d-2)\mu_q/d]}{K_q + 2\mu_q}} = \frac{\mu_p - \mu_q}{\mu_p + \dfrac{\mu_q[dK_q/2 + (d+1)(d-2)\mu_q/d]}{K_q + 2\mu_q}} \phi_p. \tag{257}$$

Following the discussion on the bulk modulus, it is helpful to consider the class of morphologies for which Eq. (256) is exact. For $d = 2$ and $d = 3$, Eq. (256) coincides with the Hashin–Shtrikman bounds on the effective shear modulus for all isotropic two-phase composites that realize the bounds (see below). Again, for such hierarchical laminates, one of the phases is always a disconnected, dispersed phase in a connected matrix phase. Equation (256) is the d-dimensional generalization of the Hashin–Shtrikman bounds on μ_e for any $d \geq 2$; it yields, for $(K_2 - K_1)(\mu_2 - \mu_1) \geq 0$ and $\mu_2 \geq \mu_1$, a lower bound for $q = 1$ and $p = 2$ and an upper bound for $q = 2$ and $p = 1$. Consequently, series (254) can be viewed as an expansion which perturbs around the optimal hierarchical laminates that achieve the Hashin–Shtrikman bounds, and therefore it is expected to have the same convergence properties as that of the series for the bulk modulus K_e. This will be demonstrated below.

If we now write down Eq. (254) through third-order terms and simplify it, we obtain

$$\phi_p \frac{\mu_{pq}}{\mu_{eq}} = 1 - \frac{D_3^{(p)}}{\phi_p}, \tag{258}$$

where

$$\frac{D_3^{(p)}}{\phi_p} = \frac{2}{(d+2)(d-1)}\mathbf{A}_s : \mathbf{B}_n^{(p)}$$

$$= \frac{2}{(d+2)(d-1)}\int\int\left[U_{ijkl}^{(q)}(\mathbf{r})U_{klij}^{(q)}(\mathbf{s}) - \frac{U_{iikl}^{(q)}(\mathbf{r})U_{klmm}^{(q)}(\mathbf{s})}{d}\right]\left[S_3^{(p)}(\mathbf{r},\mathbf{s}) - \frac{S_2^{(p)}(\mathbf{r})S_2^{(p)}(\mathbf{s})}{\phi_p}\right]d\mathbf{r}d\mathbf{s}$$

$$\frac{2\mu_q K_{pq}\mu_{pq}}{d(d-1)(K_q+2\mu_q)}\frac{M_p}{\phi_p} + \frac{\mu_q(d^2-4)(2K_q+3\mu_q)\mu_{pq}^2}{2d(d-1)(K_q+2\mu_q)^2}\frac{M_p}{\phi_p}$$
$$+\frac{1}{2d(d-1)}\left[\frac{dK_q+(d-2)\mu_q}{K_q+2\mu_q}\right]^2\mu_{pq}^2\frac{N_p}{\phi_p}. \tag{259}$$

The quantity N_p is a microstructural parameter independent of the phase moduli, given by

$$N_p = -\frac{(d+2)(5d+6)}{d^2}M_p + \frac{(d+2)^2}{\Omega_s^2}\int\int\frac{1}{r^d}\frac{1}{s^d}[d(d+2)(\mathbf{n}\cdot\mathbf{m})^4 - 3]$$
$$\times\left[S_3^{(p)}(\mathbf{r},\mathbf{s}) - \frac{S_2^{(p)}(\mathbf{r})S_2^{(p)}(\mathbf{s})}{\phi_p}\right], \tag{260}$$

where $\mathbf{n} = \mathbf{r}/|\mathbf{r}|$ and $\mathbf{m} = \mathbf{s}/|\mathbf{s}|$ are unit vectors. Similar to M_1 and M_2, N_1 and N_2 are not independent of one another; specifically, one has

$$N_1 + N_2 = (d-1)\phi_1\phi_2, \tag{261}$$

which is easily proven by writing down the expansion for μ_e through third-order in the difference in the moduli, i.e.,

$$\mu_e = \mu_q + c_1^{(p)}(\mu_p-\mu_q) + c_2^{(p)}(\mu_p-\mu_q)^2 + c_3^{(p)}(\mu_p-\mu_q)^3$$
$$+d_3^{(p)}(\mu_p-\mu_q)^2(K_p-K_q), \tag{262}$$

where, $c_1^{(p)} = \phi_p$, and

$$c_2^{(p)} = -\frac{2d\phi_p\phi_q(K_q+2\mu_q)}{\mu_q(d+2)[dK_q+2(d-1)\mu_q]},$$

$$c_3^{(p)} = \frac{4d^2(K_q+2\mu_q)^2\phi_p\phi_q^2}{\mu_1^2(d+2)^2[dK_q+2(d-1)\mu_q]^2} + \frac{2dM_q(d-2)[2K_q+3\mu_q]}{\mu_q(d+2)(d-1)[dK_q+2(d-1)\mu_q]^2}$$
$$+\frac{2d}{\mu_q^2(d-1)(d+2)^2}\left[\frac{dK_q+(d-2)\mu_q}{dK_q+2(d-1)\mu_q}\right]^2 N_p,$$

$$d_3^{(p)} = \frac{4dM_p}{(d+2)(d-1)[dK_q+2(d-1)\mu_q]^2}. \tag{263}$$

Since μ_e remains invariant under different labels of the reference phase, Eq. (261) follows immediately.

For $d = 2$ and $d = 3$, Eqs. (262) and (263) agree with the corresponding expansions of the Silnutzer (1972) and McCoy (1970) bounds on μ_e, respectively, which involve the three-point parameter ζ_p, as well as another three point parameter η_p, which also lies in the interval [0,1] (Milton, 1981a, 1982b), and is given by

$$\eta_p = \frac{N_p}{(d-1)\phi_1\phi_2}. \tag{264}$$

7.4.4 *Macroscopically Anisotropic Materials*

Since it was shown that the series expansion for macroscopically-isotropic composites may be regarded as one that perturbs around the structures that realize the isotropic Hashin–Shtrikman bounds, one may expect that the expansion for the macroscopically-anisotropic composites can be regarded as one that perturbs around the optimal structures that realize the anisotropic generalization of the Hashin–Shtrikman bounds obtained by Willis (1977); this indeed is the case.

Recall that, for statistically isotropic materials, $\mathbf{B}_2^{(p)} = \mathbf{0}$. However, the two-point tensor $\mathbf{B}_2^{(p)}$ does not vanish for statistically anisotropic materials, since the two-point probability function $S_2^{(p)}(\mathbf{r})$ depends on the distance $r = |\mathbf{r}|$ as well as the orientation of $\mathbf{r}$. Now, consider morphologies for which the n-point tensors $\mathbf{B}_n^{(p)} = \mathbf{0}$ for all $n \geq 3$. For such composites, Eq. (237) reduces exactly to

$$\phi_p^2 \mathbf{L}^{(q)} : \left[\mathbf{L}_e^{(q)}\right]^{-1} = \phi_p \mathbf{U} - \mathbf{B}_2^{(p)}. \tag{265}$$

Multiplying Eq. (265) by $\left[\mathbf{L}^{(q)}\right]^{-1}$ from the left gives

$$\begin{aligned}\phi_p^2 \left[\mathbf{L}_e^{(q)}\right]^{-1} &= \phi_p \left[\mathbf{L}^{(q)}\right]^{-1} - \left[\mathbf{L}^{(q)}\right]^{-1} : \mathbf{B}_2^{(p)}, \\ &= \phi_p \left[\mathbf{L}^{(q)}\right]^{-1} - \int_\epsilon \mathbf{H}^{(q)}(1,2)\left[S_2^{(p)}(1,2) - \phi_p^2\right] d2.\end{aligned} \tag{266}$$

Equations (266) are indeed the generalized Hashin–Shtrikman bounds for anisotropic composites derived by Willis (1977), albeit expressed in a different form than given originally by him.

7.4.5 *The Microstructural Parameter η_2*

Similar to the microstructural parameter ζ_2 (see Section 4.5.3), the parameter η_2 has also been evaluated for a variety of morphologies. In principle, η_2 depends on the volume fraction of a material's phases. Thus, let us write, in analogy with Eq. (4.82), for a two-phase material,

$$\eta_2 = \eta_2^{(0)} + \eta_2^{(1)}\phi_2 + \cdots \tag{267}$$

In general, for 3D randomly-oriented ellipsoids one has (Torquato *et al.*, 1987)

$$\eta_2^{(0)} = \begin{cases} 0, & \text{spheres,} \\ \frac{1}{4}, & \text{needles,} \\ 1, & \text{disks,} \end{cases} \tag{268}$$

while in 2D,

$$\eta_2^{(0)} = \begin{cases} 0, & \text{disks} \\ 1, & \text{needles.} \end{cases} \tag{269}$$

Some results have also been obtained for symmetric-cell materials (see Section 3.4.6). These results indicate that for the following cell shapes,

$$\eta_2 = \begin{cases} \phi_2, & d\text{-dimensional spherical cells,} \\ \frac{1}{4}\phi_1 + \frac{3}{4}\phi_2, & \text{3D disk-like cells,} \\ \phi_1, & \text{2D needle-like cells,} \end{cases} \tag{270}$$

and that $\eta_2 = \zeta_2$.

For distribution of three-dimensional, identical overlapping spheres, a good approximation is given by $\eta_2 \simeq 0.7468\phi_2$. The parameter η_2 was also evaluated exactly (Torquato *et al.*, 1987) through first order in ϕ_2 for equilibrium interpenetrable spheres, in the limit of fully-impenetrable spheres, and for 2D equilibrium distribution of impenetrable disks (cylinders). The results are

$$\eta_2 = \begin{cases} 0.48274\phi_2, & \text{fully-impenetrable spheres,} \\ \frac{56}{81}\phi_2 + 0.0428\phi_2^2, & \text{fully-impenetrable disks.} \end{cases} \tag{271}$$

Thovert *et al.* (1990) computed η_2 exactly through first order in ϕ_2 for equilibrium impenetrable spheres with two different and widely separated sizes, with the result being,

$$\eta_2 = 0.49137\phi_2. \tag{272}$$

7.4.6 Comparison with Numerical Simulation

Torquato (1998) compared his theoretical predictions derived and discussed above with the results of numerical simulations for a variety of important model composite materials that consist of well-defined shapes, such as oriented circular or elliptical cylinders, spheres or ellipsoids, in a matrix. This class of models includes certain fiber-reinforced materials, particulate-reinforced composites, and colloidal dispersions. In what follows we discuss such comparisons for both 2D and 3D systems.

7.4.6.1 Two-Dimensional materials

We first consider 2D dispersions, and therefore the ensuing results apply to fiber-reinforced materials with transverse isotropy, or to thin sheets of isotropic two-phase materials. To emphasize the planarity of the bulk modulus, we denote it by K_{e2}. In the case of fiber-reinforced dispersions, the appropriate planar bulk modulus K_{e2} is the plane-strain modulus which is related to the 3D one by the relation, $K_{e2} = K_e + \frac{1}{3}\mu_e$, where K_e is the corresponding 3D modulus. For plane stress elasticity (appropriate for a composite sheet), the plane-stress bulk modulus

K_{e2} is obtained from the following relation

$$K_{e2} = \frac{9K_e\mu_e}{3K_e + 4\mu_e},$$

where μ_e is the effective shear moduli of the corresponding 3D material.

Suppose that phases 1 and 2 represent the matrix and dispersed phases, respectively. For $d = 2$, Eqs. (248), (249) and (253) yield

$$\frac{K_{e2}}{K_1} = \frac{1 + \dfrac{\mu_1}{K_1}\kappa\phi_2 - \dfrac{2\mu_1}{K_1 + 2\mu_1}\kappa\mu\phi_1\zeta_2}{1 - \kappa\phi_2 - \dfrac{2\mu_1}{K_1 + 2\mu_1}\kappa\mu\phi_1\zeta_2}, \tag{273}$$

$$\frac{\mu_{e2}}{\mu_1} = \frac{1 + \dfrac{K_1}{K_1 + 2\mu_1}\mu\phi_2 - \dfrac{\mu_1}{K_1 + 2\mu_1}\kappa\mu\phi_1\zeta_2 - \dfrac{K_1^2}{(K_1 + 2\mu_1)^2}\mu^2\phi_1\eta_2}{1 - \mu\phi_2 - \dfrac{\mu_1}{K_1 + 2\mu_1}\kappa\mu\phi_1\zeta_2 - \dfrac{K_1^2}{(K_1 + 2\mu_1)^2}\mu^2\phi_1\eta_2}, \tag{274}$$

where

$$\kappa \equiv K_{21} = \frac{K_2 - K_1}{K_2 + \mu_1},$$

$$\mu \equiv \mu_{21} = \frac{\mu_2 - \mu_1}{\mu_2 + \dfrac{K_1}{K_1 + 2\mu_1}\mu_1}.$$

As discussed above and in Section 4.5.1, for any isotropic 2D material, the three-point parameters ζ_2 and η_2 are given by

$$\zeta_2 = \frac{4}{\pi\phi_1\phi_2}\int_0^\infty \frac{dr}{r}\int_0^\infty \frac{ds}{s}\int_0^\pi \cos(2\theta)\left[S_3^{(2)}(r, s, t) - \frac{S_2^{(2)}(r)S_2^{(2)}(s)}{\phi_2}\right]d\theta, \tag{275}$$

$$\eta_2 = \frac{16}{\pi\phi_1\phi_2}\int_0^\infty \frac{dr}{r}\int_0^\infty \frac{ds}{s}\int_0^\pi \cos(4\theta)\left[S_3^{(2)}(r, s, t) - \frac{S_2^{(2)}(r)S_2^{(2)}(s)}{\phi_2}\right]d\theta, \tag{276}$$

where θ is the angle opposite the side of the triangle of length t. The predictions of the 2D series must always lie within the most restrictive three-point upper and lower bounds. These bounds will be derived and discussed later in this chapter; for now we take them for granted and compare the predictions of the series expansions with them. Requiring the predictions of the series expansions to lie within the upper and lower bounds imposes tight restrictions on the range of the parameters ζ_2 and η_2. In most cases, such restriction can be determined only numerically, but for the present case of 2D materials one can show that, $\zeta_2 \leq 0.5$, which would then ensure that the series predictions for the bulk modulus would lie between the lower and upper bounds. It is difficult to obtain similar analytical conditions for the effective shear modulus. However, in the case of a 2D composite in which both phases are incompressible (i.e., when $K_1/\mu_1 = K_2/\mu_2 \to \infty$), it can be shown that there is no additional restriction on η_2. Thus, in this limit, for any physically realizable η_2, the predictions of Eq. (274) always lie within the most restrictive three-point bounds. Moreover, as shown below, Eq. (274) is in fact independent of ζ_2 for an incompressible composite. It is noteworthy that for a number of realistic models of 2D dispersions the parameters ζ_2 and η_2 are such that the estimates from Eqs. (273) and (274) always lie within the tightest three-point bounds.

In order to further validate Eqs. (273) and (274) as accurate approximations for the effective moduli of 2D dispersions, their predictions are compared to the results of numerical calculations. A set of such calculations was carried out by Eischen and Torquato (1993) who obtained comprehensive numerical data for elastic moduli of models of hexagonal arrays of infinitely long, aligned cylinders in a matrix, and presented data for eight different phase-material values over a wide range of inclusion volume fractions. To compare the predictions of Eqs. (273) and (274) to these data, the functions $\zeta_2(\phi_2)$ and $\eta_2(\phi_2)$ are needed. These functions were evaluated for hexagonal arrays of inclusions by McPhedran and Milton (1981) and Eischen and Torquato (1993). In Figure 7.2, the predictions of Eq. (274) for the effective transverse shear modulus μ_{e2} of hexagonal arrays of glass fibers in an epoxy matrix (with $\mu_2/\mu_1 = 22.5$, $\mu_1/K_1 = 0.3$, and $\mu_2/K_2 = 0.6$) are compared to the corresponding simulation data of Eischen and Torquato (1993);

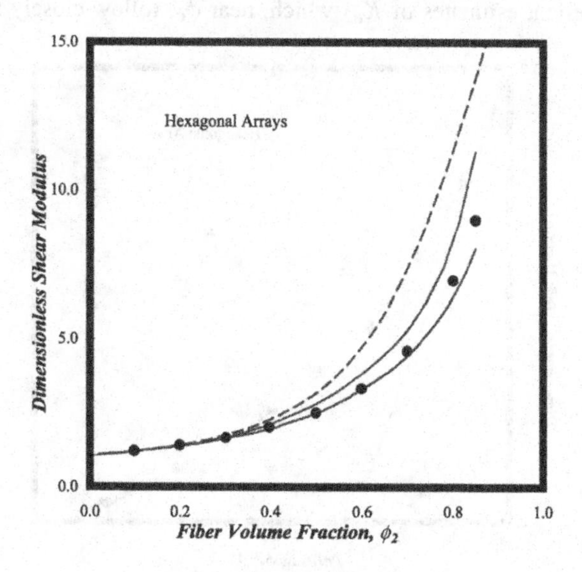

FIGURE 7.2. Dimensionless effective transverse shear modulus μ_e/μ_1 versus the fiber volume fraction ϕ_2 for hexagonal arrays of circular glass fibers in an epoxy matrix with $\mu_2/\mu_1 = 22.5, \mu_1/K_1 = 0.3$ and $\mu_2/K_2 = 0.6$. Solid circles are the simulation data, while the dashed curve shows the predictions of the EMA, Eqs. (569) and (570). The lower and upper continuous curves present the predictions of Eq. (274) and the three-point bound (419), respectively (after Torquato, 1998).

the agreement is excellent for the entire range of volume fractions. The three-point lower bound of Silnutzer (1972) (which will be derived in the next section) is very slightly below the predictions of Eq. (274), and hence is not shown. The predictions of Eq. (273) for the effective bulk modulus are equally accurate. The three-point upper bound of Gibiansky and Torquato (1995a; see below) are also shown in Figure 7.2.

Consider now a sheet (phase 1) that contains non-overlapping holes for which $K_2 = \mu_2 = 0$. As mentioned in Section 7.2.2, Day *et al.* (1992) studied the elastic moduli of hexagonal arrays of circular holes near the percolation threshold ϕ_c, and derived Eqs. (54) for the effective bulk and shear moduli. On the other hand, Eqs. (273) and (274) are simplified for this limit to

$$\frac{K_{e2}}{K_1} = \frac{\phi_1(1-2\zeta_2)}{1+\dfrac{K_1}{\mu_1}\phi_2 - 2\phi_1\zeta_2}, \tag{277}$$

$$\frac{\mu_{e2}}{\mu_1} = \frac{\phi_1(1-\zeta_2-\eta_2)}{1+\dfrac{K_1+2\mu_1}{K_1}\phi_2 - \phi_1(\zeta_2+\eta_2)}. \tag{278}$$

Figure 7.3 compares the predictions of Eq. (277) to the corresponding simulation data of Eischen and Torquato (1993) for hexagonal arrays of circular holes in a matrix for which $\mu_2/\mu_1 = K_2/K_1 = 0$ and $\mu_1/K_1 = 0.5$. Equation (277) provides excellent estimates of K_{e2} which, near ϕ_c, follow closely Eq. (54) for

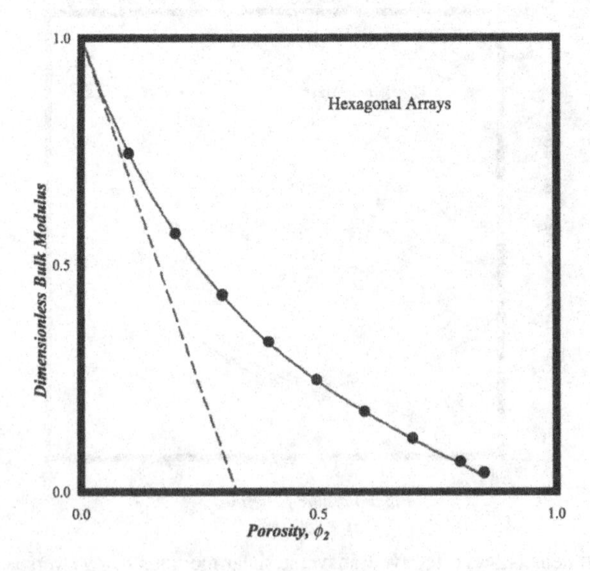

FIGURE 7.3. Dimensionless effective transverse bulk modulus K_e/K_1 versus the fiber volume fraction ϕ_2 for hexagonal arrays of circular holes in a compressible sheet, with $\mu_2/\mu_1 = K_2/K_1 = 0$, and $\mu_1/K_1 = 0.5$. Solid circles are the simulation data, solid curve shows the predictions of Eq. (277), while the dashed curve presents the EMA predictions, Eqs. (569) and (570) (after Torquato, 1998).

K_e. The three-point upper bound of Gibiansky and Torquato (1995a) is virtually indistinguishable from the predictions of Eq. (277), and of course the associated three-point lower bound is 0.

We pointed out in Section in 7.2.6 that Cherkaev *et al.* (1992) showed that the dimensionless effective Young's modulus Y_e/Y_1 of any sheet containing holes of arbitrary geometry is independent of the Poisson's ratio of the solid phase. It is straightforward to show that the above series expansions are consistent with Cherkaev *et al.*'s theorem, since using the interrelations between the elastic moduli we obtain

$$\frac{Y_e}{Y_1} = \frac{\phi_1(1-2\zeta_2)(1-\zeta_2-\eta_2)}{1+2\phi_2-2\zeta_2(1+\phi_2)+(\zeta_2+\eta_2)(2\phi_1\zeta_2-1)}, \tag{279}$$

indicating that Y_e/Y_1 is indeed independent of the Poissons' ratio, and depends only on the volume fractions and the microstructural ζ_2 and η_2.

Consider next a fiber-reinforced composite consisting of rigid fibers, i.e., one for which $\mu_2/\mu_1 = K_2/K_1 = \infty$, in a compressible matrix. For such a composite, $\kappa = \mu = 1$ and hence, according to Eqs. (273) and (274) one has

$$\frac{K_{e2}}{K_1} = \frac{1+\dfrac{\mu_1}{K_1}\phi_2 - \dfrac{2\mu_1}{K_1+2\mu_1}\phi_1\zeta_2}{1-\phi_2-\dfrac{2\mu_1}{K_1+2\mu_1}\phi_1\zeta_2}, \tag{280}$$

$$\frac{\mu_{e2}}{\mu_1} = \frac{1+\dfrac{K_1}{K_1+2\mu_1}\phi_2 - \dfrac{\mu_1}{K_1+2\mu_1}\phi_1\zeta_2 - \dfrac{K_1^2}{(K_1+2\mu_1)^2}\phi_1\eta_2}{1-\phi_2-\dfrac{\mu_1}{K_1+2\mu_1}\phi_1\zeta_2-\dfrac{K_1^2}{(K_1+2\mu_1)^2}\phi_1\eta_2}. \tag{281}$$

Figure 7.4 compares the predictions of Eq. (280) to the simulation data of Eischen and Torquato (1993) for hexagonal arrays of rigid circular fibers in a compressible matrix in which $\mu_2/\mu_1 = K_2/K_1 = \infty$ and $\mu_1/K_1 = 0.4$. The predictions are highly accurate. The three-point lower bound of Silnutzer (1972) is again indistinguishable from the predictions of Eq. (280); the associated upper bound diverges to infinity. In the case of a fiber-reinforced material consisting of rigid, non-overlapping, circular fibers that are randomly arranged in a compressible matrix (i.e., one with $\mu_2/\mu_1 = K_2/K_1 = \infty, \mu_1/K_1 = 0.4$), Torquato (1998) showed that the predictions of Eq. (280) are consistent with the upper bound of Gibiansky and Torquato (1995a).

The next model material to be considered is an incompressible fiber-reinforced composite, i.e., one with $K_1/\mu_1 = \infty$ and $K_2/\mu_2 = \infty$, for which Eq. (274) reduces to

$$\frac{\mu_{e2}}{\mu_1} = \frac{1+\mu\phi_2-\mu^2\phi_1\eta_2}{1-\mu\phi_2-\mu^2\phi_1\eta_2}, \tag{282}$$

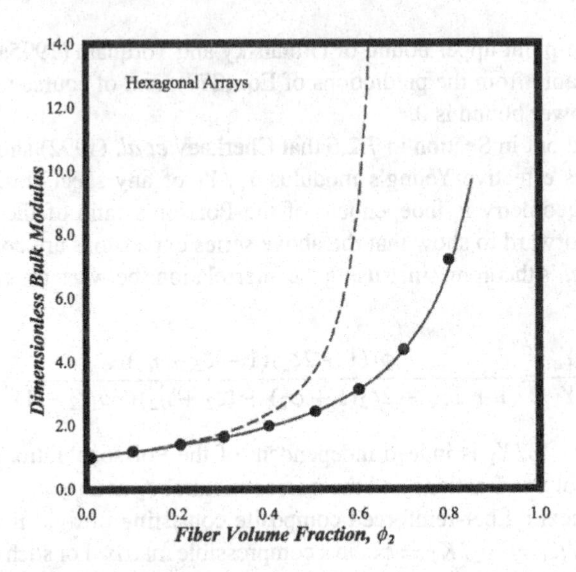

FIGURE 7.4. Dimensionless effective transverse bulk modulus K_e/K_1 versus fiber volume fraction ϕ_2 for hexagonal arrays of circular rigid fibers in a compressible sheet, with $\mu_2/\mu_1 = K_2/K_1 = \infty$ and $\mu_1/K_1 = 0.4$. Solid circles are the simulation data, solid curve shows the predictions of Eq. (280), while the dashed curve presents the EMA predictions, Eqs. (569) and (570) (after Torquato, 1998).

with, $\mu = (\mu_2 - \mu_1)/(\mu_2 + \mu_1)$. In addition, we consider the limit in which the included phase is rigid (i.e., one with $\mu_2/\mu_1 \to \infty$), which is the most difficult case to treat theoretically. For such an incompressible composite Eq. (282) yields

$$\frac{\mu_{e2}}{\mu_1} = \frac{1 + \phi_2 - \phi_1\eta_2}{1 - \phi_2 - \phi_1\eta_2}, \tag{283}$$

which, in the limit of an incompressible matrix ($K_1/\mu_1 = \infty$), is identical with Eq. (281). Figure 7.5 compares the predictions of Eq. (283) and the three-point lower bound of Silnutzer (1972) to the simulation data of Eischen and Torquato (1993) for an incompressible composite in which rigid circular fibers are arranged in an hexagonal array (with $\mu_2/\mu_1 = K_2/K_1 = \infty$, and $\mu_1/K_1 = 0$). Equation (283) provides excellent estimates of the shear modulus up to a fiber volume fraction $\phi_2 = 0.6$, but begins to deviate from the data for $\phi_2 > 0.6$.

7.4.6.2 Three-Dimensional materials

Let 1 and 2 denote the matrix and dispersed phases, respectively. For $d = 3$, series (248) and (254) yield

$$\frac{K_e}{K_1} = \frac{1 + \dfrac{4\mu_1}{3K_1}\kappa\phi_2 - \dfrac{10\mu_1}{3(K_1 + 2\mu_1)}\kappa\mu\phi_1\zeta_2}{1 - \kappa\phi_2 - \dfrac{10\mu_1}{3(K_1 + 2\mu_1)}\kappa\mu\phi_1\zeta_2}, \tag{284}$$

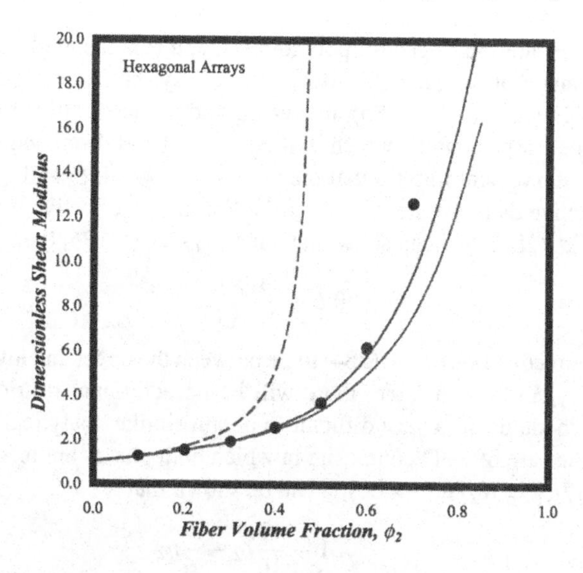

FIGURE 7.5. Dimensionless effective transverse shear modulus μ_e/μ_1 versus fiber volume fraction ϕ_2 for hexagonal arrays of rigid fibers in an incompressible matrix, with $\mu_2/\mu_1 = K_2/K_1 = \infty$ and $\mu_1/K_1 = 0.0$. Solid circles are the simulation data, while the dashed curve presents the EMA predictions, Eqs. (569) and (570). The lower and upper solid curves correspond, respectively, to the predictions of the lower bound (419) and Eq. (283) (after Torquato, 1998).

$$\frac{\mu_e}{\mu_1} = \frac{1 + \dfrac{9K_1+8\mu_1}{6(K_1+2\mu_1)}\mu\phi_2 - \dfrac{2\kappa\mu\mu_1\phi_1\zeta_2}{3(K_1+2\mu_1)} - \dfrac{\mu^2}{6}\left[\left(\dfrac{3K_1+\mu_1}{K_1+2\mu_1}\right)^2\phi_1\eta_2 + 5\dfrac{2K_1+3\mu_1}{(K_1+2\mu_1)^2}\mu_1\phi_1\zeta_2\right]}{1 - \mu\phi_2 - \dfrac{2\kappa\mu\mu_1\phi_1\zeta_2}{3(K_1+2\mu_1)} - \dfrac{\mu^2}{6}\left[\left(\dfrac{3K_1+\mu_1}{K_1+2\mu_1}\right)^2\phi_1\eta_2 + 5\dfrac{2K_1+3\mu_1}{(K_1+2\mu_1)^2}\mu_1\phi_1\zeta_2\right]}, \tag{285}$$

where

$$\kappa \equiv K_{21} = \frac{K_2 - K_1}{K_2 + \dfrac{4}{3}\mu_1},$$

$$\mu \equiv \mu_{21} = \frac{\mu_2 - \mu_1}{\mu_2 + \dfrac{\mu_1}{6}\dfrac{9K_1+8\mu_1}{K_1+2\mu_1}}.$$

For any isotropic, 3D composite, the three-point parameters ζ_2 and η_2 are defined by

$$\zeta_2 = \frac{9}{2\phi_1\phi_2}\int_0^\infty \frac{dr}{r}\int_0^\infty \frac{ds}{s}\int_{-1}^1 d(\cos\theta)P_2(\cos\theta)\left[S_3^{(2)}(r,s,t) - \frac{S_2^{(2)}(r)S_2^{(2)}(s)}{\phi_2}\right], \tag{286}$$

$$\eta_2 = \frac{5\zeta_2}{21} + \frac{150}{7\phi_1\phi_2}\int_0^\infty \frac{dr}{r}\int_0^\infty \frac{ds}{s}\int_{-1}^1 d(\cos\theta)P_4(\cos\theta)\left[S_3^{(2)}(r,s,\ell) - \frac{S_2^{(2)}(r)S_2^{(2)}(s)}{\phi_2}\right], \tag{287}$$

where P_2 and P_4 are the Legendre polynomials, and θ is the angle opposite the side of the triangle of length ℓ. Similar to the 2D systems, we require that the predictions of Eqs. (284) and (285) always lie within the most restrictive three-point upper and lower bounds, which will be derived and discussed later in this chapter. This requirement implies that the intervals in which ζ_2 and η_2 lie should be more restrictive than just the interval [0,1]. For the 3D effective bulk modulus predicted by Eq. (284), one can show that when $\mu_1/K_1 \le 0.75$, then

$$\zeta_2 \le 0.6 + \frac{8}{15}\frac{\mu_1}{K_1},$$

in order for the predictions of Eq. (284) to lie between the upper and lower bounds. However, if $\mu_1/K_1 > 0.75$, then, there will be no additional restriction on ζ_2. For the shear modulus, it is very difficult to obtain similar analytical conditions. However, in the case of a 3D composite in which both phases are incompressible (i.e., when $K_1/\mu_1 = K_2/\mu_2 \to \infty$), it can be shown that

$$\frac{1+\eta_2}{2-3\eta_2} \ge \frac{1-\frac{11}{16}\zeta_2-\frac{5}{6}\eta_2}{2(1-\zeta_2)(1+\frac{5}{16}\zeta_2-\frac{21}{16}\eta_2)}.$$

To further validate the accuracy of Eqs. (284) and (285), we compare their predictions to rigorous bounds (and, later on, to the effective-medium approximations; see Section 7.7) for ordered and disordered model microstructures. Consider first a random dispersion of identical glass spheres in an epoxy matrix, such that $\mu_2/\mu_1 = 22.5$, $K_2/K_1 = 10.0$ and $\mu_1/K_1 = 0.33$. The non-overlapping spheres are in equilibrium and thus the system has a maximum volume fraction at random-close packing when $\phi_2 \simeq 0.644$ (Rintoul and Torquato, 1996). One can compare the predictions to the rigorous bounds, since the parameters ζ_2 and η_2 have been computed for this model for various values of ϕ_2 (see Chapter 4, Table 4.2). This is illustrated in Figure 7.6 for K_e, and it is clear that the predictions of Eq. (284) lie between the very narrow bound widths formed by the upper bound of Gibiansky and Torquato (1996), obtained from cross-property relation (see Section 7.9), and the three-point lower bound of Beran and Molyneux (1966).

Consider now a composite in which the dispersed phase consists of non-overlapping cavities ($K_2 = \mu_2 = 0$). Then, Eqs. (284) and (285) reduce to

$$\frac{K_e}{K_1} = \frac{1-\phi_2-\dfrac{15K_1}{9K_1+8\mu_1}\phi_1\zeta_2}{1+\dfrac{3K_1}{4\mu_1}\phi_2-\dfrac{15K_1}{9K_1+8\mu_1}\phi_1\zeta_2}, \tag{288}$$

$$\frac{\mu_e}{\mu_1} = \frac{1-\phi_2-\dfrac{3K_1}{9(K_1+8\mu_1)}\phi_1\zeta_2-6\left(\dfrac{3K_1+\mu_1}{9K_1+8\mu_1}\right)^2\phi_1\eta_2-\dfrac{30\mu_1(2K_1+3\mu_1)}{(9K_1+8\mu_1)^2}\phi_1\zeta_2}{1+\dfrac{6(K_1+2\mu_1)}{9K_1+8\mu_1}\phi_2-\dfrac{3K_1}{9K_1+8\mu_1}\phi_1\zeta_2-6\left(\dfrac{3K_1+\mu_1}{9K_1+8\mu_1}\right)^2\phi_1\eta_2-\dfrac{30\mu_1(2K_1+3\mu_1)}{(9K_1+8\mu_1)^2}\phi_1\zeta_2}. \tag{289}$$

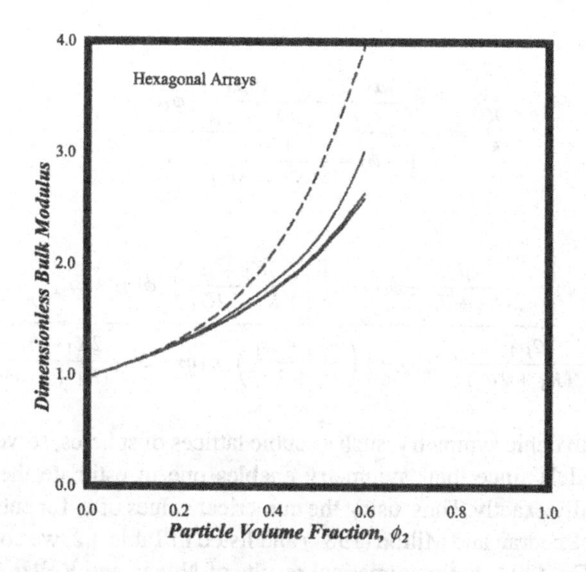

FIGURE 7.6. Dimensionless effective bulk modulus K_e/K_1 versus particle volume fraction ϕ_2 for random arrays of glass spheres in an epoxy matrix, with $\mu_2/\mu_1 = 22.5$, $\mu_1/K_1 = 0.33$ and $\mu_2/K_2 = 0.75$. Dashed curve presents the predictions of the EMA, Eqs. (569) and (570), while the middle solid curve shows the predictions of Eq. (284). The other two solid curves are the lower three-point bound (412) and the Gibiansky–Torquato cross-property upper bound, described in Section 7.9.2 (after Torquato, 1998).

For cavities in an incompressible matrix one can simplify Eqs. (288) and (289) by taking the limit, $K_1/\mu_1 \to \infty$, which leads to

$$\frac{K_e}{\mu_1} = \frac{4\phi_1}{3\phi_2}\left(1 - \frac{5}{3}\phi_1\zeta_2\right), \tag{290}$$

$$\frac{\mu_e}{\mu_1} = \frac{1 - \phi_2 - \frac{1}{3}\phi_1\zeta_2 - \frac{2}{3}\phi_1\eta_2}{1 - \frac{2}{3}\phi_2 - \frac{1}{3}\phi_1\zeta_2 - \frac{2}{3}\phi_1\eta_2}. \tag{291}$$

The effective Lamé constant λ_e is also obtained:

$$\frac{\lambda_e}{\mu_1} = \frac{4\phi_1}{3\phi_2}\left(1 - \frac{5}{3}\phi_1\zeta_2\right) - \frac{2\phi_1}{3}\left(\frac{1 - \frac{1}{3}\zeta_2 - \frac{2}{3}\eta_2}{1 - \frac{2}{3}\phi_2 - \frac{1}{3}\phi_1\zeta_2 - \frac{2}{3}\phi_1\eta_2}\right). \tag{292}$$

Interestingly, a composite consisting of spherical cavities in an incompressible matrix of shear modulus μ_1 is exactly equivalent to an incompressible liquid of shear viscosity μ containing air bubbles. The analogues of the shear modulus, bulk modulus and Lamé constant in the liquid problem are the shear viscosity, bulk viscosity and expansion viscosity, respectively.

We now consider a composite that consists of rigid inclusions, i.e., one with $\mu_2/\mu_1 = K_2/K_1 = \infty$, in a compressible matrix. In this case, $\kappa = \mu = 1$, and hence

$$\frac{K_e}{K_1} = \frac{1 + \dfrac{4\mu_1}{3K_1}\phi_2 - \dfrac{10\mu_1}{3(K_1+2\mu_1)}\phi_1\zeta_2}{1 - \phi_2 - \dfrac{10\mu_1}{3(K_1+2\mu_1)}\phi_1\zeta_2}, \tag{293}$$

$$\frac{\mu_e}{\mu_1} = \frac{1 + \dfrac{9K_1+8\mu_1}{6(K_1+2\mu_1)}\phi_2 - \dfrac{2\mu_1}{3(K_1+2\mu_1)}\phi_1\zeta_2 - \dfrac{1}{6}\left[\left(\dfrac{3K_1+\mu_1}{K_1+2G_1}\right)^2\phi_1\eta_2 + 5\mu_1\dfrac{2K_1+3\mu_1}{(K_1+2\mu_1)^2}\phi_1\zeta_2\right]}{1 - \phi_2 - \dfrac{2\mu_1}{3(K_1+2\mu_1)}\phi_1\zeta_2 - \dfrac{1}{6}\left[\left(\dfrac{3K_1+\mu_1}{K_1+2\mu_1}\right)^2\phi_1\eta_2 + 5\mu_1\dfrac{2K_1+3\mu_1}{(K_1+2\mu_1)^2}\phi_1\zeta_2\right]}. \tag{294}$$

Composites with cubic symmetry, such as cubic lattices of spheres, serve as useful benchmark models since their symmetry enables one to estimate the effective moduli essentially exactly. Thus, using the numerical values of ζ_2 for cubic lattices obtained by McPhedran and Milton (1981) and listed in Table 4.2, we compare the predictions of Eq. (293) to the numerical results of Nunan and Keller (1984) for the effective bulk moduli of rigid cubic arrays of spheres in compressible matrices. The effective stiffness tensor $\mathbf{C}_e$ of such a composite, in component form, is given by Eq. (50), from which the effective bulk modulus of such a composite is given by

$$K_e = K_1 + \mu_1(\gamma + \frac{2}{3}\alpha). \tag{295}$$

Figure 7.7 compares the predictions of Eq. (293) to the numerical data of Nunan and Keller for FCC arrays of rigid spheres in a compressible matrix in which $\mu_2/\mu_1 = K_2/K_1 = \infty$ and $\mu_1/K_1 = 0.46$. The accuracy of the predictions of Eq. (293) is remarkably good. The Beran–Molyneux three-point lower bound is virtually indistinguishable from the predictions of (293), while the associated upper bound diverges to infinity in this case.

Finally, let us consider an incompressible isotropic composite, i.e., one with $K_1/\mu_1 \to \infty$ and $K_2/\mu_2 \to \infty$, for which Eq. (285) reduces to

$$\frac{\mu_e}{\mu_1} = \frac{1 + \frac{3}{2}\mu\phi_2 - \frac{3}{2}\mu^2\phi_1\eta_2}{1 - \mu\phi_2 - \frac{3}{2}\mu^2\phi_1\eta_2}, \tag{296}$$

where, $\mu = (\mu_2 - \mu_1)(\mu_2 + 3\mu_1/2)$. If we now allow the included phase of the composite to be rigid ($\mu_2/\mu_1 \to \infty$), Eq. (296) yields

$$\frac{\mu_e}{\mu_1} = \frac{1 + \frac{3}{2}\phi_2 - \frac{3}{2}\phi_1\eta_2}{1 - \phi_2 - \frac{3}{2}\phi_1\eta_2}. \tag{297}$$

This is the most difficult system to treat theoretically, especially at high particle concentrations. Interestingly, the determination of the effective shear modulus of such a dispersion is exactly equivalent to computing the effective viscosity of the dispersion in the infinite-frequency limit. Torquato (1998) showed that the predictions of Eq. (296) are consistent with the three-point lower bound of Milton and

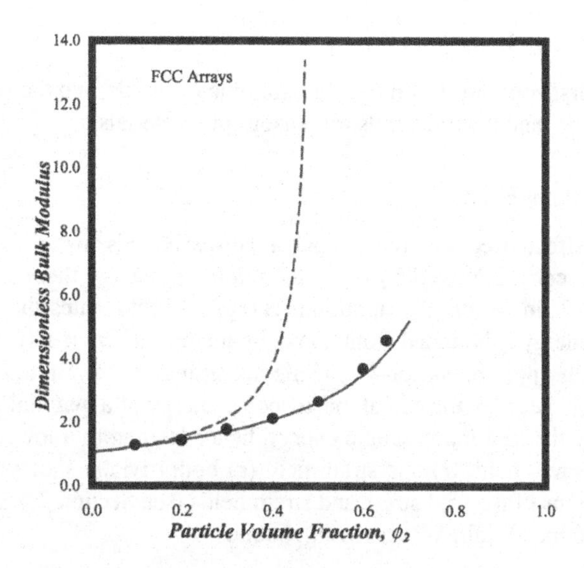

FIGURE 7.7. Dimensionless effective bulk modulus K_e/K_1 versus particle volume fraction ϕ_2 for FCC arrays of rigid spheres in a compressible matrix, with $\mu_2/\mu_1 = K_2/K_1 = \infty$ and $\mu_1/K_1 = 0.46$. Solid circles are the simulation data, dashed curve shows the predictions of the EMA, Eqs. (571) and (572), while solid curve presents the predictions of Eq. (293) (after Torquato, 1998).

Phan–Thien (1982) for the case of an incompressible composite containing a random array of rigid spheres ($\mu_2/\mu_1 = K_2/K_1 = \infty, \mu_1/K_1 = 0$) in an equilibrium arrangement.

7.5 Rigorous Bounds

In Chapter 4 we described and discussed derivation of rigorous upper and lower bounds to the effective conductivity and dielectric constant of disordered materials. Such bounds are useful for several reasons. (1) As we discussed in the last section, rigorous upper and lower bounds help one to test the merit and accuracy of a theory in the sense that, the prediction of the theory cannot violate the bounds. (2) As the accuracy of the bounds improves by incorporating in them more detailed information about the microstructure of a material, the difference between the upper and lower bounds decreases, and therefore they can provide sharper estimates of the effective property of interest. (3) As we saw in the last section, usually one of the bounds provides a relatively accurate estimate of the property of interest.

In this section we describe and discuss derivation of such bounds for the effective linear elastic moduli of heterogeneous materials. We provide some details of derivation of the bounds, since it is very similar to the procedure for deriving the bounds on the effective conductivity described in Chapter 4.

7.5.1 Isotropic Materials

As usual, we first consider the bounds for isotropic materials, and then take up the problem of upper and lower bounds for anisotropic materials.

7.5.1.1 One-Point Bounds

Similar to the effective conductivity, upper and lower bounds on the effective stiffness tensor $\mathbf{C}_e$ can be derived by using the minimum energy theorems outlined in Section 7.1.2. In words, the minimum energy theorem states that the actual macroscopic energy is bounded from above by the trial macroscopic energy, and thus leads to an upper bound on $\mathbf{C}_e$, while according to the minimum complementary energy theorem, the actual macroscopic energy of a material is bounded from below by the trial macroscopic energy, hence leading to a lower bound on $\mathbf{C}_e$. Constant stress field $\langle\sigma\rangle$ and strain field $\langle\epsilon\rangle$ both trivially satisfy the admissibility conditions of the trial stress and strain fields (see Section 7.1.2), and lead immediately to the so-called Reuss–Voigt bounds:

$$\langle \mathbf{C}^{-1}\rangle \leq \mathbf{C}_e \leq \langle \mathbf{C}\rangle. \tag{298}$$

The bounds (298) represent one-point bounds. These bounds, for macroscopically-anisotropic materials with isotropic phases, simplify to

$$d\langle K^{-1}\rangle^{-1}\mathbf{\Lambda}_h + 2\langle \mu^{-1}\rangle^{-1}\mathbf{\Lambda}_s \leq \mathbf{C}_e \leq d\langle K\rangle\mathbf{\Lambda}_h + 2\langle \mu\rangle\mathbf{\Lambda}_s, \tag{299}$$

where the tensors $\mathbf{\Lambda}_h$ and $\mathbf{\Lambda}_s$ were defined by Eq. (183). If the material is macroscopically isotropic, the bounds (299) reduce to the following one-point bounds for the elastic moduli:

$$\langle K^{-1}\rangle^{-1} \leq K_e \leq \langle K\rangle, \quad \langle \mu^{-1}\rangle^{-1} \leq \mu_e \leq \langle \mu\rangle, \quad Y_e \geq \langle Y_e^{-1}\rangle^{-1}, \tag{300}$$

which are valid for an isotropic material with $n \geq 2$ phases. The bounds (299) do not, however, lead to an upper bound for the effective Young's modulus. However, if the Poisson's ratios of all the phases are equal, then one also recovers the upper bound, $Y_e \leq \langle Y\rangle$. As in the case of the effective conductivity, the bounds (298)–(300) are not accurate.

7.5.1.2 Two-Point Bounds

Hashin and Shtrikman (1963) derived the most accurate two-point bounds on the effective elastic moduli of 3D isotropic materials, which were extended by Hashin (1965) to 2D materials. To derive these bounds, we must first state the Hashin–Shtrikman variational principle, similar to what was described in Section 4.6 for the effective conductivity and dielectric constant.

Hashin–Shtrikman Variational Principle:

Let $\mathbf{g}$ be defined by $\mathbf{g} = \epsilon - \epsilon_0$ (where ϵ_0 is the strain field of a *homogeneous comparison material*), with $\hat{\mathbf{g}} = \hat{\epsilon} - \epsilon_0$ being the associated trial difference strain field, and let $\mathbf{p} = (\mathbf{C} - \mathbf{C}_0) : \boldsymbol{\epsilon}$ [see Eq. (194)] and $\hat{\mathbf{p}}$ be the corresponding polarization fields. Then, subject to the conditions that

$$\nabla \cdot (\mathbf{C}_0 : \mathbf{g}) + \nabla \cdot \hat{\mathbf{p}} = \mathbf{0}, \tag{301}$$

$$\nabla \times (\nabla \times \mathbf{g})^{\mathrm{T}} = \mathbf{0}, \tag{302}$$

$$\langle \hat{g} \rangle = \mathbf{0}, \tag{303}$$

a hydrostatic strain field $(\epsilon_0)_{ij} = \frac{1}{d}\delta_{ij}$ imposed externally on any two-phase material results in

$$K_B = K_0 + \left\langle \hat{\mathbf{p}} : \hat{\mathbf{g}} + 2\hat{\mathbf{p}} : \epsilon_0 - \hat{\mathbf{p}} : (\mathbf{C} - \mathbf{C}_0)^{-1} : \hat{\mathbf{p}} \right\rangle, \tag{304}$$

which is an upper bound on K_e when $\mu_0 \geq \mu_1$ and a lower bound on K_e when $\mu_0 \leq \mu_2$. If, however, a deviatoric (traceless) strain field, $(\epsilon_0)_{ij} = 2^{-1/2}(\delta_{1i}\delta_{2j} + \delta_{2i}\delta_{1j})$, is imposed on the material, it will result in

$$\mu_B = \mu_0 + \left\langle \hat{\mathbf{p}} : \hat{\mathbf{g}} + 2\hat{\mathbf{p}} : \epsilon_0 - \hat{\mathbf{p}} : (\mathbf{C} - \mathbf{C}_0)^{-1} : \hat{\mathbf{p}} \right\rangle, \tag{305}$$

which is an upper bound on μ_e if $K_0 \geq K_1$ and $\mu_0 \geq \mu_1$ and a lower bound on μ_e when $K_0 \leq K_2$ and $\mu_0 \leq \mu_2$.

The original Hashin–Shtrikman bound was derived for a very large but finite sample material, subjected to a homogeneous displacement on the sample's surface (instead of requiring $\langle \hat{\mathbf{g}} \rangle = \mathbf{0}$). But because for infinite systems the effect of the boundary conditions vanishes, the above variational principle is equivalent to the original Hashin–Shtrikman principle. This variational principle has been extended to periodic media (see, for example, Nemat-Nasser and Hori, 1993), as well as anisotropic materials (Willis, 1977).

One can now derive the two-point Hashin–Shtrikman bound for a d-dimensional isotropic material. One assumes that

$$\hat{g}(\mathbf{x}) = \sum_{i=1}^{m} a_i \epsilon_i(\mathbf{x}), \tag{306}$$

$$\hat{p}(\mathbf{x}) = [\mathbf{C}(\mathbf{x}) - \mathbf{C}_0] : \sum_{i=1}^{m} a_i \epsilon_{i-1}(\mathbf{x}), \tag{307}$$

where ϵ_i is the exact ith-order solution of the strain field in the exact contrast expansion derived in Section 7.4, with $\epsilon_0 = \langle \epsilon \rangle$. If one substitutes these trial fields into the Hashin–Shtrikman variational principle and optimizes over the vector $\mathbf{a}$, one obtains a $2m$-point lower bound on the effective moduli when $\mathbf{C}_0 = \mathbf{C}_1$. Interchanging the phase subscripts then yields the corresponding $2m$-point upper bound. Hence, to derive the two-point bounds, we take $m = 1$ to obtain

$$\begin{aligned} &\phi_1 K_1 + \phi_2 K_2 - \frac{\phi_1\phi_2 d(K_2 - K_1)^2}{d(\phi_2 K_1 + \phi_1 K_2) + 2(d-1)\mu_1} \\ &\leq K_e \leq \phi_1 K_1 + \phi_2 K_2 - \frac{\phi_1\phi_2 d(K_2 - K_1)^2}{d(\phi_2 K_1 + \phi_1 K_2) + 2(d-1)\mu_2}, \end{aligned} \tag{308}$$

$$\phi_1\mu_1 + \phi_2\mu_2 - \frac{\phi_1\phi_2(\mu_2-\mu_1)^2}{\phi_2\mu_1+\phi_1\mu_2+h_1} \le \mu_e \le \phi_1\mu_1 + \phi_2\mu_2 - \frac{\phi_1\phi_2(\mu_2-\mu_1)^2}{\phi_2\mu_1+\phi_1\mu_2+h_2}, \tag{309}$$

where

$$h_i = \frac{d^2K_i + 2(d+1)(d-2)\mu_i}{2d(K_i+2\mu_i)}. \tag{310}$$

These bounds, which are exact through second order in the difference in the phase moduli, are completely similar to the bounds (4.118), the Hashin–Shtrikman bounds for the effective conductivity of isotropic two-phase materials.

The Hashin–Shtrikman bounds on K_e are optimal for the set of *all* two-phase isotropic materials with given phase volume fractions, because they can be realized exactly by certain microstructures including, as discussed in Chapter 4, singly-coated sphere assemblages and some finite-rank laminates (see Sections 4.4 and 4.5). On the other hand, although the Hashin–Shtrikman bounds on the effective shear modulus μ_e are not realized by the coated-spheres microstructures, they are still optimal because they are attained by certain finite-rank laminates (see, for example, Francfort and Murat, 1986; Milton, 1986). Milton (1986) also showed that the Hashin–Shtrikman bounds on either K_e or μ_e are attained if and only if the field in phase 1 or phase 2 is uniform. Thus, as our discussions of the bounds on the effective conductivity (see Chapter 4) and the elastic moduli indicate, attainability of any bounds by some microstructure is an important factor to their accuracy.

Walpole (1966) generalized the Hashin–Shtrikman bounds for 3D materials by removing the restrictions on the phase moduli. That is, Walpole showed that these bounds are applicable not only to the "well-ordered" materials for which $(K_2 - K_1)(\mu_2 - \mu_1) \ge 0$, which was the case considered by Hashin and Shtrikman, but also to "badly-ordered" materials for which $(K_2 - K_1)(\mu_1 - \mu_2) \ge 0$. However, we are not aware of any microstructure that can realize the bounds in the badly-ordered case.

The Hashin–Shtrikman–Walpole bounds can be generalized to d-dimensional isotropic materials with n phases. Suppose that (K_M, μ_M) and (K_m, μ_m) are the maximum and minimum bulk and shear moduli among all the phases. Then, the bounds are given by

$$\left[\sum_{i=1}^{n}\phi_i(K_m^*+K_i)^{-1}\right]^{-1} - K_m^* \le K_e \le \left[\sum_{i=1}^{n}\phi_i(K_M^*+K_i)^{-1}\right]^{-1} - K_M^*, \tag{311}$$

$$\left[\sum_{i=1}^{n}\phi_i(\mu_m^*+\mu_i)^{-1}\right]^{-1} - \mu_m^* \le \mu_e \le \left[\sum_{i=1}^{n}\phi_i(\mu_M^*+\mu_i)^{-1}\right]^{-1} - \mu_M^*, \tag{312}$$

where $K_m^* = 2d^{-1}(d-1)\mu_m$, $K_M^* = 2d^{-1}(d-1)\mu_M$, $\mu_m^* = h_m$ and $\mu_M^* = h_M$, where h_m and h_M are the minimum and maximum values of the quantity h_i defined by Eq. (310). The bounds (311) and (312) are completely similar to the bounds (4.119) for the effective conductivity of an isotropic material with $n \geq 2$ phases.

7.5.1.3 Cluster Bounds

Quintanilla and Torquato (1995) derived cluster bounds for the effective bulk and shear moduli of a two-phase material composed of equisized spheres, distributed randomly throughout a matrix. The interaction potential between the spheres was assumed to be completely arbitrary, and therefore one could consider a variety of composites, ranging from non-overlapping spheres to completely overlapping ones. Their approach is similar to what we described and discussed in Section 4.6.1.2 for deriving cluster bounds for the effective conductivity of such composites. In what follows we derive the cluster bounds for the effective bulk and shear moduli. The derivation serves another useful purpose, namely, it provides the details of the complexities that are involved in deriving such bounds.

Consider a trial strain field $\hat{\boldsymbol{\epsilon}}(\mathbf{r})$ at $\mathbf{r}$. We require the ensemble average $\langle\hat{\boldsymbol{\epsilon}}\rangle$ to be equal to the average strain field $\langle\boldsymbol{\epsilon}\rangle$. If

$$\mathcal{H}_p(\hat{\boldsymbol{\epsilon}}) = \frac{1}{2}\langle C_{ijkl}\hat{\epsilon}_{ij}\hat{\epsilon}_{kl}\rangle \tag{313}$$

is the potential energy of the system for the trial strain field, then according to the theorem of minimum energy described in Section 7.1.2, among all trial strain fields, the one that satisfies $\nabla \cdot \boldsymbol{\sigma} = \mathbf{0}$ is the one that uniquely minimizes $\mathcal{H}_p(\hat{\boldsymbol{\epsilon}})$ which, for isotropic systems, implies that

$$K_e\langle\epsilon_{ii}\rangle\langle\epsilon_{kk}\rangle + 2\mu_e\langle\tilde{\epsilon}_{ij}\rangle\langle\tilde{\epsilon}_{ij}\rangle \leq \langle K\hat{\epsilon}_{ii}\hat{\epsilon}_{kk}\rangle + 2\langle\mu\tilde{\hat{\epsilon}}_{ij}\tilde{\hat{\epsilon}}_{ij}\rangle, \tag{314}$$

where $\tilde{\epsilon}_{ij} = \epsilon_{ij} - \frac{1}{3}\epsilon_{kk}\delta_{ij}$ is the deviatoric component of the strain field $\boldsymbol{\epsilon}$ [see Eq. (15)].

We now take the trial strain field to be of the form

$$\hat{\boldsymbol{\epsilon}} = \langle\boldsymbol{\epsilon}\rangle + \lambda\boldsymbol{\epsilon}', \tag{315}$$

where λ is a parameter and $\boldsymbol{\epsilon}'$ is a perturbation chosen such that $\hat{\boldsymbol{\epsilon}}$ satisfies the requirements of a trial strain field. Setting $\langle\boldsymbol{\epsilon}\rangle = \mathbf{U}$, the unit dyadic, and minimizing over λ yields the following rigorous upper bound on K_e:

$$K_e \leq \langle K\rangle - \frac{\langle K\epsilon'_{ii}\rangle^2}{\langle K\epsilon'_{ii}\epsilon'_{kk}\rangle + 2\langle\mu\tilde{\epsilon}'_{ij}\tilde{\epsilon}'_{ij}\rangle}. \tag{316}$$

Likewise, setting $\langle\epsilon_{ij}\rangle = \delta_{i1}\delta_{j1} - \delta_{i2}\delta_{j2}$ and minimizing over λ yields an upper bound on μ_e:

$$\mu_e \leq \langle\mu\rangle - \frac{\langle\mu(\epsilon'_{11} - \epsilon'_{22})\rangle^2}{\langle K\epsilon'_{ii}\epsilon'_{kk}\rangle + 2\langle\mu\tilde{\epsilon}'_{ij}\tilde{\epsilon}'_{ij}\rangle}. \tag{317}$$

We next consider trial stress fields $\hat{\sigma}(\mathbf{r})$ that are symmetric, satisfy the equilibrium equations, and also, $\langle\hat{\sigma}\rangle = \langle\sigma\rangle$. If

$$T(\hat{\sigma}) = \frac{1}{2}\langle S_{ijkl}\hat{\sigma}_{ij}\hat{\sigma}_{kl}\rangle, \tag{318}$$

where S_{ijkl} is the compliance tensor, then, according to the theorem of minimum complementary energy described in 7.1.2, among all trial stress fields, the one that is derivable from a displacement field, $\hat{\sigma} = \mathbf{C}_e\hat{\epsilon}$, is the one that uniquely minimizes $T(\hat{\sigma})$ which, for isotropic systems, implies that

$$\frac{1}{9}K_e^{-1}\langle\sigma_{kk}\rangle^2 + \frac{1}{2}\mu_e^{-1}\langle\tilde{\sigma}_{ij}\rangle\langle\tilde{\sigma}_{ij}\rangle \leq \frac{1}{9}\langle K^{-1}\hat{\sigma}_{ii}\hat{\sigma}_{kk}\rangle + \frac{1}{2}\langle\mu^{-1}\hat{\tilde{\sigma}}_{ij}\hat{\tilde{\sigma}}_{ij}\rangle, \tag{319}$$

where $\tilde{\sigma}$ is the deviatoric component of the stress field. As before, we now take

$$\hat{\sigma} = \langle\sigma\rangle + \lambda\sigma', \tag{320}$$

set $\langle\epsilon\rangle = \mathbf{U}$, so that $\langle\sigma\rangle = 3K_1\mathbf{U}$, and minimize over λ to obtain a lower bound on K_e:

$$K_e^{-1} \leq \langle K^{-1}\rangle - \frac{2\langle K^{-1}\sigma'_{ii}\rangle^2}{2\langle K^{-1}\sigma'_{ii}\sigma'_{kk}\rangle + 9\langle\mu^{-1}\tilde{\sigma}'_{ij}\tilde{\sigma}'_{ij}\rangle}. \tag{321}$$

Likewise, setting $\langle\epsilon_{ij}\rangle = \delta_{i1}\delta_{j1} - \delta_{i2}\delta_{j2}$, so that $\sigma_{ij} = 2\mu_1(\delta_{i1}\delta_{j1} - \delta_{i2}\delta_{j2})$, and minimizing over λ yield a lower bound on μ_e:

$$\mu_e^{-1} \leq \langle\mu^{-1}\rangle - \frac{9\langle\mu^{-1}(\sigma'_{11} - \sigma'_{22})\rangle^2}{2(2\langle K^{-1}\sigma'_{ii}\sigma'_{kk}\rangle + 9\langle\mu^{-1}\tilde{\sigma}'_{ij}\tilde{\sigma}'_{ij}\rangle)}. \tag{322}$$

Thus, to obtain specific bounds, appropriate choices for ϵ' and σ' must be made. To do this, the morphology of the composite must be specified. We consider composites of volume Ω that consist of a matrix phase (phase 1) with volume fraction ϕ_1 and bulk and shear moduli K_1 and μ_1, and N possibly overlapping spherical inclusions (phase 2) of radius a, volume fraction $\phi_2 = 1 - \phi_1$, and bulk and shear moduli K_2 and μ_2. The ensembles are assumed to be statistically homogeneous. The spheres' centers, $\mathbf{r}^N = \{\mathbf{r}_1, \cdots, \mathbf{r}_N\}$, are randomly distributed and follow a probability density function $P_N(\mathbf{r}^N)$. The reduced n-particle probability density $P_n(\mathbf{r}^n)$, defined by

$$P_n(\mathbf{r}^n) = \int\cdots\int P_N(\mathbf{r}^N)\,d\mathbf{r}_{n+1}\cdots d\mathbf{r}_N,$$

is the probability of finding sphere i in a volume $d\mathbf{r}_i$ about $\mathbf{r}_i$ for $i = 1, \cdots, n$. Therefore, the probability of finding any sphere in a volume $d\mathbf{r}_i$ about $\mathbf{r}_i$ for $i = 1, \cdots, n$ is given by

$$\rho_n(\mathbf{r}^n) = \frac{N!}{(N-n)!}P_n(\mathbf{r}^n).$$

This is called the *generic* n-particle probability density function. The number density ρ and the reduced density η are given by, $\rho = \lim_{N,\Omega\to\infty} N/\Omega$, and $\eta =$

$\frac{4\pi}{3}\rho a^3$ (we use ρ here to denote the number density of the particles, instead of the usual n, so as not to confuse it with the number of particles n). Subject to these conditions, the interaction potential between the spheres is completely arbitrary. The spheres may overlap to varying degrees; an example of such an interpenetrable-sphere system is the penetrable concentric-shell (the cherry-pit) model, described in detail in Chapter 3.

As discussed in Chapters 3 and 4, the distribution function associated with finding phase i at $\mathbf{r}$ and a particular configuration of $q = n - 1$ spheres at positions $\mathbf{r}^q$ is defined as

$$G_n^{(i)}(\mathbf{r}; \mathbf{r}^q) = \frac{N!}{(N-q)!} \int \cdots \int m_i(\mathbf{r}, \mathbf{r}^N) P_N(\mathbf{r}^N)\, d\mathbf{r}_n \cdots d\mathbf{r}_N, \tag{323}$$

where

$$m_i = \begin{cases} 1, & \mathbf{r} \in D_i, \\ 0, & \text{otherwise}, \end{cases}$$

and D_i is the region of space occupied by phase i. Recall from Chapter 3 that $G_n^{(i)}(\mathbf{r}; \mathbf{r}^q)$ is called the phase point/q-particle function for phase i. It is convenient to define another set of point/q-particle distribution functions $H_n^{(i)}$ by

$$H_n^{(i)}(\mathbf{r}; \mathbf{r}^q) = G_n^{(i)}(\mathbf{r}; \mathbf{r}^q) - G_0^{(i)}(\mathbf{r})\rho_q(\mathbf{r}^q), \tag{324}$$

so that $H_n^{(i)} \to 0$ as $\mathbf{r}$ moves away from $\mathbf{r}_1, \cdots, \mathbf{r}_q$. These functions, for ensembles of spheres that are statistically homogeneous and isotropic, depend only on the relative positions of the n points. For example, $\rho_1(\mathbf{r}_1) = \rho$, $G_2^{(i)}(\mathbf{r}; \mathbf{r}_1) = G_2^{(i)}(x_1)$ and $G_3^{(i)}(\mathbf{r}; \mathbf{r}_1, \mathbf{r}_2) = G_3^{(i)}(x_1, x_2, \hat{\mathbf{x}}_1, \hat{\mathbf{x}}_2)$, where $\mathbf{x}_i = \mathbf{r} - \mathbf{r}_i$, $x_i = |\mathbf{x}_i|$, and $\hat{\mathbf{x}}_i = \mathbf{x}_i / x_i$. As discussed in Chapter 3,

$$G_2^{(2)}(x) = \rho, \quad x < a \tag{325}$$

$$H_2^{(2)}(x) = \rho\phi_1, \quad x < a. \tag{326}$$

Finally, the radial distribution function $g_2(x)$ [denoted by $C(r)$ in Chapter 3] and the total correlation function $h(x)$ are defined by, $g_2(x) = \rho_2(x)/\rho^2$ and $h(x) = g_2(x) - 1$.

To derive the cluster bounds on K_e, we use Eqs. (316) and (321) and choose

$$\epsilon'(\mathbf{r}; \mathbf{r}^N) = \sum_{i=1}^{N} \epsilon^*(\mathbf{x}_i) - \int \rho(\mathbf{x}_1)\epsilon^*(\mathbf{x}_1) d\mathbf{x}_1, \tag{327}$$

and

$$\sigma'(\mathbf{r}; \mathbf{r}^N) = \sum_{i=1}^{N} \sigma^*(\mathbf{x}_i) - \int \rho(\mathbf{x}_1)\sigma^*(\mathbf{x}_1) d\mathbf{x}_1, \tag{328}$$

where

$$\boldsymbol{\epsilon}^*(\mathbf{x}_i) = \begin{cases} -\alpha \mathbf{U}, & x_i < a, \\ \dfrac{\alpha a^3}{r^3}(3\hat{\mathbf{x}}_i\hat{\mathbf{x}}_i - \mathbf{U}), & x_i > a \end{cases} \tag{329}$$

$$\boldsymbol{\sigma}^*(\mathbf{x}_i) = \begin{cases} 4\alpha\mu_1 \mathbf{U}, & x_i < a, \\ \dfrac{2\alpha\mu_1 a^3}{r^3}(3\hat{\mathbf{x}}_i\hat{\mathbf{x}}_i - \mathbf{U}), & x_i > a \end{cases} \tag{330}$$

where, $\mathbf{x}_i = \mathbf{r} - \mathbf{r}_i$, $x_i = |\mathbf{x}_i|$, and, $\alpha = 3(K_2 - K_1)/(3K_2 + 4\mu_1)$. The ensemble averages of the various quantities are now given by

$$\langle K\epsilon'_{ii}\rangle = -12\pi\alpha(K_2 - K_1)I_1, \quad \langle K^{-1}\sigma'_{ii}\rangle = 48\pi\alpha\mu_1(K_2^{-1} - K_1^{-1})I_1, \tag{331}$$

$$\langle K\epsilon'_{ii}\epsilon'_{kk}\rangle = \alpha^2[A_1K_1 + B_1(K_2 - K_1)], \quad \langle \mu\tilde{\epsilon}'_{ij}\tilde{\epsilon}'_{ij}\rangle = \alpha^2[C_1\mu_1 + D_1(\mu_2 - \mu_1)], \tag{332}$$

$$\langle K^{-1}\sigma'_{ii}\sigma'_{kk}\rangle = 16\mu_1^2\alpha^2[A_1K_1^{-1} + B_1(K_2^{-1} - K_1^{-1})],$$

$$\langle \mu^{-1}\tilde{\sigma}'_{ii}\tilde{\sigma}'_{kk}\rangle = 4\mu_1^2\alpha^2[C_1\mu_1^{-1} + D_1(\mu_2^{-1} - \mu_1^{-1})], \tag{333}$$

where

$$A_1 = \langle \epsilon'_{ii}\epsilon'_{kk}\rangle = 9\eta + \frac{81}{2a^6}\eta^2 I_2, \quad B_1 = \langle I^{(2)}\epsilon'_{ii}\epsilon'_{kk}\rangle = \frac{27\eta}{a^3\rho}I_3 + \frac{81\eta^2}{2a^6\rho^2}I_4, \tag{334}$$

$$C_1 = \langle \tilde{\epsilon}'_{ij}\tilde{\epsilon}'_{ij}\rangle = 6\eta + 27\eta^2 I_5, \quad D_1 = \langle I^{(2)}\tilde{\epsilon}'_{ij}\tilde{\epsilon}'_{ij}\rangle = \frac{18\eta a^3}{\rho}I_6 + \frac{27\eta^2}{\rho^2}I_7, \tag{335}$$

$$I_1 = \int_0^a r^2 H_2^{(2)}(r)dr, \quad I_2 = \int_0^a r^2 dr \int_0^a s^2 ds \int_{-1}^1 h(|\mathbf{r} - \mathbf{s}|)du, \tag{336}$$

$$I_3 = \int_0^a r^2 G_2^{(2)}(r)dr, \quad I_4 = \int_0^a r^2 dr \int_0^a s^2 ds \int_{-1}^1 Q(\mathbf{r}, \mathbf{s})du, \tag{337}$$

$$I_5 = \int_a^\infty \frac{dr}{r}\int_0^\infty \frac{ds}{s}\int_{-1}^1 h(|\mathbf{r} - \mathbf{s}|)P_2(u)du, \quad I_6 = \int_a^\infty \frac{G_2^{(2)}(r)}{r^4}dr, \tag{338}$$

$$I_7 = \int_a^\infty \frac{dr}{r}\int_a^\infty \frac{ds}{s}\int_{-1}^1 Q(\mathbf{r}, \mathbf{s})P_2(u)du, \tag{339}$$

and

$$Q(\mathbf{r}, \mathbf{s}) = G_3^{(2)}(r, s, u) - \rho G_2^{(2)}(r) - \rho G_2^{(2)}(s) + \rho^2\phi_2, \tag{340}$$

where $u = (r^2 + s^2 - |\mathbf{r} - \mathbf{s}|^2)/(2rs)$.

The bounds on K_e are then given by

$$K_e \le \langle K\rangle - \frac{9\eta^2\phi_1^2(K_2 - K_1)^2}{A_1K_1 + B_1(K_2 - K_1) + 2[C_1\mu_1 + D_1(\mu_2 - \mu_1)]}, \tag{341}$$

$$K_e \geq \left\{ \langle K^{-1} \rangle - \frac{4\eta^2 \phi_1^2 (K_2^{-1} - K_1^{-1})^2}{\frac{4}{9}[A_1 K_1^{-1} + B_1 (K_2^{-1} - K_1^{-1})] + \frac{1}{2}[C_1 \mu_1^{-1} + D_1(\mu_2^{-1} - \mu_1^{-1})]} \right\}^{-1}. \tag{342}$$

To derive the cluster bounds on μ_e, one proceeds in a manner similar to that for K_e, namely, one uses (317) and (322) and again chooses Eqs. (327) and (328), where for the strain perturbation field one instead takes

$$\epsilon_{ij}^* = c_1(\delta_{i1}\delta_{j1} - \delta_{i2}\delta_{j2}), \quad r < a, \tag{343}$$

while for $r > a$

$$\epsilon_{11}^* = \frac{\partial p}{\partial x} x(x^2 - y^2) + p(3x^2 - y^2) + x\frac{\partial q}{\partial x} + q, \tag{344}$$

$$\epsilon_{12}^* = \epsilon_{21}^* = \frac{1}{2}(x^2 - y^2)\left(x\frac{\partial p}{\partial y} + y\frac{\partial p}{\partial x}\right) + \frac{1}{2}\left(x\frac{\partial q}{\partial y} - y\frac{\partial q}{\partial x}\right), \tag{345}$$

$$\epsilon_{13}^* = \epsilon_{31}^* = \frac{1}{2}(x^2 - y^2)\left(x\frac{\partial p}{\partial z} + z\frac{\partial p}{\partial x}\right) + pxz + \frac{1}{2}x\frac{\partial q}{\partial z}, \tag{346}$$

$$\epsilon_{22}^* = \frac{\partial p}{\partial y} y(x^2 - y^2) + p(x^2 - 3y^2) - y\frac{\partial q}{\partial y} - q, \tag{347}$$

$$\epsilon_{23}^* = \epsilon_{32}^* = \frac{1}{2}(x^2 - y^2)\left(y\frac{\partial p}{\partial z} + z\frac{\partial p}{\partial y}\right) - pyz - \frac{1}{2}y\frac{\partial q}{\partial z}, \tag{348}$$

and

$$\epsilon_{33}^* = \frac{\partial p}{\partial z} z(x^2 - y^2) + p(x^2 - y^2), \tag{349}$$

where

$$p(r) = \frac{5c_2}{r^7} + \frac{3K_1 + \mu_1}{\mu_1}\frac{c_3}{r^5}, \tag{350}$$

$$q(r) = -\frac{2c_2}{r^5} + \frac{2c_3}{r^3}, \tag{351}$$

$$c_1 = 6(K_1 + 2\mu_1)\gamma, \quad c_2 = -a^5(3K_1 + \mu_1)\gamma, \quad c_3 = 5a^3\mu_1\gamma, \tag{352}$$

$$\gamma = \frac{\mu_1 - \mu_2}{9\mu_1 K_1 + 6\mu_2 K_1 + 8\mu_1^2 + 12\mu_1\mu_2}. \tag{353}$$

For the stress perturbation field, one takes

$$\sigma_{ij}^* = \begin{cases} 2c_4(\delta_{i1}\delta_{j1} - \delta_{i2}\delta_{j2}), & r < a \\ (K_1 - \frac{2}{3}\mu_1)\epsilon_{kk}\delta_{ij} + 2\mu_1\epsilon_{ij}^*, & r > a, \end{cases} \tag{354}$$

where

$$c_4 = \mu_2 - \mu_1 + \mu_2 c_1 = -\mu_1(9K_1 + 8\mu_1)\gamma. \tag{355}$$

Such perturbation fields arise from the solution of the single-sphere boundary value problem described in Section 7.3.1.

One must now simplify the ensemble averages in (317) and (322). Doing so introduces the following quantities,

$$A_2 = \frac{48c_3^2}{5a^6}\eta + \frac{216c_3^2}{5a^6}\eta^2 I_5, \quad B_2 = \frac{144c_3^2}{5a^3\rho}\eta I_6 + \frac{216c_3^2}{5a^6\rho^2}\eta^2 I_7, \tag{356}$$

$$C_2 = \left\{2c_1^2 + 48\left[\frac{c_2^2}{a^1} + \frac{2c_2c_3(3K_1+\mu_1)}{5\mu_1 a^8} + \frac{c_3^2(27K_1^2+24K_1\mu_1+16\mu_1^2)}{60\mu_1^2 a^6}\right]\right\}\eta$$
$$+\left[\frac{9c_1^2}{a^6}I_2 + \frac{18c_3^2(9K_1^2+48K_1\mu_1+92\mu_1^2)}{35\mu_1^2 a^6}I_5 + \frac{72}{7a^6}I_{10}\right]\eta^2 - \frac{1}{3}A_2, \tag{357}$$

$$D_2 = \left(\frac{6c_1^2}{a^3\rho}I_3 + \frac{48}{a^3\rho}I_8\right)\eta$$
$$+\left[\frac{9c_1^2}{a^6\rho^2}I_4 + \frac{18c_3^2(9K_1^2+48K_1\mu_1+92\mu_1^2)}{35\mu_1^2 a^6\rho^2}I_7 + \frac{72}{7a^6\rho^2}I_{11}\right]\eta^2 - \frac{1}{3}B_2, \tag{358}$$

$$A_3 = 9K_1^2 A_2, \quad B_3 = 9K_1^2 B_2, \tag{359}$$

$$C_3 = \left\{8c_4^2 + 192\left[\frac{c_2^2\mu_1^2}{a^{10}} + \frac{2c_2c_3\mu_1(3K_1+\mu_1)}{5a^8} + \frac{c_3^2(3K_1^2+2K_1\mu_1+\mu_1^2)}{5a^6}\right]\right\}\eta$$
$$+\left[\frac{36}{a^6}c_4^2 I_2 + \frac{576c_3^2(9K_1^2+6K_1\mu_1+8\mu_1^2)}{35a^6}I_5 + \frac{288\mu_1^2}{7a^6}I_{10}\right]\eta^2 - \frac{1}{3}A_3, \tag{360}$$

and

$$D_3 = \left(\frac{24}{a^3\rho}c_4^2 I_3 + \frac{192}{5a^3\rho}I_9\right)\eta$$
$$+\left[\frac{36}{a^6\rho^2}c_4^2 I_4 + \frac{576c_3^2(9K_1^2+6K_1\mu_1+8\mu_1^2)}{35a^6\rho^2}I_7 + \frac{288\mu_1^2}{7a^6\rho^2}I_{11}\right]\eta^2 - \frac{1}{3}B_3. \tag{361}$$

In these equations I_1 through I_7 are given by Eqs. (336)–(339), and

$$I_8 = \int_a^\infty G_2^{(2)}(r)\left[\frac{7c_2^2}{r^8} + \frac{2c_2c_3(3K_1+\mu_1)}{\mu_1 r^6} + \frac{c_3^2(27K_1^2+24K_1\mu_1+16\mu_1^2)}{20\mu_1^2 r^4}\right]dr, \tag{362}$$

$$I_9 = \int_a^\infty G_2^{(2)}(r)\left[\frac{35c_2^2\mu_1^2}{r^8} + \frac{10c_2c_3\mu_1(3K_1+\mu_1)}{r^6} + \frac{3c_3^2(3K_1^2+2K_1\mu_1+\mu_1^2)}{r^4}\right]dr, \tag{363}$$

$$I_{10} = \int_a^\infty \frac{dr}{r^3}\left(7c_2 + \frac{3K_1+\mu_1}{\mu_1}c_3 r^2\right)\int_a^\infty \frac{ds}{s^3}\left(7c_2 + \frac{3K_1+\mu_1}{\mu_1}c_3 r^2\right)\int_{-1}^1 h(t)P_4(u)du, \tag{364}$$

and

$$I_{11} = \int_a^\infty \frac{dr}{r^3}\left(7c_2 + \frac{3K_1+\mu_1}{\mu_1}c_3 r^2\right)\int_a^\infty \frac{ds}{s^3}\left(7c_2 + \frac{3K_1+\mu_1}{\mu_1}c_3 s^2\right)\int_{-1}^1 Q(\mathbf{r},\mathbf{s})P_4(u)du, \tag{365}$$

where $t = |\mathbf{r} - \mathbf{s}|$.

Given these quantities, the effective shear modulus is bounded by

$$\mu_e \leq \langle \mu \rangle - \frac{4c_1^2\eta^2\phi_1^2(\mu_2-\mu_1)^2}{A_2K_1 + B_2(K_2-K_1) + 2[C_2\mu_1 + D_2(\mu_2-\mu_1)]}, \tag{366}$$

and

$$\mu_e \geq \left\{\langle \mu^{-1}\rangle - \frac{4c_4^2\eta^2\phi_1^2(\mu_2^{-1}-\mu_1^{-1})^2}{\frac{1}{9}[A_3K_1^{-1} + B_3(K_2^{-1}-K_1^{-1})] + \frac{1}{2}[C_3\mu_1^{-1} + D_3(\mu_2^{-1}-\mu_1^{-1})]}\right\}^{-1}. \tag{367}$$

One can now compute these general bounds for various models of composite materials. It is clear that in order to do this, the function $G_n^{(i)}$ must be specified which, as discussed in Chapters 3 and 4, depends on the microstructure of the composite and the degree of penetrability of the spheres.

Consider the case of fully-impenetrable sphere model for which the results for $G_n^{(i)}$ were already described in Chapters 3 and 4. Using these results, the followings are obtained,

$$I_2 = -\frac{2}{9}a^6, \quad I_3 = \frac{1}{3}\rho a^3, \quad I_4 = \frac{2}{9}\rho^2 a^6(\phi_2 - 2), \tag{368}$$

$$I_5 = -2/9, \quad I_6 = \frac{1}{a^3}\rho\phi_2 J_1, \quad I_7 = \frac{2}{9}\rho^2\phi_2 J_2, \tag{369}$$

where

$$J_1 = a^3 \int_{2a}^\infty \frac{r^2}{(r^2-a^2)^3} g_2(r)dr,$$

and

$$J_2 = \frac{9}{32\pi^2}\sum_{l=2}^\infty l(l-1)a^{2l-4}\int\int [g_3(r,s,t) - g(r)g(s)]\frac{P_l(u)}{r^{l+1}s^{l+1}}d\mathbf{r}d\mathbf{s},$$

where g_3 is the three-particle distribution function which is related to the function ρ_3 through, $g_3(r,s,t) = \rho_3(r,s,t)/\rho^3$. Moreover,

$$I_8 = \frac{75}{14}\rho\eta a^3(3K_1+\mu_1)^2\gamma^2\Upsilon + \frac{5}{28}\rho\eta a^3(9K_1^2 + 48K_1\mu_1 + 92\mu_1^2)\gamma^2 J_1, \tag{370}$$

$$I_9 = \frac{375}{14}\rho\eta a^3(3K_1+\mu_1)^2\mu_1\gamma^2\Upsilon + \frac{50}{7}\rho\eta a^3(9K_1^2+6K_1\mu_1+8\mu_1^2)\mu_1^2\gamma^2 J_1, \tag{371}$$

where

$$\Upsilon = \frac{1}{2\pi}\int_{2a}^{\infty} r^2 g_2(r)W(r)dr \tag{372}$$

with

$$W(r) = \frac{4}{3}\pi\left[\frac{a^3}{(r^2-a^2)^3} - \frac{14}{5}\frac{a^5}{(r^2-a^2)^4} - \frac{63}{25}\frac{a^7}{(r^2-a^2)^5} + \frac{196}{25}\frac{a^9}{(r^2-a^2)^6} + \frac{168}{25}\frac{a^{11}}{(r^2-a^2)^7}\right]. \tag{373}$$

In addition

$$I_{10} = -\frac{8}{9}(3K_1+\mu_1)^2\gamma^2 a^6, \tag{374}$$

and

$$I_{11} = 25\rho^2\eta a^6(3K_1+\mu_1)^2\gamma^2\Psi, \tag{375}$$

with

$$\Psi = \frac{1}{16\pi^2}\int g_3(r_{45}, r_{46}, r_{56})Q(r_{45}, r_{46}, r_{56})d\mathbf{r}_5 d\mathbf{r}_6, \tag{376}$$

where the origin is at $\mathbf{r}_4$, and

$$Q(r_{45}, r_{46}, r_{56}) = \frac{14}{5!}\sum_{l=4}^{\infty} l(l-1)(l-2)(l-3)\frac{2l-3}{2l-1}\frac{a^{2l}-8}{r_{45}^{l-1}r_{46}^{l-1}}$$
$$\times\left[1-\frac{2}{5}(l+1)\left(\frac{2l-1}{2l-3}\right)\left(\frac{a}{r_{45}}\right)^2\right]\left[1-\frac{2}{5}(l+1)\left(\frac{2l-1}{2l-3}\right)\left(\frac{a}{r_{46}}\right)^2\right]P_l(\cos\theta_{546})$$
$$+\frac{8}{5!}\sum_{l=3}^{\infty}\frac{l(l-1)(l-2)(11l+15)}{(2l+3)(2l-3)}\frac{a^{2l-4}}{r_{45}^{l+1}r_{46}^{l+1}}P_l(\cos\theta_{546}), \tag{377}$$

and $\cos\theta_{546} = \hat{\mathbf{r}}_{45}\cdot\hat{\mathbf{r}}_{46}$, with $\hat{\mathbf{r}}_{45} = \mathbf{r}_{45}/|\mathbf{r}_{45}|$ and $\hat{\mathbf{r}}_{46} = \mathbf{r}_{46}/|\mathbf{r}_{46}|$. Given these integrals, one then finds that for fully-impenetrable spheres,

$$A_1 = 9\phi_1\phi_2, \quad B_1 = 9\phi_1^2\phi_2, \tag{378}$$

$$C_1 = 6\phi_1\phi_2, \quad D_1 = 18\phi_2^2 J_1 + 6\phi_2^3 J_2. \tag{379}$$

The resulting bounds are then equivalent to Beran–Molyneux bounds which will be described in the next section. Note that for this composite the microstructural parameters ζ_2 is given by

$$\zeta_2 = \frac{1}{\phi_1}(3J_1\phi_2 + J_2\phi_2^2). \tag{380}$$

We now consider the bounds on the effective shear modulus μ_e of suspensions of totally-impenetrable spheres. In general, as will be discussed in the next section, the three-point bounds on μ_e are written in the following form

$$\mu_e \leq \langle \mu \rangle - \frac{6\phi_1\phi_2(\mu_2 - \mu_1)^2}{6\langle \tilde{\mu} \rangle + \Theta}, \tag{381}$$

and

$$\mu_e^{-1} \geq \left[\langle \mu^{-1} \rangle - \frac{\phi_1\phi_2(\mu_2^{-1} - \mu_1^{-1})^2}{\langle \tilde{\mu}^{-1} \rangle + 6\Xi} \right]^{-1}, \tag{382}$$

where

$$\langle b \rangle_\eta = b_1\eta_1 + b_2\eta_2, \tag{383}$$

with $\eta_2 = 1 - \eta_1$ defined by Eqs. (260) and (264) and described in detail in Section 7.4.5. For fully-impenetrable sphere model, η_2 is simplified to

$$\eta_2 = \phi_2^2(\Upsilon + \phi_2\Psi). \tag{384}$$

The required constants are then given by

$$A_2 = 240\mu_1^2\gamma^2\phi_1\phi_2, \quad B_2 = 240\mu_1^2\gamma^2\phi_1\phi_2\zeta_2, \tag{385}$$

$$C_2 = 60(3K_1^2 + 8K_1\mu_1 + 8\mu_1^2)\gamma^2\phi_1\phi_2, \tag{386}$$

$$D_2 = 72(K_1 + 2\mu_1)^2\gamma^2\phi_1\phi_2 + 60(2K_1 + 3\mu_1)\mu_1\gamma^2\phi_1\phi_2\zeta_2$$
$$+12(3K_1 + \mu_1)^2\gamma^2\phi_1\phi_2\eta_2, \tag{387}$$

$$A_3 = 2160K_1^2\mu_1^2\gamma^2\phi_1\phi_2, \quad B_3 = 2160K_1^2\mu_1^2\gamma^2\phi_1\phi_2\zeta_2, \tag{388}$$

$$C_3 = 40\mu_1^2(27K_1^2 + 48K_1\mu_1 + 32\mu_1^2)\gamma^2\phi_1\phi_2, \tag{389}$$

$$D_3 = 8\mu_1^2(9K_1 + 8\mu_1)^2\gamma^2\phi_1^2\phi_2 + 240\mu_1^3(2K_1 + 3\mu_1)\mu_1\gamma^2\phi_1\phi_2\zeta_2$$
$$+48\mu_1^2(3K_1 + \mu_1)^2\gamma^2\phi_1\phi_2\eta_2. \tag{390}$$

We then find that the cluster bounds on μ_e can also be represented in the form of (381) and (382), if we replace Θ and Ξ with

$$\Theta_1 = \frac{10\mu_1^2\langle K \rangle_\zeta + 5\mu_1(2K_1 + 3\mu_1)\langle \mu \rangle_\zeta + (3K_1 + \mu_1)^2\langle \mu \rangle_\eta}{(K_1 + 2\mu_1)^2}, \tag{391}$$

and

$$\Xi_1 = \frac{10K_1^2\langle K^{-1} \rangle_\zeta + 5\mu_1(2K_1 + 3\mu_1)\langle \mu^{-1} \rangle_\zeta + (3K_1 + \mu_1)^2\langle \mu^{-1} \rangle_\eta}{(9K_1 + 8\mu_1)^2}. \tag{392}$$

For composites with $K_1 > K_2$ and $\mu_1 > \mu_2$,

$$\Theta_1 \leq \Theta_{HS} \equiv \frac{\mu_1(9K_1 + 8\mu_1)}{K_1 + 2\mu_1}, \tag{393}$$

where Θ_{HS} refers to the corresponding expression in the Hashin–Shtrikman bounds, and therefore the cluster upper bound on μ_e is more restrictive than the Hashin–Shtrikman upper bound. Likewise, for composites with $K_1 < K_2$ and $\mu_1 < \mu_2$,

$$\Xi_1 \leq \Xi_{HS} \equiv \frac{K_1 + 2\mu_1}{\mu_1(9K_1 + 8\mu_1)}, \tag{394}$$

and therefore the cluster lower bound on μ_e is more restrictive than the Hashin–Shtrikman lower bound.

At a given volume fraction ϕ_2, both the cluster upper and lower bounds are more restrictive than the McCoy upper bound (see the next section) for some values of the elastic moduli and less restrictive for others; this is determined by the values of ζ_2 and η_2 at volume fraction ϕ_2 and the larger of K_2/K_1 and μ_2/μ_1. However, both bounds will not be as restrictive as those due to Milton and Phan–Thien, which are based on small-contrast trial fields and will be described in the next section.

For a composite with spherical voids (i.e., one with $K_2 = \mu_2 = 0$), the cluster bound is identical to the McCoy bound, is a substantial improvement upon the Hashin–Shtrikman bound, and is almost identical to the Milton–Phan–Thien bound. For an incompressible composite with impenetrable rigid spherical inclusions (i.e., one with $K_1 = K_2 = \infty$ and $\mu_2 \gg \mu_1$), the only possible non-trivial bound is the lower bound on μ_e, the determination of which for such composites is called the *Einstein problem*. Once again, the cluster bound for this case is identical to the McCoy bound, a substantial improvement on the Hashin–Shtrikman bound, but is slightly less restrictive than the Milton–Phan–Thien bound.

We now consider the fully-penetrable sphere (the Swiss-cheese) model. In this case the cluster bounds can be evaluated analytically, which is in contrast to the other three-point bounds (described below), which require numerical evaluation of ζ_2 and η_2 at a given volume fraction. One has

$$\eta = -\ln \phi_1 \tag{395}$$

$$G_2^{(2)}(x) = \begin{cases} \rho & x < a, \\ \rho\phi_2 & x > a, \end{cases} \tag{396}$$

$$h(x) = 0, \tag{397}$$

and

$$Q(\mathbf{x}_1, \mathbf{x}_2) = \begin{cases} -\rho^2\phi_1 & x_1, x_2 < a, \\ 0 & \text{otherwise}, \end{cases} \tag{398}$$

and therefore,

$$K_e \leq \langle K \rangle - \frac{3\eta\phi_1^2(K_1 - K_2)^2}{3[K_2 - \eta\phi_1(K_2 - K_1)] + 4\langle\mu\rangle}, \tag{399}$$

and

$$K_e \geq \left\{ \langle K^{-1} \rangle - \frac{4\eta\phi_1^2(K_1^{-1} - K_2^{-1})^2}{4[K_2^{-1} - \eta\phi_1(K_2^{-1} - K_1^{-1})] + 3\langle\mu^{-1}\rangle} \right\}^{-1}, \tag{400}$$

$$\mu_e \leq \langle\mu\rangle - \frac{6\eta\phi_1^2(\mu_1-\mu_2)^2}{6[\mu_2-\eta\phi_1(\mu_2-\mu_1)]+\Theta_2}, \tag{401}$$

$$\mu_e \geq \left[\langle\mu^{-1}\rangle - \frac{\eta\phi_1^2(\mu_1^{-1}-\mu_2^{-1})^2}{\mu_2^{-1}-\eta\phi_1(\mu_2^{-1}-\mu_1^{-1})+6\Xi_2}\right]^{-1}, \tag{402}$$

where

$$\Theta_2 = \frac{10\langle K\rangle\mu_1^2+\langle\mu\rangle(9K_1^2+16K_1\mu_1+16\mu_1^2)}{(K_1+2\mu_1)^2}. \tag{403}$$

and

$$\Xi_2 = \frac{10\langle k^{-1}\rangle K_1^2+\langle\mu^{-1}\rangle(9K_1^2+16K_1\mu_1+16\mu_1^2)}{(9K_1+8\mu_1)^2}. \tag{404}$$

For small ϕ_2, these bounds collapse onto those for the dilute limit described in Section 7.3.1.

If the spherical inclusions represent voids, then the lower bounds trivially become zero, while the upper bounds reduce to

$$\frac{K_e}{K_1} \leq \frac{4f\phi_1}{3\eta+4f}, \tag{405}$$

and

$$\frac{\mu_e}{\mu_1} \leq \frac{(9+8f)\phi_1}{6\eta(1+2f)+(9+8f)}, \tag{406}$$

where $f=\mu_1/K_1$. If we now consider the Einstein problem, allowing the particles to be fully penetrable, the cluster lower bound on μ_e reduces to

$$\frac{\mu_e}{\mu_1} \geq \frac{3\eta+2}{2\phi_1}, \tag{407}$$

which is not as sharp as the McCoy and Milton–Phan–Thien bounds (see the next section), but is simple, while evaluation of the other bounds requires numerical integrations.

7.5.1.4 Three- and Four-Point Bounds

In this section, we describe the most restrictive three-point bounds that are currently available. These bounds are expressed in compact forms by utilizing the so-called Y-transformation that was first described by Gibiansky and Torquato (1995b). Consider a two-phase composite material with phase volume fractions ϕ_1 and ϕ_2. In general, all the existing bounds, as well as the three-point series expansions that were described and discussed in Section 7.4, can be expressed as

$$F(a_1,a_2,\phi_1,\phi_2,y) = a_1\phi_1+a_2\phi_2 - \frac{\phi_1\phi_2(a_1-a_2)^2}{a_1\phi_2+a_2\phi_1+y}, \tag{408}$$

where a is any phase property. The Y-transformation is the inverse of the function F, as a function of the variable y, i.e.,

$$y(a_1, a_2, \phi_1, \phi_2, a_e) = -a_1\phi_2 - a_2\phi_1 + \frac{\phi_1\phi_2(a_1 - a_2)^2}{a_1\phi_1 + a_2\phi_2 - a_e}. \tag{409}$$

If we use the shortened notation, $y_a(a_1, a_2, \phi_1, \phi_2, a_e) = y(a_e)$, then it can be shown that the bounds

$$F(a_1, a_2, \phi_1, \phi_2, y_1) \le a_e \le F(a_1, a_2, \phi_1, \phi_2, y_2), \tag{410}$$

are equivalent to

$$y_1 \le y_a(a_e) \le y_2. \tag{411}$$

Beran and Molyneux (1966) derived bounds on the effective bulk modulus of 3D composites, while Silnutzer (1972) did the same for 2D materials. Their bounds were simplified by Milton (1981a,b,c,1982a,b) and, as pointed out by Gibiansky and Torquato (1995b), can be written in the following general form:

$$F\left[K_1, K_2, \phi_1, \phi_2, \frac{2}{d}(d-1)\langle\mu^{-1}\rangle_\zeta^{-1}\right] \le K_e \le F\left[K_1, K_2, \phi_1, \phi_2, \frac{2}{d}(d-1)\langle\mu\rangle_\zeta\right], \tag{412}$$

where the average is defined by, $\langle\mu\rangle_\zeta = \zeta_1\mu_1 + \zeta_2\mu_2$, and d is the dimensionality of the material. In terms of the Y-transformation, the bounds (412) can be rewritten as

$$\frac{2}{d}(d-1)\langle\mu^{-1}\rangle_\zeta^{-1} \le y_K(K_e) \le \frac{2}{d}(d-1)\langle\mu\rangle_\zeta, \tag{413}$$

which is rewritten as

$$F\left[\frac{2}{d}(d-1)\mu_1, \frac{2}{d}(d-1)\mu_2, \zeta_1, \zeta_2, 0\right]$$
$$\le y_K(K_e) \le F\left[\frac{2}{d}(d-1)\mu_1, \frac{2}{d}(d-1)\mu_2, \zeta_1, \zeta_2, \infty\right], \tag{414}$$

which, upon applying the Y-transformation again, is rewritten as

$$0 \le z_K(K_e) \le \infty, \tag{415}$$

where

$$z_K(K_e) \equiv y\left[\frac{2}{d}(d-1)\mu_1, \frac{2}{d}(d-1)\mu_2, \zeta_1, \zeta_2, y_K(K_e)\right]. \tag{416}$$

For 2D materials, these bounds were improved by Gibiansky and Torquato (1995a) who derived the following tighter bound

$$y_K(K_e) \le F(\mu_1, \mu_2, \zeta_1, \zeta_2, K_M), \tag{417}$$

or, in terms of z_K,

$$z_K(K_e) \le K_M, \tag{418}$$

with $z_K(K_e) = y[\mu_1, \mu_2, \zeta_1, \zeta_2, y_K(K_e)]$, and K_M is the maximal phase bulk modulus. Gibiansky and Torquato (1995a) showed that the most restrictive bounds on effective shear modulus of 2D materials can be written as

$$\left(2\langle K^{-1}\rangle_\zeta + \langle \mu^{-1}\rangle_\eta\right)^{-1} \leq y_\mu(\mu_e) \leq y_U^{-1}, \tag{419}$$

where the lower bound is due to Silnutzer (1972) while the upper bound was derived by Gibiansky and Torquato (1995a). Here,

$$y_U = \begin{cases} y_1, & \text{if } t \in [-K_M^{-1}, \mu_M^{-1}], \\ y_2, & \text{if } t \leq -K_M^{-1}, \\ y_3, & \text{if } t \geq \mu_M^{-1}, \end{cases} \tag{420}$$

where μ_M is the maximal phase shear modulus, and

$$y_1 = \langle 2K^{-1}\rangle_\zeta + \langle \mu^{-1}\rangle_\eta - \frac{\left[\sqrt{2\zeta_1\zeta_2(K_1^{-1} - K_2^{-1})^2} + \sqrt{\eta_1\eta_2(\mu_1^{-1} - \mu_2^{-1})^2}\right]^2}{\eta_1\mu_2^{-1} + \eta_2\mu_1^{-1} + 2\zeta_1 K_2^{-1} + 2\zeta_2 K_1^{-1}}, \tag{421}$$

$$y_2 = \left\langle \frac{1}{\mu^{-1} + K_M^{-1}} \right\rangle^{-1} + \frac{1}{K_M}, \quad y_3 = 2\left\langle \frac{1}{K^{-1} + \mu_M^{-1}} \right\rangle^{-1} - \frac{1}{\mu_M}, \tag{422}$$

$$t = \frac{\sqrt{2\zeta_1\zeta_2(K_1^{-1} - K_2^{-1})^2}(\eta_1\mu_2^{-1} + \eta_2\mu_1^{-1}) - \sqrt{\eta_1\eta_2(\mu_1^{-1} - \mu_2^{-1})^2}(\zeta_1 K_2^{-1} + \zeta_2 K_1^{-1})}{\sqrt{2\zeta_1\zeta_2(K_1^{-1} - K_2^{-1})^2} + \sqrt{\eta_1\eta_2(\mu_1^{-1} - \mu_2^{-1})^2}}. \tag{423}$$

If the material is incompressible, i.e., if $K_1/\mu_1 = K_2/\mu_2 = \infty$, then these bounds are rewritten in the following form,

$$0 \leq z_\mu(\mu_e) \leq \infty, \tag{424}$$

where $z_\mu(\mu_e) = F[\mu_1, \mu_2, \eta_1, \eta_2, y_\mu(\mu_e)]$.

For 3D composite materials, three-point bounds on the effective shear modulus were derived by McCoy (1970), which were then improved by Milton and Phan–Thien (1982), who showed that

$$\Xi \leq y_\mu(\mu_e) \leq \Theta, \tag{425}$$

where

$$\Xi = \frac{\left\langle \frac{128}{K} + \frac{99}{\mu} \right\rangle_\zeta + 45\left\langle \frac{1}{\mu} \right\rangle_\eta}{30\left\langle \frac{1}{\mu} \right\rangle_\zeta \left\langle \frac{6}{K} - \frac{1}{\mu} \right\rangle_\zeta + 6\left\langle \frac{1}{\mu} \right\rangle_\eta \left\langle \frac{2}{K} + \frac{21}{\mu} \right\rangle_\zeta}, \tag{426}$$

$$\Theta = \frac{3\langle \mu\rangle_\eta \langle 6K + 7\mu\rangle_\zeta - 5\langle \mu\rangle_\zeta^2}{6\langle 2K - \mu\rangle_\zeta + 30\langle \mu\rangle_\eta}. \tag{427}$$

In these bounds, ζ_2 can lie anywhere in [0, 1], while Milton and Phan–Thien showed that (for $d = 3$) η_2 lies in the smaller interval $[\frac{5}{21}\zeta_2, (16 + 5\zeta_2)/21]$.

In order to compare the series expansions that were described in Section 7.4 to the aforementioned three-point bounds, one must recast the series in terms of the Y-transformation. For example, Eq. (248) can be rewritten as

$$\tilde{z}_K(K_e) = \frac{1-d}{d}\frac{\mu_1\{(4+2d-2d^2)\mu_1 + d[(d+2)\zeta_2 - d]K_1\}}{[2d-(d+2)(d-1)\zeta_2]\mu_1 + dK_1}, \quad (428)$$

where $\tilde{z}_K(K_e)$ is the corresponding z-function defined by Eq. (416). Since this new form of the series expansion must satisfy the above bounds, one can obtain constraints on the numerical values of ζ_2 and η_2. For example, in order for the predictions of Eq. (284) for the effective bulk modulus of 3D materials to lie between the above upper and lower bounds, one must have, $\zeta_2 \leq 0.6 + 8\mu_1/(15K_1)$, a constraint that was in fact mentioned in Section 7.4.4.2.

Milton and Phan–Thien (1982) derived four point bounds on the elastic moduli of 3D two-phase materials. However, their bounds on the effective bulk modulus involve three different structural parameters, while those on the effective shear modulus involve eight parameters. Determination of such a large set of parameters is very difficult and impractical. Therefore, the bounds are not listed here.

7.5.2 *Anisotropic Materials*

We consider d-dimensional, macroscopically-anisotropic materials that consist of two isotropic phases. The two-point generalization of the Hashin–Shtrikman bounds on $\mathbf{C}_e$ when $\mathbf{C}_2 \geq \mathbf{C}_1$, with $\mathbf{C}_e^{(L)} \leq \mathbf{C}_e \leq \mathbf{C}_e^{(U)}$, are given by

$$\mathbf{C}_e^{(L)} = \langle \mathbf{C} \rangle - \phi_1\phi_2(\mathbf{C}_1 - \mathbf{C}_2) : \mathbf{P}^{(1)} : (\mathbf{C}_1 - \mathbf{C}_2) : \left[\mathbf{U} + \phi_1 \mathbf{P}^{(1)} : (\mathbf{C}_2 - \mathbf{C}_1)\right]^{-1}, \quad (429)$$

$$\mathbf{C}_e^{(U)} = \langle \mathbf{C} \rangle - \phi_1\phi_2(\mathbf{C}_2 - \mathbf{C}_1) : \mathbf{P}^{(2)} : (\mathbf{C}_2 - \mathbf{C}_1) : \left[\mathbf{U} + \phi_2 \mathbf{P}^{(2)} : (\mathbf{C}_1 - \mathbf{C}_2)\right]^{-1}, \quad (430)$$

where $\mathbf{P}^{(q)}$ is a two-point tensor given by

$$\mathbf{P}^{(q)} = \mathbf{A}^{(q)} - \frac{1}{\phi_p\phi_q}\int_\epsilon \mathbf{H}^{(q)}(1,2)\left[S_2^{(p)}(1,2) - \phi_p^2\right] d2. \quad (431)$$

The tensors $\mathbf{A}^{(q)}$ and $\mathbf{H}^{(q)}$ were already defined by Eqs. (204) and (205). These bounds can be generalized to the case of a macroscopically-anisotropic material with $n \geq 2$ phases (see also Nemat–Nasser and Hori, 1993).

Given only the phase volume fractions ϕ_1 and ϕ_2, bulk moduli K_1 and K_2, and shear moduli μ_1 and μ_2 of a two-phase transversely isotropic fiber-reinforced material, Hill (1964) and Hashin (1965, 1970; see also Hashin, 1980) derived the most accurate bounds on its effective transverse bulk modulus $K_e^{(t)}$, the effective axial shear modulus $\mu_e^{(a)}$, and the effective transverse shear modulus $\mu_e^{(t)}$. Silnutzer (1972) derived improved bounds on these properties that additionally depend upon ζ_2. His bounds are referred to as third-order bounds since they are exact up to third order in the difference in the phase properties. Silnutzer also derived third-order bounds on the effective transverse shear modulus $\mu_e^{(t)}$ which Milton showed can be expressed in terms of ζ_2 and η_2. The simplified forms of the Silnutzer third-order bounds (Milton, 1982a,b) for the axial shear modulus and transverse bulk modulus are given by

$$\left[\langle\mu\rangle - \frac{\phi_1\phi_2(\mu_2-\mu_1)^2}{\langle\tilde{\mu}\rangle + \langle\mu\rangle_\zeta}\right]^{-1} \le \mu_e^{(a)} \le \left[\langle 1/\mu\rangle - \frac{\phi_1\phi_2(1/\mu_2-1/\mu_1)^2}{\langle 1/\tilde{\mu}\rangle + \langle 1/\mu\rangle_\zeta}\right]^{-1}, \tag{432}$$

$$\left[\langle K^{(t)}\rangle - \frac{\phi_1\phi_2(K_2^{(t)}-K_1^{(t)})^2}{\langle \tilde{K}^{(t)}\rangle + \langle\tilde{\mu}\rangle_\zeta}\right]^{-1} \le K_e^{(t)} \le \left[\langle 1/K^{(t)}\rangle - \frac{\phi_1\phi_2(1/K_2^{(t)}-1/K_1^{(t)})^2}{\langle 1/\tilde{K}^{(t)}\rangle + \langle 1/\mu\rangle_\zeta}\right]^{-1}, \tag{433}$$

and

$$\left[\langle\mu\rangle - \frac{\phi_1\phi_2(\mu_2-\mu_1)^2}{\langle\tilde{\mu}\rangle + \langle\Theta\rangle}\right]^{-1} \le \mu_e^{(t)} \le \left[\langle 1/\mu\rangle - \frac{\phi_1\phi_2(1/\mu_2-1/\mu_1)^2}{\langle 1/\tilde{\mu}\rangle + \langle\Xi\rangle}\right]^{-1}, \tag{434}$$

with

$$\Theta = \frac{2\langle K^{(t)}\rangle_\zeta\langle\mu\rangle^2 + \langle K^{(t)}\rangle^2\langle\mu\rangle_\eta}{\langle K^{(t)} + 2\mu\rangle^2}, \tag{435}$$

$$\Xi = 2\langle 1/K^{(t)}\rangle_\zeta + \langle 1/\mu\rangle_\eta. \tag{436}$$

Here,

$$\langle b\rangle = b_1\phi_1 + b_2\phi_2, \quad \langle\tilde{b}\rangle = b_1\phi_2 + b_2\phi_2, \tag{437}$$

$$\langle b\rangle_\zeta = b_1\zeta_1 + b_2\zeta_2, \quad \langle b\rangle_\eta = b_1\eta_1 + b_2\eta_2, \tag{438}$$

where $\zeta_1 = 1-\zeta_2$ and $\eta_1 = 1-\eta_2$. The quantity $K_i^{(t)}$ is the transverse bulk modulus of phase i for transverse compression but without axial extension and may be expressed in terms of the isotropic phase moduli as $K_i^{(t)} = K_i + \mu_i/3$ ($i = 1$ and 2). Since the microstructural parameters ζ_i and η_i lie in the closed interval [0,1], the above third-order bounds always improve upon the second-order bounds of Hill and Hashin. Note that, as mentioned in Section 7.2.4, the problem of determining the effective axial shear modulus $\mu_e^{(a)}$ is mathematically equivalent to determining the effective transverse thermal or electrical conductivity (Hashin, 1970).

Fourth-order bounds on the effective transverse conductivity or axial shear modulus $\mu_e^{(a)}$ of transversely isotropic fiber-reinforced materials that depend upon the phase properties ϕ_2 and ζ_2 and an integral over the four-point probability function S_4 were derived by Milton (1981a,b,c), who demonstrated that the integral involving S_4 can be expressed in terms of ϕ_2 and ζ_2 only. The fourth-order bounds, for the case $\mu_2 \ge \mu_1$, are given by

$$\mu_L^{(a)} \le \mu_e^{(a)} \le \mu_U^{(a)}, \tag{439}$$

where

$$\mu_U^{(a)} = \mu_2 \left[\frac{(\mu_1+\mu_2)(\mu_1+\langle\mu\rangle) - \phi_2\zeta_1(\mu_2-\mu_1)^2}{(\mu_1+\mu_2)(\mu_2+\langle\tilde{\mu}\rangle) - \phi_2\zeta_1(\mu_2-\mu_1)^2} \right], \tag{440}$$

$$\mu_L^{(a)} = \mu_1 \left[\frac{(\mu_1+\mu_2)(\mu_2+\langle\mu\rangle) - \phi_1\zeta_2(\mu_2-\mu_1)^2}{(\mu_1+\mu_2)(\mu_1+\langle\tilde{\mu}\rangle) - \phi_1\zeta_2(\mu_2-\mu_1)^2} \right]. \tag{441}$$

As discussed earlier in this chapter and also in Chapters 3 and 4, computing ζ_2 and η_2 requires knowledge of the three-point probability function S_3 for infinitely long, oriented, equisized, impenetrable cylinders (disks in 2D). Consider the limit of a statistically isotropic distribution of identical, impenetrable disks (infinitely long, parallel cylinders) of radius a at an area (volume) fraction $\phi_2 = n\pi a$, where n is the number density of the disks (cylinders). Torquato and Lado (1988) found that for impenetrable disks of unit diameter one has the exact (to second order in ϕ_2) result,

$$\zeta_2 = \frac{2}{\pi\phi_1}(b_2\phi_2 + b_3\phi_2^2), \tag{442}$$

where

$$b_2 = \frac{\pi}{4}\int_1^\infty \frac{r g_2(r)}{(r^2-1/4)^2}dr, \tag{443}$$

and

$$b_3 = \sum_{n=2}^{\infty} \frac{n-1}{4^{n-3}} \int_1^\infty \frac{dr}{r^{n-1}} \int_1^r \frac{ds}{s^{n-1}} \int_0^\pi [g_3(r,s,t) - g_2(r)g_2(s)]T_n(\cos\theta)d\theta, \tag{444}$$

with $T_n(\cos\theta) = \cos(n\theta)$ being the Chebyshev polynomial of the first kind. Although the pair distribution function $g_2(r)$ is known accurately in the Percus–Yevick approximation [see Eq. (3.25), where $g_2(r)$ is denoted by $C(r)$], computing g_3 is more problematical, and hence Torquato and Lado (1988) resorted to the commonly-used superposition approximation (Hansen and McDonald, 1986) according to which

$$g_3(r_{12}, r_{13}, r_{23}) \simeq g_2(r_{12})g_2(r_{13})g_2(r_{23}). \tag{445}$$

to evaluate the integrals in Eq. (444). As discussed in Chapters 3 and 4, the exact expansion of ζ_2 through second order in ϕ_2 yields excellent agreement with the simulation results for a very wide range of volume fractions up to the disorder-order phase transition which, for disks occurs, at $\phi_2 \simeq 0.71$, indicating that cubic and higher-order terms make negligible contributions to ζ_2 (and hence to η_2), even at high densities. For disks this exact expansion is given by Eq. (4.90), which we quote here again,

$$\zeta_2 = \frac{1}{3}\phi_2 - 0.05707\phi_2^2. \tag{446}$$

Note that $\phi_2 \simeq 0.7$ corresponds to approximately 87% of the maximum random close-packing volume fraction, $\phi_{2c} \simeq 0.81$. As for η_2, Torquato and Lado (1992) derived the exact (to order ϕ_2^2) result that

$$\eta_2 = \frac{8}{\pi \phi_1}(c_2\phi_2 + c_3\phi_2^2), \tag{447}$$

where

$$c_2 = \int_1^\infty r g_2(r) W(r) dr, \tag{448}$$

where $W(r)$ is given by Eq. (373), and

$$c_3 = \sum_{n=2}^{\infty} \int_1^\infty r dr \int_1^r s ds \int_0^\pi [g_3(r,s,t) - g_2(r)g_2(s)] Q_n(r,s,\cos\theta) d\theta, \tag{449}$$

with

$$Q_n(r,s,\cos\theta) = \frac{1}{4^{n-4} r^{n-2} s^{n-2}} \left\{ \frac{n-1}{16 r^2 s^2} + (n-3) \right.$$
$$\left. \times \left[(n-2) - \frac{n(n-1)}{8r^2} \right] \left[(n-2) - \frac{n(n-1)}{8s^2} \right] \right\} T_n(\cos\theta). \tag{450}$$

Due to appearance of $g_2(r)$ and $g_3(r)$, the coefficients c_2 and c_3 depend upon the particle fraction ϕ_2.

The radial distribution function for impenetrable disks of unit diameter is known exactly through first order in ϕ_2 (Hansen and McDonald, 1986),

$$g_2(r) = U(r-1)[1 + A_2(r)\phi_2], \tag{451}$$

where

$$A_2(r) = \frac{4}{\pi} \left[\pi - 2\sin^{-1}\left(\frac{r}{2}\right) - r\left(1 - \frac{1}{4}r^2\right)^{1/2} \right] U(2-r), \tag{452}$$

and $U(x)$ is the unit step function. We then obtain

$$c_2 = \frac{7}{81}\pi + 0.1387\phi_2. \tag{453}$$

Equation (453) is exact to the order and number of significant figures indicated. The zero-density limit of g_3 is given by the expression

$$g_3(r_{12}, r_{13}, r_{23}) = U(r_{12}-1)U(r_{13}-1)U(r_{23}-1), \tag{454}$$

and hence, $c_3 = -0.3934$, which is exact to zeroth order in ϕ_2 (to the number of significant figures indicated). Putting these results together, one finally obtains the expression for η_2 given by Eq. (271) for fully-impenetrable spheres. Such results for ζ_2 and η_2 should be compared with, for example, those for the symmetric-cell material with cylindrical cells (Beran and Silnutzer, 1971) for which $\zeta_2 = \eta_2 = \phi_2$, already mentioned in Section 7.4.5.

Figure 7.8 compares the third and fourth-order bounds on the scaled effective axial shear modulus $\mu_e^{(a)}/\mu_1$ as a function of the cylinder volume fraction ϕ_2 for a composite with a shear modulus ratio $\mu_2/\mu_1 = 10$. Also shown are accurate Monte Carlo computer simulation data obtained by Kim and Torquato (1990) for

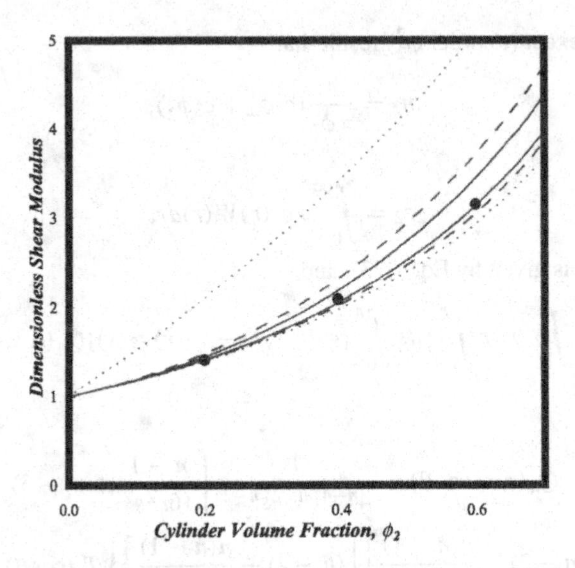

FIGURE 7.8. Dimensionless effective axial shear modulus $\mu_e^{(a)}/\mu_1$ versus the cylinder volume fraction ϕ_2 with $\mu_2/\mu_1 = 10$. Dotted curves are the two-point bounds, dashed curves are three-point bounds, while solid curves are four-point bounds. Solid circles are the simulation data (after Torquato and Lado, 1992).

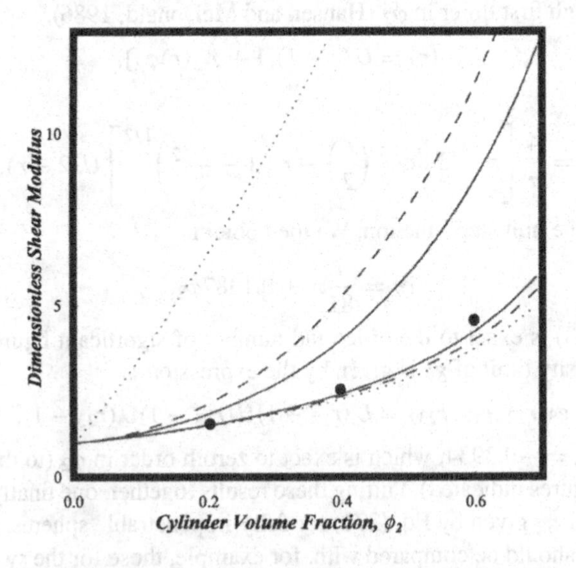

FIGURE 7.9. Same as in Figure 7.8, but with $\mu_2/\mu_1 = 50$.

the same model for the mathematically equivalent problem of determining the effective transverse conductivity of the same system. The agreement between the computer simulation data and the fourth-order lower bound is excellent. In Figure 7.9 the same bounds are shown for the case where the fibers are 50 times more rigid than the matrix ($\mu_2/\mu_1 = 50$). Not surprisingly, although the bounds widen, the fourth-order lower bound still provides an accurate estimate of the effective axial shear modulus for a wide range of cylinder volume fractions.

7.6 Multiple Scattering Method

A very different approach to calculating the effective elastic moduli of heterogeneous materials was proposed by Mondescu and Muthukumar (1999). They considered a composite material containing rigid, equisized spherical particles distributed randomly in a matrix. The spheres could overlap completely, and of particular interest was the regime in which their volume fraction was high enough for a percolation transition to occur and a sample-spanning cluster (of the spheres) to form. Since their approach to the problem is very different from what we have discussed so far, and because the predictions of the theory are in good agreement with experimental data, we describe and discuss their method.

Consider a rigid sphere of radius a inside a linear elastic, isotropic matrix of volume Ω (which is essentially infinite), on which an external force field $\mathbf{F}(\mathbf{r})$ is imposed. The system is assumed to be at steady state. Thus, the stationary equilibrium displacement field $\mathbf{u}(\mathbf{r})$ in the matrix satisfies the Lamé equation:

$$(\lambda_1 + \mu_1)\nabla(\nabla \cdot \mathbf{u}) + \mu_1 \nabla^2 \mathbf{u} + \mathbf{F} = \mathbf{0}, \tag{455}$$

where λ_1 and μ_1 are the usual Lamé constants of the matrix. Suppose that $\mathbf{R}(\theta) = \mathbf{R}_0 + \mathbf{r}_0(\theta)$ is the position of a point on the surface of the sphere, where $\mathbf{R}_0$ is the position of the center of mass of the particle, and $\mathbf{r}_0(\theta)$ is the relative coordinate of the point in the center-of-mass frame ($|\mathbf{r}_0| = a$) with θ being the angular orientation of $\mathbf{r}_0$. We also assume that the particle is fixed in the matrix, i.e., there is perfect bonding. The boundary condition at the surface of the sphere is the usual continuity of the displacement field, i.e., $\mathbf{u}_p(\theta) = \mathbf{u}_M(\theta)$, where p and M denote the particle and the matrix, respectively. The solution of the Lamé equation can be formally written in terms of a Green function:

$$\mathbf{u}(\mathbf{r}) = \int \mathbf{G}(\mathbf{r}, \mathbf{r}') \cdot \mathbf{F}(\mathbf{r}')d\mathbf{r}' + \int \mathbf{G}[\mathbf{r}, \mathbf{R}(\theta)] \cdot \mathbf{T}(\theta)d\theta, \tag{456}$$

Here, $\mathbf{T}(\theta)$ is the traction field density on the surface of the sphere, the normal at the surface pointing out of Ω and into the sphere, and $\mathbf{G}(\mathbf{r}, \mathbf{r}')$ is the (tensor) Green function given by

$$\mathbf{G}(\mathbf{r}, \mathbf{r}') = \frac{1}{16\pi\mu_1(1 - \nu_{p1})|\mathbf{r} - \mathbf{r}'|}\left[(3 - 4\nu_{p1})\mathbf{U} + \frac{(\mathbf{r} - \mathbf{r}')(\mathbf{r} - \mathbf{r}')}{|\mathbf{r} - \mathbf{r}'|^2}\right], \tag{457}$$

in which ν_{p1} is the Poisson's ratio of phase 1, the matrix. We now calculate the Fourier transform $\hat{\mathbf{G}}(\boldsymbol{\omega})$ of the Green function:

$$\hat{\mathbf{G}}(\boldsymbol{\omega}) = \frac{1}{\mu_1\omega^2}(\mathbf{U} - \hat{\boldsymbol{\omega}}\hat{\boldsymbol{\omega}}) + \frac{1}{(\lambda_1 + 2\mu_1)\omega^2}\hat{\boldsymbol{\omega}}\hat{\boldsymbol{\omega}}, \tag{458}$$

where $\hat{\omega}$ is the unit vector in the ω direction. Since the traction field associated with a rigid body rotation is zero, the integral solution (456) is valid both in the particle and in the matrix, which means, physically, that one has replaced the spherical inhomogeneity with a layer of surface forces. Averaging these forces over the matrix by an ensemble-volume averaging and applying the boundary conditions lead one to the desired effective elastic properties.

The transition from a microscopic to a macroscopic description of the material is achieved by averaging the Lamé equation, Eq. (455), and its solution, Eq. (456), over the position $\mathbf{R}_0(\theta)$ of the center of the inclusion. One then obtains

$$\langle \mathbf{u}(\mathbf{r})\rangle = \int \mathbf{G}(\mathbf{r},\mathbf{r}')\cdot\mathbf{F}(\mathbf{r}')d\mathbf{r}' + \int\int \mathbf{G}(\mathbf{r},\mathbf{r}')\cdot\mathbf{\Sigma}(\mathbf{r}',\mathbf{r}'')\cdot\langle \mathbf{u}(\mathbf{r})\rangle d\mathbf{r}'d\mathbf{r}'', \tag{459}$$

where $\mathbf{\Sigma}(\mathbf{r},\mathbf{r}')$ is called the *self-energy* of the material (named so in analogy with a similar quantity that arises in the problem of determining the effective conductivity and dielectric constant), defined by

$$\int \mathbf{\Sigma}(\mathbf{r},\mathbf{r}')\cdot\langle \mathbf{u}(\mathbf{r}')\rangle d\mathbf{r}' = \mathbf{\Sigma} * \langle \mathbf{u}\rangle|_{\mathbf{r}} = \left\langle \int \delta[\mathbf{r}-\mathbf{R}(\theta)]\mathbf{T}(\theta)d\theta \right\rangle, \tag{460}$$

where the external force field does not depend upon $\mathbf{R}_0$, the symbol $*$ combines two operations, a convolution-like integral and an inner product, and $|_{\mathbf{r}}$ indicates the final argument of the expression. The key quantity to compute is $\mathbf{\Sigma}$ which contains information on the change in the elastic properties of the matrix due to the presence of the inclusions, and is directly related to the many-body interactions among the particles.

From Eq. (459) one obtains

$$\langle \mathbf{u}(\mathbf{r})\rangle = (\mathbf{U} - \mathbf{G} * \mathbf{\Sigma})^{-1} * \mathbf{G} * \mathbf{F}|_{\mathbf{r}}. \tag{461}$$

An effective Green tensor $\mathcal{G}(\mathbf{r},\mathbf{r}')$ is now defined by

$$\mathcal{G}(\mathbf{r},\mathbf{r}') = (\mathbf{U} - \mathbf{G} * \mathbf{\Sigma})^{-1} * \mathbf{G}|_{\mathbf{r},\mathbf{r}'}, \tag{462}$$

in terms of which Eq. (461) is written in a simple form:

$$\langle \mathbf{u}(\mathbf{r})\rangle = \mathcal{G} * \mathbf{F}|_{\mathbf{r}} = \int \mathcal{G}(\mathbf{r}-\mathbf{r}')\cdot\mathbf{F}(\mathbf{r}')d\mathbf{r}'. \tag{463}$$

To obtain the change in the elastic properties (due to the presence of the inclusions), we Fourier transform Eq. (459), separate its transverse and longitudinal components by taking its inner-product with the projectors $\mathbf{U} - \hat{\omega}\hat{\omega}$ and $\hat{\omega}\hat{\omega}$ that satisfy

$$(\mathbf{U} - \hat{\omega}\hat{\omega})\cdot(\mathbf{U} - \hat{\omega}\hat{\omega}) = \mathbf{U} - \hat{\omega}\hat{\omega}, \tag{464}$$

$$\hat{\omega}\hat{\omega}\cdot(\mathbf{U} - \hat{\omega}\hat{\omega}) = \mathbf{0}, \tag{465}$$

and use the decomposition

$$\mathbf{D}(\omega) = D_{\perp}(\omega)(\mathbf{U} - \hat{\omega}\hat{\omega}) + D_{\parallel}(\omega)(\hat{\omega}\hat{\omega}) \tag{466}$$

to obtain

$$\mathcal{G}(\omega) = \frac{G_\perp(\omega)}{1 - G_\perp(\omega)\Sigma_\perp(\omega)}(\mathbf{U} - \hat{\omega}\hat{\omega}) + \frac{G_\parallel(\omega)}{1 - G_\parallel(\omega)\sum_\parallel(\omega)}\hat{\omega}\hat{\omega}$$

$$= \frac{1}{\mu_1\omega^2 - \Sigma_\perp(\omega)}(\mathbf{U} - \hat{\omega}\hat{\omega}) + \frac{1}{(\lambda_1 + 2\mu_1)\omega^2 - \Sigma_\parallel(\omega)}\hat{\omega}\hat{\omega}. \tag{467}$$

The effective elastic moduli, in the limit $|\omega| \to \mathbf{0}$, are given by

$$\frac{\mu_e - \mu_1}{\mu_1} = \frac{1}{\mu_1}\frac{\partial}{\partial\omega^2}\Sigma_\perp(\omega), \tag{468}$$

$$\frac{\lambda_e - \lambda_1}{\lambda_1} = -\frac{1}{\lambda_1}\frac{\partial}{\partial\omega^2}[\Sigma_\parallel(\omega) - 2\Sigma_\perp(\omega)], \tag{469}$$

$$K_e - K_1 = -\frac{\partial}{\partial\omega^2}\left[\Sigma_\parallel(\omega) - \frac{4}{3}\Sigma_\perp(\omega)\right]. \tag{470}$$

One must now compute the self-energy tensor $\mathbf{\Sigma}(\mathbf{r}, \mathbf{r}')$. As the sphere executes a rigid-body motion, the boundary condition can be written as

$$\mathbf{u}_0 + \boldsymbol{\varpi} \times \mathbf{r}_0(\theta) = \int \mathbf{G}[\mathbf{R}(\theta), \mathbf{r}'] \cdot \mathbf{F}(\mathbf{r}')d\mathbf{r}' + \int \mathbf{G}[\mathbf{R}(\theta), \mathbf{R}(\theta')] \cdot \mathbf{T}(\theta')d\theta, \tag{471}$$

where $\mathbf{u}_0$ and $\boldsymbol{\varpi}$ are the translational displacement of the center of mass and the angular displacement of the particle, respectively. Equation (471) can be solved for $\mathbf{T}(\theta)$. Writing

$$\int \mathbf{G}^{-1}(\theta, \theta'') \cdot \mathbf{G}[\mathbf{r}_0(\theta'') - \mathbf{r}_0(\theta')]d\theta'' = \delta(\theta - \theta')\mathbf{U}, \tag{472}$$

multiplying Eq. (469) by $\mathbf{G}^{-1}(\theta'', \theta)$, and relabeling the indices, one arrives at

$$\mathbf{T}(\theta) = -\int \mathbf{G}^{-1}(\theta, \theta') \cdot \mathbf{G}(\theta', \mathbf{r}') \cdot \mathbf{F}(\mathbf{r}')d\theta'$$

$$+ \int \mathbf{G}^{-1}(\theta, \theta') \cdot \mathbf{u}_0 \, d\theta' - \int [\mathbf{G}^{-1}(\theta, \theta') \times \mathbf{r}_0(\theta')] \cdot \boldsymbol{\varpi} d\theta'. \tag{473}$$

The new unknowns $\mathbf{u}_0$ and $\boldsymbol{\varpi}$ are eliminated by applying the constraints of force- and torque-free inclusion, namely,

$$\int \mathbf{T}(\theta)d\theta = \mathbf{0}, \tag{474}$$

$$\int \mathbf{r}_0(\theta) \times \mathbf{T}(\theta)d\theta = \mathbf{0}, \tag{475}$$

and substituting for $\mathbf{T}(\theta)$ to obtain

$$\mathbf{g}_t \cdot \mathbf{u}_0 + \mathbf{g}_{tr} \cdot \boldsymbol{\varpi} = \int\int\int \mathbf{G}^{-1}(\theta, \theta') \cdot \mathbf{G}(\theta', \mathbf{r}') \cdot \mathbf{F}(\mathbf{r}')d\mathbf{r}'d\theta d\theta', \tag{476}$$

$$\mathbf{g}_{rt} \cdot \mathbf{u}_0 + \mathbf{g}_r \cdot \boldsymbol{\varpi} = \int\int\int \mathbf{r}_0(\theta) \times \mathbf{G}^{-1}(\theta, \theta') \cdot \mathbf{G}(\theta', \mathbf{r}') \cdot \mathbf{F}(\mathbf{r}') d\mathbf{r}' d\theta d\theta', \tag{477}$$

where

$$\mathbf{g}_t = \int\int \mathbf{G}^{-1}(\theta, \theta') d\theta d\theta', \tag{478}$$

$$\mathbf{g}_{tr} = -\int\int \mathbf{G}^{-1}(\theta, \theta') \times \mathbf{r}_0(\theta') d\theta d\theta', \tag{479}$$

$$\mathbf{g}_r = -\int\int \mathbf{r}_0(\theta) \times \mathbf{G}^{-1}(\theta, \theta') \times \mathbf{r}_0(\theta') d\theta d\theta', \tag{480}$$

$$\mathbf{g}_{rt} = \int\int \mathbf{r}_0(\theta) \times \mathbf{G}^{-1}(\theta, \theta') d\theta d\theta'. \tag{481}$$

For spherical particles, $\mathbf{g}_{tr} = \mathbf{g}_{rt} = \mathbf{0}$. We can invert Eqs. (476) and (477) to obtain

$$\mathbf{u}_0 = \mathbf{g}_t^{-1} \cdot \int\int\int \mathbf{G}^{-1}(\theta, \theta') \cdot \mathbf{G}(\theta', \mathbf{r}') \cdot \mathbf{F}(\mathbf{r}') d\mathbf{r}' d\theta d\theta', \tag{482}$$

$$\boldsymbol{\varpi} = \mathbf{g}_r^{-1} \cdot \int\int\int \mathbf{r}_0(\theta) \times \mathbf{G}^{-1}(\theta, \theta') \cdot \mathbf{G}(\theta', \mathbf{r}') \cdot \mathbf{F}(\mathbf{r}') d\mathbf{r}' d\theta d\theta'. \tag{483}$$

Substituting Eqs. (482) and (483) into Eq. (473), we find that

$$\mathbf{T}(\theta) = -\int d\theta' \Big\{ \mathbf{G}^{-1}(\theta, \theta') - \int\int \mathbf{G}^{-1}(\theta, \theta_p) \cdot \mathbf{g}_t^{-1} \cdot \mathbf{G}^{-1}(\theta_q, \theta') d\theta_p d\theta_q$$
$$+ \int\int \mathbf{G}^{-1}(\theta, \theta_p) \times \mathbf{r}_0(\theta_p) \cdot \mathbf{g}_r^{-1} \cdot \mathbf{r}_0(\theta_q) \times \mathbf{G}^{-1}(\theta_q, \theta') d\theta_p d\theta_q \Big\}$$
$$\int \mathbf{G}(\theta', \mathbf{r}') \cdot \mathbf{F}(\mathbf{r}') d\mathbf{r}' \tag{484}$$

Therefore, one finally obtains

$$\mathbf{u}(\mathbf{r}) = \int \mathbf{G}(\mathbf{r}, \mathbf{r}') \cdot \mathbf{F}(\mathbf{r}') d\mathbf{r}'$$
$$+ \int\int\int \mathbf{G}(\mathbf{r}, \mathbf{r}') \cdot \mathcal{T}(\mathbf{r}', \mathbf{r}'') \cdot \mathbf{G}(\mathbf{r}'', \mathbf{r}''') \cdot \mathbf{F}(\mathbf{r}''') d\mathbf{r}' d\mathbf{r}'' d\mathbf{r}''', \tag{485}$$

which represents the governing microscopic equation of motion of the two-phase composite. The operator $\mathcal{T}$ represents the influence of the inclusions, and is given by

$$\mathcal{T}(\mathbf{r}, \mathbf{r}') = -\int\int \delta[\mathbf{r} - \mathbf{R}(\theta)] \times \mathcal{T}(\theta, \theta') \delta[\mathbf{r}' - \mathbf{R}(\theta')] d\theta d\theta', \tag{486}$$

$$\mathcal{T}(\theta, \theta') = \mathbf{G}^{-1}(\theta, \theta') - \int \mathbf{G}^{-1}(\theta, \theta_p) \cdot \mathbf{g}_t^{-1} \cdot \mathbf{G}^{-1}(\theta_q, \theta') d\theta_p d\theta_q$$
$$+ \int\int \mathbf{G}^{-1}(\theta, \theta_p) \times \mathbf{r}_0(\theta) \cdot \mathbf{g}_r^{-1} \cdot \mathbf{r}_0(\theta_q) \times \mathbf{G}^{-1}(\theta_q, \theta') d\theta_p d\theta_q. \tag{487}$$

$\mathcal{T}(\mathbf{r}, \mathbf{r}')$ depends upon the position, and the structure and geometry of the inclusion. The physical interpretation of Eq. (485) is that the displacement $\mathbf{u}$ is the result of a direct contribution by the force $\mathbf{F}$ acting at $\mathbf{r}'$ and an indirect contribution that arises from the scattering from the surface of the particle of the deformation produced by $\mathbf{F}$ at $\mathbf{r}'$.

The macroscopic description of the material's deformation is obtained by taking the ensemble-volume average of Eq. (485) which produces

$$\langle \mathbf{u}(\mathbf{r})\rangle = \mathbf{G} * \mathbf{F}|_{\mathbf{r}} + \mathbf{G} * \langle \mathbf{T}\rangle * \mathbf{G} * \mathbf{F}|_{\mathbf{r}}. \tag{488}$$

On the other hand, iteration of Eq. (459) yields

$$\langle \mathbf{u}(\mathbf{r})\rangle = \mathbf{G} * \mathbf{F}|_{\mathbf{r}} + \mathbf{G} * \boldsymbol{\Sigma} * \mathbf{G} * \mathbf{F}|_{\mathbf{r}} + \mathbf{G} * \boldsymbol{\Sigma} * \mathbf{G} * \boldsymbol{\Sigma} * \mathbf{G} * \mathbf{F}|_{\mathbf{r}} + \cdots \tag{489}$$

Representing $\boldsymbol{\Sigma}$ as a series in the number i of distinct scattering events (or the number of $\mathcal{T}$ propagators), one writes

$$\boldsymbol{\Sigma}(\mathbf{r}, \mathbf{r}') = \sum_{i=1}^{\infty} \boldsymbol{\Sigma}^{(i)}(\mathbf{r}, \mathbf{r}'), \tag{490}$$

substitutes (490) in Eq. (489), and compares the result to Eq. (488) to obtain

$$\boldsymbol{\Sigma}^{(1)}(\mathbf{r}, \mathbf{r}') = \langle \mathcal{T}(\mathbf{r}, \mathbf{r}')\rangle, \tag{491}$$

$$\boldsymbol{\Sigma}^{(2)}(\mathbf{r}, \mathbf{r}') = -\langle \mathcal{T}(\mathbf{r}, \mathbf{r}_1)\rangle * \mathbf{G}(\mathbf{r}_1, \mathbf{r}_2) * \langle \mathcal{T}(\mathbf{r}_2, \mathbf{r}')\rangle, \tag{492}$$

$$\boldsymbol{\Sigma}^{(3)}(\mathbf{r}, \mathbf{r}') = \langle \mathcal{T}\rangle * \mathbf{G} * \langle \mathcal{T}\rangle * \mathbf{G} * \langle \mathcal{T}\rangle|_{\mathbf{r},\mathbf{r}'}, \tag{493}$$

where integration in the multiple convolutions is implied.

The expression for $\boldsymbol{\Sigma}(\boldsymbol{\omega})$ is then derived by summing the series after Fourier transforming all the tensors and projecting their longitudinal and transverse parts:

$$\boldsymbol{\Sigma}(\boldsymbol{\omega}) = \frac{\mathcal{T}_{\perp}(\omega)}{1 + G_{\perp}(\omega)\mathcal{T}_{\perp}(\omega)}(\mathbf{U} - \hat{\boldsymbol{\omega}}\hat{\boldsymbol{\omega}}) + \frac{\mathcal{T}_{\parallel}(\omega)}{1 + G_{\parallel}(\omega)\mathcal{T}_{\parallel}(\omega)}\hat{\boldsymbol{\omega}}\hat{\boldsymbol{\omega}} \tag{494}$$

where $\mathcal{T}(\boldsymbol{\omega})$ is the Fourier transform of $\langle \mathcal{T}(\mathbf{r}, \mathbf{r}')\rangle$.

7.6.1 The Dilute Limit

We begin by computing the expression for $\mathcal{T}(\boldsymbol{\omega})$,

$$\begin{aligned} \mathcal{T}(\boldsymbol{\omega}) &= \frac{1}{\Omega}\int d\mathbf{R}_0 \int \mathcal{T}(\mathbf{r} - \mathbf{r}') \exp[i\boldsymbol{\omega} \cdot (\mathbf{r} - \mathbf{r}')] d(\mathbf{r} - \mathbf{r}') \\ &= \int\int \mathcal{T}(\theta, \theta') \exp\{i\boldsymbol{\omega} \cdot [\mathbf{r}_0(\theta) - \mathbf{r}_0(\theta')]\} d\theta d\theta'. \end{aligned} \tag{495}$$

Since the averaged system is isotropic, one should have, $\boldsymbol{\omega} = \omega\hat{\mathbf{z}}$. Expanding the exponential term in partial waves and the operators in spherical harmonics, one obtains

$$\mathcal{T}(\boldsymbol{\omega}) = -\frac{4\pi}{\Omega}\sum_{l=0}^{\infty}\sum_{j=0}^{\infty}(-1)^j i^{l+j}\sqrt{(2l+1)(2j+1)}\, J_l(\omega a) J_j(\omega a) \mathcal{T}_{l0:i0}, \tag{496}$$

which, in the limit $\omega \to 0$ and in the ω^2 order, reduces to

$$\mathcal{T}(\omega) = -\frac{4\pi}{\Omega}[J_0^2(\omega a)\mathcal{T}_{00;00} + 3J_1^2(\omega a)\mathcal{T}_{10;10} - 2\sqrt{5}J_0(\omega a)J_2(\omega a)\mathrm{Re}\{\mathcal{T}_{00;20}\}], \tag{497}$$

where J_n is the Bessel function of nth order, and Re{ } represents the real part of the quantity. It can be shown that $\mathcal{T}_{00;10} = 0$. We must now compute the $\mathcal{T}_{lm;l'm'}$ matrix and $\mathbf{g}_t$ and $\mathbf{g}_r$. Mondescu and Muthukumar (1999) showed that

$$\mathbf{g}_t = 4\pi\mathbf{G}_{00;00}^{-1} = \frac{24\pi a(1-\nu_{p1})\mu_1}{5-6\nu_{p1}}\mathbf{U}, \tag{498}$$

and

$$\mathbf{g}_r = 8\pi a^3\mu_1\mathbf{U}. \tag{499}$$

Introducing a vector $\mathcal{Y}_m$ by

$$\mathcal{Y}_m = \int \mathbf{r}_0(\theta)Y_{lm}^*(\theta)d\theta$$

$$= \sqrt{\frac{2\pi}{3}}a\delta_{1l}[(-\delta_{1m}+\delta_{-1m})\hat{\mathbf{x}} + i(\delta_{1m}+\delta_{-1m})\hat{\mathbf{y}} + \sqrt{2}\delta_{0m}\hat{\mathbf{z}}], \tag{500}$$

one obtains

$$\mathcal{T}_{lm;l'm'} = \mathbf{G}_{lm;l'm'}^{-1} - \mathbf{G}_{lm;00}^{-1}\cdot\mathbf{G}_{00;00}\cdot\mathbf{G}_{00;l'm'}^{-1} + \frac{1}{8\pi a^3\mu_1}$$
$$\times\sum_{i=-1}^{l}\sum_{j=-1}^{l}(\mathbf{G}_{lm;1i}^{-1}\times\mathcal{Y}_i)\cdot(\mathcal{Y}_j^*\times\mathbf{G}_{1j;l'm'}^{-1}). \tag{501}$$

Since the $\mathbf{G}_{lm;l'm'}^{-1}$ elements are diagonal in the l index, after expanding the cross-product, one obtains

$$\mathcal{T}_{00;00} = 0, \quad \mathcal{T}_{00;10} = \mathcal{T}_{10;00}^{\mathrm{T}} = 0, \quad \mathcal{T}_{00;20} = 0, \tag{502}$$

$$\mathcal{T}_{10;10} = \mathbf{G}_{10;10}^{-1} - \frac{3}{2}a\mu_1(\mathbf{U}-\hat{\mathbf{z}}\hat{\mathbf{z}})$$

$$= \frac{3}{2}a\mu_1\frac{5(1-\nu_{p1})}{4-5\nu_{p1}}(\mathbf{U}-\hat{\mathbf{z}}\hat{\mathbf{z}}) + 6a\mu_1\frac{(3-5\nu_{p1})(1-\nu_{p1})}{(4-5\nu_{p1})(1-2\nu_{p1})}\hat{\mathbf{z}}\hat{\mathbf{z}}. \tag{503}$$

Substituting these results in Eq. (497) and considering the small ω limit [so that $J_1(\omega a) \simeq (\omega a/3)$], one finds that at low ϕ_2, the volume fraction of the spheres,

$$\mathcal{T}(\omega) = -\mu_1\omega^2\left\{\frac{15(1-\nu_{p1})}{2(4-5\nu_{p1})}(\mathbf{U}-\hat{\mathbf{z}}\hat{\mathbf{z}}) + \frac{6(3-5\nu_{p1})(1-\nu_{p1})}{(4-5\nu_{p1})(1-2\nu_{p1})}\hat{\mathbf{z}}\hat{\mathbf{z}}\right\}\phi_2, \tag{504}$$

where $\phi_2 = (1/\Omega)(4\pi a^3/3)$. Therefore,

$$\frac{\mu_e-\mu_1}{\mu_1} = \frac{15(1-\nu_{p1})}{2(4-5\nu_{p1})}\phi_2, \tag{505}$$

$$\frac{\lambda_e - \lambda_1}{\lambda_1} = \frac{3(1 - \nu_{p1})}{2\nu_{p1}(4 - 5\nu_{p1})}\phi_2. \tag{506}$$

These expressions reproduce the classical result of Eshelby (1957) in the limiting case of rigid inclusions, which was discussed earlier in this chapter. Equation (505) is also in agreement with Eq. (103).

7.6.2 Nondilute Systems

To extend the model to non-dilute systems, Mondescu and Muthukumar (1999) used a type of mean-field approximation. Consider a composite with $N-1$ spherical, rigid, fully penetrable particles randomly distributed in an elastic matrix of shear modulus μ_1. Following the derivations described above, an exact multiple scattering solution for the average displacement field $\langle \mathbf{u} \rangle$ can be written as

$$\langle \mathbf{u}(\mathbf{r}) = \mathbf{G} * \mathbf{F} + \sum_{\alpha \geq 1}^{N-1} \mathbf{G} * \langle \mathcal{T}_\alpha \rangle * \mathbf{G} * \mathbf{F} + \sum_{\alpha \neq \beta}^{N-1} \mathbf{G} * \langle \mathcal{T}_\alpha * \mathbf{G} * \mathcal{T}_\beta \rangle * \mathbf{G} * \mathbf{F}$$

$$+ \sum_{\alpha \neq \beta; \beta \neq \gamma}^{N-1} \mathbf{G} * \langle \mathcal{T}_\alpha * \mathbf{G} * \mathcal{T}_\beta * \mathcal{T}_\gamma \rangle * \mathbf{G} * \mathbf{F} + \cdots, \tag{507}$$

where all convolutions are functions of $\mathbf{r}$. The Greek indices label different spheres, with $\mathcal{T}_\alpha$ being given by Eqs. (486) and (487). Thus, the self-energy $\mathbf{\Sigma}_{N-1}$ and all the elastic properties can be deduced for any types of interactions among particles, if one wishes so. Although this procedure is rigorous, it is also computationally intensive. An approximate, but simpler, way of analyzing the problem is to assume first that the two-phase composite can be represented as an effective, homogeneous medium, where the average elastic disturbances are transmitted by the effective Green function $\mathcal{G}(\mathbf{r}, \mathbf{r}')$, which is a function of some self-energy $\mathbf{\Sigma}_{N-1}$ that includes all possible sequences of elastic, long-range interactions mediated by the matrix. Adding one more particle changes the self-energy of the effective medium by $\delta \mathbf{\Sigma}_{N-1} = \mathbf{\Sigma}_{N-1}/(N-1)$. Then, provided it has a solution, the self-consistent equation for $\mathbf{\Sigma}_{N-1}$ is given by

$$\delta \mathbf{\Sigma}_{N-1}[\mathcal{G}(\mathbf{\Sigma}_{N-1})] = \frac{\mathbf{\Sigma}_{N-1}}{N-1}. \tag{508}$$

When the spheres are penetrable, one has the following exact result:

$$\frac{\mathbf{\Sigma}(\boldsymbol{\omega})}{N-1} = \frac{1}{\Omega} \mathcal{T}^e(\boldsymbol{\omega}), \tag{509}$$

where $\mathcal{T}^e(\boldsymbol{\omega})$ is the effective value of $\mathcal{T}(\boldsymbol{\omega})$ if one uses the effective $\mathcal{G}$ propagator instead of $\mathbf{G}$. Then, if $c = (N-1)/\Omega$ is the number concentration of the spheres, choosing $\boldsymbol{\omega} = \omega \hat{\mathbf{z}}$ and keeping only terms up to ω^2 order, yield

$$\mathbf{\Sigma}(\boldsymbol{\omega}) = -4\pi c [J_0^2(\omega a) \mathcal{T}^e_{00;00} - i\sqrt{3}(\mathcal{T}^e_{00;10} - \mathcal{T}^e_{10;00})$$

$$+ 3J_1^2(\omega a) \mathcal{T}^e_{10;10} - 2\sqrt{5} J_0(\omega a) J_2(\omega a) \mathrm{Re}\{\mathcal{T}^e_{00;20}\}], \tag{510}$$

which is the governing equation for $\mathbf{\Sigma}$. This implies that one must postulate a functional form for the self-energy based on the structure of $\mathbf{\Sigma}(\boldsymbol{\omega})$ deduced from Eq. (510). To select such a function, two key observations regarding the physics of the two-phase composite at large and low ωa must be kept in mind.

(1) For small ωa the composite should behave like an effective medium, where the elastic deformations are propagating freely. Thus, we expect to have, $\mathbf{\Sigma} \sim \omega^2$.
(2) Large values of ωa correspond to the region inside the particles, where the presence of other inclusions is not felt. Then, the self-energy may reach a constant value, as soon as the observation length becomes smaller than the particle diameter $2a$.

Based on these observations, the presumed limiting behavior of $\mathbf{\Sigma}(\boldsymbol{\omega})$ is given by

$$\Sigma_\perp(\omega) = \begin{cases} -\mu_1 W_\perp (\omega a)^2 & \omega a < 1/2 \\ -\mu_1 \xi_\perp^{-2} & \omega a \geq 1/2 \end{cases} \tag{511}$$

$$\Sigma_\parallel(\omega) = \begin{cases} -(\lambda_1 + 2\mu_1) W_\parallel (\omega a)^2 & \omega a < 1/2 \\ -(\lambda_1 + 2\mu_1) \xi_\parallel^{-2} & \omega a \geq 1/2 \end{cases} \tag{512}$$

where $W_\perp$, $W_\parallel$, $\xi_\perp$, and $\xi_\parallel$ are positive, unknown functions of the particle density. Continuity of $\mathbf{\Sigma}(\boldsymbol{\omega})$ at $\omega a = 1/2$ requires that

$$\xi_\perp^{-2} = \frac{1}{4} W_\perp, \quad \xi_\parallel^{-2} = \frac{1}{4} W_\parallel. \tag{513}$$

Mondescu and Muthukumar (1999) then showed that

$$a^2 W_\perp = \frac{1}{2} \frac{1}{D_1[\frac{2}{5}]} \eta, \tag{514}$$

$$a^2 W_\parallel = \frac{1 - 2\nu_{p1}}{2(1 - \nu_{p1})} \frac{D_1[\frac{4}{5}]}{D_1^2[\frac{3}{5}] - \frac{1}{5} D_1[\frac{1}{5}] f_1[1]} \eta, \tag{515}$$

where $\eta = (N-1)(4\pi a^3/3)/\Omega$ is the reduced density of the spheres, and

$$\begin{aligned} D_l[x] &= (1-x) I_{l+1/2}(y_\perp) K_{l+1/2}(y_\perp) + x \frac{1 - 2\nu_{p1}}{2(1 - \nu_{p1})} I_{l+1/2}(y_\parallel) \\ &+ \frac{2}{\pi} \left\{ A_l(1/2) \left[\frac{1-x}{1+4y_\perp^2} + \frac{x}{1+y_\parallel^2} \frac{1 - 2\nu_{p1}}{2(1 - \nu_{p1})} + \frac{x}{2(1 - \nu_{p1})} - 1 \right] \right. \\ &\left. + (1-x) y_\perp^2 F_l(1/2; y_\parallel) + x \frac{1 - 2\nu_{p1}}{2(1 - \nu_{p1})} y_\parallel^2 F_l(1/2; y_\parallel) \right\}, \end{aligned} \tag{516}$$

with I_n and K_n being the modified Bessel functions, and

$$A_l(x) = \int_0^x J_l^2(y)\, dy, \tag{517}$$

$$F_l(y,z) = \int_0^y \frac{J_l^2(x)}{x^2+z^2}\,dx, \tag{518}$$

$$f_l[1] = 5\left(D_l[\frac{1}{5}] - D_l[\frac{2}{5}]\right). \tag{519}$$

Here, $y_\perp = a/\xi_\perp$ and $y_\parallel = a/\xi_\parallel$. Mondescu and Muthukumar (1999) then derived the following equations for $y_\perp$ and $y_\parallel$:

$$y_\perp^2(\eta) = \frac{1}{8}\frac{1}{D_1[\frac{2}{5}]}\eta, \tag{520}$$

$$y_\parallel^2(\eta, \nu_{p1}) = \frac{1}{4}\frac{1-2\nu_{p1}}{2(1-\nu_{p1})}\frac{D_1[\frac{4}{5}]}{D_1[\frac{2}{5}]D_1[1]}\eta, \tag{521}$$

in terms of which the elastic moduli of the composite are predicted to be given by

$$\frac{\mu_e}{\mu_1} = 1 + 4y_\perp^2, \tag{522}$$

$$\frac{Y_e}{Y_1} = \frac{1 + 12\frac{1-\nu_{p1}}{1+\nu_{p1}}y_\parallel^2 - 8\frac{1-2\nu_{p1}}{1+\nu_{p1}}y_\perp^2}{1 + 8(1-\nu_{p1})y_\parallel^2 - 4(1-2\nu_{p1})y_\perp^2}(1+4y_\perp^2), \tag{523}$$

$$\frac{K_e}{K_1} = 1 + \frac{4}{1+\nu_{p1}}[3(1-\nu_{p1})y_\parallel^2 - 2(1-2\nu_{p1})y_\perp^2], \tag{524}$$

and the effective Poisson's ratio ν_{pe} of the material is predicted to be

$$\nu_{pe} = \frac{1}{2}\left[1 - \frac{1-2\nu_{p1}}{2(1-\nu_{p1})}\frac{1+4y_\perp^2}{1+4y_\parallel^2 - \frac{1-2\nu_{p1}}{2(1-\nu_{p1})}(1+4y_\perp^2)}\right]. \tag{525}$$

We should also recall from Chapter 3 that

$$\phi_2 = \begin{cases} \eta, & \text{impenetrable spheres,} \\ 1-\exp(-\eta), & \text{fully-penetrable spheres.} \end{cases} \tag{526}$$

7.6.3 *Comparison with the Experimental Data*

To assess the accuracy of the above theoretical predictions, comparison is made with two types of experimental data. The first type involves composites consisting of spherical glass beads in an epoxy polymer (Smith, 1976) and in polyester (Richard, 1975) matrices. The elastic properties of these composites are:

$$\text{epoxy polymer}: \begin{cases} Y_1 = 3.01\text{Gpa}, & \nu_{p1} = 0.394, \\ Y_2 = 760\text{GPa}, & \nu_{p2} = 0.23, \end{cases}$$

$$\text{polyester}: \begin{cases} Y_1 = 2.45\times 10^5\text{psi}, & \nu_{p1} = 0.444, \\ Y_2 = 10.2\times 10^6\text{psi}, & \nu_{p2} = 0.21, \end{cases}$$

where subscripts 1 and 2 denote, as usual, the matrix and the inclusions, respectively. The second type of the experiments was reported by Nishimatsu and Gurland (1960) and Doi *et al.* (1970) using an alloy consisting of tungsten carbide (WC) particles dispersed in a cobalt (Co) matrix. The elastic properties of the materials used by Nishimatsu and Gurland (1960) are:

$$Y_1 = 30 \times 10^6\text{psi}, \quad \nu_{p1} = 0.3, \quad Y_2 = 102 \times 10^6\text{psi}, \quad \nu_{p2} = 0.22,$$

and those of the materials used by Doi *et al.* (1970) are:

$$Y_1 = 2.11 \times 10^{-4}\text{kg/mm}^2, \quad \nu_{p1} = 0.31,$$
$$Y_2 = 7.14 \times 10^{-4}\text{kg/mm}^2, \quad \nu_{p2} = 0.19.$$

Figure 7.10 compares the predicted Y_e/Y_1 to the experimental data of Smith and of Richard. Figure 7.11 compares the predicted Young's modulus to the experimental data for the WC/Co composites, while Figure 7.12 makes the same comparison for the shear modulus of WC/Co composites. Recall that, in order to derive the above theoretical predictions, Mondescu and Muthukumar (1999) assumed that the inclusions are rigid, randomly distributed, equal-sized and penetrable spheres. According to Eqs. (522)–(525), the variables that control the effective elastic properties are the volume fraction ϕ_2 of the

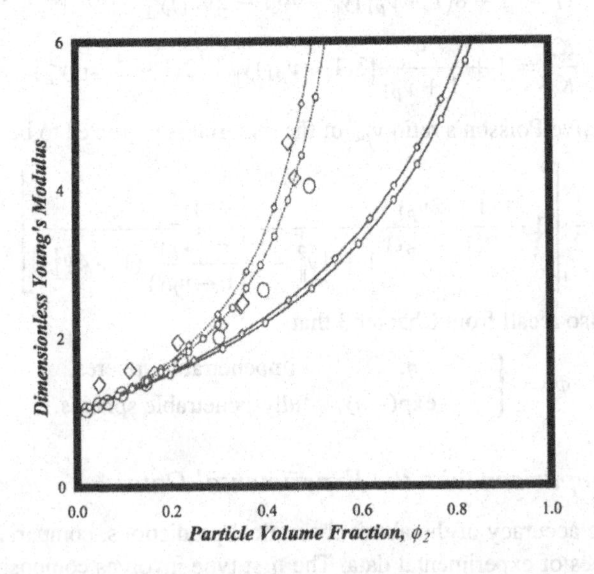

FIGURE 7.10. Dimensionless effective Young's modulus Y_e/Y_1 of composites of spherical glass beads in epoxy polymer (circles) and in polyester (diamonds). Large symbols are the experimental data, while smaller symbols joined by symbols are the theoretical predictions. The dashed curves are the theoretical predictions of Ju and Chen (1994) (who used a probabilistic pair-interaction formulation coupled with ensemble-volume averaged micromechanical field equations), while the solid curves are the predictions of multiple scattering theory (after Mondescu and Muthukumar, 1999).

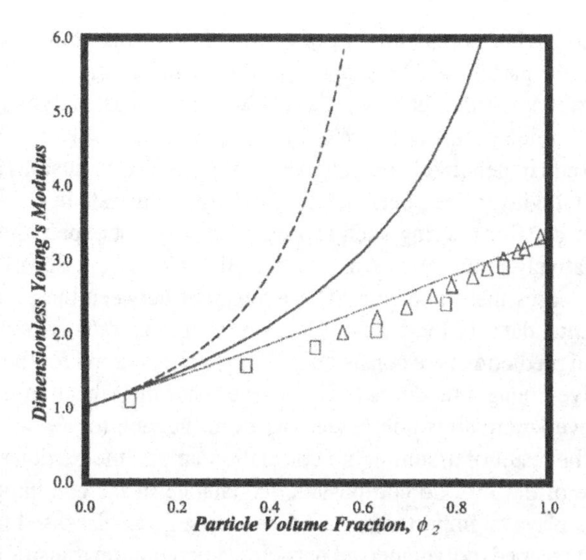

FIGURE 7.11. Dimensionless effective Young's modulus Y_e/Y_1 of the composite of WC particles dispersed in Co matrix. Squares and triangles represent, respectively, the experimental data of Nishimatsu and Gurland (1960) and Doi *et al.* (1970). Solid and dotted curves represent, respectively, the predictions of multiple scattering theory using fully- penetrable spheres and fully-impenetrable spheres models [see Eq. (526)], while dashed curve shows the predictions of Ju and Chen (1994) (after Modescu and Muthukumar, 1999).

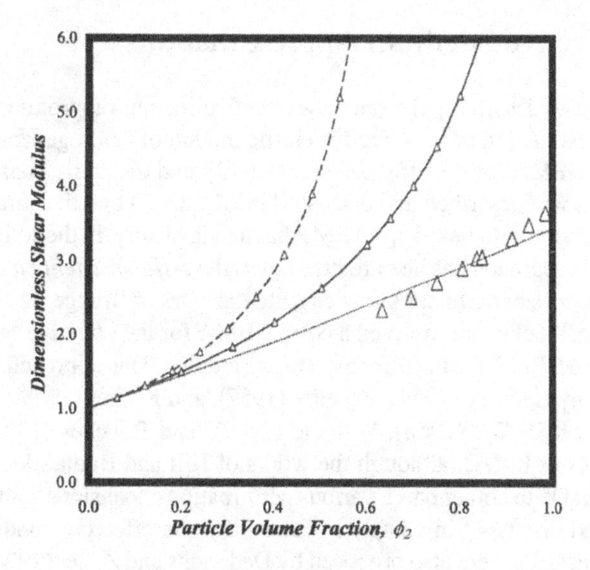

FIGURE 7.12. Same as in Figure 7.11, but for the dimensionless effective shear modulus μ_e/μ_1. Large triangles represent the experimental data.

inclusions and the Poisson's ratio ν_{p1} of the matrix. The effect of all other parameters, such as the particles' size, shapes, positions, and configurations, have been ignored. At the same time, both the glass beads in the glass/polymer system and the tungsten carbide grains in the WC/Co composites are elastic solid (although almost rigid) and impenetrable objects. Moreover, the WC inclusions are typically non-spherical, highly polydispersed, and usually agglomerate in clusters even for low values of ϕ_2. Considering such aspects, one may not expect the predictions to be quantitatively accurate, at least not for all values of ϕ_2. Surprisingly, however, in most cases there is quantitative agreement between the predictions and the experimental data. In the case of glass/epoxy and glass/polymer composites, the theoretical predictions are consistent with the general trend for the dependence of the effective Young's modulus (and also the shear modulus that is not shown) on ϕ_2. However, there are some deviations from the data for $\phi_2 > 0.4$, which is presumably the result of assuming no correlation among the particles.

In the case of the WC/Co composites, the change in the continuity of the inclusion phase plays an important role. More precisely, as discussed in Chapter 2, for irregularly-shaped, polydispersed particles, a percolation transition occurs at a volume fraction $\phi_2 \simeq 0.59$ of the inclusions, beyond which the role of the matrix and of the inclusions reverses. This is indeed the case in the experiments on the WC/Co composites, where the value of the Young's modulus varies from that of the matrix (Co) at $\phi_2 = 0$ to the value typical of WC at $\phi_2 = 1$. If the penetrability condition is assumed, then the multiple scattering method proves to be more appropriate and accurate, as indicated by the figures.

7.7 Effective-Medium Approximations

In this section we describe and discuss the effective-medium approximation (EMA) approach to estimation of the effective elastic moduli of heterogeneous materials. Similar to the EMA for the effective conductivity and dielectric constant of such materials that was described and discussed in Chapters 4 and 5, estimation of the effective elastic moduli based on an EMA has a long history. In the solid mechanics literature, this approach has been referred to as the *self-consistent method* (SCM), and its origin goes back, to our knowledge, to the work of Bruggeman (1937) who, using spherical inclusions, derived a correct EMA for the effective bulk modulus and an incorrect EMA for the effective shear modulus. This approach was revived in the 1950s by Hershey (1954), Eshelby (1957), and Kröner (1958). Hill (1965), Budiansky (1965), Wu (1966), Walpole (1969), and Boucher (1974) proposed various SCMs or EMAs, although the works of Hill and Budiansky are perhaps better known. On the other hand, various perturbation expansions, similar to those described and discussed in Chapters 4 and 5 for the effective conductivity and dielectric constants, were also proposed by Dederichs and Zeller (1973), Korringa (1973) and Gubernatis and Krumhansl (1975). These works, although referred to as multiple-scattering approaches (MSAs), are not identical with the approach of

Mondescu and Muthukumar (1999) described above, which is also referred to as a MSA. Watt *et al.* (1976) provided a comprehensive review of all of these works.

Even among the SCMs or EMAs, there are two distinct approaches. In one approach, the two phases of the material (i.e., the matrix and the inclusions) are treated asymmetrically in that, one phase is singled out to serve as the matrix. This is the approach taken by Hill (1965), Budiansky (1965), Wu (1966), Walpole (1969), and Boucher (1974). The second approach, in which the two phases are treated symmetrically, was developed by Berryman (1980a,b). The two approaches yield identical results if the inclusions are spherical. However, there are significant differences between their predictions for anisotropic inclusions, such as ellipsoids, disks, and penny-shaped cracks.

Before describing these two EMAs, certain tensors and fundamental properties must be computed. Their computation will be described in the next section, after which we describe both types of the EMAs without discussing their derivations, as they are entirely similar to those for the effective conductivity and other properties already described in Chapters 4–6. After presenting the general formulation, we will consider two particular limits. In one, the inclusions have $K = \mu = 0$, i.e., they represent holes (or exceedingly weak regions) in the matrix, and therefore the system is equivalent to the Swiss-cheese model, already described and utilized in Chapters 2–4. In the second limit, the model contains completely rigid inclusions with $K = \mu = \infty$, which we refer to it as the reinforced material.

7.7.1 Fundamental Tensors and Invariant Properties

In order to derive the EMAs, certain quantities must first be described and computed. Consider an ellipsoidal inclusion of a fixed orientation. The material is subjected to a uniform strain ϵ^0 at infinity and the ellipsoidal axes coincide with the specimen axes. In component form, the strain within the inclusion is given by

$$\epsilon_{ij}^c = \mathcal{T}_{ijkl}\epsilon_{kl}^0. \tag{527}$$

Here, the tensor $\mathcal{T}_{ijkl}$ is related to a fourth-rank tensor $\mathcal{S}_{ilmn}$ computed by Eshelby (1957) for the ellipsoidal inclusion,

$$\frac{x^2}{a^2} + \frac{y^2}{b^2} + \frac{z^2}{c^2} = 1,$$

for which the tensor $\mathcal{S}_{ilmn}$ is defined by

$$\epsilon_{ij}^c = \sum_m \sum_n \mathcal{S}_{ilmn}\epsilon_{mn}^T, \tag{528}$$

where ϵ_{ij}^c is the constrained strain and ϵ_{mn}^T is what Eshelby (1957) called the *stress-free strain*, which is the strain that the inclusion would undergo in the *absence* of the matrix. Physically, ϵ_{ij}^c is the strain produced when a surface traction is applied to the inclusion, while ϵ_{mn}^T is the strain on the outside surface which corresponds to this surface traction. Since ϵ_{mn}^c is related to uniform strains ϵ_{mn}^0 at infinity through the tensor $\mathcal{T}_{ijkl}$, ϵ_{ij}^T also has simple relations to ϵ_{mn}^0.

Eshelby (1957) showed that

$$\mathcal{S}_{1111} = qa^2 I_{aa} + \left(\frac{1}{4\pi} - \frac{1}{3}q\right) I_a,$$

$$\mathcal{S}_{1122} = qb^2 I_{ab} - \left(\frac{1}{4\pi} - \frac{1}{3}q\right) I_a, \tag{529}$$

$$\mathcal{S}_{1212} = \frac{1}{2}q(a^2+b^2)I_{ab} + \frac{1}{2}\left(\frac{1}{4\pi} - \frac{1}{3}q\right)(I_a + I_b),$$

where $q = 3/[8\pi(1-\nu_p)]$, and

$$I_a = 2\pi abc\int_0^\infty \frac{du}{(a^2+u)\Delta} = \frac{4\pi abc}{(a^2-b^2)(a^2-c^2)^{1/2}}(F-E),$$

$$I_{aa} = 2\pi abc\int_0^\infty \frac{du}{(a^2+u)^2\Delta} = \frac{4\pi}{3a^2} - I_{ab} - I_{ac}, \tag{530}$$

$$I_{ab} = \frac{2}{3}\pi abc\int_0^\infty \frac{du}{(a^2+u)(b^2+u)\Delta} = \frac{I_b - I_a}{3(a^2-b^2)},$$

with $\Delta = \sqrt{(a^2+u)(b^2+u)(c^2+u)}$. Here, $F = F(\theta, k)$ and $E = E(\theta, k)$ are elliptic integrals of the first and second kinds of amplitude $\theta = \sin^{-1}\sqrt{1-c^2/a^2}$, and modulus $k = \sqrt{(a^2-b^2)/(a^2-c^2)}$, $I_b = 4\pi - I_a - I_c$, and

$$I_c = \frac{4\pi abc}{(b^2-c^2)(a^2-c^2)^{1/2}}\left(\frac{b\sqrt{a^2-c^2}}{ac} - E\right).$$

For oblate spheroids ($a = b > c$),

$$I_a = I_b = \frac{2\pi a^2 c}{(a^2-c^2)^{3/2}}\left[\cos^{-1}\left(\frac{c}{a}\right) - \frac{c}{a}\sqrt{1-\frac{c^2}{a^2}}\right],$$

in which case, $I_{ab} = \frac{1}{3}I_{aa}$, with I_{aa} given by Eq. (480). For prolate spheroids ($b = c < a$), we have

$$I_b = I_c = \frac{2\pi ac^2}{(a^2-c^2)^{3/2}}\left[\frac{a}{c}\sqrt{\frac{a^2}{c^2} - 1} - \cosh^{-1}\left(\frac{a}{c}\right)\right].$$

The remaining coefficients are determined by simultaneous cyclic interchange of (1,2,3) and (a, b, c). Moreover, coefficients that couple an extension and a shear, such as $\mathcal{S}_{1112}$, $\mathcal{S}_{1123}$, $\mathcal{S}_{2311}, \cdots$, or one shear to another, e.g., $\mathcal{S}_{1223}$, are zero.

For a 2D inclusion, Eshelby's results can be simplified considerably (Thorpe and Sen, 1985). If we take the limit in which one of the axes of the 3D ellipsoid diverges, $c \to \infty$, while the other two major semi-axes a and b remain finite, we obtain a cylinder with an elliptical cross section in the (x, y) plane, and

$$\mathcal{S}_{1111} = \frac{b}{a+b}\left\{1 + \frac{a}{2(a+b)[1-\nu_p^{(3)}]}\right\},$$

$$\mathcal{S}_{1122} = \frac{b}{(a+b)[1-\nu_p^{(3)}]}\left[\nu_p^{(3)} - \frac{a}{2(a+b)}\right], \tag{531}$$

$$\mathcal{S}_{1212} = \frac{1}{2}\left\{1 - \frac{ab}{(a+b)^2[1-\nu_p^{(3)}]}\right\},$$

where superscript 3 indicates that the quantity should be evaluated for a 3D system. All other $\mathcal{S}_{ijkl}$ in the (x, y) plane involving only the indices 1 and 2 can be found by permuting the parameters (1,2) and (a, b). For example,

$$\mathcal{S}_{2222} = \frac{a}{a+b}\left\{1 + \frac{b}{2(a+b)[1-\nu_p^{(3)}]}\right\}. \tag{532}$$

All other $\mathcal{S}_{ijkl}$ are zero, except for

$$\mathcal{S}_{1133} = \frac{b}{a+b}\frac{\nu_p^{(3)}}{1-\nu_p^{(3)}}, \tag{533}$$

and $\mathcal{S}_{2233}$ which is determined from $\mathcal{S}_{1133}$ by interchanging 1 and 2.

Next, we proceed to determine the components of $\mathcal{T}_{ijkl}$. Wu (1966) related this tensor to $\mathcal{S}_{ijkl}$ to show that

$$\mathcal{T}_{1212} = \mathcal{T}_{2121} = \frac{1}{2(1+2A\mathcal{S}_{1212})}, \tag{534}$$

with similar expressions for $\mathcal{T}_{2323} = \mathcal{T}_{3232}$ and $\mathcal{T}_{3131} = \mathcal{T}_{1313}$. Moreover,

$$\begin{bmatrix} \mathcal{T}_{1111} & \mathcal{T}_{1122} & \mathcal{T}_{1133} \\ \mathcal{T}_{2211} & \mathcal{T}_{2222} & \mathcal{T}_{2233} \\ \mathcal{T}_{3311} & \mathcal{T}_{3322} & \mathcal{T}_{3333} \end{bmatrix}$$

$$= \begin{bmatrix} 1 + A\mathcal{S}_{1111} + B\mathcal{S}_1 & A\mathcal{S}_{1122} + B\mathcal{S}_1 & A\mathcal{S}_{1133} + B\mathcal{S}_1 \\ A\mathcal{S}_{2211} + B\mathcal{S}_2 & 1 + A\mathcal{S}_{2222} + B\mathcal{S}_2 & A\mathcal{S}_{2233} + B\mathcal{S}_2 \\ A\mathcal{S}_{3311} + B\mathcal{S}_3 & A\mathcal{S}_{3322} + B\mathcal{S}_3 & 1 + A\mathcal{S}_{3333} + B\mathcal{S}_3 \end{bmatrix}^{-1} \tag{535}$$

and all other components of $\mathcal{T}_{ijkl}$ vanish. Here,

$$A = \frac{\mu_2}{\mu_1} - 1, \tag{536}$$

$$B = \frac{1}{3}\left(\frac{K_2}{K_1} - \frac{\mu_2}{\mu_1}\right), \tag{537}$$

$$\mathcal{S}_1 = \mathcal{S}_{1111} + \mathcal{S}_{1122} + \mathcal{S}_{1133}, \tag{538}$$

with similar expressions for S_2 and S_3. As usual, K_1 and K_2, and μ_1 and μ_2 denote, respectively, the bulk and shear moduli of the matrix and the inclusion. These expressions can of course be simplified for a 2D inclusion.

The elastic properties of the 2D inclusion can also be expressed in terms of the corresponding 3D properties. Using the interrelations among the elastic moduli for 2D and 3D materials, it is straightforward to show that the relations between the 2D and 3D elastic properties are given by

$$\mu^{(2)} = \mu^{(3)}, \quad K^{(2)} = K^{(3)} + \frac{1}{3}\mu^{(3)}. \tag{539}$$

These results were derived based on the assumption that the ellipsoidal inclusion has a fixed orientation. In general, however, we must consider the case in which the inclusions are randomly and uniformly oriented. This was studied by Berryman (1980a,b) who showed that two key invariant quantities are needed when an isotropic averaging is done over all directions. These key properties are given by

$$P = \frac{1}{d}\sum_i \sum_j T_{iijj}, \tag{540}$$

$$Q = \frac{1}{c}\sum_i \sum_j \left(T_{ijij} - \frac{1}{d}T_{iijj}\right), \tag{541}$$

where $c = 2$ and 5 for $d = 2$ and 3, respectively. We are now ready to describe the symmetric and asymmetric EMAs.

7.7.2 *Symmetric Effective-Medium Approximation*

The symmetric EMA (SEMA), developed by Berryman (1980a,b), treats the host and the inclusions symmetrically, and leads to the following equations for the effective bulk and shear moduli of a n-phase material,

$$\sum_{i=1}^{n} \phi_i P^{ei}(K_e - K_i) = 0, \tag{542}$$

$$\sum_{i=1}^{n} \phi_i Q^{ei}(\mu_e - \mu_i) = 0, \tag{543}$$

where ϕ_i is the volume fraction of phase i. No distinction is made in Eqs. (542) and (543) between the different components, i.e., one does not designate one material as the matrix and the others as inclusions. The superscripts on the P and Q are intended for emphasizing that they contain $A = \mu_i/\mu_e - 1$ and $C = K_i/K_e - 1$ for inclusions with moduli K_i and μ_i which are embedded in an effective medium with effective moduli K_e and μ_e. The coefficients P^{ei} and Q^{ei} for 3D systems with various inclusion shapes were given by Berryman (1980b) and are listed in Table 7.1, where the various quantities are as follows,

$$F = \frac{\mu}{6}\frac{9K + 8\mu}{K + 2\mu}, \tag{544}$$

TABLE 7.1. The invariants P^{mi} and Q^{mi} for various 3D inclusion shapes. The quantities F, α, β and γ are defined by Eqs. (544)–(547).

Inclusion	P^{mi}	Q^{mi}
Spheres	$\dfrac{K_m+4\mu_m/3}{K_i+4\mu_m/3}$	$\dfrac{\mu_m+F_m}{\mu_i+F_m}$
Needles	$\dfrac{K_m+\mu_m+\mu_i/3}{K_i+\mu_m+\mu_i/3}$	$\dfrac{1}{5}\left(\dfrac{4\mu_m}{\mu_m+\mu_i}+2\dfrac{\mu_m+\gamma_m}{\mu_i+\gamma_m}+\dfrac{K_i+4\mu_m/3}{K_i+\mu_m+\mu_i/3}\right)$
Disks	$\dfrac{K_m+4\mu_i/3}{K_i+4\mu_i/3}$	$\dfrac{\mu_m+F_i}{\mu_i+F_i}$
Penny cracks	$\dfrac{K_m+4\mu_i/3}{K_i+4\mu_i/3+\pi\alpha\beta_m}$	$\dfrac{1}{5}\left[1+\dfrac{8\mu_m}{4\mu_i+\pi\alpha(\mu_m+2\beta_m)}+2\dfrac{K_i+2\mu_i/3+2\mu_m/3}{K_i+4\mu_i/3+\pi\alpha\beta_m}\right]$

$$\alpha = \frac{c}{a}, \tag{545}$$

$$\beta = \frac{3K+\mu}{3K+4\mu}\mu, \tag{546}$$

$$\gamma = \frac{3K+\mu}{3K+7\mu}\mu. \tag{547}$$

Consider, as an example, a 2D composite with elliptical inclusions that represent holes in the material (i.e., the generalized 2D Swiss-cheese model). In this case (Thorpe and Sen, 1985),

$$P^{ei} \to P_e = \frac{a^2+b^2}{ab(1-\nu_{pe})}, \tag{548}$$

$$Q^{ei} \to Q_e = \frac{(a+b)^2}{ab(1-\nu_{pe})}, \tag{549}$$

where ν_{pe} is the effective Poisson's ratio of the composite. If the volume fraction of the elliptical holes is $\phi_2 = 1-\phi_1$, the EMA is given by

$$\phi_1 P^{e1}(K_e - K_1) - \phi_2 P^{e0} K_e = 0, \tag{550}$$

or

$$\frac{K_e}{K_1} = \frac{1-\alpha_1/\phi_1}{1-\alpha_1}, \tag{551}$$

with $\alpha_1 = (1-P^{e1}/P^{e0})^{-1} = (1-P^{e1}/P_e)^{-1}$. Similarly, the effective shear modulus is given by

$$\frac{\mu_e}{\mu_1} = \frac{1-\alpha_2/\phi_1}{1-\alpha_2}, \tag{552}$$

with $\alpha_2 = (1-Q^{e1}/Q^{e0})^{-1} = (1-Q^{e1}/Q_e)^{-1}$. In these equations, superscript 1 refers to the component with non-zero moduli, and 0 to the void. These equations should be solved simultaneously in order to compute K_e and μ_e as functions of ϕ_1 (or ϕ_2).

As the volume fraction of the holes increases, the remaining material becomes increasingly disjointed, and eventually disconnected when its volume fraction falls below a critical value $\phi_c = \phi_{1c}$ at which K_e and μ_e vanish. In this limit one obtains,

$$\phi_c = 2\left[1 + \sqrt{\frac{2(a+b)^2}{a^2+b^2}}\right]^{-1}, \tag{553}$$

and

$$\nu_{pc} = 1 - \phi_c, \tag{554}$$

thus indicating that, at the critical volume fraction or the percolation threshold, the Poisson's ratio is independent of the bulk and shear moduli of the material, and depends only on the aspect ratio a/b, in agreement with what we discussed in Section 7.2.7. For the special case of circular inclusions ($a = b$), the SEMA predicts that at $\phi_c = 2/3$,

$$\nu_{pc} = \frac{1}{3}, \quad \frac{K_e}{\mu_e} = 2. \tag{555}$$

Similarly, for $b/a \to 0$, i.e., for needle-shaped inclusions, one obtains

$$\phi_c = 2(\sqrt{2} - 1) \simeq 0.8284, \quad \nu_{pc} = 3 - \sqrt{8} \simeq 0.1716, \quad \frac{K_e}{\mu_e} = \sqrt{2} \simeq 1.414. \tag{556}$$

Note that the universal values of ν_{pc} predicted by the SEMA are only approximate. Indeed, the point (ϕ_c, ν_{pc}) in the (ϕ_1, ν_p) plane can be thought of as a fixed point, much like the fixed points in the renormalization group theory described in Chapter 5.

7.7.3 Asymmetric Effective-Medium Approximation

In the asymmetric EMA (AEMA), one treats the matrix and inclusions asymmetrically in the sense that, a particular component is designated as the matrix. Then, the general equations of AEMA for the effective bulk and shear moduli of a n-phase composite are given by

$$\frac{1}{K_e} = \frac{1}{K_1}\left[1 + \sum_i^n (1-\phi_i) P^{ei}\left(\frac{K_1 - K_i}{K_e}\right)\right], \tag{557}$$

$$\frac{1}{\mu_e} = \frac{1}{\mu_1}\left[1 + \sum_i^n (1-\phi_i) Q^{ei}\left(\frac{\mu_1 - \mu_i}{\mu_e}\right)\right], \tag{558}$$

where material 1 has been designated as the matrix.

As the first example, consider the 2D Swiss-cheese model with elliptical inclusions, for which the AEMA predicts that

$$\frac{K_e}{K_1} = \frac{\phi_1 - \alpha_1}{1 - \alpha_1}, \tag{559}$$

$$\frac{\mu_e}{\mu_1} = \frac{p - \alpha_2}{1 - \alpha_2}, \tag{560}$$

where, $\alpha_1 = 1 - P_e^{-1} = 1 - ab(1 - \nu_{pe})/(a^2 + b^2)$, and $\alpha_2 = 1 - Q_e^{-1} = 1 - ab(1 + \nu_{pe})/[(a + b)^2]$. The percolation threshold is predicted to be given by

$$\phi_c = \left(1 + \frac{ab}{a^2 + b^2}\right)^{-1}, \tag{561}$$

at which

$$\nu_{pc} = 1 - \phi_c. \tag{562}$$

Several points are worth noting. (1) Except for circular inclusions, the percolation thresholds predicted by the SEMA and AEMA are *not* identical. (2) Both the SEMA and AEMA predict that, $\phi_c + \nu_{pc} = 1$, although the values of ϕ_c and ν_{pc} that the two approximations predict are quite different. (3) The coupled equations for K_e and μ_e *decouple* if rewritten in terms of the effective Young's modulus Y_e and ν_{pe}. This result is only true for the AEMA with elliptical holes, i.e., for the 2D Swiss-cheese model, but not so for the reinforced model or for the SEMA. (4) All the EMAs discussed here yield the correct initial slope near $\phi_1 = 1$, indicating that $dY_e/d\phi_1$ is independent of μ_i, and $d\nu_{pe}/d\phi_1$ is independent of Y_1 as $\phi_1 \to 1$, hence providing a convenient way of determining the initial slopes of K_e and μ_e based on the SEMA or AEMA. (5) All the initial slopes are infinite for needles which have $\nu_{pc} = 0$.

As the second example, consider the 2D reinforced problem, for which the AEMA yields the following equations,

$$\frac{1}{K_e} = \frac{1}{K_1}\left[1 + (1 - \phi_1)P^{e\infty}\left(\frac{K_1 - K_\infty}{K_e}\right)\right], \tag{563}$$

$$\frac{1}{\mu_e} = \frac{1}{\mu_1}\left[1 + (1 - \phi_1)Q^{e\infty}\left(\frac{\mu_1 - \mu_\infty}{\mu_e}\right)\right], \tag{564}$$

which, after taking the limits $K_\infty \to \infty$ and $\mu_\infty \to \infty$, yield

$$\frac{K_e}{K_1} = \frac{1 - \alpha_1}{\phi_1 - \alpha_1}, \tag{565}$$

$$\frac{\mu_e}{\mu_1} = \frac{1 - \alpha_2}{\phi_1 - \alpha_2}, \tag{566}$$

indicating that both moduli diverge at the percolation threshold, as expected, where

$$(1 - \alpha_1)^{-1} = \frac{2}{3 - \nu_{pe}}\left[1 + \left(\frac{1 - \nu_{pe}}{1 + \nu_{pe}}\right)\frac{(a + b)^2}{2ab}\right], \tag{567}$$

$$(1 - \alpha_2)^{-1} = \frac{1}{2}\left[\frac{(a + b)^2}{ab(3 - \nu_{pe})} + \frac{(a + b)^2}{(a + b)^2 - ab(1 + \nu_{pe})}\right]. \tag{568}$$

As another example, consider a 2D model of hexagonal array of infinitely-long, aligned cylinders in a matrix, or a thin-plate composite consisting of hexagonal arrays of disks in a matrix. The AEMAs for this model, due to Hill (1965), Budiansky (1965), Wu (1966), and others, are given by

$$\phi_1 \frac{K_e - K_1}{\mu_e + K_1} + \phi_2 \frac{K_e - K_2}{\mu_e + K_2} = 0, \tag{569}$$

$$\phi_1 \frac{\mu_e - \mu_1}{\mu_e K_e/(K_e + 2\mu_e) + \mu_1} + \phi_2 \frac{\mu_e - \mu_2}{\mu_e K_e/(K_e + 2\mu_e) + \mu_2} = 0. \tag{570}$$

The predictions of Eqs. (569) and (570) for the effective shear modulus μ_e are shown in Figure 7.2, where they are compared with the results of numerical simulations of Eischen and Torquato (1993), the three-point lower bound of Silnutzer (1972), and the truncated series expansion of Torquato (1997), Eq. (274). Figure 7.3 compares the predictions of Eqs. (569) and (570) for the effective bulk modulus of the 2D Swiss-cheese model with circular holes [i.e., in limits, $K_2 = 0$ and $\mu_2 = 0$, in Eqs. (569) and (570)], with the results of numerical simulations of Eischen and Torquato (1993) and the truncated series expansion of Torquato (1997), Eq. (273), in the limits, $K_2 = 0$ and $\mu_2 = 0$. In this case, the AEMA badly fails, except when ϕ_2, the volume fraction of the holes (i.e., the porosity of the material), is low.

As the final example, we consider a 3D random dispersion of identical spheres in a matrix. The AEMA equations (Hill, 1965; Budiansky, 1965, among others) are now given by

$$\phi_1 \frac{K_e - K_1}{4\mu_e/3 + K_1} + \phi_2 \frac{K_e - K_2}{4\mu_e/3 + K_2} = 0, \tag{571}$$

$$\phi_1 \frac{\mu_e - \mu_1}{\mu_e(9K_e + 8\mu_e)/(6K_e + 12\mu_e) + \mu_1} + \phi_2 \frac{\mu_e - \mu_2}{\mu_e(9K_e + 8\mu_e)/(6K_e + 12\mu_e) + \mu_2} = 0. \tag{572}$$

Figure 7.6 compares the predictions of these equations for the effective bulk modulus K_e with the three-point lower bound of Beran and Molyneux (1966), the upper bound of Gibiansky and Torquato (1995a), and the truncated series expansion of Torquato (1997), Eq. (284). Once again, the predictions of the AEMA are not accurate if $\phi_2 > 0.35$.

For spherical inclusions, the EMA equations can be generalized and written in a compact form. Equations (569) and (571) can be generalized to a d-dimensional material with n phases:

$$\sum_{i=1}^{n} \phi_i \frac{K_e - K_i}{K_i + 2d^{-1}(d-1)\mu_e} = 0, \tag{573}$$

while Eqs. (570) and (572) are written in the general form,

$$\sum_{i=1}^{n} \phi_i \frac{\mu_e - \mu_i}{\mu_i + h_e} = 0, \tag{574}$$

where h_e is given by Eq. (310) with $i = e$. Generally speaking, and similar to the EMA for the effective conductivity and dielectric constant (see Section 4.9), Eqs. (573) and (574) violate the bounds that improve upon the Hashin–Shtrikman bounds for realistic microstructures.

7.7.4 The Maxwell–Garnett Approximations

The Maxwell–Garnett (MG) approximation, that was described in Section 4.9.4 for the effective conductivity of heterogeneous materials, can be extended to estimating the effective elastic moduli. For spherical inclusions, the MG approximation for the effective bulk modulus is given by

$$\frac{K_e - K_1}{K_e + 2d^{-1}(d-1)\mu_1} = \phi_2 \frac{K_2 - K_1}{K_2 + 2d^{-1}(d-1)\mu_1}, \tag{575}$$

while for the effective shear modulus, the MG approximation is governed by

$$\frac{\mu_e - \mu_1}{\mu_e + h_1} = \phi_2 \frac{\mu_2 - \mu_1}{\mu_2 + h_1}, \tag{576}$$

where h_1 is defined by Eq. (310) with $i = 1$. Equations (575) and (576) are very similar to Eq. (4.262) for the effective conductivity. As in the case of the effective conductivity, the MG approximation provides reasonable estimates of the elastic moduli if the spherical inclusions are widely separated. Moreover, the MG approximation for the effective bulk modulus coincides with the optimal Hashin–Shtrikman lower bound (308) if $\mu_2 \geq \mu_1$, and the corresponding optimal upper bound (308) if $\mu_2 \leq \mu_1$. Furthermore, the MG approximation is identical to the Hashin–Shtrikman lower bound (309).

The MG approximation can be generalized to macroscopically-anisotropic materials that consist of $n-1$ different types of unidirectionally aligned isotropic inclusions of the same shape, and is given by

$$\sum_{i=1}^{n} \phi_i (\mathbf{C}_e - \mathbf{C}_1) \cdot \mathbf{T}_i^{(1)} = \mathbf{0}, \tag{577}$$

where

$$\mathbf{T}_i^{(1)} = \left[\mathbf{U} + \mathcal{S} : \mathbf{C}_1^{-1} : (\mathbf{C}_i - \mathbf{C}_1)\right]^{-1}. \tag{578}$$

Here, $\mathcal{S}$ is Eshelby's tensor for an ellipsoid in an isotropic matrix of stiffness $\mathbf{C}_1$, defined and given by Eqs. (528)–(533). The case in which the ellipsoidal inclusions are randomly distributed in the matrix can also be obtained from Eq. (577). One obtains the following equations for the effective bulk and shear moduli,

$$\sum_{i=1}^{n} \phi_i (K_e - K_i) T_i^{(1h)} = 0, \tag{579}$$

$$\sum_{i=1}^{n} \phi_i (\mu_e - \mu_i) T_i^{(1s)} = 0, \tag{580}$$

where $T_i^{(1h)}$ and $T_i^{(1s)}$ are the hydrostatic and shear contributions to $\mathbf{T}_i^{(1)}$.

7.8 Numerical Simulation

So far, we have described the theoretical methods of estimating the elastic properties of heterogeneous materials, based on the continuum models. In this section we discuss briefly the numerical simulation methods for estimating the elastic moduli.

7.8.1 Finite-Difference Methods

Early numerical calculation of the effective elastic moduli was based on the standard finite-difference (FD) discretization of the continuum equations of elasticity. Examples of such computations include the computations of Adams and Doner (1967), Chen and Cheng (1967), and Flaherty and Keller (1973) who computed the effective elastic moduli of square and hexagonal arrays of cylinders in a matrix (see Section 7.2.2). The main advantage of the FD methods is their simplicity. The equations that are obtained by the FD discretization are typically very simple, and the resulting matrix of coefficients that arise in the numerical solution of these equations is usually sparse and banded. The main disadvantages of FD methods are twofold: (1) If the boundary of the system, and/or the interface between various phases or components of a composite material is complex (e.g., curved), then, one must use highly refined FD mesh in such areas in order to ensure accuracy of the computed properties and their convergence to the true values. (2) The same sort of problem arises if the contrast between the elastic properties of the phases is large, even if the interfaces, and/or the boundary of the material, are smooth. The problem in this case is the sharp variations of the local elastic properties near the interface between the various phases of the material. Thus, once again, one must use highly refined FD mesh in order to realistically represent the material and its constituents. However, such refined meshes would require large computation time and computer memory. In addition, for large contrasts between the elastic properties of the various phases, the convergence may not even be guaranteed. In fact, the computations of Adams and Doner (1967), Chen and Cheng (1967) and Flaherty and Keller (1973) were restricted to a very limited range of material properties.

7.8.2 Boundary-Element and Finite-Element Methods

More accurate and efficient alternatives to the FD methods are based on the finite-element (FE) and boundary-element (BE) methods. There are already many books that describe these methods in great details (see, for example, Cook *et al.*, 1989; Becker, 1992), and therefore we only provide a brief discussion and describe an example.

The basis of the BE method is the integral representation of the displacement field $\mathbf{u}$ which, in component form, is given by (see, for example, Phan–Thien and Kim, 1994)

$$c_{ij}(\boldsymbol{\xi})u_i(\boldsymbol{\xi}) + \int_\Gamma T^*_{ij}(\boldsymbol{\xi},\mathbf{x})u_j(\mathbf{x})d\Gamma(\mathbf{x}) = \int_\Gamma u^*_{ij}(\boldsymbol{\xi},\mathbf{x})T_j(\mathbf{x})d\Gamma(\mathbf{x}), \qquad (581)$$

where a summation over repeated indices is implied, and $\Gamma(\boldsymbol{\xi})$ represents the bounding curve, including inclusion boundaries in a composite material, of the elastic system. Here, u_i and T_i are, respectively, the displacement and traction components *on the boundary*. Therefore, Equation (581) is a very important result as it expresses the displacement field $\mathbf{u}$ *inside* a material in terms of *boundary* displacements and tractions. Equation (581) must be satisfied at each "source" point $\boldsymbol{\xi}$ on the boundary, while the field point is at $\mathbf{x}$. In analogy with the potential theory, the integral on the left-hand side of Eq. (581) is called the double layer potential. The jump in the integral representation, from $\mathbf{u}$ to $\mathbf{0}$ across the boundary, is due to the double layer. For any Poisson's ratio ν_p, the kernel T_{ij} of the integral is strictly $O(r^{-2})$ and singular.

The integral on the right-hand side of Eq. (581) represents a displacement field generated by a distribution of surface forces of strength $T_j(\mathbf{x})d\Gamma(\mathbf{x})$, because $\mathbf{u}^*(\boldsymbol{\xi},\mathbf{x})$ is a displacement field in direction j at field point $\mathbf{x}$ due to a unit force applied in direction i at the source point $\boldsymbol{\xi}$. Likewise, $T^*_{ij}(\boldsymbol{\xi},\mathbf{x})$ is the traction in direction j at field point $\mathbf{x}$ due to a unit force applied in direction i at the source point $\boldsymbol{\xi}$. In analogy with the potential theory, this integral is called the single-layer potential, i.e., the single layer of charges distributed on the boundary is replaced by a single layer of forces. On the surface, $u^*_{ij} \sim O(1/r)$, and therefore the integral is regular, since the singularity is removable when the integral is evaluated using a local coordinate system located at the singular point, noting the fact that $d\Gamma \sim O(r^2)$. The quantities u^*_{ij} and T^*_{ij} are related to the Green function for the linear elasticity problem for isotropic materials. For example, for a 2D isotropic material and under the plane strain condition, they are given by

$$u^*_{ij}(\boldsymbol{\xi},\mathbf{x}) = \frac{1}{8\pi\mu(1-\nu_p)}\left[(3-4\nu_p)\ln\left(\frac{1}{r}\right)\delta_{ij} + \frac{r_i r_j}{r^2}\right], \tag{582}$$

$$T^*_{ij}(\boldsymbol{\xi},\mathbf{x}) = -\frac{1}{4\pi r(1-\nu_p)}\left\{\left[(1-2\nu_p)\delta_{ij} + \frac{2r_i r_j}{r^2}\right]\frac{r_i n_i}{r} - (1-2\nu_p)\frac{r_i n_j - r_j n_i}{r}\right\}, \tag{583}$$

where we recognize the $\ln(1/r)$ term as the basic feature of the 2D Green function [see Eq. (199)]. Here, μ is the transverse shear modulus, $r_i = x_i - \xi_i$, $r = \sqrt{r_i r_j}$, and n_i are (the components of) the measure numbers of the unit outward vector normal to the bounding curve Γ. Note that, in this context, μ and the Poisson's ratio ν_p are the 3D material properties. The components of the tensor c_{ij} are computed by requiring that rigid body displacements yield zero tractions.

For a composite material, the BE equation, Eq. (581), is enforced along the matrix external boundary, the inclusion external boundary, and the matrix/inclusion interfaces. Continuity of displacement and traction fields are enforced along the matrix-inclusion interfaces. If, for example, in a 2D problem there are corner points in the system, one can allow for two independent displacement components and four traction components.

Consider, as an example, a hexagonal array of circles (or infinitely-long cylinders) that was studied by Eischen and Torquato (1993) by the BE method. They

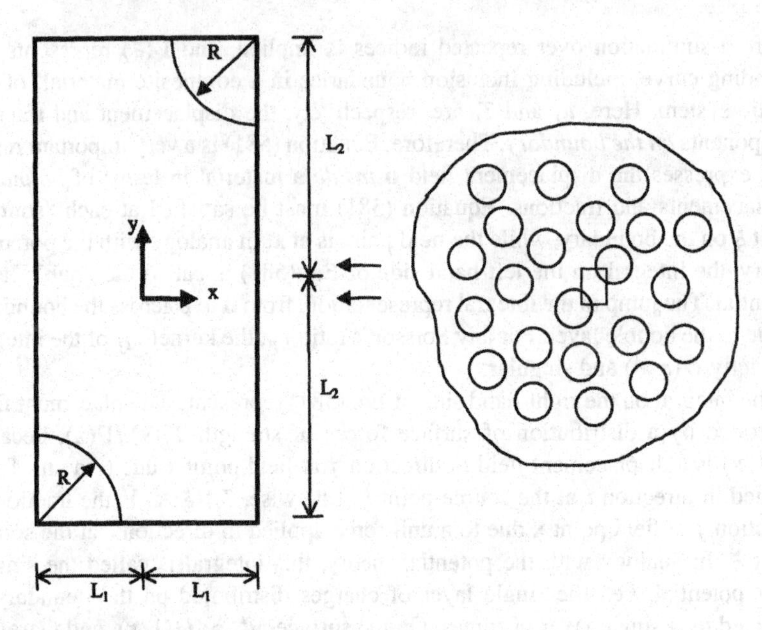

FIGURE 7.13. Unit cell for BE computations of elastic moduli of hexagonal arrays (after Eischen and Torquato, 1993).

used a rectangular unit cell shown in Figure 7.13, with dimensions $2L_1 \times 2L_2$, with $L_2 = \sqrt{3}L_1$. The origin of the Cartesian coordinates is at the center of the cell. Also shown are two quarters of circular inclusions of radius R, with a volume fraction ϕ_2 which is given by

$$\phi_2 = \frac{\pi}{2\sqrt{3}}\left(\frac{R}{L}\right)^2, \tag{584}$$

where $L^2 = L_1^2 + L_2^2$. We note that $R/L = 1$ corresponds to the critical volume fraction of the inclusions. Eischen and Torquato (1993) solved two problems with imposed average stress and strain. In the first problem, the strains were $\langle\epsilon_{11}\rangle = 0$, $\langle\epsilon_{22}\rangle = 1$, with $\langle\sigma_{12}\rangle = 0$, where $\langle\cdots\rangle$ denotes average quantities which are defined by

$$\langle\epsilon_{ij}\rangle = \frac{1}{4L_1L_2}\int_{-L_2}^{L_2}\int_{-L_1}^{L_1}\epsilon_{ij}(x_1, x_2)dx_1dx_2, \tag{585}$$

$$\langle\sigma_{ij}\rangle = \frac{1}{4L_1L_2}\int_{-L_2}^{L_2}\int_{-L_1}^{L_1}\sigma_{ij}(x_1, x_2)dx_1dx_2. \tag{586}$$

The appropriate boundary conditions are given by

$$u_2(x_1, L_2) = -u_2(x_1, -L_2) = L_2, \quad u_1(-L_1, x_2) = u_1(L_1, x_2) = 0, \tag{587}$$

$$\sigma_{12}(x_1, L_2) = \sigma_{12}(x_1, -L_2) = \sigma_{12}(-L_1, x_2) = \sigma_{12}(L_1, x_2) = 0. \tag{588}$$

The average stress $\langle\sigma_{22}\rangle$ is given by

$$\langle\sigma_{22}\rangle = \frac{1}{2L_1}\int_{-L_1}^{L_1} \sigma_{22}(x_1, L_2)dx_1, \tag{589}$$

which is related to the effective bulk and shear moduli of the composite by,

$$\langle\sigma_{22}\rangle = K_e - \mu_e. \tag{590}$$

In the second problem solved by Eischen and Torquato (1993) the imposed shear strain was $\langle\epsilon_{12}\rangle = 1/2$ with $\langle\sigma_{11}\rangle = \langle\sigma_{22}\rangle = 0$. The boundary conditions for the displacements u_1 and u_2 are the same as (587), but the stress conditions are replaced by

$$\sigma_{22}(x_1, L_2) = \sigma_{22}(x_1, -L_2) = \sigma_{11}(-L_1, x_2) = \sigma_{11}(L_1, x_2) = 0. \tag{591}$$

Moreover,

$$\langle\sigma_{12}\rangle = \mu_e = \frac{1}{2L_1}\int_{-L_1}^{L_1} \sigma_{12}(x_1, x_2)dx_1, \tag{592}$$

and therefore the effective bulk modulus K_e can be calculated using Eq. (590).

Conceptually, the FE method is somewhat similar to the BE method, in that it subdivides the system into elements. However, unlike the BE method, the resulting discrete equations govern the displacement field **u** throughout the material, as opposed to on the boundary in the BE method. Since we demonstrate in Chapter 8 how the FE discretization of the continuum equations of linear elasticity is used for generating discrete or lattice models of elastic properties of heterogeneous materials, we postpone further discussion of this method. In general, there are at least three basic issues that one must deal with when using a BE or FE method.

(1) Since the BE method involves integral equations, how to carry out numerical integration of the integral is an important issue. In particular, since as pointed above, the kernel of these integrals involves singular functions, an effective way of dealing with such singularities must be developed.

(2) The way the system and/or its boundary are discretized in the FE and BE methods is also an important issue that must be addressed carefully. For example, the integration over the entire boundary is replaced by a sum of many integrations over individual finite elements or, in the case of the BE method, a BE over the surface.

(3) In utilizing the BE or FE method, one major task is efficient solution of the system of algebraic equations which is the result of discretizing the continuum equations. The matrix of coefficients for this system is dense and fully populated. For small- or medium-size problems, a direct method of solving the matrix problem, such as the Gauss elimination method, is very efficient, as it provides the exact numerical solution of the problem (aside from the round-off errors). However, the direct methods cannot be used for large-scale problems, because the required computer memory will be too large, and therefore one must resort to iterative techniques, the most efficient of which is perhaps the

conjugate-gradient method which will be described in detail in Chapter 9 of Volume II. Iterative algorithms also lend themselves to parallel computations, a distinct advantage.

7.9 Links Between the Conductivity and Elastic Moduli

A fundamental scientific problem with great practical implications is whether various transport properties of a heterogeneous material are rigorously linked to one another. Such cross-property relations can be especially useful if one property is more easily measured than another property. Since, as emphasized throughout this book, the effective transport properties reflect certain morphological information about a material, one might to be able to extract useful information about one effective property given an accurate (experimental or theoretical) estimate of another property, even if their respective governing equations are uncoupled or seemingly unrelated. This is a fertile area of research (see, for example, Prager 1969; Berryman and Milton 1988; Torquato 1990) which has produced many interesting and useful results. The purpose of this section is to summarize the most recent and useful of such results due to Gibiansky and Torquato (1995b,1996). At the same time, their results are compared to other older predictions in order to obtain a better understanding of the progress that has been made. We begin our discussion with 2D systems and provide a somewhat detailed discussion of the results, after which we summarize the 3D results.

7.9.1 *Two-Dimensional Materials*

We wish to derive cross-property relations between the effective transverse conductivity and effective transverse elastic moduli of transversely isotropic, two-phase, fiber-reinforced composites, the phase boundaries of which are cylindrical surfaces with generators parallel to one axis. Thus, the essential problem is studying 2D two-phase, isotropic disordered materials (that is, plane elasticity and conductivity) with effective properties that are identical to the corresponding transverse properties of fiber-reinforced composites. For an isotropic medium, the material is characterized by the triples (g_1, K_1, μ_1) and (g_2, K_2, μ_2) for the two phases, where g_i, K_i and μ_i denote, respectively, the conductivity, bulk modulus and shear modulus of phase i. Similarly, the triple (g_e, K_e, μ_e) denotes the effective properties of the composite material which, of course, depend on the phase properties, phase volume fractions ϕ_1 and $\phi_2 = 1 - \phi_1$, and the material's morphology. A given fixed morphology corresponds to some point (g_e, K_e, μ_e) in a 3D $g_e - K_e - \mu_e$ space. Changing the material's morphology moves and traces out this space, and the goal is to find some bounds on this set or space. In particular, we restrict our attention to the bounds of the projection of this set onto the $g_e - K_e$ and $g_e - \mu_e$ planes. Our discussion follows closely that of Gibiansky and Torquato (1995b). Their results for the effective conductivity and bulk modulus are not restricted to only isotropic composites, but are also applicable as well to anisotropic composites with square symmetry. Moreover, as pointed out in Chapter 4, the determination of the electrical conductivity is mathematically equivalent to finding either the thermal conductivity, dielectric constant, magnetic permeability, or diffusion coefficient, and thus the cross-property relations connect the elastic moduli to any of these properties as well.

We use the notations introduced by Eqs. (408) and (409) and, for the sake of simplicity, we use the briefer notation, $F(a_1, a_2, \phi_1, \phi_2, y) = F_a(y)$. As mentioned in Section 7.5.1.4, this function is a scalar-variant of the inverse Y-transformation. Now let g_{1e} and g_{2e} denote the expressions,

$$g_{1e} = F_g(g_1), \quad g_{2e} = F_g(g_2). \tag{593}$$

Similarly, suppose that

$$K_{1e} = F_K(\mu_1), \quad K_{2e} = F_K(\mu_2), \tag{594}$$

and μ_{1e}, μ_{2e}, μ_{3e}, and μ_{4e} denote the quantities given by

$$\mu_{1e} = F_\mu[\mu_1 K_1/(K_1 + 2\mu_1)], \quad \mu_{2e} = F_\mu[\mu_2 K_2/(K_2 + 2\mu_2)], \tag{595}$$

$$\mu_{3e} = F_\mu[\mu_1 K_2/(K_2 + 2\mu_1)], \quad \mu_{4e} = F_\mu[\mu_2 K_1/(K_1 + 2\mu_2)]. \tag{596}$$

Moreover, the harmonic average of the phase bulk moduli is defined by,

$$K_h = \left(\frac{\phi_1}{K_1} + \frac{\phi_2}{K_2}\right)^{-1} = F_K(0). \tag{597}$$

Equations (593) and (594) coincide with the 2D upper and lower Hashin–Shtrikman bounds on the effective conductivity (Hashin and Shtrikman, 1962a) and effective bulk modulus (Hashin and Shtrikman, 1963) of isotropic composites, respectively, which were already described in Sections 4.6.1.1 and 7.5.1.2. Equations (595) and (596) coincide with the 2D Hashin–Shtrikman-Walpole bounds on shear modulus (Hashin and Shtrikman, 1963; Hashin 1965; Walpole 1966, 1969), which were also obtained by Hill (1964).

The bounds that Gibiansky and Torquato (1995b) derived are given by segments of hyperbolae in the $g_e - K_e$ and $g_e - \mu_e$ planes with asymptotes that are parallel to the axes $g_e = 0$ and $K_e = 0$, or $\mu_e = 0$. Thus, every such hyperbola in the $x_e - y_e$ plane can be described by the equation

$$A(x_e - x_0)(y_e - y_0) = 1, \tag{598}$$

and can be defined by three points that it passes through. We denote by

$$\text{Hyp}[(x_1, y_1), (x_2, y_2), (x_3, y_3)]$$

the segment of such hyperbola that passes through the points (x_1, y_1), (x_2, y_2) and, when extended, the point (x_3, y_3). Thus, it can be parametrically described in the

$x_e - y_e$ plane by

$$\begin{cases} x_e = \gamma x_1 + (1-\gamma)x_2 - \dfrac{\gamma(1-\gamma)(x_1 - x_2)^2}{(1-\gamma)x_1 + \gamma x_2 - x_3}, \\ y_e = \gamma y_1 + (1-\gamma)y_2 - \dfrac{\gamma(1-\gamma)(y_1 - y_2)^2}{(1-\gamma)y_1 + \gamma y_2 - y_3}, \end{cases} \tag{599}$$

where $\gamma \in [0, 1]$. The bounds that connect the effective conductivity and the elastic moduli can now be stated.

7.9.1.1 Conductivity-Bulk Modulus Bounds

To obtain cross-property bounds on the set of the pair (g_e, K_e) for any isotropic composite material at a fixed volume fraction ϕ_1, one should inscribe in the conductivity-bulk modulus plane the segments of the following four hyperbolae:

$$\text{Hyp}[(g_{1e}, K_{1e}), (g_{2e}, K_{2e}), (g_1, K_h)], \quad \text{Hyp}[(g_{1e}, K_{1e}), (g_{2e}, K_{2e}), (g_2, K_h)],$$

$$\text{Hyp}[(g_{1e}, K_{1e}), (g_{2e}, K_{2e}), (g_1, K_1)], \quad \text{Hyp}[(g_{1e}, K_{1e}), (g_{2e}, K_{2e}), (g_2, K_2)].$$

The outermost pair of these curves yield the desired bounds. This is shown in Figure 7.14 which depicts conductivity-bulk modulus bounds for the following values of the parameters:

$$\frac{g_2}{g_1} = 20, \quad \frac{K_2}{K_1} = 20, \quad \nu_{p1} = \nu_{p2} = 0.3, \quad \phi_1 = 0.2. \tag{600}$$

The corner points $A = (g_{1e}, K_{1e})$ and $B = (g_{2e}, K_{2e})$ of the set enclosed by the bounds are optimal because they correspond to assemblages of coated circles (Hashin and Shtrikman, 1963; see Section 7.5.1.2), as well as to isotropic matrix laminate composites (Francfort and Murat, 1986). The hyperbolae $\text{Hyp}[(g_{1e}, K_{1e}), (g_{2e}, K_{2e}), (g_1, K_1)]$ (curve 1 of Figure 7.14) and $\text{Hyp}[(g_{1e}, K_{1e}), (g_{2e}, K_{2*}), (g_2, K_2)]$ (curve 2 of Figure 7.14) correspond to the assemblages of doubly coated spheres, or to doubly coated matrix laminate composites (Cherkaev and Gibiansky 1992; Gibiansky and Milton 1993). Depending upon the values of the parameters, one of these curves always forms part of the bound (upper bound of Figure 7.14). Thus, this is an optimal bound because there exist composites that realize it.

In order to obtain bounds for an *arbitrary* volume fraction (of either phase) one can take the union of the sets defined by these bounds over the phase volume fractions. This union is bounded by the curves 5 and 6 of Figure 7.14. The unmarked straight line of Figure 7.14 corresponds to the upper bound of relation (Torquato, 1992),

$$\frac{K_e}{K_1} \leq \frac{g_e}{g_1}. \tag{601}$$

The bound (601) holds for d-dimensional isotropic, two-phase materials of *arbitrary* topology having *positive* phase Poisson's ratio ν_{pi}, if $K_2/K_1 \leq g_2/g_1$, and is optimal and coincides with the new bound when $g_2/g_1 = K_2/K_1$ and $\nu_{pi} = 0$

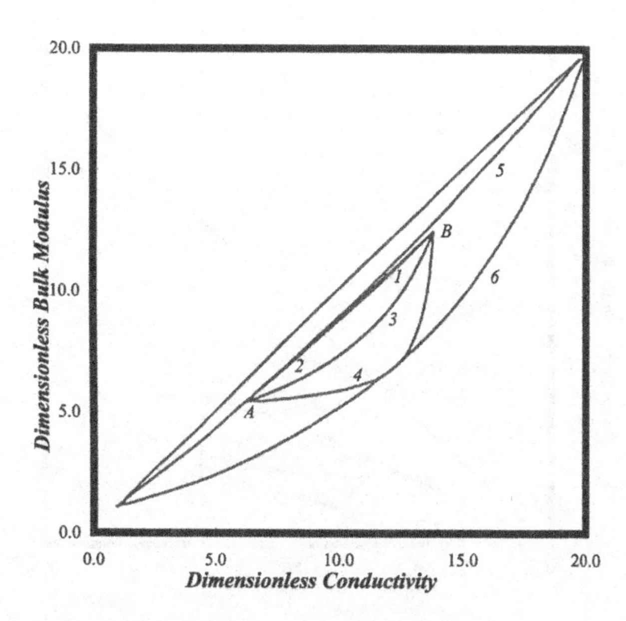

FIGURE 7.14. Cross-property bounds in the conductivity-bulk modulus plane. On the scale of the figure, curves 1 and 2 are almost indistinguishable from each other. The internal region, bounded by curves 1 and 4, represents the bounds for fixed volume fraction, while the larger region, bounded by curves 5 and 6, represents the bounds for arbitrary volume fractions. The unmarked straight line is the bound, $K_e/K_1 \leq g_e/g_1$ (after Gibiansky and Torquato, 1995b).

($i = 1$ and 2), i.e., $\mu_1/K_1 = \mu_2/K_2 = 1$. In general, the Gibiansky–Torquato volume-fraction independent upper bound is more restrictive than (601).

7.9.1.2 Conductivity-Shear Modulus Bounds

To derive cross-property bounds on the set of the pairs (g_e, μ_e) for any composite at fixed volume fraction ϕ_1, one should inscribe in the conductivity-shear modulus plane the segments of the following four hyperbolae:

$$\text{Hyp}[(g_{1e}, \mu_{1e}), (g_{2e}, \mu_{3e}), (g_1, \mu_1)], \quad \text{Hyp}[(g_{1e}, \mu_{1e}), (g_{2e}, \mu_{3e}), (g_2, \mu_2)],$$

$$\text{Hyp}[(g_{1e}, \mu_{4e}), (g_{2e}, \mu_{2e}), (g_1, \mu_1)], \quad \text{Hyp}[(g_{1e}, \mu_{4e}), (g_{2e}, \mu_{2e}), (g_2, \mu_2)],$$

and the segments of two straight lines:

$$g_e = g_{1e}, \quad \mu_e \in [\mu_{1e}, \mu_{4e}], \quad \text{and} \quad g_e = g_{2e}, \quad \mu_e \in [\mu_{2e}, \mu_{3e}].$$

The outermost of these curves yield the desired bounds. Figure 7.15 presents the conductivity-shear modulus bounds for the phase moduli and volume fractions given in (600), where the cross-property bounds in the $g_e - \mu_e$ plane are represented by a curvilinear trapezium. The straight sides AB and CD are given by the Hashin–Shtrikman bounds on the effective conductivity. The other two curvilinear sides (new bounds that are denoted as curves 1 and 4 of Figure 7.15) are the

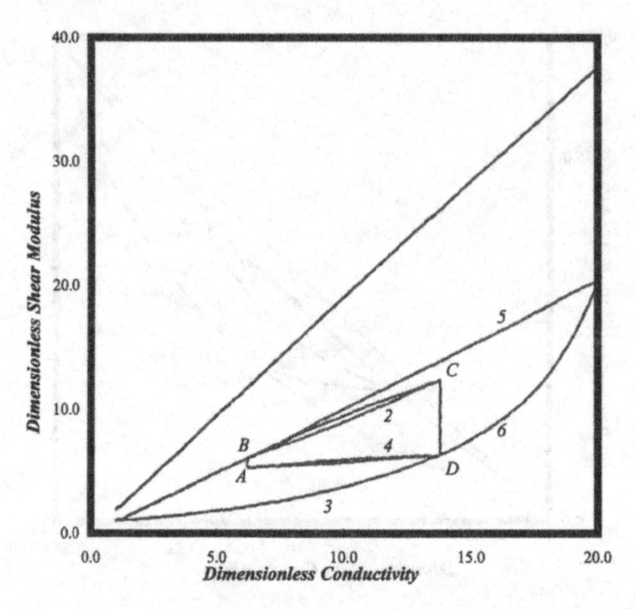

FIGURE 7.15. Cross-property bounds in the conductivity-shear modulus plane. Points B and C are joined by curve 1. The trapezium ABCD represents the bounds for fixed volume fraction, while the larger internal area, bounded by curves 5 and 6, shows the bounds for arbitrary volume fraction. The unmarked straight line is the bound (602) with $\nu_{pe} = 0$ (after Gibiansky and Torquato, 1995b).

hyperbola segments. The two corner points $A = (g_{1e}, \mu_{1e})$ and $C = (g_{2e}, \mu_{2e})$ of the set enclosed by the bounds correspond to the matrix laminate composites that realize the Hashin–Shtrikman bounds for elasticity and conductivity, and thus are optimal (Francfort and Murat, 1986). The other two points, $B = (g_{1e}, \mu_{4e})$ and $D = (g_{2e}, \mu_{3e})$, correspond to the structures that could realize the Walpole bounds on the shear modulus of a composite. Note that composites with the pair (g_e, μ_e) that lies on the vertical sides of the trapezium (see Figure 7.15) are polycrystals made of square symmetric matrix laminate composites (Cherkaev and Gibiansky, 1984).

The union of the sets defined by these bounds over the phase volume fractions yield cross-property bounds for *arbitrary* volume fraction (the set bounded by the curves 5 and 6 of Figure 7.15). The unmarked straight line of Figure 7.15 corresponds to the upper-bound (Torquato, 1992)

$$\frac{\mu_e}{K_1} \leq \frac{g_e}{g_1}\frac{1 - \nu_{pe}}{1 + \nu_{pe}}, \tag{602}$$

(where $K_2/K_1 \leq g_2/g_1$) with $\nu_{pe} = 0$ [i.e., the weaker form of (602)], which does not incorporate volume fraction information. The Gibiansky–Torquato volume-fraction independent upper bound is more restrictive than the weak form of (602) (with $\nu_{pe} = 0$).

Therefore, the effective elastic moduli and conductivities of composite materials are in fact not independent, but are connected through the morphology. However, this relation is not one-to-one because materials of different morphologies may possess the same conductivities but very different elastic moduli, and vice versa. The cross-property bounds derived by Gibiansky and Torquato indicate how the values of effective bulk or shear modulus (conductivity) of the composite are constrained assuming that the phase properties and volume fractions are fixed and the effective conductivity (bulk or shear modulus) is known.

The proof of these cross-property bounds is based on the so-called translation method (see, for example, Cherkaev and Gibiansky 1984, 1987, 1992, 1993; Francfort and Murat 1986; Milton 1990a,b; Gibiansky and Milton 1993), which bounds from below the relevant energy functional $\mathcal{H}$ that, for the present problem, has the general form

$$\mathcal{H} = \sum_{i=1}^{N} \mathcal{H}_\epsilon^*[\boldsymbol{\epsilon}_0^{(i)}] + \mathcal{H}_E^*(\mathbf{E}_0), \tag{603}$$

and is equal to the sum of the elastic and electrical energies stored in the element of periodicity of a composite which is exposed to N elastic fields and two electrical fields with fixed mean values, where $N = 1$ for the bulk modulus bounds and $N = 2$ for the shear modulus bounds. Here, $\mathcal{H}_\epsilon^*$ is the elastic energy density stored in the composite, which is given by

$$\mathcal{H}_\epsilon^*(\boldsymbol{\epsilon}_0) = \boldsymbol{\epsilon}_0 : \mathbf{C}_e : \boldsymbol{\epsilon}_0 = \inf \langle \boldsymbol{\epsilon} : \mathbf{C} : \boldsymbol{\epsilon} \rangle, \tag{604}$$

where the infimum is taken over all admissible fields, $\boldsymbol{\epsilon} = \frac{1}{2}[\nabla \mathbf{u} + (\nabla \mathbf{u})^T]$, that are the strain fields with given mean value $\boldsymbol{\epsilon}_0 = \boldsymbol{\epsilon} : \langle \boldsymbol{\epsilon} \rangle$ that satisfy compatibility conditions. For the conjugate functional of the complementary energy we have

$$\mathcal{H}_\tau^*(\boldsymbol{\tau}) = \boldsymbol{\tau}_0 : \mathbf{S}_e : \boldsymbol{\tau}_0 = \inf \langle \boldsymbol{\tau} : \mathbf{S} : \boldsymbol{\tau} \rangle, \tag{605}$$

where the effective compliance tensor $\mathbf{S}_e$ is determined as, $\mathbf{S}_e = \mathbf{C}_e^{-1}$, and the infimum is taken over stress fields with given mean value $\boldsymbol{\tau}_0 = \boldsymbol{\tau} : \langle \boldsymbol{\tau} \rangle$ that satisfy equilibrium condition $\nabla \cdot \boldsymbol{\tau} = 0$.

For the conductivity problem one uses the functionals that are sums of the energy stored by the composite, namely,

$$\mathcal{H}_E^*(\mathbf{E}_0) = \mathbf{E}_0 \cdot \hat{\mathbf{g}}_e \cdot \mathbf{E}_0 = \inf \langle \mathbf{E} \cdot \hat{\mathbf{g}} \cdot \mathbf{E} \rangle, \tag{606}$$

where the infimum is taken over all admissible fields, $\mathbf{E} = -\nabla(\varphi^{(1)}, \varphi^{(2)})$, with φ denoting the potential, $E_0 = \mathbf{E} \cdot \langle \mathbf{E} \rangle$, and $\mathbf{g}$ is the conductivity tensor. In a similar way, a functional in terms of the current density can be defined. Note that the energy functional $\mathcal{H}$ can also be written in terms of the stress and the current fields, as given above. The reason for utilizing this energy functional is that its value is equal to the energy stored by an equivalent homogeneous medium in the uniform field. The equivalent medium is characterized by the tensor of the effective properties and the uniform external field coincides with the mean value of the field in the composite. Clearly, the lower bound on the functional $\mathcal{H}$ provides bounds on the effective moduli of interest.

To better understand the functionals that yield the cross-property bounds, it is instructive to begin with functionals for the pure elasticity and pure conductivity problems before presenting those for the cross-property bounds. To obtain bounds on the bulk modulus, the composite is exposed to an external hydrostatic strain $\boldsymbol{\epsilon}_h = \epsilon_h \mathbf{d}_1$ or stress $\boldsymbol{\tau}_h = \tau_h \mathbf{d}_1$, where

$$\mathbf{d}_1 = \frac{1}{\sqrt{2}} \begin{pmatrix} 1 & 0 \\ 0 & 1 \end{pmatrix}$$

The energy of an isotropic composite under the action of such fields is proportional to the effective bulk modulus K_e according to

$$\mathcal{H}_\epsilon(\boldsymbol{\epsilon}_h) = \mathcal{H}_\epsilon^*(\boldsymbol{\epsilon}_h) = \boldsymbol{\epsilon}_h : \mathbf{C}_e : \boldsymbol{\epsilon}_h = 2K_e(\epsilon_h)^2, \tag{607}$$

or its inverse value $1/K_e$ according to

$$\mathcal{H}_\tau(\boldsymbol{\tau}_h) = \mathcal{H}_\tau^*(\boldsymbol{\tau}_h) : \mathbf{S}_e : \boldsymbol{\tau}_h = \frac{1}{2K_e}(\tau_h)^2. \tag{608}$$

The minimization (by changing the morphology) of the energy stored in the hydrostatic strain field is equivalent to the minimization of the bulk modulus. By contrast, the minimization of the energy stored in the hydrostatic stress field leads to maximization of K_e. One must express the elastic energy in terms of $\boldsymbol{\zeta} = \nabla \mathbf{u}$ instead of the strain fields $\boldsymbol{\epsilon}$, i.e.,

$$\mathcal{H}_\epsilon(\boldsymbol{\epsilon}_h) = \mathcal{H}_\zeta(\boldsymbol{\zeta}_h) = \boldsymbol{\zeta}_h : \mathbf{C}_e : \boldsymbol{\zeta}_h, \tag{609}$$

where we assume that $\boldsymbol{\epsilon}_h : \mathbf{d}_i = \boldsymbol{\zeta}_h : \mathbf{d}_i$, with $i = 1, 2, 3$ and

$$\mathbf{d}_2 = \frac{1}{\sqrt{2}} \begin{pmatrix} 1 & 0 \\ 0 & -1 \end{pmatrix}; \quad \mathbf{d}_3 = \frac{1}{\sqrt{2}} \begin{pmatrix} 0 & 1 \\ 1 & 0 \end{pmatrix}.$$

By using the same type of argument, it is clear that if $\mathbf{E} = \mathbf{E}_h = E_h \mathbf{d}_1$, then minimization of the functional

$$\mathcal{H}_E(\mathbf{E}_h) = \mathcal{H}_E^*(\mathbf{E}_h) = \mathbf{E}_h \cdot \hat{\mathbf{g}} \cdot \mathbf{E}_h = g_e(E_h)^2, \tag{610}$$

is equivalent to the minimization of the effective conductivity g_e of an isotropic composite. If $\mathbf{I} = \mathbf{I}_h = I_h \mathbf{d}_1$, then the minimization of the functional

$$\mathcal{H}_I(I_h) = \mathcal{H}_I^*(I_h) = \mathbf{I}_h \cdot \hat{\mathbf{g}}^{-1} \cdot \mathbf{I}_h = \frac{1}{g_e}(I_h)^2, \tag{611}$$

is equivalent to maximization of the effective conductivity.

In order to derive the conductivity-bulk modulus bounds, one combines the functionals for the pure cases. Then, one has a choice between stress and strain trial fields for the elasticity problem, and between the electrical and current fields for the electrostatic problem. The following functionals should then be considered for the conductivity-bulk modulus bounds:

$$\mathcal{H}_{\zeta E}(\zeta_h, \mathbf{E}_h) = \mathcal{H}^*_{\zeta}(\zeta_h) + \mathcal{H}^*_{E}(\mathbf{E}_h), \tag{612}$$

$$\mathcal{H}_{\zeta I}(\zeta_h, \mathbf{I}_h) = \mathcal{H}^*_{\zeta}(\zeta_h) + \mathcal{H}^*_{I}(\mathbf{I}_h), \tag{613}$$

$$\mathcal{H}_{\tau E}(\tau_h, \mathbf{E}_h) = \mathcal{H}^*_{\tau}(\tau_h) + \mathcal{H}^*_{E}(\mathbf{E}_h), \tag{614}$$

$$\mathcal{H}_{\tau J}(\tau_h, \mathbf{I}_h) = \mathcal{H}^*_{\tau}(\tau_h) + \mathcal{H}^*_{E}(\mathbf{I}_h). \tag{615}$$

The lower bound of each of these functionals gives some component of the boundary where the bounds are located. By using linear combinations of the four functionals, one can find the functional that moves one in any fixed direction in the $g_e - K_e$ plane. Therefore, by bounding the functionals from below one can find bounds on the set of pairs (g_e, K_e) of composites of all possible morphologies. For example, consider the functional (612). For fixed amplitudes E_h and ϵ_h, the bounds on the functional (612) show how far one can move in the direction $\mathbf{v} = -E_h^2 \mathbf{e}_g - \epsilon_h^2 \mathbf{e}_K$, where $\mathbf{e}_g$ and $\mathbf{e}_K$ are the unit vectors of the g_e-axis and K_e-axis, respectively. By changing the ratio E_h^2/ϵ_h^2 one can direct $\mathbf{v}$ at any point within the third quadrant of the $g_e - K_e$ plane, implying that the functional (612) can provide the part of the cross-property boundary that has a normal directed into the third quadrant. Similarly, the functional (613) provides the part of the boundary with a normal directed into the second quadrant of the plane $g_e - K_e$. The functional (614) corresponds to that part of the boundary that has a normal directed into the fourth quadrant of the plane (g_e, K_e), while the functional (615) provides the part of the boundary with a normal directed into the first quadrant of the plane (g_e, K_e).

In a similar manner, one can obtain bounds on the shear modulus of a composite material by bounding the energy stored in a composite exposed to shear-type trial strain or stress fields. In this way, one obtains bounds on any of the two shear moduli of the composite which, in general, is anisotropic. To ensure isotropy of the system, one must consider the response of the composite to two orthogonal shear fields. Hence, to estimate the shear modulus of an isotropic composite one should minimize the functional equal to the sum of the values of the energy stored by the medium under the action of two trial orthogonal shear strain or stress fields. The details of the proofs of these theorems are given by Gibiansky and Torquato (1995b), to whom we refer the interested reader.

7.9.1.3 Applications

Let us now discuss some specific applications of cross-property relations.

Rigid superconducting phase: Suppose that one of the phases of a material is characterized by $K_2/K_1 = \infty$, $\mu_2/\mu_1 = \infty$, and $g_2/g_1 = \infty$, i.e., phase 2 is totally rigid and superconducting. In this limit, the boundary hyperbolae degenerate into straight lines and the bounds for fixed ϕ_1 are given by

$$g_e \geq g_u, \quad K_u \leq K_e \leq K_u + \max\left[\frac{K_1 + \mu_1}{2g_1}, \frac{2\mu_2 K_2}{g_2(K_2 + \mu_2)}\right](g_e - g_u), \tag{616}$$

$$\mu_l \leq \mu_e \leq \mu_u + \max\left[\frac{K_1 + 2\mu_1}{4g_1}, \frac{\mu_2 K_2}{g_2(K_2 + \mu_2)}\right](g_e - g_u), \tag{617}$$

with

$$g_u = \frac{1 + \phi_2}{\phi_1} g_1, \quad K_u = \frac{K_1 + \mu_1 \phi_2}{\phi_1},$$

$$\mu_l = \frac{(1 + \phi_2)\mu_1 K_1 + 2\mu_1^2}{\phi_1(K_1 + 2\mu_1)}, \quad \mu_u = \frac{\phi_2 K_1 + 2\mu_1}{2\phi_1}.$$

Although it may seem unphysical that these bounds depend on the ratio of the infinite moduli of the ideal phase, we should keep in mind that a very small amount of a rigid, conducting material can yield finite effective properties. In the same manner, one obtains the cross-property bounds for an arbitrary volume fraction ϕ_1:

$$g_e \geq g_1, \quad K_1 \leq K_e \leq K_1 + \max\left[\frac{K_1 + \mu_1}{2g_1}, \frac{2\mu_2 K_2}{g_2(K_2 + \mu_2)}\right](g_e - g_1), \tag{618}$$

$$\mu_1 \leq \mu_e \leq \mu_1 + \max\left[\frac{K_1 + 2\mu_1}{4g_1}, \frac{\mu_2 K_2}{g_2(K_2 + \mu_2)}\right](g_e - g_1). \tag{619}$$

The lower bounds are trivial and the same as the Hashin–Shtrikman bounds for K_e and, for reasons discussed above, are optimal, as are the upper bounds.

Insulating phase: Now consider the opposite limit in which $K_2/K_1 = 0$, $\mu_2/\mu_1 = 0$, and $g_2/g_1 = 0$. For fixed volume fractions, the cross-property bounds are given by

$$\frac{1}{g_e} \geq \frac{1}{g_0}, \quad \frac{1}{K_e} \geq \frac{1}{K_0} + \min\left[\frac{g_1(K_1 + \mu_1)}{2\mu_1 K_1}, \frac{2g_2}{K_2 + \mu_2)}\right]\left(\frac{1}{g_e} - \frac{1}{g_0}\right), \tag{620}$$

$$\frac{1}{\mu_e} \geq \frac{1}{\mu_0} + \min\left[\frac{g_1(K_1 + \mu_1)}{\mu_1 K_1}, \frac{4g_2}{K_2 + 2\mu_2}\right]\left(\frac{1}{g_e} - \frac{1}{g_0}\right), \tag{621}$$

with

$$g_0 = \frac{\phi_1 g_1}{1 + \phi_2}, \quad K_0 = \frac{\phi_1 \mu_1 K_1}{\mu_1 + \phi_2 K_1}, \quad \mu_0 = \frac{\phi_1 \mu_1 K_1}{K_1(1 + \phi_2) + 2\phi_2 \mu_1}.$$

The corresponding bounds for arbitrary volume fractions are obtained by replacing g_0, K_0 and μ_0 with g_1, K_1 and μ_1, respectively. Gibiansky and Torquato (1995b) showed that these bounds provide excellent predictions for the effective bulk and shear moduli of various composites, if accurate estimates of the effective conductivity of the material are used.

7.9.2 Three-Dimensional Materials

We now discuss cross-property relations for 3D materials, summarizing and discussing the work of Gibiansky and Torquato (1996), and making a comparison between their theoretical predictions and the previous bounds. Similar to the 2D case, we use the notations introduced by Eqs. (408) and (409), and use again, for the sake of brevity, the notation, $F_a(y) = F(a_1, a_2, \phi_1, \phi_2, y)$. We now define the following quantities,

$$g_{1e} = F_g(2g_1), \quad g_{2e} = F_g(2g_2), \tag{622}$$

$$g_{1*} F_g(-2g_1), \quad g_{2*} = F_g(-2g_2), \tag{623}$$

$$K_{1e} = F_K\left(\frac{4}{3}\mu_1\right), \quad K_{2e} = F_K\left(\frac{4}{3}\mu_2\right). \tag{624}$$

In addition, we also define the arithmetic and harmonic averages of the conductivity,

$$g_a = \phi_1 g_1 + \phi_2 g_2 = F_g(\infty), \quad g_h = \left(\frac{\phi_1}{g_1} + \frac{\phi_2}{g_2}\right)^{-1}, \tag{625}$$

and similar averages for the bulk modulus.

The bounds derived by Gibiansky and Torquato (1996) are given by segments of hyperbolae in the $g_e - K_e$ plane with asymptotes that are parallel to the axes $g_e = 0$ and $K_e = 0$. Every such hyperbola in the (x_e, y_e) plane is described by Eq. (598) and can be parameterized by Eqs. (599). The proof of the cross-property relations discussed below are entirely similar to that of 2D systems outlined above, the details of which are given by Gibiansky and Torquato (1996).

7.9.2.1 Conductivity-Bulk Modulus Bounds

The cross-property bounds on (g_e, K_e) for any isotropic material at a fixed volume fraction ϕ_1 are obtained by inscribing in the conductivity-bulk modulus plane the following five segments of hyperbolae:

$$\text{Hyp}[(g_{1e}, K_{1e}), (g_{2e}, K_{2e}), (g_1, K_1)], \quad \text{Hyp}[(g_{1e}, K_{1e}), (g_{2e}, K_{2e}), (g_2, K_2)],$$

$$\text{Hyp}[(g_{1e}, K_{1e}), (g_{2e}, K_{2e}), (g_{1*}, K_h)], \quad \text{Hyp}[(g_{1e}, K_{1e}), (g_{2e}, K_{2e}), (g_{2*}, K_h)],$$

$$\text{Hyp}[(g_{1e}, K_{1e}), (g_{2e}, K_{2e}), (g_a, K_a)].$$

The outermost pair of these hyperbolae yields the desired bounds. This is shown schematically in Figure 7.16 for the parameters given in (600), where curves 1–5 represents the five hyperbolae.

Berryman and Milton (1988) used a different technique for deriving cross-property bounds on the pair (g_e, K_e). To derive their bounds, they used the three-point bounds described in Section 7.5.1.4 that depend on the microstructural parameters ζ_2 and η_2. By excluding these parameters from the conductivity and elastic moduli bounds, they obtained their cross-property relations which can be written in terms of the Y-transformation notation, since one can write

$$2(\zeta_1 g_1 + \zeta_2 g_2) - \frac{4\zeta_1\zeta_2(g_1 - g_2)^2}{2\zeta_2 g_1 + 2\zeta_1 g_2 + g_{\min}} \le y(g_e) \le 2(\zeta_1 g_1 + \zeta_2 g_2), \tag{626}$$

$$\left(\frac{3\zeta_1}{4\mu_1} + \frac{3\zeta_2}{4\mu_2}\right)^{-1} \le y(K_e) \le \frac{4}{3}(\zeta_1\mu_1 + \zeta_2\mu_2), \tag{627}$$

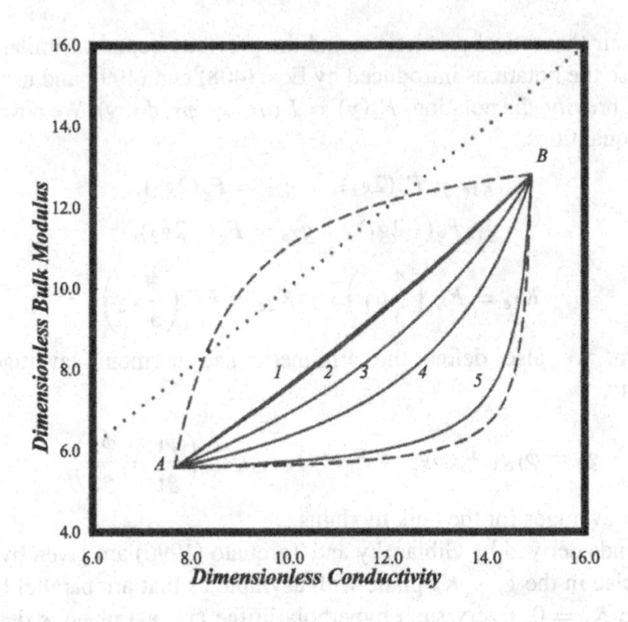

FIGURE 7.16. Cross-property bounds in the conductivity-bulk modulus plane. On the scale of the figure, curves 1 and 2 coincide. The internal region, bounded by curves 1 and 5, represents the bounds for fixed volume fraction. Dashed curves represent the Berryman–Milton bounds, while the dotted line is the bound, $K_e/K_1 \leq g_e/g_1$ (after Gibiansky and Torquato, 1996).

where ζ_1 and $\zeta_2 = 1 - \zeta_1$ are the microstructural parameters defined earlier in this chapter and in Chapters 3 and 4, and $g_{\min} = \min\{g_1, g_2\}$. The idea of Berryman and Milton (1988) was to exclude ζ_1 from these relations in order to obtain bounds on the effective bulk modulus. For a fixed ζ_1, the pair $[y(g_e), y(K_e)]$ lies within the square (626)–(627) in the $y(g_e) - y(K_e)$ plane. When ζ_1 changes within the interval [0,1], this square traces out the set that contains the pair $[y(g_e), y(K_e)]$ of any composite. The bounds of this set are traced out by certain of the corner points of the square (626)–(627). Therefore, the Berryman–Milton cross-property bounds can be obtained by inscribing in the $y(g_e) - y(K_e)$ plane the following five segments of the hyperbolae:

$$\mathrm{Hyp}\left[\left(2g_1, \frac{4}{3}\mu_1\right), \left(2g_2, \frac{4}{3}\mu_2\right), (-g_1, 0)\right], \ \mathrm{Hyp}\left[\left(2g_1, \frac{4}{3}\mu_1\right), \left(2g_2, \frac{4}{3}\mu_2\right), (-g_2, 0)\right],$$

$$\mathrm{Hyp}\left[\left(2g_1, \frac{4}{3}\mu_1\right), \left(2g_2, \frac{4}{3}\mu_2\right), (-g_1, \infty)\right], \ \mathrm{Hyp}\left[\left(2g_1, \frac{4}{3}\mu_1\right), \left(2g_2, \frac{4}{3}\mu_2\right), (-g_2, \infty)\right],$$

$$\mathrm{Hyp}\left[\left(2g_1, \frac{4}{3}\mu_1\right), \left(2g_2, \frac{4}{3}\mu_2\right), (\infty, 0)\right],$$

and the straight line that connects the points $(2g_1, \frac{4}{3}\mu_1)$ and $(2g_2, \frac{4}{3}\mu_2)$. Then, the outermost two of these curves represent the Berryman–Milton bounds, shown as dashed curves in Figure 7.16. Therefore, it is clear that the Gibiansky–Torquato

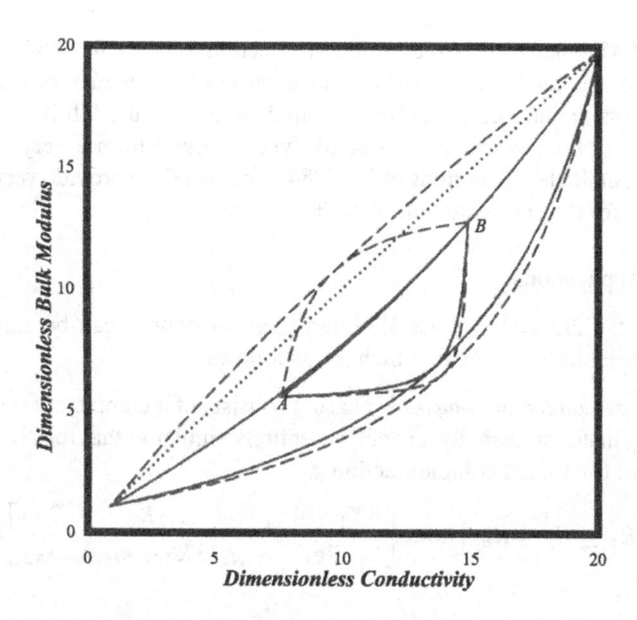

FIGURE 7.17. Cross-property bounds in the conductivity-bulk modulus plane for the composite with arbitrary (the larger external region) and fixed (the internal regions) volume fraction. Solid and dashed curves correspond, respectively, to the Gibiansky–Torquato and Berryman–Milton bounds, while the dotted line is the bound, $K_e/K_1 \leq g_e/g_1$ (after Gibiansky and Torquato, 1996).

bounds are tighter than the Berryman–Milton bounds. Also shown is the bound (601) as the straight dotted line, which is clearly a weak bound.

To obtain the cross-property bounds for arbitrary volume fractions, one should take the union of the sets defined by bounds of Figure 7.16 over the phase volume fraction. This is illustrated in Figure 7.17 for the set of parameters given in (600). The solid curves represent the Gibiansky–Torquato bounds, the dashed curves the Berryman–Milton bounds, and the dotted straight line the upper bound of (601) which coincides with the Gibiansky–Torquato bounds when $g_2/g_1 = K_2/K_1$ and the Poisson's ratios of the phases are zero, i.e., when $2\mu_1/3K_1 = 2\mu_2/3K_2 = 1$. Similar to the 2D systems discussed above, the points A= (g_{1e}, K_{1e}) and B= (g_{2e}, K_{2e}) of the set enclosed by the bounds are optimal because they correspond to the assemblages of coated spheres (Hashin and Shtrikman, 1963) described earlier, as well as to isotropic matrix laminate composites. Curves 2 and 1 of the same figure correspond to the assemblages of doubly coated spheres, or to doubly matrix laminate composites (Schulgasser, 1976). Polycrystals that consist of laminates of two original phases and made of coated cylinder geometries correspond to curve 3.

To see the usefulness of these bounds, consider a random dispersion of identical glass spheres in an epoxy matrix with $\mu_2/\mu_1 = 22.5$, $K_2/K_1 = 10.0$, and $\mu_1/K_1 = 0.33$. Figure 7.6 compares the predictions of the cross-property bounds for the effective bulk modulus with the predictions of Eq. (284), the truncated series expansion of Torquato (1997, 1998), and the three-point lower bound of

Beran and Molyneux (1966), described in Section 7.5.1.4. To obtain the cross-property bounds for the bulk modulus, estimates of the effective conductivity of the same system obtained by numerical simulations were used. It is clear that the cross-property bounds and the Beran–Molyneux bound form a very narrow region that contain the predictions of Eq. (284), and therefore provide very accurate predictions for the elastic moduli of such composites.

7.9.2.2 Applications

Similar to the 2D systems, the 3D cross-property bounds can be simplified in several special limits, some of which are as follows.

Rigid superconducting phase: If phase 2 consists of a completely rigid, superconducting material, then, by an analysis entirely similar to that for 2D materials, one obtains, for a fixed volume fraction ϕ_1,

$$g_e \geq g_u, \quad K_u \leq K_e \leq K_u + \max\left[\frac{3K_1 + 4\mu_1}{9g_1}, \frac{6\mu_2 K_2}{g_2(3K_2 + 4\mu_2)}, \frac{2\mu_2}{3g_2}\right](g_e - g_u), \tag{628}$$

with

$$g_u = \frac{1 + 2\phi_2}{\phi_1} g_1, \quad K_u = \frac{3K_1 + 4\phi_2\mu_1}{2\phi_1}.$$

To obtain the corresponding bounds for any volume fraction, one should replace g_u and K_u by g_1 and K_1.

The important point about the upper bounds of (628) is that, unlike the Hashin–Shtrikman and Berryman–Milton upper bounds, they do not diverge if g_e remains finite, since a finite g_e carries microstructural information about the material in that, it implies that the rigid phase has not formed a sample-spanning cluster, whereas the Hashin–Shtrikman and Berryman–Milton bounds lack such mictrostructural information, and therefore they diverge in this limit.

Insulating phase: If phase 2 is insulating, then, by an analysis similar to that for 2D systems, one obtains, for a fixed volume fraction ϕ_1,

$$\frac{1}{g_e} \geq \frac{1}{g_0}, \quad \frac{1}{K_e} \geq \frac{1}{K_0} + \min\left[\frac{g_1(3K_1 + 4\mu_1)}{6\mu_1 K_1}, \frac{9g_2}{3K_2 + 4\mu_2}, \frac{3g_1}{2\mu_1}\right]\left(\frac{1}{g_e} - \frac{1}{g_0}\right), \tag{629}$$

with

$$g_0 = \frac{2\phi_1 g_1}{1 + \phi_2}, \quad K_0 = \frac{4\phi_1\mu_1 K_1}{4\mu_1 + 3\phi_2 K_1}.$$

For an arbitrary volume fraction ϕ_1, one should replace g_0 and K_0 with g_1 and K_1. While the lower bound is trivial, the upper bound is tighter than the Hashin–Shtrikman bound. Gibiansky and Torquato (1996) discussed many other examples in which their cross-property bounds provided very useful estimates of the elastic moduli of the composites.

Summarizing, cross-property relations and bounds provide accurate estimates of the effective transport properties of composite materials, as they utilize (experimental or numerical) data on one type of transport property, which carry with them some information on the morphology of the material, in order to make prediction for another type of transport property.

Summary

In this chapter we described and discussed several theoretical and computational approaches to the problem of estimating the linear elastic properties of heterogeneous materials. Accurate calculation of several microstructural parameters that appear in the general formulation of rigorous upper and lower bounds on the effective elastic moduli have made such bounds progressively tighter. These bounds are very useful as long as the inclusion phase does not form a sample-spanning cluster. The exact series expansions, as well as the multiple scattering approach, provide accurate estimates of the effective elastic moduli for many composites. As in the case of the effective conductivity and dielectric constant, the effective-medium approximations provide easy-to-calculate estimates of the elastic moduli that, in many cases, are relatively accurate. Such approximations also provide simple interpolation formula for the effective transport properties. Finally, cross-property relations that utilize information on one type of transport property in order to provide estimates for other transport properties are powerful new tools for computing such characteristics of disordered material. Thus, together with the discrete models discussed in Chapter 8, we now have a reasonably deep understanding of the effective linear elastic properties of heterogeneous materials. This claim will be confirmed in Chapter 9 where we will be discussing the applications of these theoretical approaches to modeling mechanical properties of a variety of disordered materials and comparing their predictions with the relevant experimental data.

8 Rigidity and Elastic Properties: The Discrete Approach

8.0 Introduction

In this chapter the discrete models of disordered materials, which were utilized in Chapters 5 and 6 for predicting their effective conductivity, dielectric constant, and other dynamical properties, are significantly extended in order to analyze their linear elastic properties. We consider discrete models of disordered two-phase materials, the discrete counterparts of what we described in Chapter 7, and pay particular attention to elastic percolation networks (that is, networks in which each bond is an elastic element) which are examples of *vector percolation*, as opposed to the percolation models for the effective conductivity and other transport properties of disordered materials that were studied in Chapters 5 and 6, which are *scalar percolation* models. Elastic percolation networks are discrete models of disordered two-phase materials in which the contrast between the elastic properties of the two phases is large. As shown in this chapter, in at least one type of vector percolation, namely, the *rigidity percolation* model, the deformation of the material represents a linear, vector, but *non-local* transport process that possesses complex properties that are absent in many other discrete (or continuum) models of disordered materials. As shown below, the non-locality is due to the fact that, under certain conditions, formation of a sample-spanning cluster of one of the material's phases that can support transport of stress and deformation is not a function of the connectivity of the phase alone. That is, the mere fact that the cluster is sample spanning is not sufficient for it being able to support transmission of stress and strain across the material, as there are long-range correlations in the system. The existence of such long-range correlations gives rise to very interesting phenomena that are completely absent in the scalar percolation models described so far. The applications of these models to predicting the effective mechanical properties of real disordered materials will be described in Chapter 9.

In general we consider disordered lattices or networks in which each bond represents an elastic element, e.g., a spring, with an elastic constant e, the numerical value of which is selected from a probability density function $\psi(e)$. In the most general case, and similar to models of disordered two-phase materials that we considered in Chapter 7, one may consider a two-phase network in which

$$\psi(e) = ph_1(e) + (1 - p)h_2(e), \tag{1}$$

that is, the elastic constants of phases 1 and 2 are selected from probability density functions $h_1(e)$ and $h_2(e)$, respectively. The simplest case when $h_(e) = \delta(e - a)$

and $h_2(e) = \delta(e - b)$, i.e., when the network is a mixture of two types of bonds with elastic constants e_1 and e_2, is the discrete analogue of two-phase composites studied in Chapter 7.

If $h_2(e) = \delta_+(0)$,

$$\psi(e) = ph_1(e) + (1 - p)\delta_+(0), \tag{2}$$

that is, if a fraction p of the bonds have an elastic constant e which is selected from a normalized probability density function $h_1(e)$, while the elastic constant of the rest of the bonds is zero (i.e., such bonds represent voids or pores in the material), then the model is referred to as the *elastic percolation network* (EPN). Since, at this point, we have no reason to believe, *a priori*, that the percolation threshold of this system is the same as that of scalar percolation, i.e., since it is not yet clear whether the percolation threshold for the effective conductivity and elastic moduli are exactly the same, we denote the percolation threshold of the elastic networks by p_{ce} to distinguish it from p_c, the percolation threshold of scalar percolation and the effective conductivity, dielectric constant, etc. As the percolation threshold p_{ce} of the network is approached, all the elastic moduli of the network decrease and eventually vanish at $p = p_{ce}$. Similar to the effective conductivity, dielectric constant and other dynamical properties of the material that were considered in Chapters 4–6, near the percolation threshold, all the elastic moduli of the network follow the power law,

$$E_m \sim (p - p_{ce})^f, \tag{3}$$

where E_m is any of the effective elastic moduli of the network, and f is a critical exponent.

If, on the other hand, $h_1(e) = \delta(e - \infty)$,

$$\psi(e) = p\delta(e - \infty) + (1 - p)h_2(e), \tag{4}$$

that is, if a fraction p of the bonds are superrigid, while the rest of them are comparatively "soft" with a finite elastic constant that is selected from a normalized probability density function $h_2(e)$, then the model is usually called a *superelastic percolation network* (SEPN). As the percolation threshold p_{ce} is approached from below, all the elastic moduli of the network increase and eventually diverge at $p = p_{ce}$, where a sample-spanning cluster of the superrigid bonds is formed. Near p_{ce}, one has

$$E_m \sim (p_{ce} - p)^{-\chi}, \tag{5}$$

where χ is another critical exponent. It should be clear that the exponents f and χ are the elastic analogues of μ and s, the critical exponents that characterize the conductivity of conductor-insulator, and conductor-superconductor percolation composites, which were utilized extensively in Chapters 2–6. Later in this chapter, we will describe and discuss the conditions under which f and χ may be expected to be universal, i.e., to be independent of the morphology of the network *and* the constitutive equation that describes the elastic behavior of the network's bonds.

8.1 Elastic Networks in Biological Materials

Although, as we show in Section 8.4, the elastic networks that we describe in this chapter can be derived rigorously by discretizing the continuum elasticity equations, and even though Chapter 9 will describe many applications of such networks to computing and predicting the elastic properties of real materials, such as polymers and gels that do have a discrete morphology, it should be interesting to motivate the use of elastic networks by describing the structure of such networks in an important class of materials, namely, biological tissues and cells.

Two-dimensional (2D) networks arise in many situations in biological cells. Such networks are attached to the cells' plasma or nuclear membrane. Alternatively, they may be wrapped around the cells, acting as their walls. The human red blood cells contain neither a nucleus nor other cytoskeletal components (for example, microtubules). Instead, they possess only a membrane-associated cytoskeleton that is composed of tetramers of the protein spectrin. The erythrocyte cytoskeleton is highly convoluted *in vivo*, but when stretched by about a factor of about seven in area, it reveals its structure in the form of a relatively uniform 2D network with a coordination number of between four to six. The spectrin tetramers are attached, about midway along their contour length, to the plasma membrane by the protein ankyrin using, as an intermediary, a protein called band 3. Typically, 120,000 tetramers cover the 140 $\mu\mathrm{m}^2$ membrane area of a typical erythrocyte, which corresponds to a tetramer density of 800 $\mu\mathrm{m}^{-2}$. The tetramers are attached to one another at junction complexes that contain actin segments of about 35 nm long. The erythrocyte is covered by about 35,000 junction complexes (that is, nodes in the language of networks) with an average separation (bond length) of about 75 nm.

Lateral cortex of the auditory outer hair cells provides an example of a 2D biological network with low (less than six) coordination number. In such networks the nodes are connected to only four of their neighbors. In guinea pigs, for example, outer hair cells are roughly cylindrical with a diameter of about 10 μm, with the lateral cortex attached to the plasma membrane. The cortex consists of parallel filaments with a thickness of about 6 nm, that are spaced about 60 nm apart, and wound around the cylinder's axis. The filaments are crosslinked at intervals of about 30 nm by thinner filaments. Experiments indicate that the network is elastically anisotropic, as the circumferential filaments are actin while the crosslinks are spectrin.

Although these 2D materials can also be represented as solid sheets, it is better to view them as 2D networks, which are what they really are. Some key questions regarding these networks are as follows.

(1) What happens when filaments, both soft and stiff, are linked together to form an elastic network?

(2) To what extent do the network's elastic (and other) properties represent those of the individual filaments?

(3) What is the role of interactions among the filaments?
(4) How do the elastic properties of biological networks depend on stress and temperature?

It is clear that to answer these questions, proper network models must be developed, in which bending and stretching deformation of the flexible filaments are taken into account. Isotropic materials, and their network representation in terms of a triangular network, possess in-plane deformations involving only two deformation modes—compression and shear. The compression mode preserves the internal angles of the network by uniformly scaling its linear dimensions, whereas the shear mode preserves the network's area, but not its internal angles. We may characterize these two deformation modes with the area compression, or bulk, modulus K and the shear modulus μ. For a triangular network of simple Hookean springs with spring constant α, one has (see Section 8.5 for derivation of these results)

$$K_e = 2\mu_e = \frac{\sqrt{3}}{2}\alpha. \tag{6}$$

These results are applicable only for infinitesimal deformations of an *unstressed* network at zero temperature. As the temperature increases, the network may deviate from its uniform configuration, its elastic moduli change, and therefore Eqs. (6) become increasingly inaccurate. The Hookean spring description of the bonds also has its limitations. For example, at large deformations, real protein filaments cannot be extended indefinitely, whereas a Hookean spring can be. Moreover, a triangular network of Hookean springs at zero temperature can expand without limit if subjected to a tensile stress larger than $\sqrt{3}\alpha$, or collapse under moderate compression.

Three-dimensional elastic networks are also formed in biological materials, such as cells. The filaments of the cytoskeleton and extracellular matrix material form a variety of networks, the chemical and geometrical structure of which can be homogeneous or heterogeneous. To describe such networks, consider, first, a uniform straight rod of length L bent into an arc of a circle of radius R. The energy $\mathcal{H}_{\rm arc}$ required to perform this deformation is given by, $\mathcal{H}_{\rm arc} = K_f L/(2R^2)$, where K_f is called the *flexural rigidity* of the rod. Large values of K_f correspond to stiff rods. The combination $K_f/(k_B T)$, where k_B is the Boltzmann's constant and T is the temperature, is called the *persistence length* of the material. So long as the persistence length is much larger than the counter length, a filament appears relatively straight. However, if the persistence length is much smaller than the contour length, the filaments adopt more convoluted shapes.

Consider, for example, the network of microtubules that are found in the cell. The persistence length of microtubules is of the order of millimeters, so that they bend only gently on the scale of microns, and therefore do not form contorted networks. That is, they are not crosslinked, but rather extend like spikes towards the cell boundary, growing and shrinking with time. However, there exist many microtubule-associated proteins that form links between microtubules, resulting in a bundled structure. In this case, linking proteins are spaced along the microtubule like the rungs of a ladder, tying two microtubules parallel to one another. Such microtubule bundles exist in the long axon of a nerve cell, or in the whip-like flagella extending from some cells. They do not resemble the 2D regular networks of the erythrocyte cytoskeleton or the nuclear lamina described above.

The intermediate filaments of a cell form a variety of networks that may run through the cell, i.e., they may be three dimensional, or have a sheet-like, 2D structure. In some cases, the filaments terminate at a desmosome at the cell boundary; in these and other ways, extended networks of filaments bind the cells into collective structures. The principal components of the internal networks of cells are actin, intermediate filaments and microtubules. Other proteins are present in the 2D or 3D networks that surround the cells, such as the extracellular matrix. These networks tend to be composite structures with heterogeneous molecular composition and organization.

As one may expect, the engineering specifications for a biological elastic network may vary from cell to cell. For example, the elastic networks in red blood cells must be flexible enough to deform in a capillary and then recover their equilibrium shape in a reasonable time frame. Therefore, a cell, and its associated elastic network, may experience a variety of stresses during their everyday operations, and hence the network design must be able to respond to such stresses in an appropriate way. Thus, the question of how the elastic properties of these networks depend on stress or temperature is very important. In addition, we must recognize that there may or may not be permanent crosslinks between the filaments of a biological elastic network. There is a well-defined threshold for the average connectivity of the network above which the network elements are sufficiently linked to their neighbors that a continuous path can be traced along the elements from one side of the network to an opposite side, hence making clear the relevance and importance of percolation to biological elastic networks. It is only when such a continuous path exists that one can speak of the elastic properties of the network. At temperatures above zero, it is believed, but not proven yet, that the elastic moduli are non-zero above the connectivity percolation threshold p_c. However, as we demonstrate in this chapter, certain elastic networks, at least at $T = 0$, do not possess non-vanishing elastic moduli unless the their average connectivity is significantly above what is predicted by scalar percolation, described and utilized in Chapters 2–6.

8.2 Number of Elastic Moduli of a Lattice

Before embarking on describing the discrete models, it is necessary to recall (see Section 7.1) the number of independent elastic moduli that one needs for describing the elastic properties of a 2D or 3D network, since most of the networks that have been used in the numerical simulations or theoretical analyses of elastic properties of composite materials are *not*, in the sense of symmetry group of crystals, isotropic.

As is well-known, and as also described in Section 7.1, the number of independent elastic moduli of isotropic materials is two. One may measure, for example, the Young's modulus Y, the shear modulus μ, and the bulk modulus K, and any two of these moduli suffice for characterizing the elastic properties and rigidity of an isotropic network or material. The triangular network, which has been used extensively in the numerical simulations of elastic properties of materials, is an isotropic lattice. In 3D all topologically-ordered lattices are elastically anisotropic. For the cubic family of lattices, namely, the square, simple-cubic, body-centered (BCC), and face-centered (FCC) lattices, that have also been extensively used in the numerical simulations, three independent elastic moduli are needed in order to characterize their elastic behavior. In terms of the components of the elastic stiffness tensor $\mathbf{C}$ described in Chapter 7, the three main elastic moduli for a d-dimensional material are given by

$$\mu = \begin{cases} C_{44}, & \text{simple shear,} \\ \frac{1}{2}(C_{11} - C_{12}), & \text{pure shear,} \end{cases}$$

$$Y = \frac{(C_{11} - C_{12})(C_{11} + 2C_{12})}{C_{11} + C_{12}},$$

$$K = \frac{1}{d}(C_{11} + 2C_{12}).$$

The equations for the shear modulus apply when the lattice is elastically anisotropic. An example is the square lattice which can have two different shear modes, the simple shear (used in most theoretical discussions) and pure shear. This means that the Young's and shear moduli of the cubic family of lattices are *not* isotropic (Turley and Sines, 1971) in the sense that, they are direction-dependent. Thus, unless otherwise specified, whenever in this chapter we refer to Y and μ, we mean the Young's and shear moduli in the principal directions of the network. On the other hand, the bulk modulus K of the same lattices is isotropic.

8.3 Numerical Simulation and Finite-Size Scaling

Let us briefly discuss numerical simulation of linear deformation of EPNs and SEPNs and computation of their elastic moduli. To calculate the elastic properties of a given lattice, we minimize its total elastic energy $\mathcal{H}$ with respect to $\mathbf{u}_i$, i.e., we write down the equation, $\partial\mathcal{H}/\partial\mathbf{u}_i = 0$, where $\mathbf{u}_i$ is the vector displacement of site i of the lattice. Writing down this equation for every interior node of the network results in a dN simultaneous linear equations for the nodal displacements $\mathbf{u}_i$ of a d-dimensional network of N internal nodes, if no rotational motion is involved (see Section 8.11 when such rotations must also be taken into account). The boundary conditions depend on the elastic modulus to be calculated. For example, to calculate the elastic constant C_{11}, i.e., a tensile experiment, we stretch two opposite faces of the network by a given strain ϵ and impose periodic boundary conditions in the other directions. The resulting set of linear equations are solved by either Gaussian elimination or by an iterative method such as the conjugate-gradient method (see Chapter 9 of Volume II for a detailed discussion of the conjugate-gradient method). The convergence criterion for the iterative method is that for *all* sites i, $|\mathbf{u}_i^{(k)} - \mathbf{u}_i^{(k-1)}|/|\mathbf{u}_i^{(k-1)}| < \delta$, where $\mathbf{u}_i^{(k)}$ is the displacement of site i after the kth iteration, and δ is a small number of the order of $10^{-3} - 10^{-5}$. From the solution of the set of equations for the nodal displacements we calculate the elastic energy $\mathcal{H}$, and hence the elastic modulus $C_{11} = 2\mathcal{H}/\epsilon^2$.

Since throughout this chapter we will frequently speak of the critical exponents that characterize the power-law behavior of the elastic moduli near the percolation threshold, let us briefly recall from Chapter 2 how such exponents are estimated. The most accurate method for estimating most of the critical exponents is based on the finite-size scaling. Recall from Chapter 2 (see Section 2.9) that, according to the finite-size scaling theory, for any effective property P of a network of linear size L, that follows a power law in $|p - p_{ce}|$ at the percolation threshold (of the same network when $L \to \infty$), one can write

$$P(L, p_{ce}) \sim L^{-\tilde{\zeta}} \left[a_1 + a_2 h_1(L) + a_3 h_2(L)\right], \tag{7}$$

where $h_1(L)$ and $h_2(L)$ represent the non-analytical and analytical correction-to-scaling terms, respectively, which are particularly important for small to moderate values of L, and $\tilde{\zeta} = \zeta/\nu$, with ζ being the critical exponent associated with the power-law behavior of P near the percolation threshold, and ν is the critical exponent that characterizes the divergence of percolation correlation length ξ_p at $p = p_c$ [see Eq. (2.32)]. As shown later in this chapter, one may define a correlation length ξ_e for EPNs and SEPNs which also diverges at $p = p_{ce}$ according to a power law which is characterized by a critical exponent ν_e, which may be different from ν, in which case one must use ν_e rather than ν in Eq. (7). For example, an equation similar to (7) has been used for estimating the exponents f and χ by studying the elastic moduli of finite-size networks at p_{ce} (see below).

If one is interested in studying the scaling behavior of the topological properties of the elastic networks (such as the accessible fraction of the elastic bonds, and the backbone of the networks), then the functions $h_1(L)$ and $h_2(L)$ are usually combined into a single correction-to-scaling function $h(L) = a_2 L^{-\omega}$, so that Eq. (7) is rewritten as

$$P(L, p_{ce}) \sim L^{-\tilde{\zeta}} \left(a_1 + a_2 L^{-\omega}\right), \tag{8}$$

where ω is the leading correction-to-scaling exponent, and is presumably universal. On the other hand, if one is interested in investigating the scaling properties of the elastic moduli and estimating the associated exponents f and χ, then the most accurate results are obtained (Sahimi and Arbabi, 1991) if one uses

$$h_1(L) = (\ln L)^{-1}, \tag{9}$$

$$h_2(L) = L^{-1}. \tag{10}$$

8.4 Derivation of Elastic Networks from Continuum Elasticity

We begin our discussion by deriving the elastic networks, beginning with the static limit of the continuum equation of linear elasticity. The linear equation governing elastic equilibrium of a continuum is given by the usual equation,

$$\nabla \cdot \boldsymbol{\sigma} = \mathbf{0}, \tag{11}$$

where $\boldsymbol{\sigma}$ is the stress tensor defined by the usual relation,

$$\boldsymbol{\sigma} = \lambda(\nabla \cdot \mathbf{u})\mathbf{U} + \mu\left[\nabla\mathbf{u} + (\nabla\mathbf{u})^{\mathrm{T}}\right], \tag{12}$$

with μ and λ being the Lamé constants of elasticity, $\mathbf{U}$ the identity tensor, and $\mathbf{u}$ the displacement field (presumed infinitesimal) of a material point from its original position. The Galerkin finite-element approximation (see, for example, Strang and Fix, 1973) transforms the continuous medium into a discrete mesh and Eq. (11) into a discrete system of linear equations that govern the displacements of the grid points in the mesh. If ϕ is a trial function, then one forms the following equation

$$\int_{\Omega} \phi(\nabla \cdot \boldsymbol{\sigma})\, d\Omega = \mathbf{0}, \tag{13}$$

where Ω is the volume of the space over which the integration is carried out. Using the divergence theorem, Eq. (13) is converted to

$$\int_{A} (\mathbf{n} \cdot \phi\boldsymbol{\sigma})\, dS - \int_{\Omega} (\nabla\phi \cdot \boldsymbol{\sigma})\, d\Omega = \mathbf{0}, \tag{14}$$

where A is the external surface of the system, and $\mathbf{n}$ is the unit normal vector pointed outward. We can either take the stress to be zero along the surface, or fix the displacement along the surface by restricting the class of functions used to represent $\mathbf{u}$; hence the above surface integral is eliminated and will not appear anymore. In Galerkin's method $\mathbf{u}$ is represented by

$$\mathbf{u} = \sum_i \mathbf{u}_i \phi_i$$

where $\{\phi_i\}$ is a set of basis functions, and the sum is over all the basis functions in the set. Thus,

$$\boldsymbol{\sigma} = \lambda\left(\sum_i \mathbf{u}_i \nabla\phi_i\right)\mathbf{U} + \mu\left(\sum_i \mathbf{u}_i \nabla\phi_i + \sum_i \nabla\phi_i \mathbf{u}_i\right). \tag{15}$$

Hence, Eq. (14) becomes

$$\int_{\Omega} \left\{\lambda \sum_i (\mathbf{u}_i \cdot \nabla\phi_i)\nabla\phi_j + \mu\left[\sum_i (\mathbf{u}_i \cdot \nabla\phi_j)\nabla\phi_i + \sum_i (\nabla\phi_i \cdot \nabla\phi_j)\mathbf{u}_i\right]\right\} d\Omega = \mathbf{0}. \tag{16}$$

Note that, since Eq. (14) is self-adjoint, Eq. (16) can also be derived from minimizing the associated elastic energy $\mathcal{H}$ with respect to $\mathbf{u}_i$.

The basis $\{\phi_i\}$ is selected such that only a small number of the basis functions are non-zero in any particular region of the system. Those basis functions that overlap in any region are nearest neighbors. Any useful basis can represent a constant exactly; thus, if the basis is normalized, i.e., if

$$\sum_i \phi_i = 1 \tag{17}$$

then, by taking the gradient of both sides of Eq. (17) one obtains

$$\nabla\phi_k = -\sum_{i\neq k} \nabla\phi_i \tag{18}$$

Using Eq. (18) in (16) leads to an alternative form

$$\int_\Omega \sum_{i\neq j} \{\lambda\nabla\phi_j\nabla\phi_i + \mu\left[\nabla\phi_i\nabla\phi_j + (\nabla\phi_i\cdot\nabla\phi_j)\mathbf{U}\right]\cdot(\mathbf{u}_i - \mathbf{u}_j)\}\, d\Omega = \mathbf{0}, \tag{19}$$

which can be written in the compact form,

$$\sum_j \mathbf{W}_{ij}\cdot(\mathbf{u}_i - \mathbf{u}_j) = \mathbf{0}, \tag{20}$$

where the sum is over all the nearest neighbors j of i. The total region of overlap of two nearest-neighbor basis functions usually consists of one or two elements; one or both basis functions are identically zero outside this region. Hence, the surface integrals in

$$\int_\Omega \nabla\phi_i\nabla\phi_j d\Omega = \int_A \mathbf{n}\phi_i\nabla\phi_j dS - \int_\Omega \phi_i\nabla\nabla\phi_j d\Omega$$

and

$$\int_\Omega \nabla\phi_i\nabla\phi_j d\Omega = \int_A \mathbf{n}\phi_j\nabla\phi_i dS - \int_\Omega \phi_j\nabla\nabla\phi_i d\Omega$$

are both zero, so that

$$2\int_\Omega \nabla\phi_i\nabla\phi_j d\Omega = -\int_\Omega (\phi_j\nabla\nabla\phi_i + \phi_i\nabla\nabla\phi_j) d\Omega,$$

and therefore

$$\int_\Omega \nabla\phi_j\nabla\phi_i d\Omega = \int_\Omega \nabla\phi_i\nabla\phi_j d\Omega. \tag{21}$$

Thus, with the aid of Eq. (21) we obtain

$$\mathbf{W}_{ij} = \int_\Omega \left[(\lambda+\mu)\nabla\phi_i\,\nabla\phi_j + \mu(\nabla\phi_i\cdot\nabla\phi_j)\mathbf{U}\right] d\Omega. \tag{22}$$

It is then clear that $\mathbf{W}_{ij}=\mathbf{W}_{ij}^{\mathrm{T}}=\mathbf{W}_{ji}$, and thereby that Eq. (20) can be derived by minimizing the elastic energy $\mathcal{H}$ defined by

$$\mathcal{H} = \frac{1}{2}\sum_{ij}(\mathbf{u}_i - \mathbf{u}_j)\cdot\mathbf{W}_{ij}\cdot(\mathbf{u}_i - \mathbf{u}_j). \tag{23}$$

Since the sum is over all nearest-neighbor pairs ij, $\mathbf{W}_{ij} \neq \mathbf{0}$ can be thought of as forming a bond between i and j, and therefore Eq. (23) describes an elastic network in which the elastic properties of the bonds are distributed according to a statistical distribution function.

8.4.1 The Born Model

To construct the above model on a triangular lattice, which has been used extensively in the computer simulations, the simplest basis functions which are

$$\phi_1 = \frac{1}{2} + x - \frac{y}{\sqrt{3}}, \quad \phi_2 = \frac{1}{2} - x - \frac{y}{\sqrt{3}}, \quad \phi_3 = \frac{2y}{\sqrt{3}}$$

are defined on the equilateral triangle $\{(1/2, 0), (-1/2, 0), (0, \sqrt{3}/2)\}$. With such basis functions we obtain

$$\mathbf{W}_{ij} = \frac{\mu[4(1+\nu_p)\mathbf{R}_{ij}\mathbf{R}_{ji} + (1-3\nu_p)\mathbf{U}]}{4(1-\nu_p)}, \tag{24}$$

and the associated elastic energy is given by

$$\mathcal{H} = \frac{1+\nu_p}{1-\nu_p}\sum_{ij}\mu[(\mathbf{u}_i - \mathbf{u}_j)\cdot\mathbf{R}_{ij}]^2 + \frac{1-3\nu_p}{4(1-\nu_p)}\sum_{ij}(\mathbf{u}_i - \mathbf{u}_j)^2, \tag{25}$$

where ν_p is the Poisson's ratio, μ is the shear modulus, and $\mathbf{R}_{ij}$ is the unit vector along the line from i to j. The standard finite-difference discretization on the triangular mesh leads to the same equation. The first term of Eq. (25) is the energy of a network of central-force springs, i.e., springs which transmit force only in the $\mathbf{R}_{ij}$ direction, but do not transmit shear forces, whereas the second term is a contribution analogous to scalar transport (for example, the power dissipated in conduction), since $(\mathbf{u}_i - \mathbf{u}_j)^2$ represents the *magnitude* of the displacement difference $\mathbf{u}_i - \mathbf{u}_j$. This model may be considered as an analogue of a 3D solid in plane-stress with holes normal to the x-y plane, or as a 2D solid with the Poisson's ratio defined as the negative of ratio of the strain in the y-direction to that in the x-direction, when a stress is applied in the x-direction but none is applied in the y-direction. Results for a 3D solid in plane-strain can be generated from those of this model using the transformation $\nu_p' = \nu_p/(1+\nu_p)$, where ν_p' is the Poisson's ratio for the plain strain (see also Section 7.2).

8.4.2 Shortcomings of the Born Model

Equation (25), which is popularly known as the Born model (see Born and Huang, 1954), suffers from some peculiarities. For example, it is not difficult to show (although at first glance it may not be obvious) that the elastic energy $\mathcal{H}$ defined by Eq. (25) is *not* invariant with respect to arbitrary rigid body rotations, a fundamental requirement for any reasonable model of elastic properties of materials, except in the limit $\nu_p = 1/3$ which reduces the model to a network of central-force springs. When the elastic energy of a system is written in terms of an expansion in the displacement field $\mathbf{u}$, its rotational invariance is not obvious. In this situation, to demonstrate the lack of rotational invariance of the elastic energy, one substitutes an infinitesimal rotation $\boldsymbol{\omega} \times \mathbf{R}_i$ for the displacement vector $\mathbf{u}_i$, where $\mathbf{R}_i$ is the position vector of i. An arbitrary rotation of the solid should not contribute to its energy, but Eq. (25) indicates that, while the contribution of the central-force part would be zero, that of the scalar-like part would not be, and therefore $\mathcal{H}$ is not rotationally invariant. Moreover, although materials do exist that have a Poisson's ratio as high as $1/2$ (and can theoretically be as high as 1 in 2D materials), the model fails to have a strictly positive energy for $\nu_p > 1/3$ and therefore violates the thermodynamic requirement that the potential energy be a minimum at zero strain. Another example of displacements that contribute to the scalar-like portion of the energy of this model, but not to the central-force portion, arises when a significant fraction of the lattice's bonds is removed. In such a lattice a site that is connected to only one bond can have an arbitrary displacement in the direction orthogonal to the direction of the bond without affecting the central-force part of the elastic energy, as can a site which is connected to only a set of two collinear bonds.

In his original formulation of this model, Born inserted the scalar-like part of the elastic energy (25) as a substitute for the many-body, angular and bending terms (see below) that normally arise in describing the elastic properties of materials, because the expansion of such scalar two-body terms is much simpler and more convenient than expanding the many-body terms that they replace. When viewed in this way, the coefficients of the model should be treated as fitting parameters. Hence, let us rewrite Eq. (25) as

$$\mathcal{H} = \frac{1}{2}\alpha_1 \sum_{ij} [(\mathbf{u}_i - \mathbf{u}_j) \cdot \mathbf{R}_{ij}]^2 + \frac{1}{2}\alpha_2 \sum_{ij} (\mathbf{u}_i - \mathbf{u}_j)^2, \tag{26}$$

where α_1 and α_2 now represent two adjustable parameters (without any regard to their theoretical relation to the Poisson's ratio ν_p). When introduced in this way, one may use the Born model for modeling and fitting elastic properties of certain materials, a subject that will be discussed in Chapter 9.

Interestingly, as shown above, the Born model can also be derived by direct discretization of Eq. (12). At first glance, this may be surprising, since the starting point of the derivation is a well-known and well-established Eq. (11), which implies a rotationally-invariant elastic energy for the system. However, a closer inspection of the discretization process reveals the cause of this apparent contradiction. Let $\mathbf{u}_{ij} = \mathbf{u}_i - \mathbf{u}_j$. Then, $\mathbf{u}_{ij}$ has two components, a longitudinal component, $\mathbf{u}_{ij}^{\ell}$, and a transverse one, $\mathbf{u}_{ij}^{t}$. The latter component represents a *first-order rotation* $\delta\theta$ of $\mathbf{R}_{ij}$, i.e., $\delta\theta \simeq u_{ij}^{t}/R_{ij}$, where we have used the magnitude of the vectors. For a pure rotation $\delta\theta$, the distance $\mathbf{r}_{ij}$ between i and j should not change, and therefore $\delta\mathbf{r}_{ij}$, the fluctuations in $\mathbf{r}_{ij}$, must vanish *in all orders*. However, for a purely transverse deviation, $u_{ij}^{\ell} = 0$, $u_{ij}^{t} \neq 0$, and therefore, $\delta r_{ij} = (u_{ij}^{t})^2/(2R_{ij}) \neq 0$, implying that to *second-order* u_{ij}^{t} does not correctly describe the rotation. In effect, what this implies is that the discretization procedure eliminates certain first-order terms that are rotationally invariant, and gives rise to second-order terms that are not. Alexander (1998) provided a rather detailed discussion of this point.

Finally, note that, so long as $\alpha_2 > 0$, the scalar-like term of Eq. (25) or (26) is the dominating contributing factor to the elastic energy $\mathcal{H}$. This implies immediately that, although the Born model is a vector model, it behaves effectively as a scalar model and therefore, for example, the percolation threshold of the model, at which the elastic moduli vanish or diverge, is the same as that of scalar percolation models described in Chapters 2–6, and that near the percolation threshold the elastic moduli of the model follow Eqs. (3) and (5), but with $f = \mu$ and $\chi = s$, where μ and s are the critical exponents that characterize, respectively, the power-law behavior of a conductor-insulator, and a conductor-superconductor mixture near p_c. That is, the power-law behavior of the elastic moduli of the Born model is the same as that of the effective conductivity of the same model. We will return to the Born model in Chapter 9 where we describe applications of vector percolation models.

8.5 The Central-Force Network

Consider now the limit $\nu_p = 1/3$, i.e., a network of central-force, or Hookean, springs. Since the elastic materials that we wish to consider are heterogeneous, the local shear modulus (or Lamé constant) μ varies spatially. Thus, writing $\mu = e_{ij}\alpha/4$ and taking $\nu_p = 1/3$ reduce Eq. (25) to

$$\mathcal{H} = \frac{1}{2}\alpha \sum_{ij} [(\mathbf{u}_i - \mathbf{u}_j) \cdot \mathbf{R}_{ij}]^2 e_{ij} \tag{27}$$

where α is the central-force constant. We can easily compute the elastic moduli of such networks, if no percolation effect is present, i.e., no bond is broken. Consider, as an example, a uniform triangular network (in which all the e_{ij} are equal). Each spring has an unstretched length ℓ_0, and an elastic energy, $\mathcal{H} = \frac{1}{2}\alpha(\ell - \ell_0)^2$, where ℓ is the spring's length under the deformation. We consider a pure compressional mode in which each spring is stretched a small (infinitesimal) amount $\delta \equiv \ell - \ell_0$ away from its equilibrium length ℓ_0. The change in the energy is then given by

$$\Delta\mathcal{H} = \frac{3}{2}\alpha\delta^2, \tag{28}$$

where the fractor 3 is due to the fact that the number of bonds in a triangular network is three times the number of nodes. The network area per node is, $S = \frac{\sqrt{3}}{2}\ell_0^2$, and therefore the change in the energy density of the network, as a result of the deformation, is given by

$$\Delta V = \frac{\Delta\mathcal{H}}{S} = \sqrt{3}\alpha\left(\frac{\delta}{\ell_0}\right)^2. \tag{29}$$

Because the deformations are uniform, and since the displacements in the y-direction are independent of those in the x-direction, we have, $u_{xx} = u_{yy}$, and $u_{xy} = 0$. Therefore,

$$\Delta V = \frac{1}{2}K(u_{xx} + u_{yy})^2 + \mu\left[\frac{1}{2}(u_{xx} - u_{yy})^2 + 2u_{xy}^2\right] = 2K\left(\frac{\delta}{\ell_0}\right)^2, \quad (30)$$

which, when compared with Eq. (29), yields

$$K = \frac{\sqrt{3}}{2}\alpha, \quad \text{triangular network,} \quad (31)$$

which was already mentioned in Eq. (6). Note that, the first equation in (30) is easily obtained by writing the argument of the summation in Eq. (27) in a general (x, y) coordinate, and dividing the result by S, the area per node of the network.

Similar calculations can be carried out for determining the shear modulus of the network. Consider a triangle with a base that is parallel to the lines of this page. If we move the top node of the triangle by an amount δ in the x-direction (the horizontal direction), then, to first order in δ, the left spring of the triangle is lengthened by $\frac{1}{2}\delta$, while the right spring is shortened by the same amount. Thus, to lowest order in δ, the change in the elastic energy of either spring is $\frac{1}{8}\alpha\delta^2$, and since the base spring remains intact, the total energy change is $\frac{1}{2}\alpha\delta^2$, and the change in the energy density is, $(2\sqrt{3})^{-1}\alpha(\delta/\ell_0)^2$. In the simple shear mode, the y-coordinate of the nodes and the distance between nodes in the x-direction remain unchanged, and therefore, $u_{xx} = u_{yy} = 0$. On the other hand, each successive row of nodes is displaced by δ in the positive x-direction for each increase $\frac{\sqrt{3}}{2}\ell_0$ in the y-direction, and thus $u_{xy} = 3^{-1/2}\delta/\ell_0$. Hence, from the first equation of (30), we obtain, $\Delta V = (2\mu/3)(\delta/\ell_0)^2$, and therefore

$$\mu = \frac{\sqrt{3}}{4}\alpha, \quad \text{triangular network,} \quad (32)$$

also mentioned in Eq. (6). Note that Eqs. (31) and (32) imply that, for a uniform triangular network, the Poisson's ratio $\nu_p = (K_e - \mu_e)/(K_e + \mu_e) = 1/3$.

We can carry out the same type of calculations for the square network. The change in the elastic energy density ΔV as a result of deformation of a square cell is given by

$$\Delta V = \frac{1}{2}K(u_{xx} + u_{yy})^2 + \frac{1}{2}\mu_p(u_{xx} - u_{yy})^2 + 2\mu_s u_{xy}^2, \quad (33)$$

where μ_p and μ_s are the shear moduli in pure and simple shear, respectively. However, if there is no restoring force between adjacent network elements, i.e., if the energy does not depend even implicitly upon the angle between neighboring bonds, then, a square cell and its deformed configurations in simple shear have the same energy. That is, their deformation is done at no cost to the total energy of the system, and therefore the shear moduli $\mu_s = 0$, even when all the bonds of the network are connected. A similar reasoning would then yield

$$K = \mu_p = \frac{1}{2}\alpha. \tag{34}$$

Since the square network is elastically anisotropic, its Poisson's ratio is not unique and depends on how it is defined.

Calculations of similar quantities in 3D regular lattices, all of which are elastically anisotropic, is slightly more complex. The change in the elastic energy density for the cubic family of lattices is given by

$$\begin{aligned}\Delta V = \frac{1}{2}C_{11}(u_{xx} + u_{yy} + u_{zz})^2 + C_{12}(u_{xx}u_{yy} + u_{xx}u_{zz} + u_{yy}u_{zz}) \\ +2C_{44}(u_{xy}^2 + u_{xz}^2 + u_{yz}^2).\end{aligned} \tag{35}$$

If ℓ_0 is the initial length of the bonds, then, straightforward calculations indicate, for example, that

$$K = \begin{cases} \frac{1}{3\ell_0}\alpha, & \text{simple-cubic lattice,} \\ \frac{1}{\sqrt{3}\ell_0}\alpha, & \text{BCC lattice,} \\ \frac{2\sqrt{2}}{\sqrt{3}\ell_0}\alpha, & \text{FCC lattice.} \end{cases} \tag{36}$$

A simple-cubic lattice does not possess a shear modulus in simple shear. We emphasize that Eqs. (31), (32), (34) and (36) are valid at zero temperature and when the external stress is zero (infinitesimally small). Later in this chapter, we will come back to this problem when the temperature of the system and the stress are both finite.

8.6 Rigidity Percolation

If, in Eq. (27), the elastic constant e_{ij} is distributed according to the distribution (2) or (4), then one obtains an elastic percolation network (EPN) [when e_{ij} is distributed according to the distribution (2)] or a superelastic percolation network (SEPN) [when e_{ij} is distributed according to the distribution (4)]. Percolation on such networks of springs is called *rigidity percolation*, although percolation in more general elastic networks (see below) has also been referred to as such.

Such networks are of both theoretical and practical interest. In addition to the biological materials described in Section 8.1, in many engineering problems, structures composed of bars or beams connected at nodes, which are called trusses, acquire their rigidity mainly from the tensile and compressive stiffness of the beams, and these are central-force type of contributions. In contrast, those in which angular forces, e.g., covalent bonds at the molecular level, are the most important are usually referred to as frames. For example, in the absence of friction between the particles of a granular packing, which is a reasonable model of unconsolidated porous materials (such as powders), the mechanical behavior of the packing is similar to those of rigidity percolation. It is not difficult to see that rigid systems in which angular forces dominate their behavior support an applied stress as long as they are simply connected. In contrast, central-force systems require higher degrees of connectivity, and because of their importance to granular media, glasses, and gels, they must be carefully studied.

It was established in mid 1980s (Feng and Sen, 1984; Jerauld, 1985) that the bond percolation threshold p_{ce}^B of rigidity percolation is much larger than p_c^B, the corresponding threshold of scalar percolation which also represents the threshold of random resistor networks. The reason for the difference is clear: On a central-force percolation network there may be a cluster of intact springs which, although sample-spanning, is not rigid, i.e., its deformation does not cost any energy, and hence the deformation does not change the elastic energy of the system. Because of this phenomenon, it is necessary to precisely define rigidity of lattices. This is done in the next section.

8.6.1 *Static and Dynamic Rigidity and Floppiness of Networks*

Consider, as an example, a square lattice of central-force (Hookean) springs. As shown in Section 8.5, this lattice has no *shear rigidity* in the plane (in simple shear), as it can be sheared without changing the lengths of any bond (i.e., without changing the elastic energy) and cannot maintain shear stresses. We refer to such a state as *floppy*. Moreover, one can rotate the horizontal edges of all the squares belonging to the same column by one angle and the vertical edges in each row by another angle, which result in a network configuration that can also be obtained from the original square lattice by a continuous deformation of its boundaries which is not a simple shear. On the other hand, if one fixes the external boundary of this network in any one of the set of allowed shapes, then, the entire network will be completely rigid. Thus, the position of *all* the interior sites are determined by fixing the boundary. In this sense, the floppiness of the square network in central-force percolation is *macroscopic*, because its boundaries are floppy. It is also important, from a practical view point, to distinguish between macroscopic and *microscopic* or local floppiness, as the latter refers to floppiness of a small portion of the system. Thus, although a material may be macroscopically rigid, some local regions of it can still be floppy. In this case, some of the microscopic degrees of freedom remain free and are not determined by the bonding structure, even if the external boundaries are fixed.

Microscopic floppiness can appear in a trivial way when the positions of some of the sites (or atoms at the molecular level) are not completely determined by the bonds that are attached to them. Therefore, one can introduce the concept of *local rigidity*, which is the requirement that the position of every site be fixed completely by the positions of the other sites to which it is directly bonded. This is then a necessary, but not sufficient, requirement for microscopic rigidity.

Based on such considerations, one may introduce the concept of a *geometrically-rigid network* as one in which the configuration of the sites cannot be deformed without changing the lengths of at least some of the bonds. According to this definition then, the network rigidity depends only on its bonding structure, i.e., its topology. The question of whether a given network is geometrically rigid is equivalent to asking whether the configuration of a network of N sites can determine all the $\frac{1}{2}N(N-1)$ distances r_{ij} between the sites. Although this question is conceptually simple, it is, nevertheless, a quite complex problem to determine the rigidity of a given network based on geometrical criteria alone. In fact, to our knowledge, there is no known way of determining the rigidity of 3D network, based solely on geometrical criteria. We refer to networks that do not satisfy the geometrical rigidity requirements as *geometrically floppy*.

For 2D systems exact conditions relating rigidity to connectivity have existed for a long time (Laman, 1970; Lovasz and Yemini, 1982; Recski, 1992; Hendrickson, 1992). Laman's theorem states that,

A random lattice consisting of N Sites and B bonds so that 2N-B=3 is rigid if and only if there is no subset of the lattice, consisting of n sites connected by b bonds for which the relation 2n-b=3 is violated.

This condition is not difficult to understand. Each node of a 2D lattice has two degrees of freedom, which are two translations, and each bond is a constraint. In the rigidity condition, $2n - b = 3$, with the 3 indicating the fact that in 2D a rigid body has 3 degrees of freedom, which are two translations and one rotation. This constraint can be used for constructing 2D central-force percolation networks and determining whether they are rigid *without solving for the stress or strain distribution in the network*. However, this constraint is restricted to 2D systems, and its 3D version, $3N - B = 6$, is known (Hendrickson, 1992) not to be a general necessary and sufficient condition for rigidity. We will come back to this theorem and its application later in this chapter.

In practice, however, geometrical criteria are very difficult, if not impossible, to utilize for checking the rigidity of a material, since such criteria are not directly accessible to experiments. What can be verified experimentally, such as vanishing of a shear modulus, or of the velocity of sound in some directions, and the presence of local vibration modes with no harmonic restoring forces, are all related to *dynamical stability* of the material which, in turn, gives rise to the concept of *dynamical floppiness*. In turn, dynamical floppiness is related to the *stability* of the expansion of the energy $\mathcal{H}(\{\mathbf{r}\})$ of the material around reference configuration $\{\mathbf{R}\}$.

Thus, consider a network (material) of n sites (particles) and energy $\mathcal{H}(\{\mathbf{r}\})$. The reference position of the sites is the set $\{\mathbf{R}\} = \{\mathbf{R}_1, \mathbf{R}_2, \cdots, \mathbf{R}_N\}$. Then, expand the energy in the Nd dimensional space of positions $\{\mathbf{r}\} = \{\mathbf{r}_1, \mathbf{r}_2, \cdots, \mathbf{r}_N\}$ around the reference state. The problem of rigidity of the material reduces to the question of stability of this expansion. The simplest way of investigating this stability is by carrying out a linear stability analysis. Hence, consider a set of displacement deviations, $\{\mathbf{d}\} = \{\mathbf{d}_1, \mathbf{d}_2, \cdots, \mathbf{d}_N\}$, where $\mathbf{d}_i$ represents the deviation of the site or atom i from its equilibrium position $\mathbf{R}_i$. The harmonic expansion of the energy is then written as

$$\mathcal{H}_h = [\mathbf{d}]\mathbf{D}[\mathbf{d}], \tag{37}$$

where $\mathbf{D}$ is the $Nd \times Nd$ dynamical matrix for the expansion, which is defined by

$$\mathbf{D}_{ii} = [\nabla_i \cdot \nabla_i]\mathcal{H}(\{\mathbf{r}\})|_{\{\mathbf{R}\}}, \tag{38}$$

$$\mathbf{D}_{ij} = [\nabla_j \cdot \nabla_i]\mathcal{H}(\{\mathbf{r}\})|_{\{\mathbf{R}\}}. \tag{39}$$

Both $\mathbf{D}_{ii}$ and $\mathbf{D}_{ij}$ are d-dimensional matrices.

The dynamical matrix $\mathbf{D}$ has $Nd = N_1 + N_2$ eigenvalues, of which $N_1 = \frac{1}{2}d(d+1)$ correspond to the rigid-body degrees of freedom of the system, and therefore must vanish because of the macroscopic translational and rotational invariance of the system. The remaining N_2 eigenvalues λ_α are the force constants for the internal harmonic eigenmodes $\{\mathbf{d}^\alpha\}$ that satisfy, $\mathbf{D}[\mathbf{d}^\alpha] = \lambda_\alpha[\mathbf{d}^\alpha]$, with $[\mathbf{d}^\alpha] = \{\mathbf{d}_1^\alpha, \mathbf{d}_2^\alpha, \cdots, \mathbf{d}_N^\alpha\}$, where $\mathbf{d}_i^\alpha$ is the d-dimensional vector which describes the deviations of atom or site i in mode α. If the equilibrium state $\{\mathbf{R}\}$ is dynamically stable, all the force constants must be positive, $\lambda_\alpha > 0$, for all values of α. For geometrically floppy networks there are restrictions on the dynamical matrix $\mathbf{D}$ that may reduce its rank and cause some of its eigenvalues to vanish. Since such networks have free degrees of freedom, it is natural to assume that there are also free eigenmodes $\{\mathbf{d}^\beta\}$ corresponding to the free degrees of freedom. There are no restoring forces for such free modes, and therefore $\lambda_\beta = 0$ for all free modes β. This implies immediately that the equilibrium reference configuration $\{\mathbf{R}\}$ of floppy networks is not stable with respect to such free eigenmodes. In this case one has large fluctuations and anharmonic effects become important, some of which were discussed in Section 6.8.1.

Having defined precisely static and dynamical rigidity and floppiness of elastic networks, we can now resume our discussion of the properties of rigidity percolation.

8.6.2 *The Correlation Length of Rigidity Percolation*

We already mentioned that for rigidity percolation a correlation length ξ_e is defined that, similar to the correlation length ξ_p of scalar percolation, diverges as p_{ce} is approached:

$$\xi_e \sim (p - p_{ce})^{-\nu_e}. \tag{40}$$

Early computer simulations of 2D rigidity percolation (Sahimi and Goddard, 1985; Lemieux *et al.*, 1985, Arbabi and Sahimi, 1988a) yielded, $\nu_e \simeq 1.1$ for bond percolation. Moreover, these simulations, based on the finite-size scaling analysis (see Section 8.3), indicated that for rigidity percolation on the triangular network one has, $f/\nu_e \simeq 1.45$, an interesting result due to the fact that it differed from the critical exponent μ of the effective conductivity of percolation networks for which, $\mu/\nu \simeq 0.973$ (see Table 2.3). These studies did not hint at the possibility of the existence of significant corrections-to-scaling [see Section 8.2 and Eq. (7)].

However, Roux and Hansen (1988) and Hansen and Roux (1988, 1989) used a transfer-matrix method, similar to what was described in Chapter 5 (see Section 5.14.2; see also Section 8.12) for computing the effective conductivity, and studied bond percolation on the triangular network with central forces. They showed that the location of the percolation threshold p_{ce} and the scaling properties of the central-force model are extremely sensitive to whether one accounts for the effect of the correction-to-scaling. They estimated that for the triangular network, $p_{ce}^B \simeq 0.642$ and $f/\nu_e \simeq 3.0$. Although their estimate of p_{ce}^B is only 1.2% less than the earlier estimates, it apparently causes a dramatic shift in the value of f/ν_e from 1.45 to 3.0. Moreover, it appears that while the corrections-to-scaling are large at $p_{ce}^B \simeq 0.642$, they are practically non-existent at $p \simeq 0.65$, which is in agreement with the findings of the early simulations and explains why a much lower value of f/ν_e had been found by these simulations. It is thus clear that accurate estimation of the critical exponents of rigidity percolation requires very accurate estimates of the percolation thresholds, and also taking into account the effect of corrections-to-scaling. We now discuss these aspects of the problem.

8.6.3 The Force Distribution

A fruitful way of investigating an EPN is (Sahimi and Arbabi, 1989) by studying its force distribution (FD)—the distribution of the forces that the intact (uncut) bonds of the network suffer. In order to calculate the FD, one imposes a given boundary condition on the network and determines the nodal displacements $\mathbf{u}_i$ from which the total force F_i, exerted on a bond i, and thus its distribution, are calculated. Of particular interest are the moments of the FD defined by

$$M(q) = \sum_i n_{F_i} F_i^q , \tag{41}$$

where n_{F_i} is the number of bonds that suffer a force with magnitude F_i. It should be clear to the reader that the FD for EPNs is the analogue of the distribution of currents in random resistor networks described and studied in Chapter 5 (see Section 5.16). In EPNs near p_{ce}, the moments $M(q)$ follow the power law,

$$M(q) \sim (p - p_{ce})^{\tau(q)} \sim \xi_e^{-\tau(q)/\nu_e} , \tag{42}$$

where, similar to the exponents that characterize the moments of the current distribution in random resistor networks, all the $\tau(q)$ are distinct (Sahimi and Arbabi, 1989). For length scales $L \ll \xi_e$ one should replace ξ_e in Eq. (42) by L, and therefore

$$M(q) \sim Ł^{-\tilde{\tau}(q)}, \tag{43}$$

where $\tilde{\tau} = \tau/\nu_e$. One includes only non-zero values of F_i in Eq. (41), and therefore $M(0)$ is simply the total number of bonds that suffer a non-zero force, i.e., the total number of bonds in the backbone of the EPN, and hence $-\tilde{\tau}(0)$ is simply the fractal dimension D_{bb}^e of the backbone of the EPN, which is not necessarily the same as D_{bb}, the backbone of scalar percolation or random resistor networks. On the other hand, $M(1)$ is the average force that a bond suffers, and $M(2)$ is proportional (and, in some cases, is exactly equal) to the elastic moduli of the network and hence, $\tau(2) = f$. Therefore, each moment of the FD provides some insight into the properties of EPNs (or SEPNs).

8.6.4 *Determination of the Percolation Threshold*

The macroscopic floppiness of the square network in central-force models can be extended to any d-dimensional cubic network. The floppiness also implies that, for a d-dimensional simple-cubic network one has, $p_{ce}^B = p_{ce}^S = 1$, where p_{ce}^S is the site percolation threshold of the system, whereas for scalar percolation in the square network $p_c^S \simeq 0.5928$ and $p_c^B = 1/2$, and for the simple-cubic network $p_c^S \simeq 0.3116$ and $p_c^B \simeq 0.2488$ (see Tables 2.1 and 2.2). Therefore, a meaningful study of rigidity percolation is restricted to certain lattices, e.g., the triangular and BCC networks, which are regular lattices, and also random lattices discussed below. Early simulations of rigidity percolation on the triangular network (Feng and Sen, 1984; Jerauld, 1985; Sahimi and Goddard, 1985) indicated that, $p_{ce}^B \simeq 0.65$, as compared with $p_c^B = 0.347$.

Since it appears that accurate estimation of the critical exponents that characterize various properties of rigidity percolation near p_{ce} depends sensitively on precise estimates of p_{ce} itself, the important question is how to estimate p_{ce} with high precision. There are currently two methods for precise estimation of p_{ce}, one of which is applicable to both 2D and 3D networks, while the second one can be utilized only for 2D systems. Moreover, mean-field like arguments and effective-medium approximations to be described below, can also provide estimates of the percolation thresholds, although such estimates are not accurate enough for use in finite-size scaling analyses of the critical behavior of central-force networks near p_{ce}.

8.6.4.1 Moments of the Force Distribution

The first method of obtaining precise estimates of p_{ce} is based on the moments of the FD. Arbabi and Sahimi (1993) proposed that if, in a network of linear size L with a fraction p of intact springs, one calculates the ratio $r = M(2)/M(1)$ for various values of L and p, where $M(1)$ and $M(2)$ are the first two moments of the FD distribution, then at the true value of p_{ce} a plot of $\ln r$ versus $\ln L$ should be a straight line since, if one writes down an equation similar to (7) for the moments $M(q)$, then, at the true value of p_{ce} the contributions of the corrections-to-scaling to both moments should be of the same order of magnitude and would therefore cancel one another. In principle, this should be true for any ratio, $r = M(q)/M(q-1)$. However, for $q > 2$, the moments of the FD are subject to large uncertainties. This idea was tested (Arbabi and Sahimi, 1993) on the triangular network with modest network sizes, $L = 25$, 35, 45 and 55, and a few values of p, ranging from 0.636 to 0.65. It was found that a plot of $\ln r$ versus $\ln L$ is a straight line only if $0.640 < p < 0.642$; see Figure 8.1. Thus, p_{ce}^B is estimated to be

$$p_{ce}^B \simeq 0.641 \pm 0.001, \quad \text{triangular lattice,} \tag{44}$$

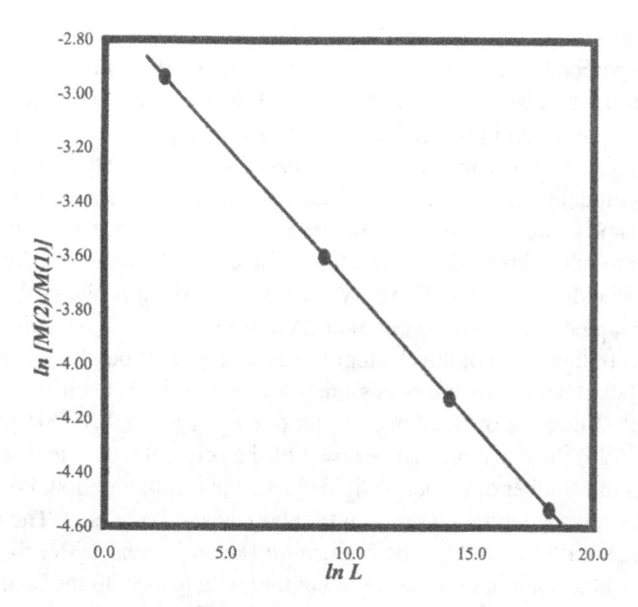

FIGURE 8.1. Dependence of the ratio $M(2)/M(1)$ of the moments of the force distribution in rigidity percolation on the linear size L of the triangular lattice at $p_{ce}^B \simeq 0.641$ (after Arbabi and Sahimi, 1993).

in excellent agreement with the estimate of Roux and Hansen mentioned above. The same method was also used (Arbabi and Sahimi, 1993) for estimating the site percolation threshold p_{ce}^S of the central-force triangular network with the result being

$$p_{ce}^S \simeq 0.713 \pm 0.002, \quad \text{triangular lattice.} \tag{45}$$

A BCC lattice of central-force springs is also rigid with respect to a transverse shear, and also a volume change, and therefore it should possess a well-defined $p_{ce} < 1$. Thus, the moment method was also used to study bond percolation on the BCC lattice with central forces (Arbabi and Sahimi, 1993). As in the case of the triangular lattice, $M(1)$ and $M(2)$ were calculated for various network sizes L in the range, $0.72 \leq p \leq 0.75$. It was found that a plot of $\ln[M(2)/M(1)]$ versus $\ln L$ for various values of p would produce a straight line if $p_{ce}^B \simeq 0.737$, but not for $p < 0.735$ or $p > 0.739$. Therefore,

$$p_{ce}^B \simeq 0.737 \pm 0.002, \quad \text{BCC lattice,} \tag{46}$$

much larger than $p_c^B \simeq 0.1795$ for scalar bond percolation on the BCC network (see Table 2.2). Note that, with pure elongational shear along the cubic axis, the BCC lattice would have no restoring elastic constant.

8.6.4.2 The Pebble Game

Estimating the rigidity percolation thresholds by the moments of its FD is a general method which is applicable to any lattice of any dimensionality. However, for 2D lattices, Laman's theorem described above can be fruitfully used for estimating the percolation threshold of central-force networks. This theorem was first used by Prunet and Blanc (1986) and Wang (1988) to count the number of rigid animals, i.e., large rigid percolation clusters below p_{ce} with radii that are *larger* than the rigidity correlation length ξ_e (see also Chapter 9 for application of lattice animals to modeling of branched polymers). Laman's theorem was also used by Moukarzel and Duxbury (1995) and Jacobs and Thorpe (1995, 1996) for determining the percolation threshold of 2D central-force, topologically-random lattices. For example (Moukarzel and Duxbury, 1995), for a given p, the probability that a site is occupied, one finds the configuration of the cluster of occupied sites and from it the configuration of the cluster of occupied (uncut) bonds. One starts with an empty lattice and adds the bonds one by one at a time. Each rigid cluster is a body with 3 degrees of freedom, so that one must generalize (Moukarzel and Duxbury, 1995) the constraint $2n - b = 3$ of the original lattice to $3n_b - b = 3$, where n_b is the number of bodies or rigid clusters in a configuration. For example, Figure 8.2 shows a configuration with two bodies and three bars. The algorithm can be made very efficient by the realization (Hendrickson, 1992) that one can determine whether a bond (or bar) is *redundant* with respect to the bonds that are already in the lattice. If the bonds are central-force (Hookean) springs, a redundant bond violates $3n_b - b = 3$, leading to internal stresses in the lattice. Thus, if a redundant spring is added to a cluster of bonds that is already *rigid* (that is, it satisfies $3n_b - b = 3$), the algorithm identifies the bonds that become internally stressed when the extra spring is added, in which case the bonds are given the same label as that of the cluster. In this way one determines whether a given central-force percolation network is rigid.

Jacobs and Thorpe (1995) used a somewhat different algorithm to implement Laman's theorem, calling it a *pebble game*. A *free* pebble is one that is on a site, while an *anchored* one covers a bond. A free pebble represents a single motion that a site can execute. If two additional free pebbles are found at a different site, then

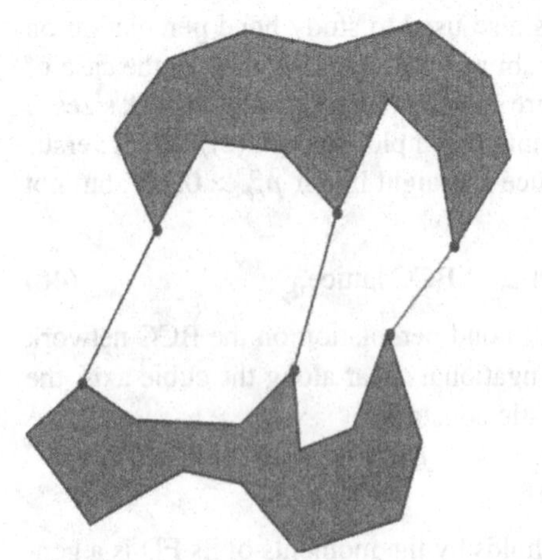

FIGURE 8.2. A rigid configuration with two bodies and three bars (after Moukarzel and Duxbury, 1995).

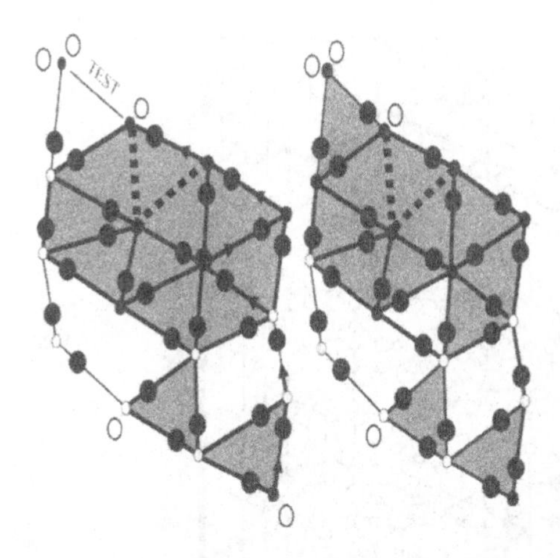

FIGURE 8.3. The pebble game in 2D. Redundant (independent) bonds are shown with dashed (solid) lines that are not (are) covered by a pebble. Large (filled, open) particles indicate (anchored, free) pebbles on (bonds, sites), while the small (filled, open) particles indicate sites belonging to (one, more than one) rigid cluster. Shaded areas are 2D rigid bodies. Left: Five free pebbles indicate five floppy modes until a new bond is added and tested for independence, while a 4th free pebble is found via the path traced by arrows. Right: The added bond is independent and hence covered (after Jacob and Thorpe, 1995).

the distance between these sites is not fixed. If we place a bond between the two sites, it will constrain their distance of separation. This independent constraint is recorded by anchoring one of the 4 free pebbles to the bond, which must always remain covered. In this way, one constructs an efficient algorithms for determining whether a cluster of Hooke's springs is rigid.

Thus, the pebble game for rigidity percolation is as follows. One starts with a network of n sites. Nearest-neighbor bonds are randomly selected and tested. The pebbles are then shuffled around the network in order to free two pebbles at each site at the ends of the test bonds. In the shuffling, an anchored pebble can be released from covering a bond by anchoring a neighboring free pebble to that bond. For example, Figure 8.3 shows an example of how pebbles are shuffled. It is always possible to free three pebbles, because they correspond to the rigid body motion of that bond. A redundant bond is identified when the search for the fourth pebble results in a closed loop back to the sites at the ends of the test bond, indicating that the distance between the incident sites is already fixed, and the redundant bond is not covered. When a redundant bond is found, the set of sites searched in the failed trial to free the fourth pebble defines an overconstrained region. If a bond is redundant, then the sites are mutually rigid. A systematic search is then undertaken for mapping out all rigid clusters after building the network. Figure 8.4 shows a section of a large network after applying the pebble game.

An important aspect of this method is that it should be used with random lattices which are called (Jacobs and Thorpe, 1995) *generic lattices*. These are lattices of a given topology but with no special symmetry. For example, one can take a regular lattice and randomly displace each site's location by a small amount, thereby eliminating the presence of connected collinear and parallel bonds, such as those that are in the triangular lattice. For a generic network built on the triangular lattice, the bond percolation threshold was estimated to be (Jacobs and Thorpe, 1995)

$$p_{ce}^{B} \simeq 0.6602 \pm 0.0003, \tag{47}$$

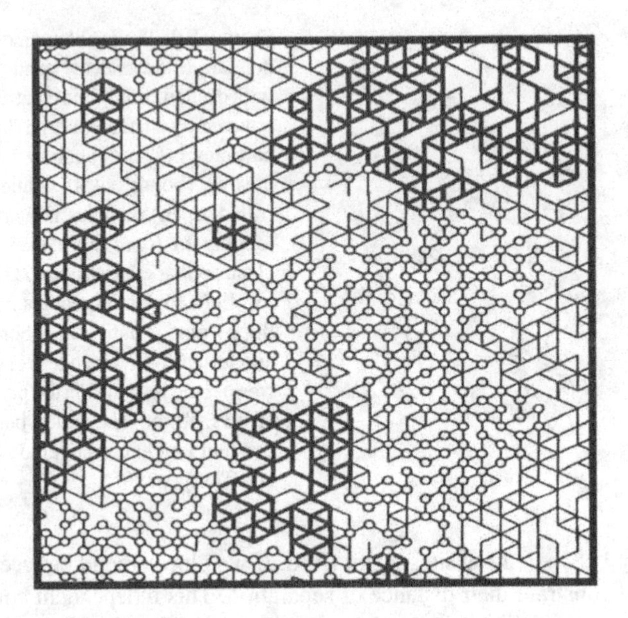

FIGURE 8.4. The structure of a generic diluted triangular network in rigidity percolation. Heavy lines are the overconstrained bonds, while open circles indicate the pivot between two or more rigid bodies (after Jacob and Thorpe, 1995).

which is somewhat larger than that of the triangular network given by (44). The advantage of this method is that it allows one to use huge lattices in the simulations. Lattices as large as 1000×1000 were used by Moukarzel and Duxbury (1995) and Jacobs and Thorpe (1995) for investigating percolation properties of central-force networks. The disadvantage of this method is that it is restricted to 2D lattices, whereas most practical applications are concerned with 3D systems.

8.6.4.3 Constraint-Counting Method

The percolation threshold of central-force networks can also be estimated by a constraint-counting argument (Feng *et al.*, 1985). Suppose that p is the fraction of uncut springs. When p is small, the central-force network consists of disconnected floppy regions, and hence has many zero-frequency modes, the number of which is given by the degrees of freedom $N_f = Nd$, minus the number of constraints $N_c = ZNp/2$, where N is the number of nodes of the lattice, and Z is its coordination number. Thus, the fraction f_0 of zero-frequency modes is given by

$$f_0 = \frac{Nd - ZNp/2}{Nd} = 1 - \frac{Zp}{2d}, \tag{48}$$

which vanishes at $p = p_{ce}^B$ and therefore

$$p_{ce}^B = \frac{2d}{Z}. \tag{49}$$

As we show shortly, Eq. (49) is identical with the bond percolation threshold that the effective-medium approximation predicts for the central-force networks. The constraint-counting method has also been used for determining the percolation threshold of a variety of other EPNs (Schwartz *et al.*, 1985). Note, however, that if the same constraint-counting arguments are used for estimating the site percolation threshold of central-force networks (Thorpe and Garboczi, 1987), it results again in Eq. (49), i.e., the same as the bond percolation threshold, and thus in this sense the constraint-counting arguments fail.

Equation (49) predicts that for the triangular network $p_{ce}^B = 2/3 \simeq 0.666$, which should be compared with the numerical estimate given by (44), a 3.7% difference. It predicts, for the BCC lattice, $p_{ce}^B = 3/4 = 0.75$, which should be compared with the numerical estimate given by (46), a 1.7% difference. Thus, as far as predicting the percolation threshold of central-force networks is concerned, *unlike* scalar percolation (e.g., resistor networks), the performance of mean-field-like arguments, such as the constraint-counting method, *improves* as the dimensionality of the system increases.

8.6.5 Mapping Between Rigidity Percolation and Resistor Networks

Let us point out that there is one central-force model with a percolation threshold which is the same as that of scalar percolation. Roux and Hansen (1987) devised a triangular lattice in which the bonds in one preferred direction are infinitely rigid. In the other two directions (that make $+60°$ and $-60°$ angles with the preferred direction), a fraction p of the bonds, selected at random, are also infinitely rigid, while the rest are Hookean springs with a finite strength. Roux and Hansen showed that the bond percolation threshold of this lattice at which the elastic moduli diverge is exactly equal to bond percolation threshold of a square lattice in scalar percolation, i.e., $p_{ce}^B = 1/2$, and, moreover, the exponent χ defined by Eq. (4) is $\chi = s \simeq 1.3$, where s is the exponent that characterizes the power-law divergence of the effective conductivity of a conductor-superconductor network at p_c.

The model of Roux and Hansen is, however, pathological. In general, the rigidity percolation thresholds are not the same as those of the scalar percolation thresholds, and therefore an interesting question is whether central-force percolation networks can be mapped onto random percolation networks or, more specifically, random resistor networks. This is in fact possible. Tang and Thorpe (1987,1988) demonstrated how this mapping can be achieved by showing that, if a central-force network with *zero natural length* is stretched onto a frame, then there is an exact mapping onto the random resistor network. More specifically, their model is as follows. On a lattice background with lattice spacing ℓ, one places between the nearest-neighbor sites Hookean springs with natural length ℓ_0 with $\ell > \ell_0$. The area of the lattice is held fixed either by a frame or by imposing periodic boundary conditions. At equilibrium, no net force is exerted on any site, but the springs around any site can be stretched, which can be considered as an internal strain. The limit $\ell_0 = 0$ corresponds to a random resistor network, while $\ell_0 = \ell$ is the usual rigidity percolation. Since the natural length of the springs is not the same as that of the lattice spacing, such networks are under tension, with the tension being supplied by the frame. Varying $\eta = \ell_0/\ell$ results in networks with percolation thresholds that vary *continuously* between p_c (corresponding to $\eta = 0$) and p_{ce} (corresponding to $\eta = 1$). However, except for the limiting cases of $\eta = 0$ and 1, the elastic properties of these networks behave *nonlinearly*, and thus are different from both random resistor and central-force networks.

8.6.6 Nature of the Phase Transition

Equations (42) and (43) are valid if the percolation transition at p_{ce} is continuous and second-order. In fact, before there was any discussion as to whether the rigidity percolation represents a second-order transition, it was *assumed* for many years that, in analogy with scalar percolation at p_c, the percolation transition at p_{ce} is also second-order and continuous. Therefore, based on this assumption, many well-known techniques for scalar percolation, such as the finite-size scaling method described above, were utilized for studying the power-law and scaling behavior of the topological properties of rigidity percolation clusters. For example, Lemieux *et al.* (1985) and Day *et al.* (1986) used a transfer-matrix approach (see Section 8.12) to study percolation properties of central-force percolation networks. Wang and Harris (1985, 1988), Marshall and Harris (1988), and Wang *et al.* (1992) used the series expansion method described in Chapters 5 (see Section 5.15.3) to estimate the exponents β_e and ν_e of rigidity percolation, where β_e is the analogue of the exponent β for scalar percolation [see Eqs. (2.29) and (2.30)], and several other exponents associated with the topology of central-force percolation clusters.

However, rigidity percolation has proven to be a far more complex problem than previously thought. As discussed earlier in this chapter, in a d-dimensional central-force lattice, no element of the structure can be moved in any direction with respect to the remaining lattice, and it cannot also be rotated along $\frac{1}{2}d(d-1)$ independent axes (Obukhov, 1995), implying that there are $\frac{1}{2}d(d+1)$ constraints on the possible motion of each element of the rigid lattice. Thus, *there are long-range, non-decaying correlations in rigidity percolation* that are absent in random scalar percolation. The existence of such long-range correlations immediately provides an important hint about the nature of phase transition in the sample-spanning rigid cluster at p_{ce}: As discussed in Chapter 2 (see Section 2.12.2), long-range, non-decaying correlations may give rise to compact clusters and first-order phase transitions (Sahimi and Mukhopadhyay, 1996; Knackstedt *et al.*, 2000), implying that rigidity percolation may represent a first-order phase transition. Using a mean-field theory, Obukhov (1995) argued that this is indeed the case, although the specific model that he claimed to be using in his analysis is not completely clear to us. Moukarzel *et al.* (1997b), using large-scale simulations, provided numerical evidence in support of Obukhov's argument. Moreover, by solving the problem on the Bethe lattices (Moukarzel *et al.*, 1997a), which corresponds to the mean-field limit of percolation at its upper critical dimension (see Chapters 2 and 5), they found that, at least in some cases, the percolation transition in the sample-spanning rigid cluster at p_{ce} is first-order. They also found that some variations of the problem are similar to bootstrap percolation described and discussed in Chapter 2 (see Section 2.12.2). As mentioned there, Sahimi and Ray (1991) had already speculated that bootstrap percolation might be relevant to describing mechanical properties of disordered lattices.

If it is true that rigidity percolation represents a first-order phase transition, then how can one interpret the power-law properties of the elastic moduli of central-force percolation networks, Eqs. (3) and (5), which are characteristics of second-order phase transitions? Although the sample-spanning central-force cluster may be compact (and, hence, the percolation transition in the cluster may be first-order), its *backbone* is *not* (Moukarzel *et al.*, 1997b). The backbone is in fact a fractal object with a well-defined fractal dimension D_{bb}^e, which Moukarzel *et al.* (1997b) estimated it to be about 1.78 for 2D central-force clusters, larger than that of scalar percolation, $D_{bb} \simeq 1.64$ (see Table 2.3). Because of this fractality, the rigidity percolation transition, *defined as the point at which the elastic moduli vanish*, is also second-order and thus the critical exponents f and χ, defined by Eqs. (3) and (5), are well-defined and non-trivial.

On the other hand, Jacobs and Thorpe (1995,1996) disputed the notion that the percolation transition in the sample-spanning central-force cluster is first order. They calculated the critical exponents ν_e and β_e, and the fractal dimensions D_f^e and D_{bb}^e, where D_f^e is the analogue of the fractal dimension D_f of the sample-spanning scalar cluster at the percolation threshold. Their estimates for 2D networks are, $\nu_e \simeq 1.21$, $\beta_e \simeq 0.175$, $D_f^e \simeq 1.86$, relatively close to (but still distinct from) those of scalar percolation (see Table 2.3), and $D_{bb}^e \simeq 1.8$, consistent with the estimate of Moukarzel *et al.* (1997b), but much larger than that of scalar percolation, $D_{bb}^e \simeq 1.64$. The numerical results of Jacobs and Thorpe (1996) also indicated that the critical exponents are the same for both bond and site rigidity percolation, consistent with the result of Knackstedt and Sahimi (1992) who reached the same conclusion using a position-space renormalization group method (see Section 8.11.9). For the most recent discussion of this issue see Moukarzel and Duxbury (1999).

Let us point out that all the results of Jacobs and Thorpe, and of Moukarzel, Duxbury, and Leath, were obtained using *random* lattices. It remains an open question whether the same conclusions are true about *regular* lattices, such as the triangular and BCC lattices.

8.6.7 Scaling Properties of the Elastic Moduli

The moments $M(q)$ of the force distribution (FD) for $q = 0 - 4$ at $p_{ce}^B \simeq 0.642$, the rigidity bond percolation threshold of the triangular network, were calculated by Hansen and Roux (1989), Sahimi and Arbabi (1989), and Arbabi and Sahimi(1993). Figure 8.5, taken from Sahimi and Arbabi (1989), presents the results, while in Table 8.1 we present the numerical estimate of $\tilde{\tau}(q)$. In particular, $-\tilde{\tau}(0) \simeq 1.62$

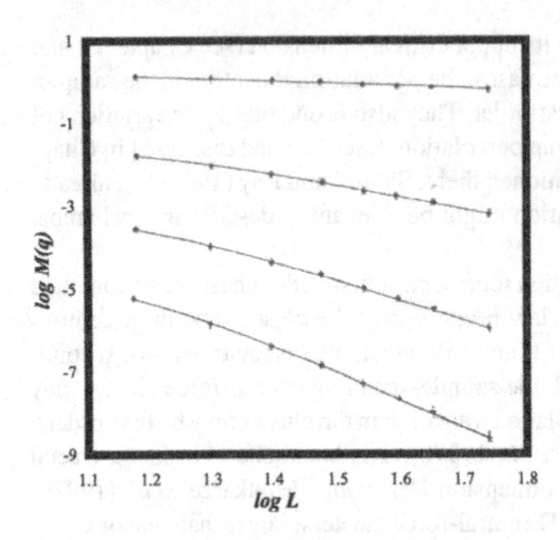

FIGURE 8.5. Dependence of the moments $M(q)$ of the force distribution in rigidity percolation on the linear size L of the triangular network at $p_{ce}^B \simeq 0.641$. The results, from top to bottom, are for $q = 1, 2, 3$, and 4 (after Sahimi and Arbabi, 1989).

TABLE 8.1. Estimates of the critical exponents $\hat{\tau}(q)$ of the moments of the force distribution in rigidity bond percolation on the triangular network at $p_{ce}^B \simeq 0.641$. Note that for random lattices, $\hat{\tau}(0) \simeq -1.78$.

q	0	1	2	3	4
$\hat{\tau}(q)$	-1.62 ± 0.06	0.86 ± 0.05	2.95 ± 0.25	4.94 ± 0.45	7.05 ± 0.75

is in agreement with the fractal dimension of the backbone of 2D scalar percolation (Rintoul and Nakanishi, 1992; Sheppard *et al.*, 1999), but lower than $D_{bb}^e \simeq 1.78$ for central-force random lattices. Note also that

$$\tilde{\tau}(2) = \frac{f}{\nu_e} \simeq 2.95\,, \tag{50}$$

and that for $q \geq 2$

$$\tilde{\tau}(q) = \tilde{\tau}(q-1) + 2\,, \tag{51}$$

that is, there is a constant gap between $\tilde{\tau}(q-1)$ and $\tilde{\tau}(q)$. We emphasize the importance of the correction-to-scaling terms $h_1(L)$ and $h_2(L)$ [see Eq. (7)], especially when L is relatively small. If such corrections are neglected, then one obtains $\tilde{\tau}(2) = f/\nu_e \simeq 2.2$, instead of 2.95. It is only by including such correction terms that one can obtain accurate and reliable estimates of the exponents.

However, the scaling exponent of the elastic moduli of rigidity site percolation on the triangular lattice appear to be quite different from that of bond percolation (Arbabi and Sahimi, 1993):

$$\frac{f}{\nu_e} \simeq 1.12 \pm 0.05, \tag{52}$$

which is much smaller than (50) for bond percolation. This difference indicates that the scaling properties of elastic moduli of central-force models in site and bond percolation on the triangular network may belong to two different universality classes.

TABLE 8.2. Estimates of the critical exponents $\hat{\tau}(q)$ of the moments of the force distribution in rigidity percolation on the BCC network at $p_{ce}^B \simeq 0.737$.

q	0	1	2	3	4
$\hat{\tau}(q)$	-2.5 ± 0.3	1.0 ± 0.1	2.1 ± 0.1	3.97 ± 0.20	5.91 ± 0.35

The FD and the scaling properties of its moments have also been calculated for rigidity bond percolation on the BCC lattice. For this lattice the second moment $M(2)$ of FD is *exactly* equal to the corresponding elastic modulus. The estimated exponents are given in Table 8.2 which indicates that

$$f \simeq 2.1 \pm 0.2. \tag{53}$$

Note that estimates of $\hat{\tau}(q)$ given in Table 8.2 satisfy the scaling relation (51). Moreover, the estimate $f \simeq 2.1$ is close to the critical exponent of conductivity of 3D random percolation networks, $\mu \simeq 2.00 \pm 0.04$ (Gingold and Lobb, 1990; see Table 2.3). However, we believe that the apparent closeness of these two exponents is coincidental, because, excepts for some pathological cases, there is no reason to believe that there is a fundamental physical basis for the equality of the critical exponent a of scalar property—the effective conductivity—and that of the elastic moduli—a vector property. On the other hand, as Table 8.2 indicates, $-\hat{\tau}(0) = D_{bb}^e \simeq 2.5$, which is essentially equal to the fractal dimension D_f of the sample-spanning scalar percolation cluster at p_c (see Table 2.3). The reason may be that the bond percolation threshold p_{ce}^B of the BCC network is so much larger than p_c^B, the corresponding percolation threshold of the scalar model, that the backbone of the elastic network is very similar to the sample-spanning cluster in the scalar problem. This is, however, speculation and, at the time of writing this book, remained to be tested further.

Superelastic percolation networks with central forces, first introduced by Sahimi and Goddard (1985), have also been studied extensively (Roux and Hansen, 1988; Hansen and Roux, 1988,1989; Burton and Lambert, 1988; Wang and Harris, 1988; Arbabi and Sahimi, 1990c,1993) by various techniques. For bond percolation on a triangular network the most accurate current estimate of the exponent χ, defined by Eq. (5), is (Arbabi and Sahimi, 1993)

$$\frac{\chi}{\nu_e} \simeq 0.92 \pm 0.02, \tag{54}$$

while for site percolation on the same network

$$\frac{\chi}{\nu_e} \simeq 1.05 \pm 0.03, \tag{55}$$

which does not agree with that of bond percolation. This again may indicate that the critical exponents of the elastic moduli in rigidity bond and site percolation in 2D may belong to two different universality classes. For rigidity bond percolation on a BCC network the estimate is (Arbabi and Sahimi, 1993)

$$\frac{\chi}{\nu_e} \simeq 0.80 \pm 0.03. \tag{56}$$

Later in this chapter we will compare these results with those of EPNs and SEPNs with stretching and bond-bending forces, which are more realistic models of disordered solids.

8.7 Green Function Formulation and Perturbation Expansion

We now formulate the problem of calculating the effective elastic moduli of the Born model (which, in the limit $\nu_p \to 1/3$, reduces to rigidity percolation) in terms of an appropriate Green function, and develop a perturbation expansion for the problem, based on which an effective-medium approximation (EMA) is derived. The Green function formulation of the problem and the derivation of the perturbation expansion are generalizations of those for diffusion and conduction developed by Sahimi *et al.* (1983a), which was described and discussed in Chapters 5 and 6. In what follows we use the same notations that we utilized in Chapters 5 and 6 in order to make transparent the similarities between the present formulation and those described in Chapters 5 and 6.

We introduce an effective-medium as a lattice in which the $\mathbf{W}_{ij}$ are replaced with $\mathbf{W}_{ij}^e$, where superscript e denotes the value of the quantity in the effective-medium. Thus, Eq. (20) becomes

$$\sum_j \mathbf{W}_{ij}^e \cdot (\mathbf{u}_j^e - \mathbf{u}_i^e) = \mathbf{0}, \tag{57}$$

where $\mathbf{u}_i^e$ is the displacement of site i in the effective-medium. Subtraction of Eq. (57) from (20) and some rearrangements yield

$$\sum_j \mathbf{W}_{ij}^e \cdot [(\mathbf{u}_j - \mathbf{u}_j^e) - (\mathbf{u}_i - \mathbf{u}_i^e)] = -\sum_j \mathbf{\Delta}_{ij} \cdot (\mathbf{u}_j - \mathbf{u}_i), \tag{58}$$

where $\mathbf{\Delta}_{ij} = \mathbf{W}_{ij} - \mathbf{W}_{ij}^e$. A *vector* Green function is now introduced by

$$\sum \mathbf{W}_{ij}^e \cdot (\mathbf{G}_{jm} - \mathbf{G}_{im}) = -\delta_{ij}, \tag{59}$$

with the aid of which the *exact* but implicit solution of Eq. (20) is obtained:

$$\mathbf{u}_i = \mathbf{u}_i^e + \sum_k \sum_j \mathbf{G}_{ij} \cdot \mathbf{\Delta}_{jk} \cdot (\mathbf{u}_j - \mathbf{u}_k). \tag{60}$$

Because the elastic moduli appear as the coefficients of $\mathbf{u}_{ij} = \mathbf{u}_i - \mathbf{u}_j$, it is more convenient to work with $\mathbf{u}_{ij}$. We thus obtain

$$\mathbf{u}_{ij} = \mathbf{u}_{ij}^e + \sum_l \sum_k (\mathbf{G}_{ik} - \mathbf{G}_{jk}) \cdot \mathbf{\Delta}_{lk} \cdot \mathbf{u}_{lk}. \tag{61}$$

Since $\mathbf{u}_{ij} = -\mathbf{u}_{ji}$, and $\mathbf{\Delta}_{lk} = \mathbf{\Delta}_{kl}$, we can rewrite Eq. (61) in a more compact form

$$\mathbf{u}_{ij} = \mathbf{u}_{ij}^e + \sum_l \sum_k \boldsymbol{\gamma}_{ijkl} \cdot \boldsymbol{\Delta}_{lk} \cdot \mathbf{u}_{lk}, \tag{62}$$

where $\boldsymbol{\gamma}_{ijkl} = (\mathbf{G}_{il} - \mathbf{G}_{jl}) - (\mathbf{G}_{ik} - \mathbf{G}_{jk})$.

8.7.1 Effective-Medium Approximation

The EMA is obtained by taking all the $\boldsymbol{\Delta}_{lk} = \mathbf{0}$, except for a finite set of bonds for which this quantity is not zero. If this set contains only one bond mk for which $\boldsymbol{\Delta}_{mk} \neq \mathbf{0}$, then Eq. (62) becomes

$$\mathbf{u}_{mk} = (\mathbf{U} - \boldsymbol{\gamma}_{mkmk} \cdot \boldsymbol{\Delta}_{mk})^{-1} \cdot \mathbf{u}_{mk}^e. \tag{63}$$

In the effective-medium approach one demands that $\langle \mathbf{u}_{mk} \rangle = \mathbf{u}_{mk}^e$, where the averaging is taken with respect to the statistical distribution of the heterogeneities. The EMA equation is finally obtained:

$$\langle (\mathbf{U} - \boldsymbol{\gamma}_{mkmk} \cdot \boldsymbol{\Delta}_{mk})^{-1} \rangle = \mathbf{U} \tag{64}$$

It remains to specify the Green functions $\mathbf{G}_{ij}$ which depend on the topology of the lattice. This can be done most conveniently by discrete Fourier transformation technique, along the lines described in Chapter 6 (see Section 6.1.4). We do not give the details of the computations, as they are entirely similar to what was discussed there. We merely note that, because the Green functions depend only on distance and orientation, we must have $\mathbf{G_{jm}} = \mathbf{G}(\mathbf{j} - \mathbf{m})$, and $\mathbf{W}_{\mathbf{jm}}^e = \mathbf{W}^e(\mathbf{j} - \mathbf{m})$. Moreover, because of translational invariance, $\mathbf{m}$ may be set to zero without loss of generality.

8.7.2 The Born Model

Consider first the Born model on a d-dimensional simple-cubic lattice. For the sake of simplicity we rewrite Eq. (24) as

$$\mathbf{W}_{ij} = a_1 \mathbf{R}_{ij} \mathbf{R}_{ij} + a_2 \mathbf{U}. \tag{65}$$

To derive the EMA we must only compute $\boldsymbol{\gamma} = \boldsymbol{\gamma}_{\mathbf{0k0k}} \cdot \mathbf{W_{0k}}$. In general, one can show that

$$\boldsymbol{\gamma}_{\mathbf{0k0k}} = -\frac{2}{(2\pi)^d} \int_{-\pi}^{\pi} \int_{-\pi}^{\pi} \int_{-\pi}^{\pi} [1 - \cos(\mathbf{k} \cdot \boldsymbol{\theta})]\, \hat{\mathbf{G}}(\boldsymbol{\theta}) d\boldsymbol{\theta}, \tag{66}$$

where $\hat{\mathbf{G}}(\boldsymbol{\theta})$ is the Fourier transform of the Green function. With $\boldsymbol{\theta} = (\theta_1, \cdots, \theta_d)$, for a d-dimensional cubic lattice one has

$$2\hat{\mathbf{G}}(\boldsymbol{\theta}) =$$

$$\begin{pmatrix} [da_2 + a_1(1 - \cos\theta_1)]^{-1} & 0 & \cdots & 0 \\ 0 & [da_2 + a_1(1 - \cos\theta_2)]^{-1} & \cdots & 0 \\ \cdot & \cdot & \cdot & \cdot \\ 0 & 0 & \cdots & [da_2 + a_1(1 - \cos\theta_d)]^{-1} \end{pmatrix}. \tag{67}$$

Without loss of generality, $\mathbf{k}$ may be taken to be $(1, 0, \cdots, 0)^{\mathrm{T}}$, as all the bonds are equivalent. Thus, we only need $\boldsymbol{\gamma}_{0k0k}$ which requires only the two quantities,

$$\begin{aligned}\gamma_{11} &= -\frac{2}{(2\pi)^d}\int\cdots\int\frac{(a_1+a_2)(1-\cos\theta_1)}{da_2+a_1(1-\cos\theta_1)}d\theta_1\\ &= -\left(1+\frac{a_2}{a_1}\right)\left[1-\frac{d}{\sqrt{d(d+2a_1/a_2)}}\right],\end{aligned} \tag{68}$$

and

$$\gamma_{kk} = -\frac{2}{(2\pi)^d}\int\cdots\int\frac{a_2(1-\cos\theta_1)}{da_2+a_1(1-\cos\theta_k)}d\theta_1 d\theta_k = -\frac{1}{\sqrt{d(d+2a_1/a_2)}}, \tag{69}$$

where both integrals are over the first Brillouin zone, $(-\pi, \pi)$. Therefore, if the effective values of a_1 and a_2 are, respectively, a_{1e} and a_{2e}, then, with a simple percolation distribution, $\psi(x) = (1-p)\delta_+(0) + p\delta(x - a_i)$ $(i = 1$ and $2)$, and $c_e = a_{e1}/a_{e2}$, one obtains

$$\frac{a_{2e}(1+1/c_e)}{a_2(1+1/c)} = \frac{p+\gamma_{11}}{1+\gamma_{11}}, \tag{70}$$

$$\frac{a_{2e}}{a_2} = \frac{p+\gamma_{22}}{1+\gamma_{22}}, \tag{71}$$

where $c = a_1/a_2$. It is then straightforward to show that the EMA, Eq. (64), predicts that the elastic moduli vanish at $p_{ce} = 1/d = 2/Z$, the same as in scalar transport derived in Chapter 5 (see Section 5.6.1). This is not surprising because, as discussed in Section 8.4.2, the presence of the scalar-like contribution to the elastic energy (25) ensures that the power-law behavior of the elastic moduli of the Born model near p_{ce} is the same as that of the effective conductivity near p_c, and that the percolation threshold of the system is p_c not p_{ce}. Moreover, if we set $a_2 = 0$ (i.e., the Poisson's ratio $\nu_p = 1/3$), we find that the EMA predicts the exact, albeit trivial, result that removal of any springs at all will cause a d-dimensional cubic lattice of central-force springs to have a zero effective spring constant, i.e., the percolation threshold of the lattice is unity.

As another example, consider the Born model on the triangular lattice. In this case, there are three distinct $\mathbf{W}_{ij}$ which correspond to three directions along three vectors with components $(1, 0)$, $(1/2, \sqrt{3}/2)$, and $(-1/2, \sqrt{3}/2)$, which can, however, be transformed into one another upon coordinate rotation. Moreover,

$$\hat{\mathbf{G}}(\boldsymbol{\theta}) = \begin{pmatrix} S_{11} & -S_{12} \\ -S_{12} & S_{22} \end{pmatrix}, \tag{72}$$

where

$$S_{11} = (5-\nu_p) - (1-3\nu_p)\cos^2\theta_1 - 4\cos\theta_1\cos\theta_2, \tag{73}$$

$$S_{22} = (7-\nu_p) - (5+\nu_p)\cos^2\theta_1 - 2(1-\nu_p)\cos\theta_1\cos\theta_2, \tag{74}$$

$$S_{12} = \sqrt{3}(1+\nu_p)\sin\theta_1\sin\theta_2. \tag{75}$$

The bond-bond Green function $\boldsymbol{\gamma}$ is then given by

$$\boldsymbol{\gamma} = -\frac{2}{(2\pi)^2}\int_{-\pi}^{\pi}\int_{-\pi}^{\pi} 2(1-\cos^2\theta_1)\hat{\mathbf{G}}(\boldsymbol{\theta})d\boldsymbol{\theta}. \tag{76}$$

Moreover, it can be shown that for the triangular lattice, $\gamma_{12} = \gamma_{21} = 0$, and therefore one needs only to compute γ_{11} and γ_{22}. In the limit $\nu_p = -1$ one recovers the random resistor network model with a percolation threshold at $p_c = 2/Z$, which is also the percolation threshold of the Born model (within the EMA), while $\nu_p = 1/3$ reduces Eq. (64) to an EMA for the elastic moduli of the triangular network with central-force springs.

8.7.3 Rigidity Percolation

If only central forces are present, then $a_2 = 0$ ($\nu_p = 1/3$), and $\mathbf{W}_{ij} = a_1\mathbf{R}_{ij}\mathbf{R}_{ij}$. Therefore, in this case

$$\gamma_{kk} = -\frac{2}{(2\pi)^d}\int\cdots\int\left\{\sum_{\mathbf{l}}[1-\cos(\mathbf{l}\cdot\boldsymbol{\theta})]\,\mathbf{ll}\right\}^{-1}\cdot[1-\cos(\mathbf{k}\cdot\boldsymbol{\theta})\mathbf{kk}]\,d\boldsymbol{\theta}. \tag{77}$$

This equation can be integrated case by case for a variety of Bravais lattices to yield the necessary Green functions for constructing the EMA. Since only γ_{11} is needed for constructing the EMA, it is not difficult to show that $\gamma_{11} = -2d/Z$, and therefore for a d-dimensional network of coordination number Z, the effective force constant α_e is given by

$$\int_0^\infty \frac{\alpha_e-\alpha}{\alpha+\alpha_e(Z/2d-1)}\psi(\alpha)d\alpha = 0, \tag{78}$$

which is very similar to Eq. (5.57) for the effective conductivity of random resistor networks. Here, $\psi(\alpha)$ is the probability density function of the force constant α. Note that, as pointed out at the beginning of this section, αe_{ij} may be interpreted as the force constant of the bond ij, so that any statistical distribution of e_{ij} may be converted to one for α. From the effective force constant α_e, one calculates the elastic moduli of the network. For example, the effective shear modulus of the triangular network is given by Eq. (32), in which α is replaced by the effective α_e.

To test the accuracy of Eq. (78), we examine its predictions in certain well-understood limits. For example, it is not difficult to see that with a percolation-type distribution, $\psi(\alpha) = (1-p)\delta_+(0) + p\delta(\alpha-\alpha_1)$, Eq. (78) predicts a bond percolation p_{ce}^B that is given by Eq. (49), which is also the prediction of constraint-counting method. According to Eq. (49), the average coordination number $\langle Z\rangle$ of a d-dimensional central-force network at the rigidity percolation threshold p_{ce} is given by, $\langle Z\rangle = p_{ce}^B Z = 2d$, independent (of course approximately) of the coordination number of the network itself. This implies that, for a network of central-force springs to be rigid, one must have

$$Z > 2d, \quad \text{macroscopic rigidity}, \tag{79}$$

hence explaining why study of central-force percolation on the simple-cubic lattices, and also the hexagonal lattice (in 2D) and the diamond lattice (in 3D), is meaningless. The inequality (79) expresses the condition for *macroscopic* rigidity of the network. It is straightforward to show that the condition for *microscopic* rigidity of central-force spring networks is

$$Z > d, \quad \text{microscopic rigidity.} \tag{80}$$

Central-force networks with $Z = 2d$ are usually called *marginally-bonded* networks. Examples include d-dimensional cubic lattices.

As usual, Eq. (78), the EMA for predicting the effective elastic constants of central-force spring networks, also predicts that the critical exponents f and χ of the elastic moduli are unity, in disagreement with the numerical estimates discussed above. However, since the predicted percolation threshold differs very little from its actual value, then, except for a very small region near p_{ce}^B, the predictions of Eq. (78) are in very good agreement with numerical simulation results (Jerauld, 1985; Feng *et al.*, 1985; Garboczi and Thorpe, 1985, 1986a,b; Thorpe and Garboczi, 1987; Pla *et al.*, 1990). A typical example of the comparison of the numerical results with the EMA predictions is shown in Figure 8.6.

Let us point out that another type of random central-force lattice models was developed and studied by Kellomäki *et al.* (1996), who generated such lattices by placing N_f 1D straight lines of length ℓ_f and zero width on a rectangle of area A. The distributions of the lines centers and of the orientation of the lines were random. Lines were bounded together at their crossing points. The dead-end lines were removed, and thus the coordination number of each node was 2, 3,

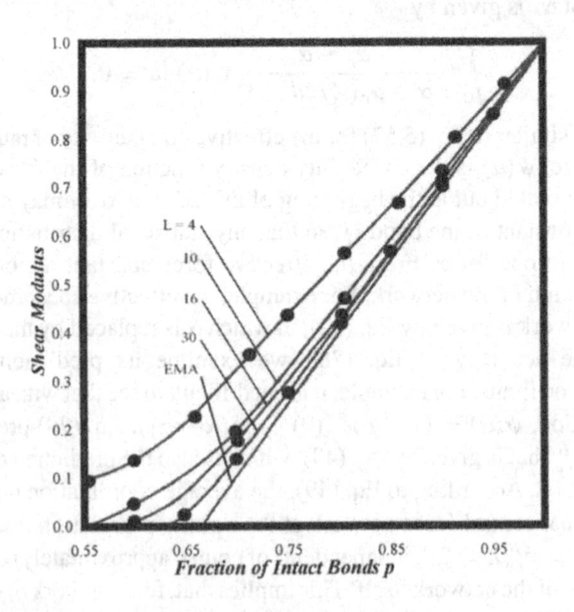

FIGURE 8.6. Dependence of the shear modulus of a triangular network, in rigidity percolation, on the fraction of p of the intact bonds. Shown are the results for several network linear sizes L, as well as the predictions of the EMA.

or 4. The lines between neighboring nodes were assumed to be Hookean springs with a given Young's modulus and cross section. The density of such networks is $\rho = N_f \ell_f / A$, and percolation (formation of the sample-spanning cluster) occurs at $\rho_c \simeq 5.71$. Kellomäki *et al.* (1996) showed that such random lattices are *not* rigid which, in view of Eq. (79) and our discussion above, is not surprising since the average coordination number of such networks is less than 4, the minimum required value for rigidity in 2D. Many other properties of such networks were also studied by Kellomäki *et al.* (1996).

8.8 The Critical Path Method

As discussed in Chapters 5 (see Section 5.13) and 6 (see Section 6.2.4), a powerful method for analyzing transport processes in heterogeneous materials is the critical path method, first proposed by Ambegaokar, Halperin and Langer (AHL, 1971). According to AHL arguments, in a material in which the local conductances are broadly distributed, the effective conductivity is approximately equal to the critical conductivity g_c, defined as the smallest conductivity needed to form a sample-spanning conducting path, when the network representing the material is built up in descending order, starting with the largest bond conductance. Applied to central-force spring networks (Garboczi, 1988a), the AHL method would predict that the effective force constant α_e is approximately equal to the critical force constant α_c, defined similarly to g_c. However, we should keep in mind that, when the AHL method is applied to central-force spring networks, the critical force constant α_c is defined as that value of α at which a sample-spanning *rigid* cluster has formed for the first time. In other words, α_c should not be taken as the value of the force constant when a cluster has formed (at p_c) and is merely connected, rather it should also be rigid which happens for the first time at p_{ce}, since for $p_c < p < p_{ce}$ such a cluster is floppy.

Therefore, suppose that the force constants are distributed uniformly between α_1 and α_2. Beginning with α_2 and working down, a fraction p of the bonds will be in place when a fraction p of the values of the force constants from α_2 to α_1 has been used. Therefore,

$$\alpha_c = \alpha_2 - p(\alpha_2 - \alpha_1) = \alpha_1 + (1 - p)(\alpha_2 - \alpha_1). \tag{81}$$

Garboczi (1988a) showed that the predictions of Eq. (81) are in good agreement with the results of computer simulations using the triangular lattice. They were also found to be very close to the predictions of the EMA, Eq. (78).

8.9 Central-Force Networks at Nonzero Temperatures and Under Stress

All the results described and discussed so far (as well as those that will be described later in this chapter for more sophisticated models of elastic networks) are applicable to unstressed networks at zero temperature. In practice, however, all the experimental measurements are carried out at temperatures above $T = 0$, and therefore it is important to understand the temperature-dependence of the elastic moduli, at least in the context of the network models. In addition, in many practical situations, the material under study is exposed to a finite stress or tension (as opposed to an infinitesimal one), and thus the role of such an external driving force in determining the elastic properties of materials must be understood. In principle, the role of the temperature can be understood by carrying out Molecular Dynamics simulations in which the material is represented as a collection of atoms and/or molecules, and Newton's equation of motion is solved for all the atoms and molecules, taking into account the effect of the interatomic interactions. This is easier said than done, as a realistic simulation may require (at the very least) hundreds of thousands of atoms, accurate representation of the interatomic interactions, and powerful computers to integrate the equation of motion for long enough times to observe the dynamic evolution of the material. Over the past few years, there has been very significant advances in all these aspects, which will be described in detail in Chapter 9 of Volume II. For now, we restrict ourselves to simple phenomenological models in order to gain a better understanding of the effect of non-zero temperatures and stresses on the mechanical properties of elastic networks. We restrict our attention to homogeneous networks, and then discuss the extension to heterogeneous networks, and in particular elastic percolation networks.

Consider, as an example, the triangular network of identical springs. At zero temperature and stress, all the springs have the same length ℓ_0. The network can be thought of as a plane of triangular plaquettes, each bounded by three springs. The length of the springs changes from ℓ_0 to ℓ_σ if a 2D isotropic tension σ is imposed on the network. We adopt the sign convention that $\sigma > 0$ corresponds to tension and $\sigma < 0$ to compression. The plaquettes, however, retain their equilateral triangular shape, albeit with a different area. The springs' length ℓ_σ is calculated by minimizing the enthalpy H of the system,

$$H = \mathcal{H}_s - \sigma S, \tag{82}$$

where $\mathcal{H}_s$ is the energy of the springs and S is the surface area of the network. As described in Section 8.5, the energy per node of the springs (of equal length ℓ) is $\frac{3}{2}\alpha(\ell - \ell_0)^2$, and the area per node is $\frac{\sqrt{3}}{2}\ell^2$, and therefore the enthalpy per node is

$$H_n = \frac{3}{2}\alpha(\ell - \ell_0)^2 - \frac{\sqrt{3}}{2}\sigma\ell^2. \tag{83}$$

Thus, ℓ_σ, obtained by solving $\partial H_n/\partial\ell = 0$, is given by

$$\ell_\sigma = \frac{\sqrt{3}\alpha\ell_0}{\sqrt{3}\alpha - \sigma}. \tag{84}$$

Equation (84) indicates that the network expands under tension and shrinks under compression. The corresponding enthalpy per node is given by

$$H_n = -\frac{3\sigma\alpha\ell_0^2}{2(\sqrt{3}\alpha - \sigma)}. \tag{85}$$

More importantly, Eq. (84) indicates that the area per node S_n of the network,

$$S_n = \frac{3\sqrt{3}\alpha^2\ell_0^2}{2(\sqrt{3}\alpha - \sigma)^2}, \tag{86}$$

diverges under tension as $\sigma \to \sqrt{3}\alpha$. This is understandable as both the energy of the springs and the term σS_n scale, at large extensions, as ℓ^2. However, these two terms have opposite signs under tension ($\sigma > 0$), and for large enough σ the tension term dominates the enthalpy, as a result of which S_n increases without bound.

In addition, Eqs. (84) and (85) indicate that the network *collapses* under compression which, however, is *not* seen in networks of equilateral plaquettes. This apparent contradiction is due to the assumption that was made for deriving Eq. (85), namely, that this equation was obtained by first solving $\partial H_n/\partial\ell = 0$. However, if shapes other than equilateral plaquettes are considered, then, Eq. (85) may not provide the global minimum of the enthalpy per node, as equilateral triangles provide the largest surface for a fixed perimeter, and therefore they may not provide the optimal shape under compression, where the term σS drives the system towards small surface areas. It is clear that the term σS would be minimum (for $\sigma < 0$) if the plaquettes have zero surface area. Plaquettes that are shaped like isosceles triangles have this property and the lowest spring energy given by

$$\mathcal{H}_s = \frac{1}{2}\alpha\left[(2\ell_1 - \ell_0)^2 + 2(\ell_1 - \ell_0)^2\right], \tag{87}$$

where a zero-surface area triangle has two short sides of length ℓ_1 and a long side of length $2\ell_1$. The minimum of H_n, as given by Eq. (87), is when $\ell_1 = \frac{2}{3}\ell_0$, and is given by, $H_n = \frac{1}{6}\alpha\ell_0^2$, implying that H_n per node of the network increases with pressure ($\sigma < 0$) according to Eq. (85) until it exceeds the enthalpy of a network of zero-surface area isosceles plaquettes ($\frac{1}{6}\alpha\ell_0^2$) at the collapse tension σ_c given by

$$\sigma_c = -\frac{\sqrt{3}}{8}\alpha. \tag{88}$$

Equation (88) implies that the spring length at which the collapse transition occurs is $\ell_\sigma/\ell_0 = 8/9$. Thus, for any $\sigma > \sigma_c$ the network will collapse.

Having determined the springs' length ℓ_σ, the elastic moduli of the network are computed by exactly the same method that we used in Section 8.5 to calculate the elastic moduli of the uniform network at zero temperature and stress. Therefore, we only present the results here:

$$K_e = \frac{1}{2}(\sqrt{3}\alpha - \sigma), \tag{89}$$

$$\mu_e = \frac{\sqrt{3}}{4}(\alpha + \sqrt{3}\sigma), \tag{90}$$

both of which reduce to Eqs. (31) and (32) in the limit $\sigma = 0$. The Poisson's ratio of the network is then give by

$$\nu_p = \frac{K_e - \mu_e}{K_e + \mu_e} = \frac{\sqrt{3}\alpha - 5\sigma}{3\sqrt{3}\alpha + \sigma}, \tag{91}$$

so that $\nu_p < 0$ if $\sigma/\alpha > \sqrt{3}/5$.

We should be cautious in using the above results, since the collapse transition actually occurs if the assumption of having springs of the same length is not made. Thus, we should expect these results to be accurate at low temperatures and/or high tensions, where the springs have more or less the same length which, however, may be different from ℓ_0. Monte Carlo simulations of Boal *et al.* (1993) indicate that Eq. (84) to be accurate at the 90% level or better for the range, $\frac{1}{8}\alpha\ell_0^2 \leq k_B T \leq \alpha\ell_0^2$.

Similar reasoning and computations can be carried out for the square network. In this case, one obtains

$$\ell_\sigma = \frac{\alpha\ell_0}{\alpha - \sigma}, \tag{92}$$

so that the surface area per node of square plaquettes increases without bound as $\sigma \rightarrow \alpha$. The collapse transition occurs under any negative tension, $\sigma < 0$, because all the square plaquettes have the same energy (see Section 8.5). As for the elastic moduli, we obtain

$$K_e = \frac{1}{2}(\alpha - \sigma), \tag{93}$$

$$\mu_p = \frac{1}{2}(\alpha + \sigma), \quad \mu_s = \sigma, \tag{94}$$

all of which reduce to Eqs. (34) in the limit $\sigma = 0$.

What happens if the networks contain percolation disorder, i.e., only a fraction $p < 1$ of the springs is intact, with the rest of them being cut? In this case, one should be able to use Eqs. (89)–(94) in which the spring constant α has been replaced by the effective value of the spring constant of the network, α_e, which can be estimated by, for example, the effective-medium approximation, Eq. (78). This approximation should be reasonably accurate so long as p is not too close to the percolation threshold p_{ce}. For p close to p_{ce} one must carry out Monte Carlo simulations in order to estimate the effective elastic moduli at low temperatures and non-zero stresses.

8.10 Shortcomings of the Central-Force Networks

One main shortcoming of central-force percolation networks is that their percolation thresholds p_{ce} are much larger than those of scalar percolation p_c. What this implies is that, for $p_c \leq p \leq p_{ce}$, where p is the fraction of the intact bonds (or sites), an EPN with only central forces is geometrically connected, but its elastic moduli are all zero, which, in most cases, is not a physically-realistic situation. If there are correlations between the intact bonds (Arbabi and Sahimi, 1988a), then the percolation threshold of a central-force EPN will be less than p_{ce}, but in order to force the percolation threshold of a correlated central-force network to be equal to p_c, one must introduce infinitely long-range correlations between the intact bonds. Another shortcoming of central-force percolation networks is that, in such models the significance of a straight-bond chain in transmitting elastic forces is ambiguous, because in a nonlinear model (such as those considered for brittle fracture of solids described in Chapter 8 of Volume II) the straight bonds could buckle under compression but not under extension, which is again unrealistic. The third shortcoming of the model is that the matrix $\mathbf{W}_{ij}$ may not be invertible for some complex configurations, hence leading to unphysical predictions. Therefore, in the rest of this chapter we consider vector percolation models that do not suffer from such shortcomings.

8.11 Elastic Percolation Networks with Bond-Bending Forces

We now describe more general vector percolation models which are free of the shortcomings of the central-force networks. It is known that microscopic *many-body interactions* (as opposed to two-body interactions in the central-force models), and in particular three-body bending and four-body twisting, have important effects on the elastic moduli of solids, especially their shear moduli. Such interactions are expected to be particularly important for weakly-bonded tenuous materials, such as certain types of aggregates and gels. Including such many-body interactions in a lattice model also reveals an important advantage of the discrete models. In a continuum model, the distinction between two-body and multibody interactions can become obscure, whereas in a lattice model the sites that interact with each other directly clearly leave a signature and distinguish themselves from any other types of interactions.

Hence, consider an EPN in which such many-body interactions exist. The simplest of such models is perhaps an EPN in which there are both central and bond-bending, or angle-changing, forces, with the latter type representing three-body interactions. One of the main advantages of such models is that their percolation threshold can be the same as that of scalar percolation, if the many-body interactions are such that any deformation of the lattice is done at some costs to its elastic energy. In general, the elastic energy of such models is given by (Kantor and Webman, 1984)

$$\mathcal{H} = \frac{1}{2}\alpha \sum_{\langle ij \rangle} [(\mathbf{u}_i - \mathbf{u}_j) \cdot \mathbf{R}_{ij}]^2 e_{ij} + \frac{1}{2}\gamma \sum_{\langle jik \rangle} (\delta\boldsymbol{\theta}_{jik})^2 e_{ij} e_{ik}, \tag{95}$$

where α and γ are the central and bond-bending force constants, respectively. Here $\langle jik \rangle$ indicates that the sum is over all triplets in which the bonds j-i and i-k form an angle with its vertex at i. The first term on the right-hand side of Eq. (95) represents the usual contribution of the central forces (see above), while the second term is due to bond-bending, or angle-changing, forces. The precise form of $\delta\boldsymbol{\theta}_{jik}$ depends on the microscopic details of the model. One may consider at least two classes of such models which we now describe.

8.11.1 The Kirkwood–Keating Model

If bending of the bonds that make an angle of 180° with one another (i.e., the collinear bonds) is not allowed, then

$$\delta\theta_{jik} = (\mathbf{u}_i - \mathbf{u}_j) \cdot \mathbf{R}_{ik} + (\mathbf{u}_i - \mathbf{u}_k) \cdot \mathbf{R}_{ij}. \tag{96}$$

We refer to this particular version of the model as the Kirkwood-Keating (KK) model. Kirkwood (1939) used this model to study vibrational properties of rod-like molecules, while Keating (1966) studied the elastic properties of covalent crystals with essentially the same model. In Keating's model, the bond-stretching term represented a repulsive interaction to perturbations from the equilibrium length of covalent bonds, while the bond-bending term represented a repulsive interaction to perturbations from the equilibrium tetrahedral angle. Neither Kirkwood not Keating studied percolation properties of this model. A similar model was used by Gazis *et al.* (1960), except that in their model there was also a central force between every site and its second-nearest-neighbor sites. All terms of the equation of motion, i.e., $\partial\mathcal{H}/\partial\mathbf{u}_i = 0$, with $\mathcal{H}$ given by Eq. (95), when written for a site i and all of its neighbors that contribute to $\mathcal{H}$, can be expressed in terms of difference operators. Using Taylor's series, such difference operators can be expressed in terms of the partial derivatives of the continuous function $\mathbf{u}(x, y, z)$. Then, if we assume that the wavelengths of deformations are much longer than the lattice constant, all but the second derivatives of $\mathbf{u}_i$ can be neglected. In the continuum limit one obtains

$$\begin{aligned}(C_{11} - C_{12} - 2C_{44}) \sum_j \frac{\partial^2 u_j}{\partial x_j^2} \mathbf{R}_j + C_{44}\,[2\nabla(\nabla \cdot \mathbf{u}) - \nabla \times (\nabla \times \mathbf{u})] \\ + C_{12}\nabla(\nabla \cdot \mathbf{u}) = \mathbf{0},\end{aligned} \tag{97}$$

where $u_j = u_x, u_y, u_z$, $x_j = x, y, z$, and $\mathbf{R}_j$ are the corresponding unit vectors in the x, y, and z directions. Here $C_{11} = \alpha/\ell$, $C_{12} = 0$, and $C_{14} = 4\gamma/\ell$ are the usual elastic constants of the lattice, with ℓ being the lattice constant. If a Hookean spring is also inserted between a site and its second-nearest neighbor (as in the model of Gazis *et al.*, 1960), then Eq. (97) would still hold but with, $C_{11} = (\alpha + 4\alpha_{sn})/\ell$, $C_{12} = 2\alpha_{sn}/\ell$ and $C_{44} = 2(2\gamma + \alpha_{sn})/\ell$, where α_{sn} is the central-force constant between a site and its second-nearest neighbors. Clearly, Eq. (97) is rotationally invariant. Therefore, similar to the Born and central-force models, the KK model can be derived by discretization of a well-defined continuum equation.

8.11.2 The Bond-Bending Model

If bending of the collinear bonds is allowed, then (Wang, 1989; Arbabi and Sahimi, 1990a)

$$\delta\theta_{jik} = \begin{cases} (\mathbf{u}_{ij} \times \mathbf{R}_{ij} - \mathbf{u}_{ik} \times \mathbf{R}_{ik}) \cdot (\mathbf{R}_{ij} \times \mathbf{R}_{ik})/|\mathbf{R}_{ij} \times \mathbf{R}_{ik}|, & \mathbf{R}_{ij} \text{ not parallel to } \mathbf{R}_{ik}, \\ |(\mathbf{u}_{ij} + \mathbf{u}_{ik}) \times \mathbf{R}_{ij}|, & \mathbf{R}_{ij} \text{ parallel to } \mathbf{R}_{ik}, \end{cases} \tag{98}$$

where, $\mathbf{u}_{ij} = \mathbf{u}_i - \mathbf{u}_j$. For *all* 2D systems, Eq. (98) is simplified to

$$\delta\boldsymbol{\theta}_{jik} = (\mathbf{u}_i - \mathbf{u}_j) \times \mathbf{R}_{ij} - (\mathbf{u}_i - \mathbf{u}_k) \times \mathbf{R}_{ik}. \tag{99}$$

We refer to this version of Eq. (95) as the bond-bending (BB) model. It is clear that, similar to the KK model, the BB model also has a well-defined continuum counterpart. For most materials to which the KK or the BB models are applicable, one has $\gamma/\alpha \leq 0.3$ (Martins and Zunger, 1984).

8.11.3 The Percolation Thresholds

Phillips and Thorpe (1985) used a constraint-counting analysis which, as described earlier, is a mean-field theory, to predict the bond percolation threshold $(p_c^B)_{kk}$ of the d-dimensional KK model in a lattice with coordination number Z. The number of constraints N_c associated with bond-stretching and BB forces is given by, $N_c = N_\alpha + N_\gamma$, where

$$N_\alpha(Z) = \frac{1}{2}Z, \quad N_\gamma(d, Z) = \frac{1}{2}(d-1)(2Z-d), \tag{100}$$

where all the sites with $Z \leq d-2$ must be eliminated from the counting. Since in a percolation network each bond is present with probability p, we must write, $N_c(p) = N_\alpha(pZ) + N_\gamma(d, pZ)$, and therefore, $(p_c^B)_{kk}Z + (d-1)[2(p_c^B)_{kk}Z - d] = 2d$ [see the inequality (79)] which, when solved for the percolation threshold, yields

$$(p_c^B)_{kk} \simeq \frac{1}{Z}\frac{d(d+1)}{2d-1}, \tag{101}$$

which should be compared to Eq. (49) for the central-force model. On the other hand, recall from Chapter 2 (see Section 2.5.2) that for scalar percolation one has (Shante and Kirkpatrick, 1971)

$$p_c^B \simeq \frac{1}{Z}\frac{d}{d-1}. \tag{102}$$

Thus, for 2D lattices, $(p_c^B)_{kk} = p_c^B \simeq 2/Z$, and simulations have confirmed that the percolation threshold of the KK model in 2D is the same as that of the scalar percolation. However, for 3D systems Eq. (101) predicts that

$$(p_c^B)_{kk} \simeq \frac{2.4}{Z}, \tag{103}$$

so that the average coordination number of the network at the percolation threshold, i.e., $\langle Z\rangle = (p_c^B)_{kk}Z$, should be about 2.4, whereas Eq. (102) yields $p_c^B \simeq 1.5/Z$; that is, in scalar percolation the average coordination number of the sample-spanning percolation cluster at p_c is about 1.5. The accuracy of Eq. (101) was confirmed by numerical simulations (He and Thorpe, 1985). Note a striking feature of Eq. (101): The percolation threshold does *not* depend on the force constants α and γ.

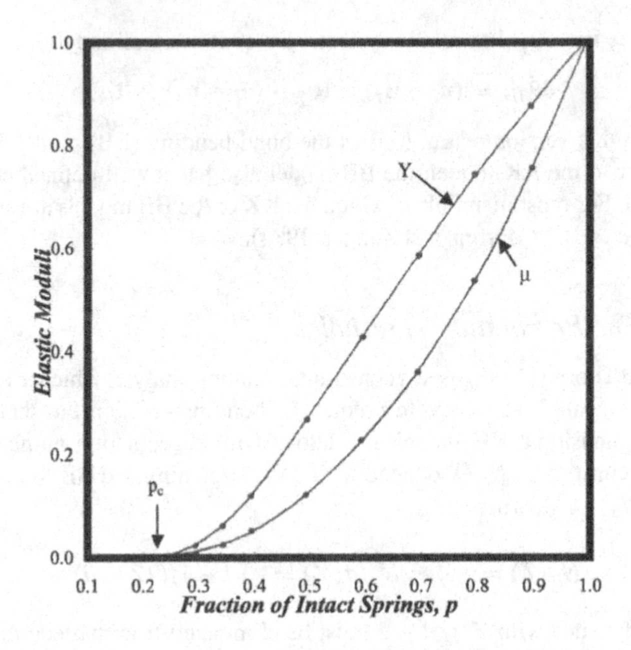

FIGURE 8.7. Dependence of the Young's modulus Y and shear modulus μ of the simple-cubic lattice, in the bond-bending model, on the fraction of p of the intact bonds in bond percolation, with $\gamma/\alpha = 1/4$.

On the other hand, the BB model in d-dimensions has the same percolation threshold as the scalar percolation if each site of the network interacts with at least $\frac{1}{2}d(d-1)$ of its nearest-neighbors. In practice, this can be easily achieved, and therefore, by using suitable three-body interactions between the sites of the lattice, one can obtain EPNs with percolation thresholds that are equal to those of scalar percolation, a highly desirable property as far as many practical applications are concerned. Hence, in what follows we restrict most of our discussions to the BB model. Figure 8.7 shows the typical dependence of the elastic moduli of the BB model in a simple-cubic network on the fraction p of the intact bonds.

8.11.4 The Force Distribution

Similar to rigidity percolation, the force distribution (FD) in the BB models has also been computed and studied (Sahimi and Arbabi, 1989,1993). Unlike the central-force model, there are some subtleties in this distribution that must be discussed. Far from p_c the contributions of the central forces totally dominate the elastic energy $\mathcal{H}$, in which case we may expect a unimodal FD, similar to that of rigidity percolation far from its percolation threshold p_{ce}. However, if we decrease γ, holding α fixed, bending of two bonds with respect to each other becomes easier, implying that the contribution to $\mathcal{H}$ of the BB forces increases. If γ/α is lowered to a small value, say of the order of 0.01, the contributions of the central and BB

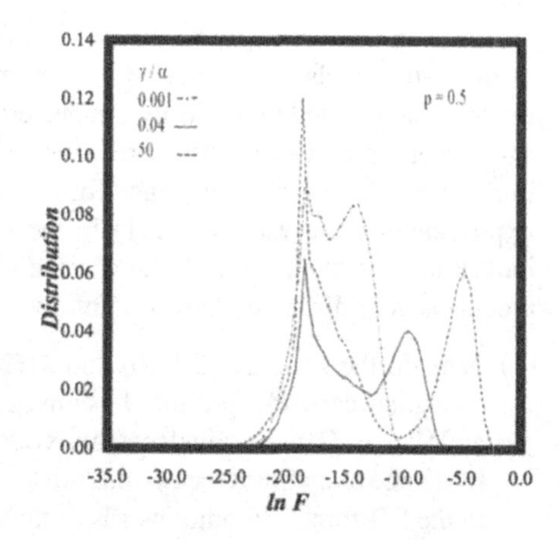

FIGURE 8.8. Distribution of the forces F in the bond-bending model in the square network at the bond percolation threshold $p_c = 1/2$, for several values of the ratio of the bending and stretching force constants (after Sahimi and Arbabi, 1993).

TABLE 8.3. Estimates of the exponents $\hat{\tau}(q)$ of the moments of the force distribution in the bond-bending model.

q	0	1	2	3	4
$\hat{\tau}(q)$ (Square)	-1.65 ± 0.07	1.08 ± 0.8	3.4 ± 0.4	7.3 ± 0.9	11.3 ± 1.5
$\tilde{\tau}(q)$ (Cubic)	-1.9 ± 0.1	2.2 ± 0.2	4.6 ± 0.5	8.5 ± 1.3	—

force become comparable, and therefore the FD takes on a distinct bimodal shape. Figure 8.8 presents the FD for the square network. The appearance of the second (smaller) maximum in the distribution, which is to the left of the larger maximum (due to central forces) is due to the BB forces. Further decrease in γ/α means that the BB contributions are so large that the central-force contributions can be neglected, and therefore we may expect the distribution to take on a unimodal shape again. At p_c and for fixed γ/α, the BB contributions are *always* larger than those of the central forces and depend only weakly on γ/α. As a result, although the FD is bimodal, the magnitude of the maximum due to the BB forces is much larger than that of the central forces. Similar results were also obtained for a simple-cubic network (Sahimi and Arbabi, 1993). The moments $M(q)$ of the force distribution in the square and cubic networks at p_c have also been studied, and their associated critical exponents have been determined (Sahimi and Arbabi, 1993); Table 8.3 presents the results. The values of $\tilde{\tau}(0)$ agree nicely with the estimates of the fractal dimension D_{bb} of the backbone of scalar percolation clusters (see Table 2.3), as they should.

8.11.5 *Comparison of the Central-Force and Bond-Bending Networks*

We now compare the central-force and the BB models, using the moments of their respective FD as the basis of the comparison. A glance at Tables 8.1 – 8.3 shows that the difference between the values of $\tilde{\tau}(q)$ for the two models is larger for larger values of q. This is due to the fact that the higher moments of the FD are affected more strongly by the fine details of the backbone, and the backbones of the two models are in fact very different. The backbone of the central-force model is dominated by multiply-connected loops, whereas that of the BB model can be well-approximated by the relatively simple node-link-blob model described in Section 2.6.3. Further evidence for the difference between the two backbones is provided by the values of $\bar{\tau}(1)$ in the two models which are quite different, indicating that the average force exerted on a bond of the backbone of the two models is very different. At this point, two questions must be addressed.

(1) Why do the scalings of $M(0)$ and $M(2)$ for rigidity bond percolation on the triangular network appear to be consistent with those of the BB model? Roux and Hansen (1989) argued that this is due to the *lever arm* effect which is caused by a force coupling at the two ends of a connected cluster, and hence it is similar to the BB forces. In other words, if the contributions of the central forces and lever arm effect are *comparable*, a central-force percolation network behaves effectively as one with BB forces. However, this argument is not rigorous.
(2) Why does rigidity site percolation not behave like rigidity bond percolation networks? If we calculate the FD for site percolation on the triangular network for $0.71 \leq p \leq 0.72$, although we still obtain a bimodal distribution (which is similar to the FD of a BB model), the magnitude of the smaller maximum, which is supposedly due to the lever arm effect, is always much smaller than the larger maximum (which is contributed by the central forces), implying that the lever arm effect for site percolation on the triangular network, if it in fact exists, is much weaker than that for bond percolation, so much so that the contribution of the central forces dominates that of the lever arm effect. This might explain why in the triangular network the scaling of elastic moduli in rigidity site percolation does not seem to be in the universality class of elastic moduli of rigidity bond percolation. Note that there appears to be no relation between the exponents of the FD of the BB model, in contrast with the central-force model for which Eq. (51) relates the exponents.

Study of the FDs of the rigidity percolation and the BB models shows that, *at most* two moments of the FD distributions of the two models, namely, $M(0)$ and $M(2)$, may have the same critical exponents. Therefore, one may consider a very general criterion for the universality of elastic percolation models: In order for two elastic percolation models belong to the same universality class, *all moments of their force distribution must have the same critical exponents*. According to this criterion then, rigidity percolation and the BB percolation models do *not* belong to the same universality class.

8.11.6 Scaling Properties

The critical exponents f and χ have been estimated for the BB models using a variety of techniques (Feng *et al.*, 1984; Feng, 1985a; Feng and Sahimi, 1985; Zabolitzky *et al.*, 1986; Bergman, 1986b; Bergman and Duering, 1986; Duering and Bergman, 1988). In 2D the most accurate estimate of f was obtained by a transfer-matrix method (Zabolitzky *et al.*, 1986), to be described below, and is given by

$$f \simeq 3.96 \pm 0.04, \tag{104}$$

while in 3D large scale Monte Carlo calculations yielded (Arbabi and Sahimi, 1988b)

$$f \simeq 3.75 \pm 0.10. \tag{105}$$

These results agree with $\tilde{\tau}(2)/\nu$ given in Table 8.3. We emphasize that, similar to rigidity percolation, the contributions of the corrections-to-scaling to the leading power law for the elastic properties of the BB model near p_c are quite large.

Two-dimensional superelastic percolation networks (Sahimi and Goddard, 1985) with BB forces have been studied extensively (Bergman, 1985, 1986b; Feng, 1985a; Bergman and Duering, 1986; Duering and Bergman, 1988). Such networks were first proposed by Sahimi and Goddard (1985) for modeling the divergence of the viscosity of a gelling solution near and below the gel point (see Chapter 9 for details). In addition to their relevance to modeling viscoelastic properties of gel polymers, SEPNs can be thought of as models of randomly-reinforced disordered materials. In analyzing the scaling properties of SEPNs, the contributions of the corrections-to-scaling have been found to be even more important than those for EPNs. The most accurate estimates of the critical exponent χ for BB models are (Arbabi and Sahimi, 1990c; Sahimi and Arbabi, 1993)

$$\chi \simeq 1.24 \pm 0.03, \quad \text{2D}, \tag{106}$$

$$\chi \simeq 0.65 \pm 0.03, \quad \text{3D}. \tag{107}$$

Both estimates are somewhat smaller than the corresponding values for the exponent s which characterizes divergence of the effective conductivity of conductor-superconductor percolation networks studied in Chapter 5.

More generally, let us rewrite Eq. (95) in a slightly different form,

$$\mathcal{H} = \frac{1}{2}\sum_{\langle ij\rangle}[(\mathbf{u}_i - \mathbf{u}_j)\cdot\mathbf{R}_{ij}]^2 k_{ij} + \frac{1}{2}\sum_{\langle jik\rangle}(\delta\theta)^2 m_{jik}, \tag{108}$$

where the meanings of k_{ij} and m_{jik} are clear. We now consider a two-component elastic network of linear size L with the two types of bonds, having (k_1, m_1) and (k_2, m_2) as their stretching and BB force constants, where $k_1 < k_2$ and $m_1 < m_2$. The fractions of the two types of bonds are $(1-p)$ and p, respectively. If

$$\tilde{m} = \frac{m_1/m_2}{L^{-(f+\chi)/\nu}}, \quad \tilde{k} = \frac{k_1/k_2}{L^{-(f'+\chi)/\nu}}.$$

then Duering and Bergman (1988) showed that, depending on the values of the elastic constants, one may have four distinct scaling regimes which are as follows,

$$\mu_e \simeq$$

$$\begin{cases} \mu_1 L^{\chi/\nu} h_1(k_1/m_1), & \tilde{k} \ll 1 \text{ and } \tilde{m} \ll 1, \\ \mu_1 h_2(k_1/m_1)(k_1/k_2)^{-\chi/(f'+\chi)}, & \tilde{k} \gg 1 \text{ and } \tilde{m} \ll 1, \\ \mu_1 h_3(k_1/m_1)(m_1/m_2)^{-\chi/(f+\chi)}, & \tilde{k} \ll 1 \text{ and } \tilde{m} \gg 1, \\ \mu_1 h_4\left[(k_1/k_2)^{f'+\chi}(m_2/m_1)^{f+\chi}, k_1/m_1\right](k_1/k_2)^{-\chi/(f'+\chi)}, & \tilde{k} \gg 1 \text{ and } \tilde{m} \gg 1. \end{cases} \tag{109}$$

Here, h_1, h_2, h_3, and h_4 are universal scaling functions, and μ_1 is the shear modulus of a homogeneous material with microscopic force constants k_1 and m_1. Observe that in the last three cases the shear modulus does not depend on the linear size L of the network. Moreover, in the fourth case, the scaling function h_4 depends on two variables, whereas for the first three cases the scaling functions depend only on a single scaling variable. Similar scaling laws can also be written down for the other elastic moduli of the network. The scaling representations (109) have been developed in analogy with those for the effective conductivity and the dielectric constant that were described in Chapter 6 (see SEction 6.5). Although, intuitively, one expects to have $f = f'$, Duering and Bergman (1988) found that, at least in 2D, $f' = \mu$, where μ is the critical exponent of the effective conductivity of the network near p_c.

8.11.7 *Relation with Scalar Percolation*

Rigorous upper and lower bounds were derived for the exponent f by Kantor and Webman (1984) and Roux and Guyon (1986) that link f with the topological exponents of scalar percolation defined and described in Chapter 2. These bounds are given by

$$1 + \nu d < f < \nu(D_{min} + d), \tag{110}$$

where D_{min} is the fractal dimension of the shortest path of a percolation cluster defined in Section 2.8. These bounds yield $11/3 \simeq 3.66 < f < 4.17$ and $3.64 < f < 3.85$ for 2D and 3D materials, respectively, and are relatively sharp.

Sahimi (1986a), and later Roux (1986), proposed that

$$f = \mu + 2\nu, \tag{111}$$

where μ is the critical exponent of the effective conductivity of percolation networks. The predictions of Eq. (111) are in excellent agreement with the numerical results given by (104) and (105), and thus Eq. (111) is likely to be an exact scaling law. The origin of Eq. (111) is clear: Elasticity introduces two extra factors of ξ_p, the correlation length of percolation, and therefore one must have

$$\frac{\text{elastic moduli}}{\text{conductivity}} \sim \xi_p^2,$$

which immediately results in Eq. (111).

On the other hand, Limat (1988a,b) argued that Eq. (111) is *not* exact because, according to him, one must include the effect of the *eccentricity* E_c of elastic percolation clusters which measures the strength of a coupling effect between displacements and rotation that tends to rigidify the loops of the cluster. He suggested instead that $f = \mu + 2\nu - \Delta_e$, where Δ_e is a supposedly new exponent which describes the power-law behavior of E_c near p_c. However, the numerical estimates of f given by (104) and (105) indicate that $\Delta_e(d = 2) \simeq 0.0$ and $\Delta_e(d = 3) \simeq 0.01$. In the mean-field approximation which becomes exact for $d \geq 6$, one has $f = 4$, $\mu = 3$, and $\nu = 1/2$ and thus, $\Delta_e = 0$. Therefore, $\Delta_e \simeq 0$ for all d, implying that either E_c is not a critical quantity near p_c at all (i.e., it does not diverge or vanish at p_c), or that near p_c one must have, $E_c \sim \ln(p - p_c)$, so that $\Delta_e = 0$.

It has also been proposed (Arbabi and Sahimi, 1990c; Sahimi and Arbabi, 1993) that

$$\chi = \nu - \frac{1}{2}\beta. \qquad (112)$$

The predictions of Eq. (112) agree with the numerical estimates of χ given by (106) and (107). Moreover, for 1D percolation systems, $\nu = 1, \beta = 0$, and hence $\chi = 1$, which is easily seen to be an exact result, and in the mean-field approximation, which is exact for $d \geq 6$, one has $\chi = 0$ which also agrees with the prediction of Eq. (112) (with $\nu = 1/2$ and $\beta = 1$). Thus, Eq. (112) may be an exact scaling relation for all dimensions $1 \leq d \leq 6$, and in particular for $d = 3$. Limat (1988c, 1989) also proposed that $\chi = s - \Delta_s$, where Δ_s is the analogue of the exponent Δ_e defined above for the superelasticity problem, and estimated that $\Delta_s \sim s/5$ and hence, $\Delta_s(d = 2) \simeq 0.26$ and $\Delta_s(d = 3) \simeq 0.14$, consistent with the numerical results given above.

8.11.8 Fixed Points of Vector Percolation: University of the Poisson's Ratio

Bergman and Kantor (1984) found that the ratio K/μ of the bulk to shear modulus of EPNs appears to approach a constant value as the percolation threshold is approached. Based on an EMA, they conjectured that $K/\mu = 4/d$ represents an exact and universal value. A universal value of K/μ at p_c would represent a type of a fixed point, much like the fixed points of renormalization group transformations described in Chapter 5 (see Section 5.11), and to be described below for the elasticity problem. Further studies and simulations of Schwartz *et al.* (1985) and Arbabi and Sahimi (1988b) indicated that, in the lattice models this ratio does appear to approach an apparently universal value, and that this type of fixed point behavior seems to hold for the ratios of other elastic moduli as well. For example, in the simple-cubic lattice and for all values of γ/α [the two force constants of the BB model; see Eq. (103)], the ratio Y/μ of the Young's and shear moduli appears to approach 4/3 (Arbabi and Sahimi, 1988b), in agreement with the conjecture of Bergman and Kantor (1984). Renormalization group calculations, to be described in the next section, also confirm the existence of such a fixed point. Moreover, in an experimental study with a randomly perforated Cu foil (Benguigui and Bergman, 1987), the ratio C_{11}/μ was also found to be universal and consistent with the theoretical prediction.

That the ratio of the elastic moduli might be universal is plausible, because as the percolation threshold of the lattice is approached, all the elastic moduli follow the same power law with the *same* critical exponent [see Eq. (3)], and therefore the ratios of the elastic moduli represent *amplitude ratios*. It is known (Aharony, 1980) that certain amplitude ratios at the percolation threshold are universal and depend only on the dimensionality of the system. Thus, it is plausible that the ratios of the elastic moduli of EPNs at the percolation threshold might also be universal. However, we must mention that Schwartz *et al.* (1985) found that in certain continuum models this apparent universality breaks down, the reason for which is presently not known.

Let us point out that, a universal value of the ratio of the elastic moduli appears to also exists in brittle fracture of materials, near the point at which a sample-spanning crack forms. The existence of such a fixed point, which has been demonstrated both by computer simulations and experiments, has important practical implications for brittle fracture of disordered materials, and will be discussed in Chapter 8 of Volume II.

8.11.9 Position-Space Renormalization Group Method

Similar to the effective conductivity of disordered networks, the elastic moduli of such networks, and in particular their power-law behavior near the percolation threshold, can be computed and studied by a position-space renormalization group (PSRG) method. This idea, which is a significant extension of the PSRG method for the effective conductivity, was first developed by Feng and Sahimi (1985) for the BB model. It was later extended to the central-force networks by Knackstedt and Sahimi (1992). More recently, Novikov *et al.* (2001) also utilized this method to compute the elastic moduli of 3D models of disordered materials.

Since an EPN with central and BB forces is described by three parameters, namely, α, γ, and p (the fraction of the intact bonds), one must also develop three RG recursion relations for the three parameters. As the percolation threshold of the BB model is the same as that of scalar percolation, the recursion relation for $p' = R(p)$ (see Section 5.11), the renormalized probability of having an intact bond in the renormalized network, does not change on passing from the scalar model to the BB model. Therefore, let us consider the 2D RG cell of Figure 5.14. Then, the recursion relation for p' is given by Eq. (5.181) which has a fixed point at $p^* = 1/2$, the exact bond percolation threshold of the square network. To develop the recursion relation for α', the renormalized stretching force constant, we fix the displacements of the left-most sites of the RG cell at $(0, \Delta)$, where Δ is a constant, hold the displacements of the right-most sites of the cell at (0,0), solve for the displacements of the interior sites (by solving the usual equation, $\partial\mathcal{H}/\partial\mathbf{u}_i=0$), and determine all the configurations of the RG cell that transmit elastic forces. In analogy with Eq. (5.192), we also write

$$p' \ln \alpha' = \sum_{i=1}^{n} a_i(p) \ln h_i(\alpha, \gamma), \tag{113}$$

where $a_i(p)$ is the probability of the spanning configuration i, $h_i(\alpha, \gamma)$ is its equivalent bond stretching force constant, n is the total number of the spanning configurations, and $\sum_i a_i(p) = p'$. Since those configurations of the RG cell that transmit elastic forces are *exactly* the same as those through which electrical current passes, the coefficients $a_i(p)$ in Eqs. (5.192) and (113) are also identical. Thus, for the RG cell of Figure 5.14 with linear size $b = 2$, we obtain

$$p' \ln \alpha' = p^5 \ln \alpha + 4p^4 q \ln \alpha \left(\frac{\alpha + 6\gamma}{2\alpha + 9\gamma} \right) + p^4 q \ln \alpha$$
$$+ (6p^3q^2 + 2p^2q^3) \ln(\alpha/2) + 2p^3q^2 \ln \left(\frac{4\alpha\gamma}{2\alpha + 9\gamma} \right), \tag{114}$$

where $q = 1 - p$.

Determination of the recursion relation for the BB force constant γ is somewhat more complex. The exterior sites of the RG cell are displaced as much as $\Delta/2$ in *each direction*, and therefore one must have symmetric cell configurations in which both of rescaled bonds are present. For the 2D RG cell of Figure 5.14 with $b = 2$, one finds that

$$p' \ln \gamma' = p^5 \ln \gamma + p^4 q \ln(5\gamma/9) + 2p^3 q^2 \ln(\gamma/3)$$
$$+ (4p^4 q + 6p^3 q^2 + 2p^2 q^3) \ln(\gamma/9). \tag{115}$$

In analogy with the conduction problem, Eqs. (114) and (115) can be thought of as the working equations for determining the renormalized force constants, and hence the effective elastic moduli of the RG cell (that is, the network).

Together with Eq. (5.181), Eqs. (114) and (115) have fixed points at $(\alpha^*, \gamma^*, p^*)$. However, if we let $r = \gamma/\alpha$ and $r' = \gamma'/\alpha'$, we obtain a single recursion relation of the form, $r' = F(p, r)$, which has two stable fixed points at $p = p^* = p_c = 1/2$, namely, $r^* = 0$ and $= \infty$, and one stable point at $r^* = 1/66843$, which is the relevant fixed point. If we linearize the recursion relation for r', we obtain an eigenvalue, $\lambda_r = \partial r'/\partial r$ evaluated at $r = r^*$ and $p = p^*$, which, for the RG cell used here, is given by $\lambda_r \simeq 0.875$. This means that, after many iterations of the RG transformation, the ratio r flows stably into the point r^*. In analogy with the problem of computing the effective conductivity described in Chapter 5 (see Section 5.11), we may write

$$\frac{f}{\nu} = -\frac{\ln \lambda_\alpha}{\ln b}, \qquad \lambda_\alpha = \frac{\partial \alpha'}{\partial \alpha} |_{\alpha=\alpha^*}, \tag{116}$$

where ν is the critical exponent of percolation correlation length, which is computed (in the PSRG approach) by Eq. (5.186). To compute λ_α one must iterate Eqs. (114) and (115) (or, equivalently, the recursion relation for r') many times to reach the true fixed points for α and γ. In the present problem, one obtains, after many iterations, $\lambda_\alpha \simeq 0.1617$, which implies that $f \simeq 3.75$, where we used $\nu \simeq 1.43$, the PSRG prediction for $b = 2$ RG cell [see the discussion after Eq. (5.186)]. This estimate is only about 5% lower than what is given by Eq. (104), hence indicating the high accuracy of the PSRG method in 2D. Calculations with a 3D RG cell of size $b = 2$ (see Figure 5.14) yields (Novikov *et al.*, 2001), $f \simeq 3.2$, which is about 30% lower than the estimate given by Eq. (105). On the other hand, the same PSRG method in 3D, but for the superelasticity problem, yields (Novikov *et al.*, 2001), $\chi \simeq 0.63$, only 3% lower than the estimate given by Eq. (107).

The PSRG method described here also provides many useful insights into the behavior of EPNs. For example, the fact that the fixed point $r^* = 1/66843$ is so small is *not* accidental, but has clear physical meaning. Recall that near p_c strongly bonded regions, i.e., the multiply-connected regions of the backbone, are connected by tenuous weak regions. Compared to these regions, the strong regions can be considered as being perfectly rigid, and therefore the elastic properties are actually controlled by the weak regions, which can be roughly approximated by tortuous chains. If $\gamma \gg \alpha$, i.e., if $r^* \gg 1$, and for any fixed chain length, it will cost less elastic energy to accomplish a displacement of the chain's ends by adjusting the length of bonds parallel to the stress, rather than bending bond angles, so that the elastic moduli of the system scale as $(p - p_c)^\mu$, i.e., the same as the effective conductivity. However, if $\gamma \ll \alpha$, i.e., if $r^* \ll 1$, then the opposite is true and the behavior of the system crosses over to the expected behavior, i.e., Eq. (3) [or Eq. (5) in the case of SEPNs]. Therefore, any reasonable PSRG method *must* yield $r^* \ll 1$, as is the case here.

Another insight that PSRG methods provide is the confirmation that the ratio of any two elastic moduli of disordered materials near p_c is a type of a RG fixed point. That is, as p_c is approached and regardless of the values of α and γ, this ratio takes on a universal value. For example, Figure 8.9 presents the ratio K/μ of the bulk and shear moduli of the simple-cubic network, calculated for various values of α/γ by the above PSRG method. It is seen that as one approaches $p = p^* \simeq 0.21$, which is the prediction of the PSRG method for the bond percolation threshold of the simple-cubic lattice with a RG cell of size $b = 2$ (see Sections 5.11 and

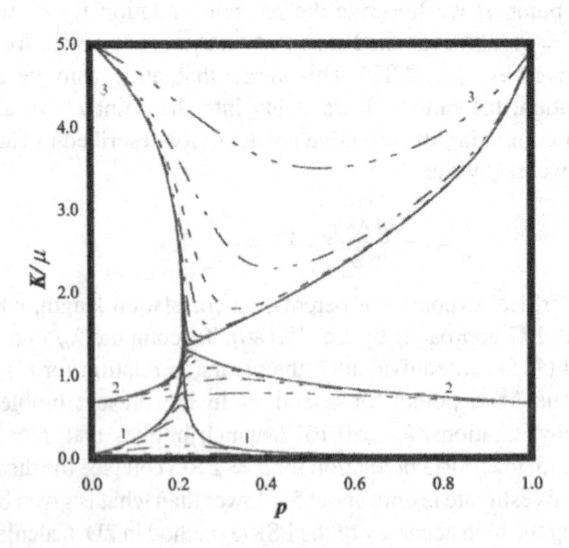

FIGURE 8.9. Ratio of bulk and shear moduli of the simple-cubic lattice versus the fraction p of the intact bonds, indicating a universal value of K/μ at p_c. Numbers on the curves indicate the initial value (at $p = 1$) of $K_1/\mu_1 = K_2/\mu_2$ which are 0.025 (1), 0.75 (2) and 5 (3) (after Novikov *et al.*, 2001).

5.12), values of K/μ for all cases approach the same value which is about 4/3, in agreement with what was discussed in the last section.

The third important insight that the PSRG approach provides is the universality of the geometrical exponents of rigidity percolation, such as the correlation length exponent ν_e defined by Eq. (40). According to the RG calculations of Knackstedt and Sahimi (1992), these exponents are the same for site, bond and correlated rigidity percolation, provided of course that the range of the correlations is short. We may then conclude that PSRG methods provide a very useful tool for computing the elastic properties of disordered materials.

8.11.10 *Effective-Medium Approximation*

The BB models have also been studied within an EMA (Schwartz *et al.*, 1985; Mall and Russel, 1987). The general form of the EMA for this case is similar to those for the continuum models in that, the equations for the effective elastic constants (for stretching and BB forces) and the moduli are coupled and nonlinear. They predicts that the bond percolation threshold of the system is at $p_c^B = 2/Z$, the same as for the scalar percolation, implying correctly that the percolation threshold of the BB model is just p_c, the scalar percolation threshold. The numerical predictions of the EMA equation for the BB models in 2D are also accurate, except very near the percolation threshold. A somewhat different approach for deriving such an EMA was developed by Böttger *et al.* (1993).

8.12 Transfer-Matrix Method

Similar to the effective conductivity, an efficient method for computing the elastic moduli of EPNs and SEPNs is by a transfer-matrix method, both for the central-force (Lemieux *et al.*, 1985; Roux and Hansen, 1988) and the BB models (Bergman, 1985; Zabolitzky *et al.*, 1986). As described in Chapter 5 (see Section 5.14.2) for the effective conductivity, the transfer-matrix method is a technique for calculating the elastic properties of percolation networks, and in particular their scaling properties near p_c, in which the properties of the network, in the shape of a long strip, are computed *exactly*. The method for computing the elastic moduli is very similar to what we described in Chapter 5 for the effective conductivity. Briefly, a symmetric compliance matrix $\mathbf{S} = (S_{ij})$ provides the displacements $\mathbf{u}_i$ that must be applied at each of the nodes at the two right-hand columns of the strip in order to produce a given set of force components $\mathbf{F}_i$:

$$\mathbf{u}_i = \sum_j \mathbf{F}_j S_{ij}.$$

Then, similar to the computation of the effective conductivity, additional bonds are added, one by one or in groups of a few (depending on the topology of the network), to the right-hand column of the strip, and the compliance matrix is recalculated at every step. Starting with $\mathbf{S} = \mathbf{0}$ is equivalent to applying a zero displacement to the left-most side of the strip. As the strip is built up, new elements of **S** appear. At the same time, older elements S_{ij} of the compliance matrix are discarded as soon as site i or j becomes an interior node of the strip, which happens when the total internal force exerted upon the site can be calculated entirely in terms of the displacements at the existing sites. Since, at equilibrium, the total internal force must vanish, one can eliminate all elements of **S** connected to that site. The compliance matrix **S** is of course much simpler for the central-force model than for the BB model. More details about this algorithm for the BB model are given by Duering and Bergman (1988).

8.13 The Beam Model

A model related to the KK and BB models is the beam model. In this model each bond of the lattice is a beam rather than a spring. The elastic behavior of each beam is governed by three material dependent constants, $a_1 = l/(YA)$, $a_2 = l/(\mu A)$, and $a_3 = l^3/(YI)$, where Y and μ are the Young's and shear moduli, A the cross section of the beam, and I the moment of inertia for flexion. Each site of the network is characterized by a displacement vector $\mathbf{u}_i$ and a rotational angle φ_i. If a site is rotated, the beams bend accordingly, and thus the local momenta are taken into account. Consider, for example, a square network. For a horizontal beam that connects sites i and j one has a longitudinal force F_{lj} acting at j given by

$$F_{lj} = \alpha_1(\mathrm{u}_{ix} - \mathrm{u}_{jx}), \tag{117}$$

a shear force

$$F_{sj} = \alpha_2(\mathrm{u}_{iy} - \mathrm{u}_{jy}) + \frac{1}{2}\alpha_2 l(\varphi_i + \varphi_j), \tag{118}$$

and a flexural torque

$$\tau_j = \frac{1}{2}\alpha_2 l(\mathrm{u}_{iy} - \mathrm{u}_{jy} + l\varphi_j) + \alpha_3 l^2(\varphi_i - \varphi_j), \tag{119}$$

where $\alpha_1 = 1/a_1$, $\alpha_2 = 1/(a_2 + a_3/12)$, and $\alpha_3 = \alpha_2(a_2/a_3 + 1/3)$. In a similar way, analogous equations for the vertical direction can be written down. In mechanical equilibrium the sum of all forces and torques acting on a site must be zero, giving rise, in the continuum limit, to the classical Cosserat equations (Cosserat and Cosserat, 1909; Toupin, 1964; Mindlin, 1964).

Elastic properties of lattices of beams were studied extensively several decades ago (Kaliski, 1963; Askar and Cakmak, 1968; Klemm and Wozniak, 1970) and also more recently (Lewinski, 1988, 1989), but these studies did not include disorder in the model. In the context of disordered lattices, and in particular elastic percolation networks, Roux and Guyon (1985) were the first to study this model. The percolation threshold of this model is of course the same as that of scalar percolation. Numerical simulations of Roux and Guyon (1985) also indicated that the scaling properties of the beam model are identical with those of the BB model.

We will come back to the beam model in Chapter 8 of Volume II where we describe lattice models of brittle fracture of materials under compression.

8.14 The Granular Model

Another discrete elastic model is the so-called *granular* or the *disk model*, which was first introduced by Schwartz *et al.* (1984) for studying vibrational properties of granular materials. In this model, the elastic energy of the system is given by

$$\mathcal{H} = \frac{1}{2}\sum_{\langle ij\rangle} e_{ij}\left\{\alpha[(\mathbf{u}_i - \mathbf{u}_j)\cdot\mathbf{R}_{ij}]^2 + \gamma\{\mathbf{R}_{ij}\times[\mathbf{R}_{ij}\times(\mathbf{u}_i - \mathbf{u}_j)] + a(\varphi_i + \varphi_j)\mathbf{z}\times\mathbf{R}_{ij}\}^2 + \delta(\varphi_i - \varphi_j)^2\right\}, \tag{120}$$

where a is the radius of the disks, δ is the elastic constant associated with the restoring torque that is necessary when two disks counterrotate, and $\mathbf{z}$ is a unit vector normal to the plane of the disks. The rest of the notations is the same as before. Figure 8.10 illustrates the physical meanings of α, γ and δ. With such couplings, any relative displacement or rotation of two neighboring disks is done at a cost to the energy $\mathcal{H}$ of the system. Therefore, the percolation threshold of this model is the same as that of scalar percolation. In terms of a lattice model, the granular model is essentially equivalent to the beam model described above. This model was studied by Feng (1985a), Schwartz *et al.* (1985), and Limat (1988a),

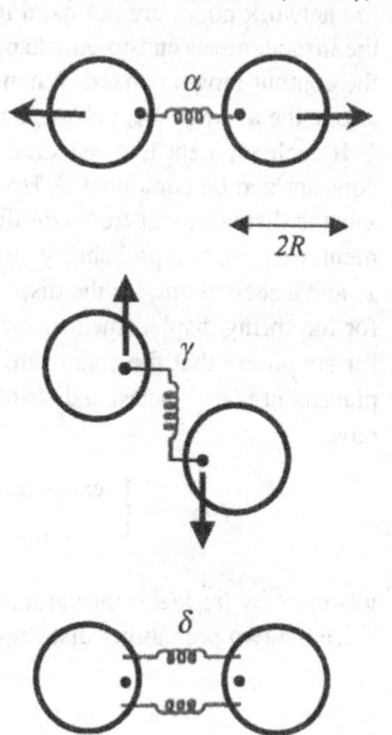

FIGURE 8.10. The physical meanings of the three elastic constants of the granular model [see Eq. (120)].

who reported that the scaling properties of the elastic moduli of the model are similar to those of the BB model.

8.15 Entropic Networks

All the network models of elastic moduli of disordered materials that have been described so far in this chapter are enthalpic models in that, the expressions for the energy of the networks are due to the enthalpy, or the potential energy, of the system. In practice, however, entropic networks—those in which the entropic part of the free energy of the system plays the leading role—are also very important. This is particularly true for polymers and gels, and also biological materials described in Section 8.1. Therefore, it is important to consider entropic networks, and the differences between them and enthalpic networks studied so far.

To construct an entropic network, flexible chains of sites (or atoms) are packed together with sufficient density that a given chain is close to its neighbors at many points along its contour length. To give the network rigidity, different chains are welded to one another at points where they are close to each other. After welding, there are n chain segments, each a disordered chain in its own right, having been cut from a larger chain, with end-to-end displacement $\mathbf{r}_{ee}$ with a probability distribution which is a generalization of Gaussian distribution, as the chains are not ideal formed by a random walk, but are in fact self-avoiding walks. Although the positions of the chains during the network formations are frozen, the network nodes are not fixed in space at non-zero temperatures, and therefore the instantaneous end-to-end displacement of a segment may change, even though the contour length is fixed. When the network deforms in response to an external stress, the average $\langle \mathbf{r}_{ee} \rangle$ changes as well.

If each segment is considered as an entropic spring, then its effective spring constant can be computed. A Hookean spring has an energy $\mathcal{H}_s = \frac{1}{2}\alpha x^2$, where x is the displacement from equilibrium. To compute the spring constant of a segment, we need two probability distributions $P(x)$, one for the spring displacement x, and a second one for the displacement of a chain. The probability distribution for the spring displacement is the Boltzmann factor, $\exp(-\mathcal{H}_s/k_BT)$. Assuming for simplicity that the chains are ideal, the probability distribution for their displacement is the Gaussian distribution. Thus, ignoring a normalization factor, we have

$$P(x) \sim \begin{cases} \exp[-\alpha x^2/(2k_BT)], & \text{for the spring,} \\ \exp(-x^2/2\sigma^2), & \text{for the ideal chain,} \end{cases} \tag{121}$$

where $\sigma^2 = \langle r_{ee}^2 \rangle / d$ is the variance of the distribution in d dimensions. Therefore, since the two probability distributions must be the same, we obtain

$$\alpha = \frac{dk_BT}{\langle r_{ee}^2 \rangle}. \tag{122}$$

Having obtained an approximate expression for the effective spring constant of the chain segments, we can now derive an approximate expression for the elastic moduli of entropic networks. Suppose, for example, that a triangular network is made of segments of the ideal chains described above. As discussed earlier in this chapter, the area per node S_n of the network is given by, $S_n \sim \frac{\sqrt{3}}{2}\langle r_{ee}^2 \rangle$, and therefore the spring constant of the chain segments can be written as, $\alpha \sim 3\sqrt{3}k_BT/(2S_n)$ (in 3D). The 2D density ρ of the chains is given by $\rho = 3/S_n$, as we have three chains per node, and therefore, $\alpha = \sqrt{3}\rho k_BT/2$. In Section 8.1 we showed that, in an enthalpic triangular network, the bulk and shear moduli are related by, $K_e = 2\mu_e$. Assuming that the same relation roughly holds for the present entropic triangular network, we obtain $K_e \sim 2\mu_e = 3\rho k_BT/4$. Similar results (but with different numerical factors) can be derived for 3D entropic networks.

This estimate for the elastic moduli of entropic networks is, of course, only an approximation because, (1) the chains are not ideal, formed by simple random walks, but are, as is well-known, self-avoiding walks, and therefore the probability distribution for the end-to-end displacements of the chains (or chain segments) is not Gaussian; (2) the repulsive interaction between the chains has been ignored, and (3) the effect of chain entanglement has been neglected. Nevertheless, computer simulations of Plischke and Joós (1998) confirmed the density-dependence of the elastic moduli of the entropic networks. Therefore,

$$\mu_e \sim \rho k_BT. \tag{123}$$

We may expect Eq. (123) to be accurate if the chain density is not too large. At large chain densities, the second and third assumptions listed above break down completely, and therefore Eq. (123) loses its accuracy. We will come back to entropic networks in Chapter 9, where we describe the application of elastic and superelastic percolation networks to describing viscoelastic properties of polymers and gels.

Summary

The geometrical and elastic properties of vector percolation models are now well-understood. An important result that has emerged is that, at least at zero temperature, the scaling properties of the elastic moduli of such models near the percolation thresholds are *not* identical with those of their effective conductivity. This difference has important practical implications that will be discussed in Chapter 9, where we describe applications of vector percolation models to modeling of mechanical properties of disordered materials and compare their predictions with the relevant experimental data.

It has also become clear that elastic percolation networks with stretching and bond-bending forces are the simplest vector percolation models that possess two fundamental properties that any reasonable model of heterogeneous materials must have: A rotationally-invariant elastic energy, and a percolation threshold that is identical with that of scalar percolation.

The contribution of enthalpy and entropy to the elastic properties of a network depends on the flexibility of its elements or bonds. If the bonds are connected to each other over distances that are much smaller than the persistence length of the corresponding material for which the elastic network is intended, then enthalpy may dominate the elasticity. In this case, the shear resistance is not much smaller than that of conventional materials. However, if the distance between the nodes of the elastic network is much larger than the persistence length of the network, then entropic effects dominate the elastic properties of the network, in which case an equation such as Eq. (123) may provide reasonable estimates of the shear modulus.

9
Rigidity and Elastic Properties of Network Glasses, Polymers, and Composite Solids: The Discrete Approach

9.0 Introduction

In Chapter 8 we described and analyzed discrete vector percolation models. These models, together with the continuum models of elastic properties described in Chapter 7, provide a fairly complete understanding of rigidity and linear elastic properties of disordered materials. At the same time, although certain properties of disordered materials, such as their vibrational density of states that was described and discussed in Section 6.6, may be approximately modeled by the discrete scalar models, such as the random resistor percolation networks, modeling and predicting their mechanical properties require, in most cases, taking into account the effect of their true vectorial nature. To predict such properties, elastic percolation networks (EPNs) and superelastic percolation networks (SEPNs), are powerful tools. The purpose of the present chapter is to describe and discuss applications of these models to predicting the rigidity and linear elastic properties of several important classes of disordered materials, and compare their predictions with the relevant experimental data.

The materials that we consider in this chapter are characterized either by the existence of a percolation threshold, at which their effective properties either vanish or diverge, or a percolation-type transition point (such as the gel point during polymerization; see Section 9.2). Our discussions in Chapter 7 should have made it clear that the rigidity and linear elastic properties of two-phase materials that are far from their percolation threshold, or those in which the contrast between the properties of the two phases is not large, are well-described and predicted by various mean-field theories, such as the effective-medium approximations (EMAs), the rigorous upper and lower bounds and other analytical approximations that were derived in Chapter 7. However, the effective properties of two-phase materials in which the contrast between the properties of the two phases is large, particularly those in which one or both phases form large clusters and is therefore near its percolation threshold, deviate greatly from the predictions of mean-field theories and other analytical approximations. It is the description of this type of two-phase materials that is best done by vector percolation, and in particular models that are based on EPNs and SEPNs. Thus, in this chapter we focus our attention mainly

TABLE 9.1. Estimates of the critical exponents of transport properties in scalar and vector percolation in d-dimensions. Values of f and χ, the exponents for the elastic moduli, for the central-force model refer to bond percolation, while those of μ and s, the conductivity exponents, are independent of the model. ν is the critical exponent of the percolation correlation length. For the central-force model ν_e, the value of which (given in Chapter 8) is different from ν, should be used.

d	μ/ν	s/ν	f/ν	χ/ν	Model
2	0.9745 ± 0.0015	0.9745 ± 0.0015	2.97 ± 0.03	0.92 ± 0.03	bond bending
	—	—	2.95 ± 0.25	0.92 ± 0.02	central force
3	2.27 ± 0.01	0.835 ± 0.005	4.3 ± 0.1	0.74 ± 0.04	bond bending
	—	—	2.1 ± 0.1	0.80 ± 0.03	central force

on this type of disordered, two-phase materials, particularly those in which one or both phases are near their percolation threshold. The region near the percolation threshold is often referred to as the critical region. It is the region in which the rigidity and linear elastic properties (as well as other effective properties of disordered materials studied in Chapters 4–6) follow universal scaling and power laws, independent of the materials' morphology. As pointed out in Chapter 5, in many materials the critical region is quite extended, in which case vector percolation and the associated universal scaling and power laws are powerful tools for describing quantitatively predicting materials' properties over much of the range of the volume fractions of their phases. At the same time, whenever necessary and/or possible, we also describe materials' properties outside the critical region in order to gain a better understanding of the strengths and weaknesses of such models. For convenience, and as a basis for comparison with the experimental data, we summarize in Table 9.1 the currently-accepted estimates of the various critical exponents for scalar and vector transport properties of materials with percolation disorder, including the central force and the bond-bending (BB) models described in Chapter 8.

Let us emphasize that our description and discussion of application of the discrete models of rigidity and elastic properties of heterogeneous materials, and in particular vector percolation, is by no means exhaustive, as the number and variety of materials to which such models may be applicable is too large. For example, many concepts that are directly or indirectly related to vector percolation and the discrete models of rigidity properties of materials, have been utilized to explain seemingly unrelated phenomena. Hammonds *et al.* (1997) showed, for example, that the existence of adsorption sites in zeolites, an important class of nanoporous catalytic materials that are used heavily in the chemical industry, is related to the existence of floppy modes that were described in Chapter 8. They demonstrated that zeolite frameworks can support large bands of rigid unit modes in the **k** (wave vector) space. Localized rigid unit modes are formed within such frameworks that do not distort the constituent tehrahedra to any significant degree. These localized modes enable cations at certain sites to pull the framework—with essentially zero cost in elastic energy—in such a way that oxygen-cation bonding distances become exactly optimal for the cation of interest. Several other such application of vector percolation and rigidity transitions are described by Thorpe and Duxbury (1999).

9.1 Network Glasses

In a pioneering work, Zachariasen (1932) introduced the notion of continuous random networks for studying network glasses, and proposed that the structure of a glass consists of an "extended three dimensional network lacking periodicity with an energy content comparable with that of the corresponding crystal." He argued that for the energy of the glass to be comparable with that of the crystal, the coordination number of the polyhedra and the manner by which they are connected must be the same in a glass and in the corresponding crystal. However, he stated that, unlike crystals, the relative orientations of adjacent polyhedra "vary within rather wide limits," leading to a lack of long-range order in glasses. Zachariasen (1932) envisaged these networks maintaining local chemical order but, by incorporating small structural disorder, having a non-crystalline topology. Warren (1934) coined the term *random networks* for such disordered materials, while Gupta and Cooper (1990) proposed the term *topologically disordered* for describing the structure of glasses as it conveys the essence of the Zachariasen's view of a glass structure. Careful diffraction experiments, from which the radial distribution function (see Chapter 3) can be determined, have confirmed the correctness of this model, and therefore it is now widely accepted.

If one examines the amorphous structure of covalent glasses, one may classify their building blocks into two groups: In one group is the bonding structure that consists of covalent bonds with densities that are of the order of $10^{22} - 10^{23}$ cm^{-3}, and can be specified by their chemical and topological structures. In the second group are defects, such as impurities, dangling bonds, and "wrong" bonds (for example, homopolar bonds in stoichiometric alloys). The density of the defects is typically two orders of magnitude less than that of the covalent bonds. Therefore, the covalent bonds are primarily responsible for such electronic properties as the band-gap energy, in contrast to gap states that are caused by the defects. Hence, topological properties play the most important role in the glassy characters, and therefore the application of the concept of percolation theory to characterizing many important properties of network glasses is natural.

However, the idea of a relation between network glasses and percolation, particularly vector percolation, was not developed for a long time after Zachariasen's early work. The reason was that, whereas the liquid-glass phase transition is the result of the system going out of a complete metastable equilibrium, most percolation models, as we have emphasized in this book, are equilibrium models [dynamical percolation models, such as the one suggested by Sahimi (1986b), are not considered in this book], and therefore it was thought for a long time that there can be no connection between the liquid-glass phase transition and the percolation model. The connection between the two transitions was finally suggested by Cohen and

Grest (1979) via the concept of *free volume*. Fox and Flory (1951) and others [for a review of the older literature on this subject see Grest and Cohen (1983)] had proposed earlier that the liquid-glass transition resulted from disappearance of the free volume of the amorphous phase at some temperature. The basic assumptions of the Cohen–Grest model were as follows.

(1) With each molecule one can associate a local volume v of molecular scale.
(2) When the local volume reaches a critical value v_c, the excess value is the free volume.
(3) Molecular transport takes place only when the total free volume, having a volume greater than some critical value, is approximately equal to the molecular volume formed by the redistribution of the free volume. The redistribution of the free volume requires no local free energy.

These assumptions are valid if each molecule is restricted to move mainly within a cell or cage defined by its nearest neighbors. There is considerable experimental evidence, in dense liquids, that indicates that this is indeed the case. If the temperature T of the system is near or above the glass transition temperature T_g, then the local free energy E_i of a cell depends only on its volume v_i. Near T_g, the function $E_i(v)$ can be approximated by (Cohen and Grest, 1979)

$$E_i(v) = \begin{cases} E_0 + \frac{1}{2}a_1(v - v_0)^2, & v \leq v_c \\ E_0 + a_2(v - v_0) + \frac{1}{2}a_1(v - v_0)^2, & v \geq v_c, \end{cases} \tag{1}$$

where a_1 and a_2 are two constants, and v_c is a critical volume. The existence of v_c enables one to classify the cells into two groups. Those with $v > v_c$ are liquid-like and have a free volume $v_f = v - v_c$, whereas cells with $v < v_c$ are solid-like. In effect, the free volume is the difference between the glass and crystalline volumes. If $f(v)$ is the probability distribution of having a cell with volume v, then the fraction p of the liquid-like cells is obviously

$$p = \int_{v_c}^{\infty} f(v)dv,$$

However, a free exchange of free volume can take place only between liquid-like cells that are nearest neighbors and have a sufficiently large number of nearest neighbor liquid-like cells, and therefore this is a correlated site percolation (Stanley, 1979). Moreover, there is a critical value $p_c(Z)$ of the probability p (where Z is the coordination number) such that, for $p > p_c$ one has a sample-spanning cluster of liquid-like cells, whereas for $p < p_c$ the liquid-like cells form only isolated clusters, and therefore the system as a whole is solid and in glassy state. This connection between percolation and liquid-glass transition made it possible to calculate the thermodynamic properties of the system. This aspect of the problem was reviewed in detail by Grest and Cohen (1983), to whose paper we refer the interested reader.

Free volume in alloy glasses is a scalar quantity that can be measured rather easily but is difficult to precisely define theoretically. It is now clear that the percolation model of Cohen and Grest (1979), which is a scalar model, is applicable to metallic glasses. What we are mainly interested in this chapter is the mechanical properties of network glasses, which we now describe and discuss.

9.1.1 Rigidity Transition

The question, "why is a glass rigid and has mechanical properties," is interesting, because we know that while in a rigid system a local disturbance results in a bulk response, the same is not true of liquids. For example, if we apply a small force to a crystal, it will lead to its collective motion, as it is energetically favorable for the crystal's atoms to maintain their *relative* positions. The same force, when applied to a liquid, generates local rearrangements of the atoms, and after a short time the liquid "forgets" its structure before the force was applied.

Unlike a liquid though, the low-temperature state of a glass depends on its history, which implies that its response to a shear deformation is not simple. If the glass contains N atoms, then the number of its possible initial low-temperature configurations grows exponentially with N. Furthermore, on short time scales, a glass exhibits complete memory of its initial configuration, when it is exposed to an applied force, whereas at much longer times, the glass "flows" and hence forgets its past, a phenomenon that is usually referred to as *aging*. Thus, a glass has certain properties that distinguishes it from a liquid.

The next question, which is important to predicting the mechanical properties of a glass, is, what distinguishes a glass from a crystal? In addition to the fact that, if a network glass and a crystal each has N atoms, their number of states is, respectively, $\exp(N)$ and $O(1)$, a crystal breaks translational symmetry and possesses long-range spatial ordering, whereas there exists no such simple long length scale associated with a glass. Moreover, the response of a crystal to a shearing force is invariant under time-translation, whereas for long measurement times a glass loses its rigidity, and therefore does not exhibit the time-invariance that the crystal possesses.

Thus, how does one characterize the rigidity transition and the existence of non-vanishing elastic moduli in a network glass? Phillips (1979,1981), in an attempt to explain the strong glass-forming propensity of some chalcogenide materials, proposed that covalent bonding in glasses can be optimized when the average coordination number $\langle Z \rangle$ of the three-dimensional (3D) covalent network in mechanical equilibrium is about 2.4. However, in order to provide a firm theoretical foundation for his proposal, Phillips invoked the scalar percolation model, but since, as discussed in Chapter 2, the average coordination number of random scalar percolation at the percolation threshold p_c is about 1.5 and not 2.4, Phillips himself rejected this idea! His idea was significantly refined by Thorpe (1983) who established the connection between this phenomenon and a vector percolation model. According to the Phillips–Thorpe idea, in covalent networks with stretching and Keating-type (BB) forces described in Section 8.11.1, for average coordination numbers $\langle Z \rangle < 2.4$, the number N_c of interatomic forces per atom that act as constraints is less than the network dimensionality $d = 3$, which is the same as the number of degrees of freedom N_d per atom. Under a shearing force such networks deform easily and possess a finite number N_0 of zero-frequency modes, $N_0 = N_d - N_c$. When $\langle Z\rangle > 2.4$, N_c is larger than N_d, and thus such networks are macroscopically rigid and possess non-vanishing elastic moduli. Hence, when $\langle Z\rangle = 2.4$, one has $N_c = N_d$ and therefore $N_0 = 0$ represents the vector percolation threshold of the system.

Two crucial points must, however, be noted here. First, the analysis that lead to $\langle Z\rangle \simeq 2.4$ is of mean-field type. However, as mentioned in Section 8.11.3, computer simulations of He and Thorpe (1985) confirmed the accuracy of this mean-field analysis. Secondly, $\langle Z\rangle \simeq 2.4$ agrees with the predictions of Eqs. (8.101) and (8.103), and thus represents the average coordination number of 3D Kirkwood–Keating models at their percolation threshold.

9.1.2 *Comparison with the Experimental Data*

Over the past several years, there have been many experimental tests of the Phillips–Thorpe (PT) proposal, using a wide variety of techniques and measuring several distinct properties of network glasses. What follows is a brief description and discussion of some these tests. The list of the papers and the data that they reported, that we discuss below, is by no means exhaustive, but represents what we view as the most illuminating tests of the PT proposal for applicability of vector percolation to rigidity transition in network glasses.

The first of such evidence was probably provided by Ota *et al.* (1978), although no mention was made of vector percolation. They reported measurements, among other things, of the bulk, shear and Young's moduli of Ge_xSe_{1-x} glasses, which are perhaps the simplest of such materials, over a wide range of the composition x. Over the composition range $0 < x < 1/3$, the coordination number of Ge and Se are 4 and 2, respectively, and thus the average coordination number of the network glass is $\langle Z\rangle = 2(x+1)$. Therefore, an average coordination number $\langle Z\rangle \simeq 2.4$ corresponds to a critical composition $x_c \simeq 0.2$. Measurements of Ota *et al.* (1978) indicated that at around $x \simeq 0.2 - 0.25$, the three elastic moduli increase significantly, thus signaling a rigidity transition.

The same type of glass was used by Bresser *et al.* (1986), who carried out experiments using such binary glasses in the composition range $0 < x < 1/3$ and ^{129}I Mössbauer emission spectroscopy, and presented strong evidence that at $x_c \simeq 0.23 \pm 0.02$ the Mössbauer-site intensity ratio I_B/I_A exhibits dramatic threshold behavior. At $x = 0$ (i.e., g–Se) only B sites are observed, and thus I_B/I_A is infinite. For $0 < x < 0.15$ the ratio I_B/I_A decreases with increasing x, reaching a minimum around $x \simeq 0.15$, but as $x \to x_c \simeq 0.23$ this ratio increases and reaches a maximum. Beyond x_c the ratio I_B/I_A decreases monotonically with x again. The molar volume V_m of Ge_xSe_{1-x} also exhibits a threshold behavior. For $0 < x < 1/3$ the molar volume decreases monotonically with x, reaching its minimum at $x_c \simeq 0.23 \pm 0.02$, beyond which it increases again monotonically as x does. Note that $x_c \simeq 0.23 \pm 0.02$ agrees reasonably well with the PT estimate, $x_c = 0.2$, which corresponds to an average coordination number of about 2.4. Similar threshold behavior for the molar volume was also reported by Sreeram *et al.* (1991) who measured V_m for several chalcogenide glasses incorporating Ge, Sb, Se, As, and Te, and observed the threshold behavior at $\langle Z \rangle \simeq 2.4$.

Further experimental evidence of the applicability of vector percolation to network glasses was provided by Halfpap and Lindsay (1986), who measured the elastic moduli C_{11} and C_{44} of the Ge-As-Se glasses over a wide range of composition. The coordination number of As is 3, and therefore if the glass has the composition $Ge_xAs_ySe_{1-x-y}$, its average coordination number is $\langle Z \rangle = 4x + 3y + 2(1 - x - y) = 2x + y + 2$. Thus, varying x and y allows one to obtain glasses over a wide range of compositions and to test the applicability of vector percolation to glass formation. Despite different local structures of the glasses that are obtained when x and y are varied, measurements of Halfpap and Lindsay (1986) indicated that at $\langle Z \rangle \simeq 2.4$ there is a dramatic increase in the elastic moduli C_{11} and C_{44}; an example is shown in Figure 9.1. More evidence for applicability of vector percolation to network glasses was provided by the experimental data of Mahadevan and Giridhar (1989), who also measured the elastic moduli C_{11} and C_{44} of various glasses formed with As, Ge, S, Sb, Se, and Te, and found essentially the same behavior that had been reported by Halfpap and Lindsay (1986).

Experiments of Tatsumisago *et al.* (1990) shed further light on this problem. They showed that the relevance of vector percolation to glass formation can be better understood by studying the *liquid-state* behavior of glass-forming systems. They measured the elastic moduli, the viscosity, the thermal-expansivity, heat

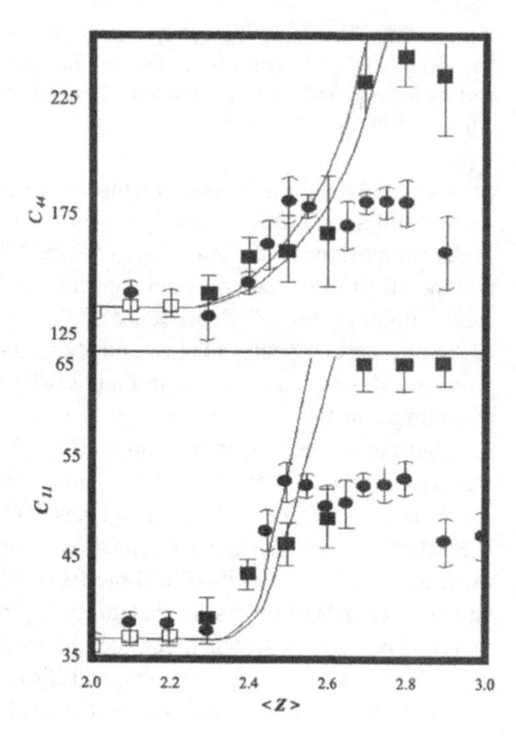

FIGURE 9.1. Threshold behavior of the elastic moduli (in kbar) of glasses versus their average coordination number $\langle Z \rangle$. Circles and squares correspond, respectively, to high- and low-As series of glasses. The theoretical curves on the left are for amorphous Ge, while those on the right are for crystalline Ge (after Halfpap and Lindsay, 1986).

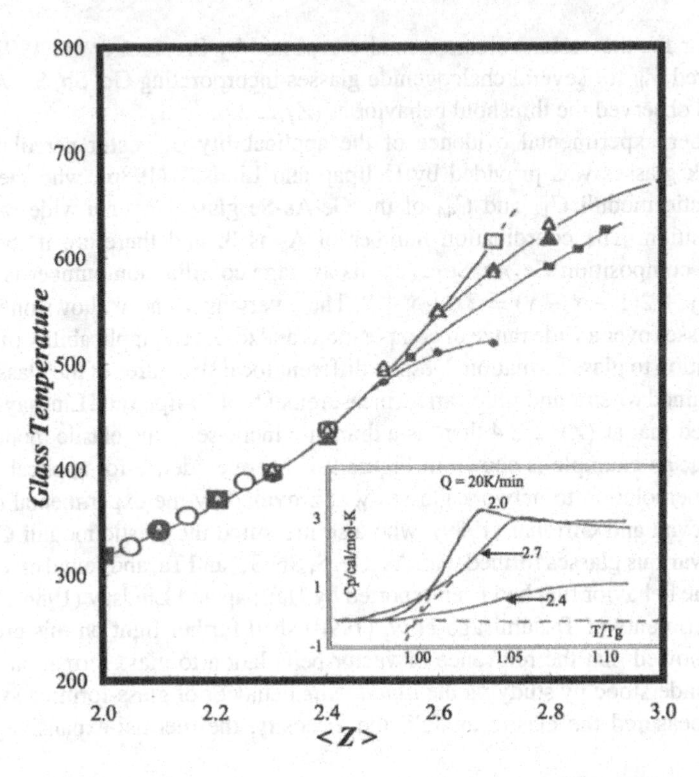

FIGURE 9.2. Glass temperature T_g versus the average coordination number $\langle Z\rangle$ of $Se_{1-x}(Ge_yAs_{1-y})_x$. Symbols are the data for various values of y. The inset shows the specific heat capacity of the glasses, with the numbers on the curves giving the value of $\langle Z\rangle$ (after Tatsumisago *et al.*, 1990).

capacity, and the glass-transition temperature T_g for various glasses formed with As, Ge, and Se and showed that T_g is a universal function of $\langle Z\rangle$, i.e., independent of the composition of the glass, up to $\langle Z\rangle \simeq 2.4 - 2.5$. It is only above this point that the effect of the composition appears; see Figure 9.2. There is, however, no sudden upturn in the $\langle Z\rangle$-dependence of T_g, but the difference between the shapes of the heat capacity curves below and above the critical value $\langle Z\rangle \simeq 2.4$ is quite pronounced; this is also shown in Figure 9.2. Other interesting features included a minimum in the activation energy at $\langle Z\rangle \simeq 2.4$ for the viscosity data, and a minimum in the heat capacity jump at $\langle Z\rangle \simeq 2.4$ and T_g. These data all indicated a structural transition at $\langle Z\rangle \simeq 2.4$. Similar data were also presented by Böhmer and Angell (1992) and by Wagner and Kasap (1996).

Further evidence for the PT transition is provided by the fact that the very small mass of H causes vibrational modes of H or H_2O to split off from the main vibrational bands of Ge-As-Se chalcogenide glasses and induces such excitations to have long lifetimes, which can then be measured with infrared hole-burning methods (see for example, Uebbing and Sievers, 1996, and references therein). Such experiments indicate linear dependence of the relaxation rates on the average coordination number $\langle Z \rangle$, with a clear change in the slope at $\langle Z \rangle \simeq 2.4 \pm 0.1$, accompanied by a small but unambiguous cusp.

Most recently, Wang *et al.* (2000) also investigated the Raman spectra described in Sections 6.6.9 and 6.7 in chalcogenide glasses, GeS_{1-x} and Ge_xSe_{1-x} with $0 \leq x \leq 0.42$, resulting in an average coordination number $2 \leq \langle Z \rangle \leq 2.84$. Their experimental data indicated that the relative degree of fragility, representing the structural relaxation of the network, is larger for $\langle Z \rangle < 2.4$ than for $\langle Z \rangle > 2.4$, thus signaling a rigidity transition at $\langle Z \rangle \simeq 2.4$.

9.1.3 Rigidity Transition at High Coordination Numbers

Although the above experimental data seem to provide strong evidence for the applicability of vector percolation to describing elastic properties of network glasses, they are not free of ambiguities. For example, Yun *et al.* (1989) reported absence of any threshold behavior at $\langle Z \rangle \simeq 2.4$ in the elastic properties Ge_xSe_{1-x}, contradicting the earlier data of Bresser *et al.* (1986) described above. Some other glasses (see below) exhibit an onset of vector percolation at values *higher* than $\langle Z \rangle \simeq 2.4$. Some of such differences may be attributed to the oxide contamination effects that adversely affect the chemical structure of the glasses. One may also argue that in real materials the van der Waals forces between unbounded neighbors can smear out the percolation transition. Let us now discuss these issues in more detail.

Feng *et al.* (1997) presented Raman scattering data for Ge_xX_{1-x} glasses ($X =$ S or Se) that indicated that there is a stiffness threshold, at a mean coordination number $\langle Z \rangle \simeq 2.46 \pm 0.01$, for the shear (transverse) optic mode. Although their reported average coordination number is only slightly larger than the predicted value of 2.4, Phillips (1999) suggested that the critical connectivity for the hydrostatic is 2.4, but is about 2.45 for the shear value, which is in agreement with the data of Feng *et al.* (1997).

Srinivasan *et al.* (1992) measured the thermal diffusivity of two sets of $Ge_xSb_ySe_{1-x-y}$, $Ge_xAs_yTe_{1-x-y}$ and $Si_xAs_yTe_{1-x-y}$ glasses, using a photoacoustic (PA) technique. In these experiments, the sample with appropriate thickness (about 150μm), in the form of a thin disc, is glued to the PA cell with silver paste which acts as a good thermal backing medium. The sample is then irradiated with intensity modulated light. The amplitude $\mathcal{A}$ of the PA signal is then recorded as a function of the modulation frequency ω. The thermal diffusion length in the sample varies with ω. From a plot of $\log \omega$ versus $\log \mathcal{A}$, a characteristic frequency ω_c is determined as the frequency at which the slope of the plot changes. The thermal diffusivity D_T is then given by, $D_T = \omega_c \ell^2$, where ℓ is the thickness of the sample. Measurements of Srinivasan *et al.* (1992) indicated that the thermal diffusivity of the three glasses increases monotonically with the average coordination number $\langle Z \rangle$, reaching a maximum at $\langle Z \rangle \simeq 2.6$, beyond which it decreases again.

The data of Srinivasan *et al.* and similar ones may be explained based on a modified picture of network glasses proposed by Tanaka (1989), who studied network glasses $Ge(Si)_xAs_yS(Se)_{1-x-y}$ and proposed, following Zallen (1983), that the dimensionality d of the network plays an important role in glasses' properties.

Zallen suggested that $d = 1, 2$ and 3, respectively, for amorphous Se, $As_2S(Se)_3$, and Si(Ge). For example, $d = 1$ for Se implies a chain-like morphology in which entangled chain molecules are held together by (weak) van der Waals intermolecular forces, whereas $d = 3$ implies 3D continuous-random networks. Tanaka (1989) found that the molar volumes for various glasses exhibit two extrema: They reach their minimum at an average coordination number $\langle Z\rangle \simeq 2.4$, which is in agreement with the data of Bresser *et al.* (1986) and Sreeram *et al.* (1991) described above, while their maximum is reached at $\langle Z\rangle \simeq 2.67$. The optical band-gap energies of the glasses also exhibited a maximum at $\langle Z\rangle \simeq 2.67$, while their elastic moduli indicate two upturns, one at $\langle Z\rangle \simeq 2.4$ and another one at $\langle Z\rangle \simeq 2.67$. Similar data had earlier been reported by Tanaka *et al.* (1986). To explain such data, the following scenario was proposed by Tanaka (1989).

(1) In glassy Se(S) with coordination number $Z = 2$, the network is roughly one dimensional.
(2) When As and/or Fe atoms are introduced into these chalcogen glasses, the 1D molecules are crosslinked, so that the structure gradually transforms for $d = 1$ to $d = 2$. At $\langle Z\rangle \simeq 2.67$ layer structures are fully evolved, and hence $d = 2$. The layer structures are segmental.
(3) With further increase in the local connectivity, the structures undergo the transition to 3D networks.

The prediction, $\langle Z\rangle \simeq 2.67$, may be explained based on a constraint-counting argument developed by Tanaka (1989). To derive his prediction by this method, let us first recall the PT constraint-counting method. The number of constraints N_c in 3D covalent glasses is $N_c = \frac{1}{2}Z + (2Z - 3)$, where $\frac{1}{2}Z$ represents the stretching (radial) constraints and $(2Z - 3)$ the number of BB valence-force constraints, with 2 corresponding to the two degrees of freedom in a spherical representation (r, θ, φ) of an atom bonded to another atom located at the origin, and the 3 being related to the system rotation around the three main axes. Since the total number N_d of degrees of freedom in 3D is 3, and because at the rigidity transition we must have $N_c = N_d$, we obtain the PT estimate $Z = \langle Z\rangle = 2.4$. Using a simple modification of this analysis, the value of the average coordination number, $\langle Z\rangle \simeq 2.67$, for the glasses that were studied by Tanaka (1989) and Srinivasan *et al.* (1992) can be predicted. In this case, the number of constraints is given by, $N_c = \frac{1}{2}Z + (Z - 1)$, so that the number of the angular constraints is reduced from $(2Z - 3)$ for fully 3D glasses to $(Z - 1)$, due to the assumption of having network glass configurations that are planar. As the number of degrees of freedom is still $N_d = 3$, and at the transition one must have, $N_c = N_d$, one obtains $Z = \langle Z\rangle = 8/3 \simeq 2.67$, in agreement with Tanaka's and Srinivasan *et al.*'s data. Therefore, it appears that in these experiments a glass is fixed stably in a 3D space, if $\langle Z\rangle \simeq 2.67$.

Liu and Löhneysen (1993) measured the low-temperature specific heat C_p of g-As_xSe_{1-x} and fitted their data to $C_p = a(x)T + b(x)T^3$. They found that both the PT rigidity transition at $\langle Z\rangle \simeq 2.4$ and the Tanaka transition at $\langle Z\rangle \simeq 2.7$ contribute to the fitted functional form, with a double maximum in $a(x)$. They also found that if they subtract the contribution of the second term of the fitted equation, then $a(x)$, which is a measure of the two-level point defects, vanishes at $\langle Z\rangle \simeq 2.4$, hence proving that, in the absence of interlayer cross-linking, one would have a crystal. One may also explain these data by the fact that the defects are all associated with chemical disorder (for example, homopolar bonds), and that the concentration of these is minimized at $x = 0.4$.

9.1.4 *Effect of Onefold-Coordinated Atoms*

The idea that network glasses must become rigid when their average coordination number is about 2.4 was refined by Boolchand and Thorpe (1994) in order to treat networks that have onefold-coordinated (OFC) atoms. Since such atoms represent, in the language of percolation theory, dangling ends, one might believe that they can be removed as in, for example, H in a–Si, a–C, and a–Ge-Si networks [see, for example, Mousseau and Thorpe (1993) and references therein]. However, it is known that the presence of H in a–Si network degrades the network's elastic moduli. Therefore, it is instructive to explicitly consider the effect of OFC atoms. This was undertaken by Boolchand and Thorpe (1994) who used a constraint-counting method of the type described above. Consider a 3D covalent network with N atoms with N_Z of them having a coordination number of Z. Following exactly the same argument that was utilized for deriving Eqs. (8.100), for $Z \geq 2$, one has $N_\alpha = Z/2$ constraints (corresponding to the stretching force of the Kirkwood–Keating model) and $N_\gamma = 2Z - 3$ constraints (corresponding to the bond-bending term of the KK model). For an atom with $Z = 1$, one has $N_\alpha = 1/2$ and $N_\gamma = 0$. Therefore, the total number of constraints is given by

$$N_c = \frac{1}{N}\left[N_1 \times \frac{1}{2} + \sum_{Z\geq 2} N_Z\left(\frac{1}{2}Z + 2Z - 3\right)\right], \tag{2}$$

and the number N_0 of the zero-frequency modes per atom is

$$\begin{aligned} N_0 = N_d - N_c &= 3 - \left(\frac{1}{2}\frac{N_1}{N} + \sum_{Z\geq 2}\frac{5}{2}Z\frac{N_Z}{N} - \sum_{Z\geq 2} 3\frac{N_Z}{N}\right) \\ &= 3 - \left(\frac{1}{2}\frac{N_1}{N} + \sum_{Z\geq 1}\frac{5}{2}\frac{ZN_Z}{N} - \frac{5}{2}\frac{N_1}{N} - \sum_{Z\geq 1} 3\frac{N_1}{N} + 3\frac{N_Z}{N}\right) \\ &= 6 - \frac{5}{2}\langle Z\rangle - \frac{N_1}{N}, \end{aligned} \tag{3}$$

where $\langle Z\rangle = \sum_{Z\geq 1} ZN_Z/N$. As the general condition for the onset of rigidity is $N_0 = 0$, we obtain

$$\langle Z\rangle = 2.4 - 0.4\frac{N_1}{N}. \tag{4}$$

Equation (4) implies that the critical average coordination number at the onset of rigidity depends on the fraction of the OFC atoms, and that their presence reduces the critical value. An example of a glass to which this model may be applicable is ternary system $Ge_xS_{1-x-y}I_y$ which is formed by a fast quench over a wide range of compositions. The Ge, Se, and I atoms possess a coordination number of 4, 2, and 1, respectively, and bond in conformity to the $8 - \mathcal{N}$ rule. For a given x, one can determine the critical value y_c of iodine concentration y at which Eq. (4) is satisfied. Since for this glass, $\langle Z \rangle = 4x + 2(1 - x - y_c) + y_c = 2(x - y_c + 1)$, one obtains,

$$y_c = \frac{1}{3}(10x - 2). \tag{5}$$

Experimental data of Dembovskii *et al.* (1971) indicated that the maximum iodine concentration at which a glass can form is $y_c \simeq 0.55$ at $x = 0.386$, whereas, given this value of x, Eq. (5) predicts that $y_c \simeq 0.62$, slightly higher than the experimental value. On the other hand, had one ignored the effect of OFC atoms, the predicted critical concentration (i.e., at $\langle Z \rangle = 2.4$) would have been completely wrong.

9.1.5 *Stress-Free Versus Stressed Transition*

On the other hand, studies of Georgiev *et al.* (2000) on bulk As_xSe_{1-x} indicated that a stress-free rigidity transition occurs at $\langle Z \rangle = 2.29 \pm 0.01$ and another transition to a stressed phase occurs at $\langle Z \rangle = 2.37 \pm 0.01$, both of which are below the value of 2.4. One possible way of explaining such data is provided by the work of Thorpe *et al.* (2000) who considered two types of models. In one, they constructed random bond networks by randomly positioning points in a plane. When such networks are very large, they contain no closed loops, and therefore are similar to the Bethe lattices (see Chapter 2). Thorpe *et al.* found that the rigidity transition in such networks is *first-order*, which is similar to the central-force percolation transition on the Bethe lattices which, as discussed in Section 8.6.6, is also first-order. The rigidity transition on this network occurred at $\langle Z \rangle = 2.3893$. In the second model, Thorpe *et al.* began with a smaller coordinated floppy network and added bonds to it. The additional bonds lead to rigid regions, and so long as they are unstressed, i.e., can have their natural length (angle) without being forced to change by their surroundings, they are retained in the network. If addition of a bond results in its being redundant and creating a stressed region, it is removed. If only central forces operate in the system, one can continue building up this network until it is entirely unstressed, in which case the percolation transition occurs at $\langle Z \rangle = 2.375 \pm 0.001$. The transition at this point is second-order. However, because of the presence of bond-bending forces, it may happen that while one part of the set of constraints associated with a given bond removes floppy modes, the other part introduces strain, and therefore there is a *stress transition* at which a single additional bond causes a large stressed region to appear, so that the associated stress percolation transition is expected to be first-order. In Thorpe *et al.*'s simulations this type of transition occurred at $\langle Z \rangle \simeq 2.392$. Although these values are not in numerical agreement with the data of Georgiev *et al.* (2000), they do point out the possibility of developing an appropriate vector percolation model for explaining their data.

Summarizing, there is little, if any, doubt that a percolation model can explain many properties of glasses. While scalar percolation can be used for explaining the behavior of metallic glasses (via the percolation of free volume) and computing their thermodynamic properties, vector percolation explains the rigidity transition that takes place in network glasses at an average coordination number $\langle Z \rangle$. Above this transition point, the elastic moduli of the material are predicted well by the EPNs based on the Kirkwood–Keating model described in Chapter 8.

9.2 Branched Polymers and Gels

Polymeric materials have wide applications in many branches of science and technology. In addition, they have many interesting, and in many cases unusual, properties that justify their study. Moreover, two key works in the early 1970s demonstrated clearly how polymers can be studied both theoretically and experimentally. de Gennes (1972) showed that there is a close connection between linear polymers (i.e., those with monomers that have functionality or coordination number $Z = 2$) and a statistical mechanical model, namely, the n-vector model. If no two polymer parts occupy the same point in space, the resulting linear polymer corresponds to the limit $n \to 0$, and the most suitable model for such polymers is the path of a self-avoiding walk in which a particle performs a random walk in space with the restriction that it never visits any point more than once. If this restriction is lifted, the polymer conformation corresponds to the path of an ordinary random walk [see Hughes (1995) for a comprehensive discussion] or the $n = -2$ limit of the n-vector model. de Gennes' discovery made it possible to apply modern methods of statistical mechanics, such as the renormalization group theory, to the study of linear polymers. On the experimental side, Cotton (1974) used small-angle neutron scattering and a labelling technique in which the hydrogen atoms along the polymer chain were replaced by deuterium, to demonstrate that it is possible to detect one polymer chain among many others in a solution. Thus, one can analyze a single polymer chain and compare the experimental results with the theoretical predictions. In fact, Cotton's data were in excellent agreement with the theoretical predictions based on the self-avoiding walk model.

The works of de Gennes and Cotton were restricted to linear polymers, i.e., those in which each monomer is connected to two neighboring monomers. However, if one uses monomers with functionality $Z > 2$, so that each monomer can be connected with more than two neighboring monomers, then at least two other classes of polymers can be obtained. If the reaction time t is relatively short and below, but close to, a characteristic time t_g, then one obtains branched polymers in the solution, usually called *sol*, that form a viscous solution. Such branched polymers are large but finite clusters of monomers. On the other hand, if the reaction time is larger than t_g, a very large solid network of connected monomers appears that is usually called a *chemical gel*, or simply a gel. The gel network has

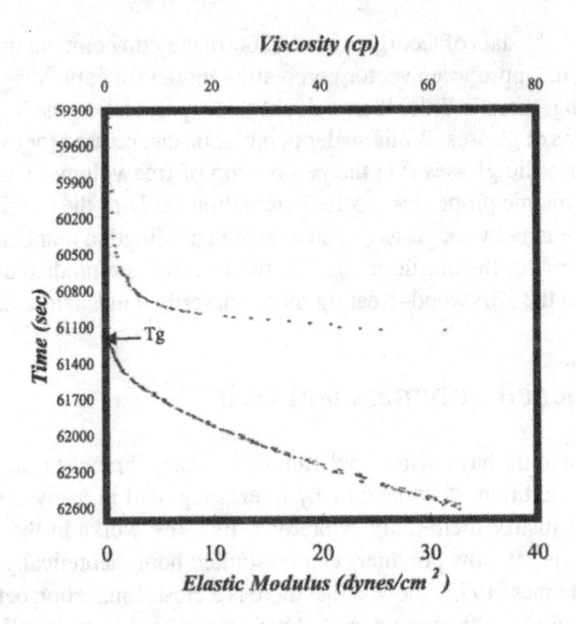

FIGURE 9.3. Threshold behavior of the viscosity of a gelling solution, and the elastic moduli of the gel polymer above the gel point, as a function of gelling time.

interesting structural, mechanical, and rheological properties which are described in this chapter. The characteristic time t_g is called the *gelation time*, and the point at which the gel network appears for the first time is called the *gel point* (GP). An example is shown in Figure 9.3. Most of us are already familiar with such sol-gel transformations in our daily lives, since we all know about milk-to-cheese transition, pudding, gelatine, etc. However, materials that contain gels, or use their specific properties, are numerous. An important example is eye humor. In addition, gels play an important role in laboratory technology (e.g., gel chromatography), in the fabrication of a wide variety of products, such as glues, cosmetics, contact lenses, etc., and in food technology. In addition, the sol-gel transition is a general phenomenon that has been utilized for producing a variety of ceramic materials (Brinker and Scherer, 1990).

Chemical reactions are responsible for the interconnectivity of the monomers in chemical gels. In general, there are three types of chemical gelation:

(1) *Polycondensation:* In this type of polymerization one begins with either bifunctional units A-A or trifunctional ones B_3, or more generally Z-functional units B_Z. In polycondensation reaction the A units are linked with the B units, with each elementary reaction being accompanied by the elimination of a molecule between units of A and B. Thus, a polymer network of the type shown in Figure 9.4 is formed in which the polymer chains are terminated by either A or B. No two units of the same class can participate in a reaction with each other, and therefore there is always exactly one bifunctional unit between polyfunctional units in the polymer network.

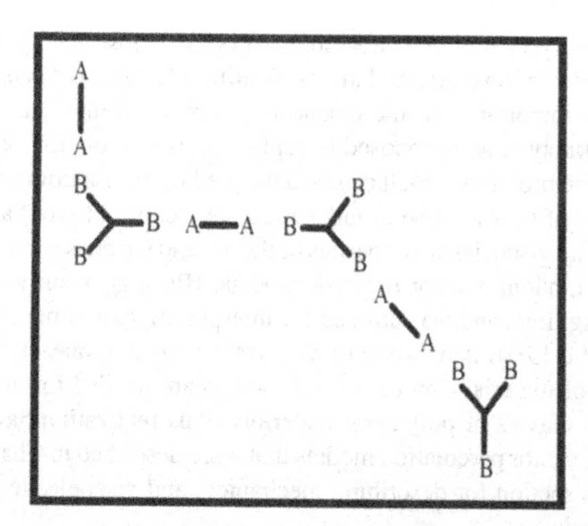

FIGURE 9.4. A polymer network formed by polycondensation.

(2) *Vulcanization:* This process starts with long linear polymer chains in a solution. The chains are then crosslinked by small units. An example is rubber, the elasticity of which is due to the introduction of S-S bonds between polyisoprene chains. One needs only a small number of bonds to crosslink the chains and form an interconnected polymer network.
(3) *Additive polymerization:* Similar to polycondensation, the initial solution contains two types of units which are A=A units that are bifunctional when the double bond opens, and B=D=B units that are quadrifunctional when the two double bonds open independently of each other. If the reaction polymerizes A=A units, one obtains A-A-A-A-··· chains, whereas reaction between the A units and the B=D=B units reticulates the network. The length of the chains between two reticulation points is not fixed, but depends crucially on the initiation process and on the relative concentrations of the bi-and quadrifunctional units.

In addition to such chemical gels, one may also have *physical gels* in which the monomers or particles are attached to each other by relatively weak and *reversible* association, or such physical processes as entanglement. A well-known example is silica aerogel, the dynamical properties of which were described and discussed in Section 6.6.9, and will further be considered below. Another example is a solution of gelatin in water below a certain critical temperature where a coil to helix transition takes place, and bonds appear to form by winding of helixes of two adjacent chains. Such physical gels can be made and also destroyed by thermal treatment.

In this section we describe modeling of morphological, mechanical and viscoelastic properties of the sol and gel phases, especially near the gel point GP. There is clearly a phase transition as one passes from the sol phase to the gel phase, and this phase transition, called *gelation*, has been described by a percolation model. As discussed in Chapter 2, modeling gel network formation was pioneered by Flory (1941) and Stockmayer (1943) whose theory is essentially equivalent to percolation on Bethe lattices. Stauffer (1976) and de Gennes (1976a) emphasized the importance of the deviations from the Bethe lattice solution of Flory and Stockmayer, and proposed to replace it by percolation on 3D lattices. This aspect of the problem, which can be described by the random scalar percolation or a variant of it, is now well-understood. de Gennes (1976a) also proposed that the elastic and viscoelastic properties of the gel and sol phases can be described by appropriate random resistor network models. His suggestion was widely accepted for a long time, and was utilized for interpreting experimental data. It was recognized in the 1980s that, while de Gennes' suggestion may be applicable to certain classes of materials, more general models are needed for at least several other important classes of polymeric materials. This realization gave rise to the development of vector percolation models that were described in Chapter 8, which are used in this section for describing mechanical and viscoelastic properties of polymers and gels.

9.2.1 Percolation Model of Polymerization and Gelation

Consider a solution of molecules or monomers with functionality $Z \geq 3$. To understand the connection with percolation, suppose that the monomers occupy the sites of a periodic lattice. With probability p, two nearest-neighbor monomers (sites) can react and form a chemical bond between them. If p is small, only small polymers are formed. As p increases, larger and larger polymers (clusters of connected monomers) with a broad size distribution are formed. This mixture of clusters of reacted monomers and the isolated unreacted monomers represents the sol phase. For $p > p_c$, where p_c is a characteristic value that depends on Z (or, the number of nearest-neighbors of a monomer of the lattice), an "infinite" cluster of reacted monomers is formed which represents the gel network described above. Near the GP the gel usually coexists with a sol such that the finite polymers are trapped in the interior of the gel. As $p \to 1$, almost all monomers react, and the sol phase disappears completely. Thus, p_c signals a *connectivity* transition: For $p > p_c$, an infinite cluster, together (possibly) with a few finite-size clusters, exist and thus the system is mainly a solid gel. The fraction of chemical bonds formed at the GP (which is related to the fraction of reacted monomers at the GP) is obviously the analogue of the bond percolation threshold p_{cb}. Thus, it should be clear that formation of branched polymers and gels is very similar to a percolation process. In reality, the monomers do not react randomly; there are usually some correlations between reaction of the monomers with one another, but such correlations do not change the essence of the main results, obtained based on the random percolation model, that are described in this chapter.

9.2.2 Morphological Properties of Branched Polymers and Gels

What is the signature of the sol-gel phase transition? Experimental studies of sol-gel transitions usually proceed by measuring the time evolution of the rheological (e.g., the viscosity) or mechanical properties (e.g., the elastic moduli) during the chemical reaction leading to gelation, assuming that the experimental parameter—time or frequency—and the theoretical one—the number of crosslinks—are linearly related in the vicinity of the GP. Rheological measurements are usually performed by using a cone and plate rheometer or by the more accurate magnetic sphere rheometer. The ranges of shear rates, deformations, and times of measurements of these devices allow the determination of the steady-state zero-shear viscosity and the steady-state linear elastic moduli up to the vicinity of the phase transition at the GP, but it has proven to be almost impossible to do such measurements *at* the GP (see below).

The correlation or connectivity length ξ of branched polymers diverges as p_c is approached according to the power law

$$\xi \sim |p - p_c|^{-\nu}, \tag{6}$$

which is the analogue of Eq. (2.32). Above the GP the correlation length of polymers can be interpreted as the mesh size of the gel network. For any length scale greater than ξ the gel network is essentially homogeneous. Below the GP the correlation length is the typical radius of the finite polymers in the sol phase. However, in this case, those polymers with radii much larger than ξ have completely different characteristics. Therefore, we discuss such polymers separately, and we refer to them as the branched polymers.

9.2.2.1 Gel Polymers

There are several important structural properties of branched polymers and gel networks that can be measured directly or indirectly. The *gel fraction* $P_g(p)$ is the fraction of the monomers that belong to the gel network. It can be measured by simply weighing the solid gel at different times during polymerization. It is obvious that $P_g(p) > 0$ only if $p > p_c$, and that P_g is the analogue of percolation fraction or percolation probability $P(p)$ defined in Section 2.6.1. Of particular interest to us is the behavior of $P_g(p)$ near p_c. It has been shown that in this region

$$P_g(p) \sim (p - p_c)^{\beta}, \tag{7}$$

which is completely similar to Eq. (2.29).

At the GP the gel network is *not* homogeneous but is a self-similar fractal object with a fractal dimension D_p that in d-dimensions is given by

$$D_p = d - \beta/\nu, \tag{8}$$

the same as Eq. (2.47). The number distribution of the polymers, i.e., the probability $Q(s, \epsilon)$ that a polymer of the sol phase contains s monomers at a distance $\epsilon = |p - p_c|$ from the GP, is clearly the analogue of n_s defined in Section 2.7. Thus, in analogy with Eq. (2.35) we can write

$$Q(s, \epsilon) \sim s^{-\tau} h_1(\epsilon s^{\sigma}), \tag{9}$$

where h_1 is a universal scaling function. Using this distribution, we define two distinct mass averages. One, the *weight-average molecular weight*, is defined by

$$M_w = \frac{\int s^2 Q(s,\epsilon)ds}{\int sQ(s,\epsilon)ds} \sim \epsilon^{-\gamma_p} \sim |p - p_c|^{-\gamma_p}, \tag{10}$$

where $\gamma_p + 2\beta = \nu d$, with d being the dimensionality of the material. In the polymer literature M_w is also called the *degree of polymerization*. The second mass average is defined by

$$M_z = \frac{\int s^3 Q(s,\epsilon)ds}{\int s^2 Q(s,\epsilon)ds} \sim \epsilon^{-1/\sigma} \sim |p - p_c|^{-1/\sigma}, \tag{11}$$

where $\sigma = (\tau - 2)/\beta$, as in percolation (see Chapter 2). Note, however, that the average $\langle M \rangle = \int sQ(s,\epsilon)ds / \int Q(s,\epsilon)ds$ does *not* diverge at the GP.

9.2.2.2 Comparison with the Experimental Data

We are now in a position to compare measurements of various polymer properties with the predictions of percolation theory. Since one of the main predictions of percolation is the existence of universal critical exponents and fractal dimensions, and because the numerical value of any polymer property, such as it average molecular weight or the location of the gel point, is not universal and depends on the morphology of the polymer, we focus here on a comparison between the measured universal exponents, such as β, γ_p and σ, and the predictions of the percolation model.

In their experiments with irradiated polystyrene solution in cyclopentane, Leibler and Schosseler (1985) coupled gel permeation chromatography and light scattering to deduce the polymer size distribution which provides a direct means of measuring the exponent τ. Figure 9.5 presents their measurements from which one obtains, $\tau \simeq 2.3 \pm 0.1$, consistent with the percolation prediction of Section 2.18 (see Table 2.3). Lapp *et al.* (1989) further checked this result by carrying out similar experiments in a system made by chemical end-linking of polydimethylsiloxane, and Patton *et al.* (1989) carried out experiments in a system in which polyester was made by bulk condensation polymerization. The measurements of both of these groups were consistent with the percolation prediction for the exponent τ. Note that since the polymer fractal dimension D_p is related to τ by

$$D_p = \frac{d}{\tau - 1}, \tag{12}$$

these data also imply that $D_p \simeq 2.5$, in agreement with the percolation prediction, $D_f \simeq 2.53$ (see Table 2.3).

Equation (10) was tested by Adam *et al.* (1987), who carried out static light scattering measurements on a polyurethane sol, and by Candau *et al.* (1985) who performed their experiments on polystyrene systems crosslinked with divinylbenzene. Figure 9.6 depicts the results of Adam *et al.* (1987) from which one obtains $\gamma_p \simeq 1.71 \pm 0.06$, only 5% smaller than the percolation prediction of 1.82. A similar estimate was reported by Candau *et al.* (1985). On the other hand, Eqs. (7)

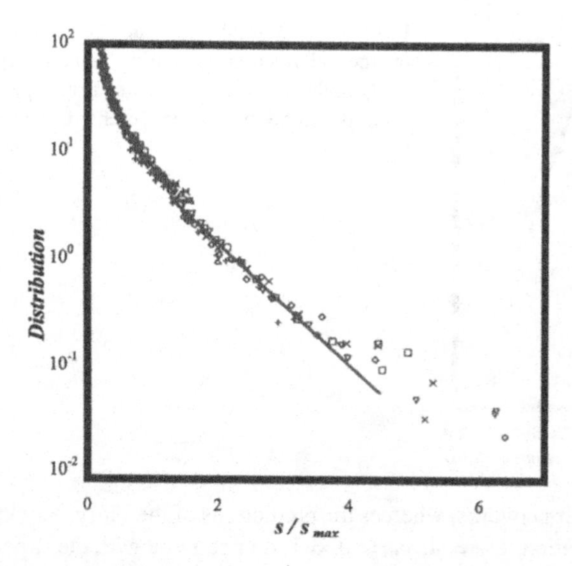

FIGURE 9.5. Normalized polymer size distribution as a function of polymer size s. Percolation theory predicts the slope to be $1-\tau \simeq -1.3 \pm 0.1$ (after Leibler and Schosseler, 1985).

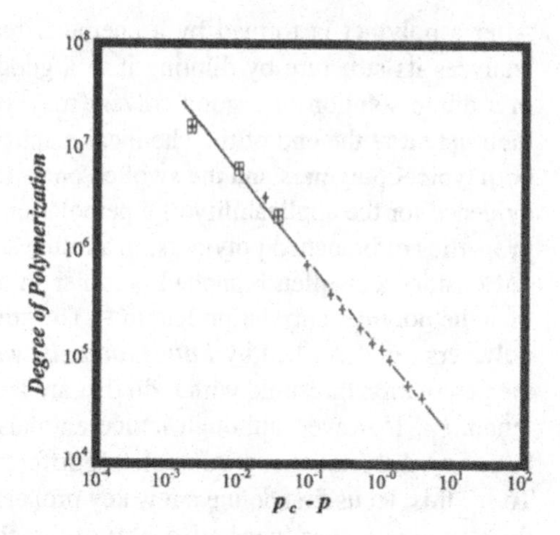

FIGURE 9.6. Dependence of degree of polymerization M_w on $p_c - p$ for a polyurethane sol. The slope of the curve is $-\gamma_p$ (after Adam *et al.*, 1987).

and (10) can be combined to yield $M_w \sim P_g^{-\gamma_p/\beta}$, and thus a plot of $\log(M_w)$ versus $\log(P_g)$ should yield an estimate of γ_p/β. Schmidt and Burchard (1981) carried out anionic copolymerization of divinylbenzene with styrene and obtained both branched polymers and gels. Light scattering was used to measure various properties of interest. When they plotted $\log(M_w)$ versus $\log(P_g)$, as shown in Figure 9.7, they obtained a straight line with the slope $\gamma_p/\beta \simeq 4.5$, in good agreement with the percolation prediction, $\gamma_p/\beta \simeq 4.44$. For more recent experiments regarding these exponents see Trappe *et al.* (1992). We conclude that 3D percolation provides a very good description of the universal morphological properties

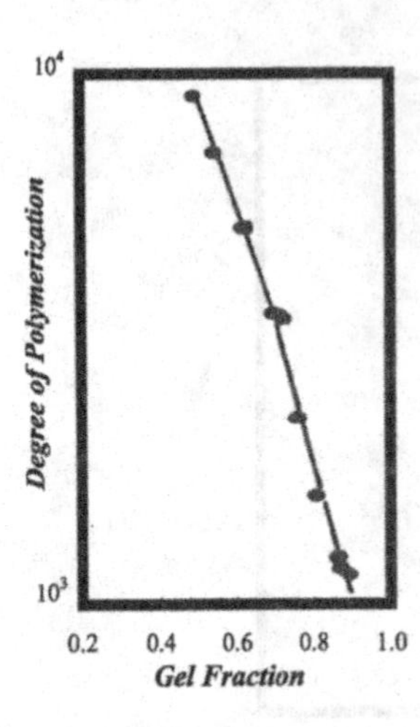

FIGURE 9.7. Dependence of degree of polymerization M_w on the gel fraction P_g during anionic copolymerization of divinylbenzene with styrene. The slope of the curve is $-\gamma_p/\beta$ (after Schmidt and Burchard, 1981).

of gel polymer networks, whereas the predictions of the Flory–Stockmayer theory of polymerization, $D_p = 4$, $\gamma_p = \beta = 1$, do not agree with the experimental data described above.

9.2.2.3 Branched Polymers

After a polymer is formed by a chemical reaction, the experimentalist usually analyzes its structure by diluting it in a good solvent. Such branched polymers in a dilute solution of a good solvent may swell and have a radius *larger* than their extent at the end of the chemical reaction. Thus, it is important to consider both typical polymers and the swollen ones. Let us now discuss the experimental evidence for the applicability of a percolation model for describing the structural properties of branched polymers, in a dilute solution of a good solvent.

Consider a swollen branched polymer in a good solvent with a radius larger than the polymer correlation length ξ. The structural properties of such branched polymers are described by *lattice animals*, which are percolation clusters below the percolation threshold with radii that are larger than the percolation correlation length ξ_p. However, although lattice animals are very large percolation cluster below p_c, their statistics are completely different from those of percolation clusters. To see this, let us first define a few key properties of lattice animals. Let $A_s(p)$ be the average number (per lattice site) of the clusters and a_{sm} be the total number of geometrically different configurations for a cluster of s sites and perimeter m. Thus, $A_s(p) = \sum_m a_{sm} p^s (1-p)^m$. For large s the asymptotic behavior of $A_s(p)$ is described by the power law

$$A_s(p) \sim \lambda_g^s s^{-\theta}, \tag{13}$$

where θ is a universal exponents, independent of the coordination number of the lattice, whereas the *growth parameter* λ_g is not universal. Moreover, for large values of s a fractal dimension D_a, defined by

$$s \sim R^{D_a}, \tag{14}$$

describes the structure of the animals or the branched polymers, where R is the radius of the lattice animal. Lubensky and Isaacson (1978) and Family and Coniglio (1980) showed that the exponents θ and D_a are not related to any of the percolation exponents defined in Chapter 2. Moreover, Parisi and Sourlas (1981) showed that

$$\theta = \frac{d-2}{D_a} + 1, \tag{15}$$

and that

$$D_a = 2, \quad d = 3. \tag{16}$$

Two other key differences distinguish lattice animals from percolation clusters. One is that the exponents θ and D_a defined above are applicable for *any* $p < p_c$ (recall that the percolation exponents are defined for $p \simeq p_c$), so long as $R \gg \xi_p$. The other difference is that the upper critical dimension for lattice animals or branched polymers, i.e., the dimension at which the mean-field approximation to the critical exponents becomes exact, is 8, two more than that of percolation.

We may also define a pair correlation function $C(r)$, i.e., the probability that two monomers or sites, separated by a distance r, belong to the same polymer or cluster. As discussed in Section 2.3.1, for a d-dimensional branched polymer and large r, we expect the correlation function to decay as

$$C(r) \sim r^{D_a - d}. \tag{17}$$

As explained in Section 2.3.2, the Fourier transform of $C(r)$ is proportional to the scattered intensity $I(q)$ in an X-ray or a neutron scattering experiment, where q is the magnitude of the scattering vector given by Eq. (2.11). By Fourier transforming of Eq. (17), one obtains Eq. (2.13) for the scattering intensity of self-similar materials, including branched polymers (lattice animals), which we quote here again,

$$I(q) \sim q^{-D_a}. \tag{18}$$

In real applications, however, polymer solutions are almost always polydispersed and contain polymers of all sizes with radii that are smaller or larger than the correlation length ξ. Thus, one must define *average* properties, where the averaging is taken over the polymer size distribution. An average polymer radius is defined by,

$$\langle R \rangle = \frac{\sum_s s^2 R(s) Q(s,\epsilon)}{\sum_s s^2 Q(s,\epsilon)}, \tag{19}$$

which, when combined with Eqs. (9) and (14), yields a relation between s and $\langle R \rangle$ (Daoud *et al.*, 1984):

$$s \sim \langle R \rangle^{D_a(3-\tau)}, \tag{20}$$

so that, in analogy with Eq. (14), an *effective fractal dimension*, $D_a^e = D_a(3-\tau)$, may be defined. In effect, Eq. (20) mixes the branched polymer exponent D_a with the percolation or gel polymer exponent τ. If a percolation description of polymerization is applicable (which the experiments described above confirm it to be the case), we should have (see Table 2.3) $\tau(d = 3) \simeq 2.18$, implying that

$$D_a^e \simeq 1.64, \quad d = 3, \tag{21}$$

indicating that the effective fractal dimension is smaller than that of a single branched polymer. Since we have defined an effective fractal dimension for a dilute polydisperse polymer solution, the scattering intensity for a solution of a polydisperse polymer should also be modified to

$$I(q) \sim q^{-D_a(3-\tau)}. \tag{22}$$

In practice, Eq. (22) is used in a scattering experiment for estimating D_e.

9.2.2.4 Comparison with the Experimental Data

Experimental evidence for Eq. (16) is actually provided through Eq. (18). Bouchaud *et al.* (1986) carried out small-angle neutron scattering experiments on a monodisperse polyurethane sample and measured the scattered intensity as a function of q. Figure 9.8 presents such data for this polymer from which one obtains, $D_a = 1.98 \pm 0.03$, in excellent agreement with Eq. (16). Bouchaud *et al.* (1986) also synthesized a natural polydisperse polyurethane sample and carried out small-angle neutron scattering on a dilute solution of it. Figure 9.8 also presents these data from which one obtains, $D_a^e \simeq 1.6 \pm 0.05$, in good agreement with the theoretical prediction given by (21). Adam *et al.* (1987) carried out static light scattering experiments in dilute polydisperse polyurethane solutions and reported that, $D_a^e \simeq 1.62 \pm 0.08$, again in good agreement with (21). Leibler and Schosseler (1985) measured the average radius of polystyrene, crosslinked by irradiation by elastic light scattering and found that, $D_a^e \simeq 1.72 \pm 0.09$, consistent

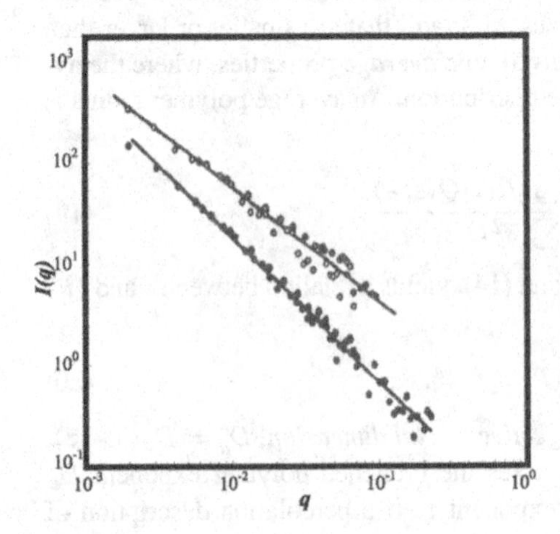

FIGURE 9.8. Small-angle neutron scattering data for branched polymers. The upper curve is for a polydisperse polymer solution with a slope −1.6. The lower curve is for a single polymer in a good dilute solvent with a slope −1.98 (after Bouchaud *et al.*, 1986).

with Eq. (21). Patton *et al.* (1989) performed both quasi-elastic and elastic light scattering experiments on branched polyesters and reported that, $D_a^e \simeq 1.52 \pm 0.1$, somewhat lower than the prediction, but still consistent with it. A different experiment was carried out by Dubois and Cabane (1989) on a silica gel, a physical gel that has a morphology more complex than the chemical gels used in the experiments described above (see Chapter 6, and also the discussion below). Despite significant differences, Dubois and Cabane reported that, $D_a^e \simeq 1.58$, quite close to the theoretical prediction, Eq. (21).

9.2.3 *Rheology of Critical Gels: Dynamic-Mechanical Experiments*

The evolution of a polymer molecular structure during the gelation process has a profound effect on molecular mobility which can be easily monitored by probing the change in viscosity and the elastic moduli. This is shown schematically in Figure 9.9. The initial ($p = 0$) liquid system has a steady shear viscosity η which increases with the extent of the reaction as the average molecular weight M_w increases. At the gel point (GP), the viscosity and the longest relaxation time $t_{\max}$ diverge. Beyond p_c, the equilibrium elastic moduli increase until they attain their highest values when the reaction is brought to completion, i.e., when $p \to 1$. All the experimental data for the elastic moduli of gel networks above, but close to, the GP indicate that they follow a power law given by

$$\text{elastic moduli} \sim (p - p_c)^z, \quad p > p_c. \tag{23}$$

On the other hand, the viscosity of the sol phase diverges as

$$\eta \sim (p_c - p)^{-k}, \quad p < p_c, \tag{24}$$

while, as shown below, the longest relaxation time diverges as

$$t_{\max} \sim |p - p_c|^{-k-z}, \quad |p - p_c| \ll 1. \tag{25}$$

In practice, it is precisely the divergence of η that signals the formation of a gel network. Due to divergence of $t_{\max}$ measurements of the viscosity and elastic

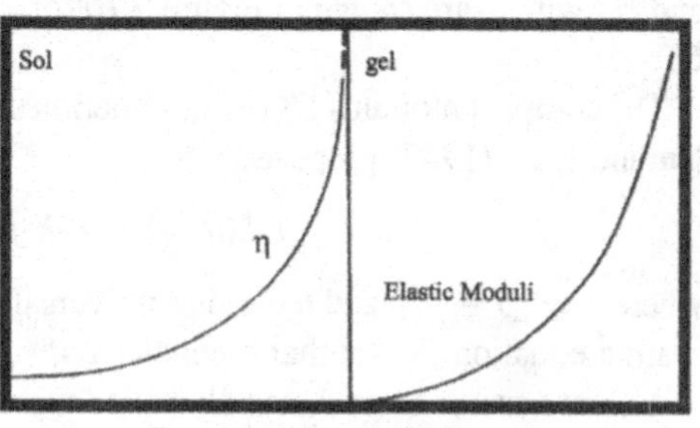

FIGURE 9.9. Typical variations of viscosity η below the gel point, and of the elastic moduli above the gel point, with the fraction of the reacted monomers.

moduli fail at p_c, since steady-state conditions cannot be reached in a finite time. Another problem is that precise measurement of the GP is often very difficult.

These problems have been partially overcome by performing *dynamic-mechanical experiments*, in which the sample is exposed to a periodically varying stress field. For example, a tensile stress is used which, at time t, is given by

$$\sigma_{zz}(t) = \sigma_{zz}^0 \exp(i\omega t). \tag{26}$$

This tensile stress results in a time-dependent longitudinal strain $\epsilon_{zz}(t)$ which varies with the frequency of the stress, but shows, in general, a phase-lag φ, so that

$$\epsilon_{zz}(t) = \epsilon_{zz}^0 \exp[-i(\omega t - \varphi)]. \tag{27}$$

We may then employ a *dynamic tensile modulus* $E^*(\omega)$, defined as

$$E^*(\omega) = \frac{\sigma_{zz}(t)}{\epsilon_{zz}(t)} = E' + iE''. \tag{28}$$

Analogous experiments can of course be carried out for other types of mechanical loading. Of particular interest are measurements for simple shear which determine the relation between the shear strain ϵ_{zx}, yielding the displacement along x per unit distance normal to the shear plane $z =$ constant, and the shear stress σ_{zx} that acts on the shear plane along x.

In any case, such dynamic-mechanical experiments measure the small amplitude oscillatory shear behavior of evolving gels. Under this condition, the gel evolution is continuous (no singularity). Even such experiments cannot entirely overcome the difficulties in the determination of the exponent k since the measurements cannot be carried out *at* the GP and in the limit of *zero frequency*. To estimate k one usually measures the frequency-dependent complex modulus $G^*(\omega) = G'(\omega) + iG''(\omega)$ at frequency ω. At the GP and for low frequencies, one has

$$G' \sim G'' \sim \omega^u, \quad p = p_c, \tag{29}$$

with

$$u = \frac{z}{z+k}, \tag{30}$$

where G' (storage modulus) and G'' (loss modulus) describe storage and dissipation in an oscillating strain field of constant amplitude. Typical variations of G' and G'' with ω are shown in Figure 9.10 for a polycondensed gel very close to the GP.

The complex modulus $G^*(\omega)$ is sometimes written as $G^* = G + i\omega\eta$, for which Durand *et al.* (1987) proposed that

$$G^*(\omega, \epsilon) \sim \epsilon^z h_2(i\omega\epsilon^{z+k}), \tag{31}$$

where $\epsilon = |p - p_c|$, and $h_2(x)$ is a universal scaling function. The significance of scaling equation (31) is that it enables one to collapse the data for all values of ϵ and ω onto a single curve, usually called the *master curve* by polymer researchers. In the low-frequency regime, we do not expect G^* to depend on ϵ, but only on ω, in

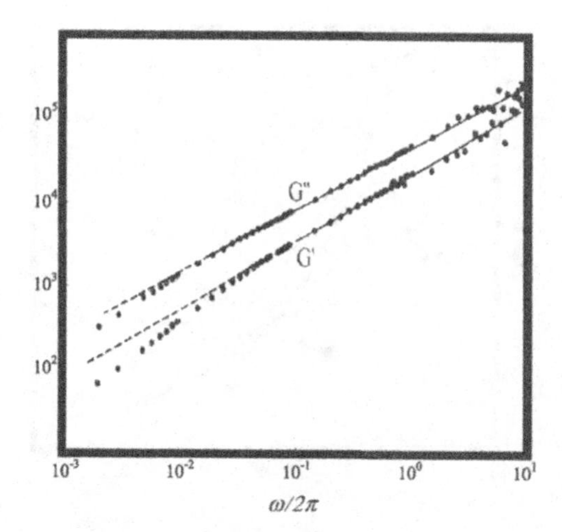

FIGURE 9.10. Frequency-dependence of the storage modulus G' and loss modulus G'' for a polycondensed gel close to the gel point (after Durand *et al.*, 1987).

which case one finds that $G^* \sim (i\omega)^u$, which is equivalent to Eq. (29). Moreover, there is a loss angle δ defined by $\tan\delta = G'/G''$. The remarkable property of δ is that at the GP it takes on a value δ_c given by

$$\delta_c = \frac{\pi}{2}(1-u) = \frac{\pi}{2}\frac{k}{z+k}, \tag{32}$$

so that, if the critical exponents z and k are universal, so will also be the loss angle δ_c.

As already mentioned above, an important problem in polymerization and gelation is accurate determination of the GP, either for avoiding it in order to prevent gelation (so that a branched polymer with specific properties can be prepared), or for making materials very close to GP, since they have unusual properties. The GP, which is the analogue of a percolation threshold, depends on the functionality of the polymer, and as we learned in Chapter 2, percolation thresholds decrease with increasing coordination numbers. Thus, polymers with crosslinks of high functionality gel very early. Holly *et al.* (1988) proposed using the loss angle δ for locating the GP. They argued that since as the GP is reached $\tan\delta$ becomes independent of the frequency [see Eq. (32)], then, if one plots $\tan\delta$ versus time at different frequencies, the intersection of the various curves should be at the GP. Figure 9.11, taken from Lin *et al.* (1991), demonstrates how this method is used for locating the GP for a physical gel.

9.2.4 The Relaxation Time Spectrum

Since the time dependence of a macroscopic relaxation process is always indicative of the underlying microscopic dynamics, one may look for kinetic equations that correctly describe the time-dependence of the observed responses of a material. In the simplest case, there is only a single characteristic time τ, the origin of which goes back to Debye who proposed it in his seminal work on the dielectric response

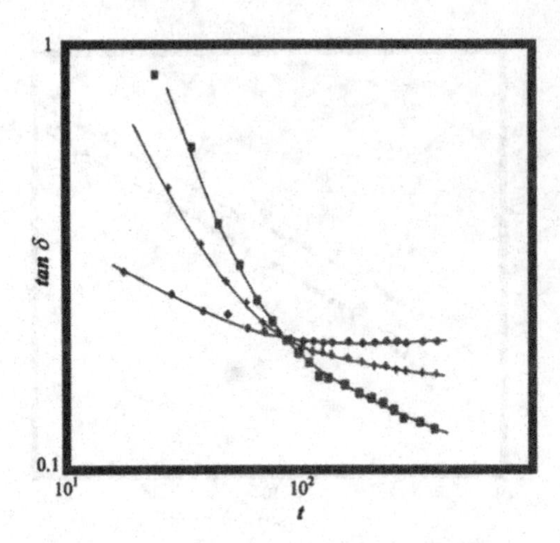

FIGURE 9.11. Determination of gel point from data for loss angle δ. Time is in minutes. The data are for frequencies 31.6 rad s^{-1} (diamonds), 1.0 rad s^{-1} (+) and 0.0316 rad s^{-1} (squares) (after Lin *et al.*, 1991).

of polar liquids. If we define a shear compliance $J(t)$ by

$$J(t) = \frac{\epsilon_{zx}}{\sigma^0_{zx}},$$

then, applying an oscillatory shear stress

$$\sigma_{zx}(t) = \sigma^0_{zx} \exp(i\omega t),$$

on a polymer means imposing on the sample an oscillatory variation for the strain $\epsilon_{zx}(t) = \Delta J \sigma_{zx}(t)$. Here, ΔJ is called the *relaxation strength* of the material. The governing equation for $\epsilon_{zx}(t)$ is then given by

$$\frac{d\epsilon_{zx}(t)}{dt} = -\frac{1}{\tau}\left[\epsilon_{zx}(t) - \Delta J \sigma^0_{zx} \exp(i\omega t)\right]. \tag{33}$$

Assuming a solution, $\epsilon_{zx}(t) = \sigma^0_{zx} J^*(\omega) \exp(i\omega t)$, where $J^*(\omega)$ is the complex shear compliance, substituting it into Eq. (33) and solving it, yield

$$J^*(\omega) = \frac{\Delta J}{1 + i\omega\tau}, \tag{34}$$

which is usually referred to as a *Debye process*.

In general, however, the dynamics of polymers and gels in the reaction bath cannot be described by a single relaxation time, rather by a statistical distribution of such characteristics times. For example, for the shear properties, we write the dynamic compliance $J^*(\omega)$ as a sum of Debye processes with relaxation times τ_i and relaxation strengths ΔJ_i, so that

$$J^*(\omega) = J_u + \sum_i \frac{\Delta J_i}{1 + i\omega\tau_i}, \tag{35}$$

which is usually represented by an integral,

$$J^*(\omega) = J_u + \int \frac{\mathcal{R}(\tau)}{1 + i\omega\tau} d\tau, \tag{36}$$

where $\mathcal{R}(\tau)$ is called the *retardation time spectrum* of the shear compliance J^*. One may also write these results in terms of the complex modulus $G^*(\omega) = 1/J^*(\omega)$, which then yields

$$G^*(\omega) = G_u - \int \frac{H(\tau)}{1 + i\omega\tau} d\tau, \tag{37}$$

where $H(t)$ is called the *relaxation time spectrum* of the complex modulus $G^*(\omega)$ [or $G(t)$] (see, for example, Ferry, 1980).

An important point is that, the observed power-law behavior of G' and G'', Eq. (29), implies a relaxation time spectrum which is self-similar (in time):

$$H(t) = \frac{G_0}{\Gamma(u)} \left(\frac{t}{t_0}\right)^{-u}, \tag{38}$$

where G_0 is the characteristic modulus, t_0 is the characteristic shortest relaxation time, and $\Gamma(x)$ is the gamma function. The modulus of a fully cross-linked polymer network is typically $10^6 - 10^7$ Pa, while the relaxation time of network strand is about $10^{-7} - 10^{-4}$ sec. This spectrum extends from the shortest time, at which strands are beginning to be probed, to the infinite relaxation time of the diverging largest polymer cluster. The parameters G_0 and t_0 are material characteristics of the gel system. Most gel systems seem to possess the same value of u the universality of which can, as explained below, be explained based on vector percolation. However, there are also gel systems which exhibit no apparent universality in the value of u.

Viscosity and elastic moduli are rheological and mechanical properties of branched polymers and gels which characterize the dynamics of the polymerization, since we may measure, directly or indirectly, the distribution of the relaxation times $H(t)$ in the reaction bath. The moments of $H(t)$ are directly related to the viscosity and the elastic moduli. Using Eq. (31), we can back-calculate $H(t)$ (Daoud, 1988):

$$H(t) \sim t^{-u} h_3(t\epsilon^{z+k}), \tag{39}$$

where h_3 is another universal scaling function. Equation (39), which indicates that in the scaling regime near the GP the relaxation time distribution is a slowly decaying power law, generalizes Eq. (38) to any value of p, the extent of the polymerization. As pointed out by Daoud (1988), two distinct average or characteristic times can now be defined. One is

$$t_1 = \frac{\int H(t)dt}{\int [H(t)/t]dt} \sim \epsilon^{-k} \sim \eta, \tag{40}$$

while the second one is given by

$$t_2 = \frac{\int tH(t)dt}{\int H(t)dt} \sim \epsilon^{-z-k}. \tag{41}$$

Note that t_2 is in fact identical with t_{max}, the longest relaxation time of the gel system; see Eq. (25). Because of the existence of the distribution of relaxation times and its scaling form given by Eq. (39), *any* relaxation property in the intermediate time or frequency range is *not* exponential, but follows a *power law*.

9.2.5 Comparison with the Experimental Data

To compare the experimental data for the scaling properties of the elasticity of gels with the predictions of vector percolation, we first divide the gels into two groups, and what follows is a brief description of each.

9.2.5.1 Physical Gels

In one group of gels are what we earlier called physical gels. Two examples are gelation of silica particles in NaCl solutions and in pure water (Gauthier–Manuel *et al.*, 1987), and silica aerogels (Woignier *et al.*, 1988). As discussed earlier, the attachment of the particles in such gels is by relatively weak association. The bond-bending (BB) forces are important in such gels since touching particles that form long chains, when deformed, roll on top of one another and this motion and the displacement of the centers of any 3 mutually-touching particles create forces that are equivalent to the BB forces. Experimental measurements (Gauthier–Manuel *et al.*, 1987; Woignier *et al.*, 1988) of the elastic moduli of such gels confirm this: The exponent z was found to be about 3.8, in excellent agreement with the critical exponent f of the 3D BB model (see Table 9.1).

More recent measurements by Devreux *et al.* (1993) indicate a crossover between the prediction of the BB model and another regime with a much smaller value of z. They measured the complex modulus G^* of silica gels formed by hydrolysis-condensation of a silicon alkoxide. For a restricted region near the gelation time, they reported $z \simeq 2.0 \pm 0.1$, close to the exponent f of the central-force percolation, whereas beyond this region they found $z \simeq 3.6 \pm 0.1$, which is close to the elasticity exponent of 3D BB model. Devreux *et al.* themselves interpreted the data for the region near the gelation time in terms of an analogy between the EPNs and random resistor networks. We will come back to this point shortly.

9.2.5.2 Chemical Gels

There are numerous experimental measurements of the elastic moduli of chemical gel networks and the associated exponent z. Examples include the measurements for hydrolyzed polyacrylamide (Allain and Salomé, 1987b,1990), tetraethylorthosilicate reactions (Hodgson and Amis, 1990; Takahashi *et al.*, 1994), gelatin solutions (Djabourov *et al.*, 1988), polycondensation of polyoxypropylated trimethylolpropane with hexamethylenediisocyanate (Durand *et al.*, 1987), and several other measurements (Gauthier–Manuel and Guyon, 1980, Adam *et al.*, 1981; Gordon and Torkington, 1981; Tokita *et al.*, 1984; Fadda *et al.*, 2001). These measurements yielded values of z in the range $1.9 - 2.4$, which do not agree with the value of the critical exponent f for the 3D BB model (see Table 9.1). In fact, if the size of a chemical gel network is large enough, the BB forces do not play an important role in determining the elastic properties of the near critical gels, and therefore the only important forces between the monomers are the central (stretching) forces. Therefore, these experimental data may be explained based on the elasticity exponent of the 3D central-force percolation, $f \simeq 2.1 \pm 0.2$.

However, a value of z in the range $1.9 - 2.4$ may also be interpreted in terms of two other models. As mentioned earlier, de Gennes (1976a) suggested that the scaling properties of the elastic moduli of (chemical) gels are in the universality class of the conductivity of scalar percolation, implying that $z = \mu \simeq 2.0$, where μ is the critical exponent of the effective conductivity of random resistor networks. On the other hand, Alexander (1984) argued that there are terms in the elastic energy of some gels and rubbers, which are under internal or external stresses, that are similar to the Born model described in Chapter 8 [see Eq. (8.25) or (8.26)]. As discussed there, the critical exponent of the elastic moduli of the Born model is equal to the conductivity exponent μ, and in particular in 3D, $f = \mu \simeq 2.0$, because near the percolation threshold p_c the contribution of the second term of the right hand side of Eq. (8.25) or (8.26), which is a purely scalar term, dominates that of the first term which is contributed by the central forces. While the data mentioned above are more or less consistent with de Gennes' hypothesis, most of them are not precise enough to distinguish between $\mu \simeq 2.0$ for 3D percolation conductivity and $f \simeq 2.1$ for the central-force percolation. However, there are also a few relatively precise sets of experimental data that seem to support de Gennes' conjecture. For example, Axelos and Kolb (1990) measured the rheological properties of pectin biopolymers that consist of randomly connected $\alpha(1-4)$ D-galacturonic acid units and their methyl esters. If the methyl ester content is low, pectin forms thermoreversible gels upon addition of cations, such as calcium. Axelos and Kolb measured the frequency dependence of the storage modulus $G'(\omega)$ and loss modulus $G''(\omega)$ [see Eq. (29)] and reported that, $z \simeq 1.93$, $k \simeq 0.82$, and $u \simeq 0.71$. Their elasticity exponent is close to the conductivity exponent $\mu \simeq 2.0$ for 3D resistor networks. Less precise data, but still supportive of de Gennes' proposal, were reported by Adam *et al.* (1997) for complex modulus of end-linked poly(dimethylsiloxane) pregel polymer clusters, quenched at different distances from the gelation threshold. They reported that $z \simeq 1.9 \pm 0.15$, consistent with the 3D value of the conductivity exponent. However, the estimated error is large enough that one can easily interpret this value of z in terms of the central-force percolation model.

At first glance, de Gennes' proposal that the critical exponent of the effective moduli of gels, a vector transport property, is equal to that of a scalar property, the effective electrical conductivity of a resistor network, may seem incorrect. However, to justify his proposal, de Gennes introduced the notion of an elastic chain between neighboring nodes or monomers, which are the analogue of quasi-one-dimensional strands that percolation clusters possess near the percolation threshold. He then argued that if such chains are elongated, then their nodes carry an extra amount of energy. If we assume that the blobs (i.e., the multiply-connected parts) in the large cluster of monomers do not contribute significantly to the elastic moduli, then one must only consider the energetics of the links or the chains. If the extra energy of such chains is larger than $k_B T$, then, as Daoud (2000) argued, one obtains de Gennes' proposal, $\mu = z$, although Daoud's analysis was a mean-field approximation, and therefore it is not clear that it would also be valid if one carries out a more general, non-mean-field, analysis.

As for Alexander's proposal, rubbers and gels differ from the Born model in several important ways, such as the presence of nonlinear terms in their elastic energy, and the possibility of negative as well as positive Born coefficients α_1 and α_2 [see Eq. (8.26)]. Therefore, as discussed in Chapter 8, while one may use the Born model to *fit* the experimental data, it is not clear that, at a fundamental level, the Born model can actually describe the elastic properties of such gels, since its elastic energy is not rotationally invariant, even though the numerical value of the critical exponent of its elastic moduli might to be close to measured values for some chemical gels.

9.2.5.3 Enthalpic Versus Entropic Elasticity

There is yet another way of rationalizing the experimental data for the scaling properties of chemical gels near the gel point. To describe this, we mention that several measurements of the elastic moduli of chemical gel networks and the associated exponent z deviate significantly from all the data described above. Examples include the measurements of Adam *et al.* (1985) for polycondensation, $z \simeq 3.3 \pm 0.5$, those of Martin *et al.* (1988) and Adolf *et al.* (1990) for gels made from 89% (by weight) of the diglycidyl ether of bisphenol A cured with 11% (by weight) of diethanolamine which yielded $z \simeq 3.3 \pm 0.3$, and the data reported by Colby *et al.* (1993) for polyester gels, which have been argued to lie in the middle of the static crossover between the Flory–Stockmayer theory and the percolation model. Colby *et al.* reported a value of $z \simeq 3.0 \pm 0.7$, which is inconsistent with both the central-force percolation and the bond-bending model, although one might argue that their reported estimate of z is agreement with $\mu = 3$, the mean-field value of the effective conductivity exponent (see Table 2.3). More recent measurements of the shear modulus of an end-linking polymer gel network by Takahashi *et al.* (1994) yielded $z \simeq 2.7$, which is again in the range of the above data.

One possible explanation for such data is that the elasticity of such gels is entropic rather than enthalpic (or energetic). Plischke and Joós (1998), Plischke *et al.* (1999), and Farago and Kantor (2000) argued that the central-force and the BB models are applicable to gels at temperature $T = 0$, and that for $T \neq 0$ there is a contribution to the shear modulus that is entropic in nature. In analogy with the physics of rubber elasticity, and as described in Section 8.15, Plischke *et al.* (1999) argued that near the percolation threshold or the gel point, the polymer network consists essentially of long chains of singly-connected particles (monomers or sites), linked to each other at various junction point. Such chains are similar to the polymer chains that are crosslinked in rubber in order to produce a rigid amorphous material. Deformation of the sample changes the distance between junctions points or crosslinks, as a result of which the entropy is generically decreased, resulting in an increase of free energy and hence a restoring force. When a sample-spanning cluster (or a large polymer network) is formed, there is a net shear restoring force, implying that the connecting chain of particles acts as a stretched spring. Molecular dynamics simulations [for description of this method see Chapter 9 of Volume II; see also Kremer (1998)] of Plischke and co-workers, and computer simulations of Farago and Kantor (2000), who used a model consisting of hard spheres in which a fraction p of the neighbors are tethered by inextensible bonds, both yielded $z \simeq \mu$.

On the other hand, del Gado *et al.* (1999) proposed a different model, also purported to be appropriate for entropic gels, in which one begins with a random collection of monomers with concentration p. Each pair of the monomers are then linked with a probability p_b to form permanent bonds. Varying p_b produces a distribution of clusters of the linked monomers, and hence gives rise to the possibility of forming a sample-spanning cluster. Monomers and the clusters then diffuse according to the bond fluctuation algorithm of Carmesin and Kremer (1988). In this algorithm, the monomers diffuse in the solution randomly but obeying the excluded-volume interaction (i.e., no two monomers can occupy the same point in space). Due to this random motion, the bonds may have to change their length in a set of allowed values, and thus they may have to bend and take on many different values of the angles between the bonds, which then gives rise to a wide variety of conformations of the polymers. The mean square displacement $\langle R^2(t)\rangle$ of the system's center of mass is then calculated which, due to the elastic potential that reduces the fluctuations proportionally to the effective elastic constant α_e, is given by

$$\langle R^2(t)\rangle \propto \alpha_e^{-1}, \tag{42}$$

and therefore the elastic constant and its power-law behavior near the percolation threshold p_c, and hence the exponent z, can be computed. Two-dimensional simulations of del Gado *et al.* (1999) yielded a value $z \simeq 2.7 \pm 0.1$, hence disagreeing with the results of Plischke and co-workers and Farago and Kantor.

How can one interpret these results? If the elasticity of these gels is due to entropic effects, then, as Daoud and Coniglio (1981) argued (see also Martin *et al.*, 1988), the elastic free energy per unit volume $\mathcal{H}$ must be given by

$$\mathcal{H} \sim \xi_p^{-d}\alpha_\ell \xi_p^2, \tag{43}$$

where ξ_p is the correlation length of percolation, and α_ℓ is the effective elastic constant of a long chain of length ξ_p connecting two nodes. Since $\alpha_\ell \sim \xi_p^{-2}$, we obtain

$$z = \nu d. \tag{44}$$

In 2D, Eq. (44) predicts that, $z \simeq 2.66$, quite different from the exponents f of the central-force and the BB models, and also the conductivity exponent $\mu \simeq 1.3$

(see Table 2.3). However, it is in agreement with the numerical simulations of del Gado *et al.* (1999). Daoud (2000) argued that Eq. (44) is valid when the energy of the chains is of the same order of magnitude as the thermal energy $k_B T$.

Equation (44) also predicts that in 3D, $z \simeq 2.64$. This value of the elasticity exponent is consistent with the experimental data of Adam *et al.* (1985), Martin *et al.* (1988), Adolf *et al.* (1990), Colby *et al.* (1993) and Takahashi *et al.* (1994) mentioned above, who all reported values of the elasticity exponent z in the range $2.7 - 3.3 \pm 0.5$. Therefore, while these experimental data may be explained by the entropic effects and Eq. (44), the numerical results of Plischke and co-workers, and those of Farago and Kantor (2000) do not agree with the prediction of Eq. (44). Indeed, while the main argument of Plischke *et al.* (1999) was that the entropic effects are important at temperatures $T \neq 0$ where one should see a crossover to the value of the conductivity exponent, all of the above experiments were carried out at finite temperatures, yet they did not yield $z \simeq \mu$.

Summarizing, the scaling properties of elastic moduli of (physical or chemical) gels may belong to *three* distinct universality classes, two of which are described by vector percolation models described and analyzed in Chapter 8, while the third class seems to be described by entropic network models, also described in Section 8.15. However, in addition to what we described above, alternative models and explanations for the experimental data for the elastic moduli of gel polymers do exist, and therefore the question of an appropriate model for describing and predicting the elastic properties of such materials is not completely solved.

9.2.5.4 Viscosity of Near-Critical Gelling Solutions

How can one explain the extensive experimental data for the scaling behavior of the viscosity η of the sol phase near the GP? To begin with, it has been proposed by Sahimi and Goddard (1985) [see also Arbabi and Sahimi (1990c) and Sahimi and Arbabi (1993)] that the scaling properties of η near the gel point is analogous to that of the shear modulus of a superelastic percolation network (SEPN) near p_c. To proceed further, we must first establish a rigorous relationship between the linear elasticity and the theory of viscous fluids, thus confirming the proposal of Sahimi and Goddard (1985).

We consider a general time-dependent system, and write down the equation of motion for a macroscopically-homogeneous material in terms of the displacement field $\mathbf{u}$ which, as described in Section 7.1, is given by

$$(\lambda + \mu)\nabla(\nabla \cdot \mathbf{u}) + \mu\nabla^2\mathbf{u} + \mathbf{F} = \rho\frac{\partial^2 \mathbf{u}}{\partial t^2}, \qquad (45)$$

where ρ is the mass density and t is the time. The rest of the notation is the same as in Chapter 7. For an *incompressible* material, i.e., one for which the bulk modulus K and the Lamé constant λ are divergent, one has the solenoidal condition,

$$\nabla \cdot \mathbf{u} = 0. \qquad (46)$$

Due to the incompressibility condition, the first term of Eq. (45) is indeterminate, and can be written in terms of the reactive hydrostatic pressure P, yielding

$$-\nabla P + \mu\nabla^2\mathbf{u} + \mathbf{F} = \rho\frac{\partial^2\mathbf{u}}{\partial t^2}. \tag{47}$$

On the other hand, let us write down the Navier-Stokes equations of motion for an incompressible and Newtonian viscous fluid,

$$-\nabla P + \eta\nabla^2\mathbf{u} + \mathbf{F} = \rho\left(\frac{\partial\mathbf{v}}{\partial t} + \mathbf{v}\cdot\nabla\mathbf{v}\right), \tag{48}$$

where $\mathbf{v}$ is the fluid velocity field, $\mathbf{F}$ is an external (body) force, and η is the fluid's dynamic viscosity. For an incompressible fluid, the continuity equation is given by

$$\nabla\cdot\mathbf{v} = 0. \tag{49}$$

For slow fluid flow, i.e., when the Reynolds number Re$\ll 1$, the nonlinear inertial term, $\mathbf{v}\cdot\nabla\mathbf{v}$, is very small and can be neglected, which means that the Navier–Stokes equations reduce to

$$-\nabla P + \eta\nabla^2\mathbf{v} + \mathbf{F} = \rho\frac{\partial\mathbf{v}}{\partial t}. \tag{50}$$

Thus, we see that, under steady-state conditions, and when the flow of the fluid is slow, the governing equations for the displacement field $\mathbf{u}$ and the velocity field $\mathbf{v}$ are exactly identical, provided that there is a one-to-one correspondence between the shear modulus μ and the dynamic viscosity η.

In addition, under certain conditions, the effective viscosity η_e of a suspension of completely rigid spheres in an incompressible fluid of viscosity η_1, under creeping (very slow) flow conditions, is related to the *steady-state* effective shear modulus μ_e of a two-phase material composed of the same completely rigid spheres in an incompressible matrix with shear modulus μ_1. In this case, the working equation is given by

$$\frac{\eta_e}{\eta_1} = \frac{\mu_e}{\mu_1}. \tag{51}$$

Equation (51) is exact when, regardless of the configuration of the particles, hydrodynamic interactions between the particles can be neglected, which is the case when the system is infinitely dilute so that the volume fraction ϕ_2 of the particles approaches zero. If the system is non-dilute, Eq. (51) would still be exact if the configurations of the particles in the flow and elasticity problems are identical.

Having established a theoretical connection between the viscosity η of a solution and the shear modulus of an appropriate two-phase material, the one-to-one correspondence between η and the shear modulus of a SEPN should be clear because, (1) η and μ both diverge at p_c (the gel point), and (2) percolation models predict accurately the morphology and elastic moduli of gel polymer networks. On the other hand, de Gennes (1979) [see also Allain *et al.* (1990)] suggested an analogy between η and the effective conductivity of a conductor-superconductor percolation networks, the effective conductivity of which diverges at p_c according to the power law, $g_e \sim (p_c - p)^{-s}$, so that $k = s$.

However, even a one-to-one correspondence between η and the shear modulus of a SEPN is not nearly enough to explain the scaling of η of the gelling solution near the gel point, because most of the experimental data indicate that the value of k is either in the range 0.6-0.9 (see, for example, Adam *et al.*, 1979, 1985; Allain and Salomé, 1987a; Durand *et al.*, 1987), or in the range 1.3-1.5 (see, for example, Djabourov *et al.*, 1988; Martin *et al.*, 1988, Martin and Wilcoxon, 1988), whereas the power-law behavior of the shear modulus of a 3D SEPN near p_c is characterized by a *unique* value of the exponent χ defined by Eq. (8.5). The reason for having two distinct values of k is (Arbabi and Sahimi, 1990c; Sahimi and Arbabi, 1993) that the dynamics of the sol solutions that yield the two distinct values of k are completely different. In one case, the solution is close to the Zimm limit in which there is little or no polymer diffusion in the reaction bath, because there are strong hydrodynamic interactions between the monomers, and also between polymers of various sizes. Hence, a SEPN, a *static* system with fixed rigid clusters, may be suitable for simulating the Zimm regime. Indeed, the estimate $\chi = \nu - \frac{1}{2}\beta \simeq 0.65$ for 3D SEPNs [where ν and β are the usual percolation exponents; see Eq. (8.112)] supports the proposal that those gelling solutions that yield a value of k is in the range 0.6-0.9, are in fact close to the Zimm regime and can be described by a static SEPN.

On the other hand, the gelling solution can also be near the Rouse regime in which there are no hydrodynamic interactions between the polymers of various sizes, and therefore the polymers can diffuse essentially freely in the reaction bath. To simulate this regime the following model was proposed (Arbabi and Sahimi, 1990c; Sahimi and Arbabi, 1993). We consider a *dynamic* SEPN in which each cluster of the totally rigid bonds represents a finite polymer. Due to randomness of percolation networks, there is of course a wide distribution of such polymers or clusters in the network. The soft bonds (those with a finite elastic constant) represent the liquid solution in which the rigid clusters move randomly, with equal probability, in any direction of the network. This simulates the diffusion of the polymers in the reaction bath. Two rigid clusters cannot overlap, but they can temporarily join and form a larger cluster, which can also be broken up again at a later time. It was shown (Arbabi and Sahimi, 1990c; Sahimi and Arbabi, 1993) that the shear modulus of this dynamic SEPN diverges with an exponent χ' given by

$$\chi' = 2\nu - \beta. \tag{52}$$

Equation (52) predicts that in 3D, $\chi' \simeq 1.35$, supporting the proposal that the gelling solutions that yield a value of k in the range 1.3-1.5 are close to the Rouse limit, and can be described by the dynamic SEPN described above. Daoud (2000) argued that, similar to the case of the elastic moduli described above, one must have two regimes. If the elastic chains carry an energy which is of the same order of magnitude as the thermal energy $k_B T$, then the exponent χ is given by Eq. (8.112). On the other hand, if the elastic chains are stretched and have an extra energy larger than $k_B T$, then one recovers Eq. (52), which had also been conjectured by de Gennes (1979), based on the analogy between the viscosity and the effective conductivity of conductor-superconductor percolation networks.

We should point out that, as in the case of the elastic moduli discussed earlier, there are some experimental data that indicate some deviations of k from χ or χ'. However, as pointed out earlier, experimental determination of k (and the elasticity exponent z) involves (see, for example, Durand *et al.*, 1987) measuring the complex modulus $G^*(\omega)$ at a series of frequencies ω. On the other hand, strictly speaking, the scaling laws for the elastic moduli of EPNs and the SEPNs are valid only in the limit of $\omega \to 0$, whereas in practice it is very difficult to reach such a limit, and therefore the measured values of k may exhibit some deviations from χ or χ'. However, we believe that such deviations are transient effects which should diminish as very low frequencies are accessed.

9.3 Mechanical Properties of Foams

Foams are a special type of material that consist of a random distribution of gas bubbles in a much smaller volume of liquid (see, for example, Weaire and Hutzler, 1999). If this mixture contains enough stabilizing surfactants, then the morphology of the material will be essentially stable and constant over time scales that can be as long as hours. Under these conditions, and due to their practical importance, understanding the rheology and mechanical properties of foams are important issues to be addressed. Consider, for example, the response of foams to an applied force. Although foams are mostly made of gas bubbles, they can support deformation and shear forces, much like a solid material. This behavior is due to the increase in the gas-liquid interface surface area, which is the result of deforming the tightly-packed gas bubbles, and the corresponding energy cost by the surface tension. If the applied force is small, then the mechanical response of foams is linear, and their shear modulus is simply the stress per unit strain, regardless of whether the experiment is carried out under controlled stress or controlled strain conditions. If the applied stress increases, the response of the system crosses over to a non-linear regime. At higher stresses, the response becomes irreversible as topological changes are induced under which a few bubbles in a finite region suddenly change their neighbors. At still higher applied stresses, the rearrangement of the bubbles becomes more frequent, but the resulting strain remains finite. Eventually, when a critical yield stress is reached, the foam starts to flow indefinitely at a non-zero strain rate by a never-ending series of neighbor-switching rearrangements. If the strain rate is low, the rearrangements are discrete avalanche-like events, but if the strain rate is high, the deformation is continuous and resembles a viscous liquid. Clearly, the transition from a non-flowing system to a flowing one is a nonlinear phenomenon, characterized by the yield stress threshold (Sahimi, 1993a). In Chapter 3 of Volume II, we will describe modeling of this transition and its many interesting features. In this section, we are mainly interested in the mechanical response of foams before they start to flow, and the relation between this phenomenon and vector percolation.

Durian (1997) developed a realistic model of foams in order to study their deformation and mechanical response. To understand his model, consider first two gas bubbles in a sea of background liquid that are brought into contact with each other. Since the ranges of van der Waals force (which is attractive and originates from the dielectric mismatch between the gas and the liquid) and of the electric double-layer force (which is repulsive and is due to the adsorbed surfactants) are typically small—less than 100 nm—compared with typical size of gas bubbles which is about 20 μm, the two bubbles experience no force until they actually touch each other. When pushed into each other, the shape of the two bubbles is distorted, and the region of their contact flattens out into a soap film with a thickness that depends on the combination of the applied, van der Waals, and double-layer forces. The formation of the soap film and the increase in the total surface area give rise to a mutually-repulsive force which is proportional to the liquid-gas surface tension, and is nearly harmonic (i.e., central-force type). Thus, Durian assumed this force to be harmonic. Then, the effective central force for bubble i is proportional to σ/R_i, where σ is the surface tension at the liquid-gas interface, and R_i is the bubble's radius. For two mutually-repulsing bubbles with their centers at $\mathbf{r}_i$ and $\mathbf{r}_j$, one has $|\mathbf{r}_i - \mathbf{r}_j| < R_i + R_j$, and the individual springs are added in series, so that the effective spring constant is $F_0/(R_i + R_j)$, where F_0 is the force constant which is of the order of, $F_0 \simeq \sigma\langle R\rangle$, with $\langle R\rangle$ being the average bubble size. Therefore, the repulsive force $\mathbf{F}_{ij}^r$ acting on the center of bubble i due to bubble j is given by

$$\mathbf{F}_{ij}^r = F_0\left[\frac{1}{|\mathbf{r}_i - \mathbf{r}_j|} - \frac{1}{R_i + R_j}\right](\mathbf{r}_i - \mathbf{r}_j). \tag{53}$$

One must also take into account the dissipation effect, since when a foam is strained at non-zero rate, energy is dissipated due to shear flow of the viscous liquid within the soap film, by flow within the adsorbed surfactant films, and by other mechanisms. The drag force generated by the dissipation is then assumed to be given by

$$\mathbf{F}_{ij}^d = -b(\mathbf{v}_i - \mathbf{v}_j), \tag{54}$$

where b is a proportionality constant, assumed to be the same for all the bubbles. Thus, writing a force balance for (the center of) each bubble and setting the result to zero, one obtains

$$\mathbf{v}_i = \langle \mathbf{v}_j\rangle + \frac{F_0}{b}\sum_j\left[\frac{1}{|\mathbf{r}_i - \mathbf{r}_j|} - \frac{1}{R_i + R_j}\right](\mathbf{r}_i - \mathbf{r}_j) + \frac{1}{b}\mathbf{F}_i^a, \tag{55}$$

where $\mathbf{F}_i^a$ is the applied force. The only relevant time scale in the system is $t_d = b\langle R\rangle/F_0$, which is set by the average bubble size and the competition between dissipation and storage of energy; the numerical values of b and F_0 are irrelevant.

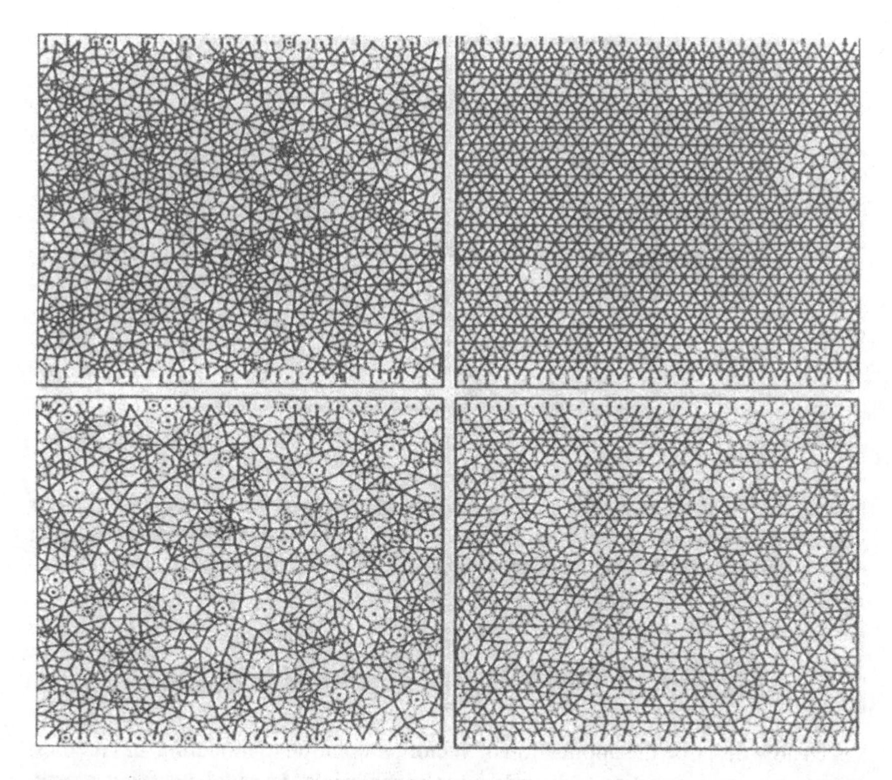

FIGURE 9.12. Equilibrium bubble configurations for four different systems. The top two and bottom two have gas fractions $\phi_2 = 1.0$ and 0.84, respectively. In the left two and right two systems the uniform bubble size distributions have widths of 0.75 and 0.1, respectively. The top and bottom edge bubbles are fixed to the horizontal plate, while periodic boundary conditions have been imposed on the left and right sides. A solid line shows a spring between the centers of two overlapping bubbles (after Durian, 1997).

By placing the bubbles on a grid, e.g., a square lattice, and solving the equation of motion (the coupled force-balance equations for all the bubbles), one can study the evolution of the system. Durian (1997) provides technical details of such simulations. Figure 9.12 presents snapshots of two different systems at two gas volume fractions ϕ_2. One has a broad bubble size distribution, while the other system has a narrow distribution. Figure 9.13 presents the shear modulus of the system as a function of the gas volume fraction. The general trends in the figure are, on one hand, entirely similar to central-force percolation and, on the other hand, to the experimental data of Mason *et al.* (1995). The percolation threshold, i.e., the critical gas volume fraction at which the shear modulus vanishes, depends weakly on the broadness of the bubble size distribution, but is around a value of 0.84, which is identical with the dense random packing of hard disks (see Chapter 3). Bolton and Weaire (1990) were in fact the first to point out the relevance of vector percolation to modeling mechanical properties of foams.

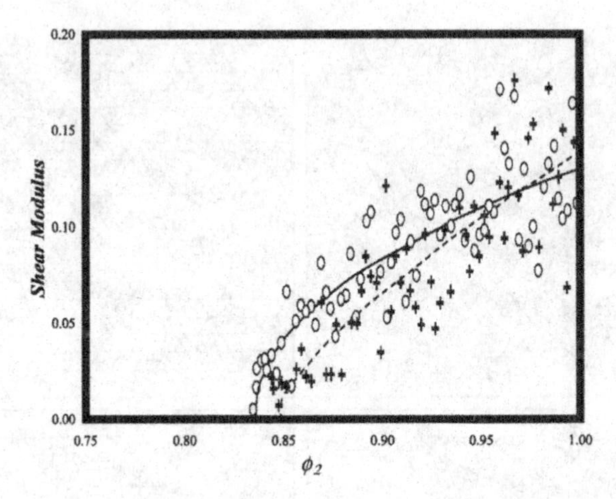

FIGURE 9.13. The shear modulus of the foam versus the gas fraction ϕ_2. The modulus vanishes at the rigidity percolation threshold. Circles and + signs denote the results for uniform bubble size distributions with widths 0.75 and 0.1, respectively (after Durian, 1997).

9.4 Mechanical Properties of Composite Solids

Let us now describe the application of vector percolation to modeling and predicting mechanical properties of disordered solid materials. As usual, we focus mostly on the behavior of materials in the vicinity of their percolation threshold, and in particular the power-law behavior of the elastic moduli, since the critical exponents that characterize such power laws are largely independent of many details of the materials' morphology, whereas the numerical value of the elastic properties, and also the percolation thresholds of materials at which an elastic modulus vanishes or diverges, are non-universal and vary greatly from one material to another. Indeed, the agreement between the measured critical exponents and the predictions of vector percolation models is the most stringent test of the applicability of such models.

9.4.1 *Porous Materials*

Deptuck *et al.* (1985) measured the Young's modulus of a sintered, submicrometer silver powder, commonly used for millikelvin and submillikelvin cryostats. The sinter remains elastic and percolating even if the volume fraction ϕ of the silver particles is as low as 0.1. They also used submicrometer copperoxide-silver powder, routinely used in heat exchangers for optimizing heat transfer to dilute ^{3}He - ^{4}He mixtures. In these composites, the silver component acts as the percolating phase with a percolation threshold ϕ_c less than 0.1. Measurements of Deptuck *et al.* (1985) indicated that, over a broad range of the volume fraction ϕ, the Young's modulus of the powder follows the power law, $Y \sim (\phi - \phi_c)^f$ with $f \simeq 3.8$, in excellent agreement with the critical exponent f of the 3D bond-bending model (see Table 9.1). Moreover, their measurements indicated that the critical region in which this power law is valid is quite broad, and hence the power law is quite useful even for quantitative estimation of the elastic moduli. Other experimental data on sintered powder samples of copper and silver were presented by Maliepaard *et al.* (1985), who reported that $f \simeq 3.6$.

Measurements of the elastic moduli of 2D porous composites were also carried out by Benguigui (1984,1986), Lobb and Forester (1987), Sofo *et al.* (1987), and Craciun *et al.* (1998). Benguigui's first experiments (1984) were done with sheets of metal in which holes had been punched; his measured exponent was, $f \simeq 3.5 \pm 0.5$, which, although consistent with the value of f for the 2D BB models, was criticized because his composite represented a *continuum* equivalent to the Swiss-cheese model described in Section 2.11, as opposed to a lattice model, and, according to Halperin *et al.* (1985) and Feng *et al.* (1987), the critical exponent f for such continua should be larger than the corresponding value for lattice systems by about 3/2 [see Eq. (2.74)]. More careful measurements by Benguigui (1986) then yielded $f \simeq 5.0 \pm 0.5$, consistent with the prediction of Halperin *et al.* (1985), $f \simeq 5.5$. Measurements of Lobb and Forrester (1987) were also carried out with metal sheets into which holes had been punched. They reported $f \simeq 4.95 \pm 1.1$, again consistent with Halperin *et al.*'s prediction. Sofo *et al.* (1987) used randomly-holed metalized Mylar and reported, $f \simeq 5.3 \pm 0.7$, in even better agreement with the prediction of Halperin *et al.* (1985).

On the other hand, Craciun *et al.* (1998) measured the elastic moduli of porous ceramics, an important class of materials that have wide applications. Their estimated exponent of the elastic moduli was, $f \simeq 2.0$, which they attributed it to the predominance of central forces in their composites. Indeed, their reported value of f agrees with that of the 3D central-force model (see Table 9.1).

Another interesting set of experiments was carried out by Allen *et al.* (1988) who measured dynamical (frequency-dependent) shear modulus of a model 2D percolation lattice, which can be thought of as the lattice analogue of a porous material, with the cut bonds representing the pore space of the material. Their measurements were made by exciting small torsional linear oscillations on a cylindrical screen as links were cut to approach p_c. Their measurements yielded $f \simeq 3.6 \pm 0.2$, relatively close to the corresponding exponent of the 2D BB model, $f \simeq 3.96$.

An interesting study of mechanical properties of two types of compressed expanded graphite was reported by Celzrd *et al.* (2001). These materials are prepared from natural graphite platelets that are inserted into sulfuric acids, which results in an intercalated compound. The resulting material is then submitted to a thermal shock by passing it through a flame which induces swelling of the platelets. The samples are then compressed by uniaxial compaction. Celzrd *et al.* (2001) measured the electrical conductivity and elastic moduli of two samples of these materials. Figure 9.14 presents their data in logarithmic scales, as functions of the apparent density of the conducting material. The electrical conductivity

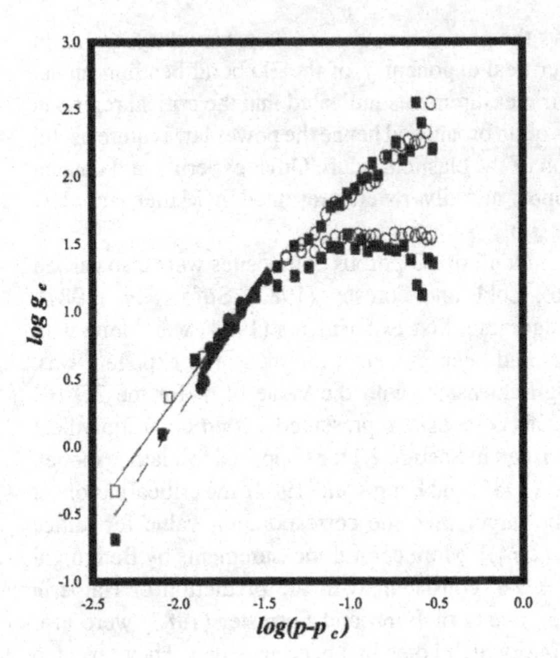

FIGURE 9.14. Electrical conductivity and elastic moduli of four cubic samples of expanded graphite. The lines show the fit of the data to power laws (after Celzard *et al.*, 2001).

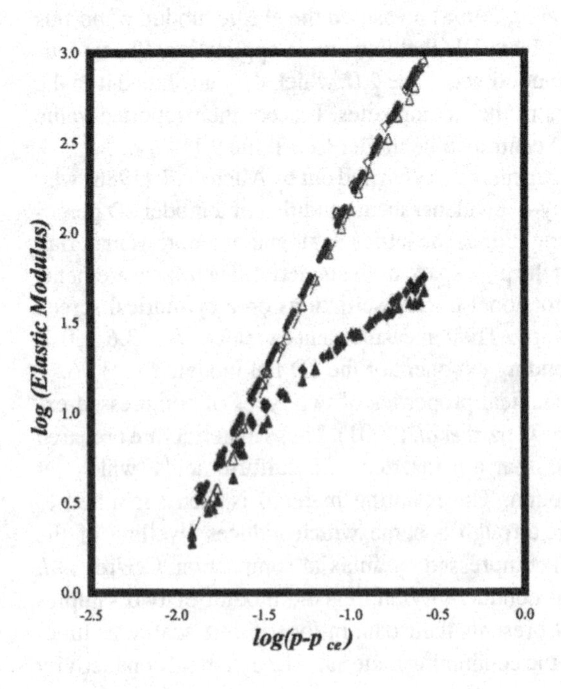

and the elastic moduli of the samples vanish at different percolation thresholds. From Eq. (8.101) and (8.102) we find that,

$$\frac{(p_c^B)_{kk}}{p_c^B} = \frac{d^2 - 1}{2d - 1},$$

where $(p_c^B)_{kk}$ is the percolation threshold of the BB Kirkwood–Keating model at which the elastic moduli vanish, p_c^B is the threshold of the scalar percolation at which the electrical conductivity vanishes, and d is the dimensionality of the sample material. Thus, for $d = 3$ one obtains, $(p_c^B)_{kk}/p_c^B = 8/5$, which was indeed confirmed by the experiments of Celzard *et al.* (2001), hence confirming the validity of vector percolation for predicting the elastic properties of composite materials, even as far as their percolation threshold for various transport properties is concerned.

9.4.2 *Superrigid Materials*

Benguigui and Ron (1993) constructed a composite of a very soft material—in their case a gel—and a very rigid material which was alumina and zirconia, and measured the Young's modulus of the material, as the volume fraction of the rigid component approached ϕ_c, the percolation threshold, or the critical volume fraction, for formation of a sample-spanning cluster of the rigid material. Their measurements indicated that the Young's modulus Y diverges at ϕ_c, and that near the percolation threshold it follows the power law, $Y \sim (\phi_c - \phi)^{-\chi}$, with $\chi \simeq 0.66 \pm 0.04$, in excellent agreement with the value of superelasticity exponent for the 3D BB model (see Table 9.1). Thus, there remains little, if any, doubt that the linear mechanical properties of disordered solids are described by vector percolation models.

9.5 Wave Speeds in Porous Materials

Elastic percolation models can also be used for calculating compressional and shear wave velocities in a porous medium, and their dependence on the porosity of the medium. The shear wave velocity V_p is given by

$$V_p = \sqrt{\frac{K + \frac{4}{3}\mu}{\rho}}, \tag{56}$$

while the compressional wave velocity is given by

$$V_s = \sqrt{\frac{\mu}{\rho}}, \tag{57}$$

where ρ is the total density, and K and μ are the usual bulk and shear modulus of the porous medium. Therefore, if we compute K and μ based on a vector percolation model of a porous medium, then the results will yield estimates of the wave speeds in the porous medium.

9.6 Elastic Properties of Composite Materials with Length Mismatch

In the absence of an induced external stress, a solid elastic material relaxes to a state of zero macroscopic stress which, however, does not necessarily imply a state of zero strain energy. The latter state is possible only if there is a single natural length throughout the solid, as in, for example, a perfectly crystalline solid. However, different regions of a material can have a mismatch between their natural lengths. An example is boron- and/or germanium-doped silicon in which the dopant atoms fill silicon-sized holes in the silicon lattice. Since boron is smaller than silicon, it induces lattice shrinkage, while germanium, which is larger than silicon, induces lattice expansion. In boron-doped silicon, hydrogen passivation of the boron acceptors causes the effective natural length of the material to increase, which is due to natural length increases at the boron sites that are caused by the formation of B-H complexes. There are many other examples of such mismatches in a wide variety of materials.

The mismatch in the natural lengths is exploited to strengthen glass using the process of chemical tempering by which larger ions are exchanged for smaller ones in the glass network, thereby inducing stored strain energy associated with compressive stress on the surface, which in turn is balanced by tensile stress in the interior, so that the macroscopic stress is still zero. The surface compression tends to deactivate the surface cracks, thereby increasing the overall tensile strength of the material by an amount roughly proportional to the surface compressive stress. Another important example is concrete, the durability of which is controlled by natural length mismatch. Concrete is produced by mixing an aggregate with portland cement and water. The aggregate can react with alkaline element constituents, leading to products that have a larger natural length than the original aggregates, hence inducing intense local compressive and tensile stresses which can cause cracking.

A simple model of a material with natural length mismatch was developed by Thorpe and Garboczi (1990), based on the central-force percolation networks, which captures all the essential features of the phenomenon. Consider a lattice in which each bond has a natural (unstretched) length L_{ij}^0 with spring constant α_{ij}. In particular, consider a two-phase composite in which a bond either has a natural length L_A^0 and a spring constant α_A with probability $1-p$, or a natural length L_B^0 with spring constant α_B with probability p. The elastic energy of the system is then given by

$$\mathcal{H} = \frac{1}{2}\sum_{ij}\alpha_{ij}\left(\mathbf{R}_i - \mathbf{R}_j - L_{ij}^0\right)^2, \tag{58}$$

where $\mathbf{R}_i$ goes to site i at the end of bond ij from some arbitrary origin. Minimizing the elastic energy yields,

$$\sum_{ij} \alpha_{ij} \left[(\mathbf{R}_i - \mathbf{R}_j) - L^0_{ij} \hat{\mathbf{R}}_{ij} \right] = 0, \tag{59}$$

where $\mathbf{R}_i$ is now the *relaxed* vector, and $\hat{\mathbf{R}}_{ij}$ is a unit vector along the *relaxed* bond direction. One may now use the Feynman–Hellman theorem (Hellman, 1937; Feynman, 1939; see also Chapter 9 of Volume II) according to which

$$\frac{\partial \langle \mathcal{H}_n \rangle}{\partial P} = \left\langle \frac{\partial \mathcal{H}_n}{\partial P} \right\rangle, \tag{60}$$

where P is any parameter of the system, and $\mathcal{H}_n = \mathcal{H}/N$, with N being the total number of the sites (or atoms) in the lattice. If we set $P = L^0_A$ or L^0_B, then Eq. (60) yields a set of two equations,

$$\alpha_A(1-p)(\langle L_A \rangle - L^0_A) = -\frac{\partial \mathcal{H}_n}{\partial L^0_A}, \tag{61}$$

$$\alpha_B p(\langle L_B \rangle - L^0_B) = -\frac{\partial \mathcal{H}_n}{\partial L^0_B}, \tag{62}$$

which involve no assumption about small displacements, but are not also very useful. If one then assumes that the displacements from a perfect lattice are small, one obtains

$$\alpha_A(1-p)(\langle L_A \rangle - L^0_A) + p\alpha_B(\langle L_B \rangle - L^0_B) = 0. \tag{63}$$

If one now combines Eqs. (61)–(63), one obtains $\partial \mathcal{H}_n/\partial L^0_A + \partial \mathcal{H}_n/\partial L^0_B = 0$, implying that $\mathcal{H}_n$ depends only on the difference $L^0_A - L^0_B$. Dimensional arguments then lead one to $\mathcal{H}_n$ actually being proportional to $(L^0_A - L^0_B)^2$, and therefore

$$\mathcal{H}_n = \frac{1}{2}(\alpha_A \alpha_B)^{1/2} p(1-p)(L^0_A - L^0_B)^2 h(p, \alpha_A, \alpha_B). \tag{64}$$

Equation (64) is valid only in the limit of small displacements. Because of the symmetry of the model in terms of exchanging A with B and p with $1-p$, we must have, $h(p, \alpha_B/\alpha_A) = h(1-p, \alpha_A/\alpha_B)$. From Eqs. (61)–(64) and the definition $\langle L \rangle = (1-p)\langle L_A \rangle + p\langle L_B \rangle$, one obtains

$$\langle L \rangle = (1-p)L^0_A + pL^0_B + \frac{2\mathcal{H}_n(\alpha_B - \alpha_A)}{\alpha_A \alpha_B (L^0_B - L^0_A)}, \tag{65}$$

$$\langle L_A \rangle = L^0_A + \frac{2\mathcal{H}_n}{\alpha_A(1-p)(L^0_B - L^0_A)}, \tag{66}$$

$$\langle L_B \rangle = L^0_B - \frac{2\mathcal{H}_n}{p\alpha_B(L^0_B - L^0_A)}. \tag{67}$$

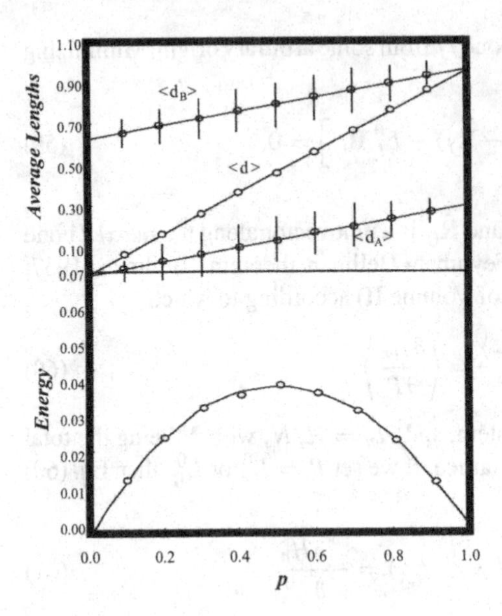

FIGURE 9.15. Average lengths $\langle d \rangle$, $\langle d_A \rangle$ and $\langle d_B \rangle$ (top) and energy per bond in units of $\alpha(L_A^0 - L_B^0)^2$ for the random triangular network with $\alpha_A = \alpha_B$. Solid curves are the theoretical predictions, while symbols are simulation results (after Thorpe and Garboczi, 1990).

The relation

$$\langle L \rangle = (1-p)L_A^0 + pL_B^0, \tag{68}$$

is known as the Vegard's law, and therefore Eqs. (65)–(67) indicate how the deviations arises from the inequality of the spring constants. Only when one has complete phase separation, i.e., when $\mathcal{H}_n = 0$, one recovers Vegard's law. The above results were all derived by Thorpe and Garboczi (1990). One may also define a *dimensionless* length d_{ij} such that, $L_{ij} = L_A^0 + d_{ij}(L_B^0 - L_A^0)$, in terms of which Eqs. (58)–(60) can be rewritten as

$$\begin{aligned} \langle d \rangle &= p, \\ \langle d_A \rangle &= p[1 - a^*(p)], \\ \langle d_B \rangle &= 1 - (1-p)[1 - a^*(p)], \end{aligned} \tag{69}$$

where $a^*(p) = \langle d_B \rangle - \langle d_A \rangle = 1 - h(p, 1)$, so that $\alpha_A = \alpha_B$. Figure 9.15 compares the theoretical predictions with the computer simulation results using a triangular lattice; the agreement is excellent. In these simulations, $L_A^0 = 1.0$ and $L_B^0 = 1.01$.

One can also develop an effective-medium approximation (EMA) for this problem (Thorpe and Garboczi, 1990). As described in Section 8.7, in the EMA approach, a central-force network is characterized by an effective spring constant α_e. If $\alpha_e = \alpha / I_W$, where I_W is a Watson integral [see Eqs. (6.130)–(6.132) for 3D examples of the Watson integral] of the lattice, then, one has the standard EMA equation for a 2D system:

$$p\frac{\alpha - \alpha_B}{\alpha_e' + \alpha_B} + (1-p)\frac{\alpha - \alpha_A}{\alpha_e' + \alpha_A} = 0, \tag{70}$$

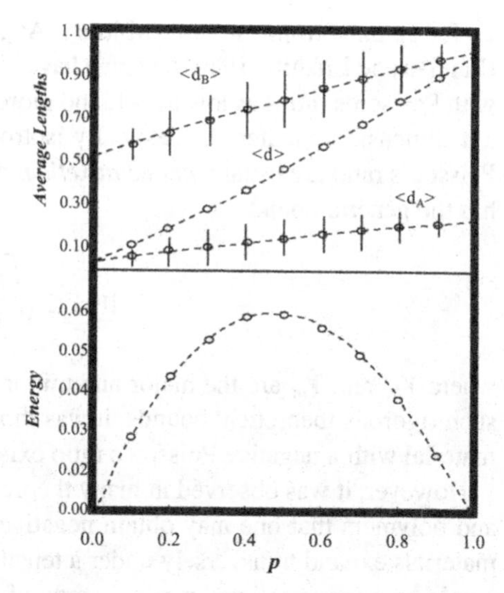

FIGURE 9.16. Same as in Figure 9.15 but with $\alpha_A = 2\alpha_B$ (after Thorpe and Garboczi, 1990).

where $\alpha'_e = \alpha_e - \alpha$. Thorpe and Garboczi (1990) showed that the EMA equations for the three main quantities are given by

$$\langle d\rangle = \frac{p\alpha_B/(\alpha'_e + \alpha_B)}{p\alpha_B/(\alpha'_e + \alpha_B) + \alpha_A(1-p)/(\alpha'_e + \alpha_A)},$$

$$\langle d_A\rangle = \frac{p\alpha_e\alpha'_e\alpha_B}{\alpha(\alpha'_e + \alpha_A)(\alpha'_e + \alpha_B)}, \tag{71}$$

$$\langle d_B\rangle = 1 - (1-p)\frac{\alpha_e\alpha'_e\alpha_A}{\alpha(\alpha'_e + \alpha_A)(\alpha'_e + \alpha_B)}.$$

Figure 9.16 compares the EMA predictions with the results of computer simulations on the triangular lattice for the case $\alpha_A = 2\alpha_B$, and $L_A^0 = 1.0$ and $L_B^0 = 1.01$. The agreement is excellent, demonstrating the accuracy of the EMA for predicting a wide variety of materials' properties.

We thus conclude that vector percolation models provide very useful tools for modeling mechanical properties of materials with natural length mismatch. Of course, in addition to the central-force networks, one may develop more sophisticated models for this phenomenon based on, for example, the bond-bending model, but the general features of the results presented here should hold for more advanced models as well.

9.7 Materials with Negative Poisson's Ratio

The Poisson's ratio ν_p is a measure of the strain in a transverse direction which results from a stress applied longitudinally. In most known materials the experimentally-measured Poisson's ratio is positive, which corresponds to materials that shrink transversely when stretched longitudinally. The definition of

ν_p for an *anisotropic* solid is arbitrary. As mentioned in Chapter 7, theoretically (Landau and Lifshitz, 1986) one may have highly compressible isotropic materials with Poisson's ratios as low as -1, and more generally, $-1 \leq \nu_p \leq (d-1)^{-1}$ for a d-dimensional material. Elastically isotropic materials that possess a negative Poisson's ratio are usually called *auxetic materials*. For anisotropic materials one has the general bound

$$|\nu_p| \leq \sqrt{\frac{Y_M}{Y_m}},$$

where Y_M and Y_m are the major and minor orthotropic Young's moduli. Despite such rigorous theoretical bounds, it was thought for a long time that no isotropic material with a negative Poisson's ratio exists.

However, it was observed in many theoretical studies of models of membranes and polymers that one may obtain negative Poisson's ratios, implying that such materials expand transversely under a tensile load applied longitudinally. For example, in a renormalization group study of a tethered membrane (Aronovitz and Lubensky, 1988) which was fluctuating about a planar mean conformation, a negative Poisson's ratio was obtained. Subsequently, Nelson and Radzihovsky (1991) showed that the negative Poisson's ratio in the model of Aronovitz and Nelson persists even if weak quenched disorder is included in the model. Boal *et al.* (1993) showed that, even for a membrane, fluctuations in the third dimension are not required for the existence of a negative Poisson's ratio, thus generalizing the result of Aronovitz and Lubensky (1988). Their simulations indicated that a 2D system subject to a square-well potential and an entropic tension can possess a negative Poisson's ratio. In these models the negative Poisson's ratio is due to entropic properties of the system. Evans (1989) constructed models by a combination of anisotropic or isotropic particles, tensile springs, and topologically-constraining rods or strings. His models all had negative Poisson's ratios. Likewise, Milton (1992) constructed a family of 2D, two-phase, composite materials with hexagonal symmetry and a Poisson's ratio arbitrarily close to -1. These composites are hierarchical laminates that were studied in Chapters 4 and 7 as relatively simple models of composite materials, and can exhibit extremal Poisson's ratio approaching -1.

Elastic percolation networks also possess negative values of the Poisson's ratio. Figure 9.17 presents ν_p for both elastic and superelastic triangular networks in the bond-bending model. While ν_p remains more or less constant in the elastic network as p, the fraction of the intact bonds, is varied, it has a rather peculiar behavior in the superelastic (randomly-reinforced) network as the fraction p of the completely rigid bonds approaches p_c: It reaches a minimum at about $p \simeq 0.05$, and then increases and achieves its maximum at about $p = 0.1$, beyond which it decreases monotonically and approaches $\nu_p \simeq -0.6$ at p_c. This peculiar behavior of ν_p has not found any explanation, since $p = 0.05$ and 0.1 are not associated with any particular topological changes in the structure of the rigid clusters in a SEPN. Note that in the EPNs considered here, all the contributions to the energy of

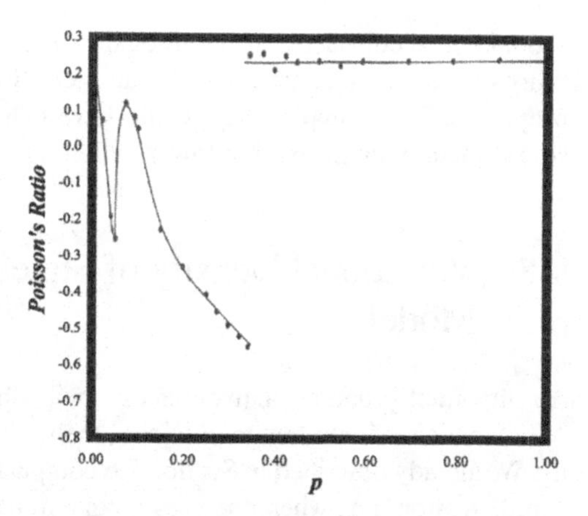

FIGURE 9.17. Poisson's ratio of the triangular network in the bond-bending model. Below $p_c \simeq 0.347$ p is the fraction of the superrigid bonds, while above p_c it is the fraction of the intact bonds with a finite elastic constant (after Sahimi and Arbabi, 1993).

the system are enthalpic, and there is no entropic effect contributing to the energy $\mathcal{H}$. In a different model, Boal *et al.* (1992) used a triangular network of Hookean springs under tension, similar to what we described in Section 8.9, and showed that such a model also exhibits negative values of the Poisson's ratio.

Since these models can possess negative values of the Poisson's ratio, and their applications encompass a large class of materials of practical importance, ranging from linear polymers to gel networks, it should be interesting to see whether real materials can also exhibit this property. Indeed, there are some manufactured foams (Lakes, 1987) which are isotropic and their Poisson's ratio is as low as -0.5, which is attributed to their microstructural geometry (see, for example, Friis *et al.*, 1988; Warren, 1990). These foams have *reentrant* (that is, non-convex) cell structures. Analogous 2D auxetic cellular materials have also been designed and fabricated (Sigmund, 1994; Larsen *et al.*, 1997; Xu *et al.*, 1999). In addition, one may achieve a Poisson's ratio of -1 by a 2D chiral honeycomb structure (Prall and Lakes, 1997).

Evans and Caddock (1989) and Caddock and Evans (1989) demonstrated that microporous, anisotropic polytetrafluoroethylene has a large negative Poisson's ratio, *regardless* of how it is defined. Alderson and Evans (1992) manufactured microporous ultra high molecular weight polyethylene, which is an anisotropic material that possesses negative Poisson's ratio as low as -1.2. Other materials with negative values of the Poisson's ratio include iron pyrites, pyrolitic graphite, hexagonal metals, and fibre-reinforced composites (Herakovich, 1984). Data from the U.S. Bureau of Reclamation indicate that there are some rocks that have slightly negative values of ν_p, such as a sample of Schist (sercite) with $\nu_p = -0.02$, Phyllite (quatoze) with $\nu_p = -0.03$, and samples of granite with $\nu_p = -0.04$ and -0.10. However, all of these materials are anisotropic for which the definition of ν_p is not unique. For such materials one can define an average or effective Poisson's ratio (Sahimi and Arbabi, 1993), where the averaging is taken with respect to all orientations in the solid. In general, however, all sensible definitions of ν_p would result in a behavior similar to Figure 9.17.

Thus, negative Poisson's ratios, as demonstrated by the EPNs, by the various models of membranes and surfaces, and by the real materials mentioned above, should not be considered as unphysical and, in fact, they point towards the possibility of designing a new class of materials with interesting, unusual, and useful properties. For example, they would have enhanced flexural rigidity and plane strain-fracture toughness, but lower bulk modulus.

9.8 Vibrational Density of States: Vector Percolation Model

An important property of materials is their vibrational density of states (DOS) $\mathcal{N}(\omega)$, which is important to obtaining their specific heat and thermal conductivity. We already described in Section 6.6 computation of the DOS within the scalar approximation, i.e., when one uses a random resistor network for computing the DOS. In this section, we describe and discuss the DOS in terms of the true vectorial nature of the vibrations. We consider heterogeneous materials with percolation-type disorder and focus our attention on the behavior of the DOS at the percolation threshold p_c, or above the threshold but at length scales $L \ll \xi_p$, where ξ_p is the correlation length of percolation. For length scales $L \gg \xi_p$, a material is macroscopically homogeneous, and thus its vibrational DOS is described by the classical laws in the phonon regime (see also below).

9.8.1 *Scaling Theory*

Recall from Chapter 6 (see Section 6.6) that vibrational states in materials with fractal morphology, such as one at its percolation threshold p_c (or above its p_c but for length scales $L \ll \xi_p$) are called *fractons* (meaning phonons on fractals!). For homogeneous materials, on the other hand, the vibrational DOS is given by the well-known relation,

$$\mathcal{N}(\omega) \sim \omega^{d-1}, \tag{72}$$

where d is the Euclidean dimensionality of the system. Equation (72) is valid at low frequencies (i.e., long wavelengths or large length scales over which the system is homogeneous) such that $\omega < \omega_{co}$, where ω_{co} is a cutoff or crossover frequency. Recall also from Chapter 6 that, according to the scaling theory of Alexander and Orbach (1982), for materials with a fractal morphology (or fractal networks in the context of the discrete models), the vibrational DOS within the scalar approximation is described by

$$\mathcal{N}(\omega) \sim \omega^{D_s-1}, \tag{73}$$

where $D_s = 2D_f/D_w$ is called the spectral (or fracton) dimension, D_f is the fractal dimension of the material, and D_w is the fractal dimension of a random walker diffusing in the fractal substrate. As we learned earlier in this book, in the context of percolation models of disordered materials, $D_f = d - \beta/\nu$ and $D_w = 2 + (\mu - \beta)/\nu$, where β, μ and ν are the usual percolation and conductivity exponents, so that

$$D_s = 2\frac{\nu d - \beta}{2\nu + \mu - \beta}. \tag{74}$$

Equations (73) and (74) are valid as long as the scalar contributions to the elastic energy of the material dominate the vectorial contributions. As discussed in Section 8.4.1, an example of such a material is one that can be described by the Born model (regardless of the model's shortcomings that we already discussed there). However, we now know that the nature of the forces that are exerted on a material, or the sites and bonds of the network that represents it, and their contributions to the elastic energy are most important. Hence, if we wish to take into account the true vectorial nature of the vibrational DOS, we are forced to introduce a new quantity D_s^e, which is called the *elastic* or *flexural spectral dimension* D_s^e, which is defined by replacing in Eq. (74) the conductivity exponent μ by f, the critical exponent of the elastic moduli, so that (Webman and Grest, 1985)

$$D_s^e = 2\frac{\nu d - \beta}{2\nu + f - \beta}. \tag{75}$$

Accordingly, we are also forced to define *two* different types of DOS for vibrations of heterogeneous materials with fractal morphology. One is $\mathcal{N}(\omega)$, which is the DOS when the scalar contributions to the elastic energy of the material dominate all other types of contributions, for which Eq. (73) is valid. The second type is what we call the *vectorial* or *elastic* DOS $\mathcal{N}_e(\omega)$ which is appropriate when the vectorial contributions, such as those of the central and bond-bending (BB) forces, dominate the elastic energy. In this case, in analogy with Eq. (73), the vibrational DOS is given by

$$\mathcal{N}_e(\omega) \sim \omega^{D_s^e - 1}. \tag{76}$$

Equations (73) and (76) are both valid for high frequencies (i.e., short wavelengths or length scales) such that $\omega > \omega_{co}$. Such vibrational states are localized as long as D_s or D_s^e is less than 2 (Rammal and Toulouse, 1983). As explained in Chapter 6, $D_s \simeq 4/3$ for all $d \geq 2$. In contrast, D_s^e varies continuously with d. Clearly, since $f > \mu$, one must have $D_s^e < D_s$, and in fact D_s^e can be *less than one* (see below).

The difference between Eqs. (73) and (76) is striking. Since $D_s \simeq 4/3$, one has

$$\mathcal{N}(\omega) \sim \omega^{1/3}, \quad \text{scalar approximation.} \tag{77}$$

If $\mathcal{N}(\omega)$ decreases with decreasing ω, then the material is mechanically stable. To understand this, recall that small frequencies imply large length scales, and if over such length scales $\mathcal{N}(\omega)$ decreases, the implication would be that there are fewer and fewer vibrational modes, i.e., the system is mechanically stable. Thus, materials (with a fractal morphology) in which the scalar contributions to their elastic energy dominate all other contributions, for which Eq. (77) is valid, are mechanically stable. On the other hand, if we use in Eq. (75), $f \simeq 3.75$ for 3D BB models (see Table 9.1), we obtain

$$D_s^e \simeq 0.87, \quad d = 3, \tag{78}$$

so that Eq. (76) becomes

$$\mathcal{N}_e(\omega) \sim \omega^{-0.13}, \quad \text{bond-bending elasticity,} \tag{79}$$

That is, as ω decreases there are larger and larger number of vibrating modes, implying that such materials *cannot become too large*, because if they do, they will lose their mechanical stability, and therefore must rearrange themselves into more stable structures. This type of DOS should be observed in materials (with a fractal morphology) in which the contributions of the BB forces dominate the elastic energy. Since for materials in which the central forces dominate, $f \simeq 2.1$ (see Table 9.1), then, the corresponding $D_s^e \simeq D_s$. On other hand, as discussed earlier in this chapter, if for entropic gels, $f \simeq \nu d$, then,

$$D_s^e = 2\frac{\nu d - \beta}{2\nu + \nu d - \beta} \simeq 1.12, \tag{80}$$

so that

$$\mathcal{N}_e \sim \omega^{0.12}, \quad \text{entropic elasticity,} \tag{81}$$

and therefore materials with entropic elasticity are also mechanically stable.

In analogy with the limit in which the scalar contributions to the elastic energy dominate all other types of contributions, one can use Eq. (76) to define a "dispersion relation" [see Eqs. (6.346), (6.348), and (6.349)]:

$$\omega \sim [L(\omega)]^{-D_f/D_s^e}. \tag{82}$$

To derive this equation, consider the vibrations of a material of linear size L. Although high-frequency modes of the vibrations are not affected by a change in the boundary conditions, the low-frequency modes will disappear from the spectrum. The crossover between the two regimes occurs at a frequency ω_L, such that

$$L \sim \Lambda(\omega_L), \tag{83}$$

where $\Lambda(\omega)$ is the wavelength. Following the same arguments that were used in Section 6.4.5.2, we consider the integrated spectral weight of the missing low-frequency modes which are lumped together in the center-of-mass degrees of freedom of the material. Since this quantity cannot depend on L and ω_L, one must have

$$[\Lambda(\omega_L)]^{D_f} \int_0^{\omega_L} \omega^{D_s^e - 1} d\omega \sim \Lambda^{D_f} \omega_L^{D_s^e} = \text{constant}, \tag{84}$$

and thus the dispersion relation for an arbitrary frequency is given by

$$\Lambda(\omega) \sim \omega^{-D_s^e/D_f}, \tag{85}$$

which is the analogue of Eq. (6.354) for scalar approximation to the DOS.

9.8.2 *Crossover Between Scalar Approximation and Vector Density of States*

Is there a crossover between the DOS described by Eqs. (73) and (76)? Feng (1985b) argued that in material with percolation disorder in which both the central and BB forces are present, another length scale L_{bb}, in addition to the percolation correlation length ξ_p, is important. If $L_{bb} \gg \xi_p$, then, at low frequencies Eq. (72) and at higher frequencies Eq. (73) govern the vibrational DOS, with a crossover at ω_{co}. However, if $L_{bb} \ll \xi_p$, then at low frequencies one is in the phonon regime until a characteristic frequency ω_{bb} is reached at which there is a crossover from Eq. (72) to Eq. (76). When the higher frequency ω_{co} is reached, there is a second crossover from Eq. (76) to (73). It is not difficult to guess that [see Eq. (82)]

$$\omega_{bb} \sim \xi_p^{-D_f/D_s^e}. \tag{86}$$

According to Feng (1985b), the characteristic length scale L_{bb} is given by

$$L_{bb} \sim \sqrt{\frac{\gamma}{\alpha}}, \tag{87}$$

where α and γ are, respectively, the usual central and BB force constants. In effect, L_{bb} is the length scale below which the motion due to central forces is energetically more favorable and above which the BB forces dominate. Large scale computer simulations of Yakubo *et al.* (1990b), to be described below, confirmed such crossover behavior. Other efforts in this direction includes those of Day *et al.* (1985), Cai and Thorpe (1989), and Monceau and Levy (1994).

9.8.3 *Large-Scale Computer Simulation*

Webman and Grest (1985) were apparently the first to calculate the vibrational DOS using vector percolation and the BB model, analyzing a system with $N \sim 10^3$ sites. They found, in agreement with Eq. (79), that the DOS was weakly divergent at low frequencies. Lam and Bao (1985) calculated the vibrational DOS of a triangular network with central forces in site percolation, and found the DOS in the fracton regime to be proportional to $\omega^{D_s^e - 1}$ with $D_s^e \simeq 0.625$. Their estimate of D_s^e should, however, be regarded with caution because of the very narrow frequency interval in the fracton regime in which they carried out their calculations. Day *et al.* (1985) calculated the vibrational DOS of the same network but with central forces and 60% BB forces. They estimated values of the percolation threshold p_{ce} and the spectral dimension D_s^e in the ranges $0.4 \le p_{ce} \le 0.405$ and $1.25 \le D_s^e \le 1.3$, which suggest that the 60%-BB model does not belong to the universality class of a full BB model. Other efforts in this direction were reviewed by Nakayama *et al.* (1994), to whom the interested reader is referred to.

Large-scale simulations of the vibrational DOS of the BB model were carried out by Yakubo *et al.* (1990b) who studied the crossover behavior of the DOS from bending to stretching fractons at the percolation threshold of a 500×500 square network. They considered the case in which the BB force constant γ was larger

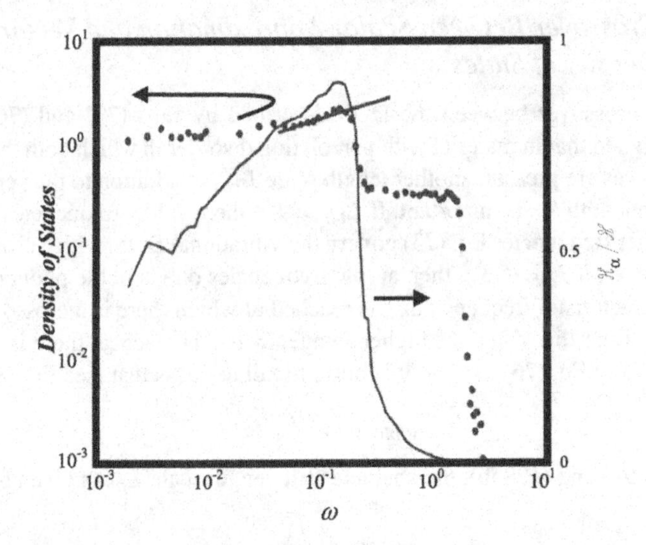

FIGURE 9.18. Calculated density of states (solid circles) at the bond percolation threshold of a square network with stretching and bond- bending forces and the force constants $\gamma = 10\alpha = 0.133$. The straight line is guide to the eye for $\mathcal{N}(\omega) \sim \omega^{1/3}$, while solid curve shows the energy ratio $\mathcal{H}_\alpha/\mathcal{H}$ and its frequency dependence (after Yakubo *et al.*, 1990b).

than the stretching one α. The calculated DOS is shown in Figure 9.18 in which $\gamma = 10\alpha = 0.133$. The network had a cutoff Debye frequency $\omega_D = 2.0784$. The steplike decrease of the states in the high-frequency regime in Figure 9.18 indicates that the stretching motions are not excited above some frequency ω_0, the value of which is determined by the stretching force constants α. For $\alpha = 0.0133$, ω_0 is estimated to be 0.2309 from the relation $\omega_0 = 2\sqrt{\alpha}$, coinciding with the observed value in Figure 9.18.

To clarify the individual contributions from the bending and stretching forces, the ratio of the contribution of each to the elastic energy was calculated as a function of the frequency. The solid curve in Figure 9.18 shows the ratio between the energy $\mathcal{H}_\alpha$ attributed to the stretching motion and the total energy $\mathcal{H}$. A crossover frequency ω_{co} is defined based on the condition, $\mathcal{H}_\alpha(\omega_{co})/\mathcal{H}(\omega_{co}) = 1/2$, leading to $\omega_{co} \sim 0.005$. The DOS in the vicinity of ω_{co} is independent of ω. The crossover region from bending to stretching fractons extends over at least two orders of magnitude in frequency, due to the fact that the ratio $\mathcal{H}_\alpha(\omega)/\mathcal{H}(\omega)$ of the energies increases logarithmically, as shown in Figure 9.18. This is in contrast with the sharp crossover from phonons to fractons for the scalar approximation described in Section 6.6. The straight line through the solid circles in Figure 9.18 has been drawn according to Eq. (77). Based on the data of Figure 9.18 in the frequency region between $\mathcal{N}(\omega) \sim \omega_{co}$ and ω_0, it is not clear that the $\omega^{1/3}$ law holds.

To clarify the contribution of bending fractons, the DOS were calculated for $\gamma/\alpha = 0.01$ and 1.0, which allow examination of the DOS for the bending-fracton

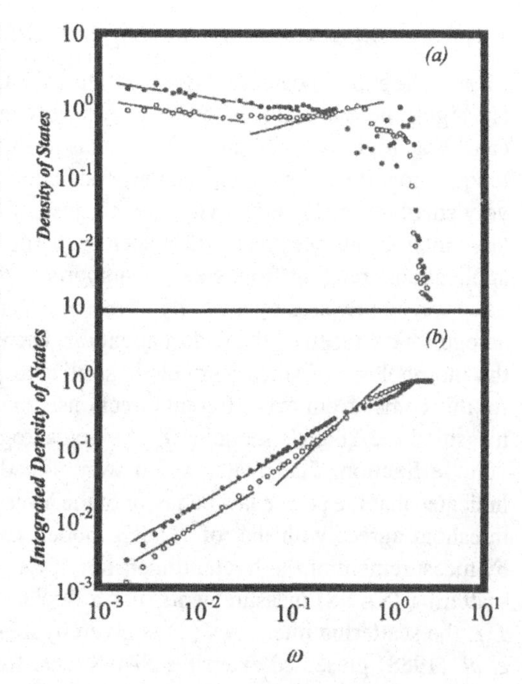

FIGURE 9.19. Density of states (a) and integrated density of states (b) at the bond percolation threshold of a square network with stretching and bond- bending forces. Filled and open circles show, respectively, the results for $\alpha = 100\gamma = 1.0$ and $\alpha = \gamma = 0.12$. The straight lines on the left sides of the figures indicate the $\omega^{D_s^e-1}$ and $\omega^{D_s^e}$ laws for the DOS and integrated DOS, while those on the right sides show the $\omega^{1/3}$ and $\omega^{4/3}$ laws for the scalar approximation (after Yakubo *et al.*, 1990b).

regime, because the stretching-fracton regime is shifted into the high-frequency region. Figures 9.19(a) and 9.19(b) show the results for the DOS and the integrated DOS, respectively, for the same network as that used for obtaining the results shown in Figure 9.18. The DOS, given by solid circles in Figure 9.19(a), weakly diverges as $\omega \to 0$, in agreement with Eq. (79). The value of D_s^e obtained by a least square fitting of the data shown in Figure 9.19 is $D_s^e \simeq 0.79$, reasonably close to (78). The straight lines on the left-hand side of the figure through the solid and open circles in Figure 9.19(a) [Figure 9.19(b)] are drawn according to $\omega^{D_s^e-1}$ $(\omega^{D_s^e})$ with $D_s^e \simeq 0.79$.

In the case of $\gamma/\alpha = 0.01$ (solid circles in Figure 9.19), the frequency ω_{co} becomes too large to distinguish the crossover frequency region, whereas for $\gamma/\alpha = 1.0$ (open circles in Figure 9.19) the crossover frequency can be estimated from evaluation of $\mathcal{H}_\alpha/\mathcal{H}$, as in the case of Figure 9.18. The frequency ω_{co} for crossover from bending to stretching fractons is close to $\omega_{bb} \sim \sqrt{\gamma/\alpha} \sim 0.1$. Note that these results do not exhibit any noticeable change in the frequency dependence of the DOS around the crossover, as shown by the open circles in the vicinity of $\omega \sim 0.1$. For comparison, the lines on the right-hand side of the figure are drawn according to the $\omega^{1/3}$ $(\omega^{4/3})$ law, indicating disagreement with the numerical data. Furthermore, it should be emphasized that the magnitudes of the DOS in the bending-fracton regimes are different for the two sets of force constants (see the difference between solid and open circles in Figure 9.19), implying that the missing modes tend to accumulate in the high-frequency region ($\omega > \omega_0$).

9.8.4 Comparison with the Experimental Data

Most of the experimental studies of vibrational DOS have involved aerogels which are highly porous solid materials with a very tenuous structure. Their porosity can be as high as 99%, and for this reason they often have very unusual and unique properties. For examples, they can be made in transparent form, they have very small thermal conductivity, and because of their large porosity they possess large internal surface area. Due to such properties, they also have a wide range of applications, ranging from catalyst supports to thermal insulators and radiators. They can be prepared by a variety of methods using different materials, but silica aerogels have received the widest attention. There has been some speculation that the morphology of silica aerogels is similar to percolation networks. Evidence for this comes from two different directions. On one hand, Woignier *et al.* (1988) measured the Young's modulus Y of silica aerogels as a function of density and volume fraction. Their data, which were already discussed in Section 9.2.3.1, indicated that the power-law behavior of the Young's modulus near the percolation threshold agrees with that of 3D BB model. Evidence has also been provided by measurement of the fractal dimension of the gel through small-angle neutron scattering (SANS) measurements. For a fractal system with a fractal dimension D_f, the scattering intensity $I(q)$ is given by an equation similar to (18). Vacher *et al.* (1988) presented extensive SANS data, from which the fractal dimension of silica aerogels was found to be, $D_f \simeq 2.45$, very close to that of 3D sample-spanning percolation clusters, $D_f \simeq 2.53$ (see Table 2.3). However, it is not yet clear whether silica aerogels do actually represent some type of a percolation structure.

The next step is to verify the scaling law for the density of states, Eq. (73) or (76). One way of doing this is by plotting the crossover or cutoff frequency ω_{co} as a function of q. Since q is inversely proportional to the length scale L, one must have

$$\omega_{co} \sim q^{D_f/D_s}, \tag{88}$$

so that such a plot should yield, $\omega_{co} \sim q^{1.88}$, if silica aerogels are percolation fractals and $D_s \simeq 4/3$. Figure 6.31 presents such a plot for a series of aerogels with various densities. The data were obtained by Brillouin scattering of visible light and indicate that, $\omega_{co} \sim q^{1.9}$, which is in very good agreement with the theoretical prediction. The vibrational density of states itself can be measured directly by incoherent neutron scattering. Vacher *et al.* (1989) used this technique to measure, $\mathcal{N}(\omega) \sim \omega^{0.85\pm0.15}$, which does not agree with either Eq. (73) or (76). This difference may be explained in various ways. One is that the material used in this study was not simply a sample-spanning percolation cluster, rather, in addition to the main gel network, many other smaller clusters (small branched polymers) could have been present, and when the solvent was removed a *distribution* of clusters (rather than a single cluster) had been obtained. Thus, the incoherent scattering was not measured from just one percolation cluster, rather from a collection of them, and hence the DOS should be written as

$$\mathcal{N}(\omega) \sim \omega^{2d/D_w - 1}, \tag{89}$$

rather than Eq. (73), because, as discussed in Chapter 2, the collection of all percolation clusters is not fractal, only the individual clusters at length scales $L \ll \xi_p$ are. Equation (89) then predicts that $\mathcal{N}(\omega) \sim \omega^{0.6}$, which is reasonably close to the experimental measurements. Vacher *et al.* (1989) gave a rather different reason for this difference. They argued that the finite clusters attach themselves to the sample-spanning cluster, making it denser, thereby increasing the *effective value* of D_s from its theoretical value of about 4/3 to a larger value of 1.85. However, this reasoning cannot give any information about the structure of the dense network, and therefore cannot be checked directly. One may also argue that energy resolution in the experiment was not sufficient to yield enough accuracy. Indeed, the experiments of Schaefer *et al.* (1990), using more accurate techniques, yielded $D_s \simeq 1.22 \pm 0.14$, which is again consistent with the theoretical expectation.

The crossover between Eqs. (73) and (76) was also studied in the experiments of Vacher *et al.* (1990), using silica aerogels. They measured the DOS of the gels, and found that at low frequencies their data can be fitted with $D_s^e \simeq 0.9$, in good agreement with the theoretical expectation (71), while at higher frequencies the data can be fitted with $D_s \simeq 1.7 \pm 0.2$, again larger than $D_s \simeq 4/3$; see Figure 9.20. Whether this discrepancy can be explained with the same type of reasoning discussed above is not clear yet.

Some other experimental studies of vibrational DOS of materials should be mentioned here. Boolchand *et al.* (1990) carried out experiments including

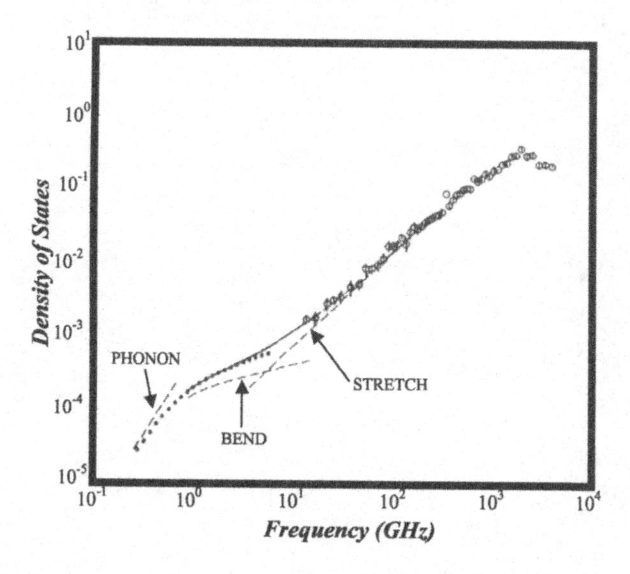

FIGURE 9.20. Crossover in the scaling of the density of states of silica aerogels. Arrows indicate the locations of the crossover frequencies (after Vacher *et al.*, 1990).

Mössbauer–Debye–Waller factors, inelastic neutron scattering, Raman scattering, and ultrasonic elastic moduli, using chalcogenide glasses that were of the general form A_xB_{1-x}. As discussed earlier in this chapter, the average coordination number of such glasses is $\langle Z\rangle = 2(x+1)$, and according to the Phillips–Thorpe theory and the vector percolation model, one should see a drastic difference in the vibrational DOS at $\langle Z\rangle \simeq 2.4$, corresponding to $x_c \simeq 0.2$. Indeed, Boolchand *et al.*'s experiments did indicate a threshold behavior at this composition. The same type of behavior was observed in the experiments of Kamitakahara *et al.* (1991), and Walter *et al.* (1988), although the threshold behavior in these experiments was not as strong as that in the experiments of Boolchand *et al.*

Summary

Vector percolation models explain qualitatively, and in many cases quantitatively, the mechanical behavior of disordered materials, ranging from branched polymers, gel polymers, and glasses, to composite solids and porous materials. In particular, such models provide a rational explanation for the power-law behavior of the mechanical properties near the percolation threshold, which mean-field theories, effective-medium approximations, and rigorous bounds of the type described in Chapter 7, fail to predict.

References

Abeles, B., H.L. Pinch, and J.I. Gittleman, "Percolation conductivity in W-Al_2O_3 granular metal films," *Phys. Rev. Lett.* **35**, 247 (1975).

Aboav, D.A., "The arrangement of grains in a polycrystal," *Metallography* **3**, 383 (1970).

Abrahams, E., P.W. Anderson, D.C. Licciardello, and T.V. Ramakrishnan, "Scaling theory of localization: Absence of quantum diffusion in two dimensions," *Phys. Rev. Lett.* **42**, 673 (1979).

Acrivos, A., and E. Chang, "A model for estimating transport quantities in two-phase materials," *Phys. Fluids* **29**, 3 (1986).

Adam, M., M. Delsanti and D. Durand, "Mechanical measurements in a reaction bath during the polycondensation reaction near the gelation threshold," *Macromolecules* **18**, 2285 (1985).

Adam, M., M. Delsanti, D. Durand, G. Hild, and J.P. Munch, "Mechanical properties near gelation threshold, comparison with classical and 3d percolation theories," *Pure Appl. Chem.* **53**, 1489 (1981).

Adam, M., M. Delsanti, J.P. Munch, and D. Durand, "Size and mass determination of clusters obtained by polycondensation near the gelation threshold," *J. Physique* **48**, 1809 (1987).

Adam, M., M. Delsanti, R. Okasha, and G. Hild, "Viscosity study in the reaction bath of the radical copolymerisation of styrene divinylbenzene," *J. Physique Lett.* **40**, L539 (1979).

Adam, M., D. Lairez, M. Karpasas, and M. Gottlieb, "Static and dynamic properties of cross-linked poly(dimethylsiloxane) pregel clusters," *Macromolecules* **30**, 5920 (1997).

Adams, D.F., and D.R. Doner, *J. Compos. Mater.* **1**, 152 (1967).

Adler, J., "Bootstrap percolation," *Physica A* **171**, 453 (1991).

Adler, J., R.G. Palmer, and M.H. Meyer, "Transmission of order in some unusual dilute systems," *Phys. Rev. Lett.* **58**, 882 (1987).

Adler, J., and D. Stauffer, "Evidence for non-universal exponents in bootstrap percolation," *J. Phys. A* **23**, L1119 (1990).

Adolf, D., J.E. Martin, and J.P. Wilcoxon, "Evolution of structure and viscoelasticity in an epoxy near the sol-gel transition," *Macromolecules* **23**, 527 (1990).

Aharony, A., "Universal critical amplitude ratios for percolation," *Phys. Rev. B* **22**, 400 (1980).

Aharony, A., S. Alexander, O. Entin-Wohlman, and R. Orbach, "Scaling approach to phonon-fracton crossover," *Phys. Rev. B* **31**, 2565 (1985).

Aharony, A., S. Alexander, O. Entin-Wohlman, and R. Orbach, "Scattering of fractons, the Ioffe-Regel criterion, and the $\frac{4}{3}$ conjecture," *Phys. Rev. Lett.* **58**, 132 (1987).

Aharony, A., O. Entin-Wohlman, and R. Orbach, in *Time Dependent Effects in Disordered Materials,* edited by T. Riste and R. Pynn (Plenum, New York, 1988), p. 233.

Aharony, A., and A.B. Harris, "Superlocalization, correlations and random walks on fractals," *Physica A* **163**, 38 (1990).

Aharony, A., A.B. Harris, and O. Entin-Wohlman, "Was superlocalization observed in carbonblack-polymer composites," *Phys. Rev. Lett.* **70**, 4160 (1993).

Aharony, A., Y. Zhang, and M.P. Sarachik, "Universal crossover in variable range hopping with Coulomb interactions," *Phys. Rev. Lett.* **68**, 3900 (1992).

Ahmed, G., and J.A. Blackman, "On theories of transport in disordered media," *J. Phys. C* **12**, 837 (1979).

Alderson, K.L., and K.E. Evans, "The fabrication of microporous polyethylene having a negative Poisson's ratio," *Polymer* **33**, 4435 (1992).

Alexander, S., "Is the elastic energy of amorphous materials rotationally invariant?" *J. Physique* **45**, 1939 (1984).

Alexander, S., "Vibrations of fractals and scattering of light from aerogels," *Phys. Rev. B* **40**, 7953 (1989).

Alexander, S., "Amorphous solids: Their structure, lattice dynamics and elasticity," *Phys. Rep.* **296**, 65 (1998).

Alexander, S., Bernasconi, J., and Orbach, R., "Low energy density of states for disordered chains," *J. Physique Colloq.* **39**(C6), 706 (1978).

Alexander, S., J. Bernasconi, W.R. Schneider, and R. Orbach, "Excitation dynamics in random one-dimensional systems," *Rev. Mod. Phys.* **53**, 175 (1981).

Alexander, S., E. Courtens, and R. Vacher, "Vibrations of fractals: dynamic scaling, correlation functions and inelastic light scattering," *Physica A* **195**, 286 (1993).

Alexander, S., O., Entin-Wohlman, and R. Orbach, "Relaxation and nonradiative decay in disordered systems. II. Two-fracton inelastic scattering," *Phys. Rev. B* **33**, 3935 (1986a).

Alexander, S., O. Entin-Wohlman, and R. Orbach, "Phonon-fracton anharmonic interactions: the thermal conductivity of amorphous materials," *Phys. Rev. B* **34**, 2726 (1986b).

Alexander, S., C. Laermans, R. Orbach, and H.M. Rosenberg, "Fracton interpretation of vibrational properties of cross-linked polymers, glasses, and irradiated quartz," *Phys. Rev. B* **28**, 4615 (1983).

Alexander, S. and Orbach, R., "Density of states of fractals: 'fractons'," *J. Physique Lett.* **43**, L625 (1982).

Alexandrowicz, Z., "Critically branched chains and percolation clusters," *Phys. Lett.* **A80**, 284 (1980).

Allain, C., L. Limat and L. Salomé, "Description of the mechanical properties of gelling polymer solutions far from gelation threshold: generalized effective-medium calculations of the superconductor-conductor site percolation problem," *Phys. Rev. A* **43**, 5412 (1991).

Allain, C., and L. Salomé, "Sol-gel transition of hydrolyzed polyacrylamide + chromium III: Rheological behavior versus cross-link concentration," *Macromolecules* **20**, 2957 (1987a).

Allain, C., and L. Salomé, "Hydrolysed polyacrylamide/Cr^3 gelation: Critical behaviour of the rheological properties at the sol-gel transition," *Polym. Commun.* **28**, 109 (1987b).

Allain, C., and L. Salomé, "Gelation of semidilute polymer solutions by ion complexation: Critical behavior of the rheological properties versus cross-link concentration," *Macromolecules* **23**, 981 (1990).

Allen, L.C., B. Golding, and W.H. Haemmerle, "Dynamic shear modulus for two-dimensional bond percolation," *Phys. Rev. B* **37**, 913 (1988).

Ambegaokar, V., B.I. Halperin, and J.S. Langer, "Hopping conductivity in disordered systems," *Phys. Rev. B* **4**, 2612 (1971).

Anderson., P.W., "Absence of diffusion in certain random lattices," *Phys. Rev.* **109**, 1492 (1958).

Andrade, J.S., A.M. Auto, Y. Kobayashi, Y. Shibusa, and K. Shirane, "Percolation conduction in vapour grown carbon fibre," *Physica A* **248**, 227 (1998).

Arbabi, S., and M. Sahimi, "Absence of universality in percolation models of disordered elastic media with central forces," *J. Phys. A* **21**, L863 (1988a).

Arbabi, S., and M. Sahimi, "Elastic properties of three-dimensional percolation networks with stretching and bond-bending forces," *Phys. Rev. B* **38**, 7173 (1988b).

Arbabi, S., and M. Sahimi, "On three-dimensional elastic percolation networks with bond-bending forces," *J. Phys. A* **23**, 2211 (1990a).

Arbabi, S., and M. Sahimi, "Critical properties of viscoelasticity of gels and elastic percolation networks," *Phys. Rev. Lett.* **65**, 725 (1990b).

Arbabi, S., and M. Sahimi, "Mechanics of disordered solids. I. Percolation on elastic networks with central forces," *Phys. Rev. B* **47**, 695 (1993).

Aronovitz, J.A., and T.C. Lubensky, "Fluctuations of solid membranes," *Phys. Rev. Lett.* **60**, 2634 (1988).

Askar, A., and A.S. Cakmak, "A structural model of a micropolar continuum," *Int. J. Eng. Sci.* **6**, 583 (1968).

Axelos, M.A.V., and M. Kolb, "Crosslinked biopolymers: Experimental evidence for scalar percolation theory," *Phys. Rev. Lett.* **64**, 1457 (1990).

Baker, G.A., *Essentials of Padé Approximants* (Academic Press, New York, 1975).

Balberg, I., "Recent developments in continuum percolation," *Philos. Mag. B* **55**, 991 (1987).

Balberg, I., and N. Bienbaum, "Cluster structure and conductivity of three-dimensional continuum systems," *Phys. Rev. A* **31**, 1222 (1985).

Balberg, I., N. Binenbaum, and S. Bozowski, "Anisotropic percolation in carbon black-polyvinylchloride composites," *Solid State Commun.* **47**, 989 (1983).

Bale, H.D., and P.W. Schmidt, "Small-angle X-ray scattering investigation of submicroscopic porosity with fractal properties," *Phys. Rev. Lett.* **53**, 596 (1984).

Barton, J.L., "La relaxation diélectrique de quelques verres ternaires silice oxyde alcalin oxyde alcalin-terreux," *Verres et Refr.* **20**, 328 (1966).

Batchelor, G.K., "Transport properties of two-phase materials with random structure," *Annul. Rev. Fluid Mech. 6*, 227 (1974).

Batchelor, G.K., and R.W. O'Brien, "Thermal or electrical conduction through a granular material," *Proc. R. Soc. Lond. A* **355**, 313 (1977).

Beck, A., O. Gelsen, P. Wang, and J. Fricke, *Rev. Phys. Appl. C* **4**, 203 (1989).

Becker, A.A., *The Boundary Element Method in Engineering* (McGraw-Hill, New York, 1992).

Bendler, J.T., and M.F. Shlesinger, "Dielectric relaxation via the Montroll-Weiss random walks of defects," in *The Wonderful World of Stochastics: A Tribute to Elliot W. Montroll*, edited by M.F. Shlesinger and G.H. Weiss (Noth Holland, Amsterdam, 1985), p. 31.

Benguigui, L., "Experimental study of the elastic properties of a percolating system," *Phys. Rev. Lett.* **53**, 2028 (1984).

Benguigui, L., "Lattice and continuum percolation transport exponents: experiments in two-dimensions," *Phys. Rev. B* **34**, 8176 (1986).

Benguigui, L., and D.J. Bergman, "On the ratio of the elastic constants near the percolation threshold of a two-dimensional perforated solid," *Europhys. Lett.* **4**, 823 (1987).

Benguigui, L., and P. Ron, "Experimental realization of superelasticity near the percolation threshold," *Phys. Rev. Lett.* **70**, 2423 (1993).

Beniot, C., E. Royer, and G. Poussigue, "The spectral moments method," *J. Phys.: Condens. Matter* **4**, 3125 (1992).

Benveniste, Y., "On the effective thermal conductivity of multiphase composites," *Z. Ang. Math. Phys.* **37**, 696 (1986).

Beran, M.J., "Use of the variational approach to determine bounds for the effective permittivity of random media," *Nuovo Cimento* **38**, 771 (1965).

Beran, M.J., *Statistical Continuum Theories* (Wiley, New York, 1968).

Beran, M.J., and J. Molyneux, "Use of classical variational principles to determine bounds for the effective bulk modulus in heterogenous media," *Q. Appl. Math.* **24**, 107 (1966).

Beran, M.J., and N. Silnutzer, "effective electrical, thermal and magnetic properties of fiber-reinforced materials," *J. Composite Mater.* **5**, 246 (1971).

Berdichevsky, *Effective Electrical, Thermal and Magnetic Properties of Fiber-Reinforced Materials* (Nauka, Moscow, The Soviet Union, 1983)

Bergman, D.J., "The dielectric constant of a composite material-a problem in classical physics," *Phys. Rep.* **43**, 377 (1978).

Bergman, D.J., "Elastic moduli near percolation: universal ratio and critical exponent," *Phys. Rev. B* **31**, 1696 (1985).

Bergman, D.J., "Elastic moduli near percolation in a two-dimensional random network od rigid and nonrigid bonds," *Phys. Rev. B* **33**, 2013 (1986b).

Bergman, D.J., and E. Duering, "Universal Poisson's ratio in a two-dimensional random network of rigid and nonrigid bonds," *Phys. Rev. B* **34**, 8199 (1986).

Bergman, D.J., E. Duering, and M. Murat, "Discrete network models for the low-field Hall effect near a percolation threshold: Theory and simulations," *J. Stat. Phys.* **58**, 1 (1990).

Bergman, and Y. Imry, "Critical behaviour of the complex dielectric constant near the percolation threshold of a heterogeneous material", *Phys. Rev. Lett.* **39**, 1222 (1977).

Bergman, D.J., and Y. Kantor, "Critical properties of an elastic fractal," *Phys. Rev. Lett.* **53**, 511 (1984).

Bergman, D.J., and D. Stroud, "Scaling theory of the low-field Hall effect near the percolation threshold," *Phys. Rev. B* **32**, 6097 (1985).

Berk, N.F., "Scattering properties of a model bicontinuous structure with a well-defined length scale," *Phys. Rev. Lett.* **58**, 2718 (1987).

Berk, N.F., "Scattering properties of the leveled-wave model of random morphologies," *Phys. Rev. A* **44**, 5069 (1991).

Berman, D., B.G. Orr, H.M. Jaeger, and A.M. Goldman, "Conductances of filled two-dimensional networks," *Phys. Rev. B* **33**, 4301 (1986).

Bernasconi, A., T. Sleator, D. Posselt, J.K. Kjems, and H.R. Ott, "Dynamic properties of silica aerogels as deduced from specific-heat and thermal-conductivity measurements," *Phys. Rev. B* **45**, 10363 (1992).

Bernasconi, J., "Conduction in anisotropic disordered systems: Effective-medium theory," *Phys. Rev. B* **9**, 4575 (1974).

Bernasconi, J., "Real-space renormalization of bond-disordered conductance lattices," *Phys. Rev. B* **18**, 2185 (1978).

Bernasconi, J., W.R. Schneider, and H.J. Wiesmann, "Some rigorous results for random planar conductance networks," *Phys. Rev. B* **16**, 5250 (1977).

Bernasconi, J., and H.J. Wiesmann, "Effective-medium theories for site-disordered resistance networks," *Phys. Rev. B* **13**, 1131 (1976).

Berryman, J.G, "Long-wavelength propagation in composite elastic media I. Spherical inclusions," *J. Acoust. Soc. Am.* **68**, 1809 (1980a).

Berryman, J.G., "Long-wavelength propagation in composite elastic media II. Ellipsoidal inclusions," *J. Acoust. Soc. Am.* **68**, 1820 (1980b).

Berryman, J.G., "Measurement of spatial correlation function using image processing technique," *J. Appl. Phys.* **57**, 2374 (1985).

Berryman, J.G., and G.W. Milton, "Micrometry of random composites and porous media," *J. Phys. D* **21**, 87 (1988).

Bhattacharya, S., J.P. Stokes, M.W. Kim, and J.S. Huang, "Percolation in an oil-continuous microemulsion," *Phys. Rev. Lett.* **55**, 1884 (1985).

Bianchi, R.F., H.P. Souza, T.J. Bonagamba, H.C. Panepucci, and R.M. Faria, "Ionic conduction and structural properties of organic-inorganic composite based on poly(propilene-glycol)," *Synth. Met.* **102**, 1186 (1999).

Black, R.D., M.B. Weissman, and F.M. Fliegel, "$1/f$ noise in metal films lacks spatial correlation," *Phys. Rev. B* **24**, 7454 (1981).

Blackman, J.A., "A theory of conductivity in disordered resistor networks," *J. Phys. C* **9**, 2049 (1976).

Bleibaum, O., H. Böttger, and V.V. Bryksin, "Effective medium theory of hopping transport," *Phys. Rev. B* **54**, 5444 (1996).

Blumenfeld, R., Y. Meir, A. Aharony, and A.B. Harris, "Resistance fluctuations in randomly diluted networks," *Phys. Rev. B* **35**, 3524 (1987).

Boal, D.H., U. Seifert, and J.C. Shillcock, "Negative Poisson ratio in two-dimensional networks under tension," *Phys. Rev. E* **48**, 4274 (1993).

Boal, D.H., U. Seifert, and A. Zilker, "Dual network model for red blood cell membranes," *Phys. Rev. Lett.* **69**, 3405 (1992).

Böhmer, R., and C.A. Angell, "Correlations of the nonexponentiality and state dependence of mechanical relaxations with bond connectivity in Ge-As-Se supercooled liquids," *Phys. Rev. B* **45**, 10091 (1992).

Bolton, F., and D. Weaire, "Rigidity loss transition in a disordered 2D froth," *Phys. Rev. Lett.* **65**, 3449 (1990).

Boolchand, P., R.N. Enzweiler, R.L. Cappelletti, W.A. Kamitakahara, Y. Cai, and M.F. Thorpe, "Vibrational thresholds in covalent networks," *Solid State Ionics* **39**, 81 (1990).

Boolchand, P., and M.F. Thorpe, "Glass-forming tendency, percolation of rigidity, and onefold-coordinated atoms in covalent networks," *Phys. Rev. B* **50**, 10366 (1994).

Born, M., and K. Huang, *Dynamical Theory of Crystal Lattices* (Clarendom Press, Oxford, 1954).

Böttger, H., and V.V. Bryksin, *Hopping Conduction in Solids* (VCH, Weinheim, Germany, 1985).

Böttger, H., T. Damker, and A. Freyberg, "Replica-trick approach to percolation networks with central and bond-bending forces," *Physica A* **199**, 219 (1993).

Bouchaud, E., M. Delsanti, M. Adam, M. Daoud, and D. Durand, "Gelation and percolation: swelling effect", *J. Physique Lett.* **47**, 539 (1986).

Boucher, S., "On the effective moduli of isotropic two-phase elastic composites," *J. Compos. Mater.* **8**, 82 (1974).

Bourret, A., "Low-density silica aerogels observed by high-resolution electron microscopy," *Europhys. Lett.* **6**, 731 (1988).

Brandt, W.W., "Use of percolation theory to estimate effective diffusion coefficients of particles migrating on various disordered lattices and in a random network structure," *J. Chem. Phys.* **63**, 5162 (1975).

Brenig, W., G. Döhler, and Wölfle, "Theory of thermally assisted electron hopping in amorphous solids," *Z. Phys.* **246**, 1 (1971).

Bresser, W., P. Boolchand, and P. Suranyi, "Rigidity percolation and molecular clustering in network glasses," *Phys. Rev. Lett.* **56**, 2493 (1986).

Brinker, C.J., and G.W. Scherer, *The Physics and Chemistry of Sol-Gel Processing* (Academic, San Diego, 1990).

Broadbent, S.R., and J.M. Hammersley, "Percolation processes. Crystals and mazes," *Proc. Camb. Philos. Soc.* **53**, 629 (1957).

Brown, R., and B. Esser, "Kinetic networks and order-statistics for hopping in disordered systems," *Philos. Mag. B* **72**, 125 (1995).

Brown, W.F., "Solid mixture permittivities," *J. Chem. Phys.* **23**, 1514 (1955).

Bruggeman, D.A.G., "Berechnung verscheidener physikalischer konstanten von heterogenen substanzen. I. Dielektrizitätskonstanten und leitfähigkeiten der mischkörper aus isotropen substanzen," *Ann. Phys.* **24**, 636 (1935).

Bruggeman, D.A.G., "Die elastischen constanten der quasiisotropen mischkörper aus isotropen substanzen," *Ann. Phys.* **29** 160 (1937).

Bryksin, V.V., "Frequency dependence of the hopping conductivity for three-dimensional systems in the framework of the effective-medium method," *Fix. Tverd. Tela* (Lennin grad) **22**, 2441 (1980) [*Sov. Phys. Solid State* **22**, 1421 (1980)].

Buchenau, U., M. Morkenbusch, G. Reichenauer, and B. Frick, "Inelastic Neutron Scattering from virgin and densified aerogels," *J. Non-Crystl. Solids* **145**, 121 (1992).

Budiansky, B., "On the elastic moduli of some heterogeneous materials," *J. Mech. Phys. Solids* **13**, 223 (1965).

Bug, A.L.R., G.S. Grest, M.H. Cohen, and I. Webman, "AC response near the percolation threshold: Transfer matrix calculation in 2D," *J. Phys. A* **19**, L323 (1986).

Bug, A.L.R., S.A. Safran, G.S. Grest, and I. Webman, "Do interactions raise or lower a percolation threshold?" *Phys. Rev. Lett.* **55**, 1896 (1985).

Bunde, A., A. Coniglio, D.C. Hong, and H.E. Stanley, "Transport in a two-component randomly composite material: scaling theory and computer simulation of termite diffusion near the superconducting limit," *J. Phys. A* **18**, L137 (1985).

Bunde, A., and H.E. Roman, "Vibrations and random walks on random fractals: anomalous behaviour and multifractality," *Philos. Mag. B* **65**, 191 (1992).

Burton, D., and C.J. Lambert, "Critical dynamics of a superelastic network," *Europhys. Lett.* **5**, 461 (1988).

Butcher, P.N., "Effective medium treatments of random simple square and simple cubic conductance networks," *J. Phys. C* **8**, L324 (1975).

Butcher, P.N., "Theory of hopping conductivity in disordered semiconductors," in *Linear and Nonlinear Electron Transport in Solids*, edited by J.T. Devreese and V.E. van Doren (New York, Plenum, 1976), p. 341.

Butcher, P.N., "Calculation of hopping transport coefficients (for impurity bands)," *Philos. Mag. B* **42**, 799 (1980).

Butcher, P.N., and J.A. McInnes, "Analytical formulae for DC hopping conductivity: r-percolation in 3D," *Philos. Mag.* **32**, 249 (1978).

Caddock, B.D., and K.E. Evans, "Microporous materials with negative Poisson's ratios. I. Microstructure and mechanical properties," *J. Phys. D* **22**, 1877 (1989).

Cahn, J.W., "Phase separation by spinodal decomposition in isotropic systems," *J. Chem. Phys.* **42**, 93 (1965).

Cai, Y., and M.F. Thorpe, "Floppy modes in network glasses," *Phys. Rev. B* **40**, 10535 (1989).

Calemczuk, R., A.M. de Goer, B. Salce, R. Maynard, and A. Zarembowitch, "Low-temperature properties of silica aerogels," *Europhys. Lett.* **3**, 1205 (1987).

Candau, S.J., M. Ankrim, J.P. Munch, P. Rempp, G. Hild, and R. Osaka, R., in *Physical Optics of Dynamical Phenomena in Macromolecular Systems* (Berlin, De Gruyter, 1985), p. 145.

Carmesin, I., and K. Kremer, "The Bond fluctuation method : A new effective algorithm for the dynamics of polymers in all spatial dimensions," *Macromolecules* **21**, 2819 (1988).

Carslaw, H.S., and J.C. Jaeger, *Conduction of Heat in Solids*, 2nd ed. (Oxford University Press, London, 1959).

Castner, T.G., N.K. Lee, G.S. Cieloszyk, and G.L. Salinger, "Dielectric anomaly and the metal-insulator transition," *Phys. Rev. Lett.* **34**, 1627 (1975).

Celzard, A., M. Krzesińska, J.F. Marêché, and S. Puricelli, "Scalar and vectorial percolation in compressed expanded graphite," *Physica A* **294**, 283 (2001).

Chalupa, J., P.L. Leath, and G.R. Reich, "Bootstrap percolation on a Bethe lattice," *J. Phys. C* **12**, L31 (1979).

Chandrasekhar, S., "Stochastic problems in physics and astronomy," *Rev. Mod. Phys.* **15**, 1 (1943).

Chayes, J.T., L. Chayes, and R. Durret, "Critical behavior of the two-dimensional first passage time," *J. Stat. Phys.* **45**, 933 (1986).

Chen, C.C., and Y.C. Chou, "Electrical-conductivity fluctuations near the percolation threshold," *Phys. Rev. Lett.* **54**, 2529 (1985).

Chen, C.H., and S. Cheng, *J. Compos. Mater.* **1**, 30 (1967).

Chen, H.-S., and A. Acrivos, "The solution of the equations of linear elasticity for an infinite region containing two spherical inclusions," *Int. J. Solids Structures* **14**, 331 (1978a).

Chen, H.-S. and A. Acrivos, "The effective elastic moduli of composite materials containing spherical inclusions at non-dilute concentrations," *Int. J. Solids Structures* **14**, 349 (1978b).

Chen, I.-G., and W.B. Johnson, "Alternating-current electrical properties of random metal-insulator composites," *J. Mater. Res.* **26**, 1565 (1991).

Chen, J., M.F. Thorpe, and L.C. Davis, "Elastic properties of rigid fiber-reinforced composites," *J. Appl. Phys.* **77**, 4349 (1995).

Chen, Z.-Y., P. Weakliem, W.M. Gelbart, and P. Meakin, "Second-order light scattering and local anisotropy of diffusion-limited aggregates and bond-percolation clusters," *Phys. Rev. Lett.* **58**, 1996 (1987).

Cheng, H., and S. Torquato, "Efective conductivity of periodic arrays of spheres with interfacial resistance," *Proc. R. Soc. Lond. A* **453**, 145 (1997).

Cherkaev, A.V., and L.V. Gibiansky, Preprint 914, *Phys.-Tech. Inst. Acad. Sci. USSR Leningrad* (1984) [English translation in *Topics in the Mathematical Theory of Composite Materials*, edited by R.V. Kohn (Birkhauser, Boston, 1996)].

Cherkaev, A.V., and L.V. Gibiansky, Preprint 1115, *Phys.-Tech. Inst. Acad. USSR Leningrad* (1987) [English translation in *Topics in the Mathematical Theory of Composite Materials*, edited by R.V. Kohn (Birkhauser, Boston, 1996)].

Cherkaev, A.V., and L.V. Gibiansky, "The exact coupled bounds for effective tensors of electrical and magnetic properties of two-component two-dimensional composites," *Proc. R. Soc. Edinb.* **122A**, 93 (1992).

Cherkaev, A.V., and L.V. Gibiansky, "Coupled estimates for the bulk and shear moduli of a two-dimensional isotropic elastic composite," *J. Mech. Phys. Solids* **41**, 937 (1993).

Cherkaev, A.V., K.A. Lurie, and G.W. Milton, "Invariant properties of the stress in the plane elasticity and equivalence classes of composites," *Proc. R. Soc. Lond. A* **438**, 519 (1992).

Chiew, Y.C., and E.D. Glandt, "Interfacial surface area in dispersions and porous media," *J. Colloid Interface Sci.* **99**, 86 (1984).

Choy, T.C., A. Alexopoulos, and M.F. Thorpe, "Dielectric function for a material containing hyperspherical inclusions to $O(c^2)$. I. Multipole expansions," *Proc. R. Soc. Lond. A* **454**, 1973 (1998).

Christensen, R.M., *Mechanics of Composite Materials* (Wiley, New York, 1979).

Christensen, R.M., and K.H. Lo, "Solutions for the effective shear properties in three phase spheres and cylinder models," *J. Mech. Phys. Solids* **27**, 315 (1979).

Clarkson, M.T., "Electrical conductivity and permittivity measurements near the percolation transition in a microemulsion. II. Interpretation," *Phys. Rev. A* **37**, 2079 (1988).

Clarkson, M.T., and S.I. Smedley, "Electrical conductivity and permittivity measurements near the percolation transition in a microemulsion. I. Experiment," *Phys. Rev. A* **37**, 2070 (1988).

Clausius, R., *Die mechanische Behandlung der Electricität* (Vieweg, Braunschweig, 1879), p. 62.

Clerc, J.P., G. Giraud, J.M. Laugier, J.M. and Luck, "The electrical conductivity of binary disordered systems, percolation clusters, fractals and related models," *Adv. Phys.* **39**, 191 (1990).

Cohen, M.H., and G.S. Grest, "Liquid-glass transition, a free volume approach," *Phys. Rev. B* **20**, 1077 (1979).

Cohen, M.H., and J. Jortner, "Effective medium theory for the Hall effect in disordered materials," *Phys. Rev. Lett.* **30**, 696 (1973).

Cohn, R.M., "The resistance of an electrical network," *Am. Math. Soc. Proc.* **1**, 316 (1950).

Colby, R.H., J.R. Gilmor, and M. Rubinstein, "Dynamics of near-critical polymer gels," *Phys. Rev. E* **48**, 3712 (1993).

Cole, K.S., and R.J. Cole, "Dispersion and absorption in dielectrics. I. Alternating current characteristics," *J. Chem. Phys.* **9**, 341 (1941).

Coniglio, A., "Thermal phase transition of the dilute s-state Potts and n-vector models at the percolation threshold," *Phys. Rev. Lett.* **46**, 250 (1981).

Coniglio, A., U. De Angelis, and A. Forlani, "Pair connectedness and cluster size," *J. Phys. A* **10**, 1123 (1977).

Conrad, H., U. Buchenau, R. Schatzler, G. Reichenauer, and J. Fricke, "Crossover in the vibrational density of states of silica aerogels studied by high-resolution neutron spectroscopy," *Phys. Rev. B* **41**, 2753 (1990).

Conwell, E.M., "Impurity band conduction in germanium and silicon," *Phys. Rev.* **103**, 51 (1956).

Cook, R.D., D.S. Malkus, and M.E. Plesha, *Concepts and Applications of Finite Element Analysis* (Wiley, New York, 1989).

Cooper, D.W., "Random-sequential-packing simulations in three dimensions for spheres," *Phys. Rev. A* **38**, 522 (1988).

Cornell, B.A., J. Middlehurst, and N.S. Parker, "Modelling the simplest form of order in biological membranes," *J. Colloid Interface Sci.* **81**, 280 (1981).

Cosserat, E., and F. Cosserat, *Théorie des Corps Déformables* (Hermann, Paris, 1909).

Cotton, J.P., *Thése, Université Paris, 6* (1974).

Courtens, E., J. Pelous, J. Phalippou, R. Vacher, and T. Woignier, "Fractal structure of base catalyzed and densified silica aerogels," *J. Non-Cryst. Solids* **95 & 96**, 1175 (1987).

Courtens, E., and R. Vacher, "Structure and dynamics of fractal aerogels," *Z. Phys. B* **68**, 355 (1987).

Courtens, E., R. Vacher, and J. Pelous, in *Fractals: Physical Origin and Properties*, edited by L. Peitronero (Plenum, London, 1990), p. 285.

Courtens, E., Vacher, R., and Stoll, E., "Fractons observed," *Physica D* **38**, 41 (1989).

Craciun, F., C. Galassi, and E. Roncari, "Experimental evidence for similar critical behavior of elastic modulus and electric conductivity in porous ceramic materials," *Europhys. Lett.* **41**, 55 (1998).

Cramer, C., and M. Buscher, "Complete conductivity spectra of fast ion conducting silver iodide/silver selenate glasses," *Solid State Ionics* **105**, 109 (1998).

Cummings, K.D., J.C. Garland, and D.B. Tanner, "Optical propeties of a small-particle composite," *Phys. Rev. B* **30**, 4170 (1984).

Dai, U., A. Palevski, and G. Deutscher, "Hall effect in a three-dimensional percolation systems", *Phys. Rev. B* **36**, 790 (1987).

Daoud, M., "Distribution of relaxation times near the gelation threshold," *J. Phys. A* **21**, L973 (1988).

Daoud, M., "Viscoelasticity near the sol-gel transition," *Macromolecules* **33**, 3019 (2000).

Daoud, M., and A. Coniglio, "Singular behaviour of the free energy in the sol-gel transition," *J. Phys. A* **14**, L301 (1981).

Daoud, M., F. Family, and G. Jannik, "Dilution and polydispersity in branched polymers," *J. Physique Lett.* **45**, 199 (1984).

Davis, V.A., and L. Schwartz, "Dielectric properties od silver-gelatin granular suspensions," *Phys. Rev. B* **33**, 6627 (1986).

Day, A.R., K.A. Snyder, E.J. Garboczi, and M.F. Thorpe, "The elastic moduli of a sheet containing circular holes," *J. Mech. Phys. Solids* **40**, 1031 (1992).

Day, A.R., R. Tremblay, and A.-M.S. Tremblay, "Spectral properties of percolating central force elastic networks," *J. Non-Cryst. Solids* **75**, 245 (1985).

Day, A.R., R.R. Tremblay, and A.-M.S. Tremblay, "Rigid backbone: A new geometry for percolation," *Phys. Rev. Lett.* **56**, 2501 (1986).

de Arcangelis, L., S. Redner, and A. Coniglio, "Anomalous voltage distribution of random resistor networks and a new model for the backbone at the percolation threshold," *Phys. Rev. B* **31**, 4725 (1985).

de Arcangelis, L., S. Redner, and A. Coniglio, "Multiscaling approach in random resistor and random superconducting networks," *Phys. Rev. B* **34**, 4656 (1986).

Debye, P., H.R. Anderson, Jr., and H. Brumberger, "Scattering by an inhomogeneous solid. II. The correlation function and its applications," *J. Appl. Phys.* **28**, 679 (1957).

Dederichs, P.H., and R. Zeller, "Variational treatment of the elastic constants of disordered materials," *Z. Phys.* **259**, 103 (1973).

de Gennes, P.G., "Exponents for excluded volume problem as derived by the Wilson method," *Phys. Lett. A* **38**, 339 (1972).

de Gennes, P.G., "On a relation between percolation theory and the elasticity of gels," *J. Physique Lett.* **37**, L1 (1976a).

de Gennes, P.G., "La percolation: Un concept unificateur," *La Recherche* **7**, 919 (1976b).

de Gennes, P.G., "Incoherent scattering near a sol gel transition," *J. Physique Lett.* **40**, L197 (1979).

de Gennes, P.G., "Suspension colloïdales dans un mélange binaire critique," *C.R. Acad. Sci.* **292**, 701 (1981).

de Goer, A.M., R. Calemczuk, B. Slace, J. Bon, E. Bonjour, and R. Maynard, "Low-temperature energy excitations and thermal properties of silica aerogels," *Phys. Rev. B* **40**, 8327 (1989).

Goodier, "Concentration of stress around spherical and cylindrical inclusions and flaws," *J. Appl. Mech.* **5**, 39 (1933).

del Gado, E., L. de Arcangelis, and A. Coniglio, "Elastic properties at the sol-gel transition," *Europhys. Lett.* **46**, 288 (1999).

Dembovskii, S.A., V.V. Kirilenko, and Ju. A. Buslaev, *Izv. Akad. Nauk USSR, Neorg, Mat.* **7**, 328 (1971).

Denteneer, P.J.H., and M.H. Ernst, "Exact results for diffusion on a disordered chain," *J. Phys. C* **16**, L961 (1983).

Deprez, N., D.S. McLachlan, and I. Sigalas, "The measurement and comparative analysis of the electrical and thermal conductivities and permeability of sintered nickel," *Physica A* **157**, 181 (1989).

Deptuck, D., J.P. Harrison, and P. Zawadzki, "Measurement of elasticity and conductivity of a three-dimensional percolation system," *Phys. Rev. Lett.* **54**, 913 (1985).

Derrida, B., R. Orbach, and K.W. Yu, "Percolation in the effective-medium approximation: Crossover between phonon and fracton excitations," *Phys. Rev. B* **29**, 6645 (1984).

Derrida, B., and Y. Pomeau, "Classical diffusion on a random chain," *Phys. Rev. Lett.* **48**, 627 (1982).

Derrida, B., and J. Vannimenus, "A transfer-matrix approach to random resistor networks," *J. Phys. A* **15**, L557 (1982).

Derrida, B., J.G. Zabolitzky, J. Vannimenus, and D. Stauffer, "A transfer matrix program to calculate the conductivity of random resistor networks," *J. Stat. Phys.* **36**, 31 (1984).

Deutscher, G., O. Entin-Wohlman, S. Fishman, and Y. Shapira, "Percolation description of granular superconductors," *Phys. Rev. B* **21**, 5041 (1980).

Deutscher, G., and M.L. Rappaport, "Critical currents of superconducting aluminium-germanium and lead-germanium thin film alloys near the metal-insulator transition," *J. Physique Lett.* **40**, L219 (1979).

Devreux, F., J.P. Boilot, F. Chaput, L. Malier, and M.A.V. Axelos, "Crossover from scalar to vectorial percolation in silica gelation," *Phys. Rev. E* **47**, 2689 (1993).

Diaz-Guilera, A., and A.-M.S. Tremblay, "Random mixtures with orientational order, and the anisotropic resistivity tensors of high$-T_c$ superconductors," *J. Appl. Phys.* **69**, 379 (1991).

Djabourov, M., J. Leblond, and P. Papon, "Gelation of aqueous gelatin solutions. II. Rheology of the sol-gel transition," *J. Phys. France* **49**, 333 (1988).

Doi, H., Y. Fujiwara, K. Miyaka and Y. Oosawa, "A systematic investigation of elastic moduli of Wc-Co alloys," *Metall. Trans.* **1**, 1417 (1970).

Doi, M., "A new variational approach to the diffusion and flow problem in porous media," *J. Phys. Soc. Japan* **40**, 567 (1976).

Doyen, P.M., "Permeability, conductivity, and pore geometry of sandstones," *J. Geophys. Res.* **93B**, 7729 (1988).

Doyle, W.T., "The Clausius-Mossotti problem for cubic arrays of spheres," *J. Appl. Phys.* **49**, 795 (1978).

Dubois, M. and B. Cabane, "Light-scattering study of the sol-gel transition in silicon tetraethoxide," *Macromolecules* **22**, 2526 (1989).

Dubrov, V.E., M.E. Levinshtein, and M.S. Shur, "Anomaly in the dielectric permeability in metal-dielectric transitions. Theory and simulation," *Sov. Phys.-JETP* **43**, 2014 (1976).

Duering, E., and D.J. Bergman, "Scaling properties of the elastic stiffness moduli of a random rigid-nonrigid network near the rigidity threshold: theory and simulations," *Phys. Rev. B* **37**, 9460 (1988).

Duering, E., and D.J. Bergman, "Current distribution on a three-dimensional, bond-diluted, random-resistor network at the percolation threshold," *J. Stat. Phys.* **60**, 363 (1990).

Dundurs, J., "Effect of elastic constants on stress in a composite under plane deformation," *J. Comp. Mater.* **1**, 310 (1967).

Durand, D., M. Delsanti, M. Adam, and J.M. Luck, "Frequency dependence of viscoelastic properties of branched polymers near gelation threshold," *Europhys. Lett.* **3**, 297 (1987).

Durian, D.J., "Bubble-scale model of foam mechanics: Melting, nonlinear behavior, and avalanche," *Phys. Rev. E* **55**, 1739 (1997).

Duval, E., A. Boukenter, T. Achibat, B. Champagnon, J. Serugetti, and J. Dumas, "Structure of silica aerogels and vibrational dynamics in fractal materials and glasses: electron microscopy and low-frequency Raman scattering," *Philos. Mag. B* **65**, 181 (1992).

Dykhne, A.M., "Conductivity of a two-dimensional two-phase system," *Sov. Phys.-JETP* **32**, 63 (1971)].

Dyre, J.C., "The random free energy barrier model for ac conduction in disordered solids," *J. Appl. Phys.* **64**, 2456 (1988).

Dyre, J.C., "Universal low-temperature ac conductivity of macroscopically disordered materials," *Phys. Rev. B* **48**, 12511 (1993).

Dyre, J.C., "Studies of ac hopping conduction at low temperatures," *Phys. Rev. B* **49**, 11709 (1994); **50**, 9692(E).

Dyre, J.C., and T.B. Schr⊘der, "Effective one-dimensionality of universal ac hopping conduction in the extreme disorder limit," *Phys. Rev. B* **54**, 14884 (1996).

Dyre, J.C., and T.B. Schr⊘der, "Universality of ac conduction in disordered solids," *Rev. Mod. Phys.* **72**, 873 (2000).

Ebrahimi, F., and M. Sahimi, "Multiresolution wavelet coarsening and analysis of transport in heterogeneous media," *Physica A 316*, 160 (2002).

Efros, A.L., and B.I. Shklovskii, "Coulomb gap and low temperature conductivity of disordered systems," *J. Phys. C* **8**, L49 (1975).

Efros, A.L., and B.I. Shklovskii, "Critical behaviour of conductivity and dielectric constant near the metal-non-metal transition threshold," *Phys. Status Solidi B* **46**, 475 (1976).

Eggarter, T.P., and M.H. Cohen, "Simple model for density of states and mobility of an electron in a gas of hard-core scatterers," *Phys. Rev. Lett.* **25**, 807 (1970).

Eischen, J.W., and S. Torquato, "Determining elastic behaviour of composites by the boundary element method," *J. Appl. Phys.* **74**, 159 (1993).

Elber, R., and M. Kaplus, "Low-frequency modes in proteins: Use of the effective-medium approximation to interpret the fractal dimension observed in electron-spin relaxation measurements," *Phys. Rev. Lett.* **56**, 394 (1986).

Elimes, A., R.A. Romer, and M. Schreiber, *Eur. Phys. J. B* **1**, 29 (1998).

Elliot, R.J., J.A. Krumhansl, and P.L. Leath, "The theory and properties of randomly disordered crystals and related physical systems," *Rev. Mod. Phys.* **46**, 465 (1974).

Elliot, S.R., "Frequency-dependent conductivity in ionically and electronically conducting amorphous solids," *Solid State Ionics* **70/71**, 27 (1994).

Entin-Wohlman, O., S. Alexander, and R. Orbach, "Inelastic extended-electron-localized-vibrational-state scattering rate," *Phys. Rev. B* **32**, 8007 (1985).

Entin-Wohlman, O., R. Orbach, and G. Polatsek, "Dynamics of tenuous structures," in *Dynamics of Disordered Materials*, edited by D. Richter, A.J. Dianoux, W. Petry, and J. Teixeira (Springer, Berlin, 1989a), p. 288.

Entin-Wohlman, O., U. Sivan, R. Blumenfeld, and Y. Meir, "Dynamic structure fractor of fractals," *Physica D* **38**, 93 (1989b).

Ertel, W., K. Frobose, and J. Jackle, "Constrained diffusion dynamics in the hard-square lattice gas at high density," *J. Chem. Phys.* **88**, 5027 (1988).

Eshelby, J.D., "The continuum theory of lattice defects," in *Progress in Solid State Physics*, edited by F. Seitz and D. Turnbull (Academic, New York, 1956), p. 79.

Eshelby, J.D., "The determination of the elastic field of an ellipsoidal inclusion and related problems," *Proc. R. Soc. Lond. A* **241**, 376 (1957).
Eshelby, J.D., "Elastic inclusions and inhomogeneities," in *Progress in Solid Mechanics*, edited by I.N. Sneddon and R. Hill (Interscience, New York, 1961), **2**, p. 89.
Essam, J.W., "Percolation and cluster size," in *Phase Transitions and Critical Phenomena*, edited by C. Domb and M.S. Green, Volume II (Academic Press, London, 1972), p. 197.
Essam, J.W., and F.M. Bhatti, "Series expansion evidence supporting the Alexander-Orbach conjecture in two dimensions," *J. Phys. A* **18**, 3577 (1985).
Essam, J.W., C.M. Place, and E.H. Sondheimer, "Self consistent calculation of the conductivity in a disordered branching network," *J. Phys. C* **7**, L258 (1974).
Eubanks, R.A., and E. Sternberg, "On the completeness of the Boussinesq-Papkovich stress functions," *J. Rational Mech. Anal.* **5**, 735 (1956).
Evans, K.E., "Tensile network microstructures exhibiting negative Poisson's ratios," *J. Phys. D* **22**, 1870 (1989).
Evans, K.E., and B.D. Caddock, "Microporous materials with negative Poisson's ratios. II. Mechanisms and interpretation," *J. Phys. D* **22**, 1883 (1989).
Every, A.G., Y. Tzou, D.P.H. Hasselman, and R. Raj, "The effect of particle size on the thermal conductivity of ZnS/diamond composites," *Acta Metall. Mater.* **40**, 123 (1992).
Eyre, D.J., and G.W. Milton, "A fast numerical scheme for computing the response of composites using grid refinement," *Euro. Phys. J.* **6**, 41 (1997).
Fadda, G.C., D. Lairez, and J. Pelta, "Critical behavior of gelation probed by the dynamics of latex spheres," *Phys. Rev. E* **63**, 061405-1 (2001).
Family, F., "Polymer statistics and universality: Principles and applications of cluster renormalization," in *Random Walks and Their Applications in the Physical and Biological Sciences*, edited by M.F. Shlesinger and B.J. West, *AIP Conference Proceedings* **109** (AIP, New York, 1984), p. 33.
Family, F., and A. Coniglio, "Cross over from percolation to random animals and compact clusters," *J. Phys. A* **13**, L403 (1980).
Family, F., and T. Vicsek (eds.), *Dynamics of Fractal Surfaces* (World Scientific, Singapore, 1991).
Farago, O., and Y. Kantor, "Entropic elasticity of two-dimensional self-avoiding percolation systems," *Phys. Rev. Lett.* **85**, 2533 (2000).
Fatt, I., "The network model of porous media I. Capillary pressure characteristics," *Trans. AIME* **207**, 144 (1956).
Feder, J., "Random sequential adsorption," *J. Theor. Biol.* **87**, 237 (1980).
Felderhof, B.U., "Bounds for the effective dielectric constant of disordered two-phase materials," *J. Phys. C* **15**, 1731 (1982).
Felderhof, B.U., and R.B. Jones, "Effective dielectric constant of dilute suspensions of sphers," *Phys. Rev. B* **39**, 5669 (1989).
Feng, S., "Percolation properties of granular elastic networks in two dimensions," *Phys. Rev. B* **32**, 510 (1985a).
Feng, S., "Crossover in spectral dimensionality of elastic percolation systems," *Phys. Rev. B* **32**, 5793 (1985b).
Feng, S., B.I. Halperin, and P.N. Sen, "Transport properties of continuum systems near the percolation threshold," *Phys. Rev. B* **35**, 197 (1987).
Feng, S., and M. Sahimi, "Position-space renormalization for elastic percolation networks with bond-bending forces," *Phys. Rev. B* **31**, 1671 (1985).
Feng, S., and P.N. Sen, "Percolation on elastic networks: new exponent and threshold," *Phys. Rev. Lett.* **52**, 216 (1984).

Feng, S., P.N. Sen, B.I. Halperin and C.J. Lobb, "Percolation on two-dimensional elastic networks with rotationally invariant bond-bending forces," *Phys. Rev. B* **30**, 5386 (1984).

Feng, S., M.F. Thorpe, and E. Garboczi, "Effective-medium theory of percolation on central-force elastic networks," *Phys. Rev. B* **31**, 276 (1985).

Feng, X., W.J. Bresser, and P. Boolchand, "Direct evidence for stiffness threshold in chalcogenide glasses," *Phys. Rev. Lett.* **78**, 4422 (1997).

Ferri, B., J. Frisken, and D.S. Cannell, "Structure of silica gels," *Phys. Rev. Lett.* **67**, 3626 (1991).

Ferry, J.D., *Viscoelastic Properties of Polymers* (New York, Wiley, 1980).

Feynman, R.P., "Forces in molecules," *Phys. Rev.* **56**, 340 (1939).

Fiegl, B., R. Kuhnert, M. Ben-Chorin, and F. Koch, "Evidence for grain boundary hopping transport in polycrystalline diamond films," *Appl. Phys. Lett.* **65**, 371 (1994).

Fisch, R., and A.B. Harris, "Critical behavior of random resistor networks near the percolation threshold," *Phys. Rev. B* **18**, 416 (1978).

Fischer, U., C. von Borczyskowski, and N. Schwentner, "Singlet-exciton transport and spatial and energetic disorder in dibenzofuran crystals," *Phys. Rev. B* **41**, 9126 (1990).

Fishchuk, I.I., "The AC conductivity and Hall effect in inhomogeneous semiconductors," *Phys. Status Solidi A* **93**, 675 (1986).

Fisher, M.E., "The theory of critical point singularities," in *Critical Phenomena*, edited by M.S. Green (Academic Press, New York, 1971), p. 1.

Fisher, M.E., and J.W. Essam, "Some cluster size and percolation problems," *J. Math. Phys.* **2**, 609 (1961).

Fitzpatrick, J.P., R.B. Malt, and F. Spaepen, "Percolation theory and the conductivity of random close packed mixtures of hard spheres," *Phys. Lett. A* **47**, 207 (1974).

Flaherty, J.E., and J.B. Keller, "Elastic Behaviour of composite media," *Commun. Pure Appl. Math.* **26**, 565 (1973).

Flory, P.J., Molecular size distribution in three dimensional polymers. I. Gelation," *J. Am. Chem. Soc.* **63**, 3083 (1941).

Fogelholm, R., "The conductivity of large percolation network samples," *J. Phys. C* **13**, L571 (1980).

Fontana, A., F. Tocca, M.P. Fontana, B. Rosi, and A.J. Dianoux, "Low-frequency dynamics in superionic borate glasses by coupled Raman and inelastic neutron scattering," *Phys. Rev. B* **41**, 3778 (1990).

Fox, T.G., and P.J. Flory, "Further studies on the melt viscosity of polyisobutylene," *J. Phys. Chem.* **55**, 221 (1951).

Francfort, G.A., and F. Murat, "Homogeneization and optimal bounds in linear elasticity," *Arch. Rat. Mech. Analy.* **94**, 307 (1986).

Frank, D.J., and C.J. Lobb, "Highly efficient algorithm for percolative transport studies in two dimensions," *Phys. Rev. B* **37**, 302 (1988).

Freltoft, T., J.K. Kjems, and D. Richter, "Density of states in fractal silica smoke-particle aggregates," *Phys. Rev. Lett.* **58**, 1212 (1987).

Fricke, J., "Aerogels," *Sci. Amer.* **258** (No. 5), 68 (1988).

Friis, E.A., R.S. Lakes, and J.B. Park, "Negative Poisson's ratio polymeric and metallic foams," *J. Mater. Sci.* **23**, 4406 (1988).

Garboczi, E.J., "Effective force constant for a central-force random network," *Phys. Rev. B* **37**, 318 (1988a).

Garboczi, E.J., and M.F. Thorpe, "Effective-medium theory of percolation on central force elastic networks. II. Further results," *Phys. Rev. B* **31** 7276, (1985).

Garboczi, E.J., and M.F. Thorpe, "Effective-medium theory of percolation on central-force elastic networks. III. The superelastic problem," *Phys. Rev. B* **33**, 3289 (1986a).
Garboczi, E.J., and M.F. Thorpe, "Density of states for random-central-force elastic networks," *Phys. Rev. B* **32**, 4513 (1986b).
Garfunkel, G.A., and M.B. Weissman, "Noise scaling in continuum percolating films," *Phys. Rev. Lett.* **55**, 296 (1985).
Garland, J.C., "Granular properties of high T_c superconductors," *Physica A* **157**, 111 (1989).
Gauthier-Manuel, B., and E. Guyon, "Critical elasticity of Polyacrylamide above its gel point," *J. Physique Lett.* **41**, L503 (1980).
Gauthier-Manuel, B., E. Guyon, S. Roux, S. Gits, and F. Lefaucheux, "Critical viscoelastic study of the gelation of silica particles," *J. Physique* **48**, 869 (1987).
Gazis, D.C., R. Herman, and R.F. Wallis, "Surface elastic waves in cubic crystals," *Phys. Rev.* **119**, 533 (1960).
Gefen, Y., A. Aharony, and S. Alexander, "Anomalous diffusion on percolation clusters," *Phys. Rev. Lett.* **50**, 77 (1983).
Georgiev, D.G., P. Boolchand, and M. Micoulaut, "Rigidity transition and molecular structure of As_xSE_{1-x} glasses," *Phys. Rev. B* **62**, R9228 (2000).
Gerber, A., and G. Deutscher, "Upper critical field of superconducting Pb films above and below the percolation threshold," *Phys. Rev. Lett.* **63**, 1184 (1989).
Gibiansky, L.V., and G.W. Milton, "On the effective viscoelastic moduli of two-phase media. I. Rigorous bounds on the complex bulk modulus," *Proc. R. Soc. Lond. A* **440**, 163 (1993).
Gibiansky, L.V., and S. Torquato, "Geometrical-parameter bounds on the effective moduli of composites," *J. Mech. Phys. Solids* **43**, 1587 (1995a).
Gibiansky, L.V., and S. Torquato, "Rigorous link between the conductivity and elastic moduli of fiber-reinforced composite materials," *Philos. Trans. R. Soc. Lond.* **343**, 243 (1995b).
Gibiansky, L.V., and S. Torquato, "Connection between the conductivity and elastic moduli of isotroppic composites," *Proc. R. Soc. Lond. A*, **452**, 253 (1996).
Gibson, L.J., and M.F. Ashby, *Cellular Solids*, 2nd ed. (Cambridge University Press, Cambridge, 1997).
Gilbert, E.N., "Random subdivisions of space into crystals," *Ann. Math. Stat.* **33**, 958 (1962).
Gingold, D.B., and C.J. Lobb, "Percolative conduction in three dimensions," *Phys. Rev. B* **42**, 8220 (1990).
Good, I.J., "The number of individuals in a cascade process," *Proc. Camb. Philos. Soc.* **45**, 360 (1949).
Gordon, M., and J.A. Torkington, *Pure Appl. Chem.* **53**, 1461 (1981).
Grabovsky, Y., G.W. Milton, and D.S. Sage, "Exact relations for effective tensors of composites: Necessary conditions and sufficient conditions," *Commun. Pure Appl. Math.* **53**, 300 (2000).
Grannan, D.M., J.C. Garland, and D.B. Tanner, "Critical behavior of the dielectric constant of a random composite near the percolation threshold," *Phys. Rev. Lett.* **46**, 375 (1981).
Granqvist, C.G., and O. Hunderi, "Conductivity of inhomogeneous materials: Effective-medium theory with dipole-dipole interaction," *Phys. Rev. B* **18**, 1554 (1978).
Greenberg, R., and W.F. Brace, "Archie's law for rocks modeled by simple networks," *J. Geophys. Res.* **74**, 2099 (1969).
Grest, G.S., and M.H. Cohen, "Percolation and the glass transition," in *Percolation Structures and Processes*, edited by G. Deutscher, R. Zallen and J. Adler, *Annals of the Israel Physical Society* **5** (Adam Hilger, Bristol, 1983), p. 187.

Grest, G.S., and I. Webman, "Vibration properties of a percolating cluster," *J. Phys. (Paris) Lett.* **45**, L1155 (1984).

Grest, G.S., I. Webman, S.A. Safran, and A.L.R. Bug, "Dynamic percolation in microemulsions," *Phys. Rev. A* **33**, 2842 (1986).

Gubernatis, T., and J.A. Krumhansl, "Macroscopic engineering preoperties of polycrystalline materials: Elastic properties,' *J. Appl. Phys.* **46**, 1875 (1975).

Gupta, P.K., and A.R. Cooper, "Topologically disordered networks of rigid polytopes," *J. Non-Cryst. Solids* **123**, 14 (1990).

Haan, S.W., and R. Zwanzig, "Series expansion in a continuum percolation problem," *J. Phys. A* **10**, 1547 (1977).

Haji-Sheikh, A., and E.M. Sparrow, "The solution of heat conduction problem by probability methods," *J. Heat Trans.*, 121 (May 1967).

Halfpap, B.L., and S.M. Lindsay, "Rigidity percolation in the germanium-arsenic-selenium alloy system," *Phys. Rev. Lett.* **57**, 847 (1986).

Halperin, B.I., S. Feng, and P.N. Sen, "Dfferences between lattice and continuum percolation transport exponents," *Phys. Rev. Lett.* **54**, 2391 (1985).

Hamilton, E.M., "Variable range hopping in a non uniform density of states," *Philos. Mag.* **26**, 1043 (1972).

Hammonds, K.D., H. Deng, V. Heine, and M.T. Dove, "How floppy modes give rise to adsorption sites in zeolites," *Phys. Rev. Lett.* **78**, 3701 (1997).

Hansen, A., and S. Roux, "Multifractality in elastic percolation," *J. Stat. Phys.* **53**, 759 (1988).

Hansen, A., and S. Roux, "Universality class of central-force percolation," *Phys. Rev. B* **40**, 749 (1989).

Hansen, J.P., and McDonald, I.R., *Theory of Simple Liquids* (Academic Press, New York, 1986).

Harris, A.B., and A. Aharony, "Anomalous diffusion, superlocalization and hopping conductivity on fractal media," *Europhys. Lett.* **4**, 1355 (1987).

Harris, A.B., and R. Fisch, "Critical behavior of random resistor networks," *Phys. Rev. Lett.* **38**, 796 (1977).

Harris, A.B., S. Kim, and T.C. Lubensky, "Epsilon expansion for the conductivity of a random resistor network," *Phys. Rev. Lett.* **53**, 743 (1984).

Harris, A.B., and S. Kirkpatrick, "Low-frequency response functions of random magnetic systems," *Phys. Rev. B* **16**, 542 (1977).

Hashin, Z., "On elastic behaviour of fiber-reinforced materials of arbitrary transverse phase geometry," *J. Mech. Phys. Solids*, **13**, 119 (1965).

Hashin, Z., "Theory of composite materials," in *Mechanics of Composite Materials* (Pergamon Press, New York, 1970), p. 126.

Hashin, Z. "Failure criteria for unidirectional fiber composites," *J. Appl. Mech.* **47**, 329 (1980).

Hashin, Z., and B.W. Rosen, "The elastic moduli of fiber-reinforced materials," *J. Apply. Mech.* **31**, 223 (1964).

Hashin, Z., and S. Shtrikman, "On some variational principles in anisotropic and nonhomogeneous elasticity," *J. Mech. Phys. Solids* **10**, 335 (1962a).

Hashin, Z., and S. Shtrikman, "A variational approach to the theory of the elastic behavior of polycrystals," *J. Mech. Phys. Solids* **10**, 343 (1962b).

Hashin, Z., and S. Shtrikman, "A variational approach to the theory of the elastic behavior of multiphase materials," *J. Mech. Phys. Solids* **11**, 127 (1963).

Haus, J.W., and K.W. Kehr, "Diffusion in regular and disordered lattices," *Phys. Rep.* **150**, 263 (1987).

Havlin, S., and D. Ben-Avraham, "Diffusion in disordered media," *Adv. Phys.* **36**, 695 (1987).

He, H., and M.F. Thorpe, "Elastic properties of glasses," *Phys. Rev. Lett.* **54**, 2107 (1985).

Heiba, A.A., M. Sahimi, L.E. Scriven, and H.T. Davis, "Percolation theory of two-phase flow in porous media," Society of Petroleum Engineers paper 11015, New Orleans, LA (1982).

Heiba, A.A., M. Sahimi, L. E. Scriven, and H. T. Davis, "Percolation theory of two-phase relative permeability," *SPE Reservoir Engineering* **7**, 123 (1992).

Heinrichs, J., and N. Kumar, "Simple exact treatment of conductance in a random Bethe lattice," *J. Phys. C* **8**, L510 (1975).

Hellmann, H., *Einführung in die Quantumchemie* (Deuticke, Leipzig, 1937).

Helsing, J., "Third-order bounds on the conductivity of a random stacking of cubes," *J. Math. Phys.* **35**, 1688 (1994).

Helsing, J., G.W. Milton, and A.B. Movchan, "Duality relations, correspondences, and numerical results for planar elastic composites," *J. Mech. Phys. Solids* **45**, 565 (1997).

Helte, A. "Fourth-order bounds on the effective conductivity for a system of fully penetrable spheres," *Proc. R. Roy. Soc. Lond. A* **445**, 247 (1994).

Henderson, S.I., T.C. Mortensen, S.M. Underwood, and W. van Megen, "Effect of particle size distribution on crystallisation and the glass transition of hard sphere colloids," *Physica A* **233**, 102 (1996).

Hendrickson, B., "Conditions for unique graph realizations," *SIAM J. Comput.* **21**, 65 (1992).

Herakovich, C.T., "Composite laminates with negative through-the-thickness Poisson's ratios," *J. Comp. Mater.* **18**, 447 (1984).

Herrmann, H.J., B. Derrida, and J. Vannimenus, "Superconductivity exponents in two- and three-dimensional percolation," *Phys. Rev. B* **30**, 4080 (1984).

Hershey, A.V., "The elasticity of an isotropic aggregate of anisotropic cubic crystals," *J. Appl. Mech.* **21**, 236 (1954).

Hetherington, J.H., and M.F. Thorpe, "The conductivity of a sheet containing inclusions with sharp corners," *Proc. R. Soc. Lond. A* **438**, 591 (1992).

Hill, R., "Elastic preoperties of reinforced solids: Some theoretical principles," *J. Mech. Phys. Solids* **11**, 357 (1963).

Hill, R., "Theory of mechanical properties of fiber-strengththened materials: I. Elastic behavior," *J. Mech. Phys. Solids* **12**, 199 (1964).

Hill, R., "Theory of mechanical properties of fiber-strenghthened materials: III. Self-consistent models," *J. Mech. Phys. Solids* **13**, 189 (1965).

Hill, R.M., "On the observation of variable range hopping," *Phys. Stat. Sol.* **A35**, K29 (1976).

Hinsen, K., and B.U. Felderhof, "Dielectric constant of a suspension of uniform spheres," *Phys. Rev. B* **46**, 12955 (1992).

Ho, F.G., and W. Strieder, "Asymptotic expansion of the porous medium, effective diffusion coefficient in the Knudsen number," *J. Chem. Phys.* **70**, 5635 (1979).

Hobson, E.W., *The Theory of the Spherical and Ellipsoidal Harmonics* (Cambridge University Press, Cambridge, 1931).

Hodgson, D.F., and E.J. Amis, "Dynamic viscoelastic characterization of sol-gel reactions," *Macromolecules* **23**, 2512 (1990).

Holly, E.E., S.K. Venkataraman, F. Chambon, and H.H. Winter, H.H., "Fourier transform mechanical spectroscopy of viscoelastic materials with transient structures," *J. Non-Newtonian Fluid Mech.* **27**, 17 (1988).

Hoover, W.G., and F.H. Ree, "Melting transition and communal entropy for hard spheres," *J. Chem. Phys.* **49**, 3609 (1968).

Hori, M., and F. Yonezawa, "Theoretical approaches to inhomogeneous transport in disordered media," *J. Phys. C* **10**, 229 (1977).

Horiguchi, T., "Lattice Green's functions for the triangular and honeycomb lattices," *J. Math. Phys.* **13**, 1411 (1972).

Hoshen, J., and R. Kopelman, "Percolation and cluster distribution. I. Cluster multiple labelling technique," *Phys. Rev. B* **14**, 3438 (1976).

Howell, F.S., R.A. Bose, P.B. Macedo, and C.T. Moynihan, "Electrical relaxation in a glass-forming molten salt," *J. Phys. Chem.* **78**, 639 (1974).

Hsu, W.Y., and T. Berzins, "Percolation and effective-medium theories for perfluorinated ionomers and polymer composites," *J. Polym. Sci. Polym. Phys. Ed.* **23**, 933 (1985).

Hughes, B.D., *Random Walks and Random Environments*, Vol. 1 (Oxford University Press, London, 1995).

Hughes, B.D., *Random Walks and Random Environments*, Vol. 2 (Oxford University Press, London, 1996), Chapter 6.

Hughes, B.D., and M. Sahimi, "Random walks on the Bethe lattices," *J. Stat. Phys.* **29**, 781 (1982).

Hui, P.M., and D. Stroud, "Theory of Faraday rotation by dilute suspension of small particles," *Appl. Phys. Lett.* **50**, 950 (1987).

Hundley, M.F. and A. Zettl, "Temperature-dependent ac conductivity of thin percolation films," *Phys. Rev. B* **38**, 10290 (1988).

Hung, C.S., and J.R. Gliessman, "The resistivity and Hall effect of Germanium at low temperatures," *Phys. Rev.* **79**, 726 (1950).

Isard, J.O., "A study of the migration loss in glass and a generalized method of calculating the rise of dielectric loss with remperature," *Proc. Inst. Electr. Eng.* **109B** (Suppl. No. 22), 440 (1961).

Ishioka, S., and M. Koiwa, "Random walks on diamond and hexagonal close packed lattices," *Philos. Mag. A* **37**, 517 (1978).

Jaccarino, V., and L.R. Walker, "Discontinuous occurence of localized moments in metals," *Phys. Rev. Lett.* **15**, 258 (1965).

Jackson, J.D., *Classical Electrodynamics*, 3rd Ed. (John Wiley & Sons, 1998).

Jacobs, D.J., and M.F. Thorpe, "Generic rigidity percolation: The pebble game," *Phys. Rev. Lett.* **75**, 4051 (1995).

Jacobs, D.J., and M.F. Thorpe, "Generic rigidity percolation in two dimensions," *Phys. Rev. E* **53**, 3682 (1996).

Jagannathan, A., and R. Orbach, "Temperature and frequency dependence of the sound velocity in vitreous silica due to scattering off localized modes," *Phys. Rev. B* **41**, 3153 (1990).

Jagannathan, A., R. Orbach, and O. Entin-Wohlman, "Thermal conductivity of amorphous materials above the plateau," *Phys. Rev. B* **39**, 13465 (1989).

Jastrzebska, M.M., S. Jussila, and H. Isotalo, "Dielectric response and a.c. conductivity of synthetic dopa-melanin polymer," *J. Mater. Sci.* **33**, 4023 (1998).

Jeffrey, D.J., "Conduction through a random suspension of spheres," *Proc. R. Soc. Lond. A* **335**, 355 (1973).

Jeffrey, D.J., "Group expansion for the bulk properties of a statistically homogeneous random suspensions," *Proc. R. Soc. Lond. A* **338**, 503 (1974).

Jerauld, G.R., *Flow and Transport in Chaotic Media: Four Case Studies*, Ph.D. Thesis, University of Minnesota (1985).

Jerauld, G.R., J.C. Hatfield, L.E. Scriven, and H.T. Davis, "Percolation and conduction on Voronoi and triangular networks: a case study in topological disorder," *J. Phys. C* **17**, 1519 (1984a).

Jerauld, G.R., L.E. Scriven, and H.T. Davis, "Percolation and conduction on the 3D Voronoi and regular networks: a second case study in topological disorder," *J. Phys. C* **17**, 3429 (1984b).

John, S., H. Sompolinsky, and M.J. Stephen, "Localisation in a disordered elastic medium near two dimensions," *Phys. Rev. B* **27**, 5592 (1983).

Joy, T., and W. Strieder, "Effective medium theory of site percolation in a random simple triangular conductance network," *J. Phys. C* **11**, L867 (1978) [*Corringendum* **12**, L53 (1979)].

Joy, T., and W. Strieder, "Effective-medium theory of the conductivity for a random-site honeycomb lattice," *J. Phys. C* **12**, L279 (1979).

Ju, J.W., and T.M. Chen, *Acta Mech.* **103**, 123 (1994).

Juretchke, H.J., R. Landauer, and J.A. Swanson, "Hall effect and conductivity in porous media," *J. Appl. Phys.* **27**, 838 (1956).

Kaliski, S., *Arch. Mech. Stosow* **19**, 33 (1963).

Kamitakahara, W.A., R.L. Cappelletti, P. Boolchand, B. Halfpap, F. Gompf, D.A. Neumann, and H. Mutka, "Vibrational densities of states and network rigidity in chalcogenide glasses," *Phys. Rev. B* **44**, 94 (1991).

Kantelhardt, J.W., and A. Bunde, "Electrons and fractons on percolation structures at iticality: Sublocalization and superlocalization," *Phys. Rev. E* **56**, 6693 (1997).

Kantor, Y., and I. Webman, "Elastic properties of random percolating systems," *Phys. Rev. Lett.* **52**, 1891 (1984).

Kapitulnik, A., and G. Deutscher, "Percolation characteristics in discontinuous thin films of Pb," *Phys. Rev. Lett.* **49**, 1444 (1982).

Kapitulnik, A., and G. Deutscher, "2D to 3D percolation crossover in the resistivity of co-evaporated Al-Ge mixture films," *J. Phys. A* **16**, L255 (1983).

Kapitulnik, A., J.W.P. Hsu, and M.R. Hahn, "Percolative properties of Al-Ge composite thin films," in *Physical Phenomena in Granular Materials*, edited by G.D. Cody, T.H. Geballe, and P. Sheng (Pittsburgh, Materials Research Society, 1990), p. 153.

Katz, A.J., and A.H. Thompson, "Quantitative prediction of permeability in porous media," *Phys. Rev. B* **34**, 8179 (1986).

Katz, A.J., and A.H. Thompson, "Prediction of rock electrical conductivity from mercury injection measurements," *J. Geophys. Res. B* **92**, 599 (1987).

Keating, P.N., "Relationship between the macroscopic and microscopic theory of crystal elasticity. I. Primitive Crystals," *Phys. Rev.* **152**, 774 (1966).

Keller, H.B., and D. Sachs, "Calculations of the conductivity of a medium containing cylindrical inclusions," *J. Appl. Phys.* **35**, 537 (1964).

Keller, J.B., "Conductivity of a medium containing a dense array of perfectly conducting spheres or cylinders or nonconducting cylinders," *J. Appl. Phys.* **34**, 991 (1963).

Keller, J.B., "A theorem on the conductivity of a composite medium," *J. Math. Phys.* **5**, 548 (1964).

Kellomäki, M., J. Åström, and J. Timonen, "Rigidity and dynamics of random spring networks," *Phys. Rev. Lett.* **77**, 2730 (1996).

Kenkre, V.M., E.W. Montroll, and M.F. Shlesinger, "Generalized master equations for continuous-time random walks," *J. Stat. Phys.* **9**, 45 (1973).

Kerstein, A.R., "Equivalence of the void percolation problem for overlapping spheres and a network problem," *J. Phys. A* **16**, 3071 (1983).

Kharadly, M.M.Z., and W. Jackson, *Proc. Elect. Eng.* **100**, 199 (1952).

Kim, I.C., and S. Torquato, "Determination of the effective conductivity of heterogenous media by Brownian motion simulation," *J. Appl. Phys.* **68**, 3892 (1990).

Kim, I.C., and S. Torquato, "Effective conductivity of suspensions of overlapping spheres," *J. Appl. Phys.* **71**, 2727 (1992).

Kim, I.C., and S. Torquato, "Effective conductivity of composites containing spheroidal inclusions: Comparison of simulation with theory," *J. Appl. Phys.* **74**, 1844 (1993).

Kimball, J.C., and L.W. Adams, Jr., "Hopping conduction and superionic conductors," *Phys. Rev. B* **18**, 5851 (1978).

Kirkpatrick, S., "Classical transport in disordered media: Scaling and effective-medium theories," *Phys. Rev. Lett.* **27**, 1722 (1971).

Kirkpatrick, S., "Percolation and conduction," *Rev. Mod. Phys.* **45**, 574 (1973).

Kirkpatrick, S., in *Proceedings of 5th International Conference on Amorphous and Liquid Semiconductor* (Taylor and Francis, London, 1974), p. 183.

Kirkpatrick, S., "Percolation thresholds in granular films - non-universality and critical current," in *Inhomogeneous superconductors - 1979*, edited by D.U. Gubser, T.L. Francavilla, J.R. Leibowitz, and S.A. Wolf, *AIP Conference Proc.* **58** (1979a), p. 79.

Kirkpatrick, S., "Models of disordered materials," in *Ill-Condensed Matter*, edited by R. Balian, R. Maynard, and G. Toulouse (North-Holland, Amsterdam, 1979b), p. 321.

Kirkpatrick, S., C.D. Gellat, Jr., and M.P. Vecchi, "Optimization by simulated annealing," *Science* **220**, 671 (1983).

Kirkwood, J.G., "The skeletal modes of vibration of long chain molecules," *J. Chem. Phys.* **7**, 506 (1939).

Kjems, J.K., "Thermal transport in fractal systems," *Physica A* **191**, 328 (1993).

Klemm, P., and Cz. Wozniak, "Dense elastic lattices of hexagonal type," *Mech. Teor. Stosow* **8**, 277 (1970).

Knackstedt, M.A., B.W. Ninham, and M. Monduzzi, "Diffusion in model disordered media," *Phys. Rev. Lett.* **75**, 653 (1995).

Knackstedt, M.A., and M. Sahimi, "On the universality of geometrical and transport exponents of rigidity percolation," *J. Stat. Phys.* **69**, 887 (1992).

Knackstedt, M.A., M. Sahimi, and A.P. Sheppard, "Invasion percolation with long-range correlations: First-order phase transitions and nonuniversal scaling properties," *Phys. Rev. E* **61**, 4920 (2000).

Knotek, M.L., M. Pollak, T.M. Donovan, and H. Kurtzman, "Thickness dependence of hopping transport in amorphous-Ge films," *Phys. Rev. Lett.* **30**, 853 (1973).

Koch, R.H., R.B. Laibowitz, E.I. Alessandrini, and J.M. Viggiano, "Resistivity-noise measurements in thin gold films near the percolation threshold," *Phys. Rev. B* **32**, 6932 (1985).

Koelman, J.M.V.A., and A. de Kuijper, "An effective medium model for the electric conductivity of an $N-$component anisotropic percolating mixture," *Physica A* **247**, 10 (1997).

Kogut, P.M., and P.L. Leath, "Bootstrap percolation transitions on real lattices", *J. Phys. C* **14**, 3187 (1981).

Kogut, P.M., and J.P. Straley, "Distribution-induced non-universality of the percolation conductivity exponents," *J. Phys. C* **12**, 2151 (1979).

Kopelman, R., S. Parus, and J. Prasad, "Fractal-like exciton kinetics in porous glasses, organic membranes, and filter papers," *Phys. Rev. Lett.* **56**, 1742 (1986).

Koplik, J., "On the effective medium theory of random linear networks," *J. Phys. C* **14**, 4821 (1981).

Koplik, J., C. Lin, and M. Vermette, "Conductivity and permeability from microgeometry," *J. Appl. Phys.* **56**, 3127 (1984).

Korringa, J., "Theory of elastic constants of heterogeneous media," *J. Meth. Phys.* **14**, 509 (1973).

Korzhenevskii, A.L., and A.A. Luzhkov "Density of phonon-fracton states of disordered solids in the vicinity of percolation phase transition," *Zh. Eksp. Teor. Fiz.* **99**, 530 (1991) [*Sov. Phys. JETP* **72**, 295 (1991)].

Koss, R.S., and D. Stroud, "Scaling behavior and surface-plasmon modes in metal-insulator composites," *Phys. Rev. B* **35**, 9004 (1987).

Kremer, K., "Numerical studies of polymer networks and gels," *Philos. Mag. B* **77**, 569 (1998).

Krohn, C.E., and A.H. Thompson, "Fractal sandstone pores: automated measurements using scanning-electron-microscope images," *Phys. Rev. B* **33**, 6366 (1986).

Kröner, E., "Berechnung der elastischen konstanten des vielkristalle aus dem konstanten des einkristalls," *Z. Phys.* **151**, 504 (1958).

Kröner, E., "Bounds for effective elastic moduli of disordered materials," *J. Mech. Phys. Solids* **25**, 137 (1977).

Kubo, R., "Statistical-mechanical theory of irreversible processes. I. General theory and simple application to magnetic and conduction problems," *J. Phys. Soc. Jpn.* **12**, 570 (1957).

Kunar, B.K., and G.P. Srivastava, "Dispersion observed in electrical properties of titanium-substituted lithium ferrites," *J. Appl. Phys.* **75**, 6115 (1994).

Kurkijärvi, J., "Conductivity in random systems. II. Finite-size-system percolation," *Phys. Rev. B* **9**, 770 (1974).

Kusy, R.P., "Influence of particle size ratio on the continuity of aggregates," *J. Appl. Phys.* **48**, 5301 (1977).

Lado, F., and S. Torquato, "Two-point probability function for distributions of oriented hard ellipsoids," *J. Chem. Phys.* **93**, 5912 (1990).

Lagar'kov, A.N., L.V. Panina, and A.K. Sarychev, "Effective magnetic permeability of composite materials near the percolation threshold," *Sov. Phys. JETP* **66**, 123 (1987) [*Zh. Eskp. Teor. Fiz.* **93**, 215 (1987)].

Laibowitz, R.B., and Y. Gefen, "Dynamic scaling near the percolation threshold in thin Au films," *Phys. Rev. Lett.* **53**, 380 (1984).

Lakes, R., "Foam structures with a negative Poisson's ratio," *Science* **235**, 1038 (1987).

Lam, P.M., and W. Bao, "Recursion method for the density of states and spectral dimension of percolation networks," *Z. Phys. B* **59**, 63 (1985).

Laman, G., "On graphs and rigidity of plane skeletal structures," *J. Eng. Math.* **4**, 331 (1970).

Lamb, W., D.M. Wood, and N.W. Aschcroft, "Long-wavelength electromagnetic propagation in heterogeneous media," *Phys. Rev. B* **21**, 2248 (1980).

Lambert, C.J., and G.D. Hughes, "Localization properties of fractons in percolating structures," *Phys. Rev. Lett.* **66**, 1074 (1991).

Landauer, R., "The electrical resistance of binary metallic mixtures," *J. Appl. Phys.* **23**, 779 (1952).

Landauer, R., "Electrical conductivity in inhomogeneous media," in *Electrical Transport and Optical Properties of Inhomogeneous Media*, edited by J.C. Garland, and D.B. Tanner, *AIP Conference Proceedings* **40** (AIP, New York, 1978), p. 2.

Lapp, A., L. Leibler, F. Schosseler, and C. Strazielle, "Scaling Behaviour of pregel sols obtained by end-linking if linear chains," *Macromolecules* **22**, 2871 (1989).

Larsen, U.D., O. Sigmund, and S. Bouwstra, "Design and fabrication of compliant micromechanisms and structures with negative Poisson's ratio," *J. Microelectromech. Sys.* **6**, 99 (1997).

Last, B.J., and D.J. Thouless, "Percolation theory and electrical conductivity," *Phys. Rev. Lett.* **27**, 1719 (1971).

Laugier, J.M., J.P. Clerc, and G. Giraud, *Proceedings of International AMSE Conference*, edited by G. Mesnard (Lyon, AMSE, 1986a).

Laugier, J.M., J.P. Clerc, G. Giraud, and J.M. Luck, "AC properties of 2D percolation networks: A transfer matrix approach," *J. Phys. A* **19**, 3153 (1986b).

Lax, M., "Molecular field in spherical mode," *Phys. Rev.* **97**, 629 (1955).

Leath, P.L., "Cluster size and boundary distribution near percolation threshold," *Phys. Rev. B* **14**, 5046 (1976).

Lee, S.B., and S. Torquato, "Porosity for the penetrable-concentric-shell model of two-phase disordered media: Computer simulation results," *J. Chem. Phys.* **89**, 3258 (1988).

Lee, S.B., and S. Torquato, "Measure of clustering in continuum percolation: Computer-simulation of the two-point cluster function," *J. Chem. Phys.* **91**, 1173 (1989).

Lee, S.-I., Y. Song, T.W. Noh, X.-D. Chen, and J.R. Gaines, J.R., "Experimental observation of nonuniversal behavior of the conductivity exponent for three-dimensional continuum percolation systems," *Phys. Rev. B* **34**, 6719 (1986).

Leibler, L., and F. Schosseler, "Gelation of polymer solutions: an experimental verification of the scaling behavior of the size distribution function," *Phys. Rev. Lett.* **55**, 1110 (1985).

Lemieux, M.A., P. Breton, and A.-M.S. Tremblay, "Unified approach to numerical transfer matrix methods for disordered systems: application to mixed crystals and to elasticity percolation," *J. Physique Lett.* **46**, L1 (1985).

Léon, C., A. Rivera, A. Várez, J. Sanz, J. Santamaria, and K.L. Ngai, "Origin of constant loss in ionic conductors," *Phys. Rev. Lett.* **86**, 1279 (12001).

Levin, V.M., "Thermal expansion coefficients of heterogeneous materials," *Mekh. Tverd. Tela* **2**, 83 (1967).

Levinshtein, M.E., M.S. Shur, and E.L. Efros, "On the relation between critical indices and percolation theory," *Sov. Phys.-JETP* **41**, 386 (1975).

Levinshtein, M.E., M.S. Shur, and E.L. Efros, "Galvanomagnetic phenomena in disordered systems. Theory and simulation," *Sov. Phys.-JETP* **42**, 1120 (1976).

Levitz, P., "Off-lattice reconstruction of porous media: critical evaluation, geometrical confinement and molecular transport," *Adv. Colloid Interface Sci.* **76-77**, 71 (1998).

Lévy, Y.-E., and B. Souillard, "Superlocalization of electrons and waves in fractal media," *Europhys. Lett.* **4**, 233 (1987).

Lewinski, T., "Dynamical tests of accuracy of Cosserat models for honeycomb gridworks," *Z. Angew. Math. Mech.* **68**, T210 (1988).

Lewinski, T., *Mech. Teor. Stosow* **22**, 389 (1989).

Lewis, F.T., "The shape of cells as a mathematical problem," *Amer. Scientist* **34**, 359 (1946).

Li, Q., C.M. Soukoulis, and E.N. Economou, "Universal behavior near the band edges for disordered systems: numerical and coherent-potential-approximation studies," *Phys. Rev. B* **37**, 8289 (1988).

Li, Q., C.M. Soukoulis, and G.S. Grest, "Vibrational properties of percolating clusters: Localization and density of states," *Phys. Rev. B* **41**, 11713 (1990).

Liang, N.T., Y. Shan, and S.-Y. Wang, "Electronic conductivity and percolation theory in aggregated films," *Phys. Rev. Lett.* **37**, 526 (1976).

Liao, K.H., and A.E. Scheidegger, "brancing-type models of flow through poroys media," *Int. Assoc. Sci. Hydr. Bull.* **14**, 137 (1969).

Limat, L., "Percolation and Cosserat elasticity: exact results on a deterministic fractal," *Phys. Rev. B* **37**, 672 (1988a).

Limat, L., "Rotationally invariant elasticity in a planar fractal network," *Phys. Rev. B* **38**, 512 (1988b).

Limat, L., "Micropolar elastic percolation: the superelastic problem," *Phys. Rev. B* **38**, 7219 (1988c).

Limat, L., "Elastic and superelastic percolation networks: imperfect duality, critical Poisson ratios, and relations between microscopic models," *Phys. Rev. B* **40**, 9253 (1989).

Lin, C., and M.H. Cohen, "Quantitative methods for microgeometric modeling," *J. Appl. Phys.* **53**, 4152 (192).

Lin, S.L., J. Mellor-Crumney, B.M. Pettitt, and G.N. Phillips, Jr., "Molecular dynamics on a distributed-memory multiprocessor," *J. Comput. Phys.* **13**, 1022 (1992).

Lin, Y.G., D.T. Mallin, J.C.W. Chien, and H.H. Winter, "Dynamical mechanical measurement of crystallization-induced gelation in thermpplastic elastomeric poly(propylene)," *Macromolecules* **24**, 850 (1991).

Liu, X., and H.v. Löhneysen, "Low-temperature thermal properties of amorphous AS_xSe_{1-x}," *Phys. Rev. B* **48**, 13486 (1993).

Lobb, C.J., and M.G. Forrester, "Measurement of nonuniversal critical behavior in a two-dimensional continuum percolating system," *Phys. Rev. B* **35**, 1899 (1987).

Lobb, C.J., and D.J. Frank, "Percolation critical exponents for conductance and critical current in two dimensions," *AIP Conference Proc.* **58**, 308 (1979).

Lobb, C.J., and D.J. Frank, "Percolative conduction and the Alexander-Orbach conjecture in two dimensions," *Phys. Rev. B* **30**, 4090 (1984).

Long, A.R., "Hopping conductivity in the intermediate frequency regimes," in *Hopping Transport in Solids,* edited by M. Pollak and B. Shklovskii (Elsevier, Amsterdam, 1991), p. 207.

Lorenz, B., I. Orgzall, and H.O. Heuer, "Universality and cluster structures in continuum models of percolation with two different radius distributions," *J. Phys. A* **26**, 4711 (1993).

Loring, R.F., and S. Mukamel, "Phonon and fracton vibrational modes in disordered harmonic structures: a self-consistent theory," *Phys. Rev. B* **34**, 6582 (1986).

Lovasz, L., and Y. Yemini, *SIAM J. Alg. Disc. Meth.* **3**, 91 (1982).

Lu, B., and S. Torquato, "n-point probability functions for a lattice model of heterogeneous media," *Phys. Rev. B* **42**, 4453 (1990).

Lu, B., and S. Torquato, "General formalism to characterize the microstructure of polydispersed random media," *Phys. Rev. A* **43**, 2078 (1991).

Lu, B., and S. Torquato, "Lineal-path function for random heterogeneous materials," *Phys. Rev. A* **45**, 922 (1992a).

Lu, B., and S. Torquato, "Lineal-path function for random heterogeneous materials. II. Effect of polydispersivity," *Phys. Rev. A* **45**, 7292 (1992b).

Lu, B., and S. Torquato, "Chord-length and free-path distribution functions for many-body systems," *J. Chem. Phys.* **98**, 6472 (1993a).

Lu, B., and S. Torquato, "Chord-length distribution function for two-phase random media," *Phys. Rev. E* **47**, 2950 (1993b).

Lubachevsky, B.D., and F.H. Stillinger, "Geometric properties of random disk packings," *J. Stat. Phys.* **60**, 561 (1990).

Lubensky, T.C., and J. Isaacson, "Field Theory for the statistics of branched polymers, gelation and vulcanization," *Phys. Rev. Lett.* **41**, 829 (1978).

Lubensky, T.C., and J. Wang, "Percolation conductivity exponent t to second order in $\epsilon = 6 - d$," *Phys. Rev. B* **33**, 4998 (1986).

Luck, J.M., "Conductivity of random resistor networks: An investigation of the accuracy of the effective-medium approximation," *Phys. Rev. B* **43**, 3933 (1991).

Macdonald, J.R., "Possible universalities in the ac frequency response of dispersed, disordered materials," *J. Non-Cryst. Solids* **210**, 70 (1997).

Mahadevan, S., and A. Giridhar, "On the chemical and mechanical thresholds of some chalcogenide glasses," *J. Non-Cryst. Solids* **110**, 118 (1989).

Mahan, G.D., "Long-wavelength absorption of cermets," *Phys. Rev. B* **38**, 9500 (1988).

Maliepaard, M.C., J.H. Page, J.P. Harrison, and R.J. Stubbs, "Ultrasonic study of the vibrational modes of sintered metal powders," *Phys. Rev. B* **32**, 6261 (1985).

Mall, S., and W.B. Russel, "Effective medium approximation for an elastic network model of flocculated suspensions," *J. Rheol.* **31**, 651 (1987).

Malliaris, A. and Turner, D.T., 1971, "Influence of particle size on the electrical resistivity of compacted mixtures of polymeric and metallic powders," *J. Appl. Phys.* **42**, 614.

Maloufi, N., A. Audouard, M. Piecuch, M. Vergant, G. Marchal, and M. Gerl, "Experimental study of the dc conductivity mechanism in amorphous Si_xSn_{1-x} alloys," *Phys. Rev. B* **37**, 8867 (1988).

Mandal, P., A. Neumann, A.G. Jansen, P. Wyder, and R. Deltour, "Temperature and magnetic-field dependence of the resistivity of carbon-black polymer composites," *Phys. Rev. B* **55**, 452 (1997).

Mandelbrot, B.B., *The Fractal Geometry of Nature* (W.H. Freeman, San Francisco, 1982).

Mandelbrot, B.B., "Self-affine fractals and fractal dimension," *Physica Scripta* **32**, 257 (1985).

Mandelbrot, B.B., "Self-affine fractal sets," in *Fractals in Physics*, edited by L. Pietronero and E. Tosatti (North-Holland, Amsterdam, 1986), pp. 3-28.

Maradudin, A.A., E.W. Montroll, G.H. Weiss, and I.P. Ipatova, *Theory of Lattice Dynamics in Harmonic Approximation*, 2nd ed. (Academic, New York, 1971).

Marshall, E.W., and A.B. Harris, "Scaling of splay and total rigidity for elastic percolation on the triangular lattice," *Phys. Rev. B* **38**, 4929 (1988).

Martin, J.E., D. Adolf, and J.P. Wilcoxon, "Viscoelasticity of near-critical gels," *Phys. Rev. Lett.* **61**, 2620 (1988).

Martin, J.E., and J.P. Wilcoxon, "Critical dynamics of the sol-gel transition," *Phys. Rev. Lett.* **61**, 373 (1988).

Martins, J.L., and A. Zunger, "Bond lengths around isovalent impurities and in semiconductor solid solutions," *Phys. Rev. B* **30**, 6217 (1984).

Mason, T.G., J. Bibette, and D.A. Weitz, "Elasticity of compressed emulsions," *Phys. Rev. Lett.* **75**, 2051 (1995).

Mattis, C.D., and J. Bardeen, "Theory of anomalous skin effect in normal and superconducting metals," *Phys. Rev.* **111**, 412 (1958).

Maxwell, J.C., *Treatise on Electricity and Magnetism* (Claredon, Oxford, 1873), p. 194.

Maxwell-Garnett, J.C., "Colours in metal glasses, in metallic films and in metallic solutions - II," *Philos. Trans. R. Soc. Lond. A* **203**, 385 (1904).

Mazzacurati, V., M. Montagna, P. Pilla, G. Viliani, G. Ruocco, and G. Signorelli, "Vibrational dynamics and Raman scattering in fractals: a numerical study," *Phys. Rev. B* **45**, 2126 (1992).

McCoy, J.J., *Recent Advances in Engineering Sciences*, Vol. 5 (Gordon and Breach, New York, 1970).

McKenzie, D.R., R.C. McPhedran, and G.H. Derrick, "The conductivity of lattices of spheres II. The body centred and face centred lattices," *Proc. R. Soc. Lond. A* **362**, 211 (1978).

McLachlan, D.S., "Measurement and analysis of a model dual-conductivity medium using a generalised effective-medium theory," *J. Phys. C* **21**, 1521 (1988).

McLachlan, D.S., I.I. Oblakova, and A.B. Pakhomov, "Dielectric-constant measurements in a system of NbC grains near the percolation threshold," *Physica A* **207**, 234 (1994).

McPhedran, R.C., and D.R. McKenzie, "The conductivity of lattices of spheres I. The simple cubic lattice," *Proc. R. Soc. Lond. A* **359**, 45 (1978).

McPhedran, R.C., and G.W. Milton, "Bounds and exact theories for the transport properties of inhomogenous media," *Appl. Phys. A* **26**, 207 (1981).

Meakin, P., *Fractals, Scaling and Growth far from Equilibrium* (Combridge University Press, Cambridge, 1998).

Meester, R., R. Roy, and A. Sarkar, "Nonuniversality and continuity of of the critical covered volume fraction in continuum percolation," *J. Stat. Phys.* **75**, 123 (1994).

Mehbod, M., P. Wyder, R. Deltour, C. Pierre, and G. Geuskens, "Temperature dependence of the resistivity in polymer-conducting-carbon-black composites," *Phys. Rev. B* **36**, 7627 (1987).

Mehrabi, A.R., and M. Sahimi, "Coarsening of heterogeneous media: Application of wavelets," *Phys. Rev. Lett.* **79**, 4385 (1997).

Meijering, J.L., "Interface area, edge length and number of verticies in crystal aggregates with random nucleation," *Phillips Res. Rep.* **8**, 270 (1953).

Meir, Y., "Universal crossover between Efros-Shklovskii and Mott variable-range-hopping regimes," *Phys. Rev. Lett.* **77**, 5265 (1996).

Mendelson, K.S., "Effective conductivity of two-phase material with cylindrical phase boundaries," *J. Appl. Phys.* **46**, 917 (1975a).

Mendelson, K.S., "A theorem on the effective conductivity of a two-dimensional heterogeneous medium," *J. Appl. Phys.* **46**, 4740 (1975b).

Mendelson, K.S., and M.H. Cohen, "The effect of grain anisotropy on the electrical properties of sedimentary rocks," *Geophysics* **47**, 257 (1982).

Meredith, R.E., and C.W. Tobias, "Resistance to potential flow through a cubical array of spheres," *J. Appl. Phys.* **31**, 1270 (1960).

Metzler, R., W.G. Glckle, and T.F. Nonnenmacher, "Fractional model equation for anomalous diffusion," *Physica A* **211**, 13 (1994).

Michels, M.A.J., "Scaling relations and the general effective-medium equation for isolator-conductor mixtures," *J. Phys.: Condens. Matter* **4**, 3961 (1992).

Michels, M.A.J., J.C.M. Brokken-Zijp, W.M. Groenewoud, and A. Knoester, "Systematic study of percolative network formation and effective electric response in low-concentration-carbon-black/polymer composites," *Physica A* **157**, 529 (1989).

Miller, A., and E. Abrahams, "Electro-optic kerr effect and polarization reversal in Deuterium doped Rochelle salt," *Phys. Rev.* **120**, 745 (1960).

Miller, C.A., and S. Torquato, "Effective conductivity of hard-sphere dispersions," *J. Appl. Phys.* **68**, 5486 (1990).

Miller, J., "Bounds for effective electrical, thermal, and magnetic properties of heterogeneous materials," *J. Math. Phys.* **10**, 1988 (1969).

Milton, G.W., "Bounds on the electromagnetic, elastic, and other properties of two-component composites," *Phys. Rev. Lett.* **46**, 542 (1981a).

Milton, G.W., "Concerning bounds on the transport and mechanical properties of multi-component composite materials," *Appl. Phys.* **A26**, 125 (1981b).

Milton, G.W., "Bounds on the elastic and transport properties of two-component composites," *J. Mech. Phys. Solids* **30**, 177 (1981c).

Milton, G.W., "Bounds on the complex permittivity of a two-component composite material," *J. Apply. Phys.* **52**, 5286 (1981d).

Milton, G.W., "Bounds on the elastic and transport properties of two-component composites," *Phys. Rev. Lett.* **46**, 542 (1982a).

Milton, G.W., "Bounds on the elastic and transport properties of two-component composites," *J. Mech. Phys. Solids* **30**, 177 (1982b).

Milton, G.W., "Correlation of the electromagnetic and elastic properties of the composites and microgeometries corresponding with effective medium approximations," in *Physics and Chemistry of Porous Media*, edited by D.L. Johnson and P.N. Sen, *AIP Conference Proceedings* **107**, 66 (1984).

Milton, G.W., "The coherent potential approximation is a realizable effective medium scheme," *Commun. Math. Phys.* **99**, 463 (1985).

Milton, G.W., "Multicomponent composites, electrical networks, and new types of continued fractions, parts I and II," *Commun. Math. Phys.* **111**, 281 (1987).

Milton, G.W., "Classical Hall effect in two-dimensional composites: a characterization of the set of realizable effective conductivity tensors," *Phys. Rev. B* **38**, 11296 (1988).

Milton, G.W., "On characterizing the set of possible effective tensors of composites: the variational method and the translation method," *Commun. Pure Appl. Math.* **43**, 63 (1990a).

Milton, G.W., in *Continuum Models and Discrete Systems*, edited by G.A. Maugin (1990b), p. 60.

Milton, G.W., "Composite materials with Poisson's ratio close to -1," *J. Mech. Phys. Solids* **40**, 1105 (1992).

Milton, G.W., R.C. McPhedran, and D.R. McKenzie, "Transport properties of arrays of intersecting cylinders," *Appl. Phys.* **25**, 23 (1981).

Milton, G.W., and N. Phan-Thien, "New bounds on the effective moduli of two-component materials," *Proc. R. Soc. Lond. A* **380**, 305 (1982).

Mindlin, R.D., "Micro-structure in linear elasticity," *Arch. Rat. Mech. Analy.* **16**, 51 (1964).

Mitchell, J.H., "On the direct determination of stress in an elastic solid, with application to the theory of plates," *Proc. Lond. Math. Soc.* **31**, 100 (1899).

Mitescu, C., M. Allain, E. Guyon, and J.P. Clerc, "Electrical conductivity of finite-size percolation networks," *J. Phys. A* **15**, 2523 (1982).

Mitescu, C., and J. Roussenq, "Une Fourmi dans labyrinthe: diffusion dans un systeme de percolation," *C. R. Acad. Sc. Paris* **283**, 999 (1976).

Mohanty, K.K., *Fluids in Porous Media: Two-Phase Distribution and Flow*, Ph.D. Thesis, University of Minnesota (1981).

Moha-Ouchane, M., J. Peyrelasse, and C. Boned, "Percolation transition in microemulsions: effect of water-surfactant ratio, temperature, and salinity," *Phys. Rev. A* **35**, 3027 (1987).

Molyneux, J.E., "Effective permittivity of a polycrystalline dielectric," *J. Math. Phys.* **11**, 1172 (1970).

Monceau, P.J.-M., and J.-C.S. Levy, "Monodimensional effects on elastic and vibrational properties of lacunary networks," *Phys. Rev. B* **49**, 1026 (1994).

Mondescu, R.P., and M. Muthukumar, "Effective elastic moduli of a composite containing rigid spheres at nondilute concentrations: A multiple scattering approach," *J. Chem. Phys.* **1123** (1999).

Montagna, M., O. Pilla, G. Viliani, V. Mazzcurati, G. Ruocco, and G. Signorelli, "Numerical study of Raman scattering from fractals," *Phys. Rev. Lett.* **65**, 1136 (1990).

Montroll, E.W., and G.H. Weiss, "Random walks on lattices. II," *J. Math. Phys.* **6**, 167 (1965).

Mossotti, O.F., Memorie di Matematica e di Fisica della Societá Italiana delle Scienze Residente in Modena, Vol. 24, pt. 2 (1850), p. 49.

Mott, N.F., "On the transition to metallic conduction in semiconductors," *Can. J. Phys.* **34**, 1356 (1956).

Mott, N.F., "Conduction in glasses containing transition metal ions," *J. Non-Cryst. Solids* **1**, 1 (1968).

Mott, N.F., "Conduction in non-crystalline materials. III. Localized states in a pseudogap and near extremities of conduction and valence bands," *Philos. Mag.* **19**, 835 (1969).

Moukarzel, C., and P.M. Duxbury, "Stressed backbone and elasticity of random central-force systems," *Phys. Rev. Lett.* **75**, 4055 (1995).

Moukarzel, C., and P.M. Duxbury, "Comparison of rigidity and connectivity percolation in two dimensions," *Phys. Rev. E* **59**, 2614 (1999).

Moukarzel, C., P.M. Duxbury, and P.L. Leath, "First-order rigidity on Cayley trees" *Phys. Rev. E* **55**, 5800 (1997a).

Moukarzel, C., P.M. Duxbury, and P.L. Leath, "Infinite-cluster geometry in central-force networks," *Phys. Rev. Lett.* **78**, 1480 (1997b).

Mousseau, N., and M.F. Thorpe, "Structural model for crystalline and amorphous Si-Ge alloys," *Phys. Rev. B* **48**, 5172 (1993).

Movaghar, B., B. Pohlmann, and W. Schirmacher, "Theory of AC and DC conductivity in disordered hopping systems," *Phys. Stat. Sol. (b)* **97**, 533 (1980a).

Movaghar, B., B. Pohlmann, and W. Schirmacher, "Random walk in disordered hopping systems," *Solid State Commun.* **34**, 451 (1980b).

Movaghar, B., and W. Schirmacher, "On the theory of hopping conductivity in disordered systems," *J. Phys. C* **14**, 859 (1981).

Movaghar, B., D. Würtz, and B. Pohlmann, "Carrier drift and trapping in one dimensional systems," *Z. Phys. B* **66**, 523 (1987).

Mukhopadhyay, S., and M. Sahimi, "Scaling behavior of permeability and conductivity anisotropy near the percolation threshold," *J. Stat. Phys.* **74**, 1301 (1994).

Mukhopadhyay, S., and M. Sahimi, "Calculation of the effective permeabilities of field-scale porous media," *Chem. Eng. Sci.* **55**, 4495 (2000).

Mura, T., *Micromechanics of Defects in Solids* (Martinus Nijhoff, Dordrecht, 1987).

Muskhelishvili, N.I., *Some Basic Problems of the Mathematical Theory of Elasticity* (P. Noordhoff, Groningen, Holland, 1953).

Nagatani, T., "Multi-bond expansion for the effective conductivity in bond-disordered resistor networks," *J. Phys. C* **14**, 4839 (1981).

Naghdi, P.M., and C.S. Hsu, "On a representation of displacements in linear elasticity in terms of three stress functions," *J. Math. Mech.* **10**, 233 (1961).

Nakajima, T., "Correlation between electrical conduction and dielectric polarization in inorganic glasses," in *Conference on Electrical Insulation and Dielectric Phenomena* (National Academy of Sciences, Washington, DC, 1972), p. 168.

Nakayama, T., *Physica A* **191**, 386 (1992).

Nakayama, T., and K. Yakubo, "Dynamic structure factor and single-length scaling for random fractals," *J. Phys. Soc. Jpn.* **61**, 2601 (1992).

Nakayama, T., K. Yakubo, and R. Orbach, "Characteristics of fractons: from specific realizations to ensemble averages," *J. Phys. Soc. Jpn.* **58**, 1891 (1989).

Nakayama, T., K. Yakubo, and R. Orbach, "Dynamical properties of fractal networks: Scaling, numerical simulations, and physical realizations," *Rev. Mod. Phys.* **66**, 381 (1994).

Namikawa, H., "Characterization of the diffusion process in oxide glasses based on the correlation between electric conduction and dielectric relaxation," *J. Non-Cryst. Solids* **18**, 173 (1975).

Nelson, D.R., and L. Radzihovsky, "Polymerized membranes with quenched random internal disorder," *Europhys. Lett.* **16**, 79 (1991).

Nemat-Nasser, S., and M. Hori, *Micromechanics: Overall Properties of Heterogeneous Materials* (North-Holland, Amsterdam, 1993).

Newman, M.E.J., and R.M. Ziff, "Efficient Monte Carlo algorithm and high-precision results for percolation," *Phys. Rev. Lett.* **85**, 4104 (2000).

Ngai, K.L., and C.T. Moynihan, "The dynamics of mobile ions in ionically conducting glasses and other materials," *Mater. Res. Soc. Bull.* **23** (11), 51 (1998).

Nicholson, D., "Capillary models for porous media. Part 2.-Sorption desorption hysteresis in three dimensional networks," *Trans. Faraday Soc.* **64**, 3416 (1968).

Nickel, B., and W.H. Butler, "Problems in strong-scattering binary alloys," *Phys. Rev. Lett.* **30**, 373 (1973).

Niklasson, G.A., "Comparison of dielectric response functions for conducting materials," *J. Appl. Phys.* **66**, 4350 (1989).

Niklasson, G.A., and C.G. Granqvist, "Optical properties and solar selectivity of coevaporated Co-Al_2O_3 composite films," *J. Appl. Phys.* **55**, 3382 (1984).

Ninham, B.W., and R.A. Sammut, "Refraction index of arrays of spheres and cylinders," *J. Theor. Biol.* **56**, 125 (1976).

Nishimatsu, C., and J. Gurland, *Trans. Am. Soc. Met.* **52**, 469 (1960).

Nitsche, L.C., and H. Brenner, "Eulerian kinematics of flow through periodic models of porous media," *Arch. Rat. Mech. Analy.* **107**, 225 (1989).

Noble, B., and J.W. Daniel, *Applied Linear Algebra*, 2nd ed. (Prentice-Hall, New Jersey, 1977).

Noh, T.W., S.G. Kaplan, and A.J. Sievers, "Observation of a far-induced sphere resonance in superconducting $La_{2-x}Sr_xCuO_{4-y}$ particles," *Phys. Rev. Lett.* **62**, 599 (1989).

Normand, J.-M., and H.J. Herrmann, "Precise numerical determination of the superconducting exponent of percolation in three dimensions," *Int. J. Mod. Phys. C* **1**, 207 (1990).

Normand, J.-M., H.J. Herrmann, and M. Hajjar, "Precise calculation of the dynamical exponent of two-dimensional percolation," *J. Stat. Phys.* **52**, 441 (1988).

Norris, A.N., P. Sheng, and A.J. Callegari, "Effective-medium theories for two-phase dielectric media," *J. Appl. Phys.* **57**, 1990 (1985).

Novikov, V.V., K.W. Wojciechowsky, D.V. Belov, and V.P. Privalko, "Elastic properties of inhomogeneous media with chaotic structure," *Phys. Rev. E* **63**, 036120 (February 2001).

Nunan, K.C., and J.B. Keller, "Effective elasticity tensor of a periodic composite," *J. Mech. Phys. Solids* **32**, 259 (1984).

Obukhov, S.P., "First order rigidity transition in random rod networks," *Phys. Rev. Lett.* **74**, 4472 (1995).

Odagaki, T., and M. Lax, "ac Hopping conductivity of a one-dimensional bond-percolation model," *Phys. Rev. Lett.* **45**, 847 (1980).

Odagaki, T., and M. Lax, "Coherent-potential approximation in the stochastic transport theory of random media," *Phys. Rev. B* **24**, 5284 (1981).

Odagaki, T., M. Lax, and A. Puri, "Hopping conduction in the d-dimensional lattice bond-percolation problem," *Phys. Rev. B* **28**, 2755 (1983).

O'Brien, R.W., "A method for the calculation of the effective transport properties of suspensions of interacting particles," *J. Fluid Mech.* **91**, 17 (1979).

Ornstein, L.S., and F. Zernike, "Accidental deviations of density and opalescence at the critical point of a single substance," *Proc. Akad. Sci. (Amsterdam)* **17**, 793 (1914).

Ortuno, M., and M. Pollak, "Hopping transport in a-Ge and a-Si," *Philos. Mag.* **47**, L93 (1983).

Ota, R., T. Yamate, N. Soga, and M. Kunugi, "Elastic properties of Ge-Se glass under pressure," *J. Non-Cryst. Solids* **29**, 67 (1978).

Ottavi, H., J.P. Clerc, G. Giraud, J. Roussenq, E. Guyon, and C.D. Mitescu, "Electrical conductivity of a mixture of conducting and insulating spheres: an application of some percolation concepts," *J. Phys. C* **11**, 1311 (1978).

Page, J.H., W.J.L. Buyers, G. Dolling, P. Gerlach, and J.P. Harrison, "Neutron inelastic scattering from fumed silica," *Phys. Rev. B* **39**, 6180 (1989).

Palevski, A., M.L. Rappaport, A. Kapitulnik, A. Fried, and G. Deutscher, "Hall coefficient and conduction in a 2D percolation system," *J. Physique Lett.* **45**, L367 (1984).

Panina, L.V., A.N. Lagarkov, A.K. Sarychev, Y.R. Smychkovich, and A.P. Vinogradov, "Effective medium theory of dielectric constant of granular materials in the presence of skin effect," in *Physical Phenomena in Granular Materials*, edited by G.D. Coby, T.H. Geballe, and P. Sheng, *MRS Symposium Proc.* **195** (1990), p. 275.

Papandreou, N., and P. Nédellec, "Interplay between microscopic and macroscopic disorder in percolating Pd films," *J. Physique. I France* **2**, 707 (1992).

Parisi, G., and N. Sourlas, "Critical behavior of branched polymers in the Lee-Yang edge singularity," *Phys. Rev. Lett.* **46**, 891 (1981).

Park, Y., A.B. Harris, and T.C. Lubensky, "Noise exponents of the random resistor network," *Phys. Rev. B* **35**, 5048 (1987).

Patton, E., J.A. Wesson, M. Rubinstein, J.E. Wilson, and L.E. Oppenheimer, "Scaling properties of branched polyesters," *Macromolecules* **22**, 1946 (1989).

Payandeh, B., "A block cluster approach to percolation," *Rivista del Nuovo Cimento* **3**, 1 (1980).

Pelous, J. R. Vacher, T. Woignier, J.L. Sauvajol, and E. Courtens, "Scaling phonon-fracton dispersion laws in fractal aerogels," *Philos. Mag. B* **59**, 65 (1989).

Percus, J.K., and G.Y. Yevick, "Analysis of classical statistical mechanics by means of collective coordinates," *Phys. Rev.* **110**, 1 (1958).

Perrins, W.T., D.R. McKenzie, and R.C. McPhedran, "Transport properties of regular arrays of cylinders," *Proc. R. Soc. Lond. A* **369**, 207 (1979a).

Perrins, W.T., R.C. McPhedran, and D.R. McKenzie, "Optical properties of dense regular cermets with relevance to selective solar absorbers," *Thin Solid Films* **57**, 321 (1979b).

Peterson, J.M., and J.J. Hermans, "The dielectric constants of nonconducting suspensions," *J. Composite Mater.* **3**, 338 (1969).

Peyrelasse, J., M. Moha-Ouchane, and C. Boned, "Dielectric relaxation and percolation phenomena in ternary microemulsions," *Phys. Rev. A* **38**, 904 (1988).

Phan-Thein, N., and S. Kim, *Microstructures in Elastic Media: Principles and Computational Methods* (Oxford University Press, Oxford, 1994).

Phillips, J.C., "Topology of covalent non-crystalline solids I: short-range order in chalcogenide alloys," *J. Non-Crystalline Solids* **34**, 153 (1979).

Phillips, J.C., "Topology of covalent non-crystalline solids II: medium-range order in chalcogenide alloys and A-Si(Ge)," *J. Non-Cryst. Solids* **43**, 37 (1981).

Philips, J.C., "Constraint theory, stiffness percolation and the rigidity transition in network glasses," in *Rigidity Theory and Applications*, edited by M.F. Thorpe and P.M. Duxbury (Kluwer Academic, New York, 1999), p. 155.

Phillips, J.C., and M.F. Thorpe, "Constraint theory, vector percolation, and glass formation," *Solid State Commun.* **53**, 699 (1985).

Pike, G.E., and C.H. Seager, "Percolation and conductivity: A computer study," *Phys. Rev. B* **10**, 1421 (1974).

Pilla, P., G. Vilani, M. Montagna, V. Mazzucurati, G. Ruocco, and G. Signorelli, "Vibrational dynamics of percolating clusters: fracton wavefunctions and Raman coupling coefficients," *Philos. Mag. B* **65**, 243 (1992).

Pla, O., R. Garcia-Molina, F. Guinea, and E. Louis, "Properties of elastic percolating networks in isotropic media with arbitrary elastic constants," *Phys. Rev. B* **41**, 11449 (1990).

Plischke, M., and B. Joós, "Entropic elasticity of diluted central force networks," *Phys. Rev. Lett.* **80**, 4907 (1998).

Plischke, M., D.C. Vernon, B. Joós, and Z. Zhou, "Entropic rigidity of randomly diluted two- and three-dimensional networks," *Phys. Rev. E* **60**, 3129 (1999).

Polatsek, G., and O. Entin-Wohlman, "Effective-medium approximation for a percolation network: The structure factor and the Ioffe-Regel criterion," *Phys. Rev. B* **37**, 7726 (1988).

Polatsek, G., O. Entin-Wohlman, and R. Orbach, "Effective-medium approximation for the dynamical excitations of percolating antiferromagnets," *Phys. Rev. B* **39**, 9353 (1989).

Ploder, D., and J.H. Van Santen, "The effective permeability of mixture of solids," *Physica* **12**, 257 (1946).

Pollak, M., "A percolation treatment of dc hopping conduction," *J. Non-Cryst. Solids* **11**, 1 (1972).

Pollak, M., "Percolation and hopping transport," in *The Metal-Non-metal Transition in Disordered Systems*, edited by L.R. Friedman and D.P. Tunstall (University of Edinburgh, 1978), p. 95.

Pollak, M., and T.H. Geballe, "Low-frequency conductivity due to hopping processes in silicon," *Phys. Rev.* **122**, 1742 (1961).

Pollak, M., and I. Riess, "Application of percolation theory to 2D-3D Heisenberg ferromagnets," *Phys. Status Solidi (b)* **69**, K15 (1975).

Posselt, D., J.K. Kjems, A. Bernasconi, T. Sleator, and H.R. Ott, "The thermal conductivity of silica aerogel in the phonon, the fracton and the particle-mode regime," *Europhys. Lett.* **16**, 59 (1991).

Postma, G.W., "Wave propagation in a stratified medium," *Geophysics* **20**, 780 (1955).

Pouliquen, O., M. Nicolas, and P.D. Weidman, "Crystallization of non-Brownian spheres under horizontal shaking," *Phys. Rev. Lett.* **79**, 3640 (1997).

Powell, M.J.D., "The volume internal to three intersecting hard spheres," *Mol. Phys.* **7**, 591 (1964).

Prager, S., "Diffusion and viscous flow in concentrated suspensions," *Physica* **29**, 129 (1963).

Prager, S., "Improved variational bounds on some bulk properties of a two-phase random medium," *J. Chem. Phys.* **50**, 4305 (1969).

Prakash, S., S. Havlin, M. Schwartz, and H.E. Stanley, "Structural and dynamical properties of long-range correlated percolation," *Phys. Rev. A* **46**, R1724 (1992).

Prall, D., and R.S. Lakes, "Properties of a chiral honeycomb with a Poisson's ratio of −1," *Int. J. Mech. Sci.* **39**, 305 (1997).

Press, W.H., S.A. Teukolsky, W.T. Vetterling, and B.H. Flannery, *Numerical Recipes*, 2nd ed. (Cambridge University Press, Cambridge, 1992).

Prunet, V., and R. Blanc, "Rigid clusters enumeration," *J. Phys. A* **19**, L1197 (1986).

Pusey, P.N., and W. van Megan, "Phase behaviour of concentrated suspensions of nearly hard colloidal spheres," *Nature* **320**, 340 (1986).

Quickenden, T.I., and G.K. Tan, "Random packing in two dimensions and the structure of monolayers," *J. Colloid Interface Sci.* **48**, 382 (1974).

Quintanilla, J., and S. Torquato, "New bounds on the elastic moduli of suspensions of spheres," *J. Appl. Phys.* **77**, 4361 (1995).

Quintanilla, J., and S. Torquato, "Lineal measures of clustering in overlapping particle systems," *Phys. Rev. E* **54**, 4027 (1996).

Quintanilla, J., S. Torquato, and R.M. Ziff, "Efficient measurement of the percolation threshold for fully penetrable disks," *J. Phys. A* **33**, L399 (2000).

Rammal, R., T.C. Lubensky, and G. Toulouse, "Superconducting networks in a magnetic field," *Phys. Rev. B* **27**, 2820 (1983).

Rammal, R., C. Tannous, and A.-M.S. Tremblay, "$1/f$ noise in random resistor networks: Fractals and percolating systems," *Phys. Rev. A* **31**, 2662 (1985a).

Rammal, R., C. Tannous, P. Breton, and A.-M.S. Tremblay, "Flicker $(1/f)$ noise in percolation networks: A new hierarchy of exponents," *Phys. Rev. Lett.* **54**, 1718 (1985b).

Rammal, R., and G. Toulouse, "Random walks on fractal structures and percolation clusters," *J. Physique Lett.* **44**, L13 (1983).

Rayleigh, Lord (John William Strutt), "On the influence of obstacles arranged in rectangular order upon the properties of the medium," *Philos. Mag.* **34**, 481 (1872).

Recski, A., "Application of combinatorics to statics - a second survery," *Disc. Math.* **108**, 183 (1992).

Redfield, D., "Observation of $\log \sigma \sim T^{-1/2}$ in three-dimensional energy-band tails," *Phys. Rev. Lett.* **30**, 1319 (1973).

Rehwald, W., H. Kiess, and B. Binggeli, "Frequency dependent conductivity in polymers and other disordered materials," *Z. Phys. B* **68**, 143 (1987).

Reichenauer, G., J. Fricke, and U. Buchenau, "Neutron scattering study of low-frequency vibrations in silica aerogels," *Europhys. Lett.* **8**, 415 (1989).

Reiss, H., and A.D. Hammerich, "Hard spheres: Scaled particle theory and exact relations on the existence and structure of the fluid/solid phase transition," *J. Phys. Chem.* **90**, 6252 (1986).

Reiss, H., H.L. Frisch, and J.L. Lebowitz, "Statistical mechanics of rigid spheres," *J. Chem. Phys.* **31**, 369 (1959).

Reńyi, A., "On a one-dimensional problem concerning random space filling," *Trans. Math. Stat. Prob.* **4**, 203 (1963).

Resnick, D.J., J.C. Garland, and R. Newrock, "Galvanomagnetic properties of randomly inhomogeneous superconductors," *Phys. Rev. Lett.* **43**, 1192 (1979).

Reynolds, P.J., H.E. Stanley, and W. Klein, "Large-cell Monte Carlo renormalization group for percolation," *Phys. Rev. B* **21**, 1223 (1980).

Richard, T.G., "The mechanical behavior of a solid microsphere filled composite," *J. Compos. Mater* **9**, 108 (1975).

Richter, D., and L. Passell, "Neutron scattering as a probe of small-particle dynamics in hydroxylated amorphous silica," *Phys. Rev. Lett.* **44**, 1593 (1980).

Rikvold, P.A., and G. Stell, "Porosity and specific surface for interpenetrable-sphere models of two-phase random media," *J. Chem. Phys.* **82**, 1014 (1985).

Rink, M., and J.R. Schopper, "Computations of network models of porous media," *Geophys. Prospect.* **16**, 277 (1968).

Rintoul, M.D., and H. Nakanishi, "A precise determination of the backbone fractal dimension on two-dimensional percolation clusters," *J. Phys. A* **25**, 945 (1992).

Rintoul, M.D., and S. Torquato, "Metastability and crystallization in hard-sphere systems," *Phys. Rev. Lett.* **77**, 4198 (1996).

Rintoul, M.D., and S. Torquato, "Precise determination of the critical threshold and exponents in a three-dimensional continuum percolation model," *J. Phys. A* **30**, L585 (1997).

Rivier, N., "Recent results on the ideal structure of glasses," *J. Physique* **43**, C9-91 (1982).

Rivier, N., and A. Lissowski, "on the correlation between sizes and shapes of cells in epithelial mosaics," *J. Phys. A* **15**, L143 (1982).

Roberts, A.P., and M.A. Knackstedt, "Structure-property correlations in model composite materials," *Phys. Rev. E* **54**, 2313 (1996).

Roberts, A.P., and M. Teubner, "Transport properties of heterogeneous materials derived from Gaussian random fields: Bounds and simulation," *Phys. Rev. E* **51**, 4141 (1995).

Rohde, M., and H. Micklitz, "Critical behavior of the Hall conductivity near the percolation threshold in granular Sn:Ar mixtures," *Phys. Rev. B* **36**, 7289 (1987).

Roling, B., "Scaling properties of the conductivity spectra of glasses and supercooled melts," *Solid State Ionics* **105**, 185 (1998).

Roman, H.E., S. Russ, and A. Bunde, "Localization and typical spatial behavior of fractons," *Phys. Rev. Lett.* **66**, 1643 (1991).

Rosen, B.W., and Z. Hashin, "Effective thermal expansion coefficients and specific heats of composite materials," *Int. J. Eng. Sci.* **8**, 157 (1970).

Roth, L.M., "Effective-medium approximation for liquid metals," *Phys. Rev. B* **9**, 2476 (1974).

Roux, S., "Relation between elastic and scalar transport exponent in percolation," *J. Phys. A* **19**, L351 (1986).

Roux, S., and E. Guyon, "Mechanical percolation: a small beam lattice study," *J. Physique Lett.* **46**, L999 (1985).

Roux, S., and E. Guyon, "Transport exponents in percolation," in *On Growth and Form*, edited by H.E. Stanley and N. Ostrowsky (Martin Nijhoff, Dordrecht, 1986), p. 273.

Roux, S., and A. Hansen, "Critical behavior of anisotropic 'superelastic' central-force percolation," *J. Phys. A* **20**, L879 (1987).

Roux, S., and A. Hansen, "Transfer-matrix study of the elastic properties of central-force percolation," *Europhys. Lett.* **6**, 301 (1988).

Royer, E., C. Beniot, and G. Poussigue, "Dynamics of percolating networks," *J. Phys.: Condens. Matter* **4**, 561 (1992).

Rozanski, S.A., G. Kermer, P. Köberle, and A. Laschewsky, "Relaxation and charge transport in mixtures of zwitterionic polymers and inorganic salts," *Macromol. Chem. Phys.* **196**, 877 (1995).

Rudman, D.A., J.J. Calabrese, and J.C. Garland, "Noise spectra of three-dimensional random metal-insulator composites," *Phys. Rev. B* **32**, 1456 (1986).

Runge, I., "On the electrical conductivity of metallic aggregates," *Z. Tech. Physik* **6**, 61 (1925).

Russel, W.B., and A. Acrivos, "On the effective moduli of composite materials: Slender rigid inclusions at dilute concentrations," *Z. Angw. Math. Phys.* **23**, 434 (1972).

Russel, W.B., D.A. Saville, and W.R. Schowalter, *Colloidal Dispersions* (Cambridge University Press, Cambridge, 1989).

Rutgers, M.A., J.H. Dunsmuir, J.Z. Xue, W.B. Russel, and P. Chaikin, "Measurement of the hard-sphere equation of state using screened charged polystyrene collids," *Phys. Rev. B* **53**, 5043 (19996).

Sahimi, M., "On the relationship between the critical exponent of percolation conductivity and the static exponents of percolation," *J. Phys. A* **17**, L601 (1984a).

Sahimi, M., "Effective-medium approximation for density of states and the spectral dimension of percolation networks," *J. Phys. C* **17**, 3957 (1984b).

Sahimi, M., "Relation between the critical exponent of elastic percolation networks and the dynamical and geometrical exponents," *J. Phys. C* **19**, L79 (1986a).

Sahimi, M., "Dynamic percolation and diffusion in disordered systems," *J. Phys. C* **19**, 1311 (1986b).

Sahimi, M., "On the determination of transport properties of disordered systems," *Chem. Eng. Commun.*. **64**, 179 (1988).

Sahimi, M., "Nonlinear transport processes in disordered media," *AIChE J.* **39**, 369 (1993a).

Sahimi, M., "Flow phenomena in rocks: From continuum models to fractals, percolation, cellular automata and simulated annealing," *Rev. Mod. Phys.* **65**, 1395 (1993b).

Sahimi, M., *Applications of Percolation Theory* (Taylor and Francis, London, 1994a).

Sahimi, M., "Long-range correlated percolation and flow and transport in heterogeneous porous media," *J. Physique I France* **4**, 1263 (1994b).

Sahimi, M., "Effect of long-range correlations on transport phenomena in disordered media," *AIChE J.* **41**, 229 (1995a).

Sahimi, M., *Flow and Transport in Porous Media and Fractured Rock* (VCH, Weinheim, Germany, 1995b).

Sahimi, M., "Non-linear and non-local transport processes in heterogeneous media: From long-range correlated percolation to fracture and materials breakdown," *Phys. Rep.* **306**, 295 (1998).

Sahimi, M., and S. Arbabi, "Force distribution, multiscaling, and fluctuations in disordered elastic media," *Phys. Rev. B* **40**, 4975 (1989).

Sahimi, M., and S. Arbabi, "On correction to scaling for two- and three-dimensional scalar and vector percolation," *J. Stat. Phys.* **62**, 453 (1991).

Sahimi, M., and S. Arbabi, "Mechanics of disordered solids. II. Percolation on elastic networks with bond-bending forces," *Phys. Rev. B* **47**, 703 (1993).

Sahimi, M., G.R. Gavalas, and T.T. Tsotsis, "Statistical and continuum models of fluid-solid reactions in porous media," *Chem. Eng. Sci.* **45**, 1443 (1990).

Sahimi, M., and J.D. Goddard, "Superelastic percolation networks and the viscosity of gels," *Phys. Rev. B* **32**, 1869 (1985).

Sahimi, M., A.A. Heiba, B.D. Hughes, L.E. Scriven, and H.T. Davis, "dispersion in flow through porous media," *Society of Petroleum Engineers Paper 10969*, New Orleans, Louisiana (1982).

Sahimi, M., B.D. Hughes, L.E. Scriven, and H.T. Davis, "Stochastic transport in disordered systems," *J. Chem. Phys.* **78**, 6849 (1983a).

Sahimi, M., B.D. Hughes, L.E. Scriven, and H.T. Davis, "Critical exponents of percolation conductivity by finite-size scaling," *J. Phys. C* **16**, L521 (1983b).

Sahimi, M., B.D. Hughes, L.E. Scriven, and H.T. Davis, "Real-space renormalization and effective-medium approximation to the percolation conduction problem," *Phys. Rev. B* **28**, 307 (1983c).

Sahimi, M., B.D. Hughes, L.E. Scriven, and H.T. Davis, "Dispersion in flow through porous media. I. One-phase flow," *Chem. Eng. Sci.* **41**, 2103 (1986).

Sahimi, M., and S. Mukhopadhyay, "Scaling properties of a percolation model with long-range correlations," *Phys. Rev. E* **54**, 3870 (1996).

Sahimi, M., and H. Rassamdana, "On position-space renormalization group approach to percolation," *J. Stat. Phys.* **78**, 1157 (1995).

Sahimi, M., and T.S. Ray, "Transport through bootstrap percolation clusters," *J. Phys. I* **1**, 685 (1991).

Sahimi, M., L.E. Scriven, and H.T. Davis, "On the improvement of the effective-medium approximation to the percolation conductivity problem," *J. Phys. C* **17**, 1941 (1984).

Sahimi, M., and D. Stauffer, "Efficient simulation of flow and transport in porous media," *Chem. Eng. Sci.* **46**, 2225 (1991).

Sahimi, M., and T.T. Tsotsis, "Transient diffusion and conduction in heterogeneous media: Beyond the classical effective-medium approximation," *Ind. Eng. Chem. Res.* **36**, 3043 (1997).

Sangani, A.S., and A. Acrivos, "The effective conductivity of a periodic array of spheres," *Proc. R. Soc. Lond. A* **386**, 263 (1983).

Sarychev, A.K., and F. Brouers, "New scaling for ac properties of percolating composite materials," *Phys. Rev. Lett.* **73**, 2895 (1994).

Sarychev, A.K., and A.P. Vinogradoff, "Drop model of infinite cluster for 2D percolation," *J. Phys. C* **14**, L487 (1981).

Schaefer, D.W., C.J. Brinker, D. Richter, B. Farago, and B. Frick, "Dynamics of weakly connected solids: silica aerogels," *Phys. Rev. Lett.* **64**, 2316 (1990).

Schaefer, D.W., and K.D. Keefer, "Structure of random porous materials: silica aerogel," *Phys. Rev. Lett.* **56**, 2199 (1986).

Schaefer, D.W., J.E. Martin, P. Wiltzius, and D.S. Cannell, "Fractal geometry of colloidal aggregates," *Phys. Rev. Lett.* **52**, 2371 (1984).

Schaefer, D.W., D. Richter, B. Farago, C.J. Brinker, C.S. Ashley, B.J. Olivier, and P. Seeger, in *Neutron Scattering for Materials Science*, edited by S.M. Shapiro, S.C. Moss, and J.D. J⊘rgensen (Materials Research Society, Pittsburgh, 1990), p. 355.

Scher, H., and M. Lax, "Stochastic transport in a disordered solid. I. Theory," *Phys. Rev. B* **7**, 4491 (1973a).

Scher, H., and M. Lax, "Stochastic transport in a disordered solid. II. Impurity conduction," *Phys. Rev. B* **7**, 4502 (1973b).

Scher, H., M.F. Shlesinger, and J.T. Bendler, "Time-scale invariance in transport and relaxation," *Physics Today* **44** (NO. 1), 26 (1991).

Scher, H., and R. Zallen, "Critical density in percolation processes," *J. Chem. Phys.* **53**, 3759 (1970).

Schmidt, M., and W. Burchard, "Critical exponents in polymers: A sol-gel study of anionically prepared styrene-divinilbenzene copolymers," *Macromolecules* **14**, 370 (1981).

Schr⊘der, T.B., and J.C. Dyre, "Scaling and universality of ac conduction in disordered solids," *Phys. Rev. Lett.* **84**, 310 (2000).

Schulgasser, K., "Relationship between single-crystal and polycrystal electrical conductivity," *J. Appl. Phys.* **47**, 1880 (1976).

Schwartz, L.M., S. Feng, M.F. Thorpe, and P.N. Sen, "Behavior of depleted elastic networks: Comparison of effective-medium and numerical calculations," *Phys. Rev. B* **32**, 4607 (1985).

Schwartz, L.M., D.L. Johnson, and S. Feng, "Vibrational modes in granular materials," *Phys. Rev. Lett.* **52**, 831 (1984).

Scott, G.D., and D.M. Kilgour, "The density of random close packing of spheres," *Brit. J. Appl. Phys.* **2**, 863 (1969).

Seaton, N.A., and E.D. Glandt, "Spatial correlation functions from computer simulations," *J. Chem. Phys.* **85**, 5262 (1986).

Sen, A.K., and S. Torquato, "Effective conductivity of anisotropic two-phase composite media," *Phys. Rev. B* **39**, 4504 (1989).

Sen, P.N., C. Scala, and M.H. Cohen, "A self-similar model for sedimentary rocks with application to the dielectric constant of fused glass beads," *Geophysics* **46**, 781 (1981).

Shahinpoor, M., *Powder Technol.* **25**, 163 (1980).

Shante, V.K.S., "Hopping conduction in quasi-one-dimensional disordered compounds," *Phys. Rev. B* **16**, 2597 (1977).

Shante, V.K.S., and S. Kirkpatrick, "An introduction to percolation theory," *Adv. Phys.* **20**, 325 (1971).

Sheng, P., "Pair-cluster theory for the dielectric constant of composite media," *Phys. Rev. B* **22**, 6364 (1980).

Sheng, P., and J. Klafter, "Hopping conductivity in granular disordered systems," *Phys. Rev. B* **27**, 2583 (1983).

Sheppard, A.P., M.A. Knackstedt, W.V. Piczzewski, and M. Sahimi, "Invasion percolation: New algorithms and universality classes," *J. Phys. A* **32**, L521 (1999).

Shklovskii, B.I., "Critical behavior of the Hall coefficient near the percolation threshold," *Sov. Phys.-JETP* **45**, 152 (1977).

Shklovskii, B.I., "Anisotropy of percolation conduction," *Phys. Stat. Sol. B* **85**, K111 (1978).

Shklovskii, B.I., and A.L. Efros, "Transition from metallic to activation conductivity in compensated semiconductors," *Sov. Phys.-JETP* **33**, 468 (1971).

Shklovskii, B.I., and A.L. Efros, *Sov. Phys.-Tech. Phys. Lett.* **1**, 83 (1975).

Shklovskii, B.I., and A.L. Efros, *Electronic Properties of Doped Semiconductors* (Springer, Berlin, 1984).

Shlesinger, M.F., "William-Watts dielectric relaxation: A fractal time stochstic process," *J. Stat. Phys.* **36**, 639 (1984).

Shlesinger, M.F., and E.W. Montroll, "Williams-Watts function of dielectric relaxation," *Proc. National Acad. Sci. USA* **81**, 1280 (1984).

Sichel, E.K., and J.I. Gittleman, "The Hall effect in granular metal films near the percolation threshold," *Solid State Commun.* **42**, 75 (1982).

Sidebottom, D.L., "Universal approach for scaling the ac conductivity in ionic glasses," *Phys. Rev. Lett.* **82**, 3653 (1999).

Siegel, R.A., and R. Langer, "A new Monte Carlo approach to diffusion in constricted porous geometries," *J. Colloid Interface Sci.* **109**, 426 (1986).

Sigmund, O., *Design of Material Structures Using Topology Optimization*, Ph.D. Thesis, Department of Solid Mechanics, Technical University of Denmark (1994).

Silnutzer, N., *Effective Constants of Statistically Homogeneous Materials*, Ph.D. dissertation, University of Pennsylvania, Philadelphia (1972).

Sinai, Ya.G., "The limiting behavior of a one-dimensional random walk in a random environment," *Theory Prob. Appl.* **27**, 256 (1982).

Sinha, S.K., T. Freltoft, and J. Kjems, "Observation of power-law correlations in silica-particle aggregates by small angle neutron scattering," in *Kinetics of Aggregation and Gelation*, edited by F. Family and D.P. Landau (North-Holland, Amsterdam, 1984), p. 87.

Sleator, T., A. Bernasconi, S. Posselt, J.K. Kjems, and H.R. Ott, "Low-temperature specific heat and thermal conductivity of silica aerogels," *Phys. Rev. Lett.* **66**, 1070 (1991).

Smith, J.C., *J. Res. Natl. Bur. Stand.* **80A**, 45 (1976).

Smith, L.N., and C.J. Lobb, "Percolation in two-dimensional conductor-insulator networks with controllable anisotropy," *Phys. Rev. B* **20**, 3653 (1979).

Smith, P.A., and S. Torquato, "Computer simulation results for bounds on the effective conductivity of composite media," *J. Appl. Phys.* **65**, 893 (1989).

Söderberg, M., P.-O. Jansson, and G. Grimvall, "Effective medium theory for resistor networks in checkerboard geometries," *J. Phys. A* **18**, L633 (1985).

Sofo, J., J. Lorenzana, and E.N. Martinez, "Critical behavior of Young's modulus for two-dimensional randomly holed metallized Mylar," *Phys. Rev. B* **36**, 3960 (1987).

Song, Y., T.W. Noh, S.-I. Lee, and J.R. Gaines, "Experimental study of the three dimensional ac conductivity and dielectric constant of a conductor-insulator composite near the percolation threshold," *Phys. Rev. B* **33**, 904 (1986).

Song, Y., R.M. Stratt, and E.A. Mason, "The equation of state of hard spheres and the approach to random closest packing," *J. Chem. Phys.* **88**, 1126 (1988).

Soven, P., "Coherent-potential model of substitutional disordered alloys," *Phys. Rev.* **156**, 809 (1967).

Spence, R., *Linear Active Networks* (Wiley, New York, 1970).

Sreeram, A.N., A.K. Varshneya, and D.R. Swiler, "Molar volume and elastic properties of multicomponent chalcogenide glasses," *J. Non-Cryst. Solids* **128**, 294 (1991).

Srinivasan, A., K.N. Madhusoodanan, and E.S.R. Gopal, "Thermal diffusivity of IV-V-VI glasses - An evidence for the existence of a mechanical threshold," *Solid State Commun.* **83**, 163 (1992).

Stachowiak, H., "Effective electric conductivity tensor of polycrystalline metals in high magnetic fields," *Physica* **45**, 481 (1970).

Stanley, H.E., "A polychromatic correlated-site percolation problem with possible relevance to the unusual behaviour of supercooled H_2O and D_2O," *J. Phys. A* **12**, L329 (1979).

Stanley, H.E., P.J. Reynolds, S. Redner, and F. Family, "Position-space renormalization for models of linear polymers, branched polymers, and gels," in *Real-Space Renormalization*, edited by T.W. Burkhardt and J.M.J. van Leeuwen (Springer, Berlin, 1982), p. 169.

Stauffer, D., "Gelation in concentrated critically branched polymer solutions. Percolation scaling theory of intramolecular bond cycles," *J. Chem. Soc. Faraday Trans. II* **72**, 1354 (1976).

Stauffer, D., J. Adler, and A. Aharony, "Universality at the three-dimensional percolation threshold," *J. Phys. A* **27**, L475 (1994).

Stauffer, D., and A. Aharony, *Introduction to Percolation Theory*, 2nd ed. (Taylor and Francis, London, 1992).

Stauffer, D., A. Coniglio, and M. Adam, "Gelation and critical phenomena," *Adv. Polymer Sci.* **44**, 103 (1982).

Stauffer, D., and L. de Arcangelis, "Dynamics and strong size effects of a bootstrap percolation problem," *Int. J. Mod. Phys. C* **7**, 739 (1996).

Stell, G., and P.A. Rikvold, "Polydispersivity in fluids and composites: Some theoretical results," *Chem. Eng. Commun.* **51**, 233 (1987).

Stinchcombe, R.B., "The branching model for percolation theory and electrical conductivity," *J. Phys. C* **6**, L1 (1973).

Stinchcombe, R.B., "Conductivity and spin-wave stiffness in disordered systems-an exactly soluble model," *J. Phys. C* **7**, 179 (1974).

Stinchcombe, R.B., and B.P. Watson, "Renormalization group approach for percolation conductivity," *J. Phys. C* **9**, 3221 (1976).

Stockmayer, W.H., "Theory of molecular size distribution and gel formation in branched-chain polymers," *J. Chem. Phys.* **11**, 45 (1943).

Stoll, E., M. Kolb, and E. Courtens, "Numerical verification of scaling for scattering from fractons," *Phys. Rev. Lett.* **68**, 2472 (1992).

Stoyan, D., W.S. Kendall, and J. Mecke, *Stochastic Geometry and its Applications*, 2nd ed. (Wiley, New York, 1995).

Straley, J.P., "Critical phenomena in resistor networks," *J. Phys. C* **9**, 783 (1976).

Straley, J.P., "Random resistor tree in an applied field," *J. Phys. C* **10**, 3009 (1977a).

Straley, J.P., "Critical exponents for the conductivity of random resistor lattices," *Phys. Rev. B* **15**, 5733 (1977b).

Straley, J.P., "Exponent theory of the Hall effect and conductivity anisotropy near the percolation threshold," *J. Phys. C* **13**, L773 (1980a).

Straley, J.P., "Conductivity anisotropy and the Hall effect in inhomogeneous conductors near the percolation threshold," *J. Phys. C* **13**, 4335 (1980b).

Straley, J.P., "Non-universal threshold behaviour of random resistor networks with anomalous distribution of conductances," *J. Phys. C* **15**, 2343 (1982).

Strang, G., and G.J. Fix, *An Analysis of the Finite Element Method* (Prentice-Hall, Englewood Cliffs, 1973).

Stroud, D., "Generalized effective-medium approach to the conductivity of an inhomogeneous material," *Phys. Rev. B* **12**, 3368 (1975).

Stroud, D., "Percolation effects and sum rules in the optical properties of composites," *Phys. Rev. B* **19**, 1783 (1979).

Stroud, D., "Classical theory of the magnetoresistance and Hall coefficient of normal-superconducting composites," *Phys. Rev. Lett.* **44**, 1708 (1980).

Stroud, D., and F.P. Pan, "Self-consistent approach to electromagnetic wave propagation in composite media: Application to model granular metals," *Phys. Rev. B* **17**, 1602 (1978).

Summerfield, S., "Effective medium theory of A.C. hopping conductivity for random-bond lattice models," *Solid State Commun.* **39**, 401 (1981).

Summerfield, S., "Universal low-frequency behaviour in the a.c. hopping conductivity of disordered systems," *Philos. Mag. B* **52**, 9 (1985).

Summerfield, S., and P.N. Butcher, "A unified equivalent-circuit approach to the theory of AC and DC hopping conductivity in disordered systems," *J. Phys. C* **15**, 7003 (1982).

Suzuki, M., "A.c. hopping conductivity in Mn-Co-Ni-Cu complex oxide semiconductors with spinel structure," *J. Phys. Chem. Solids* **41**, 1253 (1980).

Szpilka, A.M., and P. Viščor, "D.c. electrical conductivity of evaporated amorphous germanium - the low temperature limit," *Philos. Mag.* **45**, 485 (1982).

Takahashi, M., K. Yokoyama, and T. Masuda, "Dynamic viscoelasticity and critical exponents in sol-gel transition of an end-linking polymer," *J. Chem. Phys.* **101**, 798 (1994).

Tanaka, K., "Structural phase transitions in chalcogenide glasses," *Phys. Rev. B* **39**, 1270 (1989).

Tanaka, K., T. Nakagawa, and A. Odajima, "Photodarkening in glassy chalcogenide systems," *Philos. Mag. Lett. B* **54**, L3 (1986).

Tang, W., and M.F. Thorpe, "Mapping between random central-force networks and random resistor networks," *Phys. Rev. B* **36**, 3798 (1987).

Tang, W., and M.F. Thorpe, "Percolation of elastic networks under tension," *Phys. Rev. B* **37**, 5539 (1988).

Tartar, L., "Estimations fines des coefficients homogénéisés," in *Ennio De Giorgi Colloquim*, edited by P. Krée (Pitman, Boston, 1985), p. 175.

Tassopoulos, M., and D.E. Rosner, "Simulation of vapor diffusion in anisotropic particulate deposits," *Chem. Eng. Sci.* **47**, 421 (1992).

Tatsumisago, M., B.L. Halfpap, J.L. Green, S.M. Lindsay, and C.A. Angell, "Fragility of Ge-As-Se glass-forming liquids in relations to rigidity percolation, and the Kauzmann paradox," *Phys. Rev. Lett.* **64**, 1549 (1990).

Taylor, D.W., "Vibrational properties of imperfect crystals with large defects concentration," *Phys. Rev.* **156**, 1017 (1967).

Taylor, H.E., "The dielectric relaxation spectrum of glass," *Trans. Faraday Soc.* **52**, 873 (1956).

Tessler, L.R., and G. Deutscher, "Experimental investigation of two-dimensional In-Ge thin films," *Physica A* **157**, 154 (1989).

Thorpe, M.F., "Continuous deformation in random networks," *J. Non-Cryst. Solids* **57**, 355 (1983).

Thorpe, M.F., and P.M. Duxbury (eds.), *Rigidity Theory and Applications* (Kluwer Academic, New York, 1999).

Thorpe, M.F., and E.J. Garboczi, "Site percolation on central-force elastic networks," *Phys. Rev. B* **35**, 8579 (1987).

Thorpe, M.F., and E.J. Garboczi, "Elastic properties of central-force networks with bonde-length mismatch," *Phys. Rev. B* **42**, 8405 (1990).

Thorpe, M.F., D.J. Jacobs, M.V. Chubynsky, and J.C. Phillips, "Self-organization in network glases," *J. Non-Cryst. Solids* **266-269**, 859 (2000).

Thorpe, M.F., and I. Jasuik, "New results in the theory of elasticity for two-dimensional composites," *Proc. R. Soc. Lond. A* **438**, 531 (1992).

Thorpe, M.F., and P.N. Sen, "Elastic moduli of two-dimensional composite continua with elliptical inclusions," *J. Accoust. Soc. Am.* **77**, 1674 (1985).

Thovert, J.F., I.C. Kim, S. Torquato, and A. Acrivos, "Bounds on the effective properties of polydispersed suspensions of spheres: An evaluation of two relevant morphological parameters," *J. Appl. Phys.* **67**, 6088 (1990).

Tielbürger, D., R. Merz, R. Ehrenfels, and S. Hunklinger, "Thermally-activated relaxation processes in vitreous silica: An investigation by Brillouin scattering at high pressures," *Phys. Rev. B* **45**, 2750 (1992).

Timoshenko, S.P., and J.N. Goodier, *Theory of Elasticity*, 3rd ed. (McGraw-Hill, New York, 1970).

Tobochnik, J., and P.M. Chapin, "Monte Carlo simulation of hard spheres near random closest packing using spherical boundary conditions," *J. Chem. Phys.* **88**, 5824 (1988).

Tokita, M., R. Niki, and K. Hikichi, "Percolation theory and elastic modulus of gel", *J. Phys. Soc. Jpn.* **53**, 480 (1984).

Torelli, L., and A.E. Scheidegger, "Three-dimensional branching-type models of flow through porous media," *J. Hydr.* **15**, 23 (1972).

Torquato, S., "Bulk properties of two-phase disordered media. I. Cluster expansion for the effective dielectric constant of dispersions of penetrable spheres," *J. Chem. Phys.* **81**, 5079 (1984).

Torquato, S., "Electrical conductivity of two-phase disordered composite media," *J. Appl. Phys.* **58**, 3790 (1985a).

Torquato, S., "Bulk properties of two-phase disordered media. II. Effective conductivity of a dilute suspension of penetrable spheres," *J. Chem. Phys.* **83**, 4776 (1985b).

Torquato, S., "Bulk properties of two-phase disordered media. III. New bounds on the effective conductivity of dispersions of penetrable spheres," *J. Chem. Phys.* **84**, 6345 (1986a).

Torquato, S., "Microstructure characterization and bulk properties of disordered two-phase media," *J. Stat. Phys.* **45**, 843 (1986b).

Torquato, S., "Relationship between permeability and diffusion-controlled trapping constant of porous media," *Phys. Rev. Lett.* **64**, 2644 (1990).

Torquato, S., in *Macroscopic Behaviour of Heterogeneous Materials from the Microstructure*, Vol. 147, edited by S. Torquato and D. Krajcinovic (Ameican Society of Mechanical Engineers, New York, 1992), p. 53.

Torquato, S., "Mean nearest-neighbour distance in random packings of D-Dimensional hard spheres," *Phys. Rev. Lett.* **74**, 2156 (1995).

Torquato, S., "Effective stiffness tensor of composite media I. Exact series expansions," *J. Mech. Phys. Solids* **45**, 1421 (1997).

Torquato, S. "Effective stiffness tensor of composite media: II. Applications to isotropic dispersions," *J. Mech. Phys. Solids* **46**, 1411 (1998).

Torquato, S., *Random Heterogeneous Materials* (Springer-Verlag, New York, 2002).

Torquato, S., and S. Hyun, "Effective-medium approximation for composite media: Realizable single-scale dispersions," *J. Appl. Phys.* **89**, 1725 (2001).

Torquato, S., I.C. Kim, and D. Cule, "Effective conductivity, dielectric constant, and diffusion coefficients of digitized media via first-passage time equations," *J. Appl. Phys.* **85**, 1560 (1999).

Torquato, S., and F. Lado, "Characterisation of the microstructure of distributions of rigid rods and discs in a matrix," *J. Phys. A* **18**, 141 (1985).

Torquato, S., and F. Lado, "Bounds on the conductivity of a random array of cylinders," *Proc. R. Soc. Lond. A* **417**, 59 (1988).

Torquato, S., and F. Lado, "Improved bounds on the effective elastic moduli of random arrays of cylinders," *J. Appl. Mech.* **59**, 1 (1992).

Torquato, S., F. Lado, and P.A. Smith, "Bulk properties of two-phase media. IV. Mechanical properties of suspensions of penetrable spheres at nondilute concentrations," *J. Chem. Phys.* **86**, 6388 (1987).

Torquato, S., and B. Lu, "Rigorous bounds on the fluid permeability: Effect of polydispersivity in grain size," *Phys. Fluids A* **2**, 487 (1990).

Torquato, S., and B. Lu, "Chord-length distribution function for two-phase random media," *Phys. Rev. E* **47**, 2950 (1993).

Torquato, S., B. Lu, and J. Rubinstein, "Nearest-neighbour distribution functions in many-body systems," *Phys. Rev. A* **41**, 2059 (1990).

Torquato, S., and M.D. Rintoul, "Effect of the interface on the properties of composite media," *Phys. Rev. Lett.* **75**, 4067 (1995).

Torquato, S., and A.K. Sen, "Conductivity tensor of anisotropic composite media from the microstructure," *J. Appl. Phys.* **67**, 1145 (1990).

Torquato, S., and G. Stell, "Microstructure of two-phase random media. V. The n-point matrix probability functions for impenetrable spheres," *J. Chem. Phys.* **82**, 980 (1985).

Torquato, S., T.M. Truskett, and P.G. Debenedetti, "Is random close packing of spheres well defined?" *Phys. Rev. Lett.* **84**, 2064 (2000).

Toupin, R.A., "Theories of elasticity with couple-stress," *Arch. Rat. Mech. Analy.* **16**, 87 (1964).

Trappe, V., W. Richtering, and W. Burchard, "Critical behavior of anhydride cured expoxies," *J. Physique II* **2**, 1453 (1992).

Tremblay, A.-M.S., S. Feng, and P. Breton, "Exponents for $1/f$ noise near a continuum percolation threshold," *Phys. Rev. B* **33**, 2077 (1986).

Tremblay, R.R., G. Albinet, and A.-M.S. Tremblay, "Noise and crossover exponent in conductor-insulator mixtures and superconductor-conductor mixtures," *Phys. Rev. B* **45**, 755 (1992).

Troadec, J.P., D., Bideau, and E. Guyon, "Transport properties of conducting and semiconducting anisotropic mixtures," *J. Phys. C* **14**, 4807 (1981).

Truskett, T.M., S. Torquato, and P.G. Debenedetti, "Towards a quantification of disorder in materials: Distinguishing equilibrium and glassy sphere packings," *Phys. Rev. E* **62**, 993 (2000).

Tsallis, C., A. Coniglio, and S. Redner, "Break-collapse method for resistor networks and a renormalization-group application," *J. Phys. C* **16**, 4339 (1983).

Turban, L., "On the effective-medium approximation for bond-percolation conductivity," *J. Phys. C*11, 449 (1978).

Turley, J., and G. Sines, "The anisotropy of Young's modulus, shear modulus and Poisson's ratio in cubic materials," *J. Phys. D* **4**, 264 (1971).

Tyc, S., and B.I. Halperin, "Random resistor network with an exponentially wide distribution of bond conductances," *Phys. Rev. B* **39**, 877 (1989).

Uebbing, B., and A.J. Sievers, "Influence of the glass network on the vibrational energy transfer from H_2O inpurities in the Ge-As-Se glass series," *J. Non-Cryst. Solids* **203**, 159 (1996).

Underwood, E.E., *Quantitative Stereology* (Addison Wesley, New York, 1970).

Vacher, R., T. Woignier, J. Pelous, and E. Courtens, "Structure and self-similarity of silica aerogels," *Phys. Rev. B* **37**, 6500 (1988).

Vacher, R., T. Woignier, J. Pelous, G. Coddens, and E. Courtens, "The density of vibrational states of silica aerogels," *Europhys. Lett.* **8**, 161 (1989).

Vacher, R., E. Courtens, G. Coddens, A. Heidemann, Y. Tsujimi, J. Pelous, and M. Foret, "Crossovers in the density of states of fractal silica aerogels," *Phys. Rev. Lett.* **65**, 1008 (1990).

van der Marck, S.C., "Network approach to void percolation in a pack of unequal spheres," *Phys. Rev. Lett.* **77**, 1785 (1996).

van der Putten, D., J.T., Moonen, H.B. Brom, J.C.M. Brokken-Zijp, and M.A.J. Michels, "Evidence for superlocalization on a fractal network in conductive carbon-black-polymer composites," *Phys. Rev. Lett.* **69**, 494 (1992).

van Dijk, M.A., "Dielectric study of percolation phenomena in a microemulsion," *Phys. Rev. Lett.* **55**, 1003 (1985).

van Dijk, M.A., G. Castelejin, J.G.H. Joosten, and Y.K. Levine, "Percolation in oil-continuous microemulsions. A dielectric study of aerosol OT/water/isooctane," *J. Chem. Phys.* **85**, 626 (1986).

Van Siclen, C. DeW., "Walker diffusion method for calculation of transport properties of finite composite systems," *Phys. Rev. E* **65**, 026144-1 (2002).

van Staveren, M.P.J., H.B. Brom, and L.J. de Jongh, "Metal-cluster compounds and universal features of the hopping conductivity of solids," *Phys. Rep.* **208**, 1 (1991).

Viot, P., G. Tarjus, and J. Talbot, "Exact solution of a generalized ballistic-deposition model," *Phys. Rev. E* **48**, 480 (1993).

Visčor, P., and A.D. Yoffe, "Amorphous germanium prepared in ultra high vacuum and measured in situ: The DC electrical conductivity," *J. Non-Cryst. Solids* **35/36**, 409 (1982).

Visscher, W.M., and M. Bolsterli, "Random packing of equal and unequal spheres in two and three dimensions," *Nature* **239**, 504 (1972).

Viswanathan, R., and M.B. Heaney, "Direct imaging of the percolation network in a three-dimensional disordered conductor-insulator composite," *Phys. Rev. Lett.* **75**, 4433 (1995).

Volger, J., "Note on the Hall potential across an inhomogeneous conductor," *Phys. Rev.* **79**, 1023 (1950).

Voronoi, G., "Nouvelles appillications des paraméteres continus á la théorie des formes quadratiques," *J. Reine Angew. Math.* **134**, 198 (1908).

Voss, R.F., and J. Clarke, "Flicker $(1/f)$ noise: Equilibrium temperature and resistance fluctuations," *Phys. Rev. B* **13**, 556 (1976).

Voss, R.F., R.B. Laibowitz, and E.I. Allessandrini, "Fractal (scaling) clusters in thin gold films near the percolation threshold," *Phys. Rev. Lett.* **44**, 1441 (1982).

Wagner, T., and S.O. Kasap, "Glass transformation, heat capacity and structure of As_xSe_{1-x} glasses studied by modulated temperature differential scanning calorimetry experiments," *Philos. Mag. B* **74**, 667 (1996).

Walker, D., and K. Scharnberg, "Electromagnetic response of high$-T_c$ superconductors," *Phys. Rev. B* **42**, 2211 (1990).

Walpole, L.J., "On the bounds for the overall elastic moduli of inhomogenous systems-I.," *J. Mech Phys. Solids.* **14**, 151 (1966).

Walpole, L.J., "On the overall elastic moduli of composite materials," *J. Mech. Phys. Solids* **17**, 235 (1969).

Walter, U., D.L. Price, S. Susman, and K.J. Volin, "Network dynamics of chalcogenide glasses. I. Germanium diselenide," *Phys. Rev. B* **37**, 4232 (1988).

Wang, J., "Bond-diluted rigid animals for central force model," *J. Phys. A* **21**, L353 (1988).

Wang, J., "The bond-bending model in three dimensions," *J. Phys. A* **22**, L291 (1989).

Wang, J., and A.B. Harris, "Bond orientational order in the randomly diluted elastic network," *Phys. Rev. Lett.* **55**, 2459 (1985).

Wang, J., and A.B. Harris, "Splay rigidity in the anisotropic superelastic network," *Europhys. Lett.* **6**, 157 (1988).

Wang, J., A.B. Harris, and J. Adler, "Series approach to the randomly diluted elastic network," *Phys. Rev. B* **45**, 7084 (1992).

Wang, Y., M. Nakamura, O. Matsuda, and K. Murase, "Raman-spectroscopy studies on rigidity percolation and fragility on Ge-(S,Se) glasses," *J. Non-Cryst. Solids* **266-269**, 872 (2000).

Warren, B.E., "X-ray diffraction of vireous silica," *J. Am. Ceram. Soc.* **17**, 249 (1934).

Warren, T.L., "Negative Poisson's ratio in a transversely isotropic foam structure," *J. Appl. Phys.* **67**, 7591 (1990).

Watson, B.P., and P.L. Leath, "Conductivity in the two-dimensional-site percolation problem," *Phys. Rev. B***9**, 4893 (1974).

Watt, J.P., G.F. Davies, and R.J. O'Connell, "The elastic properties of composite materials," *Rev. Geophys. Space Phys.* **14**, 541 (1976).

Weaire, D., "Some remarks on the arrangement of grains in a polycrystal," *Metallography* **7**, 157 (1974).

Weaire, D., and S. Hutzler, *The Physics of Foams* (Oxford University Press, New York, 1999).

Weaire, D., and R. Phelan, "A counter-example to Kelvin's conjecture on minimal surfaces," *Philos. Mag. Lett.* **69**, 107 (1994).

Webman, I., "Effective-medium approximation for diffusion on a random lattice," *Phys. Rev. Lett.* **47**, 1496 (1981).

Webman, I., and G. Grest, "Dynamical behavior of fractal structures," *Phys. Rev. B* **31**, 1689 (1985).

Weinrib, A., "Long-range correlated percolation," *Phys. Rev. B* **29**, 387 (1984).

Weinrib, A., and B.I. Halperin, "Critical phenomena in systems with long-range-correlated quenched disorder," *Phys. Rev. B* **27**, 413 (1983).

Weissberg, H.L., "Effective diffusion coefficient in porous media," *J. Appl. Phys.* **34**, 2636 (1963).

Weissman, M.B., "$1/f$ noise and other slow, nonexponential kinetics in condensed matter," *Rev. Mod. Phys.* **60**, 537 (1988).

Widom, B., "Random squential addition of hard spheres to a volume," *J. Chem. Phys.* **44**, 3888 (1966).

Wiener, O., "Die theorie des mischk′orpers fr̈ das feld des stationären strömung," *Math.-Physichen Klasse der Königl. Sächsischen Gesellschaft der Wissenschaften* **32**, 509 (1912).

Wilkinson, D., J.S. Langer, and P.N. Sen, "Enhancement of the dielectric constant near a percolation threshold," *Phys. Rev. B* **28**, 1081 (1983).

Williams, G., and D.C. Watts, "Non-symmetrical dielectric relaxation behavior arising from a simple empirical decay function," *Trans. Faraday Soc.* **66**, 80 (1970).

Williams, G., D.C. Watts, S.B. Dev, and A.M. North, "Further considerations of non-symmetrical dielectric relaxation behavior arising from a simple empirical decay potential," *Trans. Faraday Soc.* **67**, 1323 (1971).

Williams, M.L., and H.J. Maris, "Numerical study of phonon localization in disordered systems," *Phys. Rev. B* **31**, 4508 (1985).

Willis, J.R., "Bounds and self-consistent estimates for the overall moduli of anisotropic composites," *J. Mech. Phys. Solids* **25**, 185 (1977).

Willis, J.R., "Variational and related methods for the overall properties of composites," *Adv. Appl. Mech.* **21**, 1 (1981).

Winterfeld, P.H., L.E. Scriven, and H.T. Davis, "Percolation and conductivity of random two-dimensional composites," *J. Phys. C* **14**, 2361 (1981).

Witten, T.A., and L.M. Sander, "Diffusion-limited aggregation, a kinetic critical phenomenon," *Phys. Rev. Lett.* **47**, 1400 (1981).

Woignier, T., J. Phalippou, R. Sempere, and J. Pelons, "Analysis of the elastic behaviour of silica aerogel taken as a percolating system," *J. Phys. France* **49**, 289 (1988).

Woignier, T., J. Phalippou, R. Vacher, J. Pelous, and E. Courtens, "Different kind of fractal strucures in silica aerogels," *J. Non-Cryst. Solids* **121**, 198 (1990).

Wright, D.C., D.J. Bergman, and Y. Kantor, "Resistance fluctuations in random resistor networks above and below the percolation threshold," *Phys. Rev. B* **33**, 396 (1986).

Wu, J., and D.S. McLachlan, "Percolation exponents and thresholds obtained from the nearly ideal continuum percolation system graphite-boron nitride," *Phys. Rev. B* **56**, 1236 (1997).

Wu, T.T., "The effect of inclusions shape on the elastic moduli of a two-phase material," *Int. J. Solids Struct.* **2**, 1 (1966).

Xia, T.K., P.M. Hui, and D. Stroud, "Theory of Faraday rotation in granular magnetic materials," *J. Appl. Phys.* **67**, 2736 (1990).

Xu, B., F. Arias, S.T. Brittain, X.-M. Zhao, B. Gryzbowski, S. Torquato, and G.M. Whitesides, "Making negative Poisson's ratio microstructures by soft lithography," *Advanced Materials* **11**, 1186 (1999).

Yagil, Y., M. Yosefin, D.J. Bergman, G. Deutscher, and P. Gadanne, "Scaling theory for the optical properties of semicontinuous metal films," *Phys. Rev. B* **43**, 11342 (1991).

Yakubo, K., E. Courtens, and T. Nakayama, "Missing modes in the density of states of fractal networks," *Phys. Rev. B* **42**, 1078 (1990a).

Yakubo, K., and T. Nakayama, "Fracton dynamics of percolating elastic networks: energy spectrum and localized nature," *Phys. Rev. B* **40**, 517 (1989).

Yakubo, K., T. Nakayama, and H.J. Maris, "Analysis of a new method for finding eigenmodes of very large lattice systems," *J. Phys. Soc. Jpn.* **60**, 3249 (1991).

Yakubo, K., K. Takasugi, and T. Nakayama, "Crossover behavior from vector to scalar fractons in the density of states of percolating networks," *J. Phys. Soc. Jpn.* **59**, 1909 (1990b).

Yonezawa, F., in *The Structure and Properties of Matter*, edited by T. Matsubara (Springer, Berlin, 1982), p. 43.

Yonezawa, F., and M.H. Cohen, "Granular effective medium approximation," *J. Appl. Phys.* **54**, 2985 (1983).

Young, A.P., and R.B. Stinchcombe, "A renormalization group theory for percolation problems," *J. Phys. C* **8**, L535 (1975).

Yoon, C.S., and S.-I. Lee, "Measurements of the ac conductivity and dielectric constant in a two-dimensional lattice percolation system," *Phys. Rev. B* **42**, 4594 (1990).

Yuge, Y., "Three-dimensional site percolation problem and effective-medium theory: A computer study," *J. Stat. Phys.* **16**, 339 (1977).

Yun, S.S., H. Li, R.L. Cappelletti, R.N. Enzweiler, and P. Boolchand, "Onset of rigidity in $Se_{1-x}Ge_x$ glasses: Ultrasonic elastic moduli," *Phys. Rev. B* **39**, 8702 (1989).

Zabolitzky, J.G., D.J. Bergman, and D. Stauffer, "Precision calculation of elasticity for percolation," *J. Stat. Phys.* **44**, 211 (1986).

Zachariasen, W.H., "The atomic arrangement in glass," *J. Am. Chem. Soc.* **54**, 3841 (1932).

Zallen, R., "Introduction to percolation: A model for all seasons," in *Percolation Structures and Processes*, edited by G. Deutscher, R. Zallen and J. Adler, *Annals of Israel Physical Society* **5** (Adam Hilger, Bristol, 1983), p. 3.

Zemlyanov, M.G., V.K. Malinovskii, V.N. Novikov, P.P. Parshin, and A.P. Sokolov, "A study of fractons in polymers," *Zh. Eskp. Teor. Fix.* **101**, 284 (1992) [*Sov. Phys. JETP* **74**, 151 (1992)].

Zeng, Q., and H. Li, "Diffusion equation for disordered fractal media," *Fractals* **8**, 117 (2000).

Zeng, X.C., P.M. Hui, and D. Stroud, "Numerical study of optical absorption in two-dimensional metal-insulator and normal-superconductor composites," *Phys. Rev. B* **39**, 1063 (1989).

Zhang, L., and N.A. Seaton, "Prediction of the effective diffusivity in pore networks near the percolation threshold," *AIChE J.* **38**, 1816 (1992).

Zhang, Y., P. Dai, M. Levy, and M.P. Sarachik, "Probing the Coulomb gap in insulating n-type CdSe," *Phys. Rev. Lett.* **64**, 2687 (1990).

Ziff, R.M., "Spanning probability in 2D percolation," *Phys. Rev. Lett.* **69**, 2670 (1992).

Ziman, J.M., "The localization of electrons in ordered and disordered systems. I. Percolation of classical particles," *J. Phys. C* **1**, 1532 (1968).

Zuzovsky, M., and H. Brenner, "Effective conductivities of composite materials composed of cubic arrangements of spherical particles embedded in an isotropic matrix," *J. Appl. Math. Phys. (ZAMP)* **28**, 979 (1977).

Index

Aboav–Weaire rule 89
Absorption coefficient 184
AC conductivity 339–352
effect of percolation 350
effective-medium approximation 346
scaling properties of 340–345
universality 340
Activation energy 310, 339, 341, 346, 348
Additive polymerization 601
Aerogels 614, 640
Alexander–Orbach conjecture 384, 386–389, 397
Anderson localization 335, 393
Anharmonicity 405
Animals (lattice) 551, 606
Anisotropic materials 136, 164, 187, 236, 240, 460, 488, 510
Anisotropic percolation 236, 240
Annealing schedule 100
Anomalous diffusion 305–309
Ant in a labyrinth 264
Archie's law 221
Asymmetric effective-medium approximation 510
Asymmetric hopping model 319–325
Auxetic materials 415
Average molecular weight 604
Average t-matrix approximation 209

Backbone fractal dimension 39
Beam model 582
Beran bounds 141, 146–148, 468, 486
Bethe lattice 26, 27, 96, 203–207, 316, 321
Blobs of percolation clusters 34
Body-centered cubic array of spheres 118, 421
Bohr radius 311
Boltzmann distribution 98
Boltzmann's equation 311
Bond-bending forces 569
Bond-bending models 569–581
comparison with central-force percolation 573
fixed points 577
force distribution 572
Kirkwood–Keating model 570, 592, 627
percolation threshold 571
relation to scalar percolation 576
scaling properties 574
Bond-bond Green function 211, 561
Bond percolation 27, 216, 247, 550, 571
Bootstrap percolation 50, 51, 556
Born model 541–543, 561, 615
Bose–Einstein distribution 397
Boundary-element method 514
Bounds
cluster bounds 141, 475
to effective conductivity 138–158
to dielectric constant 180
to elastic moduli 471–493
one-point bounds 138, 472
two-point bounds 139, 152, 472
three- and four-point bounds 146, 485
Box-counting method 17
Branched polymers 602, 606–609
fractal dimension 606–609
in good solvent 606, 607
in swollen state 607
Parisi-Sourlas-Family relation 607
scaling properties 606–609
size distribution 607
small-angle neutron scattering 607, 608

Carbon black materials 200, 242, 336
Cavity strain field 447–452

Cayley tree 321
Central forces 541, 543
Central-force percolation (*see* Rigidity percolation)
Chemical path 39
Chemical gels 614
Cherkaev–Lurie–Milton theorem transformation 426
Cherry-pit model 64
Chord-length distribution
  definition of 69
  identical hard spheres 77–80
  identical overlapping spheres 75–77
  link to lineal-path function 72, 73
  mean chord length 73
  polydisperse hard spheres 83
  polydisperse overlapping spheres 82
Cluster effective-medium approximation
  for effective conductivity 229
  for vibrational density of states 382
Coherent-potential approximation 230
Cohn's theorem 271
Cole–Cole representation 188
Compatibility equations 412
Compliance tensor 411
Composite materials
  bounds to conductivity (*see* Bounds)
  bounds to elastic properties (*see* Bounds)
  dilute limit 431
  exact analysis of anisotropic materials 136, 460
  exact series expansions 128, 452
  non-dilute systems 433
  two spherical inclusions 437
  with equal shear moduli 429
  with length mismatch 628
Concentric-shell model 139
Conduction
  expansion of electric field 128–132
  Green function 210, 295
  polarizability 129
  polarization in an ellipsoid 159
Conduction symmetry 108
Conductivity (*see* also Random resistor networks)
  anisotropy near percolation 240
  coated-spheres models 125
  effective-medium approximation 162–168, 230–231
  link with elastic moduli 518–531
  perturbation expansion 127–209
  upper and lower bounds (*see* Bounds)
Conjugate-gradient method 258
Constant-gap scaling 558
Constitutive nonlinearity 7
Constraint-counting method
  for Kirkwood–Keating model 571
  for rigidity percolation 554
Continuous-time random walks 295, 300–305, 337, 362
Continuum percolation 41–48
Contrast expansion 128, 447–471
Coordination number 29
Corrections to scaling 41, 265, 538, 548, 558, 575
Correlation function 21
Covalent bonding 570, 591
Critical exponents (*see* Percolation)
Critical-path method
  for electrical conductivity 255
  for hopping conductivity 325
  for rigidity percolation 565
Critical subnetwork 327
Critical volume fraction 42–44
Crossover frequency 377, 381, 386, 388
Crossover time 305, 307, 380
Cross-property relations 518–531
Cumulant approximation 342–348
Current distribution 270

Dead-end bonds 32
Degree of polymerization 604
Density-density correlation 396
Depolarization coefficient 133, 159, 160, 167, 361
Deviatoric components of stress tensor 415
Debye process 612
Dielectric constant 176–182, 352–365
  Debye model 362
  effective-medium approximation 360–362
  perturbation expansion 179, 280
  random walk model 362–365
  resistor-capacitor model 352–356
  scaling properties near percolation 365–375
  spectral representation 177–179

upper and lower bounds (*see* Bounds)
Williams–Watts formula 362
Dielectric loss 340
Diffusivity
one-dimensional systems 299–302
scaling properties 306–309
two- and three-dimensional systems 302–305
Diffusion-limited aggregates 19
Direct correlation function 21
Distribution of relaxation times 613
Drude dielectric function 183
Duality relation
for continuum composites 161
for resistor networks 207
Dundurs constants 429
Dynamic-mechanical experiments 609
Dynamic superelastic network 620
Dynamic scaling 306, 307
Dynamical effective-medium approximation 192–194, 293–305, 322–325
Dynamical floppiness 547
Dynamical matrix 548
Dynamical structure factor 398–404
effective-medium approximation 400
numerical computation 402
scaling analysis

Einstein problem 485
Effective-medium approximation
for anisotropic materials 164, 236
for bond-bending models 581
for conductor-insulator composites 216
for conductor-superconductor composites 218
for conductivity of continuous composites 162–168
for correlated composites 234
for continuum elasticity 504–513
for dielectric constant 360
for Hall conductivity 276
for materials with length mismatch 628
for multiple-coordinated networks 218
for rigidity percolation 560–565
for site percolation 231
for the Born model 561
for vibrational density of states 378–383
with energy dependence 324
Elastic constant tensor
symmetry properties of 412
Elastic density of states 634–642
Elastic moduli
cluster expansions 475
coated-inclusion models 568
definition of 411
equal phase shear moduli 429
link to conductivity 518–531
link to thermoelastic constants 430
Maxwell–Garnett approximation 513
minimum energy principles 415
one-point bounds 472
periodic arrays of spheres 421
self-consistent approximations 504–513
sheets with holes 427
strong-contrast expansions 447–471
three-point bounds 485
Elastic network
see Rigidity percolation; Bond-bending models
Elastic spectral dimension 635
Elastic symmetry 412
Elasticity
Eshelby tensor 436
expansion of strain field 447–460
Green function 449
polarizabilities 432, 452
Electrical conductivity (*see* Conductivity; Random resistor networks; Bounds)
Electromagnetic properties
effective-medium approximation 192
Entropic elasticity 584
Entropic networks 584, 585
Equilibrium hard-spheres 65
Erythrocyte cytoskeleton 534
Eshelby tensor 436
Euler's law 87
Exclusion probability 68
Extinction coefficient 190

Face-centered cubic array of spheres 118, 421
Faraday effect 188
Faraday's law 183
Fermi distribution 311
Field-theoretic approach 268
Filling factor of a lattice 43
Finite-difference methods 514

Finite-element simulation
  conductivity 514
Finite-size scaling 40
First-passage time method 170–176
Fixed point 250, 358, 577, 579
Flexural rigidity 535
Flicker noise 270
Floppy state
  dynamical 547
  geometrical 547
  macroscopic 546, 563
  microscopic 546, 564
Fluctuation-dissipation theorem 314
Foams 621–624
  mechanical properties of 621, 622
  relation to rigidity percolation 623
Fractal diffusion 305, 383, 384
Fractal dimension 17, 18
Fractional equation 309
Fractons 384–409
  crossover to phonons 385
  dispersion relation 385
  effect on sound velocity 386
  effect on thermal transport 404–409
  ensemble-averaged properties 393
  fracton dimension 380
  large-scale computer simulation 377, 386
  localization of 384
  mode patterns of 393
  phonon assisted hopping 405
Fully-impenetrable sphere model 64, 75, 81
Fully-penetrable sphere model 64, 77, 82

Gaussian random field model 91–95
Gelation 599–621
  Debye process 612
  determination of transition point 611
  dynamic-mechanical experiments 609
  dynamic superelastic network model 620
  elastic moduli 609–621
  gel fraction 603
  percolation models 602–609
  relaxation time spectrum 611–614
  retardation time spectrum 613
  Rouse regime 620
  viscosity 618–621
  Zimm regime 620
Gels
  chemical 614
  complex shear compliance 612
  dynamic shear compliance 612
  dynamic tensile modulus 610
  enthalpic 616
  entropic 616
  fluctuation algorithm for 617
  loss modulus 610
  physical gels 614
  relaxation strength 612
  storage modulus 610
Generalized master equation 295
Glass 589–599
  aging 589
  Cohen–Grest model for metallic glasses 590, 591
  constraint-counting argument 591, 592
  effect of high coordination number 595
  effect of onefold-coordinated atoms 597
  free volume 590
  mean-coordination number 591
  Phillips–Thorpe theory 571, 591
  Raman spectra 595
  stress-free versus stressed 598
Granular materials 329
Grain boundaries 89
Granular model 583
Granular superconductors 284
Green functions
  for conduction 165, 210–213
  for continuum elasticity 449
  for diffusion 295
  for the Born model 561
Green theorem 114, 119, 120, 123

Hall conductivity 275–284
Hard spheres 65
  chord-length density function 80, 83
  conductivity bounds (*See* Bounds)
  elastic moduli bounds (*See* Bounds)
  lineal-path function 79, 83
  maximally random jamming 66
  point/*q*-particle correlations 78, 79
  pore-size distribution 69
  radial distribution function 74
  random sequential addition of 59
Hashin–Shtrikman bounds (*See* Bounds)

Hashin–Shtrikman variational principles 138, 472
Hexagonal arrays of cylinders 121, 422
Hopping conductivity 309–339

Incoherent inelastic scattering 397
Indicator function 72
Interface conductivity 160
Intersection volume 74–76, 155
Inverted Swiss-cheese model 45

Kosolov constant 420

Lamé coefficient 7, 414, 539
Laminates 125, 422
Laplace equation 111
Lattice animal 606
Legendre functions 112
Lineal-path function
  definition of 69
  identical hard spheres 79
  identical overlapping spheres 77
  polydisperse overlapping spheres 82, 83
Linear chain 59
Links-nodes-blobs model 34
Localization 335–337
London penetration depth 286
Lorentz electric field 128
Loss angle 611
Loss modulus 610

Magnetoconductivity 275–282, 284
Markov random field 99
Master equation 292
Maximally random jammed state 66
Maxwell equations 182
Maxwell–Garnett approximation 168, 513
Mean chord length 73
Metal-insulator films 200–202
Metal-insulator transition 200
Microemulsions 373
Mie scattering 189
Miller–Abrahams network 311
Minimum complementary energy criterion 141, 416
Minimum energy criterion 141, 416
Minimum path 39
Mott hopping model 310, 325, 326
Multiple coordination number 218
Multiple-scattering approach
  for elastic moduli 493–501
  for optical properties 194

n-particle probability density 67, 73
n-point correlation function 21
n-point distribution function 69, 70
n-vector model
Nearest-neighbor distribution function 68
Network reduction (hopping conductivity) 311–313
Noise ($1/f$) 270
Non-self-consistent approximation 209

One-level cut model 91, 92
Ornstein–Zernike equation 44

Padé approximation 132, 268
Pair-connectedness function 32
Pair correlation function 21
Penetrable-concentric shell model 64
Percolation (definition) 26–28
  accessible fraction 32
  backbone 32
  backbone fraction 32
  clusters 28
  cluster number 32
  conductor-insulator composites 33
  conductor-superconductor composites 33
  correlation length 32
  critical exponents 35, 36
  fractal dimensions 38, 39
  mean cluster size 32
  percolation probability 32
  rigid-soft composites 34
  rigid-superrigid composites 34
  thresholds 29
  scaling properties 35–37
Percolation path approximation 350
Percus–Yevick approximation 74
Persistence length 535
Phonons 377
Phonon-assisted hopping 405
Physical gels 614
Point/n-particle distribution function 67
Poisson's ratio 415
  negative values of 631
  universality of 427, 577

Polycondensation 600
Polydispersed spheres 80–83
Polymeric materials 599
Polymer chains 599
Polymerization 600
Pore-size functions 69
Position-space renormalization 248, 266, 356, 578
Power laws 35, 36
Porod law of scattering 23

Quasi-static approximation 189

Radial distribution function
  Ornstein–Zernike formulation 44
  Percus–Yevick approximation 74
Raman scattering 642
Random-close packing fraction 66
Random resistor network (*see also* Effective-medium approximation; Percolation)
  exact result for the square lattice 207–209
  exact solution for Bethe lattices 203–207
  field-theoretic method 268
  Green function formulation 209
  Lobb–Frank–Fogelholm methods 261
  perturbation expansion 209
  series expansion 266
  transfer-matrix method 259
Random sequential addition 59
Random walk simulation 169, 263
Reconstruction models 97
Red bonds 34
Reduced network (hopping conductivity) 313
Relaxation time spectrum 613
Renormalization-group transformation
  for conductivity 251, 252
  for dielectric constant 356
  for percolation 248–251
  for bond-bending model 578
Renormalized effective-medium approximation 253
Resistance fluctuations 270
Resistor-capacitor model 352
Resistor-capacitor-inductor model 354
Retardation time spectrum 613
Reuss–Voigt bounds 472
Rigidity
  geometrical 547
  local 546
Rigidity percolation
  correlation length 548
  floppy state 552, 553
  force distribution 549, 550
  mapping to resistor networks 555
  nature of phase transition 556
  pebble game 551
  percolation threshold 550, 551
  scaling properties 557–560
  shortcomings of 568
  superelasticity 559
Rotational invariance 542, 543, 569–571

Sample-spanning cluster 28
Scattering intensity 22, 23, 396, 397, 608, 640
Second-order light scattering 24
Self-avoiding walks 599
Self-affine fractals 24–26
Self-consistent approximation (*See* Effective-medium approximation)
Self-similarity 17–20
Semiconductors 309
Series expansion 266
Sierpinsky gasket 18
Silica aerogels 395
Simulated annealing 97
Singly-connected bonds 34
Site percolation 28
Site-site Green function 210, 293
Small-angle scattering 22, 23, 396
Sol-gel transition 602
Sound velocity 386
Spatially-periodic models 61
Specific heat of superconductors 287
Spectral dimension
  for diffusion/conduction 383
  for elastic systems (flexural dimension) 635, 636
Spectral representation of dielectric function 177
Spherical harmonics 112
Square arrays of cylinders 123
Statistical self-similarity 19
Stress transition 598
Stresslet 434

Strum sequence method 378
Superconducting materials 284–290
  Bardeen–Cooper–Schrieffer theory 186, 287
  coherence length 289
  critical current 287
  differential magnetic susceptibility 289
  specific heat 287
  Type I 284
  Type II 285
  upper critical field 288, 289
Superelastic percolation networks 559, 576, 577, 618–621
Superlocalization 335
Surface-matrix correlation function 67
Surface-particle correlation function 67
Surface plasmon resonance 184
Surface-surface correlation function 67, 68
Swiss-cheese model 45
Symmetric effective-medium approximation
Symmetric hopping model 317

Taylor–Isard scaling constant 341
Tellegen's theorems 271
Tessellation models 85
Tetrakaidecahedron lattice 87
Threshold nonlinearity 8
Transfer-matrix method 259
  for resistor networks 259
  for bond-bending model 581
  for rigidity percolation 550
Two-level cut model 92–94
Two-site effective-medium approximation 323
Two-site self-consistent approximation 322

Universal scaling laws 35, 37

Variable-range hopping 335–339
Variational principles 138, 472
Vegard's law 630
Vibrational density of states 373–396, 632–640
  crossover between scalar and vector models 635
  effective-medium approximation 376–381
  fracton regime 381–384
  missing modes 387
  numerical computations 375, 384
  phonon regime 373
  scalar approximation 373
  scaling properties 381, 632
  Strum sequence method 378
  vector percolation model 632
Viscoelasticity (*see* Gelation; Gels)
Visscher–Bolsterli algorithm 83
Voronoi network 45, 90
Vulcanization 601

Wigner–Seitz cell 61
William–Watts law 362

X-ray scattering 22, 23, 120, 395, 608

Interdisciplinary Applied Mathematics

1. *Gutzwiller:* Chaos in Classical and Quantum Mechanics
2. *Wiggins:* Chaotic Transport in Dynamical Systems
3. *Joseph/Renardy:* Fundamentals of Two-Fluid Dynamics: Part I: Mathematical Theory and Applications
4. *Joseph/Renardy:* Fundamentals of Two-Fluid Dynamics: Part II: Lubricated Transport, Drops and Miscible Liquids
5. *Seydel:* Practical Bifurcation and Stability Analysis: From Equilibrium to Chaos
6. *Hornung:* Homogenization and Porous Media
7. *Simo/Hughes:* Computational Inelasticity
8. *Keener/Sneyd:* Mathematical Physiology
9. *Han/Reddy:* Plasticity: Mathematical Theory and Numerical Analysis
10. *Sastry:* Nonlinear Systems: Analysis, Stability, and Control
11. *McCarthy:* Geometric Design of Linkages
12. *Winfree:* The Geometry of Biological Time (Second Edition)
13. *Bleistein/Cohen/Stockwell:* Mathematics of Multidimensional Seismic Imaging, Migration, and Inversion
14. *Okubo/Levin:* Diffusion and Ecological Problems: Modern Perspectives (Second Edition)
15. *Logan:* Transport Modeling in Hydrogeochemical Systems
16. *Torquato:* Random Heterogeneous Materials: Microstructure and Macroscopic Properties
17. *Murray:* Mathematical Biology I: An Introduction (Third Edition)
18. *Murray:* Mathematical Biology II: Spatial Models and Biomedical Applications (Third Edition)
19. *Kimmel/Axelrod:* Branching Processes in Biology
20. *Fall/Marland/Wagner/Tyson* (Editors): Computational Cell Biology
21. *Schlick:* Molecular Modeling and Simulation: An Interdisciplinary Guide
22. *Sahimi:* Heterogeneous Materials I: Linear Transport and Optical Properties
23. *Sahimi:* Heterogeneous Materials II: Nonlinear and Breakdown Properties and Atomistic Modeling
24. *Bloch:* Nonholonomic Mechanics and Control

CPSIA information can be obtained
at www.ICGtesting.com
Printed in the USA
LVOW13*1109010418
571867LV00019B/229/P